24 Years
Previous Solved Papers

ESE 2025

UPSC ENGINEERING SERVICES EXAMINATION

Preliminary Examination

ELECTRICAL ENGINEERING

Objective Solved Papers

Volume-I

- Topicwise presentation
- Thoroughly revised & updated

Also useful for

State Engineering Services Examinations, Public Sector Examinations & Other Competitive Examinations

MADE EASY PUBLICATIONS

www.madeeasypublications.org

MADE EASY Publications Pvt. Ltd.

Corporate Office: 44-A/4, Kalu Sarai (Near Hauz Khas Metro Station), New Delhi-110016

Contact: 9021300500

E-mail: infomep@madeeasy.in

Visit us at: www.madeeasypublications.org

ESE-2025 : Preliminary Examination
Electrical Engineering : Volume-1
Topicwise Objective Solved Questions : (2001-2024)

1st Edition : 2008
2nd Edition : 2009
3rd Edition : 2010
4th Edition : 2011
5th Edition : 2012
6th Edition : 2013
7th Edition : 2014
8th Edition : 2015
9th Edition : 2016
10th Edition : 2017
11th Edition : 2018
12th Edition : 2019
13th Edition: 2020
14th Edition: 2021
15th Edition: 2022
16th Edition: 2023

17th Edition: 2024

Director's Message

B. Singh (Ex. IES)

Engineering is one of the most chosen graduating field. Taking engineering is usually a matter of interest but this eventually develops into "purpose of being an engineer" when you choose engineering services as a career option.

Train goes in tunnel we don't panic but sit still and trust the engineer, even we don't doubt on signalling system, we don't think twice crossing over a bridge reducing our travel time; every engineer has a purpose in his department which when coupled with his unique talent provides service to mankind.

I believe *"the educator must realize in the potential power of his pupil and he must employ all his art, in seeking to bring his pupil to experience this power"*. To support dreams of every engineer and to make efficient use of capabilities of aspirant, MADE EASY team has put sincere efforts in compiling all the previous years' ESE-Pre questions with accurate and detailed explanation. The objective of this book is to facilitate every aspirant in ESE preparation and so, questions are segregated chapterwise and topicwise to enable the student to do topicwise preparation and strengthen the concept as and when they are read.

I would like to acknowledge efforts of entire MADE EASY team who worked hard to solve previous years' papers with accuracy and I hope this book will stand up to the expectations of aspirants and my desire to serve student fraternity by providing best study material and quality guidance will get accomplished.

B. Singh (Ex. IES)
CMD, MADE EASY Group

Volume-I

ELECTRICAL ENGINEERING

Objective Solved Questions

of UPSC Engineering Services Examination

Contents

UNIT

I

Electrical Materials

Syllabus

Electrical engineering materials, crystal structures and defects, ceramic materials, insulating materials, magnetic materials-basics, properties and applications, ferrities, ferro-magentic materials and components, basics of solid state physics, conductors, photo-conductivity, basics of nano materials and superconductors.

Contents

OOOO

1 Introduction to Engineering Materials

1.1 Total number of electrons that can be accommodated in various electron states in a valence band of a given solid is equal to
(a) atomic number of the solid
(b) half the number of atoms in the solid
(c) the number of atoms in the solid
(d) twice the number of atoms in the solid
[ESE-2001]

1.2 What is the packing fraction of a BCC (body-centered cubic) unit cell?
(a) $\frac{\sqrt{3}\pi}{16}$ (b) $\frac{\sqrt{3}\pi}{8}$
(c) $\frac{\sqrt{3}\pi}{12}$ (d) $\frac{\sqrt{2}\pi}{8}$
[ESE-2006]

1.3 Elements can reach a stable atomic structure by
(a) losing electrons only
(b) gaining electrons only
(c) losing or gaining or sharing electrons
(d) collisions between atoms
[ESE-2006]

1.4 In a Hexagonal Close Packed (HCP) crystal structure, if 'a' and 'c' represent, respectively the short and long unit cell dimensions the (c/a) ratio should be
(a) 12.00 (b) 0.74
(c) 1.633 (d) 16.33
[ESE-2010]

1.5 Consider the following statements :
Secondary (or Molecular) bonds are
1. The attraction forces exist between atoms or molecules.
2. Stronger than primary bonds.
3. Can be divided as electrostatic bonds.
4. Weaker than primary bonds.

Which of the above statements is /are correct?
(a) 1 only (b) 2 and 3 only
(c) 1 and 4 only (d) 1, 2, 3 and 4
[ESE-2010]

1.6 **Assertion (A) :** Ionic bonds and covalent bonds are higher than metallic bonds.
Reason (R) : Ionic and covalent bonds are generally lower than other primary bonds.
(a) Both A and R are true and R is the correct explanation of A
(b) Both A and R are true but R is NOT the correct explanation of A
(c) A is true but R is false
(d) A is false but R is true
[ESE-2010]

1.7 The geometrical configuration of one molecule of C_{60}-buckminaterfullerene contains
(a) 12 hexagons and 20 pentagons of Carbon atoms
(b) 20 hexagons and 12 pentagons of Carbon atoms
(c) 20 hexagons and 20 pentagons of Carbon atoms
(d) 12 hexagons and 12 pentagons of Carbon atoms
[ESE-2011]

1.8 Einstein relation is referred between
(a) the diffusion constant and the mobility
(b) the conduction and diffusion currents
(c) the conduction and diffusion voltages
(d) none of the above
[ESE-2013]

1.9 Match **List-I (Parameter)** with **List-II (Unit)** and select the correct answer using the codes given below:

List-I
A. Boltzmann constant
B. Permeability of free space
C. Permittivity of free space
D. Mobility

List-II
1. farad/meter
2. cm^2/ volt-second
3. henry/meter
4. cm^2/second
5. electron volt/kelvin

Codes:

	A	B	C	D
(a)	5	2	1	4
(b)	1	2	5	4
(c)	5	3	1	2
(d)	1	3	5	2

[ESE-2006]

1.10 Assuming the Fermi level E_f to be independent of temperature, E_f may be defined as the level with an occupancy probability of
(a) 0% (b) 50%
(c) 75% (d) 100%

[ESE-2002]

1.11 Match **List-I** with **List-II** and select the correct answer :

List-I	List-II
A. Ga-As	1. Integrated circuit
B. Nichrome	2. Laser
C. Quartz	3. Busbar
D. Si	4. Heating element
	5. Oscillator

Codes:

	A	B	C	D
(a)	2	4	5	1
(b)	1	5	3	2
(c)	2	5	3	1
(d)	1	4	5	2

[ESE-2001]

1.12 Match **List-I** with **List-II** and select the correct answer using the code given below the lists:

List-I	List-II
A. Carbon (Diamond)	1. Conducting
B. Silicon	2. Semiconducting
C. Tin (Grey)	3. Insulating
D. Lead	

Codes:

	A	B	C	D
(a)	3	2	1	1
(c)	1	2	1	3
(c)	3	1	2	1
(d)	1	1	2	3

[ESE-2012]

1.13 Behaviour of conductors, semiconductors and insulators is explained on the basis of
(a) atomic structure
(b) molecular structure
(c) energy band structure
(d) all of the above

[ESE-2013]

1.14 What type of defect causes F-centers in a crystal?
(a) Stoichiometric defect
(b) Metal excess defect due to anion vacancies
(c) Metal excess defect due to extra cations
(d) Frenkel defect

[ESE-2017]

1.15 An atom in a crystal vibrates at a frequency, determined by
1. Crystal heat current
2. Crystal temperature
3. The stiffness of the bonds with neighbour atoms

Select the correct answer using the codes given below:
(a) 1 only (b) 2 only
(c) 3 only (d) 1, 2 and 3

[ESE-2017]

1.16 **Statement (I):** A lattice defect gets created whenever the periodicity or order of the crystal lattice gets disturbed.

Statement (II): Point defect, line defect, surface defect and volume defect create defect in lattice.
(a) Both Statement (I) and Statement (II) are individually true and Statement (II) is the correct explanation of Statement (I).
(b) Both Statement (I) and Statement (II) are individually true but Statement (II) is not the correct explanation of Statement (I).
(c) Statement (I) is true but Statement (II) is false.
(d) Statement (I) is false but Statement (II) is true.

[ESE-2017]

1.17 In Face-Centered Cubic structure (FCC), what number of atoms is present in each unit cell?
(a) 18 (b) 16
(c) 14 (d) 12

[ESE-2018]

1.18 If (n) is lattice points per unit cell of the cubic system, (N) and (M) are the Avogadro's number and atomic weight, respectively, and (ρ) is the density of the element, then the lattice constant (a) is

(a) $\left(\frac{M\rho}{nN}\right)^{1/3}$ (b) $\left(\frac{NM}{n\rho}\right)^{1/3}$

(c) $\left(\frac{nM}{N\rho}\right)^{1/3}$ (d) $\left(\frac{N\rho}{nM}\right)^{1/3}$

[ESE-2018]

1.19 **Statement (I) :** Light is capable of transferring electrons to the free-state inside a material, thus increasing the electrical conductivity of the material.

Statement (II) : The increased electrical conductivity produced by light is called photo-conductivity.

[ESE-2018]

1.20 **Statement (I) :** When a solid surface is bombarded by electrons of appreciable energy, secondary emission occurs from the surface.

Statement (II) : The major application of the secondary emission is in voltage amplification.

[ESE-2018]

1.21 Materials in which the atomic order extends uninterrupted over the entirely of the specimen; under some circumstances, they may have flat faces and regular geometric shapes, are called
(a) Anisotropy (b) Crystallography
(c) Single crystals (d) Crystal system

[ESE-2019]

1.22 The saturation flux density for Nickel having density of 8.90 g/cm^3, atomic number 58.71 and net magnetic moment per atom of 0.6 Bohr magnetons is nearly
(a) 0.82 tesla (b) 0.76 tesla
(c) 0.64 tesla (d) 0.52 tesla

[ESE-2019]

1.23 Consider the following applications:
- High temperature heat engines.
- Nuclear fusion reactors.
- Chemical processing industry.
- Aeronautical and space industry.

Which one of the materials will be used for these applications?
(a) Zironia (b) Alumina
(c) Ceramic (d) Silicon carbide

[ESE-2019]

1.24 Consider the following statements:
1. The rules for series and parallel combinations of capacitors are opposite to those for resistors.
2. The rules for series and parallel combinations of inductors are same as those for resistors.
3. An inductor is a short circuit to dc currents.

Which of the above statements are correct?
(a) 1 and 2 only (b) 1 and 3 only
(c) 2 and 3 only (d) 1, 2 and 3

[ESE-2019]

1.25 **Statement-I:** When a non-linear resistor, in series with a linear resistor, both being non-inductive, is connected to a voltage source, the current in the circuit cannot be determined by using Ohm's law.

Statement-II: If the current-voltage characteristic of the non-linear resistor is known, the current-voltage characteristics of the series circuit can be obtained by graphical solution.

[ESE-2019]

1.26 Which of the following statements regarding an atom are correct?
1. If two atoms with similar ionization potential form a bond, then this bond will most probably be either covalent or metallic.
2. When atoms with different ionization potentials form a bond, the bond will be mainly ionic.
3. If the atom or molecule already has its outer shells completely full, then the bonding between the atoms or molecules will be a secondary bond when it solidifies.

(a) 1 and 2 only (b) 1 and 3 only
(c) 2 and 3 only (d) 1, 2 and 3

[ESE-2020]

1.27 As per de Broglie's relationship, the wavelength λ related to its mass m and velocity v is

(a) $\frac{h}{mv}$ (b) $\frac{hv}{m}$

(c) $\frac{hm}{v}$ (d) $\frac{mv}{h}$

(where, h = Planck's constant)

[ESE-2020]

1.28 **Statement (I) :** In a substitutional semiconductor, atom is replaced by an occasional foreign atom. The imperfections may be deliberately controlled or created in transistor material.

Statement (II) : The lattice vacancies created when certain atoms in a semiconductor are missing are known as Schottky defects.

(a) Both Statement (I) and Statement (II) are individually true and Statement (II) is the correct explanation of Statement (I)
(b) Both Statement (I) and Statement (II) are individually true; but Statement (II) is **not** the correct explanation of Statement (I)
(c) Statement (I) is true; but Statement (II) is false
(d) Statement (I) is false; but Statement (II) is true

[ESE-2020]

1.29 What is the line energy of dislocations on BCC iron? The Burgers' vector in iron is of the 1/2 < 111 > type. The shear modulus of iron is 80.2 GN/m^2. Given that the lattice parameters of BCC iron, a = 2.87 Å.

(a) 1.40×10^{-9} J/m
(b) 3.12×10^{-9} J/m
(c) 2.476×10^{-9} J/m
(d) 6.544×10^{-9} J/m

[ESE-2021]

1.30 What is the approximate lattice constant "a" of a substance having FCC lattice, molecular weight 60.2 and density 6250 kg/m^3?

(Consider $N = 6.02 \times 10^{26}$/kg-mole)

(a) 5×10^{-10} m (b) 3×10^{-10} m
(c) 4×10^{-10} m (d) 8×10^{-10} m

[ESE-2021]

1.31 Which one of the following defects might be thought of as being formed by a cation leaving its normal position and moving into an interstitial site?

(a) Schottky defect
(b) Frenkel defect
(c) Crystallographic defect
(d) Stoichiometric defect [ESE-2022]

1.32 Which of the following classifications are correct with reference to the various types of imperfections in a semiconductor?

1. Substitutional
2. Vacancies
3. Interstitial

Select the correct answer using the code given below:

(a) 1 and 2 only (b) 2 and 3 only
(c) 1 and 3 only (d) 1, 2 and 3

[ESE-2022]

1.33 The property of periodicity in crystal structure is repeated in three dimensions. The positive ions occupy the corners of a cube as well as the centres of the cubic face. Then these crystals are called

(a) Face-centered cubic crystals
(b) Body-centered cubic crystals
(c) Halite-centered cubic crystals
(d) Rock-salt-centered cubic crystals

[ESE-2023]

1.34 What is the value of the Curie temperature of iron?

(a) About 1043 K (b) About 1555 K
(c) About 1422 K (d) About 1322 K

[ESE-2024]

1.35 Which one of the following is the correct relationship among the Curie temperatures of Fe, Ni and Co?

(a) (D/d)(Ni) Curie < (D/d) (Fe) Curie < (D/d) (Co) Curie
(b) (D/d)(Ni) Curie < (D/d) (Co) Curie < (D/d) (Fe) Curie
(c) (D/d)(Co) Curie < (D/d) (Ni) Curie < (D/d) (Fe) Curie
(d) (D/d)(Fe) Curie < (D/d) (Ni) Curie < (D/d) (Co) Curie [ESE-2024]

■■■■

Answers Introduction to Engineering Materials

1.1 (a)	1.2 (b)	1.3 (c)	1.4 (c)	1.5 (c)	1.6 (c)	1.7 (b)	1.8 (a)	1.9 (c)
1.10 (b)	1.11 (a)	1.12 (a)	1.13 (c)	1.14 (b)	1.15 (d)	1.16 (b)	1.17 (c)	1.18 (c)
1.19 (b)	1.20 (b)	1.21 (c)	1.22 (c)	1.23 (d)	1.24 (d)	1.25 (b)	1.26 (d)	1.27 (a)
1.28 (b)	1.29 (c)	1.30 (c)	1.31 (b)	1.32 (d)	1.33 (d)	1.34 (a)	1.35 (*)	

Explanations Introduction to Engineering Materials

1.1 (a)
Atomic number represents the total number of valence electrons present in various energy state in valence band.

1.2 (b)

$$\text{Packing fraction of BCC} = \frac{\sqrt{3}\pi}{8} = 0.68.$$

1.3 (c)
Elements can reach a stable state by losing or gaining or sharing of e^{-s} so as to complete its octet.
For e.g. in NaCl, Na → loses electron for completing its octet
Cl gain e^- for completing its octet
In Si csystal → one Si atom shares its outer $4e^-$ with other 4 Si atoms to complete its octet.

1.4 (c)

$$\frac{c}{a} = \sqrt{\frac{8}{3}} = 1.633$$

1.5 (c)
Secondary or molecular bonds are either ion-dipole interaction, dipole-dipole interaction or vanderwalls are weaker than primary bonds (ionic, covalent).

1.7 (b)
It contains 20-hexagons and 12-pentagons of carbon atoms.

1.8 (a)

$$\text{Einstein relation: } D = \left(\frac{KT}{e}\right)\mu$$

1.9 (c)
$K_B \rightarrow$ eV/°C
$\mu \rightarrow$ Henry/meter; $\in$ = Farad/meter
Mobility → cm^2/sec

1.10 (b)
Fermi-Dirac probability function $F(E)$ is given by

$$F(E) = \frac{1}{1+e^{(E-E_F)/kT}}$$

where, E_F = Fermi level

if $E = E_F$ then $F(E) = \frac{1}{2}$ for any value of temperature. Thus, the fermi level represents the energy state with 50% probability of being filled if no forbidden band exists.

1.11 (a)
Quartz → used for high frequency oscillation.
Si → used in integrated circuit because of two regions.
(i) at high temperature it reacts with oxygen and form SiO_2 which is a very good dielectric.
(ii) High power handling capacity.
GaAs → is a direct band gap material having larger carrier life time so suitable for designing of laser.
Nichrome → used as heating element.

1.12 (a)
In the given list Tin (grey) and lead are the conducting materials.
Carbon (diamond) is a insulating materials and silicon a semiconducting material hence option (a) is correct.

1.13 (c)
Based on energy band structure conductors, semiconductor and insualtor is classified.

1.14 (b)
F-centers are non-stoichiometric defect where electrons are trapped in anion vacancies and produced by the exposure of an alkali metal halide crystal to the alkali metal vapour.

1.15 (d)
All the above are correct.

1.22 (c)

$$\text{Atomic density} = N = \frac{\rho N_A}{A_{Ni}}\left(\frac{\text{atoms}}{m^3}\right)$$

ρ = density = $8.9\times10^6\left(\frac{g}{m^3}\right)$

N_A = Avogadro number

= 6.02×10^{23} (atoms/mole)

A_{Ni} = Atomic number of nickel

= 58.71 (g/mole)

$$N = \frac{\rho N_A}{A_{Ni}} = \frac{(8.9\times10^6)(6.02\times10^{23})}{58.7}$$

= 9.127×10^{28} (atoms/m^3)

$M = Nm$

M = Magnetization

m = Magnetic dipole moment = $0.6\ p_B$

$M_s = Nm$

$M_s = (9.127 \times 10^{28})(0.6)(9.27 \times 10^{-24})$

$M_s = 50.764 \times 10^4$ (A.m^2)

$B_s = \mu_0 M_s$

$B_s = 4\pi \times 10^{-7}(50.764 \times 10^4)$

B_s = 0.637 Tesla

1.26 (d)

All statements regarding atoms are correct.

1.27 (a)

According to de Broglie's relationship

$$\lambda = \frac{h}{p} = \frac{h}{mv}$$

λ = wavelength

h = Planck's constant

p = momentum

m = mass

v = velocity

1.28 (b)

Both statements are individually true.

1.29 (c)

Burger vector,

$$b = \frac{1}{2}\times\sqrt{3}a = 2.48 \times 10^{-10}$$

$$\text{Line energy} = \frac{1}{2}Gb^2$$

$$= \frac{1}{2}\times80.2\times(2.48\times10^{-10})^2\times10^9$$

$= 2.47 \times 10^{-9}$ J/m

1.30 (c)

$A = 60.2$

$\rho = 6250$ kg/m^3

$N_A = 6.02 \times 10^{26}$ kg-mol

$$\Rightarrow \frac{n}{V_C} = \frac{N_A\rho}{A}$$

$$= \frac{6.02\times10^{26}\times6250}{60.2}$$

$$\Rightarrow \frac{n}{V_C} = 625 \times 10^{26}$$

$$\Rightarrow V_c = \frac{n}{625\times10^{26}} = \frac{4}{625\times10^{26}} = a^3$$

$a = 4 \times 10^{-10}$ m

1.31 (b)

Frenkel defect is formed when a cation leaves its normal position and occupies an interstitial position.

1.33 (d)

Rock salt centred cubic structure is formed by interpenetration of two FCC structures. Positive and negative ions follow their respective FCC structures and their interpenetration is called Rock Salt Structure.

1.34 (a)

Material	Curie Temperature (K)
Fe	1043
Co	1404
Ni	631
Cd	289

1.35 (*)

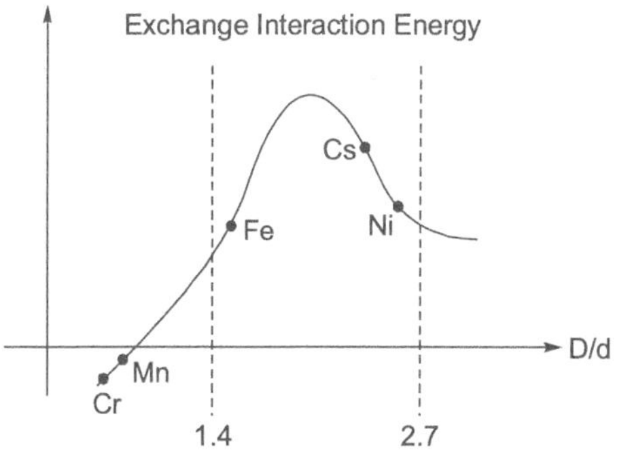

■■■■

2 Dielectric Properties of Materials

2.1 By inserting a slab of dielectric material between the plates of a parallel plate capacitor, the energy stored in the capacitor has increased three times. The dielectric constant of the material is
(a) 9 (b) 3
(c) 1/3 (d) 1/9 [ESE-2001]

2.2 When a dielectric is subjected to an alternating electric field of angular frequency 'ω', its power loss is proportional to
(a) ω (b) ω^2
(c) $1/\omega$ (d) $1/\omega^2$ [ESE-2001]

2.3 For a given dielectric, with increase in temperature the ionic polarizability
(a) increases (b) decreases
(c) remains same (d) fluctuates
[ESE-2001]

2.4 A piezoelectric crystal has an Young's modulus of 130 GPa. The uniaxial stress that must be applied to increase its polarization from 500 to 510 Cm^{-2} is
(a) 1.171 GPa (b) 1.182 GPa
(c) 2.6 GPa (d) 2.55 GPa
[ESE-2001]

2.5 On the application of the field $\vec{E}$, the modified field due to polarization $\vec{P}$ in solids and liquids having cubic symmetry is given by

(a) $\vec{E}+\dfrac{\vec{P}}{\epsilon_0}$ (b) $\vec{E}-\dfrac{\vec{P}}{\epsilon_0}$

(c) $\vec{E}+\dfrac{\vec{P}}{3\epsilon_0}$ (d) $\vec{E}-\dfrac{\vec{P}}{3\epsilon_0}$

[ESE-2002]

2.6 The complex dielectric constant of a material is given by the expression:

$$\epsilon = \epsilon' - j\epsilon''$$

If a parallel plate capacitor with area '*A*' and separation '*d*' is formed with this material as a dielectric, the loss factor will be
(a) $(A\epsilon')/(d\epsilon'')$ (b) ϵ''/ϵ'
(c) $\tan^{-1}[(A\epsilon')/(d\epsilon'')]$ (d) $\tan^{-1}[(d\epsilon'')/(A\epsilon')]$
[ESE-2002]

2.7 At optical frequencies, the major contribution to the total polarization comes from
(a) space charge polarization
(b) orientational polarization
(c) ionic polarization
(d) electronic polarization
[ESE-2002]

2.8 Which of the following is piezoelectric material?
(a) Quartz (b) Silica sand
(c) Corundum (d) Polystyrene
[ESE-2002]

2.9 Which one of the following is NOT true for Sulphur hexafluoride gas?
(a) It is electronegative in nature
(b) It has high dielectric strength
(c) It is non-toxic
(d) It is highly inflammable
[ESE-2003]

2.10 Which one of the following materials has the highest dielectric strength?
(a) Polystyrene (b) Marble
(c) Cotton (d) Transformer oil
[ESE-2003]

2.11 The losses in a dielectric subject to an alternating electric field are determined by
(a) Real part of the complex dielectric constant
(b) Imaginary part of the complex dielectric constant
(c) Absolute value of the complex dielectric constant

(d) Ratio of the magnitudes of the real and imaginary parts of the complex dielectric constant

[ESE-2003]

2.12 In a solid or liquid dielectric with externally applied electric field, as the interatomic distance increases the internal field E_i,
(a) Increases
(b) Decreases
(c) Remains unaltered
(d) Increases or decreases based on temperature

[ESE-2003]

2.13 A dielectric material has the real part of the dielectric constant (ϵ_r') as 4 and its loss tangent is 0.004. What is the complex dielectric constant (ϵ_r^*) represented by?
(a) $4 + j0.016$ (b) $4 - j0.016$
(c) $4 + j0.001$ (d) $4 - j0.001$

[ESE-2004]

2.14 Which one of the following statements is **not** correct?
(a) Vacuum can act as a dielectric material
(b) Piezoelectric materials can act as transducers
(c) Quartz crystal is a ferroelectric material
(d) The dielectric constant of dielectrics depends on the frequency of the applied field

[ESE-2004]

2.15 What are the materials which exhibit electric polarization even in the absence of an applied electric field called ?
(a) Ferromagnetic (b) Paramagnetic
(c) Ferroelectric (d) Anti-ferroelectric

[ESE-2005]

2.16 The relative dielectric constant of solid dielectrics in the alternating field is
(a) maximum at the power frequencies and decreases to unity at frequencies in the ultraviolet range.
(b) maximum at the power frequencies and decreases to zero at frequencies in the ultraviolet range.
(c) unity at the power frequencies and increases to its maximum value at frequencies in the ultraviolet range.
(d) independent of frequency variations.

[ESE-2006]

2.17 With increase in applied frequency, the dielectric loss in a material will
(a) increase (b) decrease
(c) remain constant (d) become zero

[ESE-2006]

2.18 The electronic polarisability of an inert gas atom is proportional to which one of the following?
(a) R (b) R^2
(c) R^3 (d) R^4
(Where R is the radius of the atom)

[ESE-2008]

2.19 In the case of a dielectric subjected to an alternating electric field of frequency f, the dielectric loss is proportional to which one of the following?
(a) f (b) f^2
(c) $1/f$ (d) $1/f^2$

[ESE-2008]

2.20 Which one of the following is the correct statement?
The orientational polarizability in a polyatomic gas is proportional to
(a) temperature T (b) $1/T$
(c) T^2 (d) independent of T

[ESE-2008]

2.21 Quartz and $BaTiO_3$ exhibit which of the following properties?
(a) Magnetostriction (b) Ferromagnetism
(c) Piezoelectricity (d) Ferroelectricity

[ESE-2009]

2.22 A barium titanate crystal has a thickness of 2 mm. Its voltage sensitivity is 12×10^{-3} Vm/N. It is subjected to a pressure of 0.5 MN/m^2. What is the voltage generated?
(a) 3 V (b) 6 V
(c) 5 V (d) 12 V

[ESE-2009]

2.23 Match **List-I** with **List-II** and select the correct answer using the code given below

List-I
A. Ferroelectric material
B. Anti Ferroelectric material
C. Ferrites
D. Ferro-magnetic

List-II
1. Neel temperature
2. Magnetostrictive transducers
3. Magnetocaloric effect
4. Cannot be shaped by ordinary machining process

Codes:

	A	B	C	D
(a)	2	4	1	3
(b)	3	4	1	2
(c)	2	1	4	3
(d)	3	1	4	2

[ESE-2010]

2.24 Consider the following statements
Factors affecting the dielectric loss are
1. directly proportional to the frequency of supply voltage.
2. inversely proportional to the supply frequency.
3. inversely proportional to the square of the supply voltage.
4. directly proportional to the square of the supply voltage.

Which of the above statements are correct?
(a) 1 and 3 only (b) 2 and 3 only
(c) 2 and 4 only (d) 1 and 4 only
[ESE-2010]

2.25 The property/characteristic of ferroelectric materials is
(a) Dielectric relaxation
(b) Dielectric breakdown
(c) Spontaneous polarization
(d) Spontaneous magnetization
[ESE-2010]

2.26 Consider the following statements Piezoelectric materials serve as:
1. A source of ultrasonic waves.
2. When electric field is applied, the mechanical dimensions of the substances are not at all altered.
3. Converts electrical energy to mechanical and vice versa.
4. Converts thermal energy to electrical energy.

Which of these statements is/are correct?
(a) 1 only (b) 2 and 3 only
(c) 1 and 3 only (d) 1, 2, 3 and 4
[ESE-2010]

2.27 Consider the following statements :
Piezoelectric materials
1. Crystal can be shown as electrical equivalent circuit similar to an inductor and a capacitor (Tank circuit).
2. Quartz, Rochelle salt, tourmaline.
3. Used in voltage stabilizers.
4. This exhibits the reverse effect electrostriction.

Which of these statements are correct?
(a) 1, 2 and 4 only (b) 1 and 2 only
(c) 2 and 4 only (d) 1, 2, 3 and 4
[ESE-2010]

2.28 At a measuring frequency of 10^{12} Hz, the dielectric constant of a material will be due to
(a) Electronic polarization
(b) Ionic polarization
(c) Electronic, Ionic and Orientational polarization
(d) Electronic and Ionic polarization
[ESE-2010]

2.29 High permittivity ceramic is used for capacitors of
(a) A few pF to a few hundred pF
(b) A few μF to a few hundred μF
(c) A few nF to a few hundred nF
(d) A few mF to a few hundred mF
[ESE-2011]

2.30 The following material is not used for making a piezoelectric transducer
(a) Rochelle salt
(b) Barium titanate
(c) Chromium sulphide
(d) Quartz
[ESE-2012]

2.31 **Statement (I):** Polarization is due to the application of an electric field to dielectric materials.
Statement (II): When the dipoles are created, the dielectric is said to be polarized or in a state of polarization.
(a) Both Statement (I) and Statement (II) are individually true and Statement (II) is the correct explanation of Statement (I).

(b) Both Statement (I) and Statement (II) are individually true but Statement (II) is not the correct explanation of Statement (I).
(c) Statement (I) is true but Statement (II) is false.
(d) Statement (I) is false but Statement (II) is true.

[ESE-2012]

2.32 Orientational polarization is
(a) inversely proportional to temperature and proportional to the square of the permanent dipole moment.
(b) proportional to temperature as well as to the square of the permanent dipole moment.
(c) proportional to temperature and inversely proportional to the square of the permanent dipole moment.
(d) inversely proportional to temperature as well as to the square of the dipole moment.

[ESE-2013]

2.33 Ferro-electric materials have a
(a) high dielectric constant which varies non-linearly.
(b) low dielectric constant and is non-linear.
(c) high dielectric constant which varies linearly.
(d) low dielectric constant but linear.

[ESE-2013]

2.34 In a piezoelectric crystal oscillator, the oscillation or tuning frequency is linearly proportional to the
(a) mass of the crystal
(b) square root of the mass of the crystal
(c) square of the mass of the crystal
(d) inverse of the square root of the mass of the crystal

[ESE-2013]

2.35 Which of the following are piezoelectric substances?
1. Barium Titanate
2. Lead Titanate
3. Lead Zirconate
4. Cadmium Sulphate

(a) 1, 2 and 4 (b) 1, 3 and 4
(c) 1, 2 and 3 (d) 2, 3 and 4

[ESE-2013]

2.36 Statement (I): The dielectric constant of a substance, under the influence of alternating electric fields is, in general, a 'complex' quantity.

Statement (II): The 'imaginary' part of the dielectric constant is a measure of the dielectric loss in the substance.
(a) Both Statement (I) and Statement (II) are individually true and Statement (II) is the correct explanation of Statement (I).
(b) Both Statement (I) and Statement (II) are individually true but Statement (II) is not the correct explanation of Statement (I).
(c) Statement (I) is true but Statement (II) is false.
(d) Statement (I) is false but Statement (II) is true.

[ESE-2013]

2.37 Statement (I): Electrostriction occurs due to piezoelectricity which in the reverse effect i.e. the production of polarization on application of mechanical stress if the lattice has no centre of symmetry.

Statement (II): When an electric field is applied to a substance it becomes polarized, the electrons and nuclei assume new geometric positions and the mechanical dimensions of the substance are altered.
(a) Both Statement (I) and Statement (II) are individually true and Statement (II) is the correct explanation of Statement (I).
(b) Both Statement (I) and Statement (II) are individually true but Statement (II) is not the correct explanation of Statement (I).
(c) Statement (I) is true but Statement (II) is false.
(d) Statement (I) is false but Statement (II) is true.

[ESE-2013]

2.38 The imaginary part of dielectric constant determines
(a) component of current which is in phase with the applied field
(b) component of energy absorbed per m^3
(c) amount of applied field
(d) component of voltage which is in phase with the applied field

[ESE-2014]

2.39 Enameled wires are preferred to cotton-covered wires to
(a) withstand higher temperature
(b) improve heat dissipation
(c) reduce the resistivity
(d) increase the mechanical strength

[ESE-2015]

2.40 Which one of the following materials is used for cable insulation?
(a) Phenol formaldehyde
(b) Polytetrafluoroethylene
(c) Polyvinyl chloride
(d) Acrylonitrile butadiene styrene

[ESE-2015]

2.41 A parallel plate capacitor is made of two circular plates separated by a distance of 5 mm and with a dielectric with dielectric constant of 2.2 between them. When the electric field in the dielectric is 3×10^4 V/m, the charge density of the positive plate will be, nearly
(a) 58.5×10^4 C/m^2 (b) 29.5×10^4 C/m^2
(c) 29.5×10^{-4} C/m^2 (d) 58.5×10^{-4} C/m^2

[ESE-2017]

2.42 When an alternating voltage of a given frequency is applied to a dielectric material, dissipation of energy occurs due to
1. Continual change in the orbital paths of the electrons in the atomic structure.
2. A small conduction current through the dielectric.
3. Eddy currents.

Which of the above statements are correct?
(a) 1, 2 and 3 (b) 1 and 2 only
(c) 1 and 3 only (d) 2 and 3 only

[ESE-2018]

2.43 A barium titanate crystal is inserted in a parallel plate condenser of area 10 mm × 10 mm. The plates having a separation of 2 mm, give a capacitance of 10^{-9} F. If the value of $\varepsilon_0 = 8.854 \times 10^{-12}$ Fm^{-1}, the relative dielectric constant of the crystal will be nearly
(a) 2640 (b) 2450
(c) 2260 (d) 2080

[ESE-2020]

2.44 Which one of the following statements is NOT correct regarding dielectric loss?
(a) The loss increases proportionately with the frequency of applied voltage.
(b) Presence of humidity increase the loss.
(c) Temperature rise normally decreases the loss.
(d) Voltage increase causes increased dielectric loss.

[ESE-2021]

2.45 **Statement I :** The inorganic materials are used to manufacture suspension insulators for high-voltage overhead lines and bushings on high-voltage transformers and switchgear.

Statement II : The ceramic and glass materials are formed into a series of flanged discs to decrease the creepage distance along the surface of the complete insulator.
(a) Both Statement (I) and Statement (II) is the correct explanation of Statement (I).
(b) Both Statement (I) and Statement (II) are individually true, but Statement (II) is not the correct explanation of Statement (I).
(c) Statement (I) is true, but Statement (II) is false.
(d) Statement (I) is false, but Statement (II) is true.

[ESE-2021]

2.46 A good insulating material should possess which of the following characteristics?
1. High dielectric strength
2. Low permittivity
3. Low thermal strength

Select the correct answer using the code given below:
(a) 1 and 2 only (b) 1 and 3 only
(c) 2 and 3 only (d) 1, 2 and 3

[ESE-2022]

2.47 What are the properties of good insulating material?
(a) Low dielectric strength, volume resistivity of high value and very low dissipation factor
(b) Low dielectric strength, volume resistivity of low value and very low dissipation factor
(c) Low dielectric strength, volume resistivity of low value and high dissipation factor
(d) High dielectric strength, volume resistivity of high value and very low dissipation factor

[ESE-2024]

■■■■

Answers Dielectric Properties of Materials

2.1 (b)	2.2 (a)	2.3 (c)	2.4 (c)	2.5 (c)	2.6 (b)	2.7 (d)	2.8 (a)	2.9 (d)
2.10 (a)	2.11 (b)	2.12 (b)	2.13 (b)	2.14 (c)	2.15 (c)	2.16 (a)	2.17 (a)	2.18 (c)
2.19 (a)	2.20 (b)	2.21 (c)	2.22 (d)	2.23 (c)	2.24 (d)	2.25 (c)	2.26 (c)	2.27 (d)
2.28 (d)	2.29 (b)	2.30 (c)	2.31 (a)	2.32 (a)	2.33 (a)	2.34 (d)	2.35 (c)	2.36 (a)
2.37 (d)	2.38 (b)	2.39 (a)	2.40 (c)	2.41 (*)	2.42 (b)	2.43 (c)	2.44 (c)	2.45 (b)
2.46 (a)	2.47 (d)							

Explanations Dielectric Properties of Materials

2.1 (b)

$$W_E = \frac{1}{2}CV^2 = \frac{1}{2}\int \epsilon_o \epsilon_r E^2 dv \Rightarrow W_E \propto \epsilon_r$$

2.2 (a)

$$W(\text{dielectric loss}) = \frac{\omega}{2} \cdot \epsilon_0 \cdot \epsilon_r'' E_0^2 \text{ i.e. } (\propto \omega)$$

2.3 (c)
Electronic and ionic polarizability remain constant with respect to temperature but orientational polarizability is inversely proportional to the temperature.

2.4 (c)

$$\text{Uniaxial stress } (p) = Y \cdot \frac{\Delta C}{C}$$

$$q = CV \Rightarrow \text{polarisation } P = \frac{q}{A} = \frac{C}{A} \cdot V$$

where A is area of the crystal capacitor.

$$\therefore \quad \Delta P = \left(\frac{V}{A}\right) \times \Delta C$$

$$\Rightarrow \quad \frac{\Delta P}{P} = \frac{\Delta C}{C}$$

$$\Rightarrow \text{stress } (p) = Y \cdot \frac{\Delta P}{P} = 130 \times \frac{10}{500} = 2.6 \text{ GPa}$$

2.5 (c)
In the case of solids and liquids, internal field $\vec{E}_i$ is not equal to the applied field $\vec{E}$,

(for gases $\vec{E}_i = \vec{E}$); and $\vec{E}_i = \vec{E} + \left(\frac{\gamma}{E_0}\right)\vec{P}$.

For cubic symmetry, $\gamma = \frac{1}{3}$

$$\therefore \quad \vec{E}_i = \vec{E} + \frac{\vec{P}}{3\epsilon_0}$$

2.6 (b)

$$\text{Loss factor } \tan\delta = \frac{\epsilon''}{\epsilon'}$$

2.7 (d)
The dielectric losses in the radio frequency region are usually due to dipole rotation (i.e. oriental polarisation) or to ions (i.e., ionic polarisation) jumping from one equilibrium position to another. But, the losses in the optical region, are associated with the electrons (i.e. electronic polarisabilities).

2.8 (a)
Quartz is a piezoelectric material used for high frequency oscillation

2.9 (d)
SF_6 gas is a colourless, odourless, non-toxic and non-inflammable dielectric gas.

2.10 (a)
Polystyrene has highest dielectric strength among all the given options.

2.11 (b)
The losses in a dielectric subject to an alternating electric field is loss

$= \frac{1}{2}\epsilon_0 \epsilon_0'' \omega E_0^2$, where dielectric constant

$$\epsilon_r^* = \epsilon_r' - j\epsilon_r''$$

loss $\propto \epsilon''$ i.e. proportional to absolute value of complex dielectric constant.

2.12 (b)

$$E_i = \frac{E_{\text{applied}}}{1 - \dfrac{1.2\alpha_e}{\pi \varepsilon_0 a^3}}$$

So as a ↑ denominator increases
So E_i decreases.
where a is interatomic distance.

2.13 (b)

$$\tan\delta = \frac{\epsilon_r''}{\epsilon_r'} \text{ and } \epsilon_r^* = \epsilon_r' - j\epsilon_r''$$

$$\epsilon_r'' = 4 \times 0.004 = 0.016$$

$$\Rightarrow \quad \epsilon_r = 4 - j0.016$$

2.14 (c)
Quartz crystal is a piezoelectric material.

2.15 (c)
Ferroelectric materials exhibit hysteresis effects similar to those observed in ferromagnetic materials.

2.16 (a)
At high frequencies dielectric constant is a complex quantity. Imaginary part is responsible for dielectric loss and it is high at power frequencies and decreases to unity at ultraviolet frequencies.

2.17 (a)

$$\text{Loss} = \frac{1}{2}\epsilon_0\epsilon_r''\,\omega\,E_0^2 \quad \text{so loss} \propto \omega$$

∴ Loss increases as ω increases

2.18 (c)
Electronic polarizability

$$\alpha_e = 4\pi\varepsilon_0 R^3$$

So, $$\alpha_e \propto R^3$$

2.19 (a)

$$\text{Loss} = \frac{1}{2}\epsilon_0\epsilon_r''\,\omega\,E_0^2$$

Dielectric loss $\propto \omega$

$\Rightarrow \quad \propto f$

2.20 (b)

$$\alpha_0 = \frac{\rho_p^2}{3KT} \Rightarrow \alpha_0 \propto \frac{1}{T}$$

where, ρ_p = Permanent dipole moment.

2.21 (c)
Quartz and $BaTiO_3$ both are piezoelectric material so exhibits piezoelectricity property.

2.22 (d)

$$V = gtP$$
$$= 12 \times 10^{-3} \times 2 \times 10^{-3} \times 0.5 \times 10^6$$
$$V = 12\text{ V}$$

2.24 (d)

$$W = \text{loss} = \frac{1}{2}\epsilon_0\epsilon_r''\omega\frac{V^2}{d^2}$$

$$W \propto \omega$$
$$W \propto V^2$$

2.25 (c)
Ferroelectric material shows property of spontaneous polarization.

2.26 (c)
Piezo-electric crystal get stressed when voltage is applied and vice-versa.

2.28 (d)
10^{12} Hz (Visible range) – electronic polorizotion

2.29 (b)
High permittivity ceramic is used for capacitors of a few μF to few hundred μF.

2.30 (c)
Common Piezo-electric materials are Rochelle salt, ammonium dihydrogen phosphate, lithium sulphate Barium titanate, quartz and ceramics.

2.31 (a)
Application of an electric field causes relative displacement of these charges, leading to the creation of dipoles and hence polarization.

2.32 (a)
Orientation polarization

$$P = \frac{Np_p^2 E}{3KT}$$

2.33 (a)
Ferroelectric material shows spontaneous polarization and electric hysteresis. This shows that high dielectric strength and varies non-linearly.

2.34 (d)

Since, $\omega \propto \frac{1}{\sqrt{LC}}$

either series or parallel.

Piezoelectric in electrical equivalent in series

$$\omega \propto \frac{1}{\sqrt{m}}$$

2.39 (a)

Enameled wires are preferred to cotton-covered wires to withstand higher temperature.

2.41 (*)

Given parallel plate capacitor of area A and distance between plates is 5 mm. $\in_r = 2.2$ and $E = 3 \times 10^4$ V/m.

Since, $C = \frac{\varepsilon A}{d}$

and $Q = CV = C\ Ed$

$$Q = \frac{\varepsilon A}{d} \cdot E \cdot d$$

Charge density

$$= \frac{Q}{A} = \varepsilon E$$

$$= 2.2 \times 8.8 \times 10^{-12} \text{ F/m} \times 3 \times 10^4 \text{ V/m}$$

$$= 58.08 \times 10^{-8} \text{ C/m}^2$$

2.43 (c)

$$C = \frac{\in_r \in_0 A}{d}$$

$$\Rightarrow \quad \in_r = \frac{c \cdot d}{\in_0 A}$$

$$\Rightarrow \quad \in_r = \frac{10^{-9} \times 2 \times 10^{-3}}{8.854 \times 10^{-12} \times 100 \times 10^{-6}}$$

$$= 2258.8$$

Nearest option (c).

2.44 (c)

$$W = 2\pi f V^2 \in_0 \in_r \tan\delta$$

Additionally, W increases with increase in temperature and humidity.

2.45 (b)

The correct reason for statement-I is that in inorganic materials (specially ceramics), the bonding is predominantly ionic, which gives very high insulation resistance and negligible leakage current, with high mechanical strength.

2.46 (a)

A good insulating material should have-

- High dielectric strength
- Low permittivity
- High thermal strength

2.47 (d)

- High dielectric strength
- Volume resistivity of high value
- Very low dissipation factor

■■■■

3 Magnetic Properties of Materials

3.1 A large value of the exchange interaction energy in a ferromagnetic material implies
(a) Large saturation magnetization
(b) High Curie temperature
(c) High melting point
(d) Large diamagnetic susceptibility

[ESE-2001]

3.2 Which one of the following pairs is NOT correctly matched ?
(a) Copper : Diamagnetic
(b) Sodium : Anti ferromagnetic
(c) Iron : Ferromagnetic
(d) Ferrite : Ferrimagnetic

[ESE-2001]

3.3 Ferromagnetic behaviour is shown by those transition metals where the ratio of the atomic diameter to 3d orbital diameter is
(a) in the range of 0.5 to 1
(b) in the range of 1 to 1.5
(c) in the range of 1.5 to 2
(d) greater than 2

[ESE-2001]

3.4 Ferrites can be considered as mixed oxides of metals A and B having inverse spinel structure. Their formula can be written as
(a) ABO_2 (b) A_2BO_2
(c) AB_2O_3 (d) AB_2O_4

[ESE-2001]

3.5 Match **List-I** (Magnetic materials) with **List-II** (Main applications) and select the correct answer:

List-I
A. Ni-Zn ferrite
B. Co-Sm alloy
C. Yttrium Iron Garnet
D. Mg-Zn ferrite

List-II
1. Recording head
2. Permanent magnets
3. Audio and TV transformers
4. Memory cores
5. Microwave isolators

Codes:

	A	B	C	D
(a)	3	4	5	2
(b)	1	2	3	4
(c)	3	2	5	4
(d)	1	4	3	2

[ESE-2001]

3.6 Match **List-I (Type of the material)** with **List-II (Name of the material)** and select the correct answer:

List-I	List-II
A. Ferroelectric material	1. Permalloy
B. Piezoelectric material	2. $BaTiO_3$ ceramic
C. Soft magnetic material	3. KH_2PO_4
D. Hard magnetic material	4. Tungsten steel

Codes:

	A	B	C	D
(a)	3	2	1	4
(b)	1	4	3	2
(c)	3	4	1	2
(d)	1	2	3	4

[ESE-2002]

3.7 Which one of the following materials can not be used for permanent magnets?
(a) Alnico (b) Barium Ferrite
(c) Carbon-Steel (d) Iron-Cobalt alloy

[ESE-2002]

3.8 Upto about 4% silica is added in iron to be used as a soft magnetic material. The major reason for this is to

(a) increase permeability of the material
(b) increase electrical resistivity of the material
(c) increase the coercive force
(d) increase the saturation flux density

[ESE-2002]

3.9 Susceptibility of a diamagnetic material is
1. Negative
2. Positive
3. Dependent on the temperature
4. Independent of the temperature

Select the correct answer using the codes given below:
(a) 1 and 3 (b) 2 and 4
(c) 1 and 4 (d) 2 and 3

[ESE-2003]

3.10 Match **List-I** (Magnetic Materials) with **List-II** (Applications) and select the correct answer:

	List-I		List-II
A.	Silicon Steel	1.	Current transformer
B.	Ferrite	2.	Power transformer
C.	Alnico	3.	Permanent magnet
		4.	High frequency transformer

Codes:

	A	B	C
(a)	1	2	3
(b)	1	4	2
(c)	2	1	4
(d)	2	4	3

[ESE-2003]

3.11 Match **List-I** with **List-II** and select the correct answer using the codes given below:

	List-I		List-II
A.	Larmor frequency	1.	$\chi = C/(T-\theta)$
B.	Bohr magneton	2.	$B = \mu_0(H+M)$
C.	Magnetic induction	3.	$eh/4\pi$ m
D.	Curie-Weiss law	4.	$eB/2$ m

Codes:

	A	B	C	D
(a)	2	1	4	3
(b)	2	3	4	1
(c)	4	1	2	3
(d)	4	3	2	1

[ESE-2004]

3.12 Which one of the following is the temperature below which certain material are anti-ferromagnetic and above which they are paramagnetic?
(a) Curie temperature
(b) Neel temperature
(c) Transition temperature
(d) Weiss temperature

[ESE-2004]

3.13 Match **List-I (Type of the material)** with **List-II (Name of the material)** and select the correct answer using the codes given below:

	List-I		List-II
A.	Ferroelectric	1.	Rochelle salt
B.	Soft magnetic	2.	Alnico
C.	Hard magnetic	3.	Permalloy
D.	Semiconductor	4.	Ga As

Codes:

	A	B	C	D
(a)	3	1	2	4
(b)	1	3	2	4
(c)	3	1	4	2
(d)	1	3	4	2

[ESE-2004]

3.14 Which one of the following statements is correct?
A ferrite core has lower specific eddy current loss compared to an iron core because the iron core has
(a) higher electrical resistance
(b) lower electrical resistance
(c) higher permeability
(d) lower permeability

[ESE-2004]

3.15 **Assertion (A):** Soft magnetic material is used in making electromagnets.
Reason (R): Soft magnetic materials have a high coercive field.
(a) Both A and R are true and R is the correct explanation of A
(b) Both A and R are true but R is NOT the correct explanation of A
(c) A is true but R is false
(d) A is false but R is true

[ESE-2004]

3.16 Consider the following statements regarding magnetic materials:

1. Relative permeability of water is 0.99999 and that of oxygen is 1.00002; hence water is diamagnetic and oxygen is paramagnetic material.
2. Ferrimagnetic material has no eddy current loss.
3. Permalloy and Alnico are two examples of hard magnetic materials.
4. The magnetisation and applied electric field in ferro-magnetic materials are related non-linearly.

Which of the statements given above are correct?

(a) 1, 2 and 3 (b) 2, 3 and 4
(c) 1, 3 and 4 (d) 1, 2 and 4

[ESE-2005]

3.17 The hysteresis loop for the material of the core of a transformer should be

(a) Short and narrow (b) Tall and narrow
(c) Short and wide (d) Tall and wide

[ESE-2005]

3.18 All magnetic materials lose their magnetic properties when

(a) Cooled to at low temperature
(b) Heated to at high temperature
(c) Kept in an aluminium box
(d) Kept in vacuum

[ESE-2005]

3.19 Which one of the following is *not* a permanent magnetic material ?

(a) Chromium steel (b) Silicon iron
(c) Cobalt steel (d) Alnico

[ESE-2005]

3.20 Above the Curie temperature, ferro-magnetic materials behave like

(a) Paramagnetic (b) Diamagnetic
(c) Anti-ferromagnetic (d) Ferrimagnetic

[ESE-2005]

3.21 Match **List-I** (Magnetic Material) with **List-II** (Order of Susceptibility) and select the correct answer using the codes given below:

List-I	List-II
A. Diamagnetic	1. $\approx 10^{+5}$
B. Paramagnetic	2. $\approx 10^{-5}$
C. Ferromagnetic	3. $\approx -10^{-5}$

Codes:

	A	B	C
(a)	1	3	2
(b)	1	2	3
(c)	3	2	1
(d)	3	1	2

[ESE-2005]

3.22 Metallic copper is a

(a) Paramagnetic substance
(b) Diamagnetic substance
(c) Ferromagnetic substance
(d) Ferrimagnetic substance

[ESE-2005]

3.23 The magnetic field required to reduce the residual magnetisation to zero is called

(a) Retentivity
(b) Coercivity
(c) Hysteresis
(d) Saturation magnetisation

[ESE-2005]

3.24 Bohr magneton is unit of

(a) Magnetic energy
(b) Permanent dipole moment due to spin
(c) Polarisability
(d) Hysteresis loss

[ESE-2005]

3.25 Magnetostriction is a phenomenon of

(a) generation of electricity in ferro-magnetic materials
(b) generation of magnetism in conductors
(c) change in permeability of ferro-magnetic materials during magnetisation
(d) change in physical dimensions of ferro-magnetic materials during magnetisation

[ESE-2005]

3.26 Which among the following magnetic materials has the highest energy-product to make it a permanent magnet?

(a) Alnico (b) Ferrite
(c) Samarium Cobalt (d) Cobalt-Iron alloy

[ESE-2006]

3.27 When the temperature of a magnetic material is raised above the Curie point, it becomes

(a) diamagnetic (b) paramagnetic

(c) ferromagnetic (d) ferrimagnetic

[ESE-2006]

3.28 Soft iron is used in the manufacture of electro-magnets because of its

(a) high saturation magnetisation only

(b) low retentivity only

(c) low coercive field only

(d) high saturation magnetisation, low reten-tivity and low coercive field

[ESE-2006]

3.29 With increase in temperature, magnetic susceptibility of a ferromagnetic material will

(a) increase

(b) decrease

(c) increase initially and then decrease

(d) remain constant

[ESE-2006]

3.30 Match **List-I** (Material) with **List-II** (Curie Temperature) and select the correct answer using the code given below the lists:

List-I	List-II
A. Fe	**1.** 783 K
B. $MnOFe_2O_3$	**2.** 523 K
C. $MgOFe_2O_3$	**3.** 863 K
D. $NiOFe_2O_3$	**4.** 1043 K

Codes:

	A	B	C	D
(a)	2	3	4	1
(b)	4	1	2	3
(c)	2	1	4	3
(d)	4	3	2	1

[ESE-2006]

3.31 Match **List-I** with **List-II** and select the correct answer using the code given below the lists:

List-I

A. No eddy current loss

B. Small hysteresis loss

C. Large hysteresis loss

List-II

1. Ferrimagnetic material

2. Soft magnetic material

3. Hard magnetic material

4. Non-ferrous material

Codes:

	A	B	C
(a)	2	1	3
(b)	2	3	4
(c)	1	3	4
(d)	1	2	3

[ESE-2006]

3.32 In which one of the following magnetic materials, is the net magnetic moment zero?

(a) Paramagnetic (b) Ferromagnetic

(c) Ferrimagnetic (d) Anti-ferromagnetic

[ESE-2007]

3.33 Which one of the following is the correct statement?

YIG and YAG are two types of crystals used extensively in technology and are

(a) non-magnetic and magnetic, respectively

(b) magnetic and non-magnetic, respectively

(c) both magnetic

(d) both non-magnetic

[ESE-2007]

3.34 What is a material with equal, anti-parallel atomic magnetic moments, known as?

(a) Ferrimagnetic (b) Ferrite

(c) Ferromagnetic (d) Anti-ferromagnetic

[ESE-2007]

3.35 A coil wound on a magnetic core is excited from an a.c. source. The source voltage and its frequency are both doubled. What will be the eddy current loss in the core?

(a) Four times of the original value

(b) Double of the original value

(c) Same as the original value

(d) Half of the original value

[ESE-2007]

3.36 Consider the following statements:

Assertion (A): Ferrites are useful at very high frequencies.

Reason (R): Ferrites have high permeability and high resistivity.

Of these statements:
(a) Both A and R are true and R is the correct explanation of A
(b) Both A and R are true but R is not the correct explanation of A
(c) A is true but R is false
(d) A is false but R is true

[ESE-2008]

3.37 Which one of the following is the correct statement?
During the process of magnetisation of ferromagnetic material, the magnetic domains
(a) only expand
(b) rotate first and then expand
(c) expand first and then rotate
(d) neither rotate nor expand

[ESE-2008]

3.38 Match **List-I** (Material) with **List-II** (Application) and select the correct answer using the code given below the lists:

List-I	List-II
A. Silicon steel	1. High frequency transformers
B. Ferrites	2. Permanent magnets
C. Alnico	3. Current transformers
	4. Power transformers

Codes:

	A	B	C
(a)	1	2	4
(b)	4	3	2
(c)	3	4	1
(d)	4	1	2

[ESE-2008]

3.39 Materials which lack permanent magnetic dipoles are known as:
(a) Paramagnetic (b) Diamagnetic
(c) Ferromagnetic (d) Ferrimagnetic

[ESE-2008]

3.40 Why is the core of the transformer built up of laminations?
(a) To reduce eddy current loss
(b) For convenience of fabrication
(c) No specific advantage
(d) For increasing the permeability

[ESE-2008]

3.41 Magnetically hard materials do not possess which of the following characteristics?
(a) High retentivity
(b) High coercivity
(c) Strong magnetic reluctance
(d) Zero differential permeability

[ESE-2009]

3.42 On which of the following factors does hysteresis loss not depend?
(a) Magnetic field intensity
(b) Frequency of the field
(c) Volume of the material
(d) Neel temperature

[ESE-2009]

3.43 When a ferromagnetic substance is magnetized, there are small changes in dimensions. The phenomenon is called
(a) Hysteresis (b) Magnetostriction
(c) Diamagnetism (d) Dipolar relaxation

[ESE-2010]

3.44 Consider the following statements:
The coercive force can be increased by
1. Adding Cobalt because it is ferro-magnetic material.
2. Adding Gold because it is a diamagnetic material.
3. Adding Super alloy.
4. Space charge polarizing.

Which of the above statements is/are correct?
(a) 1, 2, 3 and 4 (b) 1 only
(c) 2 only (d) 1 and 3 only

[ESE-2010]

3.45 Temperature below which certain materials are anti-ferromagnetic is called
(a) Curie temperature (b) Neel temperature
(c) Wein temperature (d) Debye temperature

[ESE-2010]

3.46 Consider the following statements:
Electrets are the materials which are
1. having permanent electric moments
2. electromagnets
3. very similar to permanent magnet materials
4. similar to anti-ferroelectric materials

Which of these statements is/are correct?

(a) 2 only (b) 1 and 3 only
(c) 2 and 3 only (d) 1, 2, 3 and 4

[ESE-2010]

3.47 Consider the following statements
Magnetic susceptibility

1. depends on the nature of the magnetic material.
2. is not dependent on the relative permeability of the medium.
3. cannot be determined by measuring the force exerted on a magnetic material when placed in a magnetic field.
4. can be determined from M - H curve.

Which of these statements is/are correct?

(a) 1, 2, 3 and 4 (b) 1 only
(c) 1 and 4 only (d) 2 only

[ESE-2010]

3.48 Consider the following statements referring to the magnetization :

1. In solenoid magnetization is due to a surface current distribution.
2. Magnetization has its origin in circulating current.
3. The solenoid dipole is represented by an infinitesimal current loop.
4. The magnetization is entirely solenoidal and divergent.

Which of the above statements is/are correct?

(a) 1, 2 and 3 only (b) 2, 3 and 4 only
(c) 3 only (d) 2, 3 and 4

[ESE-2010]

3.49 **Assertion (A) :** Magnetic cores are generally used in main memory of a digital computer.

Reason (R) : Magnetic cores are slow and volatile.

(a) Both A and R are true and R is the correct explanation of A
(b) Both A and R are true but R is NOT the correct explanation of A
(c) A is true but R is false
(d) A is false but R is true

[ESE-2010]

3.50 **Assertion (A) :** The spins within a magnetic domain are aligned permanently below Curie temperature in a ferromagnetic material.

Reason (R) : Ferromagnetic material is magnetic only when the domains are aligned by an external field.

(a) Both A and R are true and R is the correct explanation of A
(b) Both A and R are true but R is NOT the correct explanation of A
(c) A is true but R is false
(d) A is false but R is true

[ESE-2010]

3.51 Heating a permanent magnet results in the loss of magnetic behaviour because

(a) The atoms start vibrating
(b) The magnetic dipoles start vibrating
(c) The magnetic dipoles start realigning
(d) The atoms start conducting

[ESE-2011]

3.52 Diamagnetic materials posses

(a) Permanent dipoles
(b) Induced dipoles
(c) Both permanent and induced dipoles
(d) No dipoles

[ESE-2011]

3.53 Paramagnetic susceptibility of a material

(a) Increase linearly with temperature
(b) Decreases linearly with temperature
(c) Increases linearly with $(1/T)$
(d) Decrease linearly with $(1/T)$

[ESE-2011]

3.54 The magnetic domains, during the process of magnetization of ferromagnetic materials,

(a) Only expand
(b) Rotate first and then expand
(c) Expand first and then rotate
(d) Neither rotate nor expand

[ESE-2011]

3.55 If the domain walls in a magnetic material can easily be moved, then the material displays

(a) High flux density
(b) High permeability
(c) Permanent magnetic behaviour
(d) High permittivity

[ESE-2011]

3.56 Magnetic field of Earth has no vertical component at
(a) Magnetic poles (b) Magnetic equator
(c) Lattitude 45° (d) Longitude 45°
[ESE-2011]

3.57 Match **List-I** with **List-II** and select the correct answer using the code given below the lists:

List-I	List-II
A. Magnetic induction	1. Bohr magneton
B. Magnetic field	2. Tesla
C. Magnetic moment	3. Henry/metre
D. Permeability	4. Ampere/metre

Codes:

	A	B	C	D
(a)	2	1	4	3
(b)	3	1	4	2
(c)	2	4	1	3
(d)	3	4	1	2

[ESE-2011]

3.58 Soft iron is used to manufacture electro-magnets because it has
(a) High retentivity (b) High coercive field
(c) Low retentivity (d) Low coercive field
[ESE-2011]

3.59 Consider the following statements with regards to soft iron:
1. It is a magnetic material.
2. It conducts electricity.
3. It is an alloy of iron and copper.
4. It is used to make permanent magnets.

Which of these statements are correct?
(a) 1 and 2 (b) 2 and 3
(c) 3 and 4 (d) 1 and 4
[ESE-2011]

3.60 A permeable substance is one
(a) which is strong magnetic
(b) which is weak magnetic
(c) which is good conductor
(d) through which magnetic lines of force can pass easily
[ESE-2011]

3.61 High-frequency transformer cores are generally made from
(a) Mu-metal (b) Mone-metal
(c) ferrites (d) cobalt
[ESE-2012]

3.62 **Statement (I):** Alnico magnet alloys have the highest energy per unit of cost or volume of any permanent magnetic material commercially available.
Statement (II): They are very hard and brittle, therefore they cannot be machined and have to be cast and finished by grinding.
(a) Both Statement (I) and Statement (II) are individually true and Statement (II) is the correct explanation of Statement (I).
(b) Both Statement (I) and Statement (II) are individually true but Statement (II) is not the correct explanation of Statement (I).
(c) Statement (I) is true but Statement (II) is false.
(d) Statement (I) is false but Statement (II) is true.
[ESE-2012]

3.63 The presence of one of the following materials, in iron or steel for use as a magnetic material, tends to reduce the hysteresis loss
(a) Carbon (b) Sulphur
(c) Phosphorus (d) Silicon
[ESE-2012]

3.64 Match **List-I** with **List-II** and select the correct answer using the code given below the lists:

List-I
A. Antiferromagnetic
B. Ferrimagnetism
C. Diamagnetic
D. Ferromagnetic

List-II
1. Permanent magnetic dipoles
2. Dipoles interact or line up in parallel
3. Neighbouring magnetic moments are aligned anti-parallel with equal magnitudes
4. Neighbouring magnetic moments are aligned anti-parallel with unequal magnitudes

Codes:

	A	B	C	D
(a)	4	3	1	2
(b)	2	3	1	4
(c)	4	1	3	2
(d)	2	1	3	4

[ESE-2012]

3.65 The dependence of B (flux density) on H (magnetic field intensity) for different types of materials is

(a) 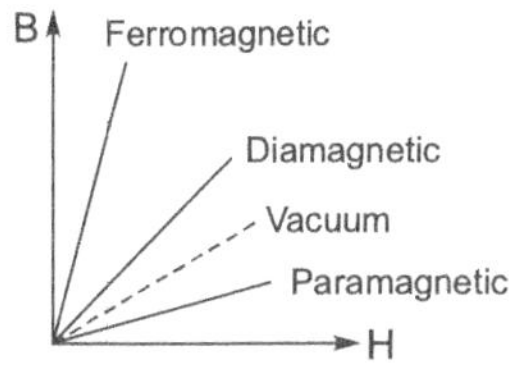

(b) 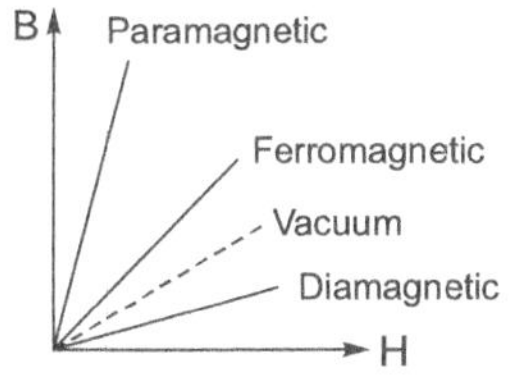

(c) 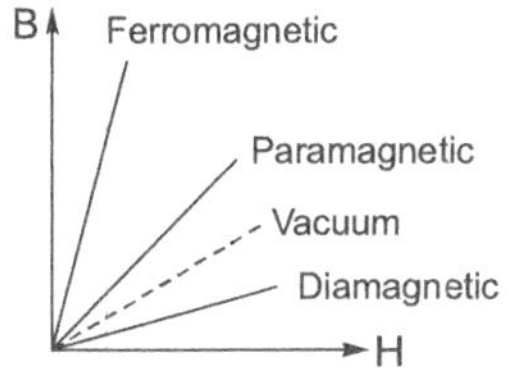

(d) 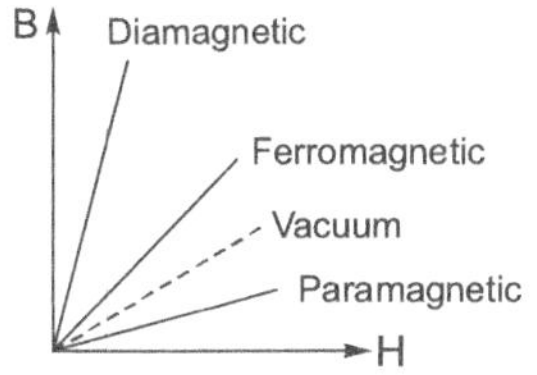

[ESE-2012]

3.66 Consider the following statements regarding magnetic materials:

1. A diamagnetic material has no permanent dipole.
2. Paramagnetic material has anti parallel orientation of equal moments with neighbouring dipoles.
3. Ferrimagnetic material has anti parallel orientation of unequal moments between neighbouring dipoles.
4. Anti ferromagnetic material has negligible interaction between neighbouring dipoles.

Which of these statements are correct?

(a) 1 and 2 (b) 3 and 4
(c) 2 and 4 (d) 1 and 3

[ESE-2013]

3.67 Consider the following statements regarding hysteresis loops of hard and soft magnetic materials:

1. Hysteresis loss of hard magnetic material will be less than that of soft material.
2. Coercivity of hard material will be greater than that of soft material.
3. Retentivity of the two materials will always be equal.

Which of these statements are correct?

(a) 1, 2 and 3 (b) 2 only
(c) 3 only (d) 1 and 3 only

[ESE-2013]

3.68 In the magnetic core the electromotive forces (emf) induced in accordance with Faraday's law of electromagnetic induction give rise to

(a) Eddy current (b) Excitation current
(c) Armature current (d) Field current

[ESE-2013]

3.69 Some magnetic materials may be classified on the basis of

1. Susceptibility
2. Saturation
3. Spin arrangement
4. Nature of hysteresis loop
5. Domain structure
6. Critical temperature above which it behaves as a paramagnetic material.

Which of these can be used to distinguish between ferri and ferromagnetic materials?

(a) 1, 3 and 4 only (b) 2, 3 and 6 only
(c) 3, 4 and 5 only (d) 1, 2, 3, 4, 5 and 6

[ESE-2013]

3.70 Magnetism is mainly due to only electron spin around their own axis in case of

(a) diamagnetic materials
(b) paramagnetic materials
(c) ferromagnetic materials
(d) paramagnetic and diamagnetic materials

[ESE-2013]

3.71 For paramagnetic materials, the relative permeability is

(a) less than unity but magnetic susceptibility is relatively small and positive
(b) greater than unity and magnetic susceptibility is relatively small but positive
(c) equal to unity and magnetic susceptibility is large but positive
(d) less than unity but magnetic susceptibility is relatively large and positive

[ESE-2013]

3.72 Permalloy and Mumetal are examples of
(a) Silicon and Iron alloys
(b) Nickel and Iron alloys
(c) Cobalt and Iron alloys
(d) Permanent magnet materials

[ESE-2013]

3.73 When a Ferromagnetic substance is magnetized, the phenomenon of 'magnetostriction' causes
(a) increase in the body temperature
(b) change in the permeability of the substance
(c) small changes in its dimensions
(d) decrease in the saturation flux-density

[ESE-2013]

3.74 The resistivity of 'Ferrites' is very much higher than that of the Ferromagnetic materials, because
(a) Ferrites are chemical compounds and the electrons in them are subject to the restraint of valence forces
(b) Ferrites have a low eddy current loss
(c) Ferrites have a non-homogeneous molecular structure
(d) Ferrites have varying flux-density inside the core

[ESE-2013]

3.75 When the temperature exceeds the transition temperature, a ferromagnetic material becomes similar to
(a) anti-ferromagnetic material
(b) diamagnetic material
(c) ferrimagnetic material
(d) paramagnetic material

[ESE-2013]

3.76 Statement (I): The magnetic moments of diamagnetic materials are mainly due to the orbital angular momentum of the electrons.

Statement (II): A steady current flowing in the orbit produces a magnetic field equivalent to that set up by a dipole perpendicular to the plane of orbit (Ampere's law).
(a) Both Statement (I) and Statement (II) are individually true and Statement (II) is the correct explanation of Statement (I).
(b) Both Statement (I) and Statement (II) are individually true but Statement (II) is not the correct explanation of Statement (I).
(c) Statement (I) is true but Statement (II) is false.
(d) Statement (I) is false but Statement (II) is true.

[ESE-2014]

3.77 Statement (I): Soft magnetic materials are not used in the construction of permanent magnets.

Statement (II): Soft magnetic materials have narrow hysteresis loop, low retentivity and low coercivity.
(a) Both Statement (I) and Statement (II) are individually true and Statement (II) is the correct explanation of Statement (I).
(b) Both Statement (I) and Statement (II) are individually true but Statement (II) is not the correct explanation of Statement (I).
(c) Statement (I) is true but Statement (II) is false.
(d) Statement (I) is false but Statement (II) is true.

[ESE-2014]

3.78 Curie law of paramagnetism (with χ = susceptibility, B = flux density and C = a constant) is
(a) $\chi = CT$ (b) $\chi = \frac{CB}{T}$
(c) $\chi = \frac{CT^2}{2B}$ (d) $\chi = \frac{C}{T}$

[ESE-2014]

3.79 If the magnetic susceptibility of a specimen is small and positive, the specimen is
(a) diamagnetic (b) paramagnetic
(c) ferromagnetic (d) non-magnetic

[ESE-2014]

3.80 Manganese ferrite is a 1 : 1 mixture of
(a) Manganese nitride and iron oxide
(b) Manganese oxide and iron oxide
(c) Manganese nitride and iron sulphide
(d) Manganese oxide and iron sulphide

[ESE-2014]

3.81 When a ferromagnetic substance is magnetized, small changes in dimensions occur. Such a phenomenon is known as

(a) Magnetic hysteresis
(b) Magnetic expansion
(c) Magnetostriction
(d) Magneto-colorisation

[ESE-2014]

3.82 In ferromagnetic, antiferromagnetic and ferrimagnetic materials, the atomic thermal motions counteract the coupling forces between the adjacent atomic dipole moments, thereby causing
(a) some dipole misalignment regardless of whether an external field is present
(b) increase in dipole alignment regardless of whether an external field is present
(c) no effect on dipole alignment
(d) atoms tend to de-randomize the direction of moments

[ESE-2014]

3.83 For high-speed reading and storing of information in a computer, the material used is
(a) ferrite
(b) piezoelectric
(c) pyroelectric
(d) ferromagnetic above 768 °C

[ESE-2015]

3.84 The temperature above which an anti-ferromagnetic material becomes paramagnetic is called
(a) peak temperature (b) Neel temperature
(c) critical temperature (d) Weiss temperature

[ESE-2015]

3.85 Magnetic materials which may be readily magnetized in either direction are
(a) soft magnetic materials
(b) hard magnetic materials
(c) high eddy current loss materials
(d) high hysteresis loss materials

[ESE-2015]

3.86 Consider the following statements regarding a ferromagnetic material :
1. Below the ferromagnetic Curie temperature, the ferromagnetic materials exhibit hysteresis effect.
2. The coercive force is the field required to reduce the flux density to zero.

Which of the above statements is /are correct?
(a) Both 1 and 2 (b) Neither 1 nor 2
(c) 1 only (d) 2 only

[ESE-2015]

3.87 Permeance is inversely related to
(a) resistance (b) conductance
(c) reluctance (d) capacitance

[ESE-2016]

3.88 Consider the following statements regarding an ideal core material:
1. It has very high permeability.
2. It loses all its magnetism when there is no current flow.
3. It does not saturate easily.

Which of the above statements are correct?
(a) 1 and 2 only (b) 1 and 3 only
(c) 2 and 3 only (d) 1, 2 and 3

[ESE-2016]

3.89 The relative permeability of a medium is equal to (with M = magnetization of the medium and H = magnetic field strength)
(a) $1+\frac{M}{H}$ (b) $1-\frac{M}{H}$
(c) $1+\sqrt{\frac{M}{H}}$ (d) $1-\sqrt{\frac{M}{H}}$

[ESE-2016]

3.90 Permanent magnet loses the magnetic behaviour when heated because of
1. atomic vibration
2. dipole vibration
3. realignment of dipoles

Which of the above are correct?
(a) 1 and 2 only (b) 1 and 3 only
(c) 1, 2 and 3 (d) 2 and 3 only

[ESE-2016]

3.91 The magnetic field required to reduce the residual magnetization to zero is called
(a) retentivity (b) coercivity
(c) hysteresis (d) saturation

[ESE-2016]

3.92 Consider the following statements:
1. Both ferromagnetic and ferrimagnetic materials have domain structures; each domain has randomly oriented magnetic moments when no external field is applied.

2. Both ferromagnetic and ferrimagnetic materials make those domains that have favourable orientation to the applied field grow in size.
3. The net magnetic moment in ferromagnetic material is higher than that in ferrimagnetic material.
4. The net magnetic moment in ferrimagnetic material is higher than that in ferromagnetic material.

Which of the above statements are correct?

(a) 1 and 4 only (b) 1, 2 and 4
(c) 2 and 4 only (d) 1, 2 and 3

[ESE-2016]

3.93 Which of the following exhibit electric hysteresis?

(a) Ferromagnetic materials only
(b) Ferroelectric materials only
(c) Ferrielectric materials only
(d) Both ferroelectric and ferrielectric materials

[ESE-2007]

3.94 The complex permeability and resulting wave losses are due to

(a) free electrons or ion oscillation and dipole relaxation.
(b) free electrons oscillation and relaxation of free space charge.
(c) bound electrons oscillation and relaxation of free space charge.
(d) bound electrons or ion oscillation and dipole relaxation.

[ESE-2015]

3.95 **Statement (I):** Hard magnetic materials are used for making permanent magnets.

Statement (II): Hard magnetic materials have relatively small and narrow hysteresis loop.

(a) Both Statement (I) and Statement (II) are individually true and Statement (II) is the correct explanation of Statement (I).
(b) Both Statement (I) and Statement (II) are individually true but Statement (II) is not the correct explanation of Statement (I).
(c) Statement (I) is true but Statement (II) is false.
(d) Statement (I) is false but Statement (II) is true.

[ESE-2014]

3.96 For high speed reading and storing of information in a computer, the core shall be of

(a) Ferrite
(b) Piezoelectric
(c) Pyroelectric
(d) Ferromagnetic above 768°C

[ESE-2017]

3.97 Soft magnetic materials should have

(a) large saturation magnetization and large permeability.
(b) low saturation magnetization and large permeability.
(c) large saturation magnetization and low permeability.
(d) low saturation magnetization and low permeability.

[ESE-2017]

3.98 Orbital magnetic moment of an electron, in an atom, is of the order of

(a) 0.1 Bohr magneton
(b) 1.0 Bohr magneton
(c) 10 Bohr magneton
(d) 100 Bohr magneton

[ESE-2017]

3.99 When the temperature of a ferromagnetic material exceeds the Curie temperature, it behaves similar to a

(a) Diamagnetic material
(b) Ferrimagnetic material
(c) Paramagnetic material
(d) Antiferromagnetic material

[ESE-2017]

3.100 The total magnetic moment

1. is called saturation magnetization.
2. depends on the number of magnetic dipoles per unit volume, the instant electric current and the area of the current loop.

Which of the above statements is/are correct?

(a) 1 only (b) 2 only
(c) Both 1 and 2 (d) Neither 1 nor 2

[ESE-2017]

3.101 Susceptibility of a diamagnetic material is

1. Negative
2. Positive

3. Dependent on the temperature
4. Independent of the temperature

Select the correct answer using the codes given below:

(a) 1 and 3 only (b) 2 and 3 only
(c) 1 and 4 only (d) 2 and 4 only

[ESE-2017]

3.102 Consider the following statements:

1. The susceptibility of diamagnetic materials is small and negative.
2. The susceptibility of para and antiferromagnetic materials is small but positive.
3. The susceptibility has a finite value for free space or air.

Which of the above statements are correct?

(a) 1 and 2 only (b) 1 and 3 only
(c) 2 and 3 only (d) 1, 2 and 3

[ESE-2017]

3.103 The phenomenon of magnetostriction occurs when a ferromagnetic substance is magnetized resulting in

(a) heating
(b) small changes in the dimensions
(c) small changes in its crystal structure
(d) some change in its mechanical properties

[ESE-2017]

3.104 The saturation magnetization for Fe_3O_4, given that each cubic unit cell contains 8 Fe^{2+} and 16Fe^{3+} ions, where Bohr magneton is 9.274×10^{-24} A.m^2 and that the unit cell edge length is 0.839 mm, will be

(a) 1.25×10^5 A/m (b) 5×10^5 A/m
(c) 10×10^5 A/m (d) 20×10^5 A/m

[ESE-2019]

3.105 The Temperature at which iron ceases to be ferromagnetic and becomes paramagnetic is

(a) Curie-Weiss point
(b) Thermo-magnetic point
(c) Ferro-paramagnetic point
(d) Curie point

[ESE-2019]

3.106 **Statement-I:** Soft iron does not retain magnetism permanently.

Statement-II: Soft iron has no retentivity.

(a) Both Statement (I) and Statement (II) are individually true and Statement (II) is the correct explanation of Statement (I).
(b) Both Statement (I) and Statement (II) are individually true but Statement (II) is not the correct explanation of Statement (I).
(c) Statement (I) is true but Statement (II) is false.
(d) Statement (I) is false but Statement (II) is true.

[ESE-2019]

3.107 **Statement-I:** Soft magnetic materials, both metallic and ceramic are used for making transformers core, whereas, hard magnetic materials both metallic and ceramic are used for making permanent magnets.

Statement-II: Magnetic materials, both metallic and ceramic are classified as soft or hard according to the magnetic hysteresis loop being narrow or broad.

(a) Both Statement (I) and Statement (II) are individually true and Statement (II) is the correct explanation of Statement (I).
(b) Both Statement (I) and Statement (II) are individually true but Statement (II) is not the correct explanation of Statement (I).
(c) Statement (I) is true but Statement (II) is false.
(d) Statement (I) is false but Statement (II) is true.

[ESE-2019]

3.108 A transformer core is wound with a coil carrying an alternating current at a frequency of 50 Hz. The hysteresis loop has an area of 60000 units when the axes are drawn in units of 10^{-4} Wb m^{-2} and 10^2 Am^{-1}. If the magnetization is uniform throughout the core volume of 0.01 m^3, the power loss due to hysteresis will be

(a) 300 W (b) 350 W
(c) 400 W (d) 450 W

[ESE-2020]

3.109 When ferromagnetic or ferrimagnetic materials are magnetized, the direction of magnetization in any domain will be rotated from its preferential

direction. This will show an anisotropic behaviour. On removal of the magnetizing force, the total magnetization will in general have a non-zero value. This behaviour is due to
(a) Crystal anisotropic
(b) Stress anisotropic
(c) Shape anisotropic
(d) Crystal, stress and shape anisotropic

[ESE-2020]

3.110 The paramagnetic susceptibility varies inversely with the absolute temperature for ordinary fields and temperatures. It is given by the relation

$$\chi = \frac{C}{T}$$

The relation is known as
(a) Phenomenon of magnetostriction
(b) Curie law of paramagnetism
(c) Hall Effect
(d) Diamagnetism

[ESE-2020]

3.111 If the interaction between the atomic permanent dipole moments is zero or negligible and the individual dipole moments are oriented at random, the material will be a
(a) Ferromagnetic material
(b) Ferrimagnetic material
(c) Paramagnetic material
(d) Antiferromagnetic material

[ESE-2020]

3.112 The magnetic moments of diamagnetic materials are mainly due to
(a) Electron spin angular momentum
(b) Nuclear spin angular momentum
(c) Orbital angular momentum of the electrons
(d) Centrifugal angular momentum

[ESE-2020]

3.113 The ideal core should exhibit very high permeability in case of transformers and inductor core due to magnetic saturation caused by DC imbalance condition that can be minimized by
(a) Low permeability core only
(b) High permeability core only
(c) Low and high permeability combination core
(d) No permeability core

[ESE-2020]

3.114 What is the loss of energy per hour in a ferromagnetic specimen when it is subjected to 50 Hz magnetisation, if the specimen weighs 50 kg, area of hysteresis loop is 200 joules per m^3 and density of iron is 7500 kg/m^3?
(a) 2.4×10^5 J (b) 3.9×10^4 J
(c) 2.9×10^4 J (d) 4.5×10^5 J

[ESE-2021]

3.115 In a ferromagnetic material, the losses due to hysteresis are
(a) directly proportional to the supply frequency.
(b) inversely proportional to the supply frequency.
(c) inversely proportional to square of the supply frequency.
(d) directly proportional to square of the supply frequency.

[ESE-2021]

3.116 The spontaneous magnetization is the most important characteristic of
(a) paramagnetic materials.
(b) ferromagnetic materials.
(c) diamagnetic materials.
(d) permalloy

[ESE-2021]

3.117 **Statement I :** Ferromagnetic properties are confined almost entirely to iron, nickel and cobalt and their alloys.

Statement II : By substituting the basic ferrite of magnetite with other divalent oxides, a wide range of ferrimagnetic compounds with useful properties can be produced, which have high electrical resistivity to minimise eddy currents at high frequencies.
(a) Both Statement (I) and Statement (II) is the correct explanation of Statement (I).
(b) Both Statement (I) and Statement (II) are individually true, but Statement (II) is not the correct explanation of Statement (I).
(c) Statement (I) is true, but Statement (II) is false.
(d) Statement (I) is false, but Statement (II) is true.

[ESE-2021]

3.118 The temperature above which the ferromagnetic materials lose their magnetic properties is called

(a) saturation point (b) breakdown point
(c) Curie point (d) peak point

[ESE-2022]

3.119 What is the energy lost per hour in a specimen of iron subjected to magnetization at 50 c/s, if the specimen weighs 50 kg and the hysteresis loop is equivalent in area to 250 J/m^3 and the density of iron is 7500 kg/m^3?

(a) 3×10^5 J (b) 4×10^5 J
(c) 5×10^5 J (d) 6×10^5 J

[ESE-2022]

3.120 Most substances are not magnets, because

(a) they do not have sufficient energy to produce magnetic behavior
(b) their electrons do not move truly
(c) the electrons usually pair up with their spins opposite to each other, so that their fields cancel each other
(d) their electrons strongly bind to the nucleus as they have more number of protons than electrons

[ESE-2022]

3.121 Whenever a particle has angular momentum, it will contribute to permanent dipole moment. Which one of the following does not contribute to the angular momentum of an atom?

(a) Orbital angular momentum of electron
(b) Proton spin angular momentum
(c) Electron spin angular momentum
(d) Nuclear spin angular momentum

[ESE-2022]

3.122 The classical theory of the diamagnetism of the bonded electrons in a free atom was elaborated by

(a) Langevin diamagnetism
(b) Larmor diamagnetism
(c) Lorentz diamagnetism
(d) Thomson diamagnetism

[ESE-2023]

3.123 Which one of the following is the element exhibiting antiferromagnetism at room temperature?

(a) Iron (b) Nickel
(c) Chromium (d) Cobalt

[ESE-2023]

3.124 What are the important characteristics that must be satisfied for the materials used to build permanent magnets?

(a) Low permeability, high coercive force and high Curie temperature
(b) Low permeability, high coercive force and low Curie temperature
(c) High permeability, high coercive force and high Curie temperature
(d) High permeability, low coercive force and low Curie temperature

[ESE-2024]

3.125 What are the important characteristics that must be satisfied for the materials used to build core of the transformer?

(a) High permeability, low hysteresis and high eddy current losses
(b) High permeability, low hysteresis and eddy current losses
(c) High permeability, high hysteresis and eddy current losses
(d) Low permeability, low coercive force and low Curie temperature

[ESE-2024]

■■■■

Answers Magnetic Properties of Materials

3.1	(a)	3.2	(b)	3.3	(c)	3.4	(d)	3.5	(c)	3.6	(a)	3.7	(b)	3.8	(b)	3.9	(c)
3.10	(d)	3.11	(d)	3.12	(b)	3.13	(b)	3.14	(b)	3.15	(c)	3.16	(d)	3.17	(b)	3.18	(b)
3.19	(b)	3.20	(a)	3.21	(c)	3.22	(b)	3.23	(b)	3.24	(b)	3.25	(d)	3.26	(a)	3.27	(b)
3.28	(c)	3.29	(b)	3.30	(b)	3.31	(d)	3.32	(d)	3.33	(b)	3.34	(d)	3.35	(a)	3.36	(a)
3.37	(c)	3.38	(d)	3.39	(b)	3.40	(a)	3.41	(d)	3.42	(d)	3.43	(b)	3.44	(b)	3.45	(b)
3.46	(c)	3.47	(c)	3.48	(b)	3.49	(c)	3.50	(c)	3.51	(b, c)	3.52	(b)	3.53	(c)	3.54	(c)
3.55	(b)	3.56	(b)	3.57	(c)	3.58	(d)	3.59	(a)	3.60	(d)	3.61	(c)	3.62	(b)	3.63	(d)
3.64	(*)	3.65	(c)	3.66	(d)	3.67	(b)	3.68	(a)	3.69	(a)	3.70	(c)	3.71	(b)	3.72	(b)
3.73	(c)	3.74	(a)	3.75	(d)	3.76	(a)	3.77	(a)	3.78	(d)	3.79	(b)	3.80	(b)	3.81	(c)
3.82	(b)	3.83	(a)	3.84	(b)	3.85	(a)	3.86	(a)	3.87	(c)	3.88	(d)	3.89	(a)	3.90	(c)
3.91	(b)	3.92	(d)	3.93	(b)	3.94	(d)	3.95	(c)	3.96	(a)	3.97	(a)	3.98	(b)	3.99	(c)
3.100	(b)	3.101	(c)	3.102	(a)	3.103	(b)	3.104	(b)	3.105	(d)	3.106	(c)	3.107	(a)	3.108	(a)
3.109	(d)	3.110	(b)	3.111	(c)	3.112	(c)	3.113	(b)	3.114	(a)	3.115	(a)	3.116	(b)	3.117	(b)
3.118	(c)	3.119	(a)	3.120	(c)	3.121	(b)	3.122	(a)	3.123	(c)	3.124	(a)	3.125	(b)		

Explanations Magnetic Properties of Materials

3.1 (a)

Higher the exchange interaction energy, higher will be area under B-H curve i.e. high saturation magnetization (because value of B at saturation will be very high).

3.2 (b)

Copper → Cu is a diamagnetic material
Iron → Ferromagnetic material
Ferrite → Ferrimagnetic material
Na → Is not a antiferromagnetic material actually it is not a magnetic material but NaCl is a diamagnetic material.

3.3 (c)

Ferromagnetic behaviour is shown by those transition metals where the ratio of atomic diameter to 3d-orbital diameter is in between 1.5 to 2.

3.4 (d)

Chemical formula of simple ferrites may be written as

$$Me^{2+}Fe_2^{3+}O_4^{2-} \quad \text{i.e. } AB_2O_4$$

where Me^{2+} → represents a variety of divalent metallic ions such as

$Fe^{2+}, CO^{2+}, Mn^{2+}, Zn^{2+}, Cd^{2+}, Mg^{2+}$ etc. formula may be written as mixture of MeO and Fe_2O_3.

3.5 (c)

Ni – Zn ferrite → used in audio and TV Transformers
Co – Sm alloy → used as permanent magnets (because it exhibits the property of high coercive force and high resistivity).
Yttrium Iron Garnet → used as microwave isolator (It is a magnetic material).
Mg – Zn ferrite → used as memory cores.

3.6 (a)

Ferroelectric material → e.g. KH_2PO_4
Piezoelectric material → $BaTiO_3$, Ceramic
Soft magnetic material → e.g. permalloy
Hard magnetic material → e.g. Tungsten steel.

3.7 (b)

The dc resistivity of ferrites is many orders of ten higher than that of iron; consequently it prevents penetration of magnetic flux into the material. Further, coercive force (H_c) for a high stability permanent magnet may be as high as 10^6 ampere–m^{-1} as compared to ferrites having as low as 1 ampere–m^{-1}.

3.8 (b)
Electrical resistivity has to be high for ferrites, 4% Si is added to Fe which is ferromagnetic material. Further upon adding 4% Si, coercive force and B_{sat} both decreases. Permeability is increased by grain oriented silicon steel.

3.9 (c)
χ_m is negative for diamagnetic material and as long as the electronic structure of the material is independent of temperature, the diamagnetic susceptibility is also essentially independent of temperature.

3.10 (d)
Silicon steel → used in power transformer.
Ferrite → used in high frequency transformer.
Alnico → It is hard ferrite used as permanent magnet.

3.11 (d)

$$\text{Larmour frequency} = \frac{eB}{2m}$$

m → mass of e^-

$$\text{Bohr magnetron} = \frac{eh}{4\pi m}$$

m = mass of e^-,
h = Planck's constant

Magnetic induction,

$$B = \mu_0(H + M)$$

M = magnetization.

$$\text{Curie-Weiss law,} \quad \chi = \frac{C}{T - \theta}$$

3.12 (b)
The Neel temperature plays a similar role in antiferromagnetic materials as does the curie temperature in ferromagnetic materials.

3.13 (b)
Ferroelectric material → e.g.. Rochelle salt
Soft magnetic material → e.g. Permalloy
Hard magnetic material → e.g.. Alnico
Semiconductor → GaAs (direct band gap)

3.14 (b)
The dc resistivity of ferrites is many orders of ten higher than that of iron, consequently the eddy current problem preventing penetration of magnetic flux into the material is much less severe in ferrites than in iron.

3.15 (c)
Hard magnetic materials have a high coercive field.

3.16 (d)
Permeability of water $\mu_{rw} = 0.99999 < 1$
Permeability of oxygen $\mu_{ro} = 1.00002 > 1$
so, water is diamagnetic material .
and oxygen is paramagnetic material. So 1 is right. Permalloy is soft alnico is hard so statement (3) is wrong.

3.17 (b)
It should have high $B_{saturation}$ and low coercive field H_C.

3.18 (b)

$$\chi(\text{susceptibility}) \propto \frac{1}{\text{Temperature}}$$

as T increases, χ decreases, so these losses their magnetic properties.

3.19 (b)
4% Si - Fe is a soft magnetic material having coercive field $H_C = 40$ amp-m^{-1}.

3.20 (a)
Above curie temperature ferromagnetic material loses their property of ferro-magnetism as χ_m magnetic susceptibility decreases and becomes paramagnetic material. ($\chi_m \propto 1/T$).

3.21 (c)
Diamagnetic material → Magnetic susceptibility is –ve and small $\simeq -10^{-5}$
Paramagnetic material → +ve and small $\simeq 10^{-5}$
Ferromagnetic material → +ve and large $\approx 10^5$

3.22 (b)
Metallic copper is diamagnetic substance having negative value of magnetic susceptibility.

3.23 (b)
Magnetic field required (applied) in reverse direction to reduce residual magnetization (spontaneous magnetization) is called coercive magnetic field and this phenomenon is called coercivity.

3.24 (b)

Bohr magnetron = $\dfrac{eh}{4\pi m}$ is unit of permanent dipole moment due to spin of e^{-s} in orbital.

3.25 (d)
Magnetostriction is phenomenon of change in dimension of ferromagnetic material due to magnetization Reverse of this effect is called as villari effect.

3.26 (a)
In magnetic materials Alnico having maximum energy product used for making permanent magnet.

3.27 (b)
In the region above the ferromagnetic curie temperature, the behaviour of a ferromagnetic material is some what similar to that of a paramagnetic material. Below the ferromagnetic curie temperature, ferromagnetic materials exhibit the well-known hysteresis in the *B* versus *H* curves.

3.28 (c)
Magnetic materials which have low coercive field are known as soft materials. Permanent magnets are made of hard materials which have a broad hysteresis loop (large coercive force), so that they are not subject to self-demagnetisation.

3.29 (b)

$$\chi = \frac{C}{(T-\theta)}$$

3.30 (b)

	Actual values
Fe → 1043 K	1043 K
$MnOFe_2O_3$ → 863 K	573 K
$MgOFe_2O_3$ → 523 K	713 K
$NiOFe_2O_3$ → 783 K	863 K

3.31 (d)
Ferrimagnetic material such as ferrites do not have eddy current loss.

3.32 (d)
In Antiferromagnetic material the dipole moment of adjacent atoms are align in opposite direction and equal in magnitude so net magnetization is zero.

3.33 (b)
In YIG, I represent Iron (Yttrium Iron Garnet)
In YAG, A represent Aluminium
(Yttrium Aluminium garnet).
I → Iron is magnetic
Al → Non magnetic

3.34 (d)
Dipole moment for antiferromagnetic

↑↓↑↓↑↓↑↓↑↓↑↓ equal and antiparallel.

3.35 (a)

$$P_e = K_e B^2 f^2$$

$$B = \frac{V}{f} = \text{const. if both doubled}$$

So, $P_e = K_e V^2$

So, $P_e \propto V^2$

for $V' = 2\,V$

$\therefore$ $P_e' = 4\,P_e$

3.36 (a)
High permeability and resistivity are main cause of application at high frequency for ferrites.

3.37 (c)
On application of field firstly domain wall expand and then rotates.

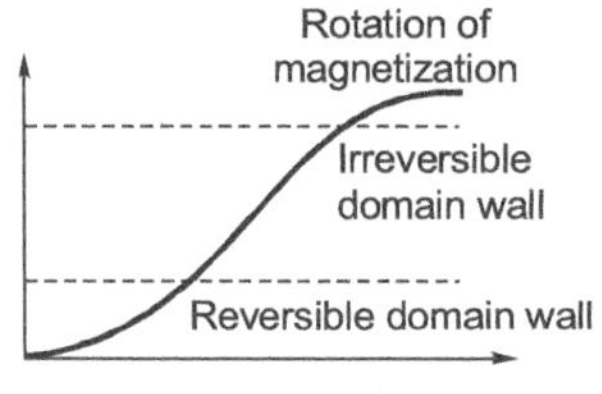

3.38 (d)
Silicon steel → Used in power transformer.
Ferrites → Used in high frequency transformer.
Alnico → Permanent magnets.

3.39 (b)
Diamagnetic material does not have permanent dipole moment in the absence of magnetic field. Paramagnetic material: It has permanent dipole moment but all atom have dipole moment in random direction so net dipole moment is zero in absence of magnetic field.

3.40 (a)
Lamination is provided to reduce eddy current losses.

$$P_e = k_e B^2 f^2 t^2$$

3.41 (d)
Magnetically hard materials possess high retentivity and high coercivity.

3.42 (d)
Neel temperature is a transition temperature above which Antiferromagnetic material becomes

paramagnetic, It has no deal with hysteresis loss.

$$P_h \propto K_h f B^n V$$

3.43 (b)

There are small changes in dimensions when ferromagnetic substance magnetized this phenomena is called magnetostriction.

3.44 (b)

Coercive force can be increased by adding hard magnetic material like cobalt.

3.45 (b)

Below Neel temperature magnetic material exhibit anti-ferromagnetic and above neel temperature behaves as paramagnetic.

3.46 (c)

Electrets are electrical analogy of electromagnets very similar to permanent magnet materials.

3.47 (c)

Magnetic susceptibility depends on

1. Nature of material
2. Permeability of medium

3.49 (c)

Magnetic core made of ferromagnetic metal or ferrimagnetic which is non-volatile since used for recording heads.

3.50 (c)

Ferromagnetic material is also magnetic if no external field applied.

3.51 (b, c)

Heating causes vibration of magnetic dipoles, so their alignment in one particular direction is lesser so loss of magnetic behaviour.

3.52 (b)

Diamagnetic materials possess only induced dipole, in absence of magnetic field there is no dipoles and dipole moment is zero.

3.53 (c)

For paramagnetic material,

Magnetic susceptibility

$$\chi_m = \frac{C}{T-\theta}$$

θ is curie temperature

$$\chi_m \propto \frac{1}{T}$$

so increases with $1/T$

3.54 (c)

During process of magnetization of ferromagnetic materials magnetic domains first expand and then rotate.

3.55 (b)

Domain walls in magnetic material can be easily moved in case of ferromagnetic materials which has high value of permeability.

3.56 (b)

At magnetic equator, magnetic field of Earth has only horizontal component (tangential component) and no vertical component (normal component) as earth is a conductor.

3.57 (c)

Magnetic field H has unit A/m. $B-4$.

Magnetic field → Bohr magnetron, $C-1$.

Magnetic induction → Tesla or Weber/m^2 $A-2$.

Permeability → H/m → $D-3$.

3.58 (d)

Soft iron is used to manufacture electromagnets because it has low cohesive field so they can response easily to alternating field.

3.59 (a)

Soft iron is a magnetic material. It is used to make electromagnets not permanent magnets so 4 is wrong.

Statement 1 is right, option (a) is correct.

3.60 (d)

A permeable substance means having a high value of permeability means through which magnetic lines of force can pass easily.

3.61 (c)

Ferrites with useful magnetic properties have d.c. resistivity of many orders of ten higher than in iron, and are used for frequencies upto microwave in transformer cores.

3.62 (b)

Alnico magnet alloys have the highest energy per unit of cost or volume of any permanent magnet material commercially available. They are usually characterised by a higher coercivity, a higher energy, and a lower retentivity then the magnet-steel type.

3.63 (d)

On addition of silicon in iron or steel the hysteresis loop becomes narrow therefore hysteresis loss decreases.

3.64 (*)

- Antiferromagnetic → ↑↓↑↓
- Ferromagnetic → ↑↑↑↑
- Ferrimagnetism → ↑↓↑↓

Correct answer is

A	B	C	D
3	4	2	1

3.65 (c)

$$B = \mu H$$

$$\Rightarrow \quad \mu_r = \frac{B}{\mu_o H}$$

For

paramagnetic $\rightarrow \chi_m \sim 10^{-5}$ (positive)

diametric $\rightarrow \chi_m \sim 10^{-5}$ (negative)

Ferromagnetic $\rightarrow \chi_m \rightarrow \infty$ (very large)

The B-H curve for different types of materials is

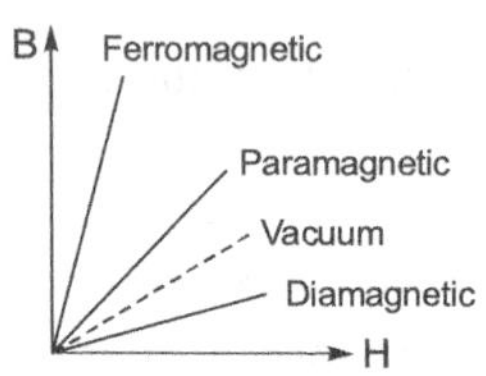

Hence option (c) is correct.

3.66 (d)

Diamagnetic material has no permanent dipole. It has induced dipole. Also in ferrimagnetic material has antiparallel orientation of unequal moments between neighbouring dipoles.

3.67 (b)

Hard magnetic materials have high hysteresis loss and high coercivity.

3.68 (a)

Current comes due to Faraday's law of electromagnetic induction is termed as eddy current.

3.70 (c)

In case of ferromagnetic materials magnetization is due to electron spin around their own axis.

3.71 (b)

For paramagnetic materials

$$\chi_m \approx 10^{-5}$$

and $\mu_r = \chi_m + 1$

i.e. $\chi_m > 0$

and $\mu_r > 1$

3.72 (b)

Permalloy – 80% Ni + 20% Fe

Mumetal – Ni-Fe alloy

Both are soft magnetic material

3.73 (c)

Just like electrostriction, in magnetostriction also small change in dimension occurred.

3.79 (b)

$\chi \sim -10^{5}$ (diamagnetic)

$\sim +10^{-5}$ (paramagnetic)

$\sim \infty$ (ferromagnetic & ferrimagnetic)

3.80 (b)

Manganese ferrite is a 1 : 1 mixture of manganese oxide and iron oxide.

Note: Ferrite is mixture of MO + Fe_2O_3

(M = Mn, CO, Ni etc.)

3.83 (a)

Ferrite in used in computer memory.

3.84 (b)

Anti ferromagnetic material $\xrightarrow[\text{Temp.}]{\text{Neel}}$ paramagnetic

Ferromagnetic material $\xrightarrow[\text{Temp.}]{\text{Curie}}$ paramagnetic

3.85 (a)

Magnetic materials which may be readily magnetized in either direction are soft magnetic materials also used in electromagnets.

3.86 (a)

Because above curie temperature, the ferromagnetic materials behave as paramagnetic.

3.88 (d)

⇒ An ideal core material should have high permeability which causes the magnetic field lines to be concentrated in the core material.

⇒ The magnetic field is often created by a coil of wire around the core that carries a current. So if there is no current then it should looses all its magnetism.

⇒ Core saturation means the core flux stops increasing with increase in magnetizing current, which is undersirable. Hence for proper operation the core should not saturate easily.

3.89 (a)

We have susceptibility,

$$\chi_M = \frac{M}{H} = \mu_r - 1$$

so relative permeability,

$$\mu_r = 1 + \frac{M}{H}$$

3.90 (c)

⇒ As a permanent magnet is heated, the electron spins (like tiny magnets) become more likely to be in high energy state, which leads to atomic vibration. That means that are less lined up so the total magnetism is reduced.

⇒ Heating means providing extra thermal energy because of which it becomes easy for domain walls (the boundaries between regions that are lined up pointing different directions) to slide around. Hence the domain walls will rearrange so that they reduce the large scale field energy by pointing different directions, hence permanent magnet losses magnetic behaviour.

⇒ As we further heat the individual dipole spin (dipole vibration) within domain become more likely to point opposite to their neighbour which results in losing permanent magnetic behaviour.

3.91 (b)

The magnetic field required to reduce the residual magnetization (or spontaneous magnetization) to zero value is applied in reverse direction and is called coercive field.

3.92 (d)

In ferrimagnetic materials dipoles are oriented in opposite direction but having different magnitude as shown below:

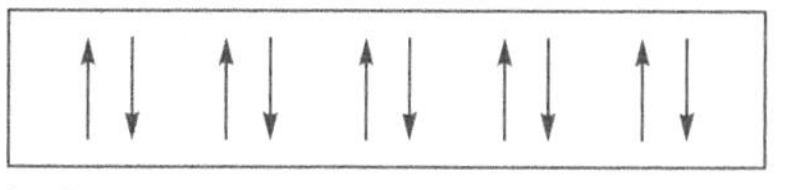

While in ferromagnetic materials in a domain all the dipoles are oriented in same direction as shown below:

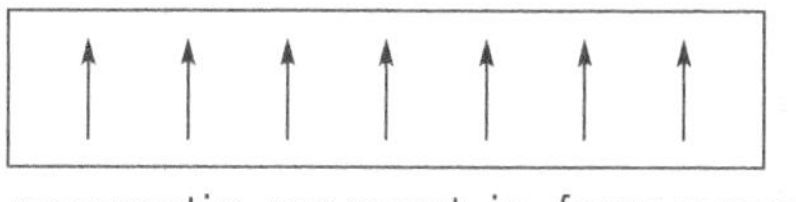

So net magnetic moment in ferromagnetic materials is higher than that in ferrimagnetic materials. Hence statement 4 is wrong, so answer will be option (d).

3.93 (b)

There is no ferrielectric material and ferroelectric exhibit hysteresis, as like ferromagnetic and ferrimagnetic.

3.95 (c)

Properties of Hard Magnets:

- High value of residual flux density
- High permeability
- High coercive force
- High curie temperature
- Board hysteresis loop

Hard magnets retain considerable amount of magnetic energy when magnetic field is removed and it is difficult to demagnetize these materials, once they get magnetized.

3.96 (a)

Ferrites with rectangular hysteresis loop have application in magnetic memory core in computers. The switching time for these ferrites are about 1 msec. Commonly used ferrites are Manganese ferrites, Manganese copper ferrites, Lithium nickel ferrite.

3.97 (a)

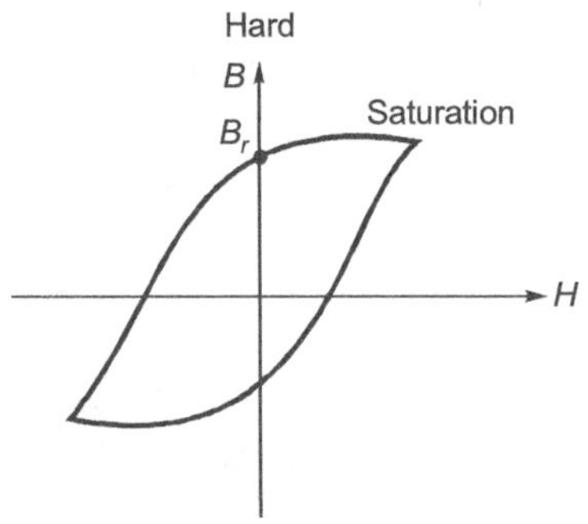

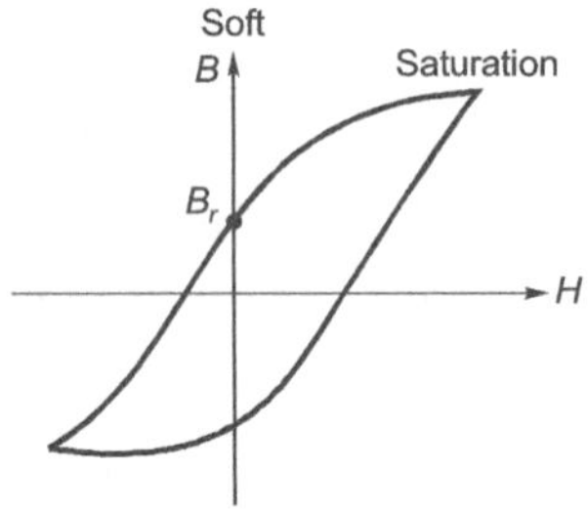

The reminant magnetic field should be less in soft with large permeability. So from large saturation magnetic field it should come to very less reminant magnetic field. So answer should be (a).

3.98 (b)

Orbital magnetic moment of an electron,

$$\bar{\mu}_m = -\frac{e}{2m}\bar{M}_a$$

$\bar{M}_a$ is angular momentum of electron = $h/2\pi$

$$\mu_m = \frac{e}{2m}\cdot\frac{h}{2\pi} = 1 \text{ Bohr magneton}$$

3.99 (c)

For temperature θ_f (Curie temperature), ferromagnetic material behaves as paramagnetic material.

For $T > \theta_f$, $\chi_m \rightarrow$ susceptibility decided by Curie Weiss law.

$$\chi_m = \mu_r - 1 = \frac{C}{T-\theta} \quad \text{for } T > \theta_f$$

3.100 (b)

$$\vec{P} = N\vec{m}$$

where, N = number of dipoles per unit volume.

3.101 (c)

Magnetic susceptibility, $\chi_m = \mu_r - 1$ is negative for diamagnetic materials. It range is -10^6 to -10^{-5}. Negative value implies that diamagnetic material acquire magnetization opposite to applied field. Susceptibility is independent of temperature.

3.102 (a)

$$\chi = \mu_r - 1 = 1 - 1 = 0 \quad (\text{for air } \mu_r = 1)$$

3.103 (b)

Due to magnetostriction there is a change in the dimension for a ferromagnetic substance when it is magnetized.

3.104 (b)

- Each Fe_3O_4 molecule has $4p_B$ magnetic dipole moment,

$$\vec{m} = 4\vec{p}_B$$

- Density of Fe_3O_4 molecules

$$N = \frac{\text{Number of molecules}}{\text{Volume}}$$

$$N = \frac{8}{(0.839\times10^{-9})^3}\left(\frac{1}{m^3}\right)$$

- $\vec{M}$ = Magnetization;

$$\vec{M} = N\vec{m}$$

$$\vec{M} = N4\vec{p}_B$$

$$M = \frac{8}{(0.839)^3\times10^{-27}}(4)(9.274\times10^{-24})$$

$$M = 5.024\times10^5 \text{ (A/m)}$$

3.108 (a)

Hysteresis loss = Area of Hysteresis loop × Frequency × Volume of core

= (60000) ($10^{-4}\times10^2$) (50) (0.01) = 300 W

3.109 (d)

Crystal, shape, stress anisotropic nature is reason behind non-zero value of total magnetization.

3.110 (b)

Curie law for paramagnetic materials,

$$\chi = \frac{C}{T}$$

χ = susceptibility;
C = Curie constant;
T = temperature in K

3.111 (c)

Paramagnetic materials contain randomly oriented permanent dipoles. There is no interaction among the dipoles, hence net permanent dipole moment is zero in the absence of field.

3.112 (c)

Diamagnetism is a very weak form of magnetism that is non-permanent and persists only while an external field is being applied. It is induced by a change in the orbital motion of electrons due to an applied magnetic field.

3.113 (b)

Magnetic saturation caused by DC imbalance condition that can be minimized by high permeability core only.

3.114 (a)

$$\text{Density} = \frac{\text{Mass}}{\text{Volume}} \text{ or } V = \frac{M}{\rho}$$

Energy Loss Per Second (ELPS)
= Volume × Energy loss per unit volume × 50
Loss of energy per hour

$$= \text{ELPS} \times 60 \times 60$$
$$= \frac{50}{7500} \times 200 \times 50 \times 60 \times 60$$
$$= 2.4 \times 10^5 \text{ J}$$

3.115 (a)

$$P_{HYS} = \eta B^n f \text{ Watt/m}^3$$

3.116 (b)

Spontaneous magnetizaton is the appearance of an ordered spin state (magnetization) at zero applied magnetic field in a ferromagnetic or ferrimagnetic material below a critical point called the Curie temperature or Tc. When heated to temperatures above Tc, ferromagnetic materials become paramagnetic and their magnetic behaviour is dominated by spin waves, and spontaneous magnetization vanishes away.

3.118 (c)

Above curie temperature ferromagnetic materials become paramagnetic.

3.119 (a)

The energy loss

$$= \frac{250}{7500} \times 50 \times 50 \times 60 \times 60$$
$$= 3 \times 10^5 \text{ J}$$

3.120 (c)

Because of the pairing net electron spin dipole moments get cancelled and hence material becomes nonmagnetic.

3.121 (b)

Permanent magnetic dipole moment of an atom is due to :

- Orbital angular momentum of electron
- Electron spin angular momentum.
- Nuclear spin angular momentum.

3.122 (a)

Theory of diamagnetism was worked out by **Paul Langevin**. Negative magnetism present in an applied field even though the material is composed of atoms that have no net magnetic moment.

3.123 (c)

At room temperature, Iron, Nickel and Cobalt are ferromagnetic.

■■■■

4 Conductive Materials

4.1 Consider the following properties pertaining to an alloy used as a precision resistor:

1. Uniform resistance
2. Stable resistance
3. Zero or low temperature coefficient of resistivity

Which of these properties are desirable?

(a) 1, 2 and 3 (b) 1 and 2
(c) 1 and 3 (d) 2 and 3

[ESE-2001]

4.2 When temperature of a conductor is approaching zero Kelvin, the mean free path of the free electrons in the conductor is proportional to

(a) T (b) T^3
(c) $(1/T)^{1/3}$ (d) $1/T^3$

[ESE-2001]

4.3 In a conductor of size 20 mm × 10 mm × 10 mm, the wavelength of the slowest moving free electron is

(a) 5 mm (b) 10 mm
(c) 20 mm (d) 40 mm

[ESE-2001]

4.4 The conductivity of a superconducting material on being subject to critical magnetic field changes to

(a) normal state
(b) unstable state
(c) temperature-independent state
(d) temperature-dependent state

[ESE-2001]

4.5 Which of the following has the greatest mobility?

(a) Positive ion (b) Negative ion
(c) Electron (d) Hole

[ESE-2002]

4.6 Which of the following insulating materials has the least affinity for moisture?

(a) Cotton (b) Paper
(c) Asbestos (d) Mica

[ESE-2002]

4.7 **Assertion (A):** Glazing is done on ceramic insulators to make the surface smooth and non-absorbent.

Reason (R): Moisture from the atmosphere can collect on the surface discontinuities on a ceramic and result in electrical breakdown.

(a) Both A and R are true and R is the correct explanation of A
(b) Both A and R are true but R is not a correct explanation of A
(c) A is true but R is false
(d) A is false but R is true

[ESE-2003]

4.8 According to Wiedemann-Franz law the ratio of thermal conductivity to electrical conductivity of a conductor is

(a) Independent of temperature
(b) Directly proportional to temperature
(c) Inversely proportional to temperature
(d) inversely proportional to square of temperature

[ESE-2003]

4.9 The average drift velocity V_d of electrons in a metal is related to electric field E and collision time T as

(a) $V_d = Q_e ET/m_e$ (b) $V_d = m_e Q_e T$
(c) $V_d = m_e Q_e \mathrm{T}/2E$ (d) $V_d = Q_e ET/2m_e$

[ESE-2003]

4.10 Which one of the following statements is correct? Superconductors are popularly used for

(a) generating very strong magnetic field
(b) reducing i^2R losses
(c) generating electrostatic field
(d) generating regions free from magnetic field

[ESE-2004]

4.11 On which of the following factors does the value of critical current density in a superconductor depend ?
(a) Temperature
(b) Applied magnetic field
(c) Temperature and applied magnetic field
(d) Silsbee's rule

[ESE-2004]

4.12 Match **List-I** (Materials) with **List-II** (Equation/rule) and select the correct answer using the codes given below:

List-I	**List-II**
A. Dielectric material	1. Debye equation
B. Ferromagnetic material	2. Curie Weiss law
C. Conductors	3. Matthiessen's rule
D. Superconductor	4. Meissner effect

Codes:

	A	B	C	D
(a)	3	4	1	2
(b)	1	4	3	2
(c)	3	2	1	4
(d)	1	2	3	4

[ESE-2004]

4.13 **Assertion (A):** An electric or mechanical input makes the flux jump from one superconductor to another, generating large direct currents.
Reason (R): The electric resistivity of superconductors depends upon the magnetic field.
(a) Both A and R are true and R is the correct explanation of A
(b) Both A and R are true but R is not a correct explanation of A
(c) A is true but R is false
(d) A is false but R is true

[ESE-2004]

4.14 **Assertion (A):** Cryotron is a switch constructed out of superconducting materials.
Reason (R): Switching action is characteristic of superconductors only.
(a) Both A and R are true and R is the correct explanation of A
(b) Both A and R are true but R is not a correct explanation of A
(c) A is true but R is false
(d) A is false but R is true

[ESE-2004]

4.15 The magnetic field at which a super-conductor remains in its superconducting state at a temperature less than the transition temperature is
(a) Zero
(b) Greater than the critical field corresponding to the given temperature
(c) Less than the critical field corresponding to the given temperature
(d) Equal to the critical field corresponding to the transition temperature

[ESE-2005]

4.16 In metals, resistivity is composed of two parts: one part is characteristic of the particular substance. The other part is due to
(a) Applied voltage
(b) Crystal imperfections
(c) Applied magnetic field
(d) Supplied thermal energy

[ESE-2005]

4.17 Match **List-I** (Insulating Material) with **List-II** (Application) and select the correct answer using the codes given below:

List-I
A. Steatite
B. Rutile (Titanium dioxide)
C. Barium titanate
D. Teflon

List-II
1. Ceramic capacitor
2. Piezoelectric application
3. Insulating materials for machine windings
4. High frequency insulator

Codes:

	A	B	C	D
(a)	4	3	2	1
(b)	2	1	4	3
(c)	4	1	2	3
(d)	2	3	4	1

[ESE-2005]

4.18 Match **List-I** (Insulator) with **List-II** (Application) and select the correct answer using the codes given below:

	List-I		List-II
A.	Mica	1.	Bushings
B.	Polystyrene	2.	Electric wires and cables
C.	Porcelain	3.	Low voltage capacitors
D.	Silicon rubber	4.	Iron
		5.	Radio cabinets

Codes:

	A	B	C	D
(a)	4	5	1	2
(b)	2	1	3	4
(c)	4	1	3	2
(d)	2	5	1	4

[ESE-2005]

4.19 Superconductivity is destroyed
(a) At high temperature
(b) At high magnetic field
(c) In presence of magnetic impurities
(d) In all the above cases

[ESE-2005]

4.20 What is the magnetic susceptibility χ of an ideal superconductor?
(a) 1 (b) –1
(c) 0 (d) Infinite

[ESE-2006]

4.21 **Assertion (A)** : Superconductivity of a superconducting material can be destroyed by application of an external magnetic field.
Reason (R) : If the applied magnetic field is greater than the critical magnetic field at a given temperature which is more than the transition temperature, super-conductivity can be destroyed.
(a) Both A and R are true and R is the correct explanation of A
(b) Both A and R are true but R is NOT the correct explanation of A
(c) A is true but R is false
(d) A is false but R is true

[ESE-2006]

4.22 Consider the following:
Applications of low permittivity ceramics include which of the following?
1. Suspension insulators for high voltage lines.
2. Superconductors.
3. Pin insulators for low voltage lines.

Select the correct answer using the code given below:
(a) 1 and 2 (b) 1 and 3
(c) 2 and 3 (d) 1, 2 and 3

[ESE-2007]

4.23 Match **List-I** (Material) with **List-II** (Properties) and select the correct answer using the code given below the lists:

List-I
A. Superconductor at very low temperatures
B. Ferric chloride
C. Diamond
D. Manganese oxide

List-II
1. Susceptibility +ve
2. Very small –ve susceptibility
3. Very high –ve susceptibility
4. Susceptibility inversely proportional to $T-\theta$

(Here T is temperature of the material in Kelvin, θ is a constant for a material)

Codes:

	A	B	C	D
(a)	2	4	3	1
(b)	3	1	2	4
(c)	2	1	3	4
(d)	3	4	2	1

[ESE-2007]

4.24 Consider the following statements in connection with superconductors:
1. Transition temperature of a super-conductor can be reduced by the application of a magnetic field.
2. Transition from the superconducting to normal state under the influence of a magnetic field is irreversible.

Which of these statements is/are correct?
(a) 1 only (b) 2 only
(c) Both 1 and 2 (d) Neither 1 nor 2

[ESE-2007]

4.25 Which one of the following is the correct statement?
Superconducting metal in super-conducting state has relative permeability of
(a) more than one (b) one
(c) zero (d) negative

[ESE-2007]

4.26 Two materials having temperature coefficient of 0.004 and 0.0004 respectively are joined in series. The overall temperature coefficient is approximately

(a) 0.08 (b) 0.04
(c) 0.001 (d) 0.0001

[ESE-2008]

4.27 Match **List-I** with **List-II** and select the correct answer using the code given below the lists

	List-I		List-II
A.	Enamel covering	1.	Laminations
B.	Insulation	2.	Wires
C.	Fibrous Materials	3.	Machines
D.	Empire cloth	4.	Transformers

Codes:

	A	B	C	D
(a)	2	4	1	3
(b)	3	4	1	2
(c)	2	1	4	3
(d)	3	1	4	2

[ESE-2010]

4.28 Consider the following statements:
Characteristics of a good insulating material are

1. Should give uniform electric and thermal properties.
2. High permittivity.
3. Low dissipation factor.
4. Low insulating resistance.

Which of the above statements are correct?

(a) 1 and 4 only (b) 2 and 4 only
(c) 1 and 3 only (d) 1, 2, 3 and 4

[ESE-2010]

4.29 Consider the following statements about superconductors :

1. The temperature at which the conductor becomes a super conductor is called transition temperature.
2. Superconductors repel magnetic flux lines.
3. All superconductors are paramagnetic materials.
4. Superconductors become normal when placed in a magnetic field of certain critical value.

Which of the above statements are correct?

(a) 1 and 2 only (b) 2 and 4 only
(c) 1, 2 and 4 only (d) 1, 2, 3 and 4

[ESE-2010]

4.30 Match **List-I** with **List-II** and select the correct answer using the code given below the Lists:

	List-I		List-II
A.	Precision work	1.	Graphite
B.	Rheostat	2.	Nichrome
C.	Heating devices	3.	Constantan
D.	Brushes	4.	Magnesium

Codes:

	A	B	C	D
(a)	4	3	2	1
(b)	1	3	2	4
(c)	4	2	3	1
(d)	1	2	3	4

[ESE-2010]

4.31 Match **List-I** with **List-II** and select the correct answer using the code given below the Lists:

List-I

A. High conductivity materials
B. High resistivity materials
C. Metal for lightly loaded contacts
D. Materials for bimetallic strip

List-II

1. Tungsten, Carbon
2. Platinum, Molybednum
3. Aluminium, Copper
4. Iron, Nickel, Constantan

Codes:

	A	B	C	D
(a)	3	1	2	4
(b)	4	1	2	3
(c)	3	2	1	4
(d)	4	2	1	3

[ESE-2010]

4.32 Consider the following statements with regard to manufacture of a standard resistor :

1. The material should be of high resistivity and low temperature coefficient.
2. Resistors are shielded against magnetic field.
3. Nickel-chromium is best suited for resistance of high value.

Which of these statements is/are correct?

(a) 1 and 2 only (b) 1 only
(c) 1 and 3 only (d) 3 only

[ESE-2010]

4.33 According to free electron theory, electron in a metal are subjected to
(a) Constant potential
(b) Sinusoidal potential
(c) Square wave potential
(d) Non-periodic potential [ESE-2011]

4.34 Materials in superconducting state have the property of
(a) Absorbing magnetic field
(b) Repelling magnetic field
(c) Absorbing electric field
(d) Repelling electric field [ESE-2011]

4.35 A superconductor may be used for generating
(a) Voltage (b) Pressure
(c) Temperature (d) Magnetic field
[ESE-2011]

4.36 As temperature falls below the transition temperature, the value of critical magnetic field of a superconductor
(a) Remains unchanged
(b) Increases
(c) Decreases
(d) First increases, reaches a peak and then decreases
[ESE-2011]

4.37 The energy gap of a superconductor
(a) is independent of temperature
(b) increases with temperature
(c) is maximum at a critical temperature
(d) is minimum at a critical temperature
[ESE-2011]

4.38 Which of the following properties is NOT correct for a superconductor in its superconducting stage?
(a) Its resistivity is zero.
(b) Magnetic flux density inside the conductor is zero.
(c) Its relative permeability is unity.
(d) Its magnetic susceptibility is negative.
[ESE-2011]

4.39 The commercial thermopiles are formed by
(a) Series of Si-Al thermocouples in an IC by doping Ai layers on p-type Si on n-type Si epitaxial layers
(b) Series of Cu-W thermocouples strips
(c) Piezoelectric material strips piled together
(d) Series of bismuth-telluride couples
[ESE-2011]

4.40 Consider a metallic conductor of length L m and a constant cross-sectional area of Am^2. A steady potential difference of V volts is applied between the ends of the conductor. The drift velocity of the free electrons is γ m/s. The mobility of the electrons is defined as $u = \frac{\gamma}{V/L}$ m. If the number of free electrons per m^3 is N and each carries a charge of Coulomb, the resistance R of the conductor is
(a) $\frac{L}{NAue}$ ohm (b) $\frac{LN}{Aue}$ ohm
(c) $\frac{Lu}{NAe}$ ohm (d) $\frac{Lue}{NA}$ ohm
[ESE-2012]

4.41 The resistivity of hard drawn copper at 20°C is 1.9×10^{-6} Ω cm. The resistivity of annealed copper compared to hard drawn copper is
(a) lesser (b) slightly larger
(c) same (d) much larger
[ESE-2016]

4.42 The electrical resistivity of many metals and alloys drops suddenly to zero when they are cooled to a low temperature (i.e., nearly equal to liquid helium temperature). Such materials (metals and alloys) are known as
(a) piezoelectric materials
(b) diamagnetic materials
(c) superconductors
(d) high-energy hard magnetic materials
[ESE-2016]

4.43 The conductivity of insulating materials (a very small value) is called as
(a) residual conductivity
(b) dielectric conductivity
(c) ionic conductivity
(d) bipolar conductivity
[ESE-2016]

4.44 When a very small amount of higher conducting metal is added to a conductor, its conductivity will

(a) increases
(b) decreases
(c) remain the same
(d) increases or decreases depending on the impurity

[ESE-2016]

4.45 An electrically balanced atom has 30 protons in its nucleus and 2 electrons in its outermost shell. The material made of such atom is
(a) a conductor
(b) an insulator
(c) a semiconductor
(d) a superconductor

[ESE-2016]

4.46 **Statement (I):** The electrical conductivity of a solid solution alloy drops off rapidly with increased alloy content.
Statement (II): A solid solution has a less regular structure than a pure metal.
(a) Both Statement (I) and Statement (II) are individually true and Statement (II) is the correct explanation of Statement (I).
(b) Both Statement (I) and Statement (II) are individually true but Statement (II) is not the correct explanation of Statement (I)
(c) Statement (I) is true but Statement (II) is false
(d) Statement (I) is false but Statement (II) is true

[ESE-2016]

4.47 The temperature coefficient of resistance of a wire is 0.0008/°C. If the resistance of the wire is 8 ohm at 0°C, what is the resistance at 100°C?
(a) 8.64 Ohm (b) 8.08 Ohm
(c) 7.92 Ohm (d) 7.20 Ohm

[ESE-2007]

4.48 Consider the following characterizing parameters of a material:
1. Magnetic permeability
2. Electron relaxation time
3. Electron effective mass
4. Energy band gap

In case of metals, increase in one of the above parameter decreases its conductivity, while increase in another increases the conductivity. These are respectively.
(a) 1 and 3 (b) 3 and 2
(c) 4 and 3 (d) 1 and 2

[ESE-2013]

4.49 **Statement (I)** : Cold-working of a conductor material results in a decrease in the electrical conductivity of the metal because the localized strains interfere with the electron movement.
Statement (II) : Subsequent annealing of the conductor material restores the electrical conductivity by establishing greater regularity in the crystal lattice.
(a) Both Statement (I) and Statement (II) are individually true and Statement (II) is the correct explanation of Statement (I).
(b) Both Statement (I) and Statement (II) are individually true but Statement (II) is not the correct explanation of Statement (I)
(c) Statement (I) is true but Statement (II) is false
(d) Statement (I) is false but Statement (II) is true

[ESE-2015]

4.50 For which one of the following materials, is the Hall coefficient zero?
(a) Metal
(b) Insulator
(c) Intrinsic semiconductor
(d) Alloy [ESE-2006]

4.51 The following data are obtained by measurement on gold:
Density = 19.32 gm/cc
Resistivity = 2.42 μΩ-cm
Atomic weight = 197.2
The mobility of electrons in gold is
(a) 4.39×10^{-3} m²/ V-sec
(b) 4.39×10^{-2} m²/ V-sec
(c) 4.39 m²/ V-sec
(d) 4.39×10^{3} m²/ V-sec [ESE-2010]

4.52 The Fermi energy E_F of a metal is proportional to (n is the number of free electrons per unit volume of the metal) as
(a) n^2 (b) $n^{1/2}$
(c) $n^{2/3}$ (d) $n^{3/2}$ [ESE-2011]

4.53 The temperature coefficient of a resistance of a doped semiconductor is
(a) Always positive
(b) Always negative
(c) Zero
(d) Positive or negative depending on the level of doping

[ESE-2011]

4.54 What happens to the resistance of a conductor if its length is increased three times and diameter is halved?
(a) Resistance remains the same
(b) Resistance is increased 3 times
(c) Resistance is increased 6 times
(d) Resistance is increased 12 times
[ESE-2015]

4.55 If the diameter of a copper wire is increased by two times keeping its terminal voltage same, then the drift velocity will
(a) become twice
(b) become half
(c) become four times
(d) remain unchanged
[ESE-2015]

4.56 Which effect is the converse of Peltier effect?
(a) Seebeck effect (b) Thomson effect
(c) Hall effect (d) Joule effect
[ESE-2015]

4.57 A copper conductor has a resistance of 15.5 Ω at 0°C. What is the percentage conductivity at 16°C (at nearest unit value) assuming the temperature coefficient of copper at 0.00428 per °C at 0°C?
(a) 54% (b) 68%
(c) 94% (d) 98% [ESE-2017]

4.58 At temperature above a limiting value, the energy of lattice vibrations, in a conductor, increases linearly with temperature so that resistivity increases linearly with temperature. In this region, this limiting value of temperature is called
(a) Bernoulli temperature
(b) Curie temperature
(c) Debye temperature
(d) Neel temperature [ESE-2017]

4.59 What is the correct sequence of the following materials in ascending order of their resistivity?
1. Iron 2. Silver
3. Constantan 4. Mica
5. Aluminium
Select the correct answer using the codes given below:
(a) 2, 5, 1, 3 and 4 (b) 4, 5, 3, 1 and 2
(c) 2, 3, 1, 5 and 4 (d) 4, 5, 1, 3 and 2
[ESE-2017]

4.60 An electric kettle is marked 500 W, 230 V and is found to take 15 minutes to bring 1 kg of water at 15°C to 100°C. If the specific heat of water is 4200 J/kg/°C, the heat efficiency of the kettle will be
(a) 87.3% (b) 83.6%
(c) 79.3% (d) 75.6%
[ESE-2019]

4.61 The standard resistor is a coil of wire of some alloy having the properties of
Which of the above statements are correct?
(a) Low electrical resistivity and high temperature coefficient of resistance.
(b) High electrical resistivity and high temperature coefficient of resistance.
(c) Low electrical resistivity and low temperature coefficient of resistance.
(d) High electrical resistivity and low temperature coefficient of resistance.
[ESE-2019]

4.62 Light is capable of transferring electrons to the free-state inside a material thus increasing the electrical conductivity of the material. When the energy imparted to the electrons is quite large, the latter may be emitted from the material into the surrounding medium. This phenomenon is known as
(a) Photoemissive effect
(b) Photovoltaic effect
(c) Photoconductivity effect
(d) Photo absorptive effect [ESE-2020]

4.63 Which of the following statements is/are correct?
1. Conductor contain a large number of electrons in the conduction band at room temperature. No energy gaps exist and the valence and conduction bands overlap.
2. A semiconductor is a material in which the energy gap is so large that practically no electron can be given enough energy to jump this gap.
3. An insulator is a solid with an energy gap small enough for electrons to cross rather easily from the valence band to the conduction band.
(a) 1 only (b) 2 only
(c) 3 only (d) 1, 2 and 3
[ESE-2020]

4.64 A heater element is made of nichrome wire having resistivity equal to 100×10^{-8} Ωm. The diameter of the wire is 0.4 mm. The length of the wire required to get a resistance of 40 Ω and 1000 W is
(a) 5.0 metres (b) 4.5 metres
(c) 5.5 metres (d) 4.0 metres
[ESE-2021]

4.65 What is the temperature coefficient of resistance of material used in a resistor if the resistance at 25°C is 50 Ω and at 70°C is 57.2 Ω?
(a) $\frac{1}{168.5}$ (b) $\frac{1}{378.5}$
(c) $\frac{1}{463.5}$ (d) $\frac{1}{287.5}$ [ESE-2021]

4.66 What is the supply frequency for the case hardening of shaft having specific resistivity of 5×10^{-5} Ω-cm and the relative permeability equal to 1 for depth of heating 2.5 mm?
(a) 20.24 kHz (b) 26.24 kHz
(c) 15.62 kHz (d) 32.15 kHz
[ESE-2021]

4.67 What is the approximate diameter of a copper wire of length 100 m, if it is to be used as winding material in a transformer such that the resistance of the whole winding is 2 Ω? (Take resistivity of copper as 1.7×10^{-8} Ω-m)
(a) 0.25 mm (b) 1.05 mm
(c) 2.25 mm (d) 3.05 mm
[ESE-2022]

4.68 Which one of the following quantities is a material parameter?
(a) Peridynamic lattice
(b) Electrical elongation
(c) Thermal elongation
(d) Electrical conductivity [ESE-2023]

4.69 The electrical conductivity depends on
(a) the volume concentration of the holes and on their mobility.
(b) the volume concentration of the electrons of conduction and on their mobility.
(c) concentration of both electrons and protons.
(d) concentration of electrons only.
[ESE-2023]

4.70 What is the order of resistivity of conductive material?
(a) 10^{-6} Ω-m to 10^{-8} Ω-m
(b) 10^{6} Ω-m to 10^{8} Ω-m
(c) 10^{-10} Ω-m to 10^{8} Ω-m
(d) 10^{-9} Ω-m to 10^{12} Ω-m
[ESE-2024]

■■■■

Answers Conductive Materials

4.1 (d)	4.2 (d)	4.3 (c)	4.4 (a)	4.5 (c)	4.6 (c)	4.7 (a)	4.8 (b)	4.9 (a)
4.10 (a)	4.11 (d)	4.12 (d)	4.13 (d)	4.14 (c)	4.15 (c)	4.16 (b)	4.17 (c)	4.18 (d)
4.19 (d)	4.20 (b)	4.21 (c)	4.22 (b)	4.23 (c)	4.24 (c)	4.25 (c)	4.26 (c)	4.27 (b)
4.28 (c)	4.29 (c)	4.30 (a)	4.31 (a)	4.32 (c)	4.33 (a)	4.34 (b)	4.35 (d)	4.36 (b)
4.37 (d)	4.38 (c)	4.39 (a)	4.40 (a)	4.41 (a)	4.42 (c)	4.43 (c)	4.44 (b)	4.45 (a)
4.46 (a)	4.47 (a)	4.48 (b)	4.49 (b)	4.50 (a)	4.51 (a)	4.52 (c)	4.53 (d)	4.54 (d)
4.55 (d)	4.56 (a)	4.57 (c)	4.58 (c)	4.59 (a)	4.60 (c)	4.61 (d)	4.62 (a)	4.63 (a)
4.64 (a)	4.65 (d)	4.66 (a)	4.67 (b)	4.68 (d)	4.69 (b)	4.70 (a)		

Explanations Conductive Materials

4.1 (d)

For precision resistor: It's resistance should not change with temperature i.e. zero or low temperature coefficient of resistivity.

Resistance should be stable i.e. it should not change with ambient condition.

4.2 (d)

At low temperatures mean free path ($\lambda = v\tau_c$)

$\tau_c \rightarrow$ time between two collision (successive)

$v \rightarrow$ Velocity of electron (e^-)

and it is $\lambda \propto \frac{1}{T^3}$ at low temperatures.

4.3 (c)

$$\lambda \propto \frac{1}{\text{Energy of the electron}};$$

slow moving electron has low energy.

4.4 (a)

The transition temperature of a superconductor can be reduced by the application of a magnetic field. Suppose a superconductor has a temperature $T < T_C$. If a magnetic field H is applied, the material remains superconducting until a critical field H_C is reached such that for $H > H_C$, the material is in the normal state.

4.5 (c)

Positive ion and negative ion have lowest mobility almost immobile and mobility of e^- is greater than that of hole.

4.6 (c)

Asbestos shows least affinity for moisture.

4.7 (a)

Both are correct and R is also the reason for A. Glazing is always done to make surface smooth and absorbent because if surface will not be smooth moisture will be collected from atmosphere on surface discontinuities and results in electrical breakdown.

4.8 (b)

$$\frac{K}{\sigma} = \frac{\pi^2 K_B}{3e^2} \cdot T = LT$$

where, $K \rightarrow$ thermal conductivity

$L \rightarrow$ Lorentz number

$\sigma \rightarrow$ electrical conductivity

$K_B \rightarrow$ Boltzmann constant

4.9 (a)

Drift velocity in metals is given by

$$v_d = \frac{qE}{m}\tau_c$$

$q \rightarrow$ Charge on e^-

$E \rightarrow$ Electric field applied

$m \rightarrow$ Mass of e^-

$\tau_c \rightarrow$ Avg. collision time

4.10 (a)

Super conductors due to their property of perfect diamagnetism used for generating very strong magnetic field.

4.11 (d)

Superconductivity in a long circular wire of radius r may be destroyed when the current I exceeds the value I_C, which at the surface of the wire would produce the critical magnetic field H_C.

Hence, the critical current $I_C = 2\pi r H_C$, this is referred to as Silsbee's rule. This rule prevents the use of superconductors as coils for the production of strong magnetic fields.

4.12 (d)

Ionic polar solids (dielectric material) follow debye equation which is generalization of CM equation.

$$\Pi = \frac{NA\alpha_e}{3\epsilon_0} = \frac{\epsilon_r - 1}{\epsilon_{r2} + 2} \cdot \frac{M}{\rho}$$

where, $N \rightarrow$ Number dipoles per unit volume

$A \rightarrow$ Avogadro Number

$\alpha_e \rightarrow$ electronic polarizability

$M \rightarrow$ Molecular weight

$\rho \rightarrow$ density (kg/m³)

Ferromagnetic material follow curie – weiss law above a temperature called as curie temperature.

$$\boxed{\chi_m = \frac{C}{T - \theta}}$$

Conductors follows Matthiessen's rule superconductors shows meissner effect i.e. expulsion of magnetic flux under super-conducting phase.

4.13 (d)
If a magnetic field H is applied, the material remains superconducting until a critical field H_C is reached such that for $H > H_C$, the material is in the normal state.

4.14 (c)
R is false here because switching action is not only the characteristic of super-conducting material but also exhibited by conductors and semiconductors (like tunnel diode, PIN diode).

4.15 (c)
If a magnetic field H is applied, the material remains superconducting until a critical field H_C is reached such that for $H > H_C$, the material is in the normal state. at a certain temperature as temperature ↑ require H will be less.

4.16 (b)
In metal resistivity is due to particular substance i.e. thermal component and other is residual component due to impurities and crystal imperfections.

4.17 (c)
Steatite → High frequency insulator Rutile [Titanium dioxide – TiO_2] Ceramic capacitor.
Barium Titanate → Piezoelectric material above curie temperature.
Teflon → Insulating materials for machine windings.

4.18 (d)
Mica → Electric wires and cables.
Polystyrene → Very high dielectric strength used in bushings.
Silicon rubber → Mixed with iron to increase resistivity.

4.20 (b)
$\chi = \mu_r - 1$, for superconductor $\mu_r = 0$ which is referred as perfect diamagnetism.

4.21 (c)
Temperature should be less than the transition temperature.

4.22 (b)
Low permittivity ceramics cannot be super conductor.

4.23 (c)
Superconductor is diamagnetic so very small negative susceptibility. Ferric chloride is paramagnetic so susceptibility is positive. MnO is Antiferro so follow curie weiss law after Neel temperature.

4.24 (c)
If applied magnetic field is more than critical field transition temperature will be reduced.
Super conductor are used as a Magnetic Switch which can be on & off by application of temperature or Magnetic field.

4.25 (c)
Exhibits perfect diamagnetism and field exclusion law (Meissner effect)

$$\chi_m = \mu_r - 1 \Rightarrow \chi_m = -1$$

$$\therefore \quad \mu_r = 0$$

4.26 (c)
Here temperature coefficient means resistance temperature coefficient so on connection in series overall coefficient will be average of two or lies within two. Series connected $\alpha_1 < \alpha_{eq} < \alpha_2$.
Nearest average value of two given coefficient is 0.001.

4.28 (c)
Good insulating material possess high insulating resistance. So, (c) is correct.

4.29 (c)

$$B = \mu_0(H + M)$$

Super conductor repel the magnetic flux

so, $$B = 0$$

$$0 = \mu_0(H + M)$$

$$M = -H$$

$$M = \chi_m H$$

so, $$\chi_m = -1$$

Condition for perfect diamagnetic

4.33 (a)
According to free electron theory electrons in metal are subjected to constant potential.

4.34 (b)
In superconductivity phase material show perfect diamagnetism i.e. $\mu_r = 0 \Rightarrow X_m = -1$ since χ_m is –ve so they repel magnetic field.

4.35 (d)
A superconductor may be used for generation of magnetic field.

4.36 (b)

Critical field required for destroying superconductivity is zero above transition temperature as above transition temperature material is in normal state. And below transition temperature critical field required is

$$H_C = H_0\left[1-\left(\frac{T}{T_C}\right)^2\right]$$

H_0 critical field required at $T = 0°K$ (max.)
as T fall below T_C, H_C increases.

4.37 (d)

Energy gap of superconductor is minimum at critical temperature as it inters in infinite conductive stage.

4.38 (c)

For a superconductor, material should exhibit perfect diamagnetism i.e. $\mu_r = 0$.

4.39 (a)

The commercial thermopiles are formed by series of Si-A*l* thermocouples in an IC by doping A*l* layers on p-type Si and on n-type Si epitaxial layers.

4.40 (a)

We know,

$$i = neAv_d$$

v_d is drift velocity,

$$i = neA\cdot u\cdot\frac{V}{L}$$

$$\Rightarrow \quad \frac{V}{i} = \frac{L}{neAu} = \text{Resistance}$$

4.41 (a)

The mechanical treatment such as cold working produces localized strain in the material which results in the increase in resistivity of material. A hard drawn copper wire thus has lower conductivity (i.e., higher resistivity) than annealed copper.

4.42 (c)

Superconductors are the materials whose resistivity becomes very small or zero below a critical temperature.

4.44 (b)

When we add a metal with another metal this makes an alloy. Conductivity of alloy is less than the conductivity of metal as alloy has less regular structure than metal.

4.45 (a)

There are 30 protons in electrically balanced atom of material, hence atomic number of material is 30. So this material is Z_n (having atomic number 30, and 2 electrons in its outermost shell), which is a conductor.

4.46 (a)

Alloy has less regular structure than a metal because of which conductivity of alloy decreases with increase in alloy content and resistivity of alloy increases in comparison to metal.

4.47 (a)

$$R = R_0(1+\alpha\,\Delta t)$$
$$= 8(1 + 0.0008 \times 100) = 8.64\ \Omega$$

4.48 (b)

$$\sigma = \frac{ne^2\tau_c}{m}$$

4.49 (b)

Cold-working and annealing are different process.

4.51 (a)

$$\text{Molar volume} = \frac{\text{Atomic weight}}{\text{Density}}$$
$$= \frac{197.2}{19.32}\times 10^{-6}\ \text{m}^3 = 10.207\times 10^{-6}\ \text{m}^3$$
$$n = \frac{\text{Avagodro's number}}{\text{Molar volume}}$$
$$= \frac{6.022\times 10^{23}}{10.207\times 10^{-6}} = 5.9\times 20^{28}$$
$$\mu = \frac{\sigma}{ne} = \frac{1}{\rho ne}$$
$$= \frac{1}{2.42\times 10^{-8}\times 1.6\times 10^{-19}\times 5.9\times 10^{28}}$$
$$= 4.39\times 10^{-3}\ \text{m}^2/\text{V-sec}$$

4.52 (c)

Fermi energy E_F of a metal is proportional to

$$E_F \propto n^{2/3}$$

Note:
$$E_F = \frac{(nC)^2}{8mC^2}\left(\frac{3}{\pi}\right)^{2/3} n^{2/3}$$
$$E_F \propto n^{2/3}$$

4.53 (d)

For a doped SC, below room temperature conductivity increases with temperature (i.e. resistance decreases with temperature) and above room temp. Conductivity decreases and resistance increases, so temperature coefficient can be +ve or –ve.

4.54 (d)

Resistance, $R = \rho\frac{l}{A} = \rho\cdot\frac{l}{\frac{\pi}{4}d^2}$

$\because$ $l' = 3l$ and $d' = d/2$

New resistance,

$$R' = \rho\cdot\frac{l'}{\frac{\pi}{4}\cdot d'^2} = \rho\cdot\frac{3l}{\frac{\pi}{4}\times\frac{d^2}{4}}$$

$$= \rho\cdot\frac{12\cdot l}{\frac{\pi}{4}\cdot d^2} = 12R$$

4.55 (d)

$$V_d = \mu E$$

μ dependent upon temperature

So, V_d dependent upon Electric field.

4.56 (a)

Seebeck effect is converse of peltier effect.

4.57 (c)

Resistivity = $\rho = \rho_0(1 + \alpha t)$

$$\text{Resistivity, } \sigma = \frac{1}{\rho} = \frac{\frac{1}{\rho_0(1+\alpha t)}}{\frac{1}{\rho_0}}\times 100$$

$$= \frac{100}{1+\alpha t}\% = \frac{100}{1.06848} = 93.59\%$$

Approximately equal to 94%.

4.58 (c)

As the temperature of a metal is raised, the electrical resistivity increases. At temperature above Debye temperature, the energy of the lattice vibrations increases linearly with T and consequently the resistivity increases linearly with T in this region.

4.59 (a)

Silver: 1.59×10^{-8} (Ω-m)

Aluminium: 2.65×10^{-8} (Ω-m)

Iron: 9.71×10^{-8} (Ω-m)

Constantan: 49×10^{-8} (Ω-m)

Mica: 1×10^{13} (Ω-m)

4.60 (c)

$$\text{Heat} = mC\Delta t$$

m - Mass

C - Specific heat

Δt - Change in temperature

$$\text{Output heat} = 1\times 10^3 \times 4.2 \times (100 - 15) = 397000 \text{ Joule}$$

$$\text{Output power, } P = \frac{\text{Output heat}}{\text{Time}} = \frac{357000}{15\times 60} = 396.67 \text{ Watt}$$

Input power = 500 Watt

$$\text{Heat efficiency} = \frac{396.67}{500}\times 100 = 79.3\%$$

4.62 (a)

Photoemission of electrons occurs when the energy of light incident upon a surface is used to enable electrons to leave that surface.

4.63 (a)

Insulators have very large energy gap while semiconductors have small energy gap.

4.64 (a)

$$R = \frac{\rho l}{A} \Rightarrow l = \frac{RA}{\rho}$$

$\rho = 100 \times 10^{-8}$ Ω-m

$d = 0.4 \times 10^{-3}$ m

$R = 40\ \Omega$

$$l = \frac{RA}{\rho} = \frac{R\times\left(\frac{\pi d^2}{4}\right)}{\rho}$$

$$= \frac{40\times 1.256\times 10^{-7}}{100\times 10^{-8}}$$

$l = 5.024$ m

4.65 (d)

$$R_T = R_o(1 + \alpha T)$$

$$\frac{R_{25}}{R_{70}} = \frac{50}{57.2} = \frac{R_o(1+\alpha\times 25)}{R_o(1+\alpha\times 70)}$$

$\Rightarrow$ $50 + \alpha\times 50\times 70 = 57.2 + \alpha\times 25\times 572$

$\Rightarrow$ $\alpha(3500 - 1430) = 7.2$

$$\alpha = \frac{1}{287.5}$$

4.66 (a)

Given resistivity,

$\rho = 5\times 10^{-5}\times 10^{-2} = 5\times 10^{-7}$ Ω-m

$\mu_r = 1$; Skin depth, $\delta = 2.5\times 10^{-3}$ m

$$\delta = \frac{1}{\sqrt{\pi f\mu\sigma}}$$

Squaring both sides in above equation,

$$\pi f \mu \sigma = \frac{1}{\delta^2}$$

As conductivity, $\sigma = \frac{1}{\rho}$

or $$f = \frac{\rho}{\delta^2 \cdot \pi \times \mu_0}$$

$$= \frac{5\times10^{-7}}{(2.5\times10^{-3})^2\times\pi\times4\pi\times10^{-7}}$$

$$= 20.26 \text{ kHz}$$

4.67 (b)

$$l = 100 \text{ m}$$

$$R = 2\ \Omega$$

$$\rho = 1.7\times10^{-8}\ \Omega\text{-m}$$

$$R = \frac{\rho l}{A}$$

$$\Rightarrow \quad A = \frac{\rho l}{A} = \frac{1.7\times10^{-8}\times100}{2}$$

$$\Rightarrow \quad A = \frac{\pi d^2}{4} = \frac{1.7\times10^{-6}}{2}$$

$$d = \sqrt{\frac{2\times1.7\times10^{-6}}{\pi}}$$

$$\Rightarrow \quad d = 1.04\times10^{-3} \text{ m}$$

4.68 (d)

Electrical conductivity is a material parameter.

4.69 (b)

$$\sigma = \eta q \mu_n$$

4.70 (a)

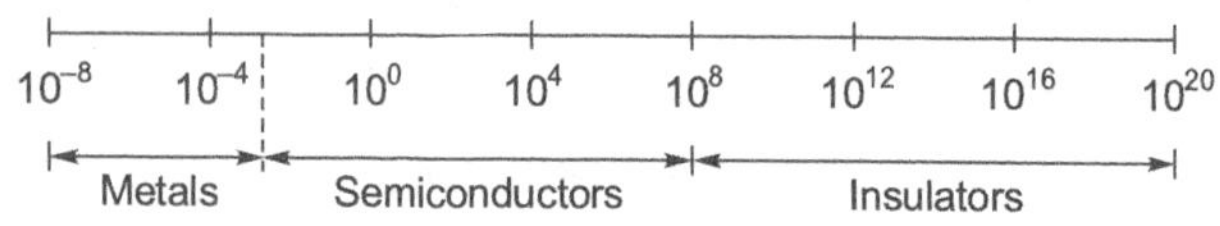

Fig. : Resistivity scale for materials found in nature

5 Semiconductor Materials

5.1 For a semiconductor, which one of the following statements is NOT correct ?

(a) When an electron and a hole recombine, energy must be liberated.
(b) Electrons in the conduction band can acquire a net acceleration from a field because there are empty energy levels available.
(c) An electron in the valence band cannot be accelerated by the field unless there are empty energy levels available.
(d) Holes cannot be accelerated by the field unless there are empty energy levels available.

[ESE-2001]

5.2 Consider the following statement:
Extrinsic semiconductors show high electrical conductivity because the impurities are

1. of high conductivity
2. highly mobile
3. highly charged

Which of these statements are **NOT** correct ?

(a) 1 and 2 (b) 2 and 3
(c) 1 and 3 (d) 1, 2 and 3

[ESE-2001]

5.3 Consider the following energy-momentum (E-k) curves labelled as 1, 2, 3 and 4 of different semiconductors:

1. 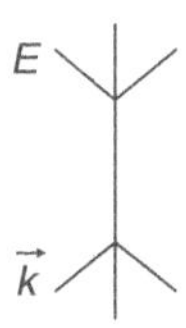

2. 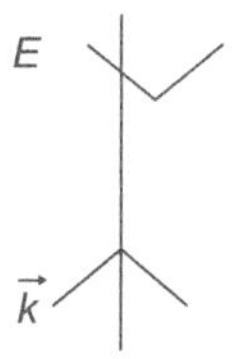

3. 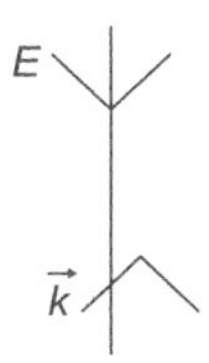

4. 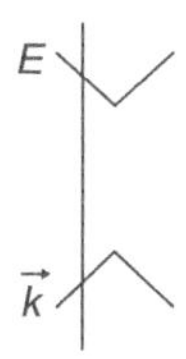

Sets of direct and indirect gap curves are respectively.

(a) (1, 2) and (3, 4) (b) (1, 4) and (2, 3)
(c) (1, 3) and (2, 4) (d) (2, 3) and (1, 4)

[ESE-2001]

5.4 Match **List-I (p-n junction devices)** with **List-II (Application)** and select the correct answer:

List-I	**List-II**
A. Zener Diode	**1.** Fast switching circuits
B. Pin Diode	**2.** Microwave switches
C. Tunnel Diode	**3.** Local oscillators for radars
D. Varactor Diode	**4.** Frequency converters
	5. Voltage regulators

Codes:

	A	B	C	D
(a)	5	2	1	4
(b)	1	2	5	4
(c)	5	3	1	2
(d)	1	3	5	2

[ESE-2001]

5.5 **Assertion (A):** A doped semiconductor will behave as a perfect insulator at zero Kelvin but its electrical conductivity will rise if the temperature is slowly increased above zero Kelvin.
Reason (R): The rise in electrical conductivity is mainly, due to increased ionization as the temperature is raised above zero Kelvin.

(a) Both A and R are true and R is the correct explanation of A
(b) Both A and R are true but R is NOT the correct explanation of A
(c) A is true but R is false
(d) A is false but R is true

[ESE-2002]

5.6 **Assertion (A):** Energy released when an electron jumps from a higher to a lower level is usually in the form of photons.

Reason (R): Energy released when an electron jumps from a higher to a lower level is absorbed by the nucleus.

(a) Both A and R are true and R is the correct explanation of A
(b) Both A and R are true but R is NOT the correct explanation of A
(c) A is true but R is false
(d) A is false but R is true

[ESE-2003]

5.7 With an increase in temperature, the Fermi level in an intrinsic semiconductor

(a) moves closer to the conduction band edge
(b) moves closer to the valence band edge
(c) moves into the conduction band
(d) remains at the centre of the forbidden gap

[ESE-2003]

5.8 The development of barrier potential in the depletion zone of a P-N junction is consequent to

(a) Diffusion of majority carriers across junction
(b) Drift of minority carriers across junction
(c) Generation of minority carriers due to thermal energy
(d) Initial flow of conduction current

[ESE-2003]

5.9 The current flow in a certain PN junction at room temperature 300 K is 2×10^{-7} A when a large reverse bias voltage is applied. The current flowing when a forward bias of 0.1 V is applied will be

(a) $2\times10^{-7}\exp\left[\frac{1.6\times10^{-19}\times0.1}{1.38\times10^{-23}\times300}\right]$

(b) $2\times10^{-7}\times\left[\frac{-1.6\times10^{-19}\times0.1}{1.38\times10^{-23}\times300}\right]$

(c) $2\times10^{-7}\exp\left[\frac{-1.38\times10^{-23}\times300}{1.6\times10^{-19}\times0.1}\right]$

(d) $2\times10^{-7}\times\left[\frac{1.38\times10^{-23}\times300}{1.6\times10^{-19}\times0.1}\right]$

[ESE-2003]

5.10 On which of the following factors does the electrical conductivity of a semiconductor depend ?

1. Carrier concentration
2. Carrier mobility
3. Sign of the carrier

Select the correct answer using the codes given below:

(a) 1 and 2 (b) 1 and 3
(c) 2 and 3 (d) 1, 2 and 3

[ESE-2003]

5.11 An intrinsic semiconductor at a temperature of absolute zero behaves like an insulator because of

(a) Non-availability of free electrons
(b) Non-recombination of electrons with holes
(c) Low drift velocity of free electrons
(d) Low (almost zero) electron energy

[ESE-2003]

5.12 Consider the following statements:

E_f^n and E_f^p are the energies of the Fermi levels on the n and p sides of p-n junction diode, respectively. They will vary with applied bias as follows:

1. $E_f^n = E_f^p$ with no bias applied
2. E_f^n increases and E_f^p decreases with forward bias.
3. E_f^n decreases and E_f^p increases with reverse bias.
4. E_f^n decreases and E_f^p increases with forward bias.
5. E_f^n increases and E_f^p decreases with reverse bias.

Select the correct answer using the codes given below:

(a) 1, 4 and 5 (b) 2 and 3
(c) 4 and 5 (d) 1, 2 and 3

[ESE-2003]

5.13 The forbidden energy gap in silicon at 300 K is

(a) 1.41 eV (b) 1.1 eV
(c) 0.785 eV (d) 0.72 eV

[ESE-2003]

5.14 In which one of the following ways can the Hall voltage across an impurity semiconductor crystal be increased?
(a) By increasing the thickness of the crystal
(b) By increasing the concentration of impurity atoms in the crystal
(c) By increasing the width of the crystal
(d) By increasing the current flowing through the crystal

[ESE-2004]

5.15 Which one of the following statements is correct ? The Hall coefficient of an intrinsic semiconductor is
(a) positive (b) negative
(c) zero (d) infinite

[ESE-2004]

5.16 **Assertion (A):** An intrinsic semi-conductor has low electrical conductivity. Hence, it is normally not used in the device fabrication without doping.
Reason (R): Its electrical resistivity can be increased by raising its temperature.
(a) Both A and R are true and R is the correct explanation of A
(b) Both A and R are true but R is NOT the correct explanation of A
(c) A is true but R is false
(d) A is false but R is true

[ESE-2004]

5.17 Which one of the following statement is correct? If the Fermi level lies midway between the conduction and valence bands, then the semiconductor is
(a) intrinsic (b) extrinsic
(c) p-type (d) n-type

[ESE-2004]

5.18 The electrical conductivity of a semiconductor increases with increase in temperature because
(a) the mobility of the carriers increases
(b) the carrier concentration increases
(c) both carrier concentration and mobility increase
(d) thermal energy of electrons increases

[ESE-2005]

5.19 Fermi level in a p-type semiconductor lies close to
(a) the top of the valence band
(b) the bottom of the valence band
(c) the top of the conduction band
(d) the bottom of the conduction band

[ESE-2005]

5.20 Hall Effect can be used
(a) to find type of semiconductor (whether p or n type)
(b) to find carrier concentration
(c) to measure conductivity
(d) all of the above

[ESE-2005]

5.21 Width of energy bands depends on which of the following?
(a) Temperature
(b) Pressure
(c) Relative freedom of electrons in the crystal
(d) Mass of atom in the material

[ESE-2006]

5.22 Which one of the following statements is ***not*** true for a p-type semiconductor?
(a) Conduction is by the movement of holes in the valence band
(b) Holes constitute the majority charge carriers
(c) Fermi level lies closer to the conduction band
(d) It is formed by adding an acceptor impurity to germanium

[ESE-2006]

5.23 Principle of Hall effect is used in the construction of which one of the following?
(a) Ammeter (b) Voltmeter
(c) Galvanometer (d) Gaussmeter

[ESE-2006]

5.24 **Assertion (A):** A p-type GaAs sample can be converted to a semi-insulating substrate by doping it with chromium.
Reason (R): Chromium increases the band gap of GaAs.
(a) Both A and R are true and R is the correct explanation of A

(b) Both A and R are true but R is NOT the correct explanation of A
(c) A is true but R is false
(d) A is false but R is true

[ESE-2006]

5.25 In an intrinsic semiconductor, the number of electrons is equal to the number of holes at which temperature?

(a) 0 K (b) 0°C
(c) High temperature (d) All temperatures

[ESE-2006]

5.26 The values of radiative and non-radiative lifetime of minority carriers in a semi-conductor are 50 ns and 100 ns, respectively. What is the effective lifetime?

(a) 12.2 ns (b) 150 ns
(c) 33.3 ns (d) 75 ns

[ESE-2006]

5.27 III-V alloy semiconductor crystallizes in what form?

(a) Simple cubic structure
(b) Body-centered cubic structure
(c) Zinc blende structure
(d) Wurtzite structure

[ESE-2006]

5.28 The measurement of which one of the following will reveal the sign of the charge carriers?

(a) Conductivity (b) Mobility
(c) Hall coefficient (d) Diffusion constant

[ESE-2007]

5.29 Consider the following statements:

Assertion (A): At absolute zero degree Kelvin temperature the semiconductor materials behave as insulators.

Reason (R): At absolute zero degree Kelvin temperature, there is no energy available to generate the current carriers in the semiconductor.

Of these statements:

(a) Both A and R are true and R is the correct explanation of A
(b) Both A and R are true but R is not the correct explanation of A
(c) A is true but R is false
(d) A is false but R is true

[ESE-2007]

5.30 Which one of the following is the Fermi function $f(E)$?

(a) $\frac{1}{1+e^{(E-E_F)/(kT)}}$ (b) $\frac{1}{1-e^{(E-E_F)/(kT)}}$

(c) $\frac{1}{1+e^{(E_F-E)/(kT)}}$ (d) $\frac{1}{1-e^{(E_F-E)/(kT)}}$

[ESE-2007]

5.31 What is the output voltage produced across the crystal by Hall effect, due to?

(a) Drop across the crystal due to current passing through it
(b) Movement of charge carriers towards one end
(c) Induced voltage by the applied magnetic field
(d) Induced voltage by the applied electric field

[ESE-2007]

5.32 Tunnel diode is commonly used as which one of the following?

(a) Logic memory storage device
(b) FM modulator
(c) Voltage regulator
(d) AM modulator

[ESE-2007]

5.33 Consider the following statements about p-n junctions:

1. p-n junction behaves as a capacitor when forward biased.
2. p-n junction has p and n type semiconductors with depletion layer in between.
3. p-n junction has a wider depletion layer as compared to a Zener diode.

Which of these statements are correct?

(a) 1, 2 and 3 (b) 1 and 2
(c) 2 and 3 (d) 1 and 3

[ESE-2008]

5.34 Which one of the following is the correct statement?

When a p-n junction is reverse-biased, then

(a) holes and electrons move away from the junction
(b) depletion region decreases
(c) barrier breaks down
(d) minority carriers are not affected

[ESE-2008]

5.35 Consider the following statements about semiconductors:

1. The forbidden energy gap in semiconductors lies between the valence band and conduction band.
2. The forbidden energy gap in germanium at 0 K is about 0.72 eV.
3. Semiconductors have negative temperature coefficient of resistivity.

Which of these statements are correct?

(a) 1, 2 and 3 (b) 1 and 2
(c) 2 and 3 (d) 1 and 3

[ESE-2008]

5.36 Fermi level in a p-type semiconductor lies close to which one of the following?

(a) The top of the valence band
(b) The bottom of the valence band
(c) The top of the conduction band
(d) The bottom of the conduction band

[ESE-2008]

5.37 Which of the following is/are primary standard(s) for voltage maintained by National Standards Laboratories?

1. Zener diode with 1.5 V output at 20°C.
2. Nickel cadmium rechargeable cell of 1.25 V output at 20°C.

Select the correct answer using the code given below:

(a) 1 only (b) 2 only
(c) Both 1 and 2 (d) Neither 1 nor 2

[ESE-2008]

5.38 Which one of the following is the correct statement? The type of majority charge carriers in a semiconductor can be found by

(a) Hall effect
(b) Piezoelectric effect
(c) Photoelectric effect
(d) Meissner effect [ESE-2008]

5.39 For which one of the following materials, is the Hall coefficient zero?

(a) Insulator
(b) Intrinsic semiconductor
(c) Metal
(d) Non-metal

[ESE-2009]

5.40 Consider the following statements:

In a Hall effect experiment, the sign of Hall voltage will change if

1. Direction of applied field is changed.
2. Direction of applied magnetic field is changed.
3. Direction of both applied electric and magnetic fields are changed.
4. Direction of current is changed.

Which of these statements is/are correct?

(a) 1, 2 and 3 (b) 3 only
(c) 1, 2 and 4 (d) 3 and 4

[ESE-2009]

5.41 In semiconductor strain gauges, what happens when a tensile strain is applied?

(a) Resistance increases in N-type of materials
(b) Resistance increases in P-type of materials
(c) Resistance increases in both P and N-type of materials
(d) Resistance decreases in both P and N-type of materials

[ESE-2009]

5.42 For intrinsic GaAs, the room-temperature electrical conductivity is 10^{-6} (ohm-m)$^{-1}$, the electron and hole mobilities are, respectively, 0.85 and 0.04 m^2/V-s. What is the intrinsic carrier concentration n_i at the room temperature?

(a) 10^{21} m^{-3} (b) 10^{-20} m^{-3}
(c) $7.0 \times 10^{+12}$ m^{-3} (d) 7.0×10^{-20} m^{-3}

[ESE-2009]

5.43 Consider the following statements:

In an n-type semiconductor

1. Fermi level lies below the donor level at room temperature (T).
2. Fermi level lies above the donor level as $T \rightarrow 0$.
3. Fermi level lies in valence band.
4. Fermi level remains invariant with temperature.

Which of these statements is/are correct?

(a) 1 only (b) 1 and 2 only
(c) 2, 3 and 4 (d) 1, 2 and 3

[ESE-2009]

5.44 For photoconductors with equal electron and hole mobilities and perfect ohmic contacts at the ends, an increase in intensity of optical illumination results in

(a) a change in open circuit voltage
(b) a change in short circuit current
(c) decrease in resistance
(d) increase in resistance

[ESE-2010]

5.45 A semiconductor has a band gap of 2 eV. The wavelength of radiation emitted from the semiconductor when electrons and holes recombine is

(a) 625 nm (b) 625 μm
(c) 625 mm (d) 625 cm

[ESE-2010]

5.46 Match **List-I** with **List-II** and select the correct answer using the code given below the Lists:

List-I
A. Metals
B. Semiconductors
C. Insulators
D. Ferroelectric crystals

List-II
1. Are in spontaneously polarize state
2. Finite forbidden gap
3. Smaller forbidden gap
4. Partially filled bands

Codes:

	A	B	C	D
(a)	4	3	2	1
(b)	1	3	2	4
(c)	4	2	3	1
(d)	1	2	3	4

[ESE-2010]

5.47 A magnetic field $\bar{B}$ of $2T$ is normal to a copper strip 0.5 mm thick carrying an electron current of 40 A. If electron density is 10.0×10^{28} per cubic metre, the voltage across the strip in micro volt is

(a) 40 (b) 30
(c) 20 (d) 10

[ESE-2010]

5.48 Increase in the applied reverse voltage to a p-n junction results in increase in the

(a) Depletion width
(b) Barrier height
(c) Depletion width and barrier height
(d) Junction temperature

[ESE-2010]

5.49 **Assertion (A) :** Intrinsic semiconductors show negative Hall coefficient.

Reason (R) : The number of electrons and holes are equal in an intrinsic semiconductor.

(a) Both A and R are true and R is the correct explanation of A
(b) Both A and R are true but R is NOT the correct explanation of A
(c) A is true but R is false
(d) A is false but R is true

[ESE-2010]

5.50 The Fermi function for an electron is $f(E)$, where E is energy Then, the Fermi function for a hole is

(a) $f(E)$ (b) $1 - f(E)$
(c) $1/f(E)$ (d) $1 + f(E)$

[ESE-2010]

5.51 The measurement of Hall coefficient of a semiconductor with one type of charge carrier gives the information about

(a) sign of charge carrier
(b) density of charge carrier
(c) both sign and density of charge carrier
(d) mass of the charge carrier

[ESE-2010]

5.52 The Fermi level in an n-type semi-conductor at zero degree Kelvin lies

(a) Below the donor level
(b) Half-way between the conduction band and the donor level
(c) Half-way between the conduction band and the valence level
(d) Close to the valence band

[ESE-2011]

5.53 Controlled addition of group III element to an elemental semiconductor results in the formation of

(a) Intrinsic semiconductor
(b) n-type semiconductor
(c) p-type semiconductor
(d) Degenerate semiconductor

[ESE-2011]

5.54 If w is the width of the depletion region in a p-n junction, the transition capacitance is

(a) w (b) w^2
(c) $1/w$ (d) $1/w^2$

[ESE-2011]

5.55 The current flow in a semiconductor is due to
1. Drift current
2. Displacement current
3. Diffusion current
(a) 1, 2 and 3 (b) 1 and 2 only
(c) 1 and 3 only (d) 2 and 3 only
[ESE-2011]

5.56 The drift velocity of electron in Silicon
(a) is proportional to electric filed for all values of electric field.
(b) is independent of electric field.
(c) increases at lower values and decreases at higher values of electric field.
(d) increases linearly with electric field at low values and gradually saturates at higher values of electric field.
[ESE-2013]

5.57 Consider the following statements:
1. Fermi level in a p-type semiconductor lies close to the top of the valence bond.
2. The forbidden energy in Germanium at 0°K is exactly 0.75 eV.
3. When a p-n junction is reverse biased, then electrons and holes move away from the junction.

Which of these statements are correct?
(a) 1, 2 and 3 (b) 1 and 2 only
(c) 2 and 3 only (d) 1 and 3 only
[ESE-2013]

5.58 **Statement (I):** By measuring the Hall-effect voltage, one can determine:
(a) The strength of the field in terms of the current, or
(b) The current in terms of the field.

Statement (II): In the case of a semiconductor, the mobility of the carriers can be determined by using (b), so long as only one kind of carrier is present.
(a) Both Statement (I) and Statement (II) are individually true and Statement (II) is the correct explanation of Statement (I).
(b) Both Statement (I) and Statement (II) are individually true but Statement (II) is not the correct explanation of Statement (I).
(c) Statement (I) is true but Statement (II) is false.
(d) Statement (I) is false but Statement (II) is true.
[ESE-2013]

5.59 The electrical conductivity of semiconductor increases with increase in temperature because
(a) the carrier concentration increases
(b) the mobility of carrier increases
(c) both carrier concentration and mobility increases
(d) the band gap decreases
[ESE-2014]

5.60 The Hall Effect voltage in intrinsic silicon
(a) is positive
(b) is zero
(c) is negative
(d) changes its sign based on application of magnetic field
[ESE-2014]

5.61 Most outstanding property of indium antimonide is
(a) a very wide range gap
(b) high resistivity at room temperature
(c) high carrier mobility
(d) very low conductivity at room temperature
[ESE-2014]

5.62 **Statement (I):** A semi-conductor is not capable of sustaining the movement of free negative electrons, but capable of positive charges or holes.

Statement (II): The positive charges cannot be positrons because to liberate the latter, an energy as high as one million *eV* or more would be required.
(a) Both Statement (I) and Statement (II) are individually true and Statement (II) is the correct explanation of Statement (I).
(b) Both Statement (I) and Statement (II) are individually true but Statement (II) is not the correct explanation of Statement (I).
(c) Statement (I) is true but Statement (II) is false.
(d) Statement (I) is false but Statement (II) is true.
[ESE-2014]

5.63 Which of the following semiconducting component is used in photoconductive devices?
(a) Caesium antimode
(b) Barium oxide
(c) Lead sulphide
(d) Zinc oxide [ESE-2014]

5.64 Heat conduction in a semiconductor takes place
(a) by the mobility of the carriers
(b) due to energy gap between conduction band and valency band
(c) by the holes and thermal vibrations of atoms
(d) by the electrons and thermal vibrations of atoms

[ESE-2015]

5.65 The resistivity of intrinsic germanium at 30 °C is 0.46 Ω-m. What is the intrinsic carrier density at 30 °C if the electron mobility is 0.38 m^2/V-s and the hole mobility is 0.18 m^2/V-s?
(a) $9.2 \times 10^5/m^3$ (b) $2.77 \times 10^3/m^3$
(c) $2.43 \times 10^{19}/m^3$ (d) $8.9 \times 10^{12}/m^3$

[ESE-2015]

5.66 For the intrinsic gallium-arsenide, the conductivity at room temperature 25 °C is 10^{-6} (ohm-m)$^{-1}$, the electron and hole mobilities are 0.85 m^2/V-s and 0.04 m^2/V-s respectively. What is the intrinsic carrier concentration at the room temperature?
(a) $7.0 \times 10^{12}\ m^{-3}$ (b) $7.0 \times 10^{-12}\ m^{-3}$
(c) $7.0 \times 10^{-12}\ m^3$ (d) $7.0 \times 10^{12}\ m^3$

[ESE-2015]

5.67 If μ_I and μ_p represent the impurity scattering and photon scattering limited values of mobility of a semiconductor, the overall mobility would be
(a) $\mu_I + \mu_p$ (b) $\frac{\mu_I + \mu_p}{2}$
(c) $\sqrt{\mu_I \mu_p}$ (d) $\frac{\mu_I \mu_p}{\mu_I + \mu_p}$

[ESE-2015]

5.68 An electrical breakdown of a *p-n* junction occurs if
(a) forward voltage increases up to the rating
(b) reverse voltage increases beyond the rating
(c) forward voltage decreases below the rating
(d) reverse voltage decreases below the rating

[ESE-2015]

5.69 Which one of the following properties is not observed in the carbon nanotubes?
(a) High stiffness and strengths
(b) Low densities
(c) Unusual electrical property
(d) Non-ductile

[ESE-2015]

5.70 The impact ionization phenomenon in semiconductor may be viewed as the reverse process of
(a) radiative recombination
(b) Auger recombination
(c) surface recombination
(d) Shockley-Read-Hall recombination

[ESE-2015]

5.71 The number of electrons excited into the conduction band from valence band (with ΔE = forbidden energy gap and k = Boltzmann's constant) is proportional to
(a) $\exp\left(\frac{\Delta E}{kT}\right)$ (b) $\exp\left(\frac{2\Delta E}{kT}\right)$
(c) $\exp\left(-\frac{\Delta E}{kT}\right)$ (d) $\exp\left(-\frac{2\Delta E}{kT}\right)$

[ESE-2016]

5.72 An intrinsic semiconductor has equal number of electrons and holes in it. This is due to
(a) doping (b) free electrons
(c) thermal energy (d) valence electrons

[ESE-2016]

5.73 The temperature coefficient of resistance of a doped semiconductor is
(a) always positive
(b) always negative
(c) zero
(d) positive or negative depending upon the level of doping

[ESE-2016]

5.74 In the slice processing of an integrated circuit
(a) components are formed in the areas where silicon dioxide remains.
(b) components are formed in the areas where silicon dioxide has been removed.
(c) the diffusing elements diffuse through silicon dioxide.
(d) only on diffusion process is used.

[ESE-2016]

5.75 The resistivity of intrinsic germanium at 30° is 0.46 Ω-m. What is the intrinsic carrier density n_i at 30°C, taking the electron mobility m_n as 0.38 m^2/V-s and hole mobility μ_p as 0.18 m^2/V-s?
(a) $2.4 \times 10^{19}/m^3$ (b) $4.2 \times 10^{19}/m^3$
(c) $2.4 \times 10^{10}/m^3$ (d) $4.2 \times 10^{10}/m^3$

[ESE-2017]

5.76 For intrinsic gallium arsenide conductivity at room temperature is 10^{-6} $(\Omega\text{-m})^{-1}$, the electron and hole mobilities are, respectively 0.85 and 0.04 $m^2/V\text{-s}$. The intrinsic carrier concentration n at room temperature is

(a) 7.0×10^{12} m^{-3} (b) 0.7×10^{12} m^{-3}
(c) 7.0×10^{-12} m^{-3} (d) 0.7×10^{-12} m^{-3}

[ESE-2017]

5.77 Consider the following statements regarding electrical conductivity σ:

1. It increases with temperature in semi-conductors.
2. It increases with temperature is exponential.
3. It increases in metals and their alloys, linearly with temperature.

Which of the above statements are correct?

(a) 1 and 2 only (b) 1 and 3 only
(c) 2 and 3 only (d) 1, 2 and 3

[ESE-2017]

5.78 Photoconductivity is a characteristic of semiconductors. When light falls on certain semiconductors, it

(a) sets free electrons from some of the atoms, increasing the conductivity.
(b) ejects electrons into space.
(c) establishes a potential difference creating a source of emf.
(d) produces heat raising the temperature.

[ESE-2017]

5.79 **Statement (I) :** The conductivity of an intrinsic semi-conductor increases exponentially with temperature.

Statement (II) : As the temperature rises, more and more covalent bonds are broken resulting in more electron hole pairs.

[ESE-2018]

5.80 Fick's laws refer to

(a) Finding whether a semiconductor in *n* or *p* type
(b) Diffusion
(c) Crystal imperfections
(d) Electric breakdown

[ESE-2019]

5.81 A magnetic field applied perpendicular to the direction of motion of a charged particle exerts a force on the particle perpendicular to both the magnetic field and the direction of motion of the particle. This phenomenon results in

(a) Flux effect
(b) Hall effect
(c) Magnetic field effect
(d) Field effect

[ESE-2019]

5.82 Which of the following statements regarding superconducting materials are correct, when a large number of metals becomes superconducting below a temperature?

1. The resistivity ρ of the superconductor is zero.
2. The magnetic flux density B vanishes through the substance.
3. Ferromagnetic and Antiferromagnetic metals are good examples of superconducting materials.

(a) 1, 2 and 3 (b) 1 and 3 only
(c) 1 and 2 only (d) 2 and 3 only

[ESE-2020]

5.83 Consider the following statements related to Hall effect :

1. Hall effect is used to determine whether a semiconductor is *n*-type or *p*-type.
2. To find the carrier concentration.
3. Measuring the conductivity (σ), the mobility (μ) can be calculated.
4. The Hall effect has been incorporated into a magnetic field meter.

Which of the above statements are correct?

(a) 1 and 2 only (b) 2, 3 and 4 only
(c) 1, 3 and 4 only (d) 1, 2, 3 and 4

[ESE-2021]

5.84 Electro-optic devices that emit radiations of different colours are called

(a) Opto-couplings
(b) Photoresistors
(c) Light emitting diodes
(d) Void generators

[ESE-2023]

5.85 Which one of the following corresponds to the variation of the rotation frequency of the electron when the external field is applied?

(a) Langevin frequency
(b) Lorentz frequency
(c) Thomson frequency
(d) Larmor frequency [ESE-2023]

5.86 Which one of the following agents is an intrinsic agent that influences conductivity in electro insulating materials?
(a) Humidity
(b) The nature of the dielectric
(c) Temperature
(d) Pressure [ESE-2023]

5.87 A semiconductor indium phosphide material behaves as an insulator at
(a) high temperature
(b) medium temperature
(c) normal room temperature
(d) very low temperature [ESE-2023]

5.88 Which one of the following statements is correct related to extrinsic and degenerate semi-conductors?
(a) Lightly and moderately doped semi-conductor is referred to as extrinsic and when it is doped to such high levels that it acts more like a conductor than a semiconductor, it is referred to as degenerate.
(b) Heavily doped semiconductor is referred to as extrinsic and when it is doped to such high levels that it acts more like a conductor than a semiconductor, it is referred to as degenerate.
c) Moderately doped semiconductor is referred to as extrinsic and when it is doped to such low levels that it acts more like a conductor than a semiconductor, it is referred to as degenerate.
(d) Lightly doped semiconductor is referred to as extrinsic and when it is doped to such low levels that it acts more like a conductor than a semiconductor, it is referred to as degenerate.
[ESE-2024]

■■■■

Answers Semiconductor Materials

5.1 (d)	5.2 (a)	5.3 (b)	5.4 (a)	5.5 (d)	5.6 (b)	5.7 (d)	5.8 (a)	5.9 (a)
5.10 (a)	5.11 (a)	5.12 (d)	5.13 (b)	5.14 (d)	5.15 (d)	5.16 (c)	5.17 (a)	5.18 (b)
5.19 (a)	5.20 (d)	5.21 (a)	5.22 (c)	5.23 (d)	5.24 (*)	5.25 (d)	5.26 (c)	5.27 (c)
5.28 (c)	5.29 (a)	5.30 (a)	5.31 (b)	5.32 (b)	5.33 (a)	5.34 (d)	5.35 (d)	5.36 (a)
5.37 (d)	5.38 (a)	5.39 (c)	5.40 (c)	5.41 (b)	5.42 (c)	5.43 (b)	5.44 (c)	5.45 (a)
5.46 (a)	5.47 (d)	5.48 (c)	5.49 (b)	5.50 (b)	5.51 (c)	5.52 (c)	5.53 (c)	5.54 (c)
5.55 (c)	5.56 (d)	5.57 (d)	5.58 (a)	5.59 (c)	5.60 (c)	5.61 (c)	5.62 (d)	5.63 (c)
5.64 (d)	5.65 (c)	5.66 (a)	5.67 (d)	5.68 (b)	5.69 (d)	5.70 (b)	5.71 (c)	5.72 (c)
5.73 (d)	5.74 (b)	5.75 (a)	5.76. (a)	5.77 (a)	5.78 (a)	5.79. (a)	5.80 (b)	5.81. (b)
5.82 (c)	5.83 (d)	5.84 (c)	5.85 (d)	5.86 (b)	5.87 (d)	5.88 (a)		

Explanations Semiconductor Materials

5.1 **(d)**
Electrons are accelerated not holes which are not a physical charge carriers.

5.2 **(a)**
Impurities are not of high conductivity as P(pentavalent atom) n-type impurity is not a very good conductor.
Impurity atoms are also not exhibit high mobility normally immobile.

5.3 **(b)**
In case of 1 and 4 momentum of e^- and hole is same and in opposite direction i.e. direct band gap curve.

In case of 2 and 3 momentum are not equal and not in opposite direction so indirect band gap material curve.

5.4 (a)
Zener diode → used as voltage regulator in power supply circuits.
Pin diode → used as microwave switches
Tunnel diode → fastest switch (switching time is 10^{-12} sec).
Varactor diode → Value of capacitor changes with applied reverse biase voltage so $f = \frac{1}{\sqrt{LC}}$ also change so corresponding input voltage we get corresponding output frequency.

5.5 (d)
Doped SC behave like insulator at 0 K but not perfect. An undoped semiconductor will behave as a perfect insulator at 0 K.

5.6 (b)
Both A and R are correct but R is not explanation, both are individually correct.
A → In case of direct band gap material when e^- falls from conduction band to valence, energy is released in the form of photon.
R → In case of indirect band gap material when e^- falls from conduction band to valence band, energy is released in the form heat.

5.7 (d)
Fermi level in an intrinsic semiconductor is independent of temperature.

5.8 (a)
Initially when junction is formed majority carriers started diffusing, e^{-s} from n to p and holes from p to n, when they cross the junction, due to recombination after sometime near the junction mobile carriers concentration becomes zero and a field is created from n to p that does not allow the diffusion of majority carriers and a depletion region is formed near the junction.

5.9 (a)

$$I_0 = 2 \times 10^{-7} \text{ A} \qquad \left(\therefore I \simeq I_0 e^{\frac{V}{\eta V_T}}\right)$$

$\eta = 1$ for Si,

$$V_T = \frac{kT}{q} = \frac{1.38 \times 10^{-23} \times 300}{1.6 \times 10^{-19}}$$

$$\therefore \quad I = 2 \times 10^{-7} \exp \frac{1.6 \times 10^{-19} \times 0.1}{1.38 \times 10^{-23} \times 300}$$

5.10 (a)

$$\sigma = ne(\mu_e + \mu_h)$$

i.e. $\sigma \propto n(\mu_e + \mu_h)$

5.11 (a)
Because of non-availability e^{-s} because thermal energy is not sufficient to excite the e^{-s} from valence band to conduction band.

5.12 (d)
band diagrams p-n junction diode with no bias, F.B. and R.B. is as

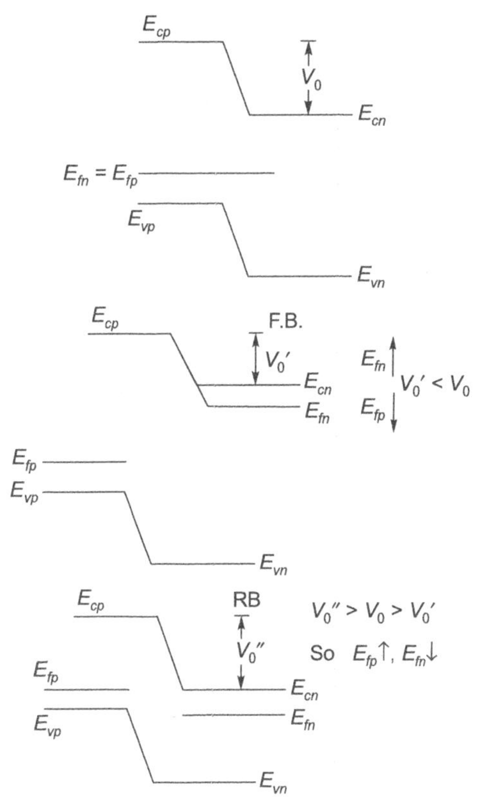

5.13 (b)
For Si, $E_G = 1.21 - 3.60 \times 10^{-4} T$
at 300°K, $E_G = 1.21 - 0.108 = 1.1 \text{ eV}$

5.14 (d)

$$V_H = -\frac{B_Z \cdot I_x}{e \cdot n_e \cdot d}$$

where d is thickness of the crystal. So $V_H \propto I_x$

5.15 (d)

$$R_H = \frac{1}{e \cdot n_h} = -\frac{1}{e \cdot n_e}$$

In intrinsic semiconductor carrier concentration $n_e = 0$. So, $R_H = \infty$.

5.16 (c)
By raising temperature electrical conductivity would increase and resistivity would decrease.

5.17 (a)
In an intrinsic semiconductor, all energy levels below a certain value W_F are filled, all those above W_F being empty; W_F is called the fermi level of the electrons.

5.18 (b)
With increase in temperature more covalent bonds are ruptured so more free electrons are available for conduction. Mobility of the carriers decreases with increase in temperature.

Note: $n_i \propto T^{3/2}$; $\sigma = \frac{n_i e^2 \tau}{m}$; $\sigma \propto n_i$

5.19 (a)
Fermi level in p-type SC lies closed to the top of valence i.e.. Just above the top of valence band.

5.20 (d)
Hall effect is used to find out the type of SC. by checking the polarity of induced Hall voltage V_H.
Used to find out concentration of charge carriers by knowing the Hall coefficient value R_H.
Also can find conductivity by formula $\mu = \frac{8}{3\pi}\sigma R_H$.

5.21 (a)
For example, energy gap (E_G) variation of silicon with temperature (T) is $E_G(T) = 1.21 - 3.6 \times 10^{-4}\, T$.

5.22 (c)
Fermi level lies closer to the valence band in p-type SC.

5.25 (d)
At absolute zero, silicon and germanium are insulators, simply because there are no mobile charge carriers available.
At $t = 0°C = 273$ K
At high temperature also,

$$n = p = n_i$$

5.26 (c)

$$\frac{1}{\tau} = \frac{1}{\tau_r} + \frac{1}{\tau_{nr}} = \frac{1}{100} + \frac{1}{50}$$

$$\tau = 33.3 \text{ nsec}$$

5.28 (c)

Hall coefficient $R_H = \frac{1}{\rho}$

where $\rho \rightarrow$ charge density
For $e^{-s} \rightarrow \rho = ne^-$ i.e. –ve
For holes $\rightarrow \rho = p \times 1.6 \times 10^{-19}$ i.e. +ve
So sign of R_h decide the type of charge carriers.

5.29 (a)
Both are correct and R is also the reason.
By Fermidirac function,

$$F(E) = \frac{1}{1 + (e)^{\frac{E - E_f}{KT}}} = 0 \text{ for } T = 0 \text{ K}$$

So there is no free e^{-s} in energy state so as no holes in valence band.

5.30 (a)
Fermi function i.e. probability of finding a e^- in a available energy state E is given by

$$f(E) = \frac{1}{1 + \exp\left[(E - E_F)/kT\right]}$$

$K \rightarrow$ Boltzman constant $\left(\frac{ev}{K}\right)$

$T \rightarrow$ Temp. in Kelvin
$E_F \rightarrow$ Fermi level in eV

5.31 (b)
Due to magnetic field (transverse) moving charge carriers feels a force ⊥ to direction of both magnetic and electric field and move towards one end of crystal and due to this a electric field is induced and due to this a voltage is also induced.

5.32 (b)
Tunnel diodes are used in FM modulator circuit.

5.33 (a)
PN junction also behaves as a capacitor when reverse biased but value of capacitor is less.
Zener diode is heavily doped so thin depletion layer.

5.34 (d)
When p-n junction is reverse biased majority carriers i.e. holes on p-side and e^{-s} on n-side moves away from junction but minority carrier i.e. e^{-s} on p-side and holes on n-side cross the junction but concentration of minority carrier is not affected so as reverse saturation current which only depend on minority carrier connection.

5.35 (d)
Energy gap in Germanium is 0.785 eV at 0 K but at room temperature it is 0.72 so point number 2 is wrong.

5.36 (a)
In p-type SC fermi level lie just above the top of the valence band and in n-type SC just below the lowest level of conduction bond.

5.37 (d)
Zener diode is used as secondary standard.

5.39 (c)

$$R_H = \frac{1}{ne}$$

For metal $n\uparrow$ (very high)

$\therefore \quad R_H = 0$

5.40 (c)
Hall voltage (induced voltage) is given by

$$V_H = \frac{BI}{\rho W}$$

$\rho \rightarrow$ Charge density
$W \rightarrow$ Width of sample
$B \rightarrow$ Magnetic field applied
$I \rightarrow$ Current flowing through sample
So if direction of B is changed sign of V_H will be changed and if direction of I is changed keeping B as it is the direction of V_H will be changed.

5.41 (b)
In a SC strain gauge when tensile stress is applied the Resistance increases in p-type SC while resistance decreases in N-Type SC and reverse will be happen in case of compressive stress.

5.42 (c)

$$\sigma = en_i(\mu_n + \mu_p)$$

$$10^{-6} = n_i[1{\cdot}6 \times 10^{-19}[0.85 + 0.04]]$$

$$\therefore \quad n_i = 7{\cdot}0 \times 10^{12} m^{-3}$$

5.43 (b)
In n-type SC
Fermi level lies below donor ionization level at room temp in n-type SC fermi level E_F is given by

$$E_F = E_C - KT \ln\frac{N_C}{N_D}$$

At 0 K,
$E_F = E_C$ since E_C is above donor level
So, E_F is also above donor level
$E_F \rightarrow$ depends on temperature.

5.44 (c)
When intensity of optical illumination increases conductivity of photoconductor increases.

5.45 (a)

$$\lambda(\mu m) = \frac{1.24}{E_g(eV)} = \frac{1.24}{2}\ \mu m$$

$$\lambda = 620 \text{ nm}$$

5.47 (d)

$\bar{B} = 2\,T$
$W = 0.5$ mm
$I = 40$ A
$n = 10 \times 10^{28}/m^3$

$$V_H = \frac{BI}{\rho W} = \frac{BI}{nqW}$$

$$= \frac{2 \times 40}{10^{29} \times 1.6 \times 10^{-19} \times 0.5 \times 10^{-3}}$$

$$= 10\,\mu V$$

5.48 (c)
By increasing the reverse bias of p-n junction depletion width will increase so barrier height will also increase.

5.51 (c)
Hall coefficient gives information about
1. Type of semiconductor
2. Carrier concentration

5.52 (c)
Is wrong as answer for this is fermi level coincide with conduction band E_C at 0°K for n-type semiconductor.

5.53 (c)
3rd group element like B are accepters so addition of these to SC like Si or Ge result in p-type SC.

5.54 (c)

Transition capacitance

$$C = \frac{\in \cdot A}{W}$$

W is width of depletion region.

A area (cross-sectional area) of sample or SC of which diode is made up of

So, $$C \propto \frac{1}{W}$$

5.55 (c)

In SC current flow due to drift and diffusion, displacement current flows in a dielectric or (conductor not perfect) under alternative field.

5.56 (d)

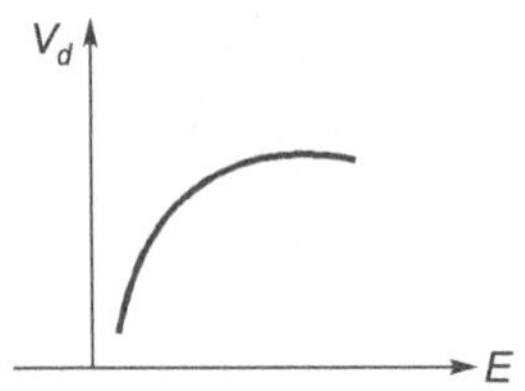

5.57 (d)

Forbidden energy in germanium at 0°K is 0.785 eV not 0.75 eV. So, option (d) is correct.

5.59 (c)

The electric conductivity of a semiconductor increases with increase in temperature because both carrier concentration and mobility increases. The expression for conductivity of any intrinsic semiconductor is

$$\sigma = \left\{e(\mu_n + \mu_p)\sqrt{A_0}\right\}T^{3/2}\exp\left(-\frac{E_{GO}}{2kT}\right)$$

where, $n = p = n_i$

$$n_i = \sqrt{A_0}\, T^{3/2}\exp\left(-\frac{E_{GO}}{2kT}\right)$$

5.60 (c)

The hall effect voltage is negative in intrinsic semiconductor since $\mu_e > \mu_h$.

5.64 (d)

Heat conduction in a semiconductor takes place by the electrons and thermal vibrations of atoms.

5.65 (c)

$$\sigma = ne(\mu_e + \mu_n)$$

$$\Rightarrow \frac{1}{0.46} = n_i \times 1.6 \times 10^{-19} \times (0.38 + 0.18)$$

$$\Rightarrow n_i = \frac{10^{19}}{0.46 \times 1.6 \times 0.56}$$

$$= 2.43 \times 10^{19}/\text{m}^3$$

5.66 (a)

$$\sigma = ne(\mu_e + \mu_n)$$

$$\Rightarrow 10^{-6} = n_i \times 1.6 \times 10^{-19} \times (0.85 + 0.04)$$

$$\Rightarrow n_i = \frac{10^{-6} \times 10^{19}}{1.6(0.85 + 0.04)}$$

$$= \frac{10^{13}}{1.6(0.85 + 0.04)} = 7.0 \times 10^{12}\ \text{m}^{-3}$$

5.67 (d)

$$\text{Overall mobility} = \frac{\mu_I \mu_P}{\mu_I + \mu_P}$$

5.68 (b)

An electrical breakdown of a p-n junction occurs if reverse voltage increases beyond the rating.

5.70 (b)

The impact ionization phenomenon in semiconductor may be viewed as the reverse process of Auger recombination.

5.71 (c)

$$n = N_c e^{-(E_C - E_F)/kT}$$

$$= N_c e^{-(E_C - E_V) + (E_F - E_V)/kT}$$

$$= N_c e^{-\Delta E/kT} . e^{(E_F - E_V)/kT}$$

$$n \propto e^{-\Delta E/kT}$$

5.72 (c)

In an intrinsic semiconductor if we increase temperature then few of the electrons in valence band acquire sufficient energy so that they overcome forbidden gap and reach into the conduction band; and a corresponding hole is generated in valence band. Hence because of the thermal energy equal number electrons and holes are generated in an intrinsic semiconductor.

5.73 (d)

For a normally doped semiconductor temperature coefficient of resistance is negative. But a heavily doped semiconductor (i.e. degenerate semiconductor) has metal like properties, hence it has positive temperature coefficient of resistance.

So the temperature coefficient of resistance for a doped semiconductor can be positive or negative depending upon level of doping.

5.74 (b)
In the slice processing of an integrated circuit the components are formed in selective areas known as "windows" where silicon dioxide has been etched.

5.75 (a)

$$P_i = 0.46\ \Omega\text{-m}$$
$$n_i = ?$$
$$\mu_n = 0.38\ \text{m}^2/\text{V-sec}$$
$$\mu_P = 0.18\ \text{m}^2/\text{V-sec}$$
$$\sigma_i = n_i(\mu_n + \mu_p)q$$
$$\sigma_i = \frac{1}{\rho_i} = \frac{1}{0.46} = 2.17(\Omega - \text{m})^{-1}$$
$$2.17 = n_i(0.38 + 0.18)1.6 \times 10^{-19}$$
$$= n_i(0.896 \times 10^{-19})$$
$$n_i = \frac{2.17}{0.896 \times 10^{-19}}$$
$$= 2.4 \times 10^{19}/\text{m}^3$$

5.76 (a)

$$\sigma_i = 10^{-6}\ (\Omega\text{-m})^{-1}$$
$$\mu_n = 0.85\ \text{m}^2/\text{v sec}$$
$$\mu_p = 0.04\ \text{m}^2/\ \text{v sec}$$
$$n_i = ?$$
$$\sigma_i = n_i(\mu_n + \mu_p)q$$
$$10^{-6} = n_i(0.85 + 0.04) \times 1.6 \times 10^{-19}$$
$$n_i = \frac{10^{-6}}{1.424 \times 10^{-19}} = \frac{1}{1.424} \times 10^{13}$$
$$= 0.702 \times 10^{13}$$
$$= 7.0 \times 10^{12}\ \text{m}^{-3}$$

5.77 (a)
Electrical conductivity σ

(i) It increase with temperature in semi-conductors (NTC).

(ii) It increase with temperature is exponential ($\sigma \propto n_i$),

$$n_i = A_0 T^{3/2} e^{-Eg_0/2kT}$$

5.78 (a)
Emission on some electrons from the atoms increases conductivity of semiconductor material.

5.79 (a)
Statement (I) is correct.
Statement (II) is correct.
Statement (II) is a reason for statement (I).

5.82 (c)
Conditions for superconductivity:
1. Resistivity should be equal to zero.
2. Perfect diamagnetism (magnetic flux inside the material should be zero).

5.83 (d)
Applications of Hall Effect : By using Hall Effect, we can determine :
1. The type of semiconductor
2. Mobility and conductivity
3. Carrier concentration
4. Magnetic flux density (using magnetofield meter)
5. Measurement of displacement
6. Measurement of power in EM waves
7. If one input signal is applied in the form of current when other input signal is applied in the form of magnetic field then the induced hall voltage is directly proportional to the product of two input signals. Hence, the hall experiment can be used as hall multiplier.

5.84 (c)
LEDs emit radiations of different colours depending upon bandgap of materials.

5.85 (d)
Larmour angular frequency is defined as the change in angular frequency of the electron due to applied flux density.

$$\omega_L = \frac{qB}{2\ m}$$

5.86 (b)
Humidity, temperature and pressure are extrinsic agents.

5.87 (d)
At 0 K (very low temperature)
Semiconductor behaves as a perfect insulator.

5.88 (a)
Extrinsic can be lightly and moderately doped. Degenerate semiconductors can be heavily doped and they behave like conductors.

■■■■

6 Superconductors

6.1 The chief deterrent to the widespread application of superconducting materials is

(a) very difficult to form, machine or cast
(b) the difficulty in attaining and maintaining extremely low temperature
(c) the poor strength-to-weight ratio
(d) the lower oxidation rate at elevated temperatures

[ESE-2015]

6.2 The evidence for the importance of electron-phonon interaction in superconductors comes from

(a) Meissner effect
(b) Josephson effect
(c) isotope effect
(d) flux quantization experiments [ESE-2015]

6.3 **Statement (I) :** A large number of metals become 'superconducting' below a temperature which is characteristic of the particular metal.

Statement (II) : Superconducting compounds and alloys should necessarily have components which are themselves superconducting.

(a) Both Statement (I) and Statement (II) are individually true and Statement (II) is the correct explanation of Statement (I).
(b) Both Statement (I) and Statement (II) are individually true but Statement (II) is not the correct explanation of Statement (I)
(c) Statement (I) is true but Statement (II) is false
(d) Statement (I) is false but Statement (II) is true

[ESE-2015]

6.4 Superconductivity in a material can be destroyed by

(a) increasing the temperature above a certain limit
(b) applying a magnetic field above a certain limit
(c) passing a current above a certain limit
(d) decreasing the temperature to a point below the critical temperature

Which of the above are correct?

(a) 1 and 2 only (b) 2 and 3 only
(c) 1, 2 and 3 only (d) 1, 2, 3 and 4

[ESE-2016]

6.5 Which of the following statements is **NOT** true?

(a) Superconductors show perfect diamagnetism.
(b) Superconductors have almost zero resistivity.
(c) The external magnetic field has no effect on superconductors.
(d) Entropy increases on going from superconducting state to normal state.

[ESE-2002]

6.6 The transition temperature of Mercury at which it becomes superconductive is

(a) 4.12° F (b) 4.12°C
(c) 4.12 K (d) 41.2 K [ESE-2002]

6.7 **Assertion (A):** Superconductors cannot be used as coils for production of strong magnetic fields.
Reason (R): Superconductivity in a wire may be destroyed if the current in the wire exceeds a critical value.

(a) Both A and R are individually true and R is the correct explanation of A
(b) Both A and R are individually true but R is not the correct explanation of A
(c) A is true but R is false
(d) A is false but R is true [ESE-2011]

6.8 It is possible to destroy the super-conductivity of a material by applying

(a) a strong magnetic field
(b) a temperature much below the transition temperature
(c) a strong electric field
(d) a pressure below that of the atmosphere

[ESE-2012]

6.9 In general, for a superconductor, which of the following statements is true?

(a) A superconductor is a perfect paramagnetic material with the magnetic susceptibility equals to positive unity.
(b) A superconductor is a perfect diamagnetic material with the magnetic susceptibility equals to negative one.
(c) A superconductor is a perfect ferromagnetic material with the magnetic susceptibility equals to positive one.
(d) A superconductor is a perfect piezoelectric material with the magnetic susceptibility equals to negative unity.

[ESE-2013]

6.10 **Statement (I):** A large number of metals become superconducting below a certain temperature which is characteristic of the particular metal.

Statement (II): Superconducting compounds and alloy must have components which are themselves superconducting.

(a) Both Statement (I) and Statement (II) are individually true and Statement (II) is the correct explanation of Statement (I).
(b) Both Statement (I) and Statement (II) are individually true but Statement (II) is not the correct explanation of Statement (I).
(c) Statement (I) is true but Statement (II) is false.
(d) Statement (I) is false but Statement (II) is true.

[ESE-2013]

6.11 Which of the following are associated with soft superconductors?

1. Silsbee's rule 2. Meissner effect
3. Faraday rotation 4. Curie-Weiss law

(a) 2, 3 and 4 (b) 1 and 3 only
(c) 1 and 3 only (d) 2 and 3 only

[ESE-2014]

6.12 Materials, whose resistivity at very low temperature plunges from a finite value to zero and remains there upon further cooling, are known as

(a) ferromagnetic materials
(b) high energy hard magnetic materials
(c) superconductors
(d) ferrimagnetic materials

[ESE-2014]

6.13 In a superconductor, if the temperature decreases below is critical temperature, the value of critical magnetic field will

(a) increases
(b) decreases
(c) not changes
(d) increases or decreases depending on the superconductor material

[ESE-2014]

6.14 Consider the following statements:

1. Superconductors exhibit normal conductivity behaviour above a transition temperature T_C.
2. Superconductors lose their super-conducting nature in an external magnetic field, provide the external magnetic field is above a critical value.
3. High T_C superconductors have T_C values in the range 1 to 10 K.

Which of the above statements are correct?

(a) 1 and 2 only (b) 1 and 3 only
(c) 2 and 3 only (d) 1, 2 and 3

[ESE-2017]

6.15 Consider the following statements:

1. Nano means 10^{-9} so that nano materials have an order of dimension higher than the size of atom and come in the form of rods, tubes, spheres or even thin sheets/films.
2. Nano materials have enhanced or changed structural property.
3. Nano elements lend themselves to mechanical processing like rolling, twisting, positioning.
4. Nano elements show important electrical, magnetic and optical characteristics that are useful in electrical industry.

Which of the above statements are correct?

(a) 1, 2 and 3 only (b) 1, 2 and 3 only
(c) 3 and 4 only (d) 1, 2 and 4 only

[ESE-2017]

6.16 The lead material works as superconductor at a temperature of T_c = 7.26 K. If the constant characteristics of the lead material at 0 K is $H_0 = 8 \times 10^5$ A/m, then what is the magnetic field in the lead at 5 K?

(a) $\frac{8\times10^5}{\pi}$ A/m (b) $\frac{4\times10^5}{4\pi}$ A/m

(c) $\frac{10^5}{2\pi}$ A/m (d) $4\pi \times 10^5$ A/m

[ESE-2021]

6.17 Which one of the following statements is not correct regarding superconductivity of material?
(a) Superconducting compounds and , alloys do not necessarily have compounds which are themselves superconducting.
(b) The metals which are very good conductors at room temperature do not exhibit superconducting properties.
(c) The metals and compounds which are superconducting are rather bad conductors at ordinary temperature.
(d) Monovalent metals, ferromagnetic and anti-ferromagnetic metals are superconducting.
[ESE-2022]

6.18 There are some metals and chemical compounds whose resistivity becomes zero when their temperature is brought near 0 K (-273 °C). Such metals or compounds are said to have attained
(a) piezoelectricity (b) superconductivity
(c) semiconductivity(d) electromagnetism
[ESE-2022]

6.19 The phenomenon where superconductivity in a material is suppressed by exceeding the maximum current the material can conduct (J_c) or the maximum magnetic field it can withstand (H_c) is called
(a) Sinter (b) Quench
(c) Stripes (d) Proximity effect
[ESE-2023]

6.20 Which one of the following is correct related to Type-1 superconductors along with the critical transition temperature (T_c)?
(a) Lead (Pb) of 4.15 K, Mercury (Hg) of 7.196 K and Aluminium (Al) of 1.175 K
(b) Lead (Pb) of 7.196 K, Mercury (Hg) of 4.15 K and Aluminium (Al) of 1.175 K
(c) Lead (Pb) of 1.175 K, Mercury (Hg) of 4.15 K and Aluminium (Al) of 7.196 K
(d) Lead (Pb) of 7.196 K, Mercury (Hg) of 1.175 K and Aluminium (Al) of 4.15 K
[ESE-2024]

6.21 Organic superconductors are composed of
(a) both an electron donor (the planar organic molecule) and an electron acceptor (a non-organic anion)
(b) an electron donor (the planar organic molecule) only
(c) an electron acceptor (a non-organic anion) only
(d) both an electron donor (a non- organic anion) and an electron acceptor (the planar organic molecule)
[ESE-2024]

■■■■

Answers Superconductors

6.1 (b) 6.2 (c) 6.3 (c) 6.4 (c) 6.5 (c) 6.6 (c) 6.7 (d) 6.8 (a) 6.9 (b)
6.10 (c) 6.11 (c) 6.12 (c) 6.13 (a) 6.14 (a) 6.15 (b) 6.16 (*) 6.17 (d) 6.18 (b)
6.19 (b) 6.20 (b) 6.21 (d)

Explanations Superconductors

6.1 (b)

The chief deterrent to the widespread application of superconducting material is the difficulty in attaining and maintaining extremely low temperature.

6.2 (c)

The evidence of the importance of electron-phonon interaction in super-conductors comes from Isotope effect.

6.3 (c)

Superconductivity mainly depends upon temperature and magnetic field.

6.4 (c)

Superconductivity can be destroyed:

1. by increasing the temperature above a critical temperature (T_C).
2. by applying magnetic field greater than the critical magnetic field (H_C).
3. by passing a current greater than the critical current (I_C) as per Silsbee rule.

6.5 (c)

The transition temperature of a super conductor can be reduced by the application of a magnetic field.

6.6 (c)

$T_C = 4.12°$ K with $H_C = 0$.

6.8 (a)

Effect of magnetic field on super-conductor.

$$H_C = H_o\left[1-\left(\frac{T}{T_c}\right)^2\right]$$

Here H_o is the critical field at absolute zero and T_C is the transition temperature. Hence it is possible to destroy super-conductivity by the application of sufficiently strong magnetic field.

6.9 (b)

$$B = \mu_0(H + M)$$

In case of superconductors

$B = 0,\quad H = -M$

$\Rightarrow \quad -XH = H$

$\Rightarrow \quad X = -L$

So, material is diamagnetic.

6.13 (a)

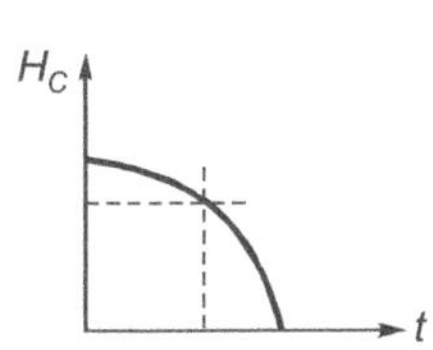

6.14 (a)

- For $T > T_C$ material is in normal state.
- For $T \leq T_C$ material is superconductor.

6.15 (b)

Only options (1), (2) and (3) are correct.

6.16 (*)

$$H_c = H_o\left[1-\left(\frac{T}{T_c}\right)^2\right]$$

$$H_c = 8\times10^5\left(1-\left(\frac{5}{7.26}\right)^2\right)$$

$$= 4.2 \times 10^5 \text{ A/m}$$

6.17 (d)

Fe is ferromagnetic but it does not exhibit superconductivity.

6.18 (b)

Superconductivity is the state of material when its resistivity becomes zero below certain critical temperature

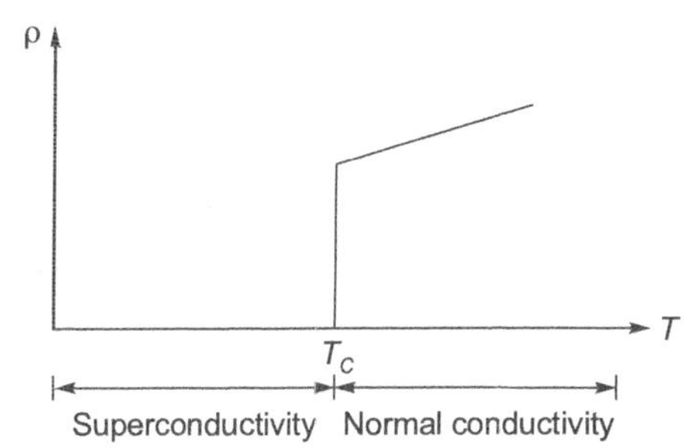

6.19 (b)

		Transition Temperature
Lead (Pb)	–	7.2 K
Mercury (Hg)	–	4.15 K
Aluminium (Al)	–	1.17 K

■■■■

7 Miscellaneous

7.1 Fermion particles obey
(a) Maxwell-Boltzmann statistics
(b) Bose-Einstein statistics
(c) Pauli's exclusion principle
(d) Heisenberg's uncertainty principle

[ESE-2015]

7.2 de Broglie wavelength associated with a material particle is
(a) inversely proportional to its energy
(b) directly proportional to its momentum
(c) directly proportional to its energy
(d) inversely proportional to its momentum

[ESE-2002]

7.3 **Assertion (A):** Lepidolite type mica is unsuitable for electric insulation.
Reason (R): Lepidolite is hard and brittle.
(a) Both A and R are true and R is the correct explanation of A
(b) Both A and R are true but R is not a correct explanation of A
(c) A is true but R is false
(d) A is false but R is true

[ESE-2003]

7.4 Lead
1. is not used to form cable sheaths.
2. is least affected by sea water.
3. has good malleable and ductile properties
4. will not alloy with many other metals
(a) 1 and 2 are correct
(b) 2 and 3 are correct
(c) 3 and 4 are correct
(d) 1 and 4 are correct

[ESE-2011]

7.5 Match **List-I** with **List-II** and select the correct answer using the code given below the lists:

List-I
A. Porcelain
B. Steatite
C. Mica
D. Rutile

List-II
1. Used for high frequency applications
2. Used in capacitors to be operated at high frequencies
3. Used for insulators
4. Releases water when heated

Codes:

	A	B	C	D
(a)	3	1	4	2
(b)	1	2	4	3
(c)	3	4	2	1
(d)	1	4	2	3

[ESE-2011]

7.6 Consider the following properties of insulators:
1. Insulation resistance
2. Dielectric breakdown strength
3. Dielectric losses
4. Permittivity

Some of these properties are for DC use and some are for AC use.
The correct combination is

	DC use	AC use
(a)	1, 2 and 3	1, 2 and 4
(b)	1 and 3	1, 2 and 3
(c)	2 and 4	1 and 2
(d)	1 and 2	1, 2, 3 and 4

[ESE-2012]

7.7 Refractive index of a slice glass can be reduced by doping it with tiny amount of
(a) GeO_2
(b) B_2O_3
(c) P_2O_5
(d) Al_2O_3

[ESE-2015]

7.8 Which one of the following statements is correct for four-point probe method of determining resistivity?
(a) The sample must be extrinsic.
(b) The current source is connected to the two inner probes.
(c) One probe point must inject minority carriers.
(d) Current flow only in a small area of the sample.
[ESE-2003]

7.9 **Assertion (A):** An insulator has a high value of resistivity, and with increasing temperature the value of its resistivity decreases exponentially.
Reason (R): With increasing temperature, the value of the energy band gap decreases.
(a) Both A and R are true and R is the correct explanation of A
(b) Both A and R are true but R is NOT the correct explanation of A
(c) A is true but R is false
(d) A is false but R is true [ESE-2005]

7.10 Phonons (Quanta of lattice vibration) obey
(a) Maxwell distribution
(b) Maxwell-Boltzman distribution
(c) Fermi-Dirac distribution
(d) Bose-Einstein distribution [ESE-2011]

7.11 Consider the following statements:
1. The critical magnetic field of a superconductor is maximum at absolute zero.
2. Transition temperature of a superconductor is sensitive to its structure.
3. The critical magnetic field of a superconductor is zero at its critical temperature.
4. Superconductors show very high conductivity below the critical temperature.

Which of the above statements are correct?
(a) 1, 2 and 3 only (b) 1, 2 and 4 only
(c) 2, 3 and 4 only (d) 1, 3 and 4 only
[ESE-2017]

7.12 Super conductivity is a material property associated with
(a) changing shape by stretching
(b) stretching without breaking
(c) a loss of thermal resistive
(d) a loss of electrical resistance
[ESE-2017]

7.13 **Statement (I):** A superconductor is a perfect diamagnetic material.
Statement (II): A superconductor is a perfect conductor.
(a) Both Statement (I) and Statement (II) are individually true and Statement (II) is the correct explanation of Statement (I).
(b) Both Statement (I) and Statement (II) are individually true but Statement (II) is not the correct explanation of Statement (I).
(c) Statement (I) is true but Statement (II) is false.
(d) Statement (I) is false but Statement (II) is true. [ESE-2017]

7.14 **Statement (I):** Superconducting compounds and alloys must have components which are themselves superconducting.
Statement (II): Metals and compounds which are superconducting are rather bad conductors at ordinary temperatures.
(a) Both Statement (I) and Statement (II) are individually true and Statement (II) is the correct explanation of Statement (I).
(b) Both Statement (I) and Statement (II) are individually true but Statement (II) is not the correct explanation of Statement (I).
(c) Statement (I) is true but Statement (II) is false.
(d) Statement (I) is false but Statement (II) is true. [ESE-2017]

7.15 **Statement (I):** An electrolytic capacitor consists of two electrodes immersed in an electrolyte with a chemical film on one of the electrodes acting as the dielectric.
Statement (II): The electrolytic capacitor may be operated with any one of the electrodes as anode positive with respect to the other. [ESE-2018]

7.16 Which material possesses the following properties?
- Shining white colour with lustre
- Soft, malleable and can be drawn into wires
- Poor in conductivity and tensile strength
- Used in making alloys with lead and copper
- Used for fuses and cable sheathing

(a) Silver (b) Tin
(c) Nickel (d) Aluminium
[ESE-2019]

7.17 Consider the following applications of the materials:

- Bismuth strontium calcium copper oxide used as a high temperature superconductor.
- Boron carbide used in helicopter and tank armour.
- Uranium oxide used as fuel in nuclear reactor.
- Bricks used for construction.

The materials used in these applications can be classified as :

(a) Ceramic (b) Constantan
(c) Manganin (d) Tantalum [ESE-2019]

7.18 With reference to nano materials, the prefix nano stands for

(a) Nano centimeter
(b) Nanometer
(c) Nano micrometer
(d) Nano millimeter [ESE-2019]

7.19 The machine used for the preparation of nano particles of alumina is

(a) Attrition mill
(b) Grinding machine
(c) Vending machine
(d) Welding machine [ESE-2019]

7.20 Which one of the following statements is NOT correct regarding tungsten, which is used as filament material?

(a) It has highest melting point amongst all metals.
(b) It can be drawn into very thin wires.
(c) It has very high tensile strength in its thinnest form.
(d) It becomes brittle at high temperature.

[ESE-2021]

7.21 Nichrome is an alloy of

(a) Manganese 2.5%, Nickel 81% to 84%, Chromium 14% to 17% and a little percentage of Iron.
(b) Manganese 2.0%, Nickel 78% to 81%, Chromium 17% to 20% and a little percentage of Iron.
(c) Manganese 1.5%, Nickel 75% to 78%, Chromium 20% to 23% and a little percentage of Iron.
(d) Manganese 0.5%, Nickel 72% to 75%, Chromium 23% to 26% and a little percentage of Iron.

[ESE-2021]

7.22 According to Bragg's law, if an electromagnetic wave is diffracted when it is passed through a series of small slits spaced a distance d apart and order of diffraction is n and if the angle between the diffracted beam and incident beam is 2θ, then the relationship between the wavelength λ and this angle of diffraction is

(a) $n\lambda = 2d \sin \theta$ (b) $\lambda = 2dn \sin \theta$
(c) $\lambda d = n \sin 2\theta$ (d) $n\lambda = d \sin 2\theta$

[ESE-2022]

7.23 A blue lamp emits light of mean wavelength of $4500\dot{A}$. The lamp is rated at 150 W and 8% of the energy appears as emitted light. How many photons are emitted by the lamp per second?

(Take $h = 6.625 \times 10^{-34}$ J-sec and $c = 3 \times 10^{8}$ m/sec)

(a) 10.55×10^{18} (b) 13.62×10^{18}
(c) 27.15×10^{18} (d) 33.25×10^{18}

[ESE-2022]

7.24 **Statement (I)** : Aluminium oxidizes quickly in normal atmospheric conditions and acquires a thin film of oxide Al_2O_3.

Statement (II) : The high melting point of aluminium oxide coating and the rapidity with which a freely exposed aluminium surface becomes oxidized, make soldering difficult through conventional means.

[ESE-2022]

7.25 **Statement (I) :** When the carrier is generated by a crystal oscillator, the frequency is fixed by the crystal.

Statement (II) : The equivalent circuit of a crystal is an L-C-R circuit with both series and parallel resonant points.

[ESE-2022]

7.26 A small region of an impure silicon crystal with dimensions 1.25×10^{-6} m $\times 10^{-3}$ m $\times 10^{-3}$ m has only the ions (with charge $+1.6 \times 10^{-19}$ C) present with a volume density of $10^{25}/m^3$. The rest of the crystal volume contains equal densities of electrons (with charge -1.6×10^{-19} C) and positive ions. The net total charge of the crystal is

(a) 3×10^{-6} C (b) 1×10^{-6} C
(c) 1.5×10^{-6} C (d) 2×10^{-6} C

[ESE-2023]

7.27 Which one of the following is the process of heating a material to just below its melting point?
(a) Sinter
(b) Squid
(c) Josephson junction
(d) Stripes

[ESE-2023]

7.28 Which of the following are the mixtures of materials used for the sliding contacts which appear between the trolley and the trolley cable in the electrical traction of high braking power?
(a) Silver, Electrolytic copper, Lead with Zinc and Aluminium
(b) Silver, Copper, Bronze with Beryllium and Cadmium
(c) Silver, Copper, Lead with Zinc and Gold
(d) Silver, Electrolytic copper, Bronze with Zinc and Gold

[ESE-2023]

7.29 What is the property of an isotropic material that must be satisfied?
(a) It has the same electric properties for the various directions of appliance of the electric field
(b) It has the different electric properties for the various directions of appliance of the electric field
(c) It has the different electric properties for the same direction of appliance of the electric field
(d) It has the same electric properties for the opposite directions of appliance of the electric field

[ESE-2024]

■■■■

Answers Miscellaneous

7.1 (c)	7.2 (d)	7.3 (a)	7.4 (b)	7.5 (c)	7.6 (d)	7.7 (b)	7.8 (b)	7.9 (d)
7.10 (d)	7.11 (d)	7.12 (d)	7.13 (b)	7.14 (d)	7.15 (c)	7.16 (b)	7.17 (a)	7.18 (b)
7.19 (a)	7.20 (d)	7.21 (c)	7.22 (a)	7.23 (c)	7.24 (b)	7.25 (b)	7.26 (d)	7.27 (a)
7.28 (a)	7.29 (a)							

Explanations Miscellaneous

7.1 (c)
Fermion particles obey Pauli's exclusion principle.

7.2 (d)
Debroglie wave length is

$$\lambda \propto \frac{1}{\text{Momentum}}$$

$$\lambda = \frac{h}{p} \quad \text{where } p = mv$$

7.3 (a)
Both are correct

A → Lepidolite type mica is unsuitable for electric isolation because it is hard and brittle for electric isolation material should be soft and ductile.

7.4 (b)
Lead has good malleable and ductile properties and less affected by sea water.

7.5 (c)
Porcelain: is a ceramic material and used for insulator.
Mica: used in capacitors to be operated at high frequencies. So, option (c) is correct.

7.6 (d)
Out of the given four properties, dielectric losses and permittivity are for AC use only. Hence option (d) is correct.

7.8 (b)
Current source is connected to the two inner probes and voltmeter is connected to outer two probes to measure voltage.

7.9 (d)
A is wrong because with increase in temperature resistivity increases and conductivity decreases.

Note: $P \propto T$ (low temperature),
$P \propto T^5$ (high temperature)

7.10 (d)
Phonons obey Bose-Einstein distribution.

7.11 (d)
Transition temperature of a superconductor is non-sensitive to its structure.

7.12 (d)
Super conductivity is a material property associated with a loss of electrical resistance.

7.13 (b)
Both statements are correct but statement (II) is not reason for statement (I).

7.20 (d)
Tungsten is brittle at room temperature and it becomes somewhat ductile at elevated temperatures.

7.22 (a)
Bragg's law $n\lambda = 2d \sin \theta$

7.23 (c)
Energy of emitted photon,

$$E = \frac{hc}{\lambda}$$

$$= \frac{6.625 \times 10^{-34} \times 3 \times 10^8}{4500 \times 10^{-10}} = 4.42 \times 10^{-19} \text{ J}$$

Energy emitted by lamp
= 8% of (150 W × 1 sec) = 8% of 150 J
($\because$ time, t = 1 sec)

$$E_{\text{total}} = \frac{8 \times 150}{100} \text{J/s} = 12 \text{ J/s}$$

$\therefore$ No. of photon emitted

$$\frac{E_{\text{total}}}{E} = \frac{12}{4.42 \times 10^{-19}}$$
$$= 2.72 \times 10^{19} = 27.2 \times 10^{18}$$

7.25 (b)
Statement I : When carrier is generated by crystal oscillator, the frequency is fixed by crystal : frequency depends on crystal. (True)
Statement II : The equivalent circuit of crystal is an LCR circuit with both series and parallel resonant points. (True).

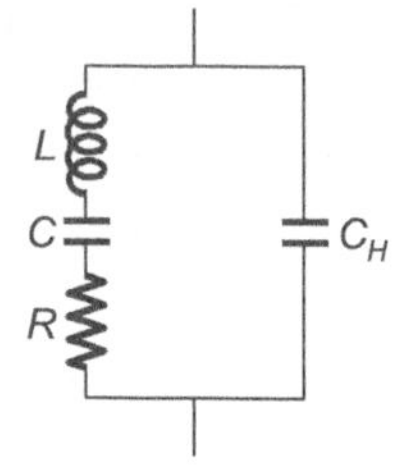

7.26 (d)
Net total charge = Volume of crystals having only ions × Volume density of ions × Charge on ions = $1.6 \times 10^{-19} \times 10^{25} \times 1.25 \times 10^{-6} \times 10^{-3} \times 10^{-3} = 2 \times 10^{-6}$ C

7.27 (a)
Sintering is the process of compacting and forming a solid mass of material by pressure or heat without melting it to the point of liquefaction.

7.28 (a)
Copper and aluminium alloy is used.

7.29 (a)
Isotrope materials exhibit same electric properties in all directions of applied electric field.

■■■■

UNIT II

Electric Circuits

Syllabus

Circuit elements, network graph, KCL, KVL, Node and Mesh analysis, ideal current and voltage sources, Thevenin's Norton's, Superposition and Maximum power transfer theorems, transient response of DC and AC networks, Sinusoidal steady state analysis, Basic filter concepts, Two-port networks, Three-phase circuits, Magnetically coupled circuits.

Contents

OOOO

1 Basics

1.1 The current in the given circuit with a dependent voltage source is

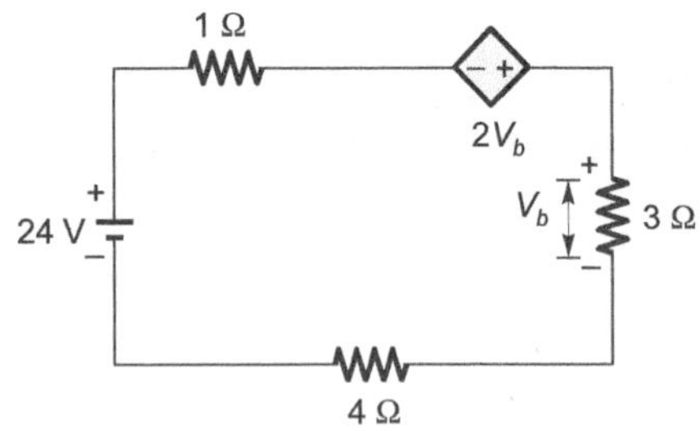

(a) 10 A (b) 12 A
(c) 14 A (d) 16 A

[ESE-2001]

1.2 The value of resistance 'R' shown in the given figure is

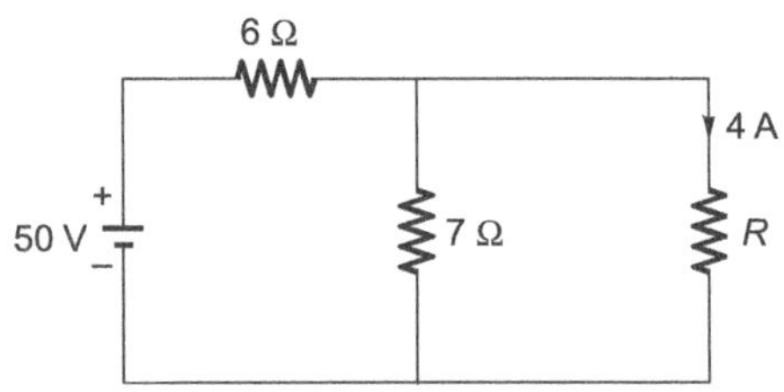

(a) 3.5 Ω (b) 2.5 Ω
(c) 1 Ω (d) 4.5 Ω

[ESE-2001]

1.3 **Assertion (A):** Equivalent network obtained from Δ-Y transformation relationships in general is valid only for one frequency.
Reason (R): The impedances involved in Δ-Y vary with frequency.
(a) Both A and R are true and R is the correct explanation of A
(b) Both A and R are true but R is NOT the correct explanation of A
(c) A is true but R is false
(d) A is false but R is true

[ESE-2001]

1.4 The figure shows a network in which the diode is an ideal one.

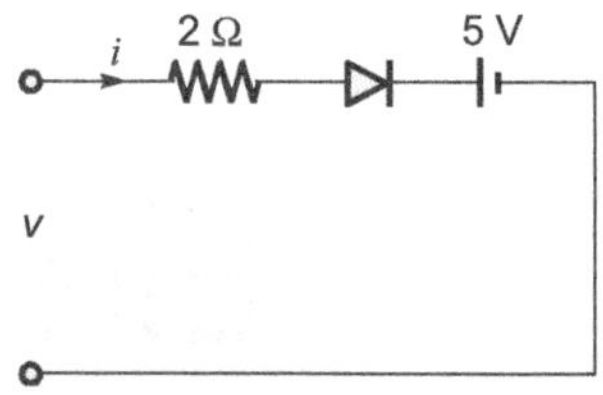

the terminal v-i characteristics of the network is given by

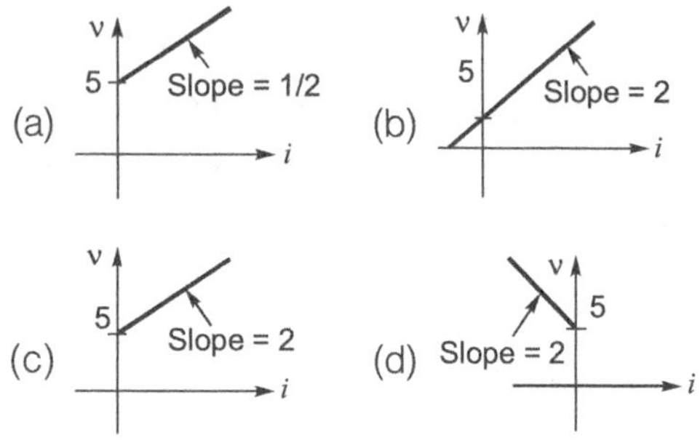

[ESE-2002]

1.5 The v-i characteristic of an element is shown in the figure given below. The element is

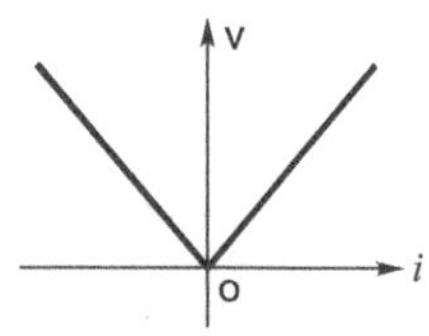

(a) non-linear, active, non-bilateral
(b) linear, active, non-bilateral
(c) non-linear, passive, non-bilateral
(d) non-linear, active, bilateral

[ESE-2002]

1.6 The current through 120 ohm resistor in the circuit shown in the figure below is

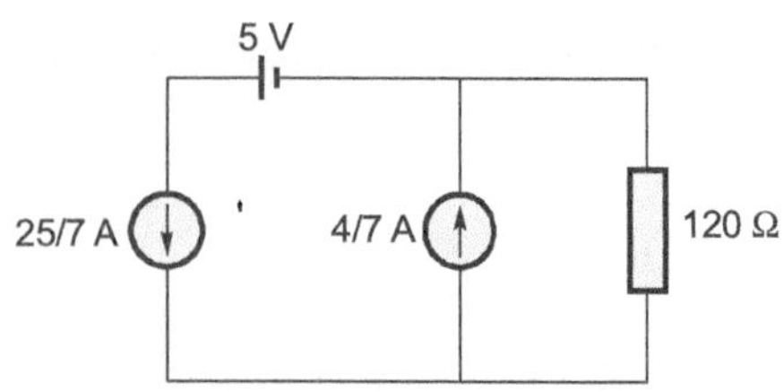

(a) 1 A (b) 2 A
(c) 3 A (d) 4 A [ESE-2002]

1.7 For the circuit given in figure below the power delivered by the 2 volt source is given by

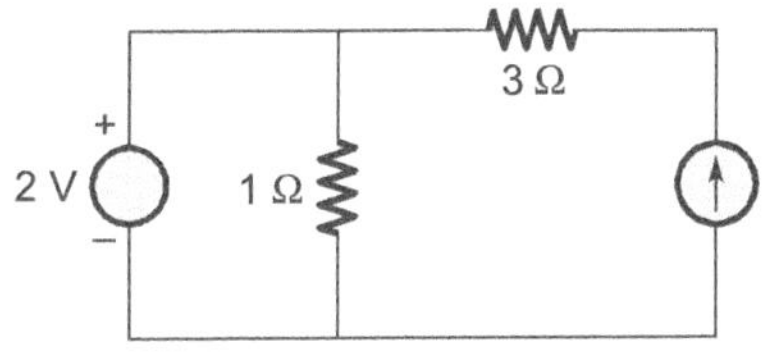

(a) 4 W (b) 2 W
(c) –2 W (d) –4 W [ESE-2002]

1.8 **Assertion (A):** Node-voltage analysis of networks is a method that uses Kirchhoff's current law to obtain a set of simultaneous equations that, when solved, will provide information concerning the magnitudes and phase angles of the voltages across each branch.

Reason (R): The ideal generator maintains a constant voltage amplitude and wave-shape regardless of the amount of current it supplies to the circuit.

(a) Both A and R are true and R is the correct explanation of A
(b) Both A and R are true but R is NOT the correct explanation of A
(c) A is true but R is false
(d) A is false but R is true

[ESE-2002]

1.9 **Assertion (A):** All networks made up of passive, linear time invariant elements are reciprocal.

Reason (R): Passivity and time-invariance of elements do not guarantee reciprocity of the network.

(a) Both A and R are true and R is the correct explanation of A
(b) Both A and R are true but R is NOT the correct explanation of A
(c) A is true but R is false
(d) A is false but R is true

[ESE-2002]

1.10 The incandescent bulbs rated respectively as P_1 and P_2 for operation at a specified mains voltage are connected in series across the mains as shown in the above figure. Then the total power supplied by the mains to the two bulbs are

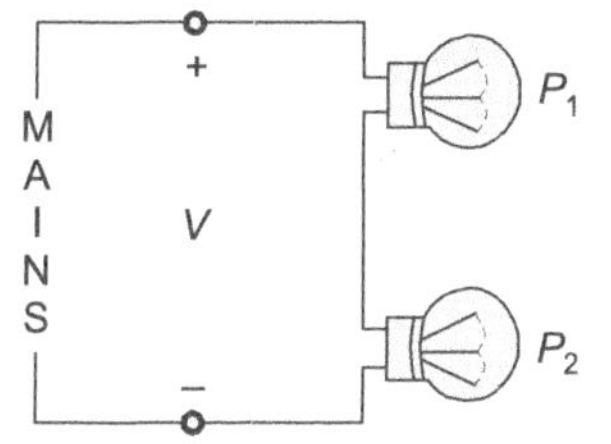

(a) $\dfrac{P_1 P_2}{P_1 + P_2}$ (b) $\sqrt{P_1^2 + P_2^2}$
(c) $(P_1 + P_2)$ (d) $\sqrt{P_1 \times P_2}$

[ESE-2003]

1.11 A certain network consists of a large number of ideal linear resistances, one of which is designated as R and two constant ideal sources. The power consumed by R is P_1 when only the first source is active, and P_2 when only the second source is active. If both sources are active simultaneously, then the power consumed by R is

(a) $P_1 \pm P_2$ (b) $\sqrt{P_1} \pm \sqrt{P_2}$
(c) $(\sqrt{P_1} \pm \sqrt{P_2})^2$ (d) $(P_1 \pm P_2)^2$

[ESE-2003]

1.12 For the circuit shown below, the value of V_s is 0 when I = 4 A. The value of I when V_s = 16 V, is

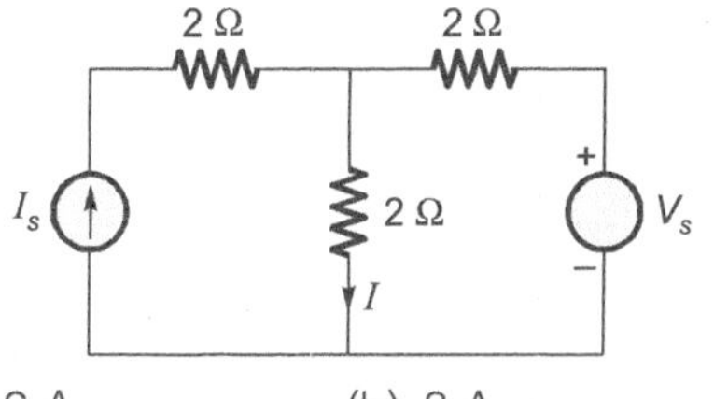

(a) 6 A (b) 8 A
(c) 10 A (d) 12 A

[ESE-2003]

1.13 In a network made up of linear resistors and ideal voltage sources, values of all resistors are doubled. Then the voltage across each resistor is

(a) Doubled
(b) Halved
(c) Decreases four times
(d) Not changed

[ESE-2003]

1.14 Consider the circuit as shown below which has a current-dependent current source. The value V_2/V_1 is

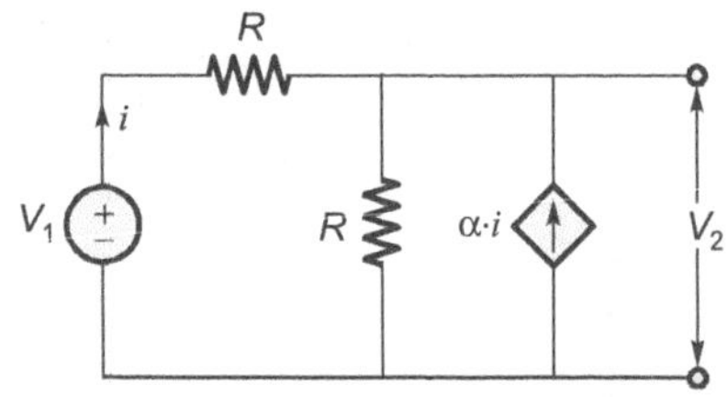

(a) 1 (b) 2

(c) $\frac{1+\alpha}{2+\alpha}$ (d) $\frac{\alpha}{2+\alpha}$

[ESE-2003]

1.15 Consider the following circuit:

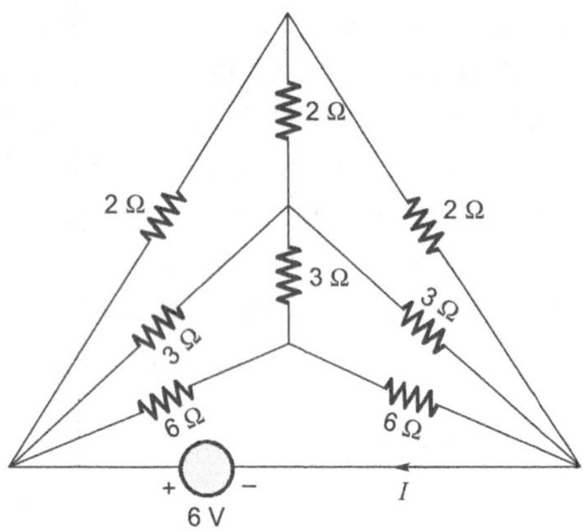

What is the value of the current-I in the above circuit?

(a) 1 A (b) 2 A

(c) 3 A (d) 4 A

[ESE-2004]

1.16 Consider the following circuit:

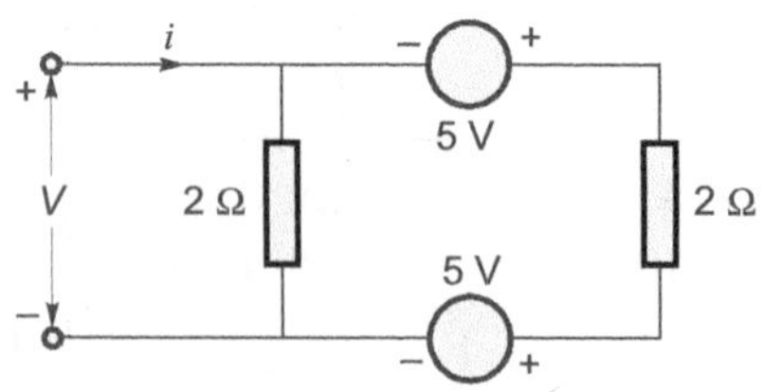

Which one of the following statements is correct? The circuit shown above is

(a) passive and linear

(b) active and linear

(c) passive and non-linear

(d) active and non-linear

[ESE-2004]

1.17 Three parallel resistive branches are connected across a d. c supply. What will be the ratio of the branch currents $I_1 : I_2 : I_3$ if the branch resistances are in the ratio $R_1 : R_2 : R_3 :: 2 : 4 : 6$?

(a) 3 : 2 : 6 (b) 2 : 4 : 6

(c) 6 : 3 : 2 (d) 6 : 2 : 4

[ESE-2004]

1.18 For the circuit shown below, what is the voltage across the current source I_s?

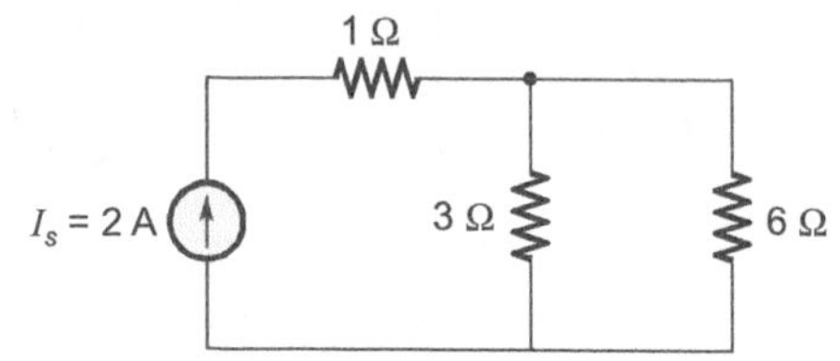

(a) 0 (b) 2 V

(c) 3 V (d) 6 V

[ESE-2005]

1.19 For the circuit shown below, the value of R is adjusted, so as to make the current in R_L equal to zero. What is the value of R?

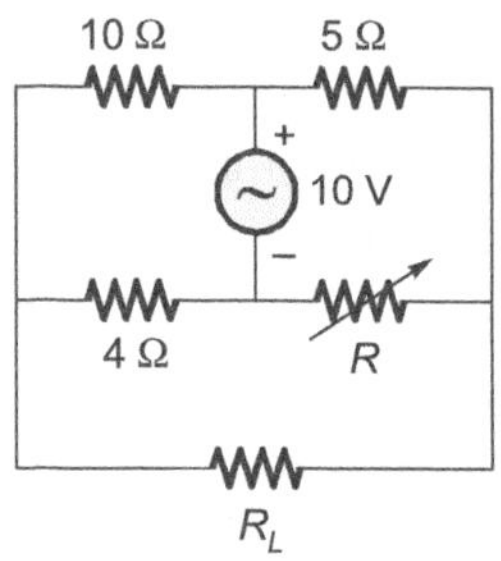

(a) 1 Ω (b) 2 Ω

(c) 3 Ω (d) 4 Ω [ESE-2005]

1.20 In the circuit shown below, what is the voltage across 5 Ω resistor?

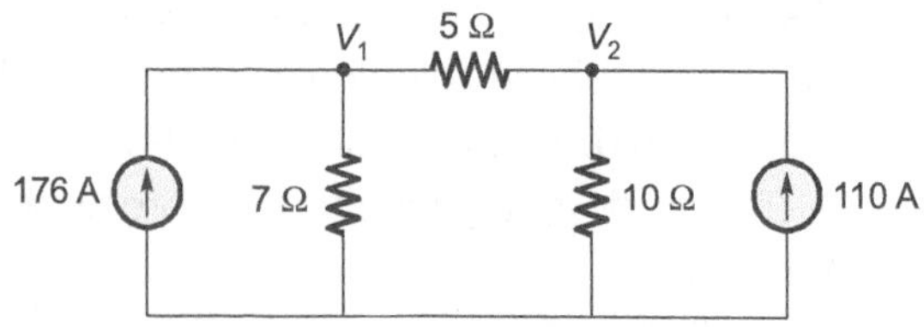

(a) –30 V (b) 30 V

(c) 1250 V (d) –1250 V

[ESE-2006]

1.21 Consider the circuit in the below figure. What is the power delivered by the 24 V source?

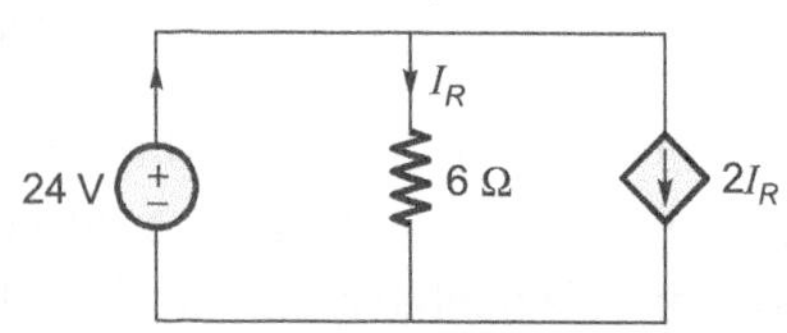

(a) 96 W (b) 144 W

(c) 192 W (d) 288 W [ESE-2007]

1.22 Consider the following statements on mesh and nodal analysis:

1. Networks that contain many series-connected elements, voltage sources or meshes having common current sources (super meshes) are more suitable for mesh analysis than for nodal analysis.
2. Networks with parallel connected elements, current sources or nodes connected by voltage sources are more suitable for nodal analysis than mesh analysis.
3. A circuit with fewer nodes than meshes is better analyzed using mesh analysis, while a circuit with fewer meshes, than nodes is better analyzed using nodal analysis.

Which of the statements given above are correct?

(a) 1 and 2 (b) 2 and 3
(c) 1 and 3 (d) 1, 2 and 3

[ESE-2007]

1.23 In the circuit shown below, what is the value of the current-I?

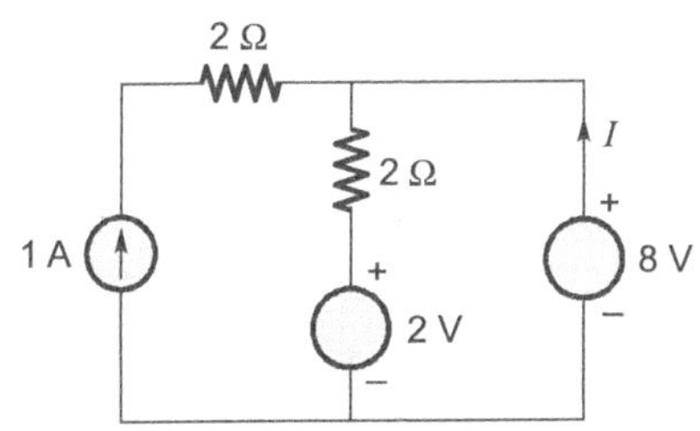

(a) 1 A (b) 2 A
(c) 3 A (d) 4 A

[ESE-2008]

1.24 If the voltage V across 10 Ω resistance is 10 V, what is the voltage E of the voltage source in the circuit shown below?

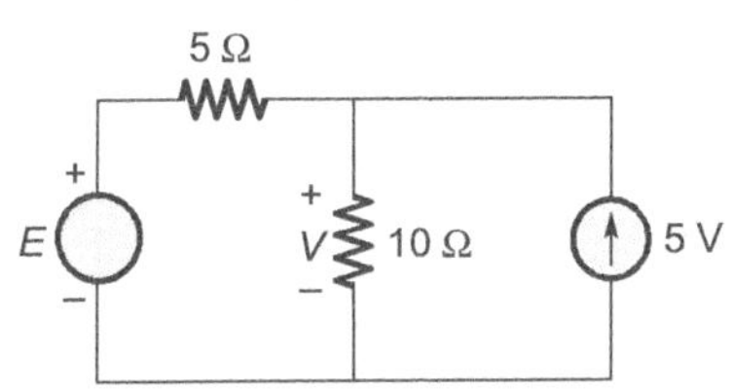

(a) −50 V (b) −10 V
(c) +10 V (d) +50 V

[ESE-2008]

1.25 What is the value of the current I in the circuit shown below?

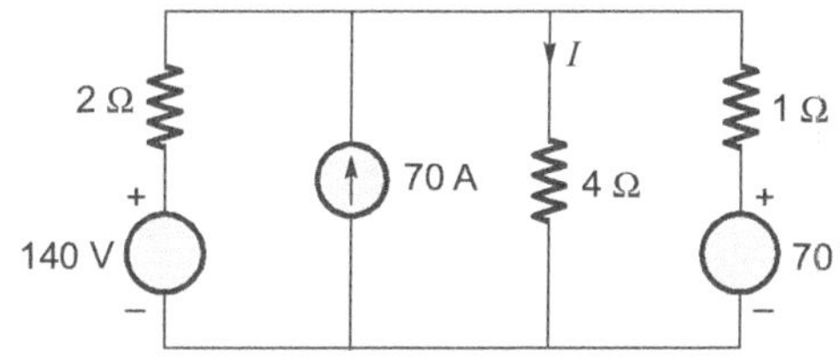

(a) 20 A (b) 25 A
(c) 30 A (d) 36 A

[ESE-2008]

1.26 For the network shown in the figure below, what is the voltage across the current source?

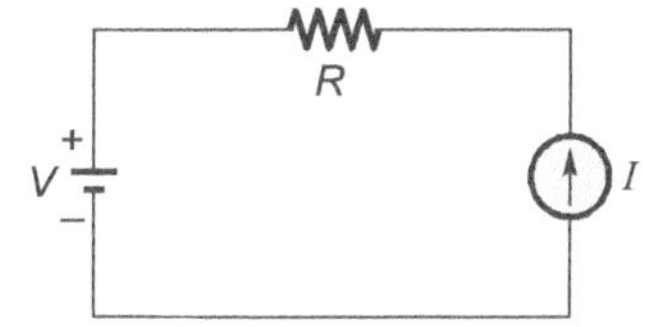

(a) $V - RI$ (b) $V + RI$
(c) Zero (d) $RI - V$

[ESE-2008]

1.27 For the circuit shown, what is the voltage V if the source voltage is reduced by 50%?

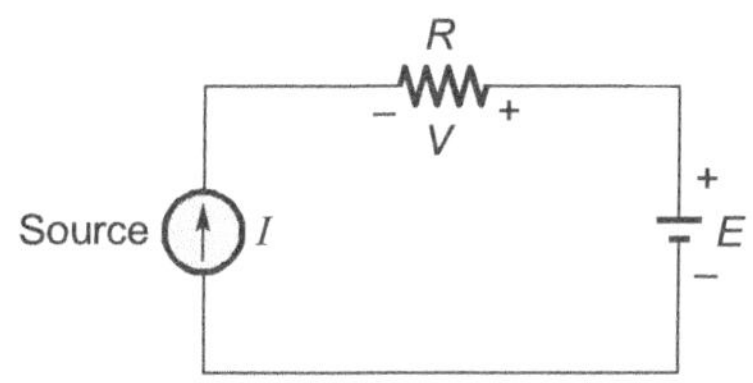

(a) $IR + E$ (b) $E - IR$
(c) $2\,IR - (E/2)$ (d) $(E/2) - IR$

[ESE-2008]

1.28 What is the current through the 2 Ω resistance for the circuit as shown below?

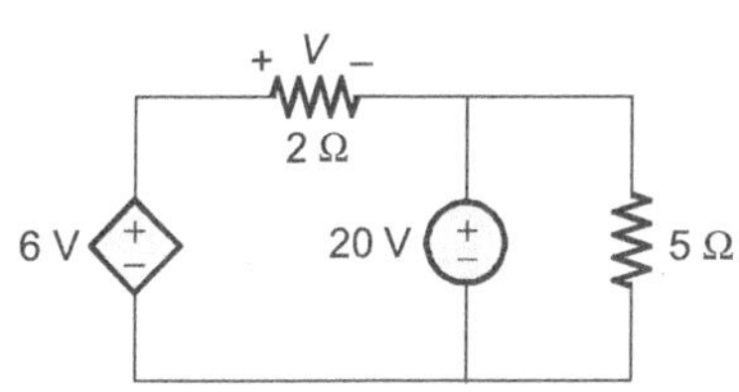

(a) 5 A (b) 4 A
(c) 3 A (d) 2 A [ESE-2009]

1.29 What is the voltage across the current source for the below shown circuit?

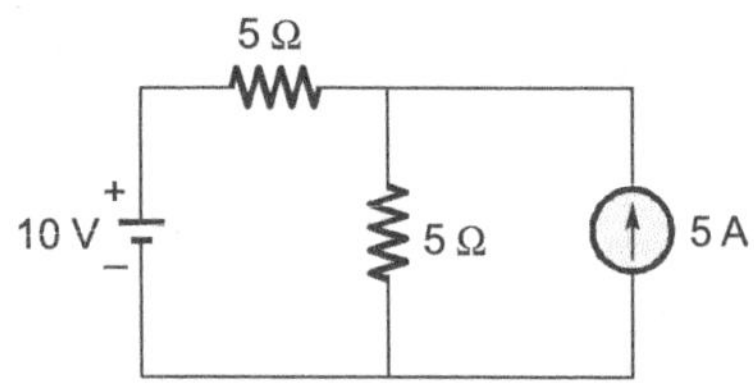

(a) 5.0 V (b) 7.5 V
(c) 12.5 V (d) 17.5 V

[ESE-2009]

1.30 What is the value of I for the below shown circuit, if $V = 2$ volts?

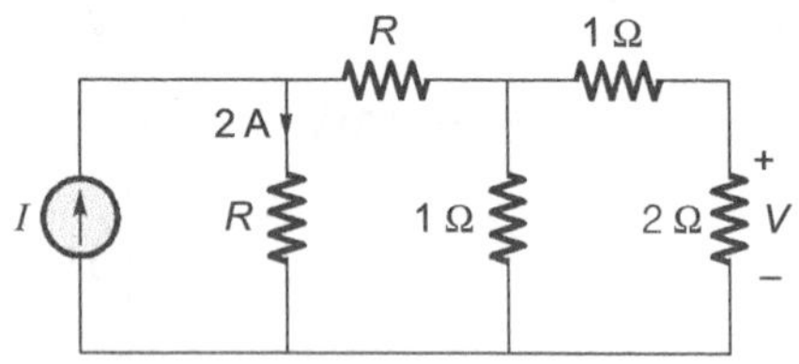

(a) 2 A (b) 4 A
(c) 6 A (d) 8 A

[ESE-2009]

1.31 Find the voltage of the node A with respect to 'O' for the circuit as shown in below.

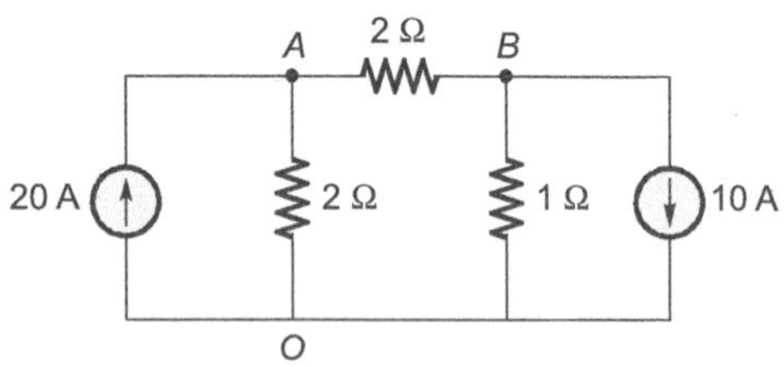

(a) 40 V (b) 20 V
(c) 50 V (d) 60 V

[ESE-2009]

1.32 For the circuit as shown below, what is the value of I?

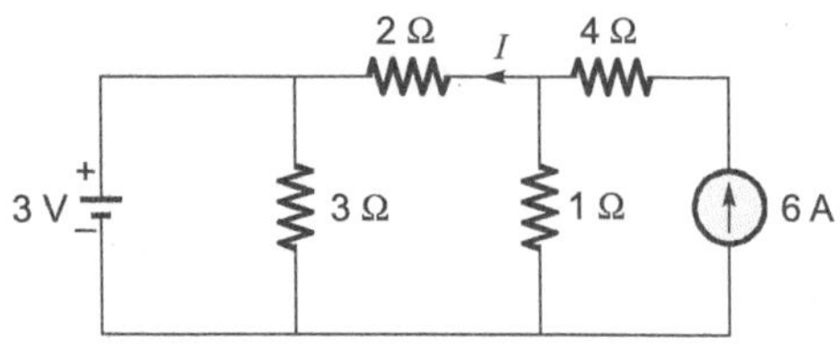

(a) 4 A (b) 3 A
(c) 2 A (d) 1 A

[ESE-2009]

1.33 The current waveform as shown below, is applied in a pure resistor of 10 Ω. What is the power dissipated in the resistor?

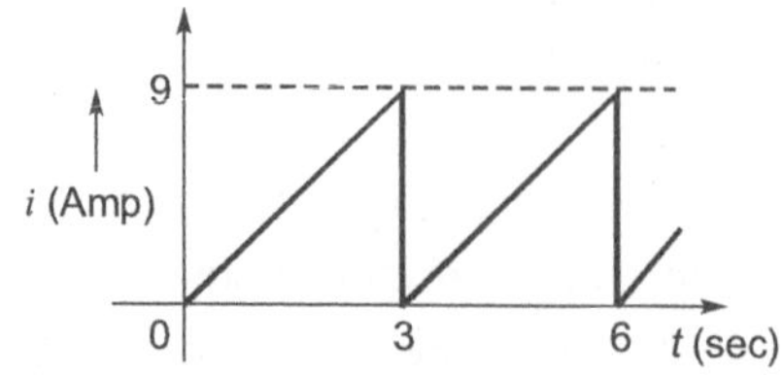

(a) 270 W (b) 135 W
(c) 52 W (d) 7 W

[ESE-2009]

1.34 Consider the following equations with respect to the below network:

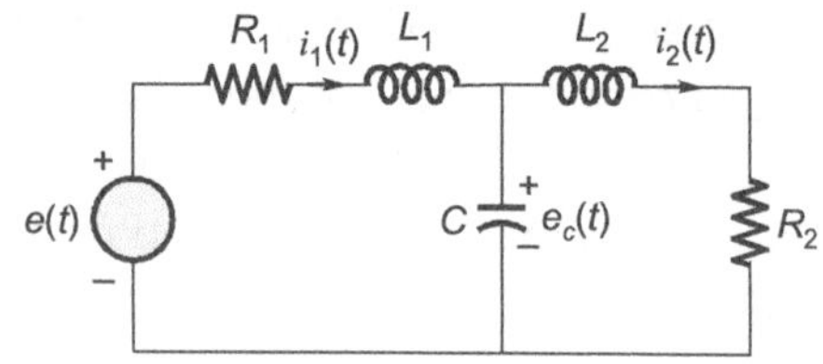

1. $L_1 \dfrac{di_1(t)}{dt} = R_1 i_1(t) - e_c(t) + e(t)$

2. $L_1 \dfrac{di_1(t)}{dt} = -R_1 i_1(t) - e_c(t) + e(t)$

3. $L_2 \dfrac{di_2(t)}{dt} = -R_2 i_2(t) + e_c(t)$

4. $C \dfrac{de_c(t)}{dt} = i_1(t) - i_2(t)$

Which of the above statements is/are correct?

(a) 1 only (b) 2, 3 and 4
(c) 1, 3 and 4 (d) 1, 2 and 4

[ESE-2009]

1.35 For the circuit shown in the figure below, the voltage across the 1 ohm resistor is given by

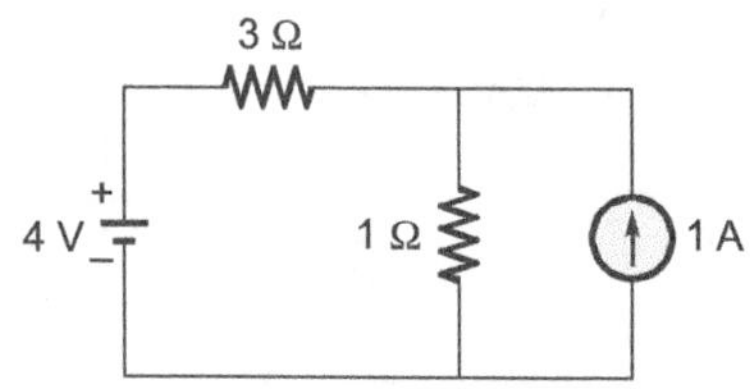

(a) 7/4 V (b) 5/4 V
(c) 7/3 V (d) 2/3 V

[ESE-2010]

1.36 The currents I_1 and I_2 in the below circuit are respectively.

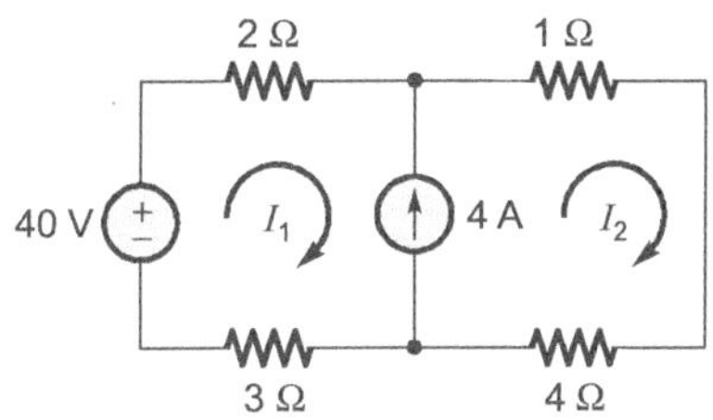

(a) 4 A; 4 A (b) 3 A; 5 A
(c) 2 A; 6 A (d) 6 A; 2 A

[ESE-2010]

1.37 The currents I_1 and I_2 in the below circuit are respectively

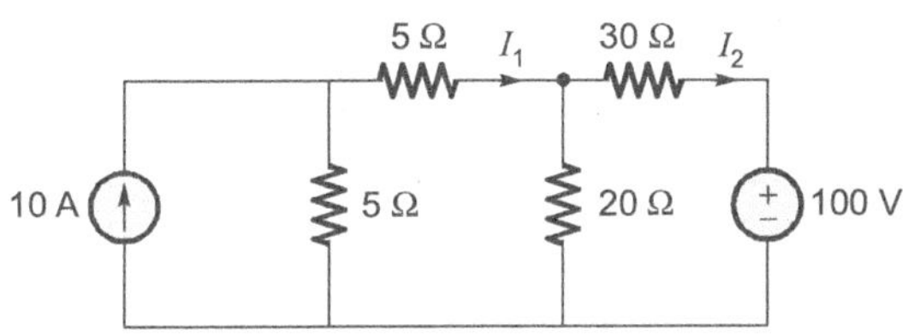

(a) 1.818 A; –0.4545 A
(b) 2.451 A; –1.568 A
(c) 0.4545 A; –1.818 A
(d) 1.56 A; –2.45 A

[ESE-2010]

1.38 If the power dissipated in the circuit shown below is 8 W, then the value of E will be

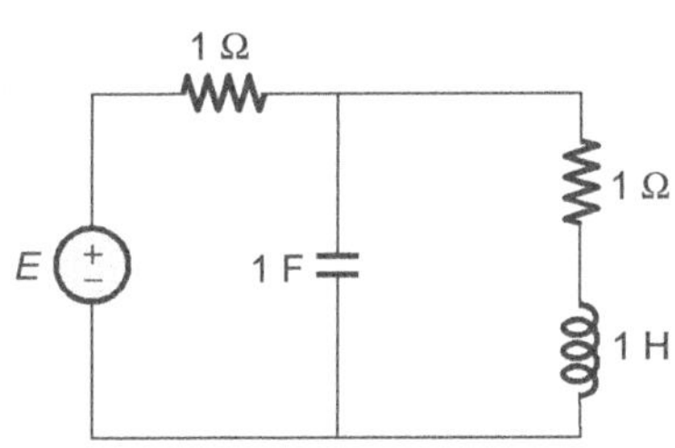

(a) 2 V (b) 4 V
(c) 8 V (d) 16 V

[ESE-2010]

1.39 The currents I_x and V_x in the below circuit are respectively

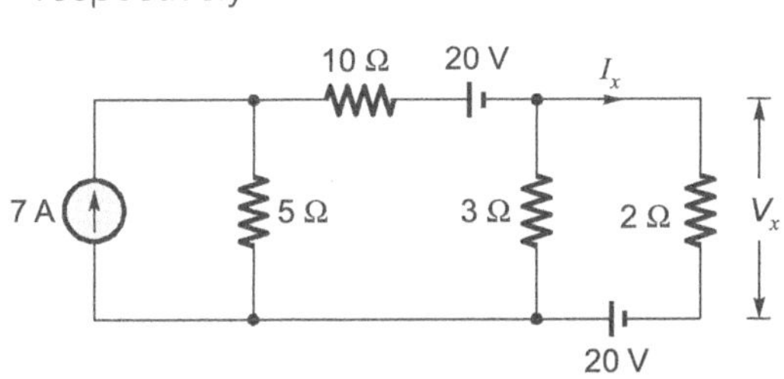

(a) 5 A; 10 V (b) 10 A; 20 V
(c) 6 A; 12 V (d) 4 A; 8 V

[ESE-2010]

1.40 For the circuit shown in figure below, the value of current, I is

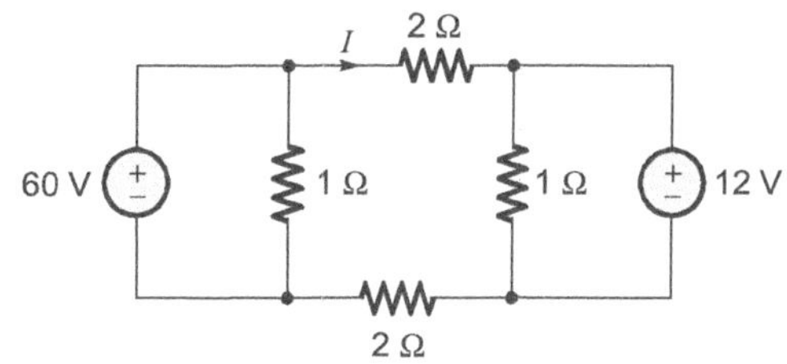

(a) 2 A (b) 3 A
(c) 6 A (d) 12 A

[ESE-2010]

1.41 In the circuit shown, the current i_1 is

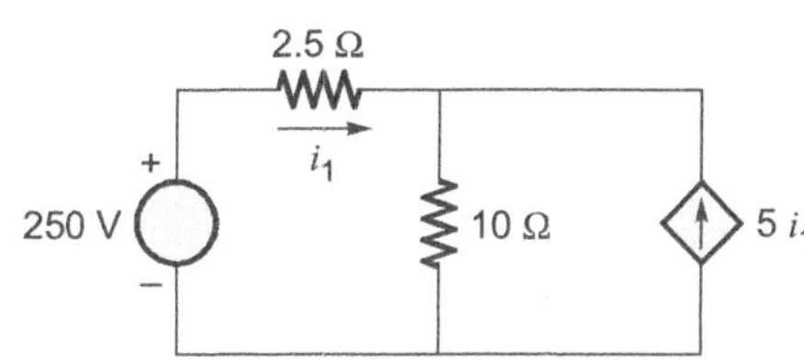

(a) 4 A (b) 2 A
(c) 4.76 A (d) 20 A

[ESE-2011]

1.42 When KCL is applied at the super node in the below circuit, the current equation in terms of node voltages V_1 and V_2 is

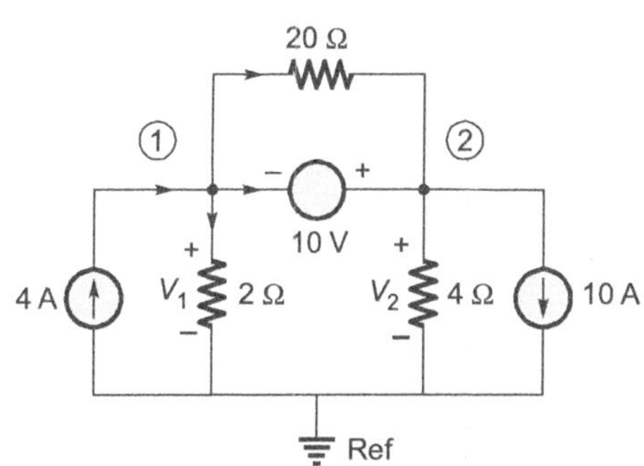

(a) $-6 = \frac{V_1}{2} + \frac{V_2}{4}$ (b) $4 = \frac{V_1 - V_2}{2} + \frac{V_1 - V_2}{20}$

(c) $4 = \frac{V_1}{2} + \frac{V_1 - V_2}{20}$ (d) $4 = \frac{V_1}{2} + \frac{V_2}{4}$

[ESE-2011]

1.43 The node voltage V in the circuit is

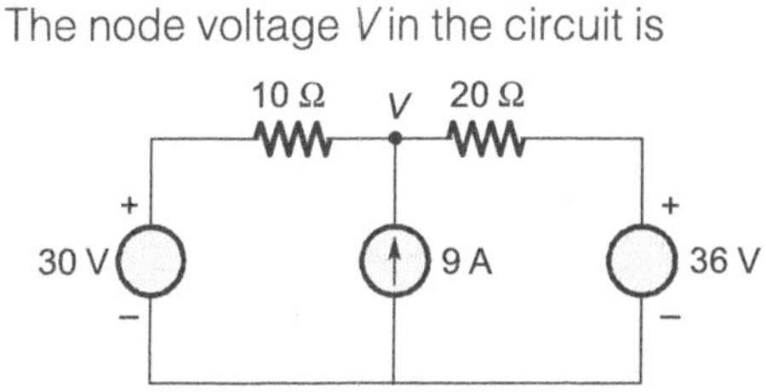

(a) 6 V (b) 30 V
(c) 36 V (d) 92 V

[ESE-2011]

1.44 The voltage V_x across the 2 Ω resistance in the circuit is

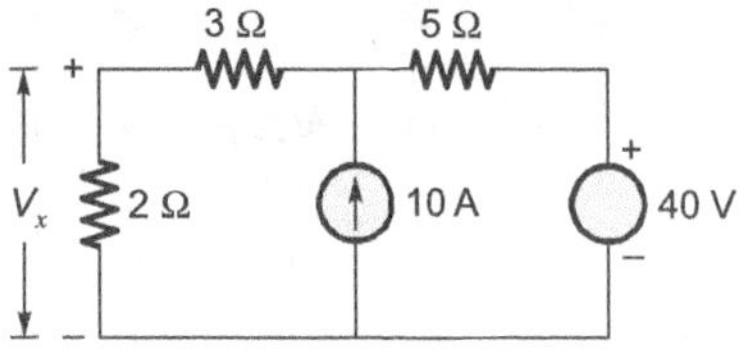

(a) 16 V (b) 60 V
(c) 18 V (d) 10 V

[ESE-2011]

1.45 Three parallel branches of resistors are connected across a d.c. source as shown in the figure. What is $I_1 : I_2 : I_3$?

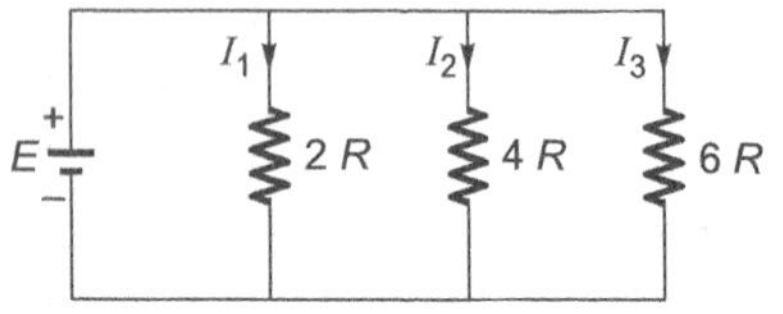

(a) 3 : 2 : 6 (b) 2 : 4 : 6
(c) 6 : 3 : 2 (d) 6 : 2 : 4

[ESE-2011]

1.46 The following are the results of tests conducted on the below star-connected load:

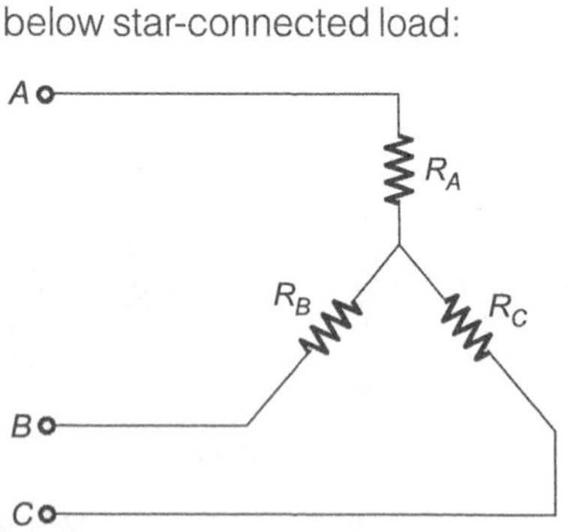

The resistance between *B* and *C* with *A* open : 22 Ω

The resistance between *C* and *A* with *B* open : 18 Ω

The individual resistance of R_A, R_B and R_C are respectively,

(a) 8 Ω, 14 Ω and 4 Ω
(b) 10 Ω, 2 Ω and 8 Ω
(c) 4 Ω, 8 Ω and 14 Ω
(d) 6 Ω, 6 Ω and 8 Ω

[ESE-2011]

1.47 Match **List-I** with **List-II** and select the correct answer using the code given below the lists:

List-I
A. Resistance
B. Inductance
C. Loop current
D. Number of loops

List-II
1. Number of node pairs
2. Conductance
3. Capacitance
4. Node pair voltage

Codes:

	A	B	C	D
(a)	1	3	4	2
(b)	2	3	4	1
(c)	1	4	3	2
(d)	2	4	3	1

[ESE-2012]

1.48 The number of independent KVL and KCL equations for a network with n-nodes and *l* links are respectively

(a) *l* and *n* (b) *l* and *n* – 1
(c) *n* – 1 and *l* (d) *n* – 1 and *l* – 1

[ESE-2012]

1.49 A triangular pyramid, built up of six wires whose resistances are shown in the figure, is fed from a 1 V battery at the terminals *A* and *B*. The current through the branch *DB* is

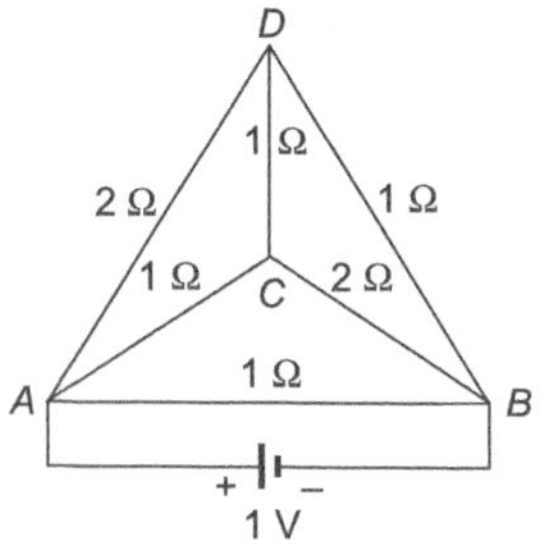

(a) $\frac{1}{7}$ A (b) $\frac{2}{7}$ A
(c) $\frac{3}{7}$ A (d) $\frac{4}{7}$ A

[ESE-2012]

1.50 For a series and a parallel circuit, the equivalent total value of certain parameter *X* is given by

$$X_e = X_1 + X_2 + X_3 + X_4 + \cdots + X_n$$

where X_i is the ith value of the parameter and X_e is the equivalent value, and *n* is the number of elements.

The parameter X can be

(a) resistance (b) current
(c) voltage (d) power

[ESE-2012]

1.51 Consider the following statements:
Any element connected in

1. series with a voltage source is redundant.
2. parallel with a voltage source is redundant.
3. series with a current source is redundant.
4. parallel with a current source is redundant.

The correct statements are

(a) 1 and 3 (b) 2 and 3
(c) 3 and 4 (d) 1 and 2

[ESE-2012]

1.52 The power dissipated in the controlled source of the network shown below is

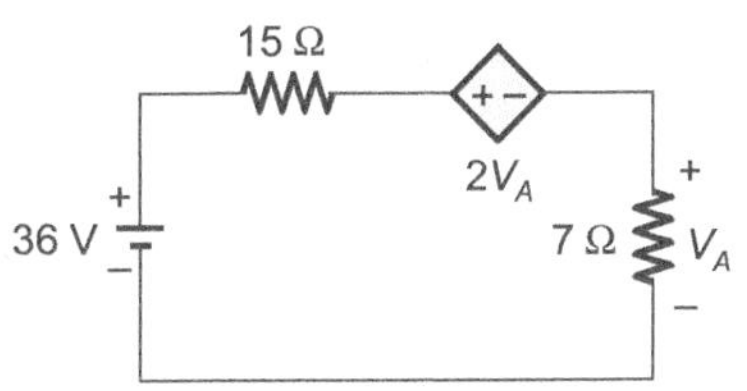

(a) 36 W (b) 15 W
(c) 07 W (d) 14 W

[ESE-2013]

1.53 A 3-phase distribution network is as shown below. The resistance across the terminals *RB*, *BY* and *RY* are

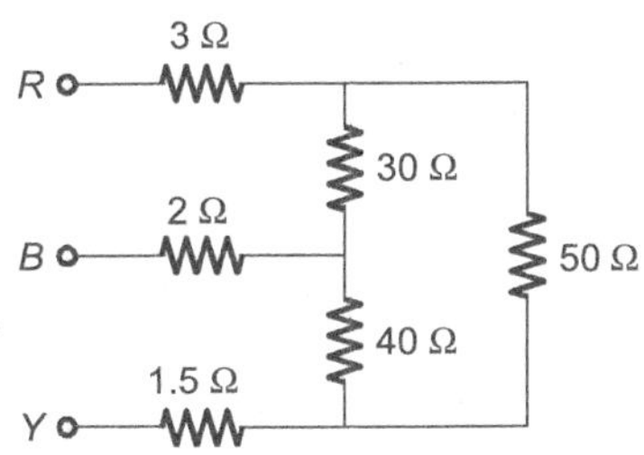

(a) 25.5 Ω, 31.2 Ω and 33.6 Ω
(b) 27.5 Ω, 30.17 Ω and 33.67 Ω
(c) 28.5 Ω, 32.3 Ω and 34.5 Ω
(d) 21.2 Ω, 42.3 Ω and 45.6 Ω

[ESE-2013]

1.54 A network *N* consists of resistors, independent voltage and current sources. The value of its determinant based on the loop analysis:

1. cannot be negative
2. cannot be zero
3. is independent of the values of voltage and current sources.
4. dependent on the values of the resistances and the voltage and current sources

(a) 1, 2 and 3 (b) 1, 2 and 4
(c) 1, 3 and 4 (d) 2, 3 and 4

[ESE-2013]

1.55 The number of 2 μF, 300 V capacitors needed to obtain a capacitance value of 2 μF rated for 1200 V is

(a) 16 (b) 12
(c) 10 (d) 8

[ESE-2013]

1.56 For the circuit shown below the value of *r* connected between *C* and *D* is such that the equivalent resistance of the circuit by looking into circuit through terminals *A* and *B* is *r* only. Then the value of *r* is

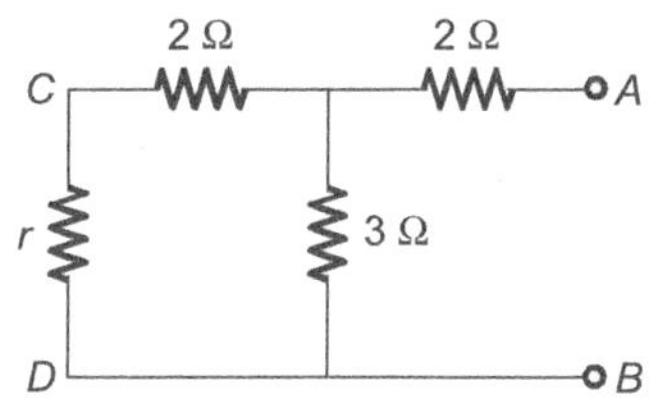

(a) 2 Ω (b) 4 Ω
(c) 3 Ω (d) 6 Ω

[ESE-2013]

1.57 If an ideal voltage source and ideal current source are connected in series, the combination

(a) has the same properties as a current source alone.
(b) has the same properties as a voltage source alone.
(c) has the same properties as the source which has a higher value.
(d) results in the branch being redundant.

[ESE-2013]

1.58 A conductor of diameter *d*, length *l* consumes a power of W when a current *I* flows through it. What will be the power consumed if *d* is doubled, *l* is halved and current is tripled?

(a) 18 W (b) 36 W
(c) 48 W (d) 72 W

[ESE-2014]

1.59 Consider the following statements:
Any element connected in
1. Series with an ideal current source is redundant.
2. Parallel with an ideal current source is redundant.
3. Series with an ideal voltage source is redundant.
4. Parallel with an ideal voltage source is redundant.

Which of the above statements is/are correct?
(a) 1 only (b) 2 and 3
(c) 2 only (d) 1 and 4

[ESE-2014]

1.60 Consider the following two types of non-identical sources:
1. Voltage source $e_1(t)$ and $e_2(t)$
2. Current sources i_1 and $i_2(t)$

Regarding the mode of their connection in a circuit.
(a) 1 cannot be connected in parallel, and 2 cannot be connected in series.
(b) 1 cannot be connected in series, and 2 cannot be connected in parallel.
(c) Both 1 and 2 cannot be connected in parallel.
(d) Both 1 and 2 cannot be connected in parallel

[ESE-2014]

1.61 The voltage and current waveforms for an element are shown in the figures.

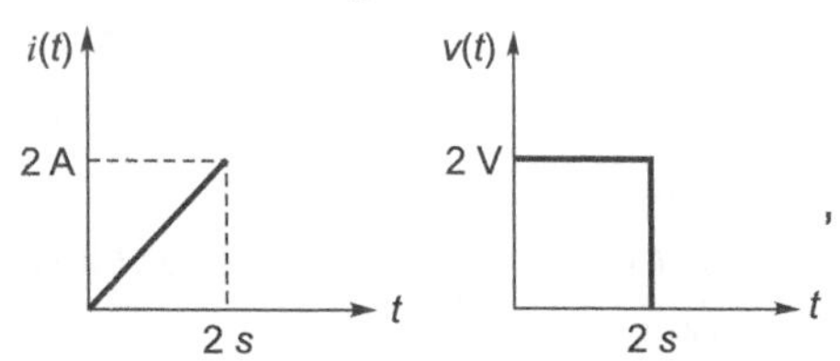

The circuit element and its value are
(a) capacitor and 2 F
(b) inductor and 1 H
(c) inductor and 2 H
(d) resistor and 1 Ω

[ESE-2014]

1.62 n resistors each of resistance R when connected in series offer an equivalent resistance of 50 Ω and when reconnected in parallel the effective resistance is 2 Ω. The value of R is
(a) 2.5 Ω (b) 5 Ω
(c) 7.5 Ω (d) 10 Ω

[ESE-2016]

1.63 The potential difference V_{AB} in the circuit is

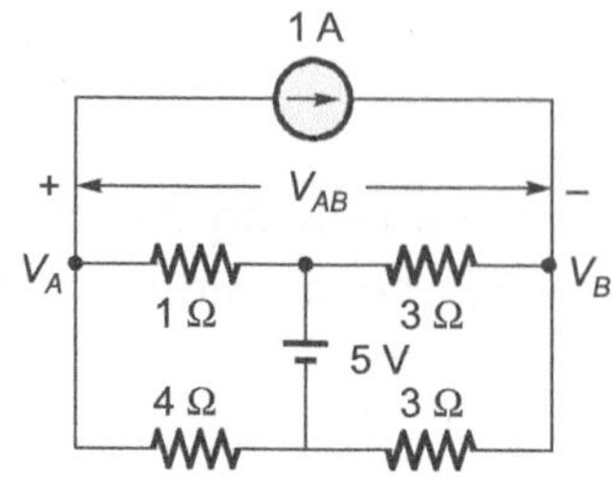

(a) 0.8 V (b) −0.8 V
(c) 1.8 V (d) −1.8 V

[ESE-2016]

1.64 Two bulbs of 100 W/250 V and 150 W/250 V are connected in series across a supply of 250 V. The power consumed by the circuit is
(a) 30 W (b) 60 W
(c) 100 W (d) 250 W

[ESE-2016]

1.65 Three 30 Ω resistors are connected in parallel across an ideal 40 V source. What would be the equivalent resistance seen by the load connected across this circuit?
(a) 0 Ω (b) 10 Ω
(c) 20 Ω (d) 30 Ω

[ESE-2016]

1.66 In a parallel resistive circuit, opening a branch results in
1. increase in total resistance
2. decrease in total power
3. no changes in total voltage and branch voltage

Which of the above is/are correct?
(a) 1 only (b) 2 only
(c) 3 only (d) 1, 2 and 3

[ESE-2016]

1.67 In nodal analysis, the preferred reference node is a node that is connected to
1. ground
2. many parts of the network
3. the highest voltage source

Which of the above is/are correct?
(a) 1 only (b) 2 only
(c) 3 only (d) 1, 2 and 3

[ESE-2016]

1.68 The mesh-current method
1. works with both planar and non-planar circuits.
2. uses Kirchhoffs voltage law.

Which of the above is/are correct?
(a) 1 only (b) 2 only
(c) Both 1 and 2 (d) Neither 1 nor 2
[ESE-2016]

1.69 A short in any type of circuit (series, parallel or combination) causes the total circuit
1. resistance to decrease
2. power to decrease
3. current to increase
4. voltage to increase

Which of the above are correct?
(a) 2 and 3 (b) 2 and 4
(c) 1 and 4 (d) 1 and 3
[ESE-2016]

1.70 The voltage and current waveforms for an element are shown in the figure.

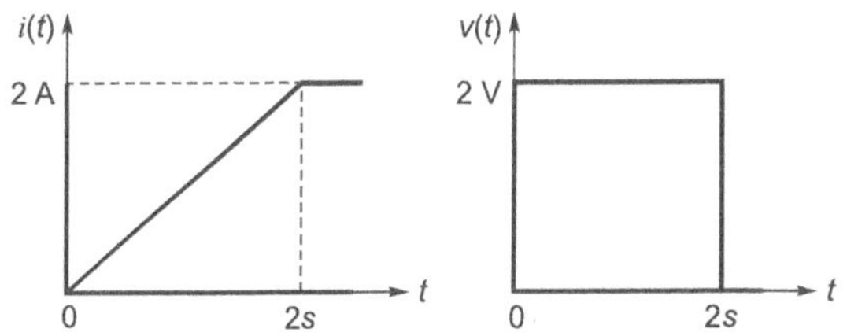

The circuit element and its value are
(a) Capacitor, 2 F (b) Inductor, 2 H
(c) Capacitor, 0.5 F (d) Inductor, 0.5 H
[ESE-2017]

1.71 What is the current through the 8 Ω resistance connected across terminals, *M* and *N* in the circuit?

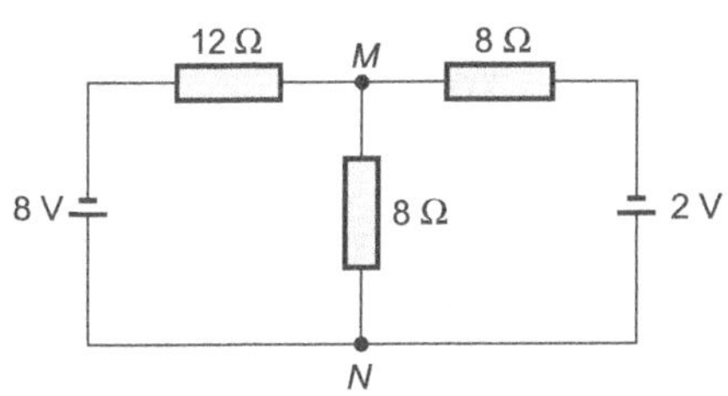

(a) 0.34 A from *M* to *N*
(b) 0.29 A from *M* to *N*
(c) 0.29 A from *N* to *M*
(d) 0.34 A from *N* to *M* [ESE-2017]

1.72 What is the potential drop across the 80 Ω resistor in the figure?

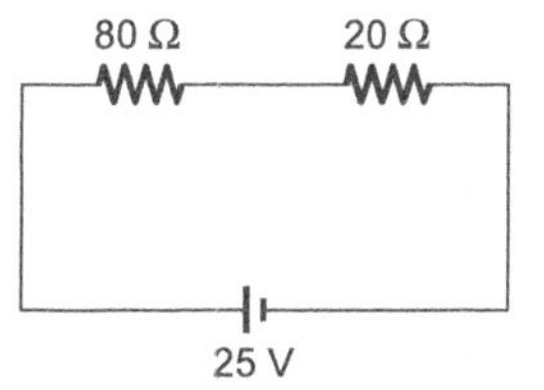

(a) 20 V (b) 15 V
(c) 10 V (d) 5 V [ESE-2017]

1.73 Kirchhoff's current law is applicable to
1. Closed loops in a circuit.
2. Junction in a circuit
3. Magnetic circuits

Which of the above statements is/are correct?
(a) 1 only (b) 2 only
(c) 3 only (d) 1, 2 and 3
[ESE-2017]

1.74 Which of the following are satisfied in a nonlinear network?
1. Associative 2. Superposition
3. Homogeneity 4. Bilaterality

Select the correct answer using the codes given below:
(a) 1 and 3 only (b) 1 and 4 only
(c) 2 and 3 only (d) 2 and 4 only
[ESE-2017]

1.75 What is the current through the 5 Ω resistance in the circuit shown?

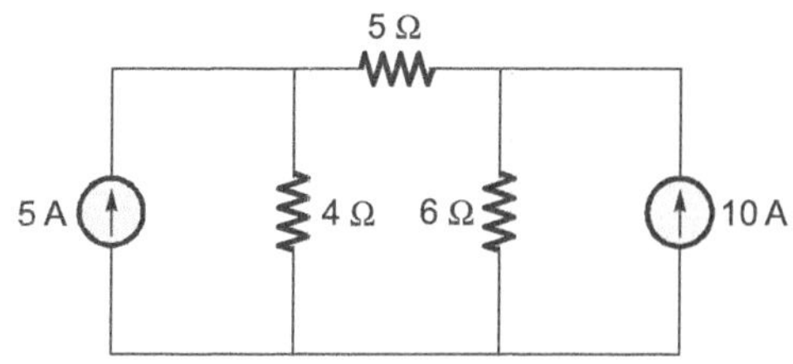

(a) 5.33 A (b) 4.66 A
(c) 2.66 A (d) 1.33 A
[ESE-2017]

1.76 The voltage and current characteristic of an element is as shown in figure. The nature and value of the element are

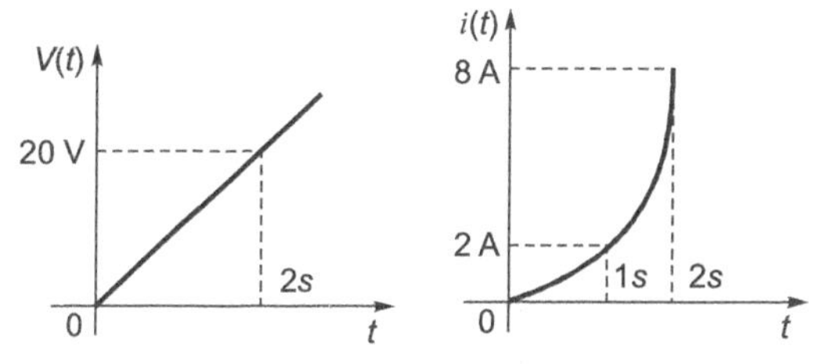

(a) Capacitor of 3.3 μF
(b) Inductor of 2.5 H
(c) Capacitor of 6.7 μF
(d) Inductor of 5.0 H [ESE-2018]

1.77 In the circuit shown, what value of R will result in $I = 4$ A?

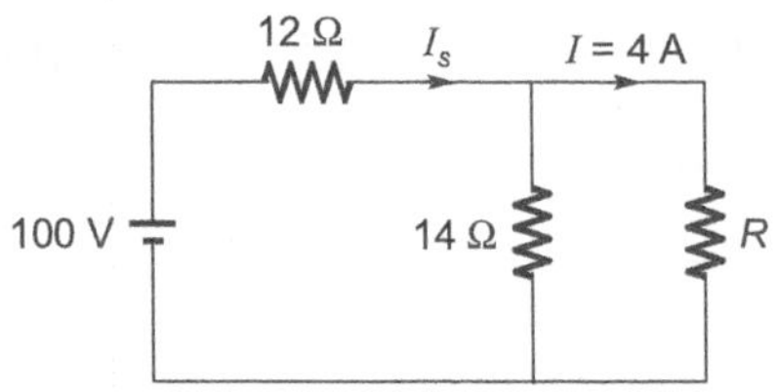

(a) 9 Ω (b) 7 Ω
(c) 5.5 Ω (d) 3.5 Ω

[ESE-2018]

1.78 In the circuit as shown, the currents I_1, I_2 and I_3 through three resistors are, respectively.

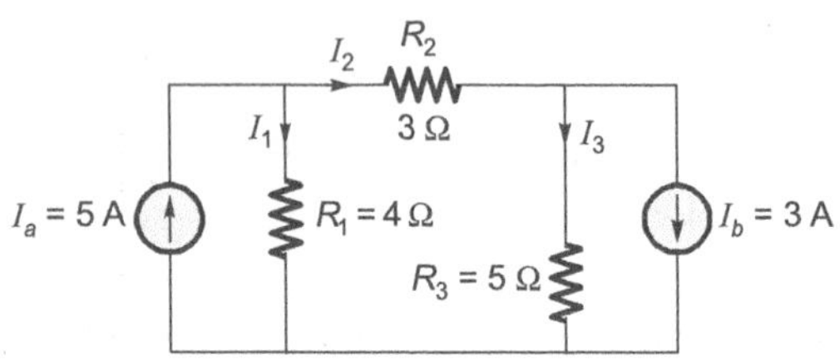

(a) 2.08 A, 2.92 A and –0.08 A
(b) 3.08 A, 2.5 A and –0.06 A
(c) 2.08 A, 2.5 A and –0.08 A
(d) 3.08 A, 2.92 A and –0.06 A

[ESE-2018]

1.79 The $v - i$ relationship for a circuit containing R and C and a battery of voltage E, all in series is

1. $\frac{1}{C}\int i\,dt + iR = E$
2. $\frac{1}{RC}i + \frac{di}{dt} = 0$
3. $\frac{1}{C}i + R\int i\,dt = E$

Which of the above relationships are correct?
(a) 1 and 2 only (b) 1 and 3 only
(c) 2 and 3 only (d) 1, 2 and 3

[ESE-2018]

1.80 A triangular wave voltage, as shown in figure, is applied across the terminals of a 0.5F pure capacitor at time $t = 0$. The corresponding current-wave is

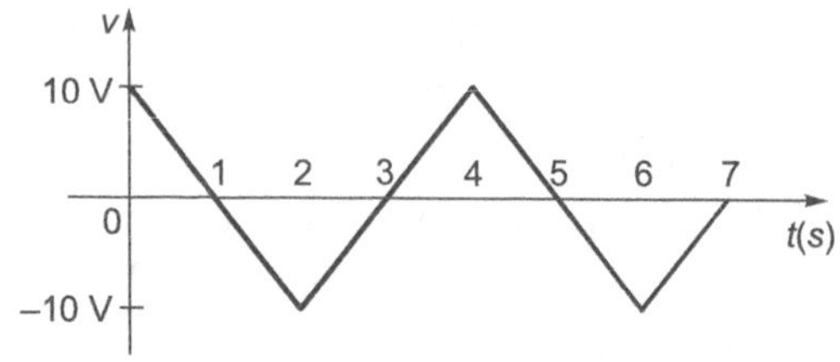

(a)
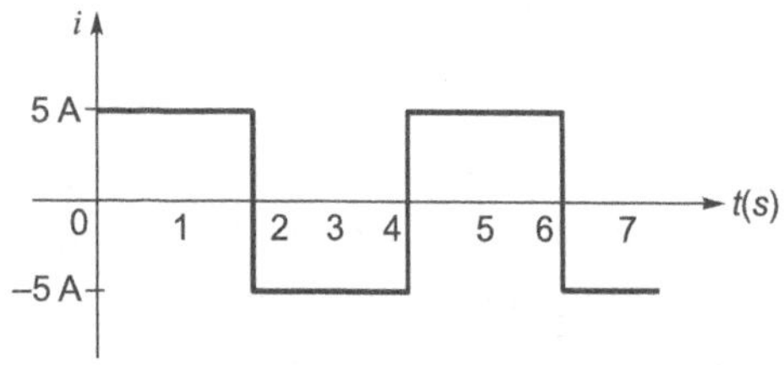

(b)
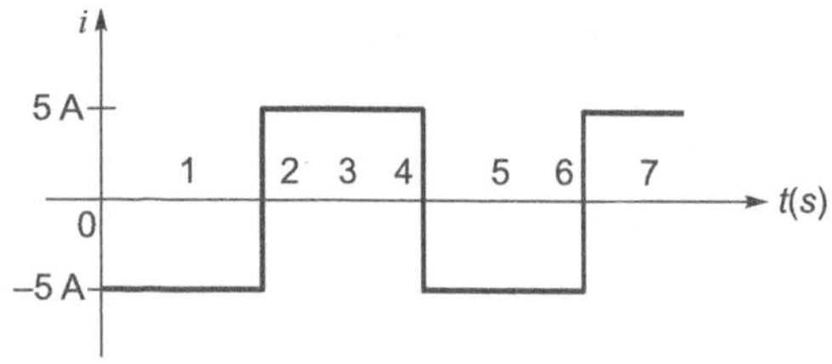

(c)
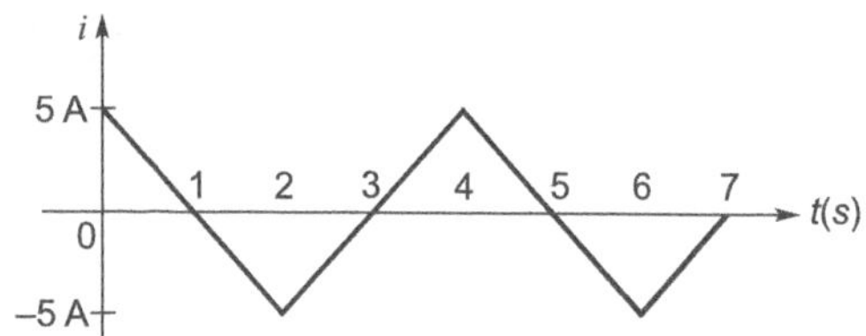

(d)
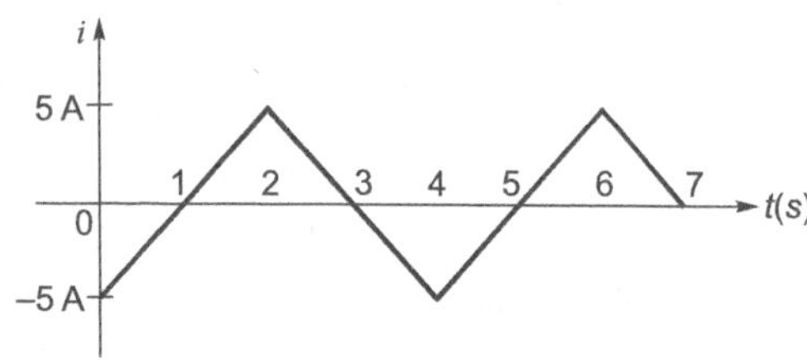

[ESE-2018]

1.81 For the network shown in the figure, the current flowing through the 5 Ω resistance will be

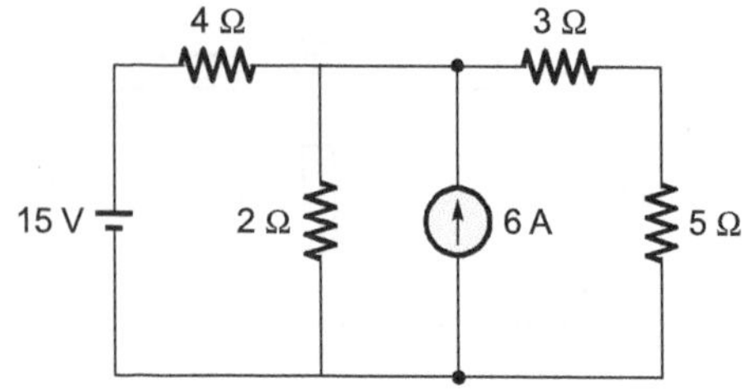

(a) $\frac{37}{25}$ A (b) $\frac{40}{28}$ A

(c) $\frac{39}{28}$ A (d) $\frac{41}{28}$ A

[ESE-2018]

1.82 The current in the 1 Ω resistor in the network as shown is

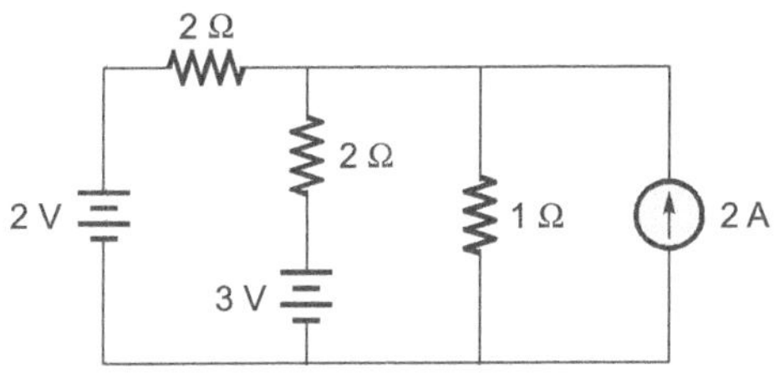

(a) 2.00 A (b) 2.25 A
(c) 2.50 A (d) 2.75 A

[ESE-2018]

1.83 A pulse of +10 V in magnitude and 2*s* in duration is applied to the terminals of a lossless inductor of 1.0 H. The current through the inductor would

(a) be a pulse of +20 A for the duration of 2*s*.
(b) be a pulse of –20 A for the duration of 2*s*.
(c) increase linearly from zero to 20 A in 2*s*, and in the positive direction and from thereon, it remains constant at +20 A.
(d) increase linearly from zero to –20 A in 2*s*, and in the negative direction and from thereon, it remains constant at –20 A.

[ESE-2018]

1.84 If the voltage across an element in a circuit is linearly proportional to the current through it, then it is a

(a) Capacitor (b) Transformer
(c) Resistor (d) Inductor

[ESE-2019]

1.85 The impedance of a parallel circuit is $(10 - j30)\ \Omega$ at 1 MHz. The values of circuit elements will be

(a) 10 Ω and 6.4 mH (b) 100 Ω and 4.7 nF
(c) 10 Ω and 4.7 mH (d) 100 Ω and 6.4 nF

[ESE-2019]

1.86 The inductance of an air-cored coil is proportional to

1. The square of the number of turns.
2. The diameter of the coil.
3. A form factor, F, dependent on the ratio of coil radius to coil length plus winding depth.

Which of the above statements are correct?

(a) 1 and 2 only (b) 1 and 3 only
(c) 2 and 3 only (d) 1, 2 and 3

[ESE-2020]

1.87 A voltage source-series resistance combination is equivalent to a current source-parallel resistance combination if and only if their

1. Respective open-circuit voltages are equal.
2. Respective short-circuit current are equal.
3. Resistance remains same in both cases.

Which of the above statements are correct?

(a) 1 and 2 only (b) 1 and 3 only
(c) 2 and 3 only (d) 1, 2 and 3

[ESE-2020]

1.88 The resistance *R* of a conductor is

(a) $\frac{EA}{Jl}$ (b) $\frac{EJ}{Al}$
(c) $\frac{El}{JA}$ (d) $\frac{JA}{El}$

where, E = Electrical field intensity
A = Cross-sectional area
J = Current density
l = Length of conductor

[ESE-2020]

1.89 Which of the following statements are correct for an ideal constant voltage source?

1. Its output voltage remains absolutely constant whatever the change in load current.
2. It possesses zero internal resistance so that internal voltage drop in the source is zero.
3. Output voltage provided by the source would remain constant irrespective of the amount of current drawn from it.
4. Output voltage provided by the source varies with the amount of current drawn from it.

(a) 1, 2 and 4 only (b) 1, 3 and 4 only
(c) 2, 3 and 4 only (d) 1, 2 and 3 only

[ESE-2020]

1.90 An uncharged capacitor of 0.01 F is charged first by a current of 2 mA for 30 s and then by a current of 4 mA for 30 s. The final voltage in it will be

(a) 12 V (b) 18 V
(c) 24 V (d) 30 V

[ESE-2020]

1.91 An energy source forces a constant current of 2 A for 10 s to flow through a light bulb. If 2.3 kJ is given off in the form of light and heat energy, what is the voltage drop across the bulb?

(a) 120 V (b) 115 V
(c) 110 V (d) 105 V

[ESE-2021]

1.92 What is the value of V_o in the given circuit?

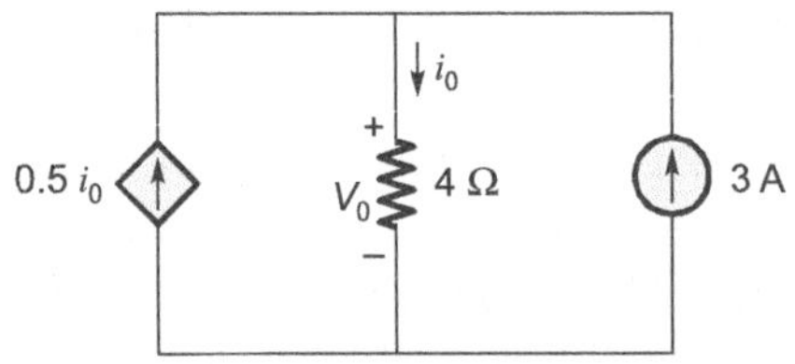

(a) 12 V (b) 6 V
(c) 24 V (d) 15 V [ESE-2021]

1.93 Three light bulbs are connected to a 9 V battery as shown in the figure. What are the values of the resistance of 20 W, 15 W, 10 W bulbs respectively?

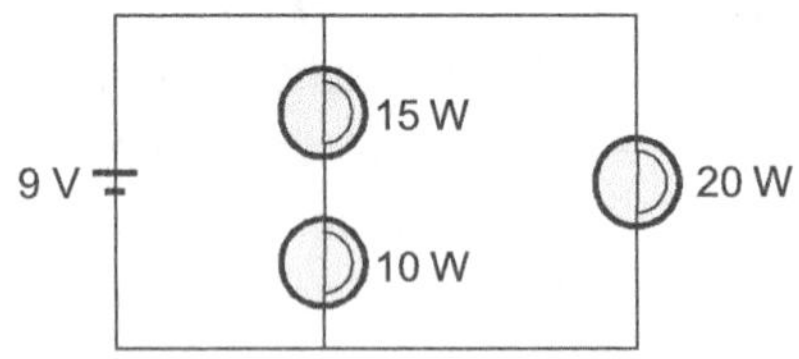

(a) 4.05 Ω, 1.945 Ω, 1.297 Ω
(b) 6.02 Ω, 3.762 Ω, 3.162 Ω
(c) 7.672 Ω, 4.887 Ω, 4.223 Ω
(d) 8.345 Ω, 6.893 Ω, 5.634 Ω [ESE-2021]

1.94 For the circuit in figure, the values of i_1 and i_4 are respectively,

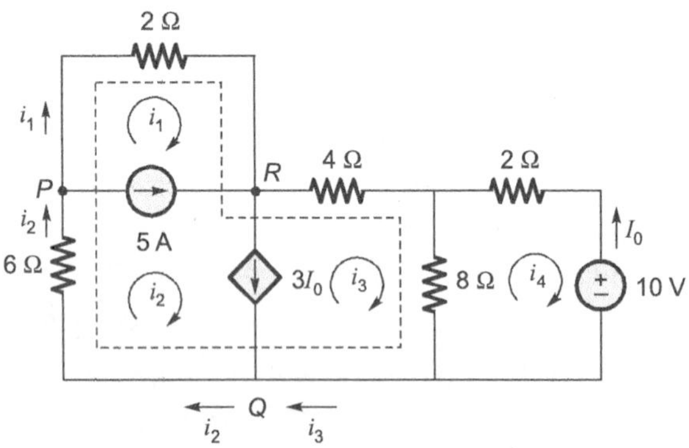

(a) –7.5 A, 2.143 A (b) –2.5 A, 3.93 A
(c) 3.4 A, –6.5 A (d) 7.1 A, –3.4 A
[ESE-2021]

1.95 What is C_{eq} for the given circuit?

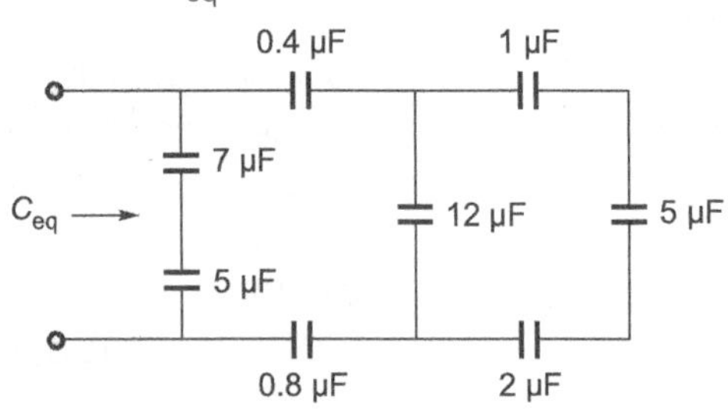

(a) 6.18 μF (b) 3.18 μF
(c) 8.23 μF (d) 12.67 μF [ESE-2022]

1.96 Consider the following statements with respect to Kirchhoff's laws for a circuit comprising of resistances and independent sources:

1. The number of independent element volt-ampere equations is equal to the number of resistances.
2. The number of independent KVL equations is equal to one more than the number of nodes.
3. The number of independent KVL equations is equal to the number of independent loops.

Which of the above statements is /are not correct?

(a) 1 only (b) 2 only
(c) 2 and 3 only (d) 1, 2 and 3
[ESE-2022]

1.97 What is the voltage v in the circuit diagram?

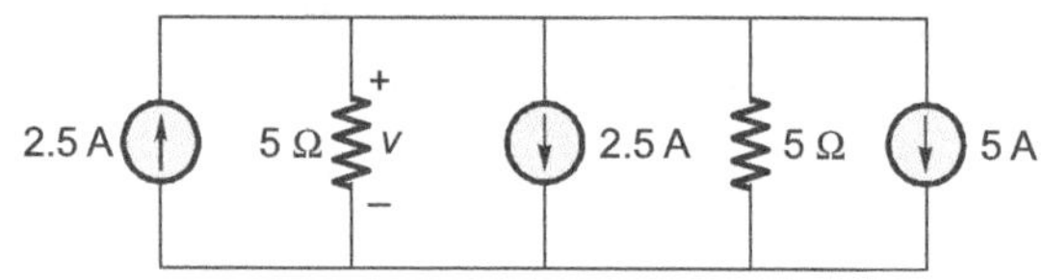

(a) 7.5 V (b) 16.5 V
(c) 12.5 V (d) 14.4 V [ESE-2022]

1.98 Choke coils are made of iron core, because
(a) it has less loss in iron cores
(b) large-valued flux densities can be produced in iron cores
(c) it is easily available in the market
(d) it has laminated core with high resistance
[ESE-2022]

1.99 What are the unknown currents i_5 and i_3 respectively for the circuit showni n the figure below?

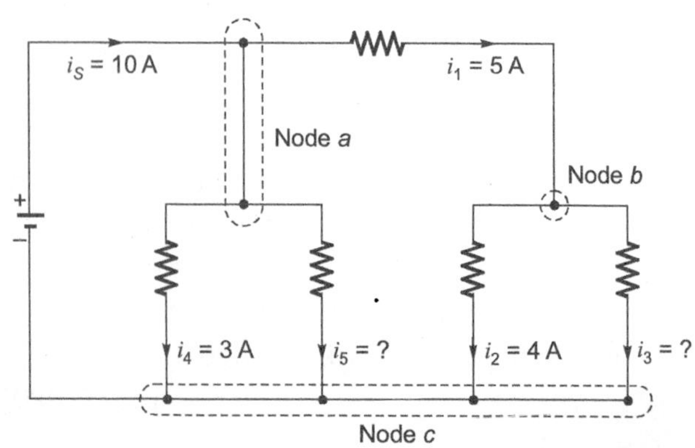

(a) 2 A and 1 A (b) 1 A and 2 A
(c) 18 A and 9 A (d) 9 A and 18 A

[ESE-2023]

1.100 What is the current delivered by the 10 V source in the circuit shown below?

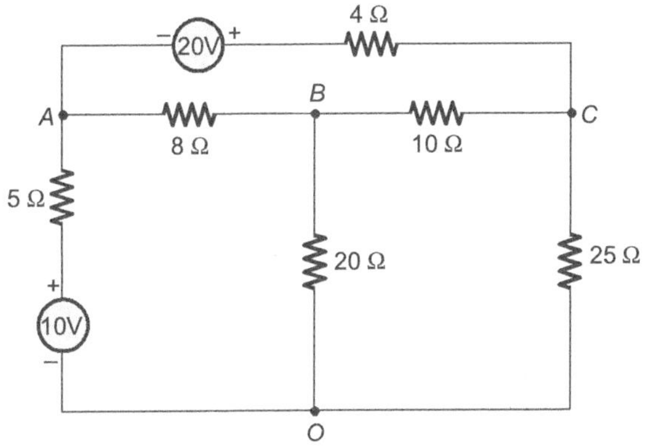

(a) 2.132 A (b) 3.132 a
(c) 1.132 A (d) 0.532 A

[ESE-2023]

1.101 What is the value of V_{AB} in the network shown in the figure?

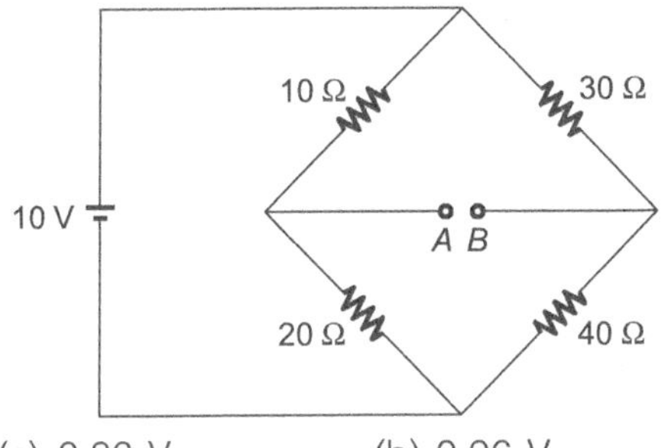

(a) 0.86 V (b) 0.96 V
(c) 0.66 V (d) 0.76 V

[ESE-2024]

1.102 Five resistances of 10 Ω each are connected between terminals *A* and *B* as shown in the figure. What is the total resistance between terminals *A* and *B*?

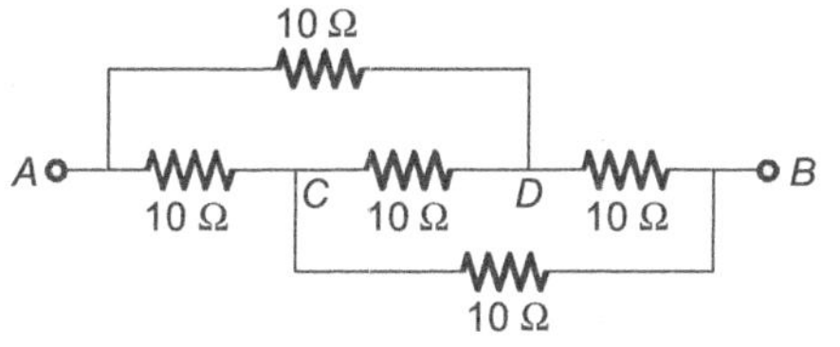

(a) 5 Ω (b) 10 Ω
(c) 15 Ω (d) 20 Ω

[ESE-2024]

1.103 An initially uncharged 1 mF capacitor has the current shown in the figure. What is the voltage across it at t = 2 ms?

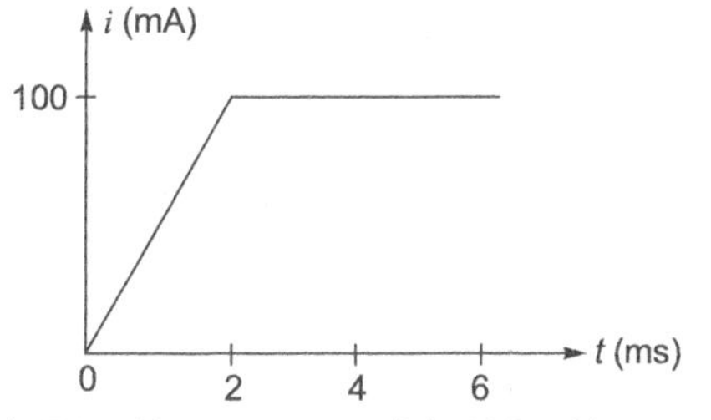

(a) 50 mV (b) 100 mV
(c) 200 mV (d) 250 mV

[ESE-2024]

■■■■

Answers Basics

1.1 (b)	1.2 (a)	1.3 (d)	1.4 (c)	1.5 (a)	1.6 (c)	1.7 (b)	1.8 (b)	1.9 (c)
1.10 (a)	1.11 (c)	1.12 (b)	1.13 (d)	1.14 (c)	1.15 (c)	1.16 (a)	1.17 (c)	1.18 (d)
1.19 (b)	1.20 (b)	1.21 (d)	1.22 (a)	1.23 (b)	1.24 (b)	1.25 (c)	1.26 (b)	1.27 (*)
1.28 (d)	1.29 (d)	1.30 (c)	1.31 (b)	1.32 (d)	1.33 (a)	1.34 (b)	1.35 (a)	1.36 (c)
1.37 (c)	1.38 (b)	1.39 (a)	1.40 (d)	1.41 (a)	1.42 (a)	1.43 (d)	1.44 (c)	1.45 (c)
1.46 (c)	1.47 (b)	1.48 (b)	1.49 (c)	1.50 (d)	1.51 (b)	1.52 (d)	1.53 (b)	1.54 (a)
1.55 (a)	1.56 (b)	1.57 (a)	1.58 (*)	1.59 (d)	1.60 (a)	1.61 (c)	1.62 (d)	1.63 (b)
1.64 (b)	1.65 (a)	1.66 (d)	1.67 (b)	1.68 (b)	1.69 (d)	1.70 (b)	1.71 (d)	1.72 (a)
1.73 (b)	1.74 (b)	1.75 (c)	1.76 (b)	1.77 (b)	1.78 (a)	1.79 (a)	1.80 (b)	1.81 (c)
1.82 (b)	1.83 (c)	1.84 (c)	1.85 (b)	1.86 (b)	1.87 (d)	1.88 (c)	1.89 (d)	1.90 (b)
1.91 (b)	1.92 (c)	1.93 (a)	1.94 (a)	1.95 (b)	1.96 (b)	1.97 (c)	1.98 (b)	1.99 (a)
1.100 (c)	1.101 (b)	1.102 (b)	1.103 (b)					

Explanations Basics

1.1 (b)

Applying KVL in the loop

$24 - 1I + 2V_b - V_b - 4I = 0$

where, $V_b = 3I$

$\Rightarrow \quad 24 - 5I + V_b = 0$

$\Rightarrow \quad 24 - 5I + 3I = 0$

$I = 12 \text{ A}$

1.2 (a)

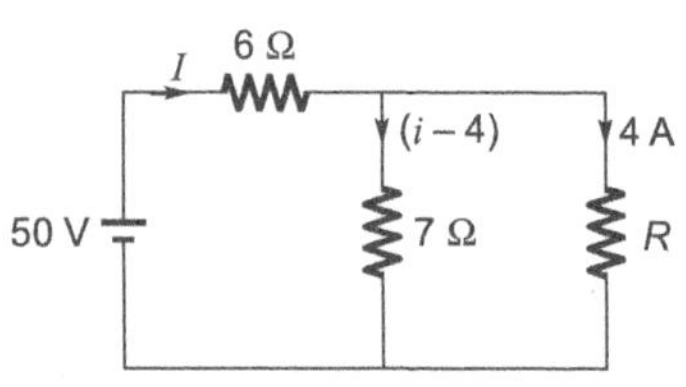

By applying KVL in 1st loop

$50 = 6i + 7(i - 4)$

$\Rightarrow \quad 13i = 78$

$\Rightarrow \quad i = 6 \text{ A}$

Now, by applying KVL in 2nd loop

$7 \times 2 = 4 \times R$

$R = 3.5\ \Omega$

1.4 (c)

By applying KVL

The terminal v-i characteristic is $v = 2i + 5$

1.6 (c)

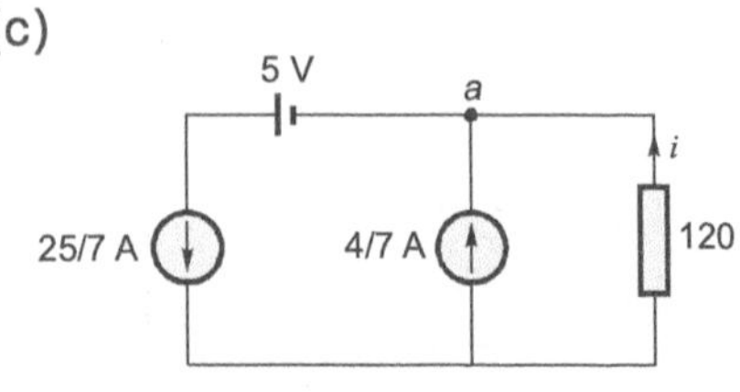

By applying KCL at node a

$$i = \frac{25}{7} - \frac{4}{7} = 3 \text{ A}$$

Voltage source in series with constant current source will behave like short circuit.

1.7 (b)

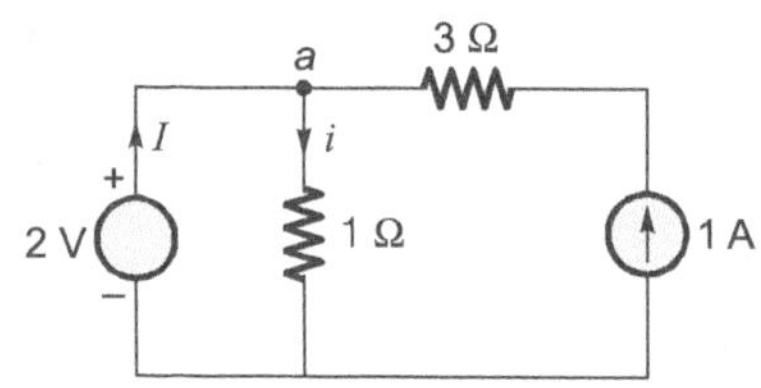

From figure, $i = \dfrac{2}{1} = 2 \text{ Amp}$

By applying KCL at node a

$I = i - 1 = 2 - 1 = 1 \text{ Amp}$

i.e. the current delivered by voltage source = 1 amp.

$\therefore$ power delivered by voltage source

$= 2 \times 1 = 2 \text{ W}$

1.9 (c)

Other requirements for reciprocity

(i) Network should have only single-source.

(ii) Networks should not have dependent sources.

1.10 (a)

$$R_1 = \frac{V^2}{P_1} \text{ and } R_2 = \frac{V^2}{P_2}$$

Bulbs are connected in series

$$R_{eq} = R_1 + R_2$$

$$= V^2\left[\frac{1}{P_1}+\frac{1}{P_2}\right] = V^2\left[\frac{P_1+P_2}{P_1P_2}\right]$$

$$\text{Total power} = \frac{V^2}{R_{eq}} = \frac{P_1P_2}{P_1+P_2}$$

1.11 (c)

$$i_1 = \sqrt{\frac{P_1}{R}} \text{ and } i_2 = \sqrt{\frac{P_2}{R}}$$

when both source are active

$$i = i_1 \pm i_2 = \sqrt{\frac{P_1}{R}} \pm \sqrt{\frac{P_2}{R}}$$

$$\text{Total power} = i^2R$$

$$= \left(\sqrt{\frac{P_1}{R}} \pm \sqrt{\frac{P_2}{R}}\right)^2 R$$

$$= (\sqrt{P_1} \pm \sqrt{P_2})^2$$

1.12 (b)

When, $V_s = 0$

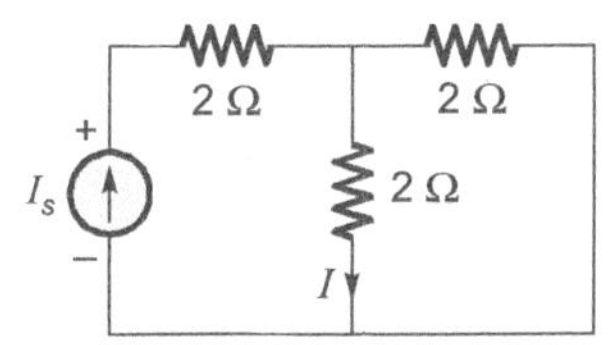

$$I = \frac{I_s}{2}$$

$$\Rightarrow \quad I_s = 2I = 2 \times 4 = 8 \text{ A}$$

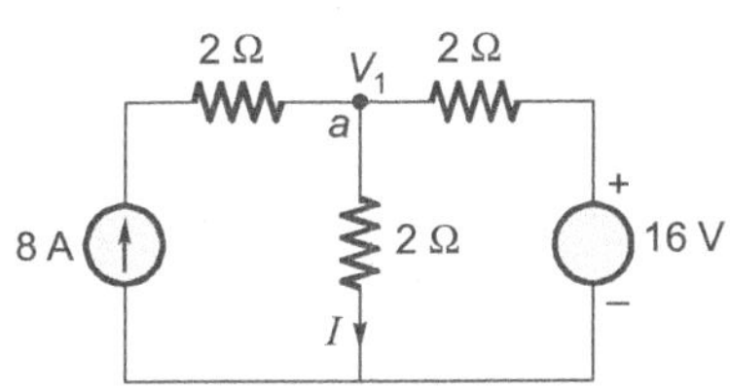

When $V_s = 16$ V

Applying KCL node a,

$$-8+\frac{V_1}{2}+\frac{V_1-16}{2} = 0$$

$$V_1 = 16 \text{ V}$$

$$I = \frac{V_1}{2} = \frac{16}{2} = 8 \text{ A}$$

Alternate solution:

By superposition,

$V_s = 0, I_s = 8$ A, $I_1 = 4$ A,

When, $V_s = 16$ V, $I_s = 0, I_2 = 4$

So, $I = 4 + 4 = 8$ A

1.13 (d)

Ideal voltage source keeps the terminal voltage constant so accordingly current will change and the voltage across each resistor is unchanged following superposition principle.

1.14 (c)

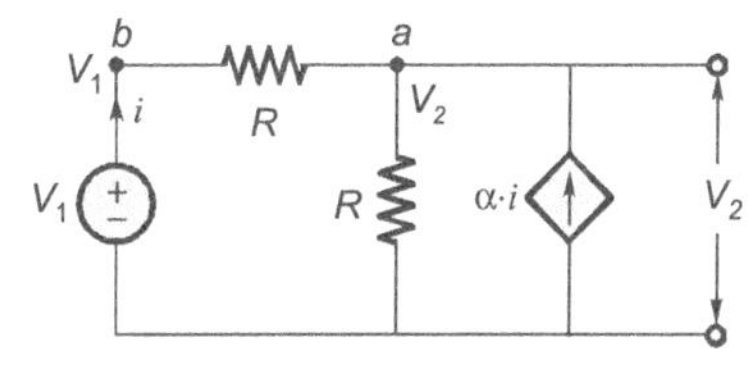

$$\frac{V_1 - V_2}{R} = i \qquad \ldots(i)$$

By applying KCL at node a

$$(1+\alpha)i = \frac{V_2}{R} \qquad \ldots(ii)$$

From equation (i),

$$(1+\alpha)\left(\frac{V_1 - V_2}{R}\right) = \frac{V_2}{R}$$

$$\therefore \quad \left(\frac{V_1}{V_2} - 1\right) = \frac{1}{1+\alpha}$$

$$\frac{V_1}{V_2} = \frac{2+\alpha}{1+\alpha}$$

$$\therefore \quad \frac{V_2}{V_1} = \frac{1+\alpha}{2+\alpha}$$

1.15 (c)

Circuit is symmetrical, so voltage at nodes V_A, V_B and V_C are equal and no current flows between these nodes.

OR

DCEB and *DBEA* are two balanced Wheatstone's bridges hence these will not be any current in branch *CB* and *BA*.

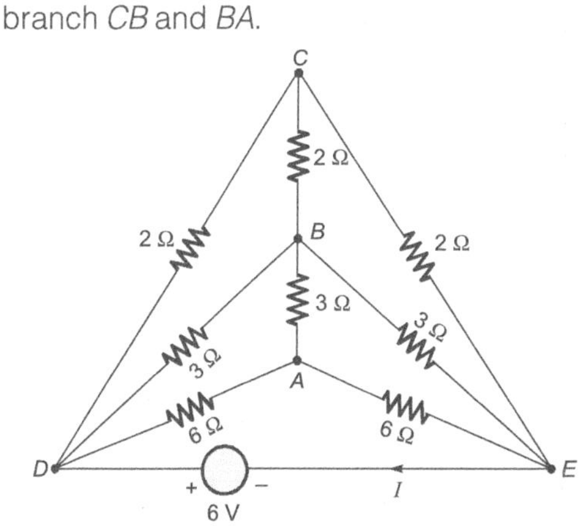

$$I = \frac{6}{2+2} + \frac{6}{3+3} + \frac{6}{6+6} = 3\text{ A}$$

1.16 (a)

In the given circuit there are only resistor/inductor or capacitor and a voltage source, of $V = 5 - 5 = 0$ V. As these are passive elements and follow the superposition theorem. The circuit is passive and linear.

1.17 (c)

As all the resistive branches are in parallel.

$\therefore$ voltage across each branch will be constant and will be equal to the dc supply voltage

$\therefore \quad V = IR = \text{constant}$

$$\Rightarrow \quad I \propto \frac{1}{R}$$

$$I_1 : I_2 : I_3 = \frac{1}{R_1} : \frac{1}{R_2} : \frac{1}{R_3}$$

$$= R_2R_3 : R_1R_3 : R_1R_2$$
$$= 24 : 12 : 8$$
$$= 6 : 3 : 2$$

1.18 (d)

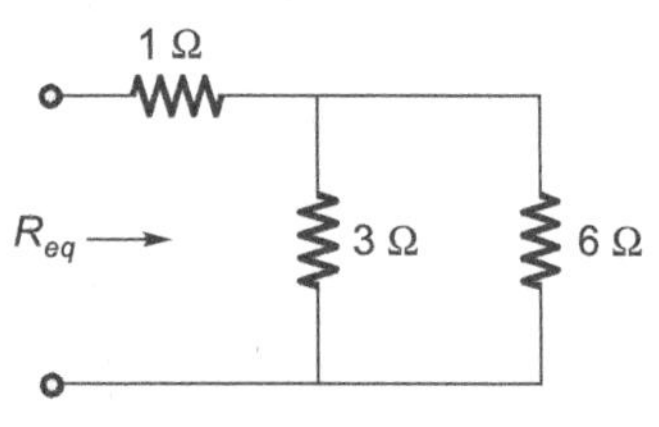

$R_{eq} = 1 + (3 \,||\, 6) = 3\ \Omega$

Voltage across current source

$I_sR_{eq} = 2 \times 3 = 6$ V

1.19 (b)

Apply Wheatstone's bridge balance equation,

$$\frac{4}{R} = \frac{10}{5} \Rightarrow R = 2\ \Omega$$

1.20 (b)

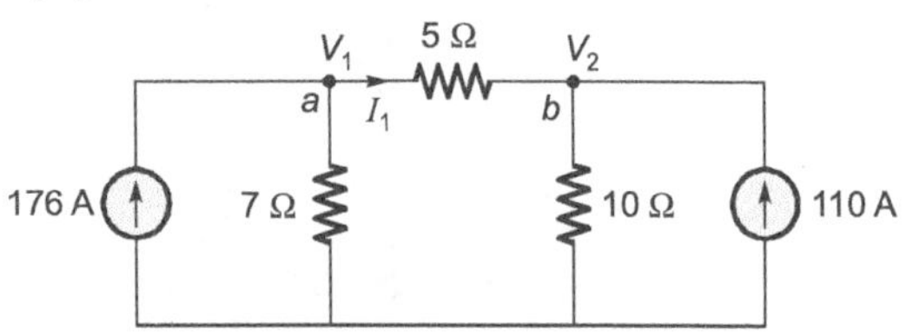

From the superposition theorem,

$$I_1 \text{ due to 176 A source} = \frac{7}{22} \times 176 = 56\text{ A}$$

From node a to node b,

$$I_2 \text{ due to 110 A source} = \frac{10}{22} \times 110 = 50\text{ A}$$

From node b to node a

$\therefore$ Voltage across 5 Ω resistor $= (V_1 - V_2)$

$= (56 - 50)\, 5 = 30$ V

1.21 (d)

$$I_R = \frac{24}{6} = 4\text{ A}$$

Current delivered by the voltage source.

$$I = I_R + 2I_R = 3I_R = 3 \times 4 = 12\text{ A}$$

Power delivered by the voltage source

$= 24 \times 12 = 288$ W

1.22 (a)

Fewer nodes will give lesser nodal equation than mesh equations hence better analyzed by node analysis. So point 3 is not correct.

1.23 (b)

$$\text{Applying KCL,} \quad -1 + \frac{8-2}{2} - I = 0 \Rightarrow I = 2\text{ A}$$

1.24 (b)

Applying KCL,

$$-5 + \frac{V}{10} + \frac{V-E}{5} = 0$$

$$\Rightarrow \quad -5 + \frac{10}{10} + \frac{10-E}{5} = 0$$

$$\Rightarrow \quad E = -10\text{ V}$$

1.25 (c)

Assuming voltage of the node V

Applying KCL,

$$\frac{V-140}{2} - 70 + \frac{V}{4} + \frac{V-70}{1} = 0$$

$$\Rightarrow \quad V = 120\text{ V}$$

$$I = \frac{V}{4} = \frac{120}{4} = 30\text{ A}$$

1.26 (b)

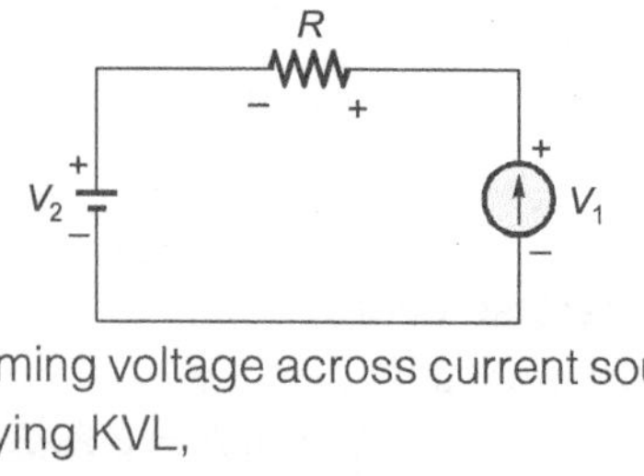

Assuming voltage across current source is V_1

Applying KVL,

$$V + IR - V_1 = 0$$

$$\Rightarrow \quad V_1 = V + IR$$

1.27 (*)

By KVL Source voltage,

$$V_S = E + IR$$

So new source voltage $= \dfrac{E + IR}{2}$

Hence again by KVL,

$$E - V - \frac{E + IR}{2} = 0$$

So, $$V = \frac{E - IR}{2}$$

1.28 (d)

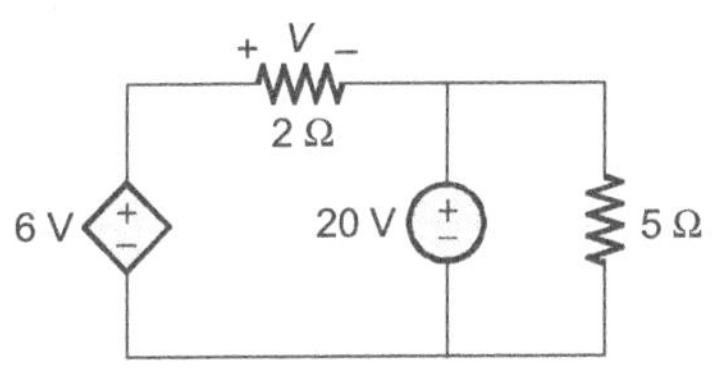

$$6\text{ V} - V - 20 = 0$$

$$V = 4\text{ V}$$

Current through 2 Ω resistor $= \dfrac{V}{2} = \dfrac{4}{2} = 2\text{ A}$

1.29 (d)

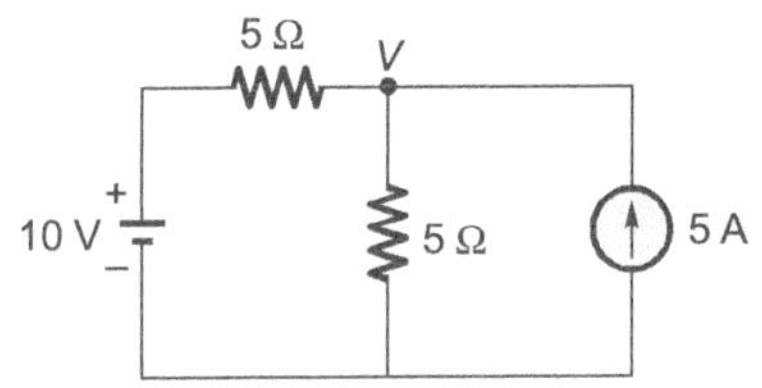

Applying Nodal Analysis,

$$\frac{10 - V}{5} + 5 = \frac{V}{5}$$

$$\therefore \quad 10 - V + 25 = V$$

$$\therefore \quad V = 17.5\text{ V}$$

1.30 (c)

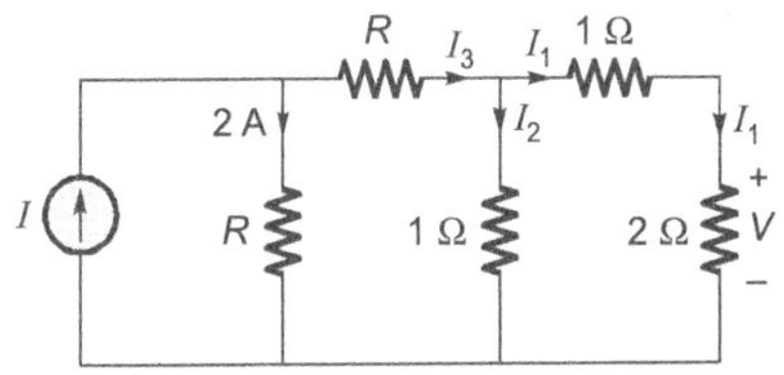

$$V = 2\text{ V}$$

$$\therefore \quad I_1 = \frac{V}{2} = 1\text{ A}$$

and $$I_1 = \left(\frac{1}{1+3}\right) I_3$$

$$\frac{1}{I} = \frac{1}{4} I_3$$

$$\therefore \quad I_3 = 4\text{ A}$$

and $$I = 4 + 2 = 6\text{ A}$$

1.31 (b)

Applying Nodal Analysis

$$20 = \frac{V_A}{2} + \frac{V_A - V_B}{2}$$

$$40 = 2V_A - V_B \quad \ldots(i)$$

and $$\frac{V_A + V_B}{2} = \frac{V_B}{1} + 10$$

$$V_A - V_B = 2V_B + 20$$

$$\therefore \quad V_A - 3V_B = 20 \quad \ldots(ii)$$

Solving (*i*) and (*ii*),

$$V_A = 20\text{ V}$$

1.32 (d)

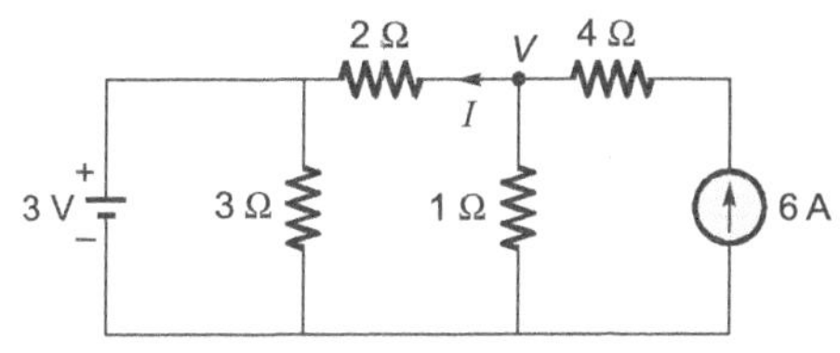

Applying Nodal Analysis,

$$6 = \frac{V-3}{2} + \frac{V}{1}$$

$$12 = V - 3 + 2V$$

$$\therefore \quad \boxed{V = 5\text{ V}}$$

$$I = \frac{5-3}{2} = 1\text{ A}$$

Alternate solution:

In the above figure 3 Ω and 4 Ω resistor are redundant so the circuit can be transformed into

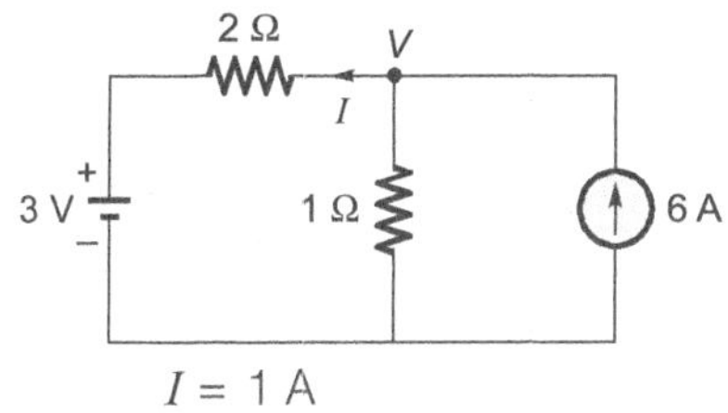

so, $$I = 1\text{ A}$$

1.33 (a)

$$i_{rms}^2 = \frac{1}{T}\int_0^T (i_{avg})^2 \cdot dt$$

$$\therefore \quad i_{avg} = \left(\frac{9-0}{3-0}\right) \times t = 3t$$

$$\therefore \qquad i_{rms}^2 = \frac{1}{3}\int_0^3 (3t)^2 \times dt$$

$$= \frac{1}{3}\times 9 \times \left[\frac{t^3}{3}\right]_0^3 = 27$$

and $\qquad P = i_{rms}^2 \times R$

$$= 27 \times 10 = 270 \text{ W}$$

1.34 (b)

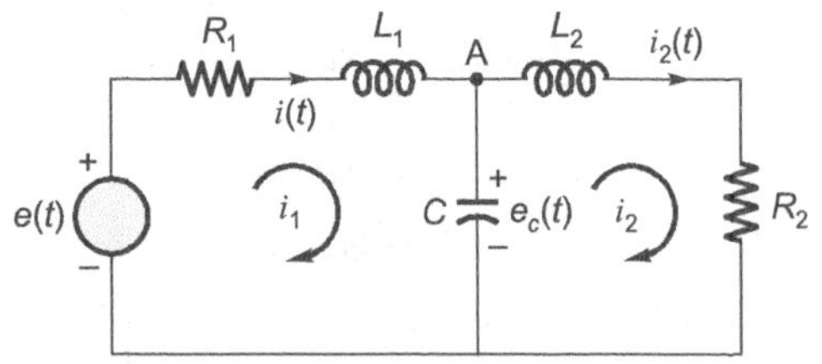

Applying Nodal at Point A

$$i_1(t) = \frac{Cde_c(t)}{dt} + i_2(t)$$

$$\frac{Cde_c(t)}{dt} = i_1(t) - i_2(t)$$

Applying KVL in loop 1

$$L_2\frac{di_2(t)}{dt} = -R_2\, i_2(t) + e_c(t)$$

Applying KVL in loop 2

$$L_1\frac{di_1(t)}{dt} = -R_1 i_1(t) - e_c(t) + e(t)$$

1.35 (a)

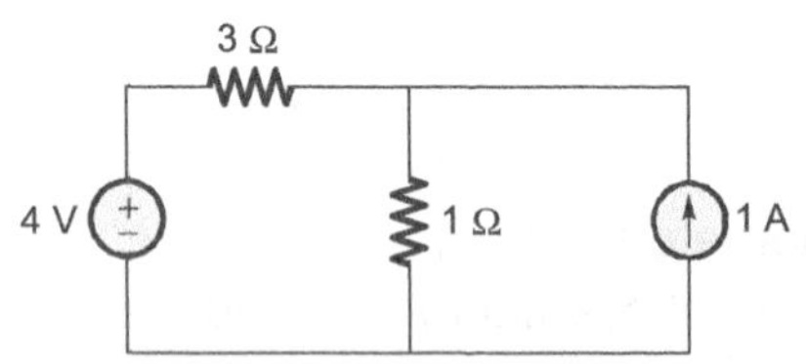

1. By O.C. the 1 A current source.

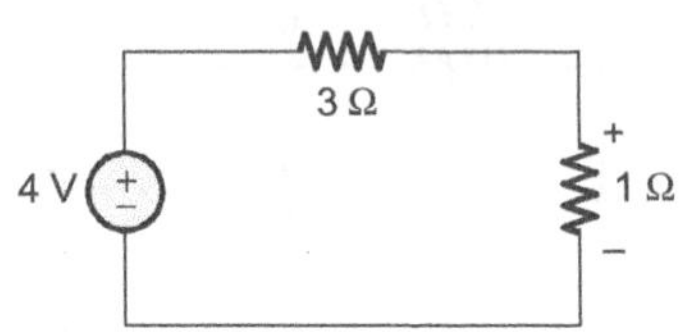

$$V'_{1\Omega} = \frac{4\times 1}{4} = 1\text{V}$$

2. By S.C. the voltage source.

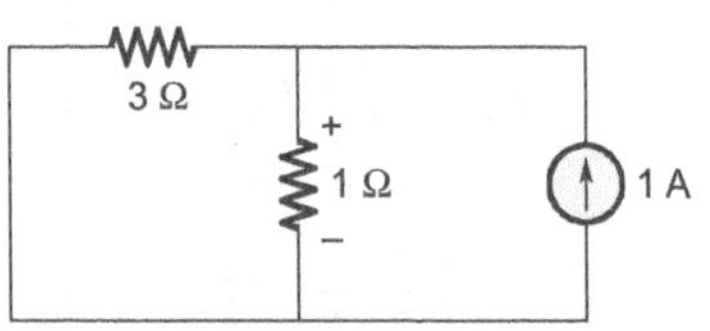

$$V''_{1\Omega} = \frac{3\times 1}{4} = \frac{3}{4}\text{V}$$

By superposition theorem,

$$V_{1\Omega}^{\text{total}} = V'_{1\Omega} + V''_{1\Omega} = 1 + \frac{3}{4} = \frac{7}{4}\text{V}$$

1.36 (c)

By open circuiting the 4 A current source.

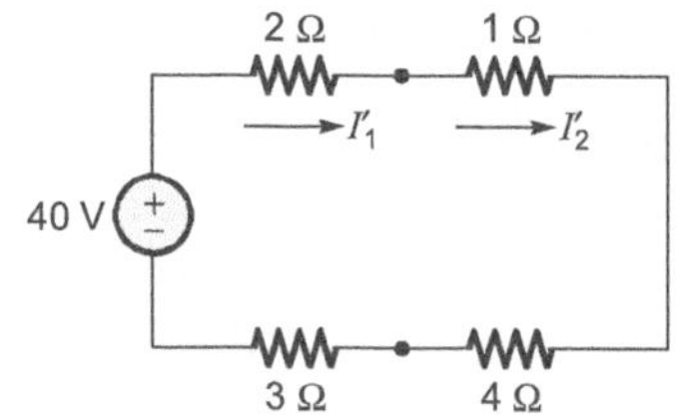

$$I'_1 = I'_2 = 4 \text{ A}$$

By short-circuiting 40 V voltage source.

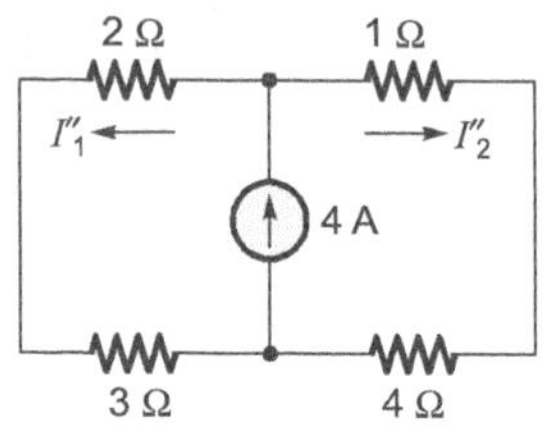

$$I''_1 = I''_2 = 2 \text{ A}$$

By superposition theorem

$$I_1 = I'_1 - I''_1 = 2 \text{ A}$$

$$I_2 = I'_2 - I''_2 = 6 \text{ A}$$

1.37 (c)

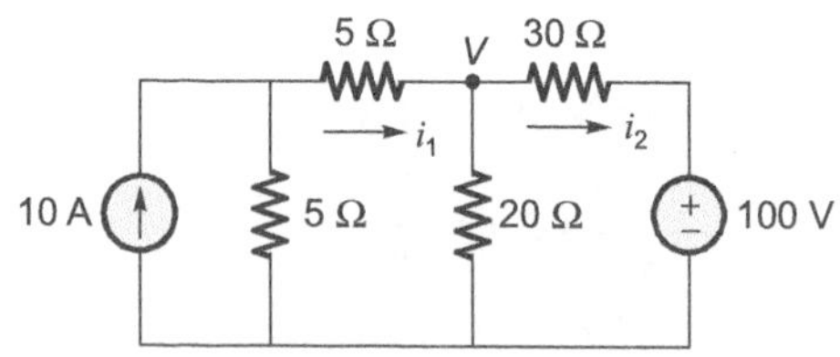

By applying source transformation,

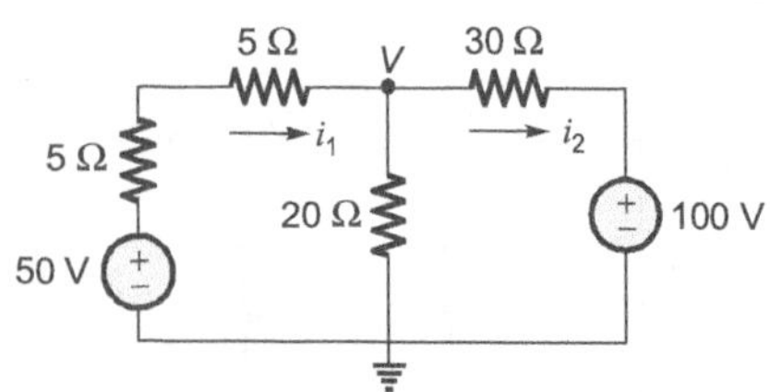

By applying KCL

$$\frac{V-50}{10} + \frac{V}{20} + \frac{V-100}{30} = 0$$

$$\frac{11}{60}V = 8.33$$

$$V = 45.45 \text{ V}$$

$$i_1 = \frac{50 - 45.45}{10} = 0.4545 \text{ A}$$

$$i_2 = \frac{45.45 - 100}{30} = -1.818 \text{ A}$$

1.38 (b)

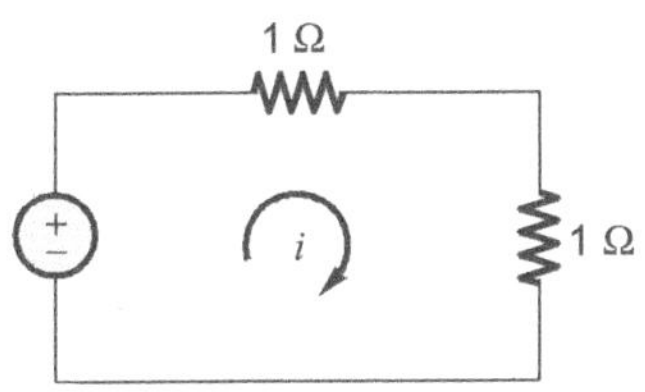

$$i = \frac{E}{2}$$

Power dissipated,

$$P = \left(\frac{E}{2}\right)^2 \times 2 = 8$$

$$E^2 = 16$$

$$E = 4 \text{ V}$$

1.39 (a)

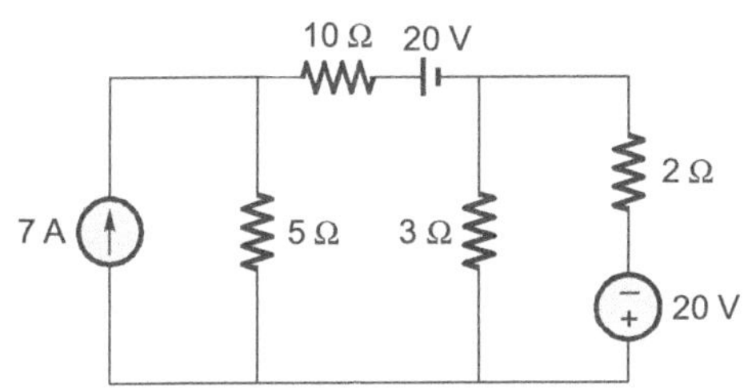

By applying source transformation,

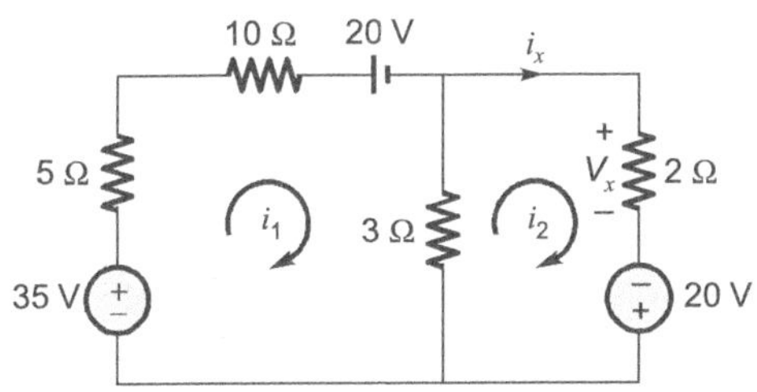

$$35 - 20 = 18i_1 - 3i_2$$

$$15 = 18i_1 - 3i_2 \quad ...(i)$$

$$20 = 5i_2 - 3i_1 \quad ...(ii)$$

By equation (i) and (ii),

$$i_1 = 1.66 \text{ A}$$

$$i_2 = 5 \text{ A}$$

so, $\quad i_x = 5 \text{ A}$

$$V_x = 10 \text{ V}$$

1.40 (d)

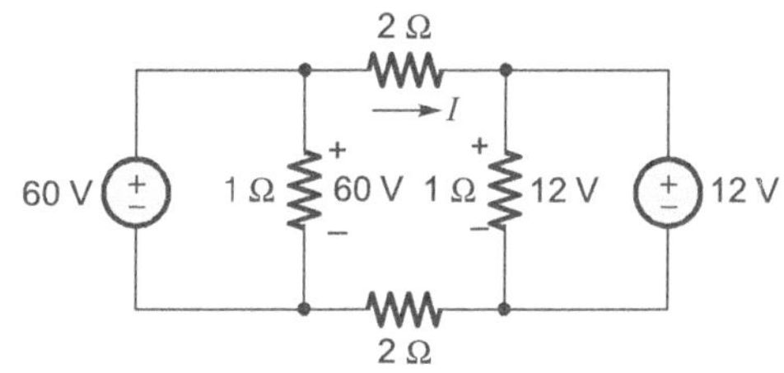

$$-60 + 2I + 12 + 2I = 0$$

$$48 = 4I \Rightarrow I = 12 \text{ A}$$

1.41 (a)

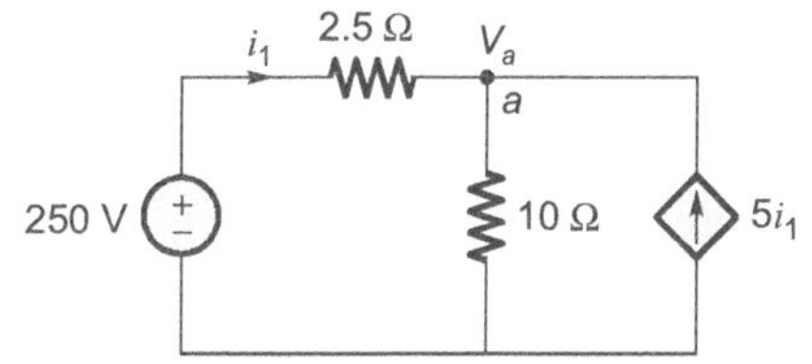

Applying KCL at node a.

$$-i_1 - 5i_1 + \frac{V_a}{10} = 0$$

$$\Rightarrow \quad V_a = 60\, i_1$$

$$i_1 = \frac{250 - V_a}{2.5} = \frac{250 - 60i_1}{2.56}$$

$$\Rightarrow \quad i_1 = \frac{250}{62.5} = 4 \text{ A}$$

Hence, option (a) is correct.

1.42 (a)

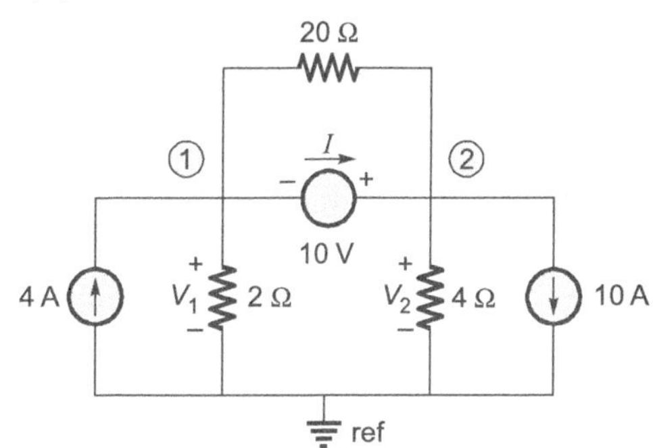

Let current through 10 V source is I.

Applying KCL at node 1,

$$-4 + \frac{V_1}{2} + \frac{V_1 - V_2}{20} + I = 0 \quad ...(i)$$

Applying KCL at node 2,

$$-I + \frac{V_2}{4} + \frac{V_2 - V_1}{20} + 10 = 0 \quad ...(ii)$$

Adding equation (i) and (ii) we get,

$$-6 = \frac{V_1}{2} + \frac{V_2}{4}$$

Hence, option (a) is correct.

1.43 (d)

Applying KCL,

$$\frac{V-30}{10}-9+\frac{V-36}{20}=0$$

$$V = 92 \text{ V}$$

Hence, option (d) is correct.

1.44 (c)

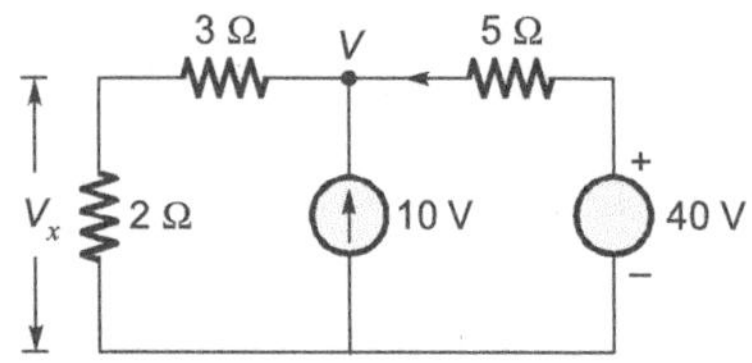

Applying KCL,

$$\frac{V}{5}+\frac{V-40}{5}-10 = 0$$

$$V = 45 \text{ V}$$

$$V_x = \frac{V}{5}\times 2 = \frac{45}{5}\times 2 = 18 \text{ V}$$

Hence, option (c) is correct.

1.45 (c)

$$I_1 = \frac{E}{2R};\ I_2 = \frac{E}{4R};\ I_3 = \frac{E}{6R}$$

$$I_1 : I_2 : I_3 = \frac{E}{2R}:\frac{E}{4R}:\frac{E}{6R} = 6:3:2$$

1.46 (c)

The resistance between A and B with C open: 12 Ω

$$R_A + R_B = 12\ \Omega \quad \ldots(i)$$

The resistance between B and C with A open: 22 Ω

$$R_B + R_C = 22\ \Omega \quad \ldots(ii)$$

The resistance between C and A with B open: 18 Ω

$$R_A + R_C = 18\ \Omega \quad \ldots(iii)$$

On solving equation (*i*), (*ii*) and (*iii*) we get,

$R_A = 4\ \Omega$, $R_B = 8\ \Omega$ and $R_C = 14\ \Omega$

Hence, option (c) is correct.

1.47 (b)

List-I and List-II are dual of each other.

1.48 (b)

In any linear, planer network number of independent KVL equations are equal to number of links and KCL equations are $(n-1)$.

1.49 (c)

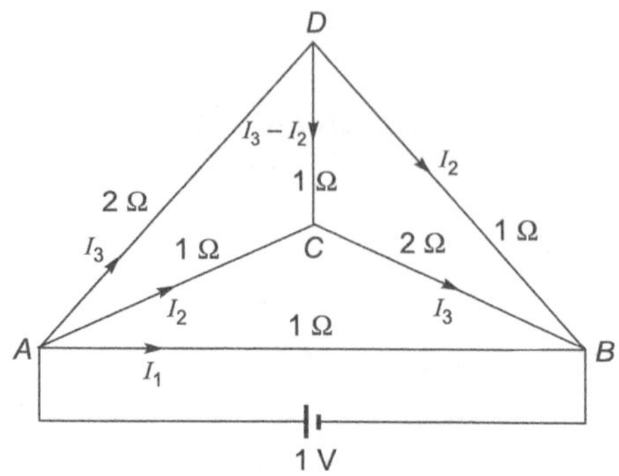

$$1 = 2I_3 + I_2 \quad \ldots(i)$$

$$2I_3 + I_3 - I_2 - I_2 = 0 \quad \ldots(ii)$$

$$\Rightarrow \quad I_3 = \frac{2}{3}I_2$$

From equation (*i*),

$$1 = 2\times\frac{2}{3}I_2 + I_2$$

$$1 = \frac{7}{3}I_2 \Rightarrow I_2 = \frac{3}{7}\text{ A}$$

1.50 (d)

As the circuit is a combination of series and parallel circuit and we know in series circuit current is same and in parallel voltage is same therefore X can not be current/voltage/resistance.

According to Tellegen's theorem the total instantaneous power given out by different sources always equals to the power consumed by various passive elements in various branches of the network at that time.

Therefore option (d) is correct.

1.51 (b)

Any element connected in parallel with a voltage source is redundant because a voltage source have a constant value of voltage across it and similarly the element connected in series with a current source is redundant because a current source supply a constant current to a network.

1.52 (d)

Let current flowing through the circuit = I

So that, $V_A = 7I$

∴ Using KVL,

$$36 - 15I - 2V_A - V_A = 0$$

$$36 - 15I - 3\times(7I) = 0$$

$$\Rightarrow \quad I = 1\text{ A}$$

and $\qquad V_A = 7\text{ V}$

$\therefore$ Power dissipated in the controlled source

$= 2\,V_A I = 14\text{ W}$

1.53 (b)

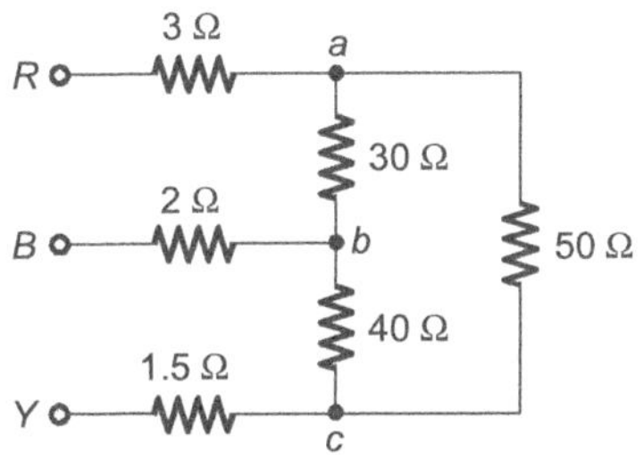

Using suitable star delta conversion

We have,

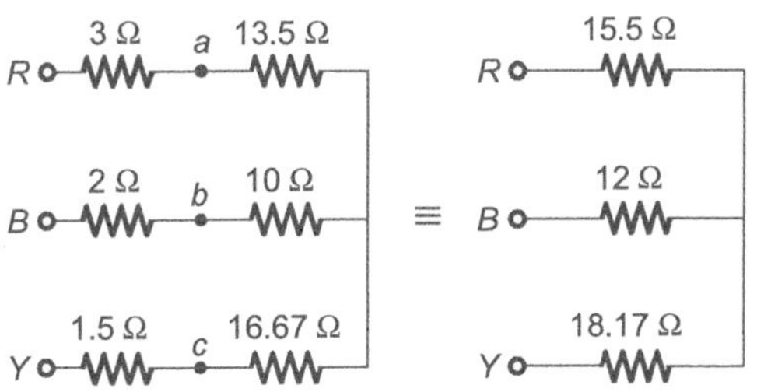

Now taking 2 phases at a time and grounding the third phase,

We have,

$$Z_{RB} = 27.5\ \Omega$$
$$Z_{BY} = 30.17\ \Omega$$
$$Z_{RY} = 33.67\ \Omega$$

1.54 (a)

Hint: Determinant of any network is independent of voltage and current sources.

1.55 (a)

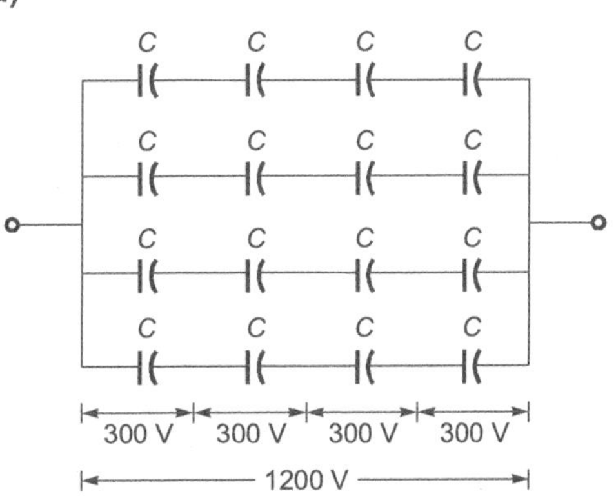

$$C = 2\,\mu\text{F}$$

$$C_{\text{overall}} = \frac{C}{4}+\frac{C}{4}+\frac{C}{4}+\frac{C}{4} = 2\,\mu\text{F}$$

Number of capacitor = 16

1.56 (b)

$$r = \frac{(r+2)\times 3}{r+2+3}+2$$

On solving, $\quad r = 4\ \Omega$

1.57 (a)

Whenever an ideal voltage source and ideal current source are connected in series, they will behave like an ideal current source alone and if they are connected in parallel, they will have an ideal voltage source alone.

1.58 (*)

$$P = I^2R = I^2 \times P\frac{l}{A}$$

$$P \propto I^2 \times \frac{l}{d^2}$$

$$\frac{w}{x} = \frac{P_1}{P_2} = \frac{I^2 \times \frac{l}{d^2}}{(3I)^2 \times \frac{l/2}{(2d)^2}}$$

$$= \frac{I^2 \times \frac{l}{d^2}}{9I^2 \times \frac{l}{2} \times \frac{1}{4d^2}} = \frac{8}{9}$$

$$x = \frac{9}{8}w$$

$\therefore$ No option is matching.

1.60 (a)

Voltage sources of different values cannot be connected in parallel, because voltage across a parallel paths will be equal.

The current sources of different values cannot be connected in series, because current through a series circuit path is same.

1.61 (c)

$$V = L\frac{di}{dt} = \frac{Ld(t)}{dt}$$

$$V = L = \text{constant} = 2\text{ V}$$

So, $\quad L = 2\text{ H}$

1.62 (d)

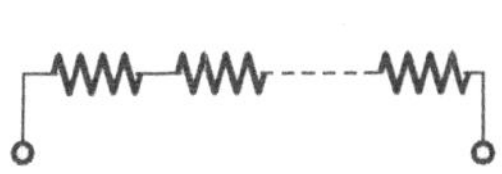

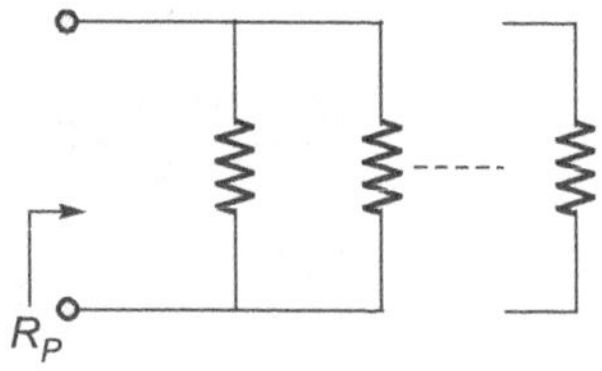

$$R_S = nR$$

$$R_P = \frac{R}{n}$$

$$\Rightarrow \quad n \cdot R = 50$$

$$2 = \frac{R}{n}$$

$$R = \frac{50}{n} \quad ...(1)$$

$$\Rightarrow \quad R = 2n \quad ...(2)$$

Multiplying equation (1) and (2),

$$R^2 = \frac{50}{n} \times 2n = 100$$

$$\Rightarrow \quad R = \sqrt{100}$$
$$= 10\ \Omega$$

1.63 (b)

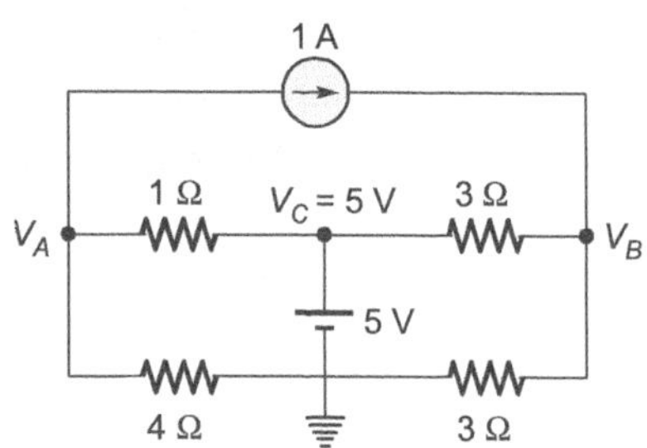

$$V_C - 0 = 5 \Rightarrow V_C = 5\text{ V}$$

At node A:

$$1 + \frac{V_A - 5}{1} + \frac{V_A}{4} = 0$$

$$5\ V_A = 16$$

$$\Rightarrow \quad V_A = \frac{16}{5}\text{ V}$$

At node B:

$$-1 + \frac{V_B - 5}{3} + \frac{V_B}{3} = 0 \Rightarrow 2V_B = 8$$

$$\therefore \quad V_B = 4\text{ V}$$

$$\therefore \quad V_{AB} = V_A - V_B$$

$$= \frac{16}{5} - 4$$

$$= -0.8\text{ V}$$

1.64 (b)

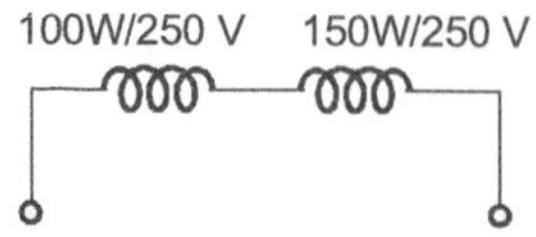

$$R_1 = \frac{V^2}{P_1} = \frac{250 \times 250}{100} = 625\ \Omega$$

$$R_2 = \frac{V^2}{P_2}$$

$$= \frac{250 \times 250}{150} = 416.67\ \Omega$$

$$R_{eq} = R_1 + R_2 = 1041.67\ \Omega$$

$$P = \frac{V^2}{R_{eq}} = \frac{250 \times 250}{1041.66} \simeq 60\text{ W}$$

1.65 (a)

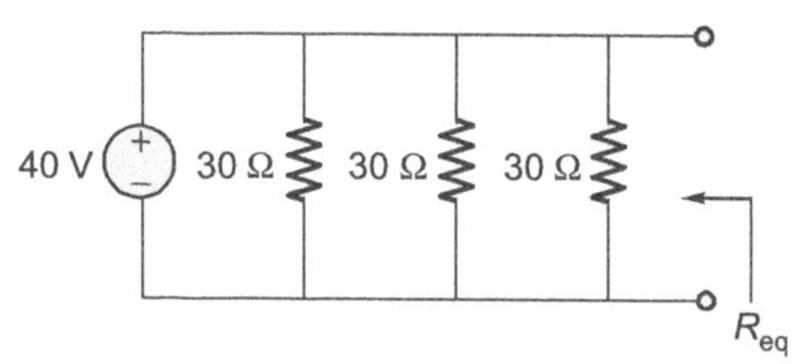

Replacing voltage source by a short circuit

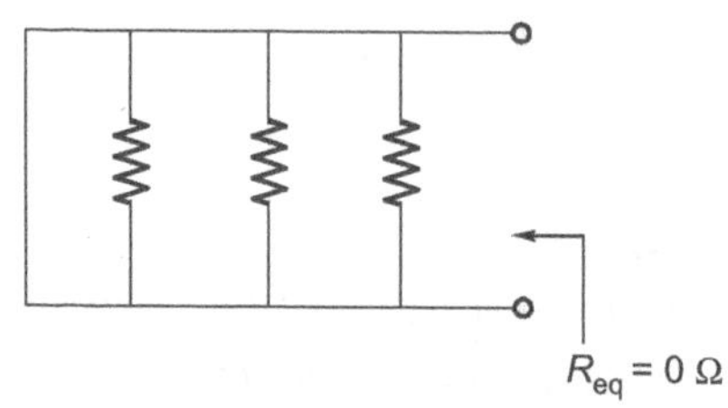

1.66 (d)

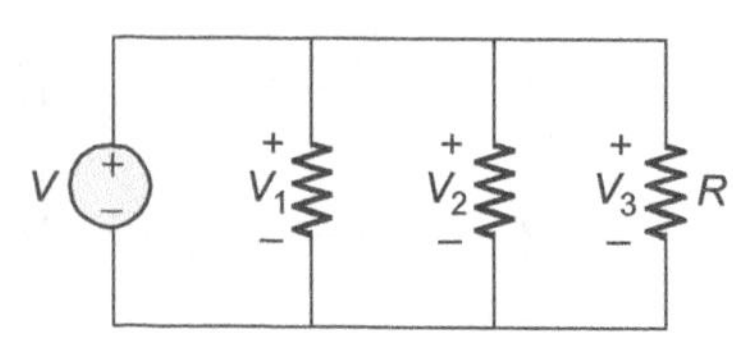

$$R_{eq} = \frac{R}{3}$$

$$P = \frac{V^2}{R_{eq}} = \frac{V^2}{R/3} = \frac{3\,V^2}{R}$$

$$V_1 = V_2 = V_3 = V$$

$$R_{eq} = \frac{R}{2}$$

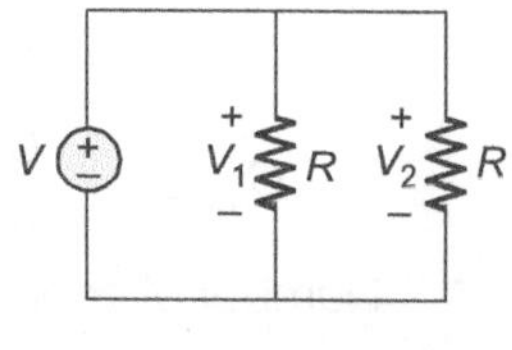

$$P = \frac{V^2}{R/2} = \frac{2V^2}{R}$$

$$V_1 = V_2 = V$$

As we can see from above circuits,

- Total resistance increases.
- Total power decreases.
- Total voltage and branch voltage remain unaffected.

We have taken equal resistances even unequal values give the same analysis.

1.67 (b)

We always prefer to take that node as the reference node that is connected to many parts of the network.

1.68 (b)

Mesh analysis is valid only for planar networks and for its application we apply KVL.

1.69 (d)

A short in any type of circuit causes the total circuit resistance to decrease and thereby the current to increase.

1.70 (b)

From the given waveforms,

$$\text{slope} = \frac{di}{dt} = \frac{2}{2} = 1 \text{ A/sec}$$

$$v = L\frac{di}{dt}$$

$$2 = L(1)$$

$$L = 2 \text{ H}$$

1.71 (d)

From the given network,

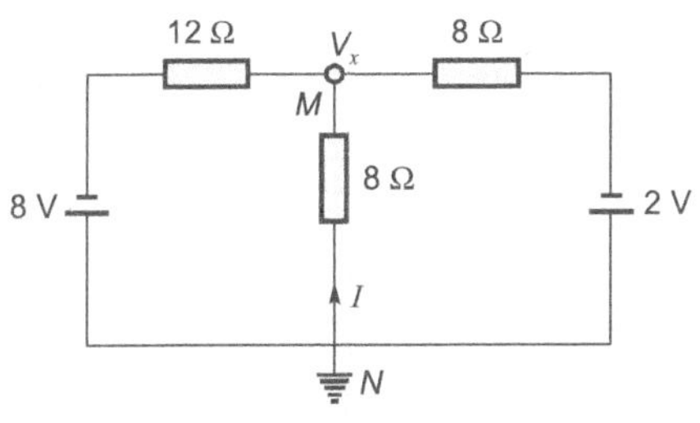

Applying KCL at M,

$$\frac{V_x+8}{12} + \frac{V_x}{8} + \frac{V_x+2}{8} = 0$$

$$2V_x + 16 + 3V_x + 3V_x + 6 = 0$$

$$8V_x = -22$$

$$\Rightarrow \quad V_x = -\frac{22}{8} \text{ V}$$

$$I = \frac{0-V_x}{8} = \frac{0-\left(-\frac{22}{8}\right)}{8} = \frac{22}{64} \text{ A}$$

$$= 0.343 \text{ A from } N \text{ to } M$$

1.72 (a)

$$V_{80\,\Omega} = \frac{25}{80+20} \times 80 = \frac{25}{100} \times 80$$

$$= \frac{1}{4} \times 80 = 20 \text{ V}$$

1.73 (b)

Kirchhoff's current law states that current entering the node = current leaving the node. Thus it is applicable at node only.

1.74 (b)

The network is linear if it satisfy superposition and homogeneity. Thus non linear network will not satisfy superposition and homogeneity.

1.75 (c)

Applying source transformations,

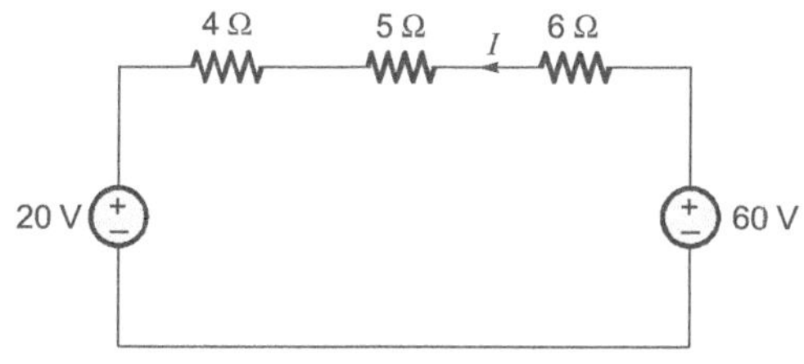

$$I = \frac{60-20}{15} = \frac{40}{15} = 2.66 \text{ A}$$

1.76 (b)

$$V(t) = 10t$$

$$i = \frac{1}{L}\int i\,dt$$

$$i = \frac{1}{L}\int 10t\,dt$$

$$i = \frac{10t^2}{L2}$$

From plot at $t = 1$ sec $i = 2$ A

$$2 = \frac{(10)}{2L} \Rightarrow L = \frac{10}{4} = 2.54$$

$\therefore$ Inductor of 2.5 H

1.77 (b)

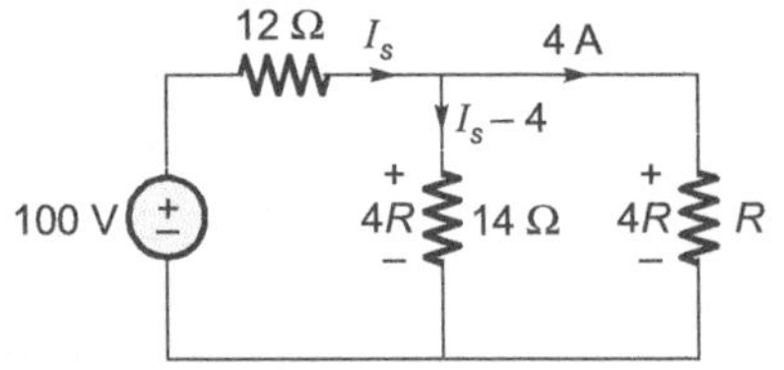

Write KVL in first loop,

$$100 = 12\,I_s + 14\,(I_s - 4)$$

$$I_s = 6\text{ A}$$

The voltage across 14 Ω is

$$(I_s - 4) \times 14 = 28\text{ V}$$

$$R = \frac{28}{4} = 7\,\Omega$$

1.78 (a)

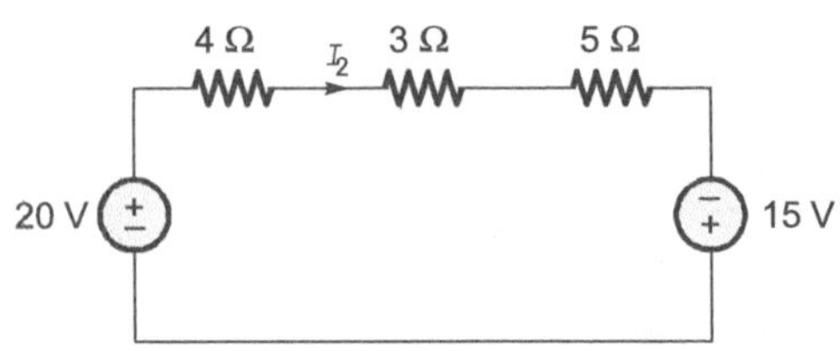

Using source transformation:

$$I_2 = \frac{35}{12} = 2.92\text{ A}$$

$$I_2 = 2.92\text{ A}$$

$$5 = I_1 + I_2$$

$\Rightarrow$ $I_1 = 5 - 2.92$

$\Rightarrow$ $I_1 = 2.08\text{ A}$

$\Rightarrow$ $I_2 = I_3 + 3$

$$I_3 = 2.92 - 3$$

$$= -0.08\text{A}$$

$\Rightarrow$ $I_3 = -0.08\text{A}$

1.79 (a)

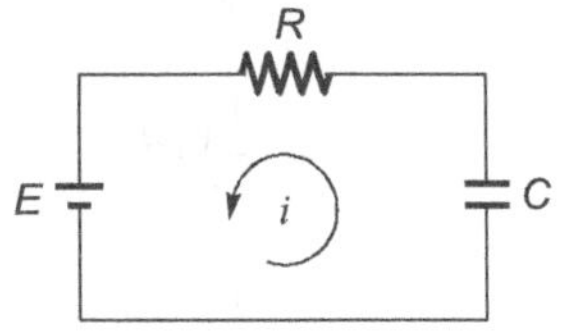

KVL:

$$E = iR + \frac{1}{C}\int i\,dt \quad \text{...(i)}$$

Differentiating equation (i),

$\therefore$ $0 = R\dfrac{di}{dt} + \dfrac{i}{C}$

$\Rightarrow$ $\dfrac{di}{dt} + \dfrac{i}{RC} = 0 \quad \text{...(ii)}$

1.80 (b)

$$m = \frac{10-(-10)}{0-2} = \frac{20}{-2} = -10$$

$$m = -10$$

$V(t) = -10t + c \qquad [0 < t < 2]$

At, $t = 1 \quad V(t) = 0$

$$0 = -10 + C$$

$\Rightarrow$ $C = 10$

$\therefore$ $V(t) = -10t + 10 \qquad [0 < t < 2\,]$

As, $i = C\dfrac{dv}{dt}$

$\Rightarrow$ $i = (0.5)\dfrac{d}{dt}[-10t + 10]$

$i = (0.5)\,[-10] \Rightarrow i = -5\text{ A}$

$\therefore$ $i = -5\text{ A} \qquad [0 < t < 2]$

1.81 (c)

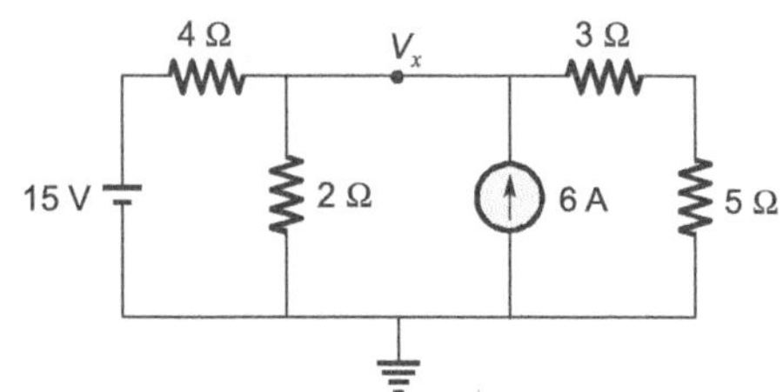

$$\frac{V_x - 15}{4} + \frac{V_x}{2} + \frac{V_x}{8} = 6$$

$$\frac{7}{8}V_x = \frac{39}{4}$$

$\Rightarrow$ $V_x = \dfrac{39 \times 2}{7}$

$\therefore$ $I = \dfrac{V_x}{8} = \dfrac{39 \times 2}{7 \times 8} = \dfrac{39}{28}\text{ A}$

1.82 (b)

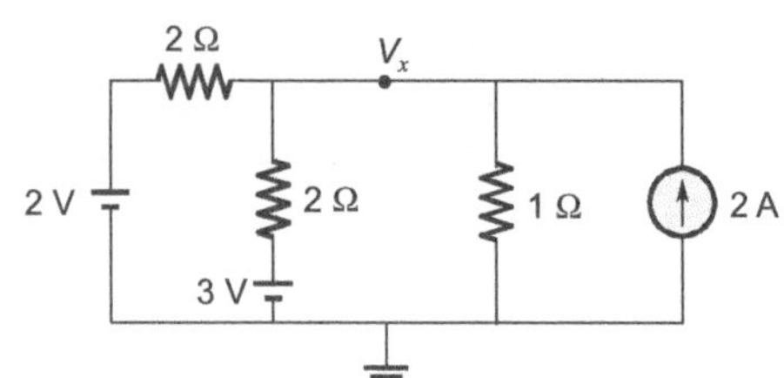

$$\frac{V_x - 2}{2} + \frac{V_x - 3}{2} + \frac{V_x}{1} = 2$$

$$4V_x = 9$$

$\Rightarrow$ $V_x = \dfrac{9}{4}$

$$I = \frac{V_x}{1} = \frac{9}{4} = 2.25\text{ A}$$

1.83 (c)

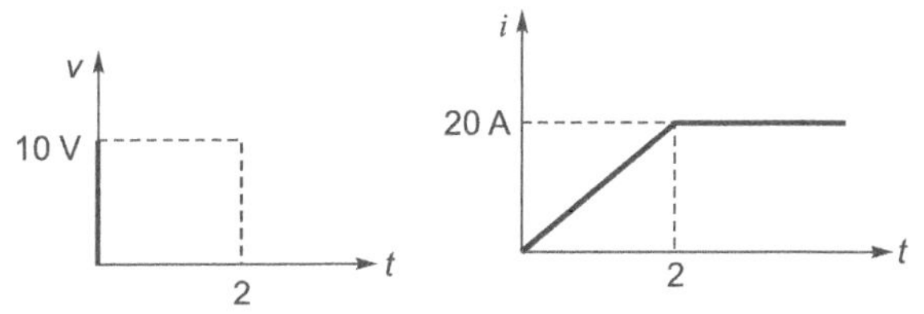

$$i = \frac{1}{L}\int v\,dt$$

As, $L = 1$ H

$$i = \int_0^t v\,dt = 10t$$

1.84 (c)

$V \propto I$ from Ohm's law

$V = RI$

R is resistor.

1.85 (b)

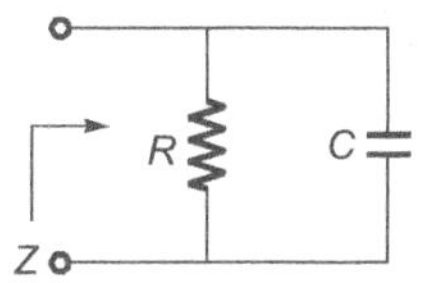

$$Z = 10 - j30$$

$$Y = \frac{1}{10 - j30} \times \frac{10 + j30}{10 + j30}$$

$$= \frac{10}{1000} + \frac{j30}{1000}$$

$$= \frac{1}{100} + j30 \times 10^{-3} = \frac{1}{R} + \frac{1}{-jX_C}$$

$$R = 100\ \Omega,\ \frac{1}{X_C} = 30 \times 10^{-3}$$

$$2\pi fC = 30 \times 10^{-3}$$

$$2\pi \times 1 \times 10^6 \times C = 30 \times 10^{-3}$$

$$C = 4.7\text{ nF}$$

1.86 (b)

$$L = \frac{N^2 \mu a}{l} = \frac{N^2 \mu \pi}{l}\frac{D^2}{4}$$

$$L \propto N^2$$

$$L \propto D^2$$

So second statement is incorrect.

1.87 (d)

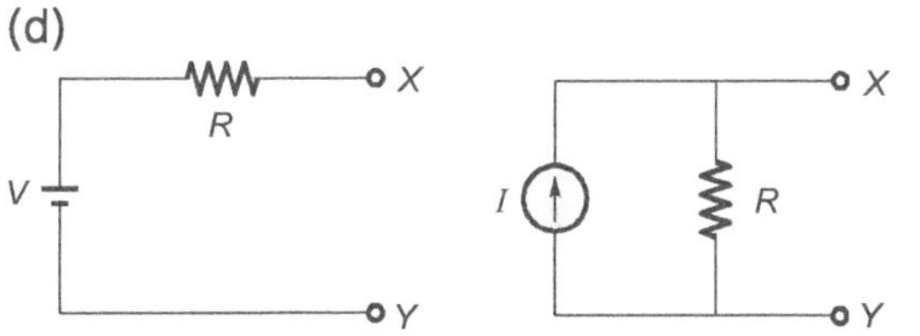

According to source transformation all the given statement is correct.

1.88 (c)

As, $J = \sigma E$

$$J = \frac{1}{\rho}E$$

$$\rho = \frac{E}{J}$$

$$\therefore \quad \rho\frac{l}{A} = \frac{E}{J}\frac{l}{A}$$

As, $R = \dfrac{\rho l}{A} \Rightarrow R = \dfrac{El}{JA}$

1.89 (d)

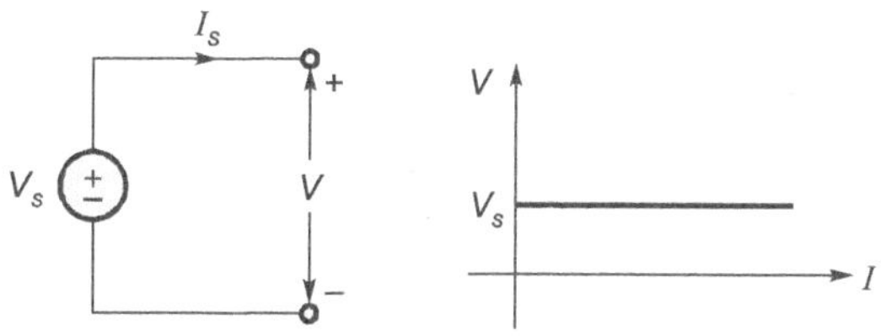

(Ideal voltage source)

By the V-I characteristic, it is clear statement 1, 2 and 3 are correct and statement 4 is not correct.

1.90 (b)

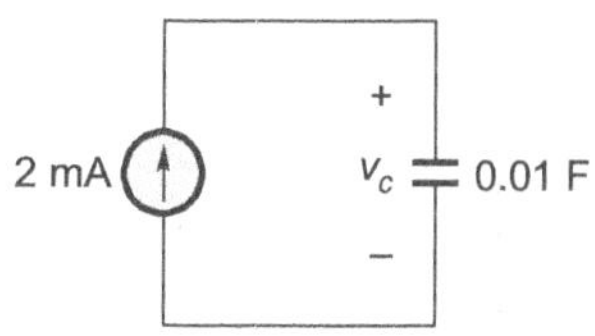

$$V_C = \frac{1}{0.01}\int_0^{30} (2 \times 10^{-3})\,dt + 0$$

$$= 100 \times 2 \times 10^{-3}(t)\Big|_0^{30}$$

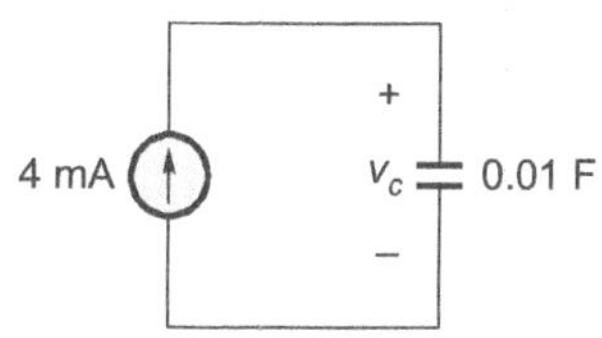

$$V_C = 6 + \frac{1}{0.01}\int_0^{30} (4 \times 10^{-3})\,dt$$

$$= 6 + \left[100 \times 4 \times 10^{-3} \times (t)\Big|_0^{30}\right]$$

$$= 6 + 12 = 18\text{ V}$$

1.91 (b)

$$H = I^2Rt$$

$$2.3 \times 10^3 = 2^2 \times R \times 10$$

$$R = 57.5\ \Omega$$

$$V = IR$$

$$V = 115\text{ V}$$

1.92 (c)

Apply KCL,

$$3 + 0.5i_o = i_o$$

$$i_o = 6\text{ A}$$

By Ohm's law,

$$V_o = 6 \times 4 = 24\text{ V}$$

1.93 (a)

$$P_{20} = \frac{V^2}{R_{20}}$$

$$R_{20} = \frac{9\times9}{20} = 4.05\ \Omega$$

In series combination,

$$P = I^2R,\ P \propto R$$

$$V = IR,\ V \propto R$$

$$\therefore \quad V \propto P$$

$$\therefore \quad V_X = \frac{P_X \cdot V}{P_X + P_Y}$$

$$V_{10} = \frac{9\times10}{10+15} = 3.6\ \text{V}$$

$$V_{15} = 9 - 3.6 = 5.4\ \text{V}$$

$$P_{15} = \frac{V_{15}^2}{R_{15}}$$

$$\Rightarrow \quad R_{15} = \frac{(5.4)^2}{15} = 1.945\ \Omega$$

$$P_{10} = \frac{V_{10}^2}{R_{10}}$$

$$\Rightarrow \quad R_{10} = \frac{(3.6)^2}{10} = 1.297\ \Omega$$

1.94 (a)

Write KVL in 4th loop

$$10i_4 - 8i_3 = -10 \quad ...(1)$$

Substitute $i_4 = -I_o$ in eq. (1)

$$10I_o + 8i_3 = 10$$

$$I_o = 1 - 0.8i_3$$

Write KVL equation for 3 loops at a time (supermesh indicated by dotted lines)

$$2i_1 + 6i_2 + 12i_3 - 8i_4 = 0$$

$$i_4 = -I_0$$

$$2i_1 + 6i_2 + 12i_3 + 8I_0 = 0 \quad ...(2)$$

Substitute I_o value in eqn. (2)

$$2i_1 + 6i_2 + 5.6i_3 = -8 \quad ...(3)$$

KCL at node P:

$$i_2 - i_1 = 5 \quad ...(4)$$

KCL at node R:

$$5 + i_1 = 3I_0 + i_3$$

Substitute i_1 value from eqn. (4)

$$i_2 = 3I_0 + i_3$$

$$i_2 - i_3 = 3I_0 \quad ...(5)$$

Sub. I_0 value in eq. (5)

$$i_2 - i_3 = 3(1 - 0.8i_3)$$

$$i_2 + 1.4i_3 = 3 \quad ...(6)$$

Solve eq. (3), (4) and (6),

$$i_1 = -7.5\ \text{A},$$

$$i_3 = 3.928\ \text{A}$$

$$I_0 = -2.143\ \text{A}$$

$$\therefore \quad i_4 = 2.143\ \text{A}$$

1.95 (b)

Given circuit can be simplified as,

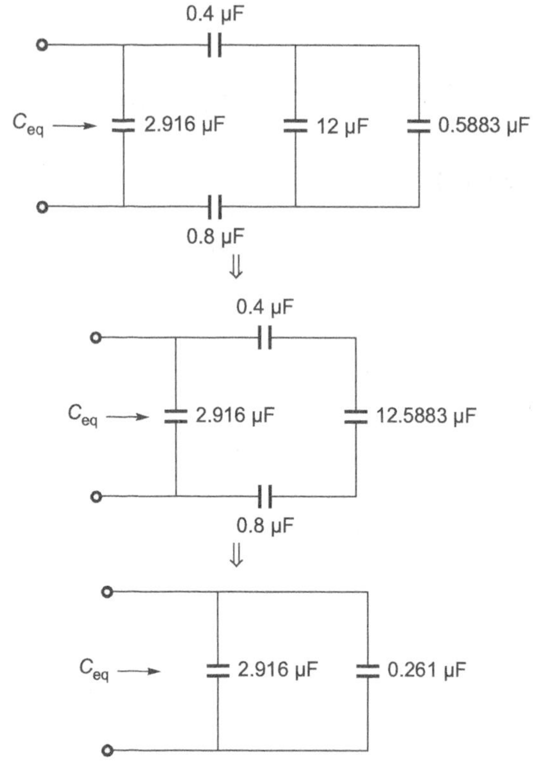

$$\therefore \quad C_{eq} = 3.177\ \mu\text{F} = 3.18\ \mu\text{F}$$

(b) option is correct.

1.97 (c)

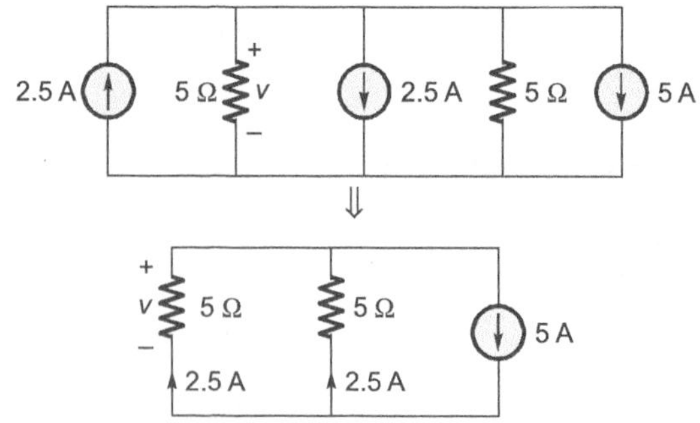

$$V = -2.5 \times 5 = -12.5$$

$$|V| = 12.5$$

1.98 (b)

For iron core, reluctance is low, flux value is high and hence flux density also high.

∴ (b) option is correct.

1.99 (a)

Given : $i_s = 10$ A

$i_1 = 5$ A

$i_2 + i_3 = 5$ A

$i_3 = 5 - i_2$

$= 5 - 4 = 1$ A

Similarly, $i_4 + i_5 = 5$

$i_5 = 5 - 3 = 2$ A

1.100 (c)

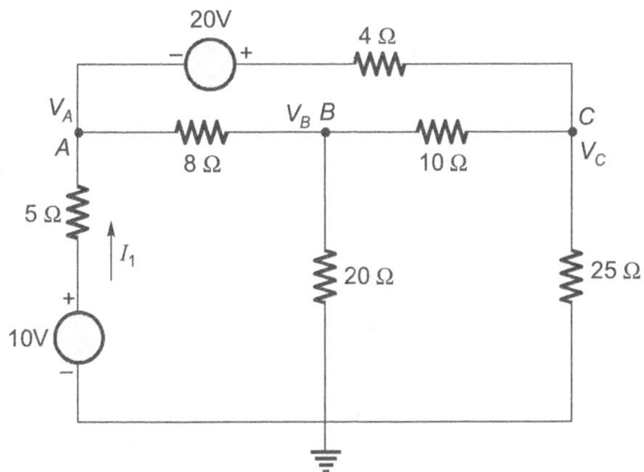

KCL at node (A)

$$\frac{V_A - 10}{5} + \frac{V_A - V_B}{8} + \frac{V_A + 20 - V_C}{4} = 0$$

$$\frac{23}{40}V_A - \frac{V_B}{8} - \frac{V_C}{4} = -3 \quad \text{...(1)}$$

KCL at node (B)

$$\frac{V_B - V_C}{10} + \frac{V_B}{20} + \frac{V_B - V_A}{8} = 0$$

$$\frac{-V_A}{8} + \frac{11}{40}V_B - \frac{V_C}{10} = 0 \quad \text{...(2)}$$

KCL at node (C)

$$\frac{V_C}{25} + \frac{V_C - V_B}{10} + \frac{V_C - 20 - V_A}{4} = 0$$

$$\frac{-V_A}{4} + \frac{-V_B}{10} + \frac{39}{100}V_B = 5 \quad \text{...(3)}$$

Solving this, $V_A = 4.34$ V

So, $I_1 = \dfrac{10 - V_A}{5} = 1.132$

1.101 (b)

$$V_{AB} = V_A - V_B = 10\times\frac{20}{30} - 10\times\frac{40}{70}$$

$$V_{AB} = \frac{20}{3} - \frac{40}{7}$$

$$V_{AB} = 0.96 \text{ V}$$

1.102 (b)

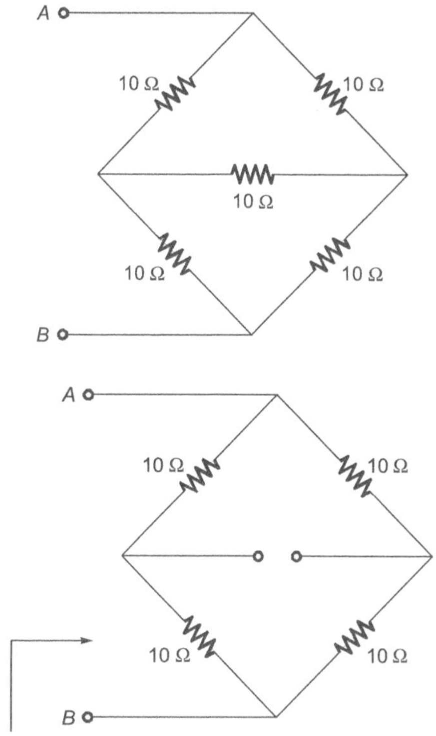

$R_{AB} = (20 \parallel 20) = 10\ \Omega$

1.103 (b)

$$i(t) = \frac{100\times10^{-3}t}{2\times10^{-3}}$$

$$i(t) = 50t;\ 0 < t \le 2$$

$$V_c(t) = \frac{1}{C}\int_0^{2\times10^{-3}} i(t)dt$$

$$V_c = \frac{1}{1\times10^{-3}}\int_0^{2\times10^{-3}} 50t\,dt$$

$$= \frac{1}{1\times10^{-3}}\times50\times\left[\frac{(2\times10^{-3})^2}{2}\right]$$

$$= 100 \text{ mV}$$

V_c at 2 msec is 100 mV.

■■■■

2 Network Theorems and Steady State Analysis

2.1 **Assertion (A):** Norton's theorem is applied to a network for which no equivalent Thevenin's network exists.

Reason (R): Norton's Theorem enables one to calculate quickly current and voltage in a particular branch of interest in a complicated network.

(a) Both A and R are true and R is the correct explanation of A
(b) Both A and R are true but R is NOT the correct explanation of A
(c) A is true but R is false
(d) A is false but R is true

[ESE-2001]

2.2 For the circuit shown in the given figure, when the voltage E is 10 V, the current I is 1 A. If the applied voltage across terminal C-D is 100 V, the short circuit current flowing through the terminal A-B will be

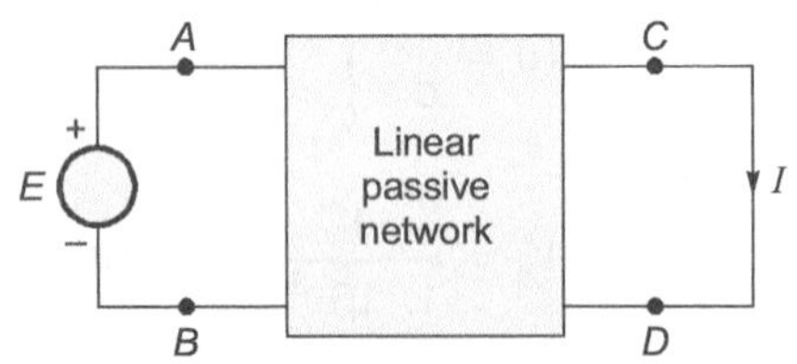

(a) 0.1 A (b) 1 A
(c) 10 A (d) 100 A

[ESE-2001]

2.3 For the circuit shown in the given figure the current I is given by

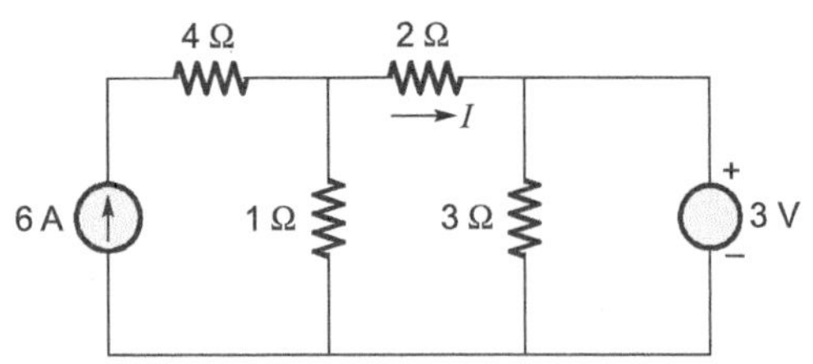

(a) 3 A (b) 2 A
(c) 1 A (d) zero [ESE-2001]

2.4 The Thevenin's equivalent resistance R_{Th} for the given network is

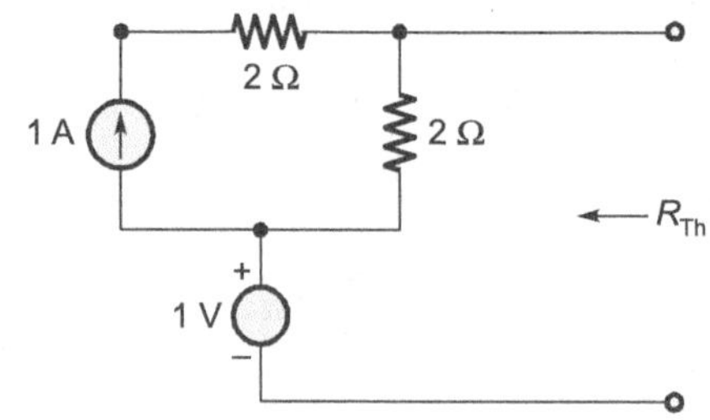

(a) 1 Ω (b) 2 Ω
(c) 4 Ω (d) infinity

[ESE-2001]

2.5 The value of V in the circuit shown in the given figure is

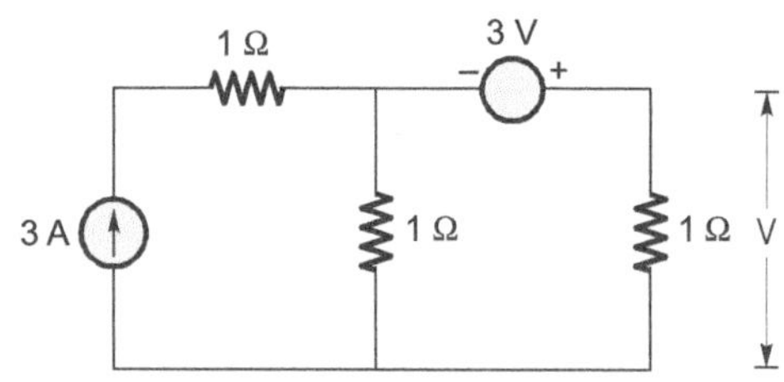

(a) 1 V (b) 2 V
(c) 3 V (d) 4 V [ESE-2001]

2.6 The Norton's equivalent of circuit shown in Figure-I is drawn in the circuit shown in Figure-II. The value of I_{SC} and R_{eq} in Figure-II are respectively

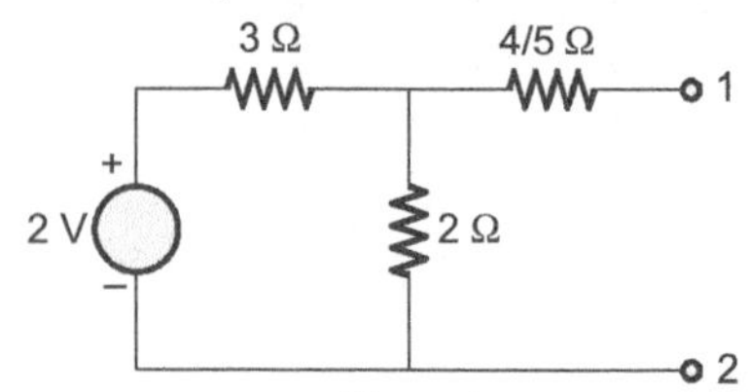

Figure I

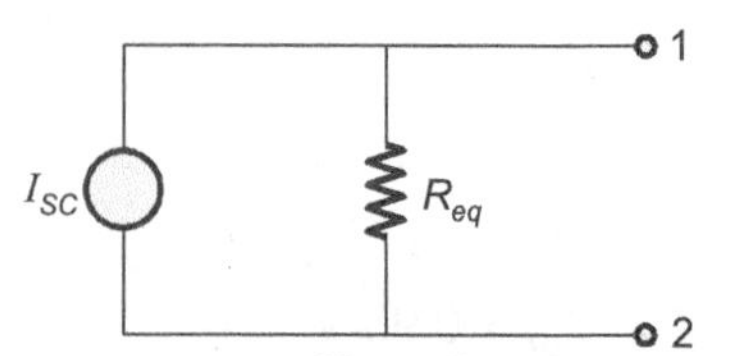

Figure II

(a) 5/2 A and 2 Ω (b) 2/5 A and 1 Ω
(c) 4/5 A and 12/5 Ω (d) 2/5 A and 2 Ω

[ESE-2001]

2.7 Two current waveforms as shown in the Figure-I and Figure-II, are passed through identical resistors of 1 Ω. The ratio of heat produced in these resistors in a given time by current of Figure-I to Figure-II is

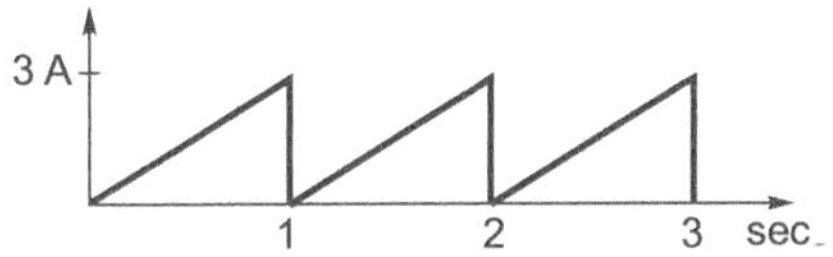

Figure-I

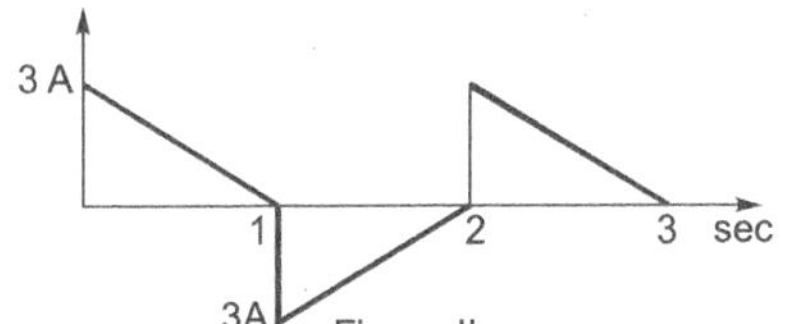

Figure-II

(a) 2 : 1 (b) 1 : 2
(c) 1 : 1 (d) $1:\sqrt{2}$

[ESE-2001]

2.8 The resistance seen from the terminals A and B of the device whose characteristic is shown in the figure below is

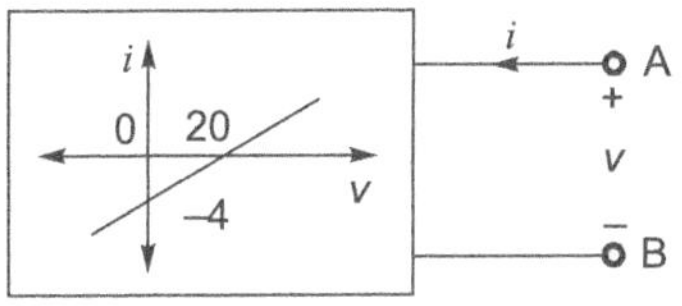

(a) –5 Ω (b) –1/5 Ω
(c) 1/5 Ω (d) 5 Ω

[ESE-2002]

2.9 If the combined generator and line impedance is (5 + *j*10) ohm, then for the maximum power transfers to a load impedance from a generator of constant generated voltage, the load impedance is given by which one of the following

(a) (5 + *j*10) Ω (b) (5 – *j*10) Ω
(c) (5 + *j*5) Ω (d) 5 Ω

[ESE-2002]

2.10 Superposition theorem is not applicable for

(a) voltage calculation
(b) bilateral elements
(c) power calculation
(d) passive elements

[ESE-2002]

2.11 In the given RC circuit, the current $i(t) = 2\cos 5000t$ A.

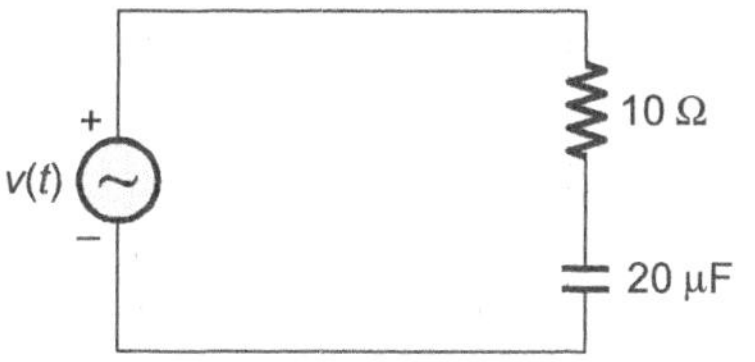

The applied voltage $v(t)$ is

(a) 28.28 cos (5000*t* – 45°) V
(b) 28.28 cos (5000*t* + 45°) V
(c) 28.28 sin (5000*t* – 45°) V
(d) 28.28 sin (5000*t* + 45°) V

[ESE-2002]

2.12 In the circuit shown in the given figure,

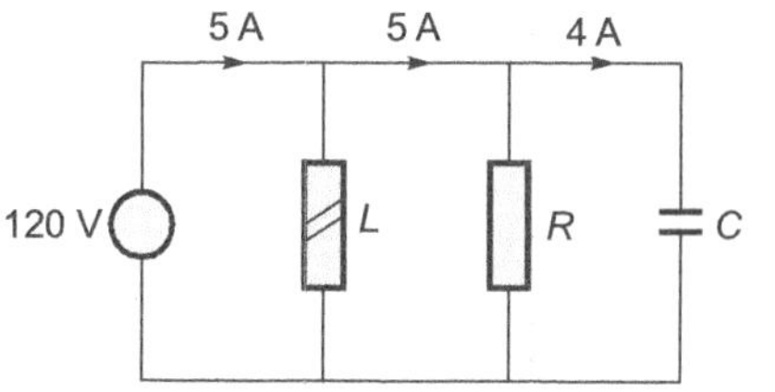

The current through the inductor *L* is

(a) 0 A (b) 3 A
(c) 4 A (d) 8 A

[ESE-2002]

2.13 Three currents i_1, i_2 and i_3 are approaching a node. If $i_1 = 10\sin(400t + 60^0)$ A, and $i_2 = 10\sin(400t - 60^0)$ A, then i_3 is

(a) 0 A
(b) 10 (sin 400*t*) A
(c) –10 (sin 400*t*) A
(d) $-5\sqrt{3}(3\sin 400t)$ A

[ESE-2002]

2.14 The black box as shown in the circuit below contains resistors and independent sources. For *R* = 0 and 2, the value of current *I* is 3 and 1.5 respectively. The value of *I* for *R* = 1 will be

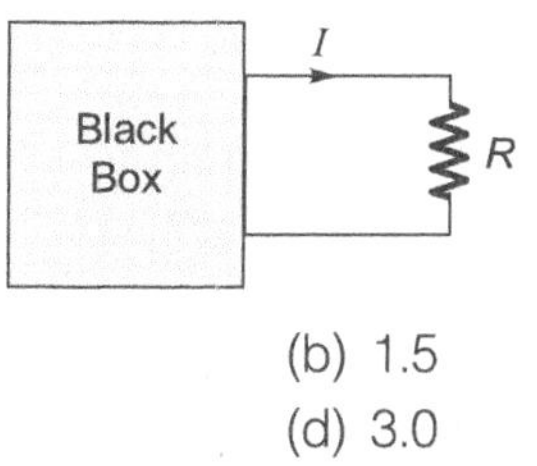

(a) 0.5 (b) 1.5
(c) 2.0 (d) 3.0

[ESE-2003]

2.15 The linear network as shown below has only resistors. If I_1 = 8 A and I_2 = 12 A; V is found to be 80 V. V = 0 when I_1 = –8 A and I_2 = 4 A. Then the value of V when $I_1 = I_2$ = 10 A, is

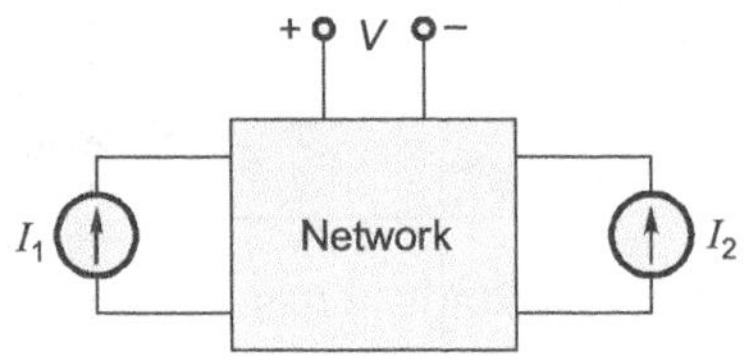

(a) 25 V (b) 50 V
(c) 75 V (d) 100 V

[ESE-2003]

2.16 The total impedance of $Z(j\omega)$ of the circuit shown below is

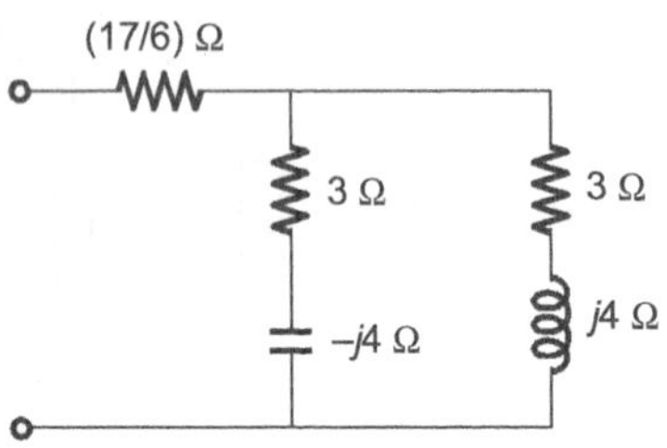

(a) (6 + j0) Ω (b) (7 + j0) Ω
(c) (0 + j8) Ω (d) (6 + j8) Ω

[ESE-2003]

2.17 A 10 mH inductor carries a sinusoidal current of 1 A r.m.s. at a frequency of 50 Hz. The average power dissipated by the inductor is

(a) 0 W (b) 0.25 W
(c) 0.5 W (d) 1.0 W

[ESE-2003]

2.18 Consider the following circuit:

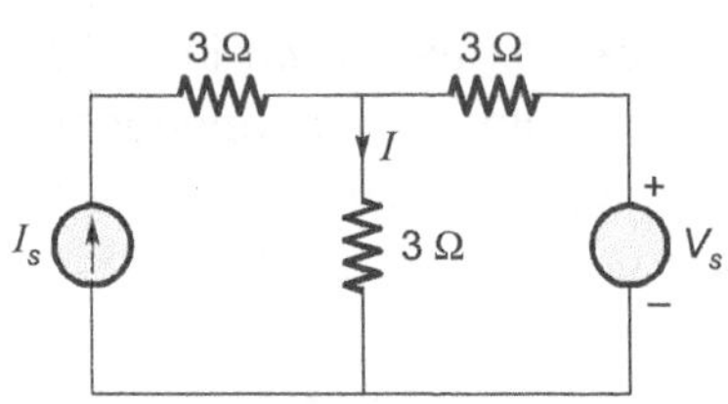

In the above circuit, when V_s = 3 V, I = 4 A, what is the value of I when V_s = 12 V ?

(a) 5.5 A (b) 10 A
(c) 15 A (d) 20 A

[ESE-2004]

2.19 Consider the following circuit:

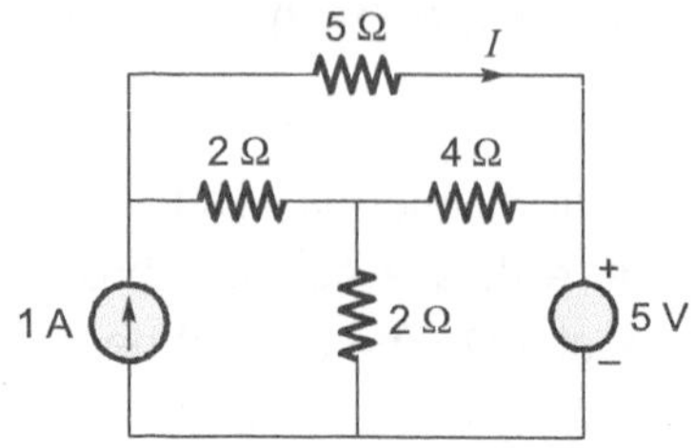

What is the value of current I in the 5 Ω resistor in the above circuit ?

(a) 0 A (b) 2 A
(c) 3 A (d) 4 A

[ESE-2004]

2.20 A voltage source having an internal impedance of 8 + j6 ohms supplies power to a resistive load. What should be the load resistance for maximum power transferred to it?

(a) 8 ohms (b) 6 ohms
(c) 10 ohms (d) $\sqrt{10}$ ohms

[ESE-2004]

2.21 Consider the circuits A and B given below:

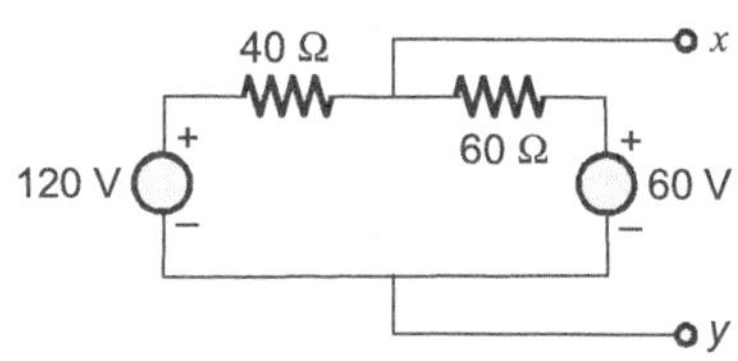

(Circuit A)

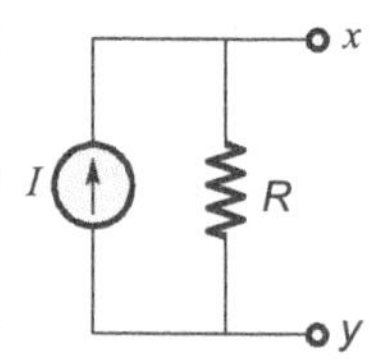

(Circuit B)

For what values respectively of I and R, the circuit B is equivalent to circuit A?

(a) 3 A, 40 Ω (b) 4 A, 24 Ω
(c) 1 A, 60 Ω (d) 2 A, 100 Ω

[ESE-2004]

2.22 Consider the following circuit:

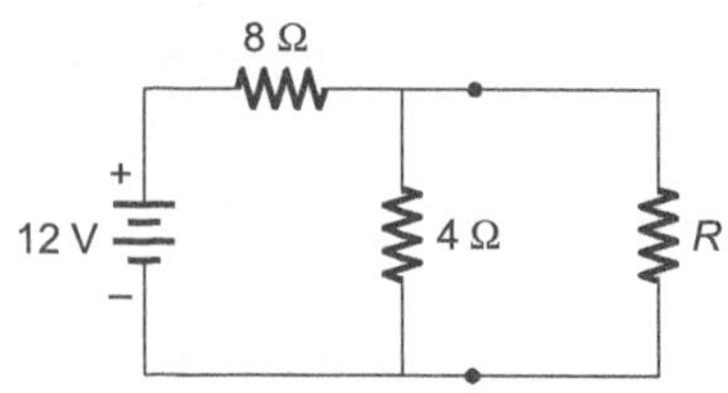

What should be the value of resistance R, in the above circuit if it has to absorb the maximum power from the source ?

(a) 8/3 ohms (b) 3/8 ohms
(c) 4 ohms (d) 8 ohms

[ESE-2004]

2.23 **Assertion (A) :** If Thevenin's equivalent of a circuit is known, its Norton equivalent is also known.
Reason (R) : Norton's equivalent is reciprocal of Thevenin's equivalent.
(a) Both A and R are true and R is the correct explanation of A
(b) Both A and R are true but R is NOT the correct explanation of A
(c) A is true but R is false
(d) A is false but R is true
[ESE-2004]

2.24 Consider the following circuit:

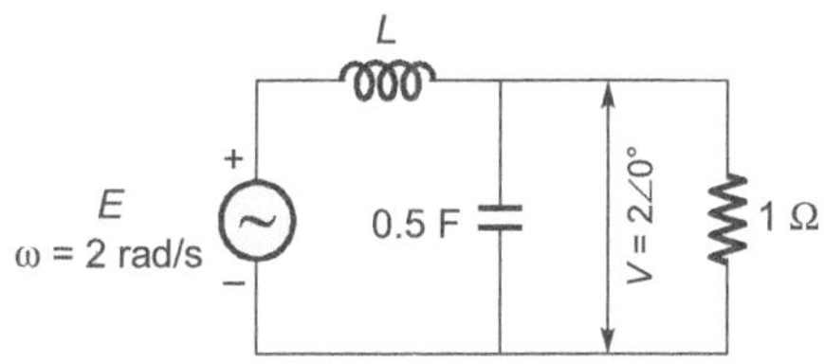

For the circuit shown above, by how much the voltage across the inductor leads the voltage across the capacitor?
(a) 45° (b) 90°
(c) 135° (d) 180° [ESE-2004]

2.25 Match **List-I (Property of Network)** with **List-II (Relevant Theorem)** and select the correct answer using the code given below the Lists:

List-I	List-II
A. Linearity	1. Superposition Theorem
B. Structure	2. Norton's Theorem
C. Equivalent Circuit	3. Tellegen's Theorem
D. Bilateral	4. Millman's Theorem

Codes:

	A	B	C	D
(a)	2	4	1	3
(b)	1	3	2	4
(c)	2	3	1	4
(d)	1	4	2	3

[ESE-2005]

2.26 In the circuit given below, I = 1 A for I_s = 0. What is the value of I for I_s = 2 A?

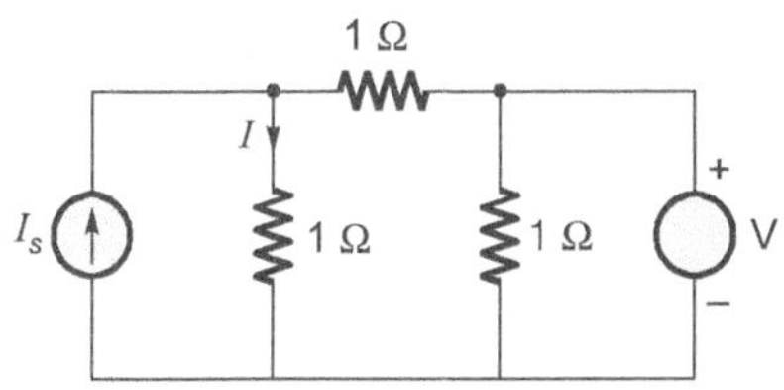

(a) 7 A (b) 4 A
(c) 3 A (d) 2 A
[ESE-2005]

2.27 The black-box, N contains resistors and independent sources. If I = 3 A and 1.5 A for R = 0 and 2 Ω, respectively, then what is the value of I for R = 1 Ω?

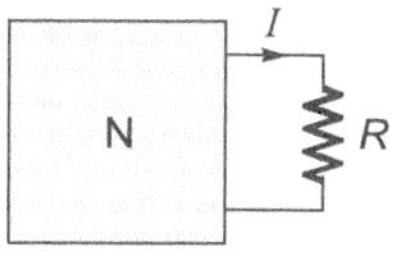

(a) 1 A (b) 2 A
(c) 3 A (d) 4 A
[ESE-2005]

2.28 For the a.c. circuit given below, what is the value of I ?

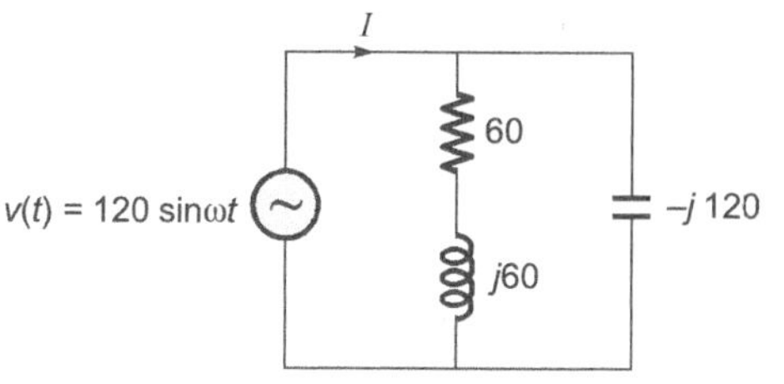

(a) 1 + j1 (b) 1 + j0
(c) 0 – j1 (d) 0 + j0
[ESE-2005]

2.29 In the circuit shown in the figure below, for what value of C will the current-I be in phase with the sinusoidal source voltage V_s = sin2t?

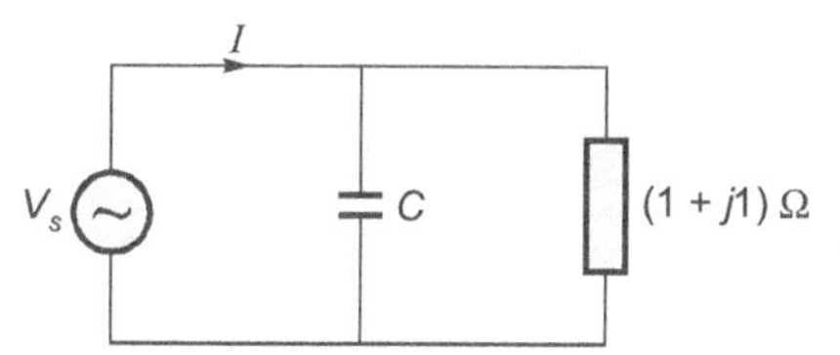

(a) $\frac{1}{4}$F (b) $\frac{1}{2}$F
(c) $\frac{1}{\sqrt{2}}$F (d) 1F [ESE-2006]

2.30 In the network shown below, what is the current I in the direction shown?

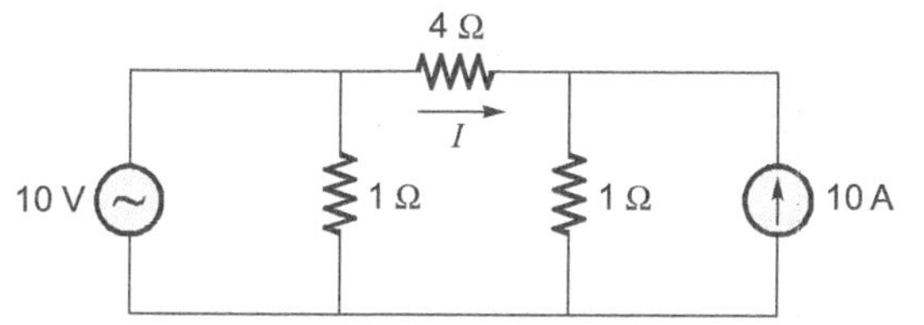

(a) 0 (b) 1/3 A
(c) 5/6 A (d) 4 A

[ESE-2006]

2.31 For the network shown as below, when $I = 0$, $V = 20$ V and when $R = 0$, $I = 10$ A. If now $R = 3\ \Omega$, what is the value of the current I?

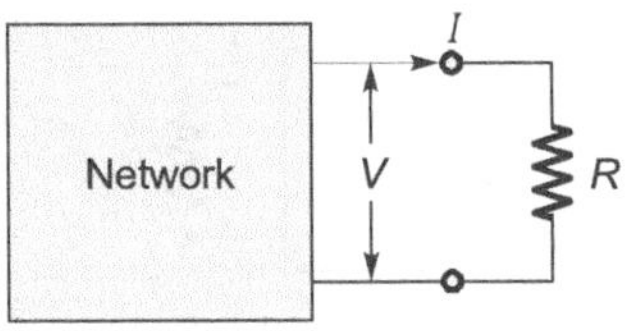

(a) 6.67 A (b) 6.0 A
(c) 4.0 A (d) 10 A [ESE-2006]

2.32 Norton equivalent to the network N to the left of AB is a current source $I_N = 4$ A from B to A, $R_N = 2\ \Omega$. The current through R when it is connected across $AB = 2$ A. What is the value of resistance R?

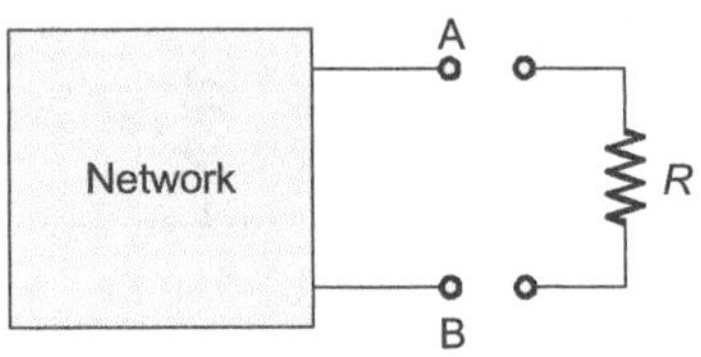

(a) 1 Ω (b) 2 Ω
(c) 3 Ω (d) 4 Ω

[ESE-2006]

2.33 In the circuit shown in the below figure,

$e_1(t) = \sqrt{3}\cos(\omega t + 30°)$ and

$e_2(t) = \sqrt{3}\cos(\omega t + 60°)$. What is the voltage $v(t)$ across the 1 Ω grounded resistor?

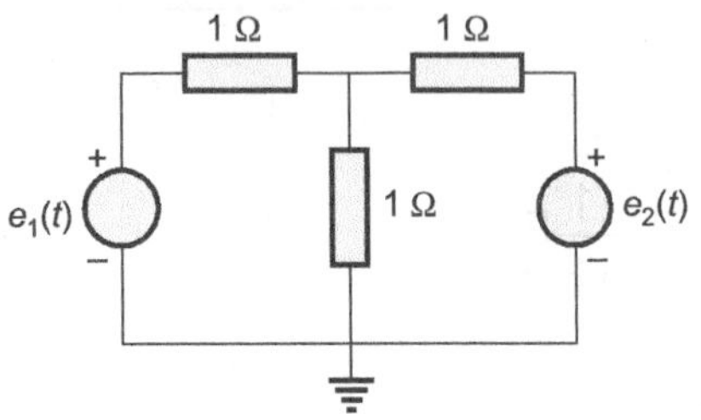

(a) $\{\cos \omega t\}$ V
(b) $\{\sin(\omega t + 30°) + \cos(\omega t + 60°)\}$ V
(c) $\{1\angle -90°\}$ V
(d) $\{j\,1\}$ V

[ESE-2006]

2.34 For the circuit shown below, what is the value of I?

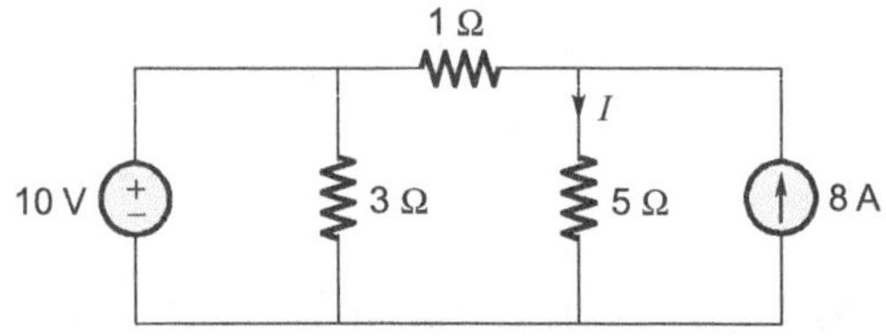

(a) 10 A (b) 6 A
(c) 3.7 A (d) 3 A [ESE-2007]

2.35 **Assertion (A):** A four terminal passive linear network will always have an equivalent T-network.
Reason (R): The Thevenin's theorem is applicable in cases of transfer functions also.
(a) Both A and R are true and R is the correct explanation of A
(b) Both A and R are true but R is not the correct explanation of A
(c) A is true but R is false
(d) A is false but R is true

[ESE-2007]

2.36 According to maximum power transfer theorem, when is the maximum power absorbed by one network from another network?
(a) The impedance of one of the network is half that of the outer.
(b) The impedance of one is the complex conjugate of the other.
(c) The impedance of one is equal to that of the other.
(d) Only the resistive parts of both are equal.

[ESE-2008]

2.37 For the circuit shown below, the black box contains resistors and independent sources only. The current I is 3 A and 1.5 A for $R = 0$ and 2 Ω, respectively. For $R = 1\ \Omega$, what is the current?

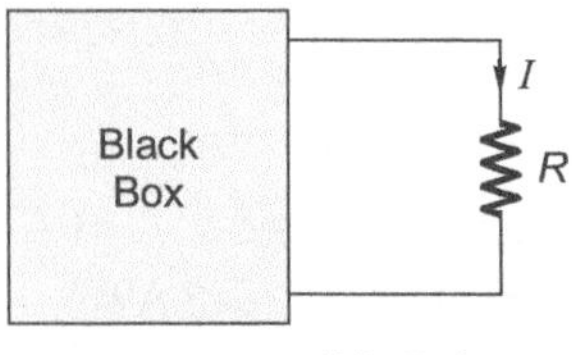

(a) 1 A (b) 2 A
(c) 3 A (d) 4 A

[ESE-2008]

2.38 Voltage source is represented as shown in the figure given below.

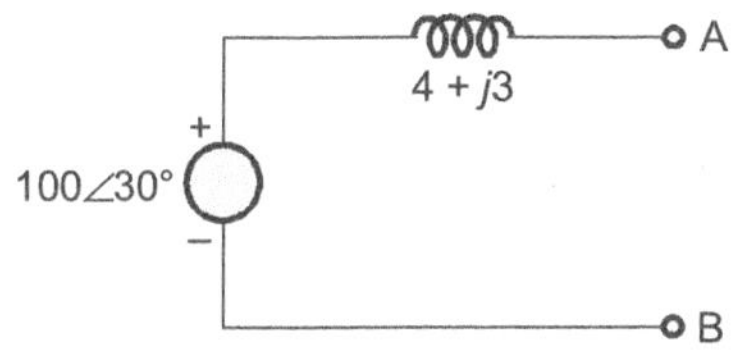

What is the equivalent current source?

(a)
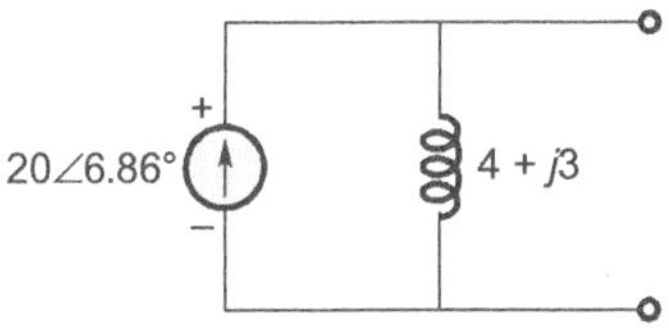

(b)
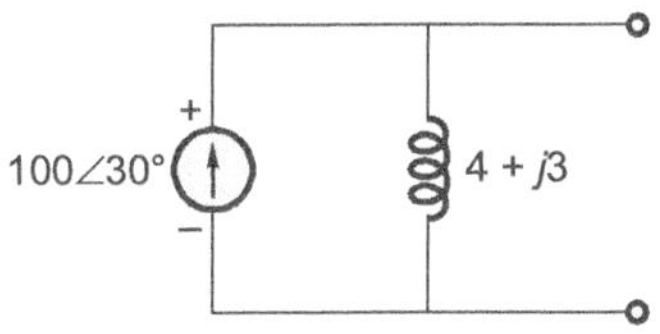

(c)
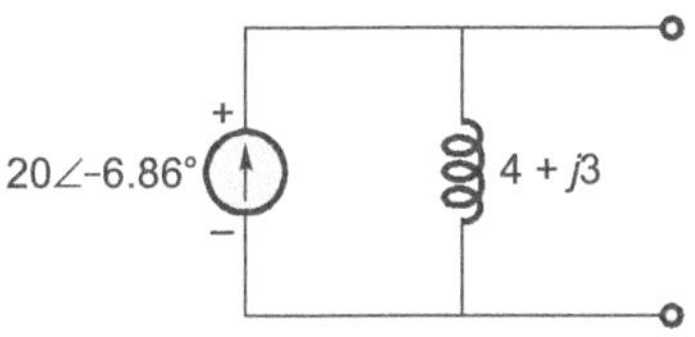

(d)
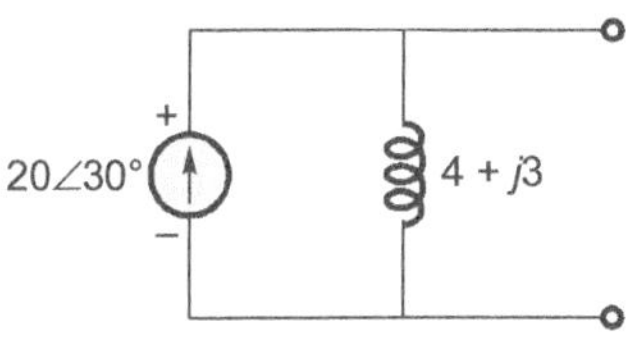

[ESE-2008]

2.39 In the circuit shown below, when is the power absorbed by the 1 Ω resistor maximum?

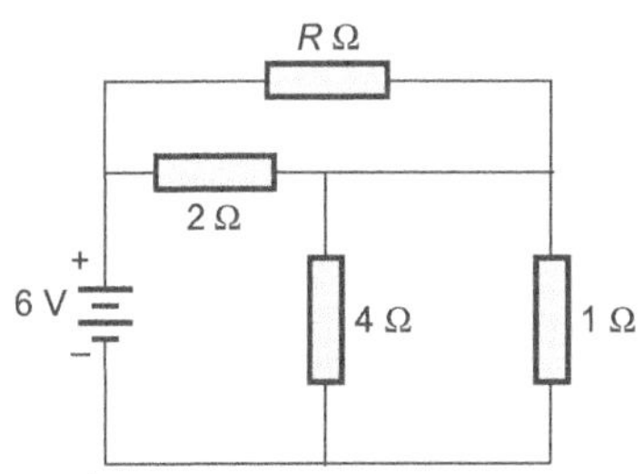

(a) $R = 0$ (b) $R = 2\ \Omega$
(c) $R = 4\ \Omega$ (d) $R = \infty$

[ESE-2008]

2.40 For the network shown below, if the current $i(t) = \sqrt{2}\sin(\omega t - 30°)$, then what is the value of R?

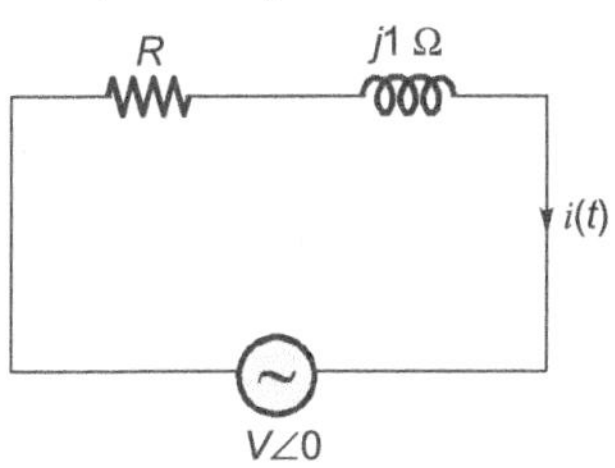

(a) 1 Ω (b) 3 Ω
(c) $\sqrt{3}\ \Omega$ (d) $3\sqrt{3}\ \Omega$

[ESE-2008]

2.41 In a two element series network, the voltage and current respectively are given as,
$v(t) = 50\sin(314t) + 50\sin(942t)$ V
$i(t) = 10\sin(314t + 60°) + 8\sin(942t + 45°)$ A, then the power factor of the network is approximately
(a) 0.9 (b) 0.6
(c) 0.3 (d) 0.1

[ESE-2008]

2.42 A 10 μF capacitor is fed from an a.c. voltage source containing a fundamental and a third harmonic of strength one-third of fundamental. The third harmonic current flowing through the capacitor expressed as percentage of the fundamental under steady-state condition will be
(a) 150% (b) 100%
(c) 50% (d) 33%

[ESE-2008]

2.43 A voltage of $V = 100\angle 30°$ is applied to an impedance $Z = 3 + j4$. Which one of the following is the power triangle?

(a)
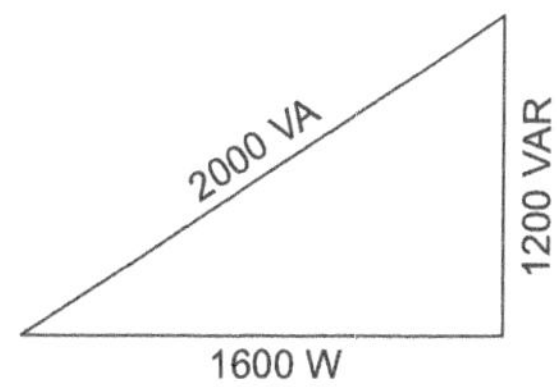

(b)
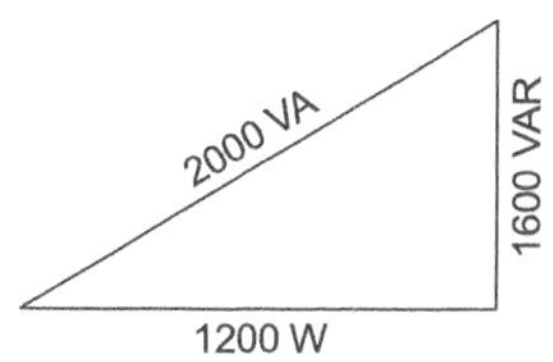

(c)
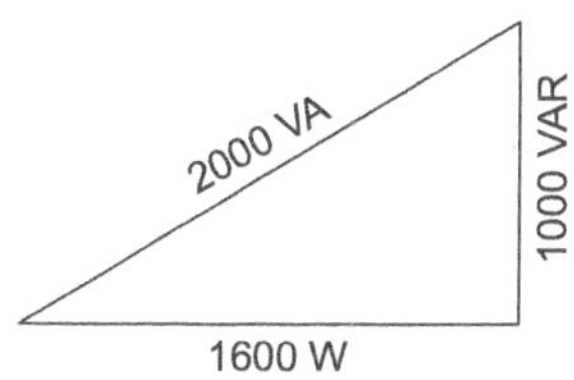

(d)
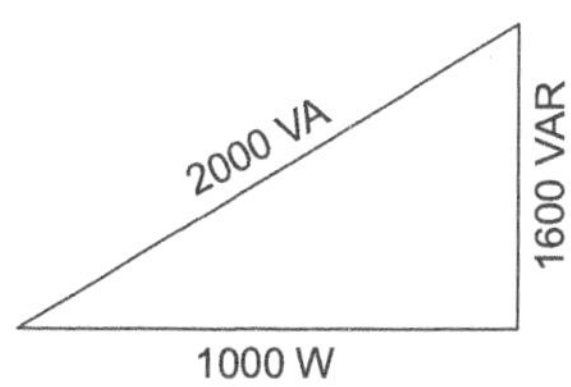

[ESE-2008]

2.44 A human nerve cell has an open circuit voltage of 80 mV and it can deliver a current of 5 nA through a 6 M ohm load. What is the maximum power available from the cell?

(a) 0.16 nW (b) 16 mW
(c) 1.6 W (d) 16 pW

[ESE-2009]

2.45 For the circuit as shown below, what are the values of the Norton's equivalent current and conductance between AB terminals?

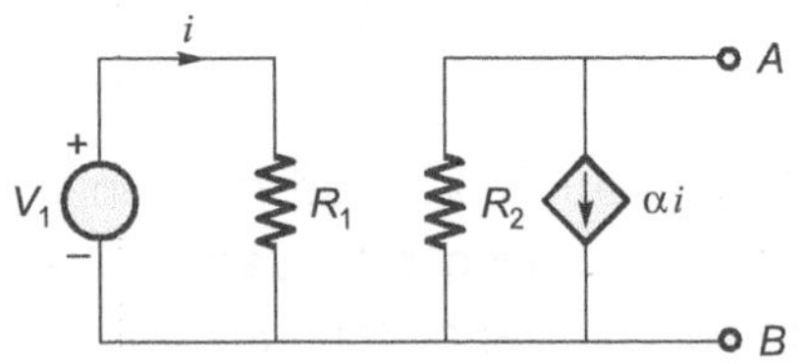

(a) $-\alpha\frac{V_1}{R_1}$ and $G=\frac{1}{R_2}$

(b) $\alpha\frac{V_1}{R_2}$ and $G=\frac{1}{R_1}$

(c) $\alpha\frac{V_1}{R_2}$ and $G=\frac{1}{R_2}$

(d) $-\alpha\frac{V_1}{R_1}$ and $G=\frac{1}{R_1}$

[ESE-2009]

2.46 For the circuit as shown in Fig. (a), the current through the ammeter is 4∠–45° Amps. What is the current in the ammeter for the circuit in Fig. (b)?

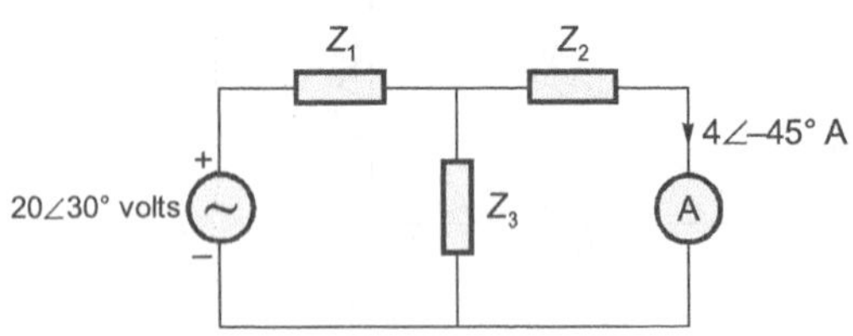

Fig. (a)

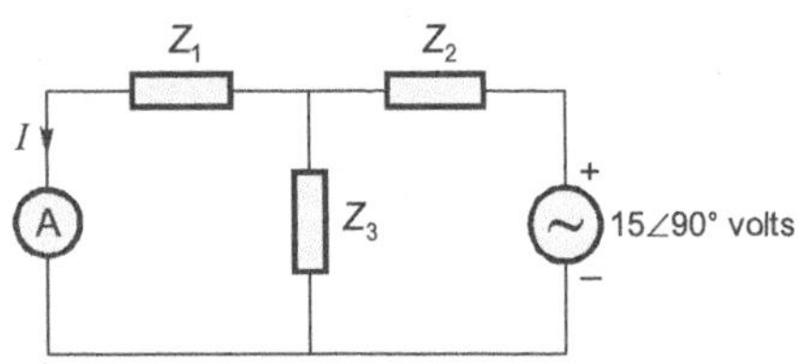

Fig. (b)

(a) 3 ∠15° Amps (b) 2 ∠30° Amps
(c) 4 ∠45° Amps (d) 5 ∠–90° Amps

[ESE-2009]

2.47 In the below shown circuit, if V = 3 volts for E = 1 volt, I = 0; and V = 2 volts for I = 2 A and E = 0. When E = 1 volt and I is replaced by a resistor of 2 ohm, then what is the value of V?

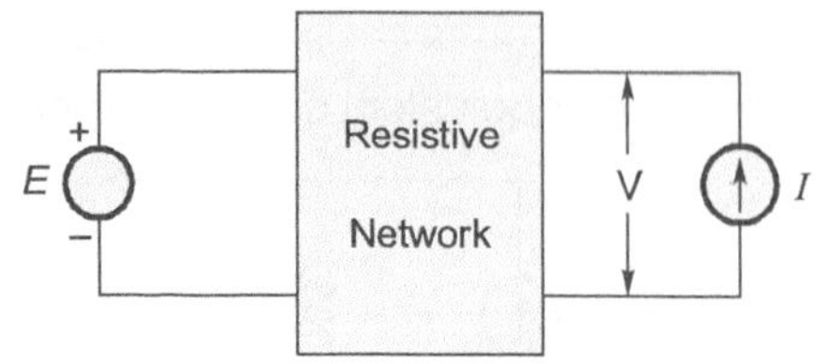

(a) 2 volts (b) 4 volts
(c) 6 volts (d) 8 volts

[ESE-2009]

2.48 Replace the below shown circuit by a single voltage source in series with an impedance.

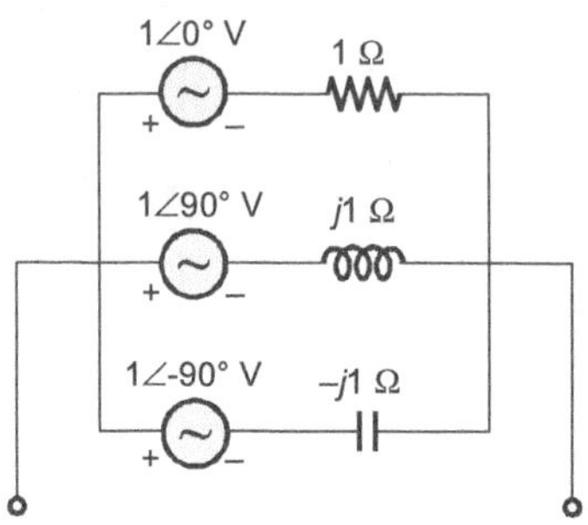

(a) 2 V, 1 Ω (b) 1 V, 3 Ω
(c) 3 V, 1 Ω (d) 2 V, 3 Ω

[ESE-2009]

2.49 For the circuit as shown below, if $E = E_1$ and I is removed, then V = 5 volts. If E = 0 and I = 1 A, then V = 5 volts. For $E = E_1$ and I replaced by a resistor of 5 Ω, what is the value of V in volts ?

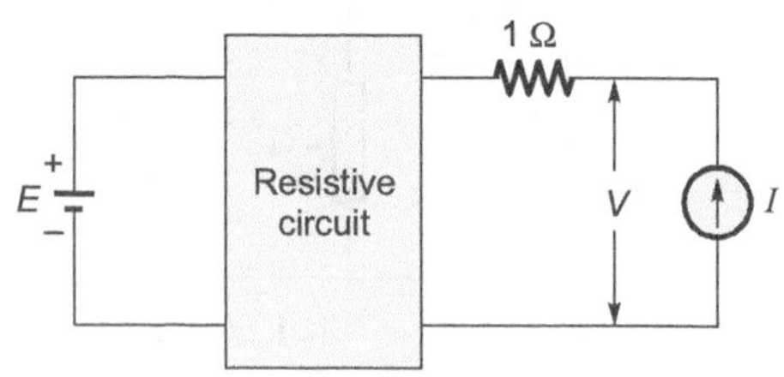

(a) 5.0 (b) 2.5
(c) 7.5 (d) 3.5

[ESE-2009]

2.50 Consider the following with respect to the circuit as shown below:

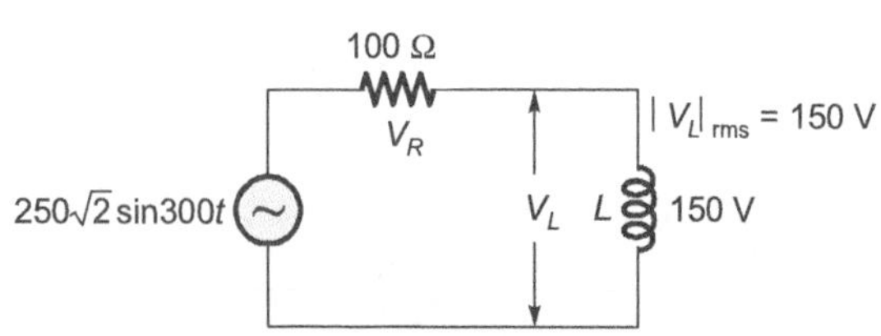

1. $V_R = 100\sqrt{2}$ V
2. $|I|_{rms} = 2$A
3. $L = 0.25$ H

Which of the above statements is/are correct?

(a) 1 only (b) 2 and 3
(c) 1 and 3 (d) 1 and 2

[ESE-2009]

2.51 For the below given circuit, if supply frequency, $\omega = 2$ rad/sec and $V_2 = 2\angle 0°$ volts, then what is the lead angle of V_L with V_2?

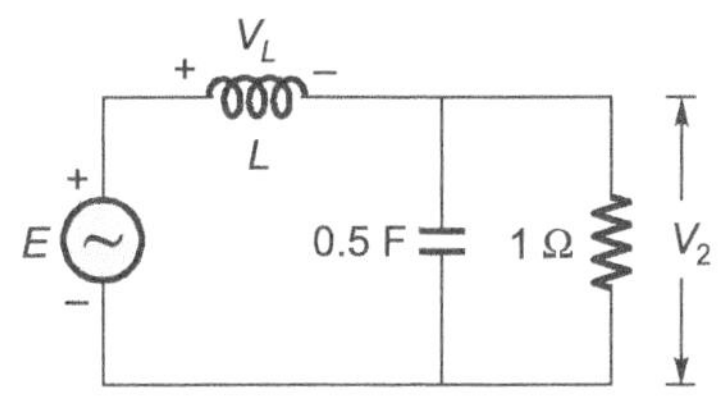

(a) 15° (b) 45°
(c) 90° (d) 135°

[ESE-2009]

2.52 For the circuit as shown below, if the current leads the applied voltage by $\tan^{-1} 2$, what is the resistance value in ohm?

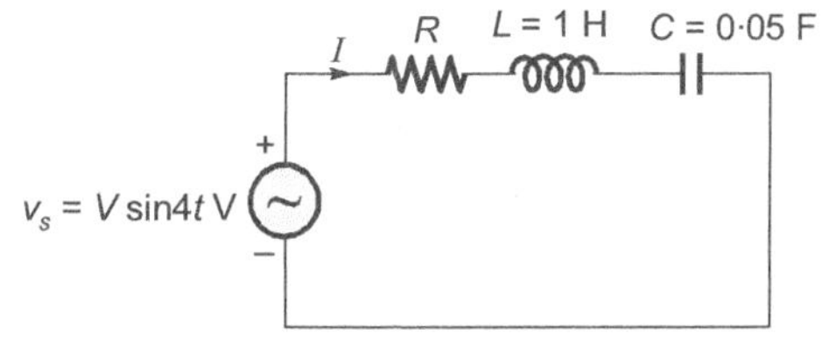

(a) 0.5 (b) 1.0
(c) 2.0 (d) 9.5

[ESE-2009]

2.53 For the AC circuit as shown below, if the rms voltage across the resistor is 120 V, what is the value of the inductor?

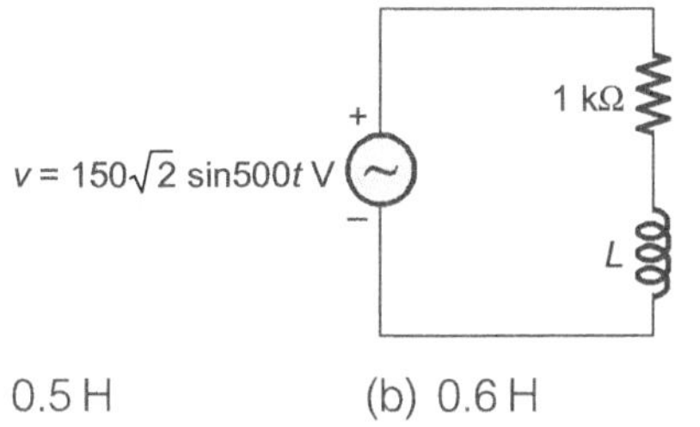

(a) 0.5 H (b) 0.6 H
(c) 1.0 H (d) 1.5 H

[ESE-2009]

2.54 Consider the following statement:

The transfer impedances of a 2-port network remain constant when the position of excitation and response are interchanged if the network

1. is linear
2. contains bilateral elements
3. has high impedance
4. is resonant

Which of these statements is/are correct?

(a) 1 and 2 only (b) 1, 3 and 4 only
(c) 2 only (d) 1, 2, 3 and 4

[ESE-2010]

2.55 A load is connected to an active network. At the terminals to which the load is connected, R_n = 10 Ω and V_{th} = 60 V. Then the maximum power supplied to the load is

(a) 360 W (b) 90 W
(c) 60 W (d) 10 W

[ESE-2010]

2.56 Applying Norton's Theorem, the Norton's equivalent circuit to the left of the terminals *a* and *b* in the below circuit is having equivalent current source (I_N) and equivalent resistance (R_N) as

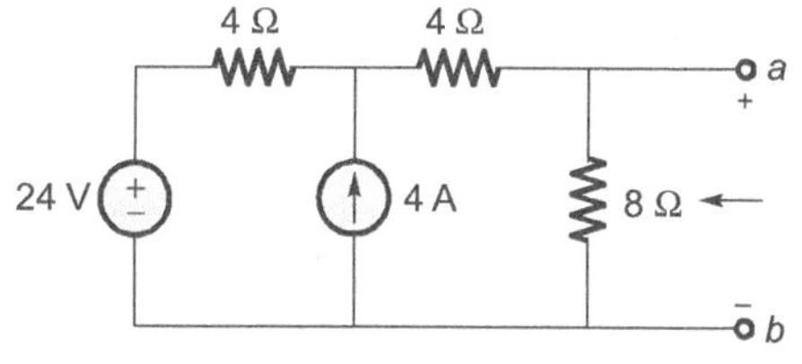

(a) I_N = 5 A; R_N = 4 Ω
(b) I_N = 4 A; R_N = 60 Ω
(c) I_N = 9 A; R_N = 1.6 Ω
(d) I_N = 4 A; R_N = 3.0 Ω

[ESE-2010]

2.57 The maximum power that can be transferred in the load Z_L in the below circuit is

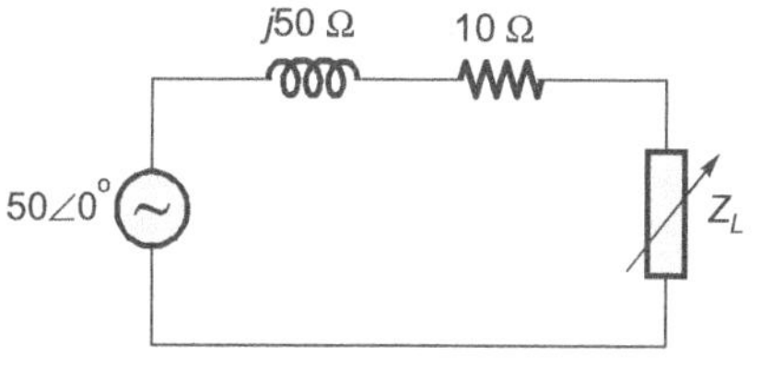

(a) 12.25 W (b) 62.5 W
(c) 24.5 W (d) 500 W

[ESE-2010]

2.58 The Thevenin's equivalent of the circuit shown in the figure below is

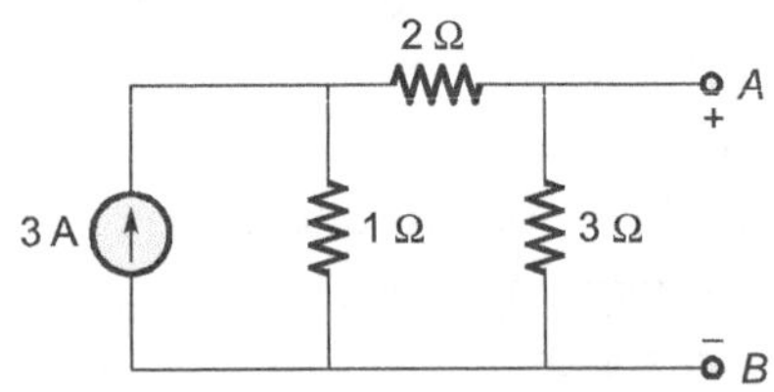

(a) 0.75 V, 1.5 Ω (b) 1.5 V, 1.5 Ω
(c) 1.5 V, 0.75 Ω (d) 5.0 V, 1.5 Ω

[ESE-2010]

2.59 For the circuit shown in the figure below, I is

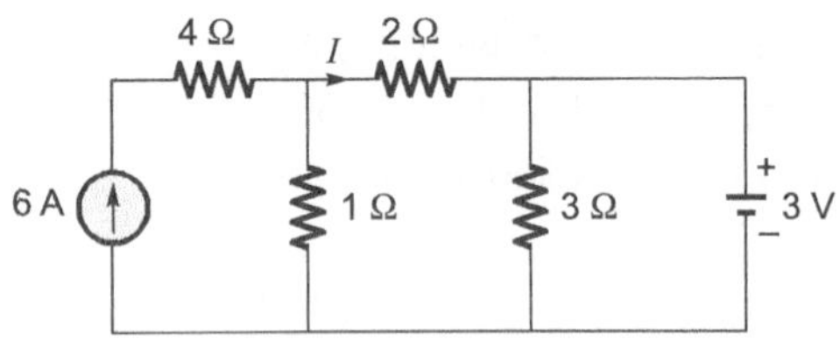

(a) 0 A (b) 1 A
(c) 2 A (d) 3 A

[ESE-2010]

2.60 For an a.c. circuit, if
$v(t) = 160 \sin(\omega t + 10°)$ and
$i(t) = 5 \sin(\omega t - 200°)$, the reactive power absorbed by the circuit is

(a) 100 VARs (b) 200 VARs
(c) 300 VARs (d) 400 VARs

[ESE-2010]

2.61 Match **List-I** with **List-II** and select the correct answer using the code given below the lists:

List-I	List-II
A. Equivalent circuit	1. Superposition
B. Linearity	2. Norton's
C. Bilateral	3. Tellegen's
D. Structure	4. Reciprocity

Codes:

	A	B	C	D
(a)	3	1	4	2
(b)	2	1	4	3
(c)	3	4	1	2
(d)	2	4	1	3

[ESE-2011]

2.62 The Norton equivalent between A and B for the circuit is

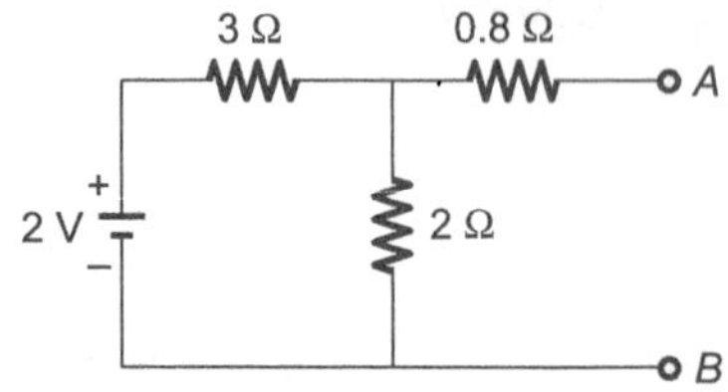

(a) 2 A and 2.5 Ω (b) 0.5 A and 1 Ω
(c) 1 A and 2 Ω (d) 0.4 A and 2 Ω

[ESE-2011]

2.63 Thevenin equivalent circuit to the left of the terminals a and b in the circuit, has equivalent voltage source V_{Th} and equivalent resistance R_{Th}, respectively, as

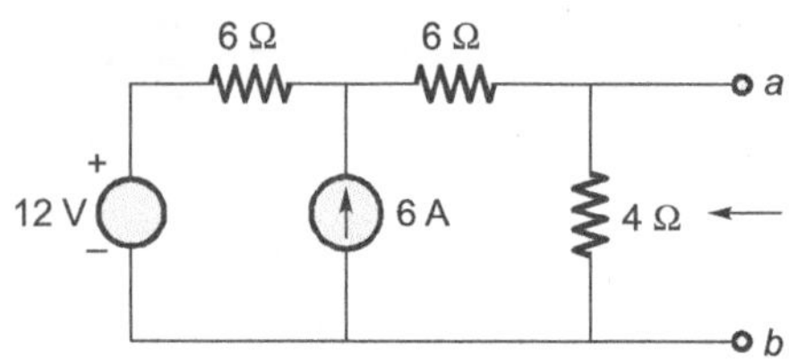

(a) 12 V and 16 Ω (b) 20 V and 4 Ω
(c) 12 V and 12 Ω (d) 12 V and 3 Ω

[ESE-2011]

2.64 The maximum power will be transferred from a voltage source to a load when

(a) the source impedance is half that of the load impedance.
(b) the source impedance is equal to that of the load impedance.
(c) the source impedance is twice that of the load impedance.
(d) both source and load impedance must be zero.

[ESE-2012]

2.65 The output resistance of the circuit at port AB is

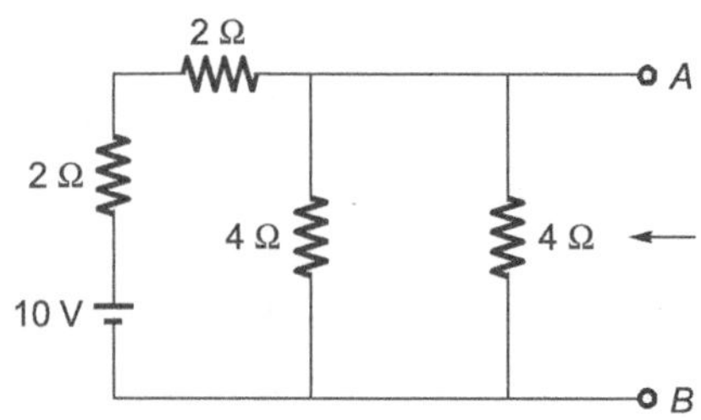

(a) 1 Ω (b) 1.2 Ω
(c) 1.33 Ω (d) 1.5 Ω

[ESE-2012]

2.66 In the circuit shown below, for different values of R, the values of V and I are given, other elements remaining the same

When $R = \infty$, $V = 5$ V

When $R = 0$, $I = 2.5$ A

When $R = 3\ \Omega$, the value of V is given by

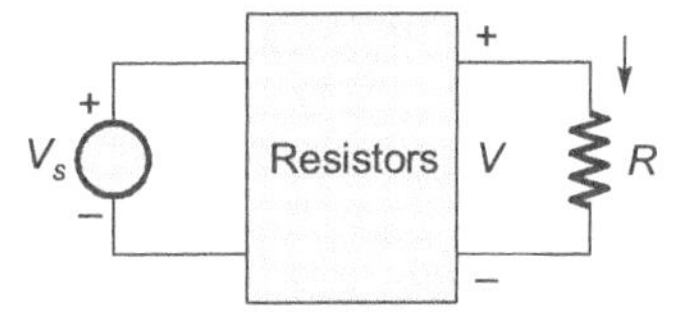

(a) 1 V (b) 5 V
(c) 3 V (d) 2 V

[ESE-2013]

2.67 In the circuit shown below, the maximum power absorbed by the load resistance R_L is

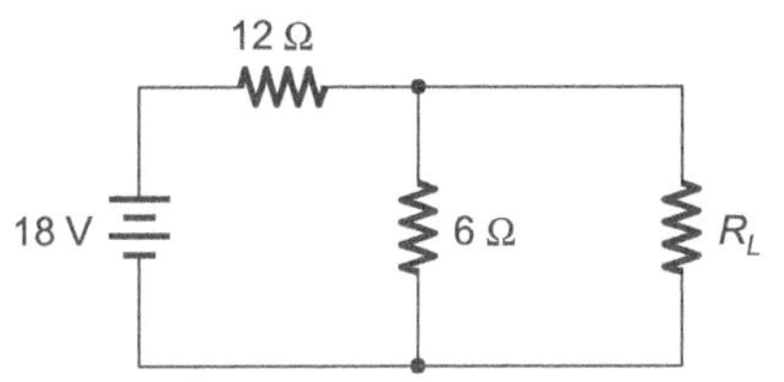

(a) 1.5 W (b) 2.25 W
(c) 2.5 W (d) 5 W

[ESE-2013]

2.68 The voltage-current relationship feeding the network N is shown in the below figure. The Thevenin's equivalent of network N will have V_{Th} and R_{Th} as

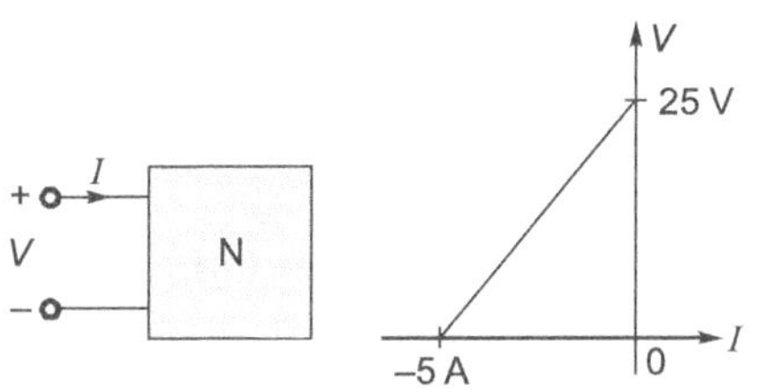

(a) 5 V and 25 Ω (b) –25 V and 5 Ω
(c) 25 V and –5 Ω (d) 25 V and 5 Ω

[ESE-2013]

2.69 An AC source of voltage E_s and an internal impedance of $Z_s = (R_s + jX_s)$ is connected to a load of impedance $Z_L = (R_L + jX_L)$. Consider the following conditions in this regard:

1. $X_L = X_s$, if only X_L is varied.
2. $X_L = -X_s$ if only X_L is varied.
3. $R_L = \sqrt{\left[R_s^2 + (X_s + X_L)^2\right]}$, if only R_L is varied.
4. $|Z_L| = |Z_S|$, if the magnitude alone of Z_L is varied, keeping the phase angle fixed.

The valid conditions for maximum power transfer from the source to the load are

(a) 1, 2, 3 and 4 (b) 1, 2 and 3 only
(c) 1 and 4 only (d) 2, 3 and 4 only

[ESE-2013]

2.70 In the network shown below $V_s = 4\cos 2t$. The value of C is so chosen that the circuit impedance is maximum. Then I_1 leads I_2 by

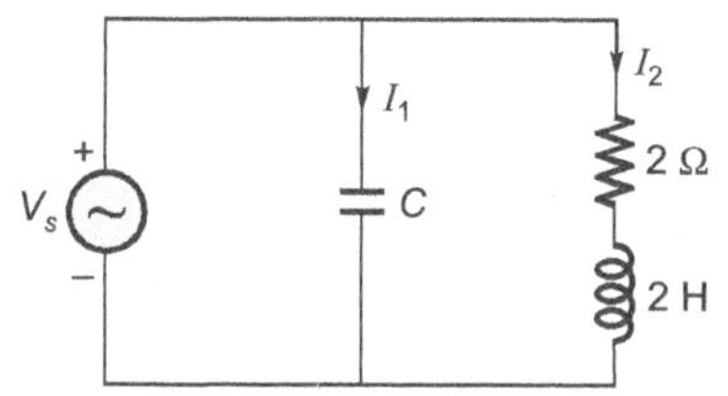

(a) 45° (b) 90°
(c) 0° (d) 135°

[ESE-2013]

2.71 The resistance of a 1 kW electric heater when energized by a 230 V 1-phase AC is

(a) 52.9 Ω (b) 230 Ω
(c) 1000 Ω (d) 4.2 Ω

[ESE-2013]

2.72 When a source is delivering maximum power to a load, the efficiency of the circuit is always

(a) 50%
(b) 75%
(c) 100%
(d) Depends on the circuit parameters

[ESE-2014]

2.73 A battery charger can drive a current of 5 A into a 1 Ω resistance connected at its output terminals. If it is able to charge an ideal 2 V battery at 7 A rate, then Thevenin's equivalent will be

(a) 7.5 V in series with 0.5 Ω
(b) 12.5 V in series with 1.5 Ω
(c) 7.5 V in parallel with 0.5 Ω
(d) 12.5 V in parallel with 1.5 Ω

[ESE-2014]

2.74 As shown in the figure, 1 Ω resistance is connected across a source that has a load line $v + i = 100$. The current through the resistance is

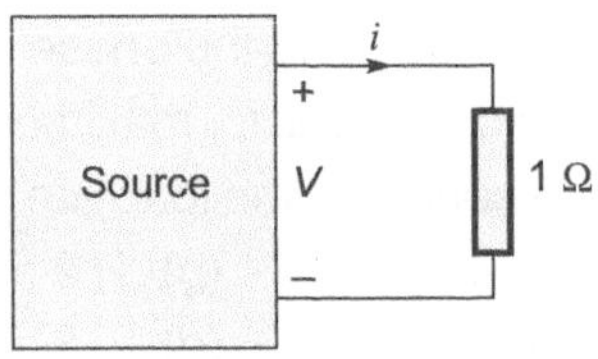

(a) 25 A (b) 50 A
(c) 100 A (d) 200 A

[ESE-2014]

2.75 A single-phase ac voltage source has 200 V rms and a system connected consumes an active power of 300 W. What is the reactive power consumed by the system if 2.5 A rms current is drawn?

(a) 100 VAR (b) 200 VAR
(c) 300 VAR (d) 400 VAR

[ESE-2014]

2.76 For a series R-C circuit, the power factor corresponding to maximum power is

(a) 0.5 lag (b) 0.5 lead
(c) 0.707 lag (d) 0.707 lead

[ESE-2014]

2.77 The power consumed by a coil is 300 W when connected to a 30 V dc source and 108 W when connected to a 30 V ac source. The reactance of the coil is

(a) 3 Ω (b) 4 Ω
(c) 5 Ω (d) 6.67 Ω

[ESE-2014]

2.78 The Thevenin equivalent voltage and resistance across *AB* shown in the figure respectively are

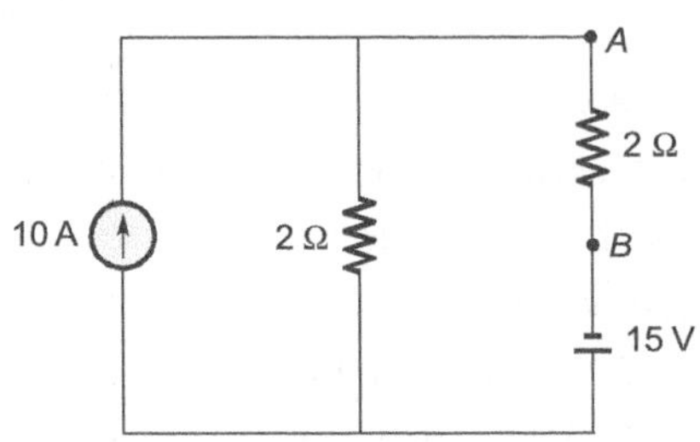

(a) 5 V and 5 Ω (b) 25 V and 3 Ω
(c) 35 V and 2 Ω (d) 25 V and 5 Ω

[ESE-2015]

2.79 The theorem which states that in any linear, non-linear, passive, active, time-variant and time-invariant network, the summation of instantaneous powers is zero will be called as

(a) Tellegen's theorem
(b) compensation theorem
(c) reciprocity theorem
(d) superposition theorem [ESE-2015]

2.80 The power factor of a circuit in which voltage and current waves are non-sinusoidal is defined as

(a) it is the cosine of the angle of phase difference between the voltage and current waves.
(b) it is the cosine of the angle of phase difference between the two complex waves.
(c) it is the cosine of the angle of phase difference between two equivalent sine waves having respectively r.m.s. values equal to those of the voltage and current in the circuit.
(d) it is the sine of the angle of phase difference between the two complex waves.

[ESE-2015]

2.81 Which of the following can produce maximum induced voltage?

(a) 1 A d.c. current
(b) 50 A d.c. current
(c) 1 A, 60 Hz a.c. current
(d) 1 A, 490 Hz a.c. current [ESE-2015]

2.82 An iron-cored choke coil has an equivalent resistance of 5 Ω. It draws 10 A when the applied voltage is 240 V, 50 Hz. Its inductance and power factor respectively are

(a) 7.5 mH and 0.1 (lag)
(b) 747 mH and 0.1 (lag)
(c) 74.7 mH and 0.208 (lag)
(d) 7.5 mH and 0.208 (lag)

[ESE-2015]

2.83 A voltage of 100 V is applied to an impedance of $Z = (3 + j4)$ Ω. What are the values of active power, reactive power and volt-amperes respectively?

(a) 1200 W, 1200 VAR and 2000 VA
(b) 1200 W, 1600 VAR and 2200 VA
(c) 1200 W, 1600 VAR and 2000 VA
(d) 1600 W, 1200 VAR and 2200 VA

[ESE-2015]

2.84 The voltage across an impedance Z is $100\angle 15°$ V and the current through Z is $20\angle -45°$ A. The active and the reactive powers in Z respectively are
(a) 1000 W and 1732 VAR
(b) 500 W and 1732 VAR
(c) 1000 W and 6000 VAR
(d) 500 W and 6000 VAR [ESE-2015]

2.85 An a.c. source of 200 V r.m.s. supplies an active power of 1200 W and a reactive power of 1600 VAR to a load. The r.m.s. current and the power factor of the load respectively are
(a) 10 A and 0.6 (b) 8 A and 0.8
(c) 10 A and 0.8 (d) 8 A and 0.6
[ESE-2015]

2.86 Consider two nodes A and B connected by an impedance of $j5\ \Omega$. If the voltages at nodes A and B are $100\angle 30°$ V and $100\angle 0°$ V respectively, the real power that can be transferred from node A to B is
(a) 1120 W (b) 2000 W
(c) 2769 W (d) 276.9 W
[ESE-2015]

2.87 Thevenin's equivalent of a circuit, operating at $\omega = 5$ rad/s, has

$$V_{OC} = 3.71\angle -15.9°\ \text{V}$$
$$Z_O = 2.38 - j0.667\ \Omega$$

At this frequency, the minimal realization of the Thevenin's impedance will have
(a) a resistor, a capacitor and an inductor
(b) a resistor and a capacitor
(c) a resistor and an inductor
(d) a capacitor and an inductor
[ESE-2016]

2.88 Thevenin's equivalents of the network in Fig. (i) are 10 V and 2 Ω. If a resistance of 3 Ω is connected across terminals AB as shown in Fig. (ii), what are Thevenin's equivalents?

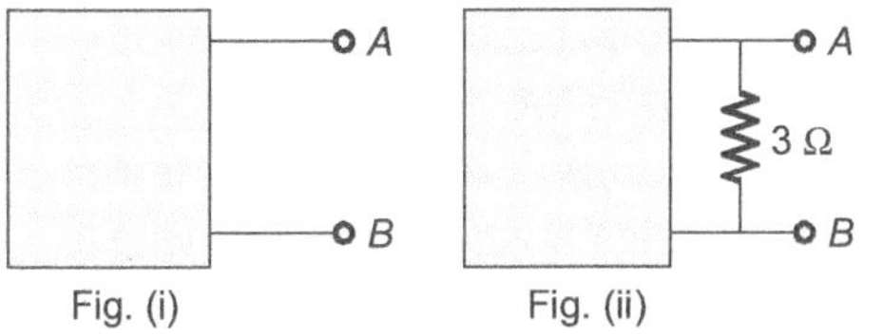

Fig. (i) Fig. (ii)

(a) 10 V and 1.2 Ω (b) 6 V and 1.2 Ω
(c) 10 V and 5.2 Ω (d) 6 V and 5.2 Ω
[ESE-2016]

2.89 A voltage source, connected to a load, has an e.m.f. of 10 V and an impedance of $(500 + j100)\ \Omega$. The maximum power that can be transferred to the load is
(a) 0.2 W (b) 0.1 W
(c) 0.05 W (d) 0.01 W
[ESE-2016]

2.90 Reciprocity theorem is applicable to a network
1. containing R, L and C elements
2. which is initially not a relaxed system
3. having both dependent and independent sources

Which of the above is/are correct?
(a) 1 only (b) 1 and 2 only
(c) 2 and 3 only (d) 1, 2 and 3
[ESE-2016]

2.91 If a constant current generator of 5 A, shunted by its own resistance of 1 Ω, delivers maximum power P in watts to its load of R_L Ω, then the voltage across the current generator and P are
(a) 5 V and 6.25 (b) 2.5 V and 12.5
(c) 5 V and 12.5 (d) 2.5 V and 6.25
[ESE-2016]

2.92 An ideal transformer is rated 220/110 V. A source of 10 V and internal impedance of 2 Ω is connected to the primary. The power transferred to a load Z_L connected across the secondary would be a maximum, when $|Z_L|$ is
(a) 4 Ω (b) 2 Ω
(c) 1 Ω (d) 0.5 Ω
[ESE-2016]

2.93 For a series R-L circuit

$$i(t) = \sqrt{2}\sin(\omega t - 45°)$$

If $\omega L = 1\ \Omega$, the value of R is
(a) 1 Ω (b) 3 Ω
(c) $\sqrt{3}\ \Omega$ (d) $3\sqrt{3}\ \Omega$
[ESE-2016]

2.94 Consider the following values for the circuit shown below:

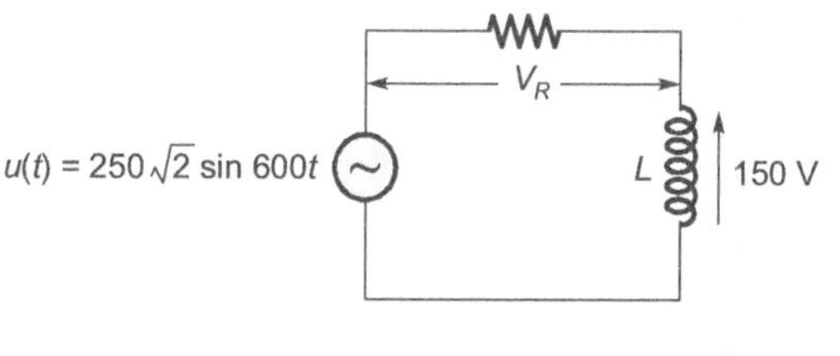

1. $V_R = 100\sqrt{2}$ V
2. $I = 2$A
3. $L = 0.25$ H

Which of the above values are correct?
(a) 2 and 3 only (b) 1 and 2 only
(c) 1 and 3 only (d) 1, 2 and 3

[ESE-2016]

2.95 For the circuit shown, Thevenin's open circuit voltage V_{OC} and Thevenin's equivalent resistance R_{eq} at terminals $A - B$ are, respectively,

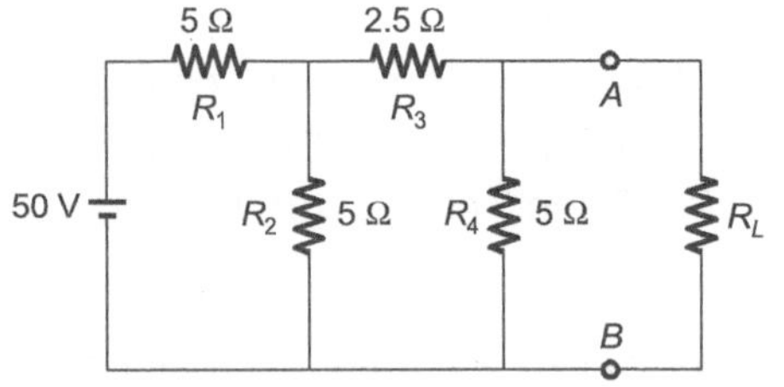

(a) 6.25 V and 2.5 Ω
(b) 12.5 V and 5 Ω
(c) 6.25 V and 5 Ω
(d) 12.5 V and 2.5 Ω

[ESE-2017]

2.96 What is the value of resistance R which will allow maximum power dissipation in the circuit?

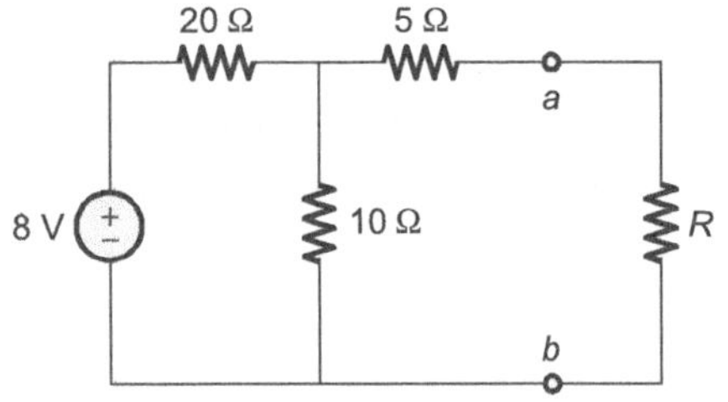

(a) 11.66 Ω (b) 10.33 Ω
(c) 8.33 Ω (d) 7.66 Ω

[ESE-2017]

2.97 Consider the following statements:
1. Network theorems are not derivable from Kirchhoff's law.
2. To get the Norton current, one has to short the current source.
3. Thevenin's theorem is suitable for a circuit involving voltage sources and series connections.

Which of the above statements is/are correct?
(a) 1, 2 and 3 (b) 1 only
(c) 2 only (d) 3 only

[ESE-2017]

2.98 What are the Thevenin's equivalent voltage V_{TH} and resistance R_{TH} between the terminals A and B of the circuit?

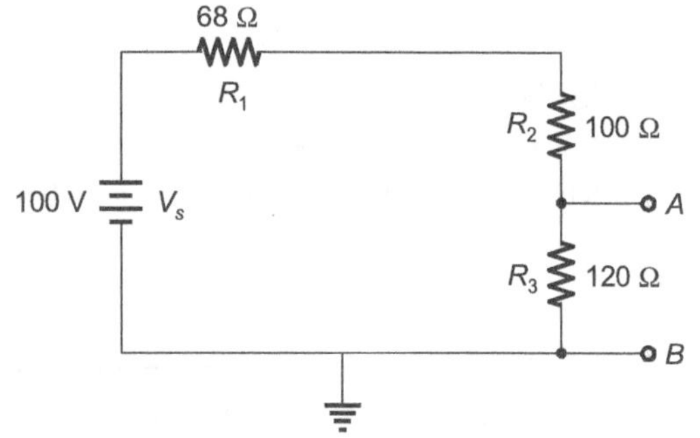

(a) 4.16 V and 120 Ω
(b) 41.67 V and 120 Ω
(c) 4.16 V and 70 Ω
(d) 41.67 V and 70 Ω [ESE-2017]

2.99 Two resistance of 5 Ω and 10 Ω and an inductor L are connected in series across a 50 cosωt voltage source. If the power consumed by the 5 Ω resistor is 10 W, the power factor of the circuit is
(a) 1.0 (b) 0.8
(c) 0.6 (d) 0.4 [ESE-2017]

2.100 A two element series circuit is connected across an AC source given by

$$e = 200\sqrt{2}\sin(314t + 20)\text{ V}$$

The current is then found to be

$$i = 10\sqrt{2}\cos(314t - 25)\text{ A}$$

Then parameters of the circuit are
(a) $R = 20$ Ω and $C = 160$ μF
(b) $R = 14.14$ Ω and $C = 225$ μF
(c) $L = 45$ mH and $C = 225$ μF
(d) $L = 45$ mH and $C = 160$ μF [ESE-2017]

2.101 Consider the following statements for Norton's theorem:
1. Short the branch resistance through which current is to be calculated
2. Obtain the current through this short-circuited branch, using any of the network simplification techniques
3. Develop Norton's equivalent circuit by connecting current source I_N with the resistance R_N in series with it.

Which of the above statements are correct?
(a) 1, 2 and 3 (b) 1 and 3 only
(c) 1 and 2 only (d) 2 and 3 only

[ESE-2018]

2.102 The circuit as shown in figure is connected to a load Z_L across $X - X$. For a maximum power transfer to the load, Z_L should be

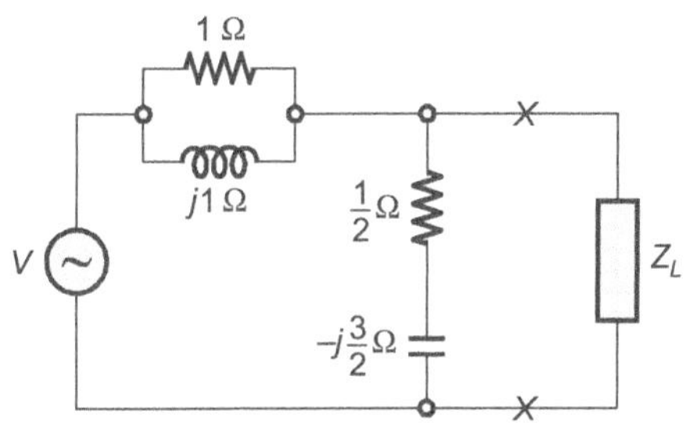

(a) $\frac{3-j}{4}\,\Omega$ (b) $\frac{3+j}{4}\,\Omega$

(c) $\frac{3j}{4}\,\Omega$ (d) $\frac{-3-j}{4}\,\Omega$

[ESE-2018]

2.103 Statement (I) : Two ideal current sources with currents I_1 and I_2 cannot be connected in series.

Statement (II) : Superposition theorem cannot be applied to current sources when one terminal of each of these sources is connected to a common node.

(a) Both Statement (I) and Statement (II) is the correct explanation of Statement (I).

(b) Both Statement (I) and Statement (II) are individually true, but Statement (II) is not the correct explanation of Statement (I).

(c) Statement (I) is true, but Statement (II) is false.

(d) Statement (I) is false, but Statement (II) is true.

[ESE-2018]

2.104 Thevenin's equivalent circuit consists of

(a) Current source and series impedance

(b) Voltage source and series impedance

(c) voltage source and shunt impedance

(d) Current source and shunt impedance

[ESE-2019]

2.105 When the voltage sources are replaced with short circuits and current sources are replaced with open circuits, leaving dependent sources in the circuit, the theorem applied is

(a) Superposition (b) Thevenin

(c) Norton (d) Millman

[ESE-2019]

2.106 The maximum power is delivered from a source to a load when the source resistance is

(a) greater than the load resistance

(b) equal to zero

(c) less than the load resistance

(d) equal to the load resistance

[ESE-2019]

2.107 A network delivers maximum power to the load resistance when it is

(a) greater than Norton's equivalent resistance of the network.

(b) equal to Thevenin's equivalent resistance of the network.

(c) less than source resistance.

(d) less than Norton's equivalent resistance of the network.

[ESE-2019]

2.108 A current of 10 A flows in a circuit with a 30° angle of lag when the applied voltage is 100 V. What are the values of resistance and reactance in the circuit respectively?

(a) 8.66 Ω and 5 Ω (b) 5 Ω and 8.66 Ω

(c) 6.66 Ω and 4 Ω (d) 4 Ω and 6.66 Ω

[ESE-2021]

2.109 Consider the following statements regarding reciprocity theorem :

1. In any passive linear bilateral network, if the single voltage source V_x in branch x produces the current response I_y in branch y, then the removal of the voltage source from branch x and its insertion in branch y will produce the current response I_y in branch x.
2. The interchange of an ideal voltage source and an ideal ammeter in any passive, linear, bilateral circuit will not change the ammeter reading.
3. The interchange of an ideal current source and an ideal voltmeter in any passive linear bilateral circuit will change the voltmeter reading.

Which of the above statements are correct?

(a) 1 and 3 only (b) 1 and 2 only

(c) 2 and 3 only (d) 1, 2 and 3

[ESE-2021]

2.110 What is the input impedance of the circuit, if the circuit operates at ω = 50 rad/s?

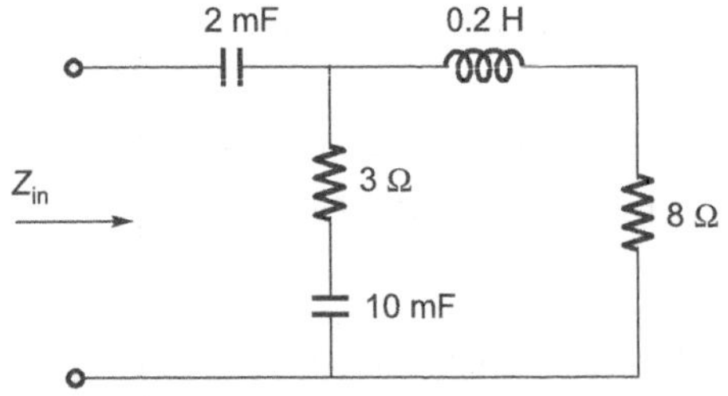

(a) (5.63 – j5.94)Ω (b) (3.22 – j11.07)Ω
(c) (4.54 + j6.79)Ω (d) (6.86 + j13.54)Ω

[ESE-2021]

2.111 The Thevenin voltage and resistance across the terminal *a-b* of the circuit in the figure respectively are

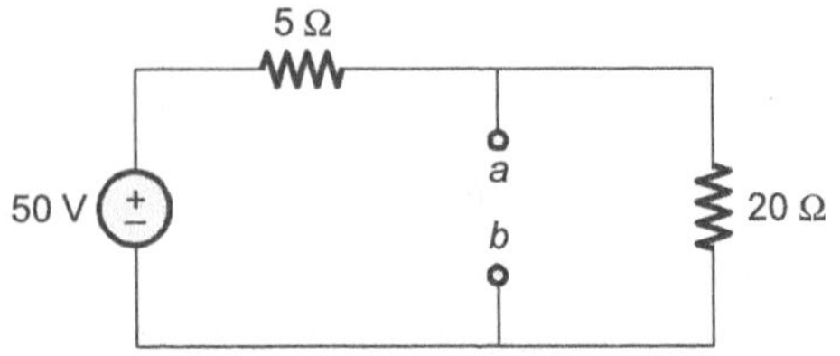

(a) 40 V, 4 Ω (b) 20 V, 8 Ω
(c) 40 V, 8 Ω (d) 20 V, 4 Ω

[ESE-2022]

2.112 The current wave form is shown in the figure below. What is the average value over the time interval of 1 to 7 seconds?

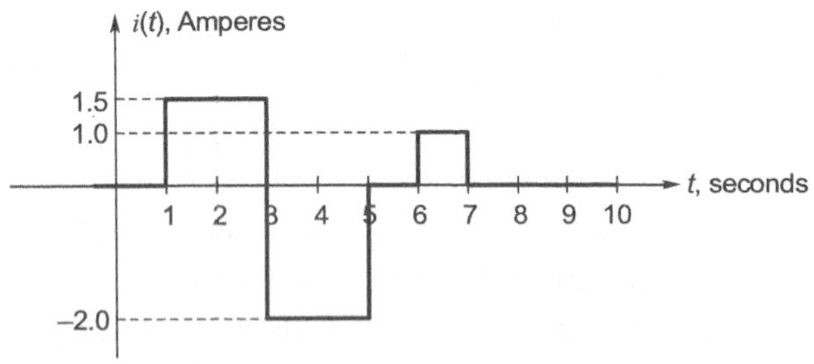

(a) 5.5 A (b) 1.5 A
(c) 0 A (d) 2.5 A

[ESE-2023]

2.113 The following circuit shown in the figure has a voltage source and a dependent current source. What is the Thevenin equivalent resistance at terminals *a* – *b*?

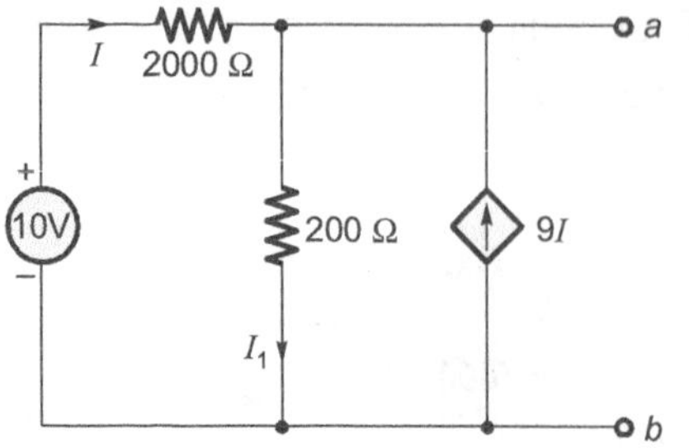

(a) 10 Ω (b) 20 Ω
(c) 100 Ω (d) 200 Ω

[ESE-2023]

2.114 An experimental circuit as shown in the figure below has the variable resistor R_L which is adjusted to the value of the load resistor as 10 Ω. What is the maximum load power?

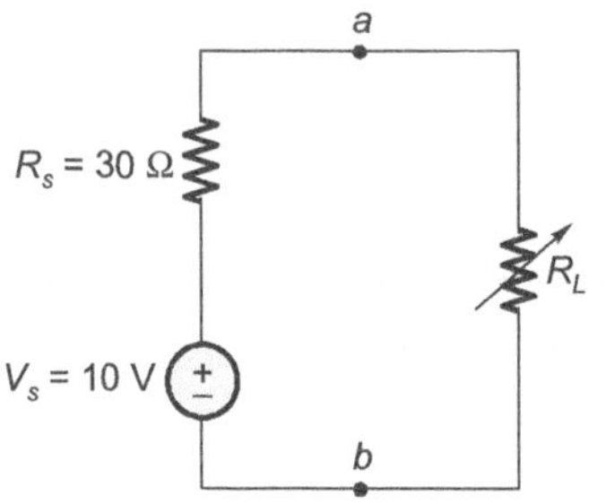

(a) 981 mW (b) 816 mW
(c) 733 mW (d) 625 mW

[ESE-2023]

2.115 A circuit composed with series combination of sinusoidal voltage source $V_s = 3\cos(100t - 3°)$ V, a 500 Ω resistor, a 30 mH inductor and an unknown impedance. What is the value of unknown impedance (Z_L) if the voltage source delivers maximum average power?

(a) 500 – j30 Ω (b) 500 + j30 Ω
(c) 500 – j3 Ω (d) 500 + j3 Ω

[ESE-2023]

2.116 The Thevenin equivalent circuit voltage and resistance for the given circuit between terminals *a* and *b* respectively

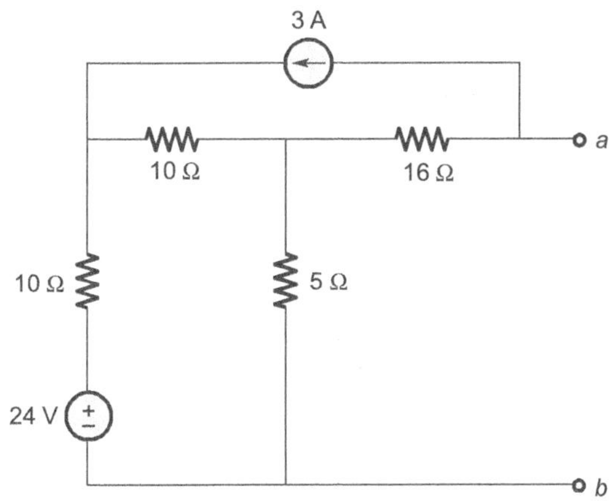

(a) 49.2 V, 10 Ω (b) –49.2 V, 15 Ω
(c) –49.2 V, 20 Ω (d) 49.2 V, 20 Ω

[ESE-2024]

2.117 What is the maximum power that can be delivered to the variable resistor R in the circuit?

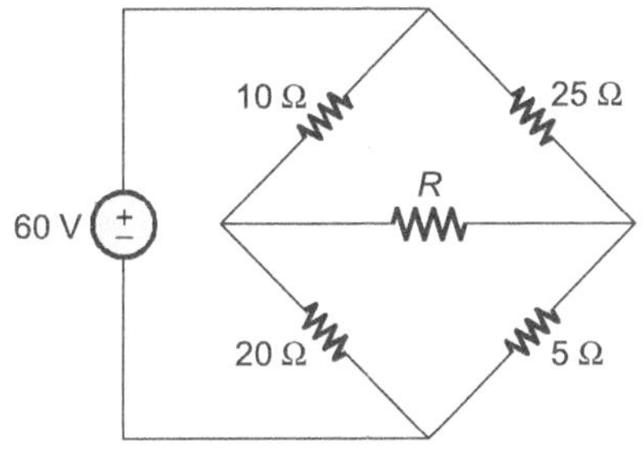

(a) 15.77 W (b) 18.77 W
(c) 19.77 W (d) 20.77 W

[ESE-2024]

2.118 What is the value of voltage V_o shown in the circuit?

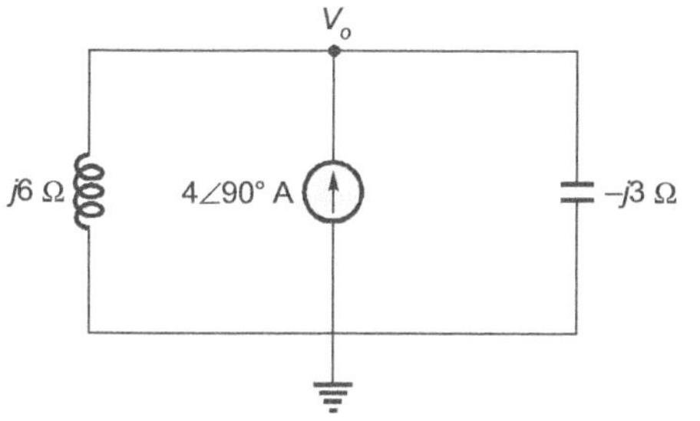

(a) –8 V (b) 8 V
(c) 24 V (d) –24 V

[ESE-2024]

■■■■

Answers Network Theorems and Steady State Analysis

2.1	(d)	2.2	(c)	2.3	(c)	2.4	(b)	2.5	(c)	2.6	(d)	2.7	(c)	2.8	(d)	2.9	(b)
2.10	(c)	2.11	(a)	2.12	(d)	2.13	(c)	2.14	(c)	2.15	(c)	2.16	(b)	2.17	(a)	2.18	(a)
2.19	(a)	2.20	(c)	2.21	(b)	2.22	(a)	2.23	(c)	2.24	(c)	2.25	(b)	2.26	(d)	2.27	(b)
2.28	(b)	2.29	(a)	2.30	(a)	2.31	(c)	2.32	(b)	2.33	(a)	2.34	(d)	2.35	(b)	2.36	(b)
2.37	(b)	2.38	(c)	2.39	(c)	2.40	(c)	2.41	(b)	2.42	(b)	2.43	(b)	2.44	(a)	2.45	(a)
2.46	(a)	2.47	(a)	2.48	(c)	2.49	(b)	2.50	(b)	2.51	(d)	2.52	(a)	2.53	(d)	2.54	(a)
2.55	(b)	2.56	(a)	2.57	(b)	2.58	(b)	2.59	(b)	2.60	(b)	2.61	(b)	2.62	(d)	2.63	(d)
2.64	(b)	2.65	(c)	2.66	(c)	2.67	(b)	2.68	(d)	2.69	(d)	2.70	(d)	2.71	(a)	2.72	(a)
2.73	(b)	2.74	(b)	2.75	(d)	2.76	(d)	2.77	(b)	2.78	(c)	2.79	(b)	2.80	(c)	2.81	(d)
2.82	(c)	2.83	(c)	2.84	(a)	2.85	(a)	2.86	(*)	2.87	(b)	2.88	(b)	2.89	(c)	2.90	(a)
2.91	(d)	2.92	(d)	2.93	(a)	2.94	(*)	2.95	(d)	2.96	(a)	2.97	(d)	2.98	(d)	2.99	(c)
2.100	(b)	2.101	(c)	2.102	(a)	2.103	(c)	2.104	(b)	2.105	(a)	2.106	(b)	2.107	(b)	2.108	(a)
2.109	(b)	2.110	(b)	2.111	(a)	2.112	(c)	2.113	(c)	2.114	(d)	2.115	(c)	2.116	(c)	2.117	(d)
2.118	(c)																

Explanations Network Theorems and Steady State Analysis

2.1 (d)

Both theorems are dual i.e., they follow duality principle.

2.2 (c)

According to reciprocity theorem, the ratio of excitation to response is constant when the position of excitation and response are interchanged.

$$\frac{E_{AB}}{I_{CD}} = \frac{E_{CD}}{I_{AB}}$$

$$\Rightarrow \quad I_{AB} = \frac{E_{CD} \times I_{CD}}{E_{AB}} = \frac{100 \times 1}{10} = 10 \text{ A}$$

2.3 (c)

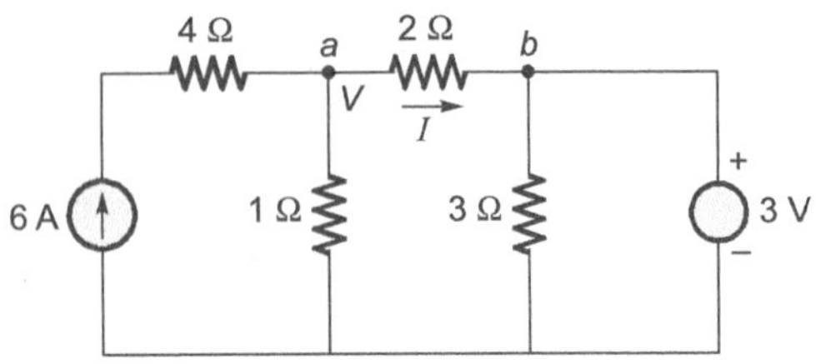

Applying KCL at node a

$$-6 + \frac{V}{1} + \frac{V-3}{2} = 0$$

$$-12 + 2V + V - 3 = 0$$

$$3V = 15 \Rightarrow V = 5 \text{ volts}$$

$$I = \frac{V-3}{2} = \frac{5-3}{2} = 1 \text{ A}$$

2.4 (b)

For finding R_{Th}, voltage source is short circuited and current source is open circuited.

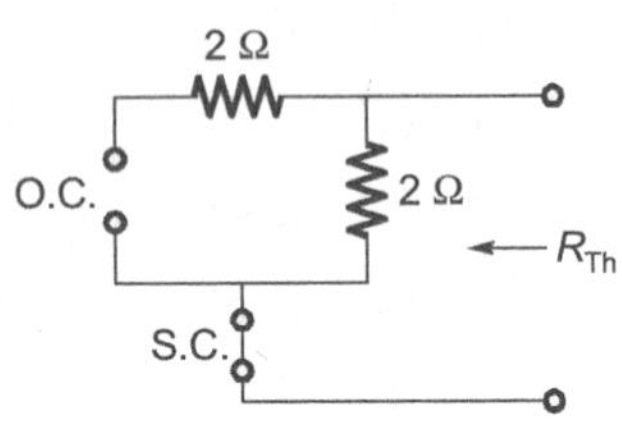

$$\therefore \quad R_{Th} = 2\ \Omega$$

2.5 (c)

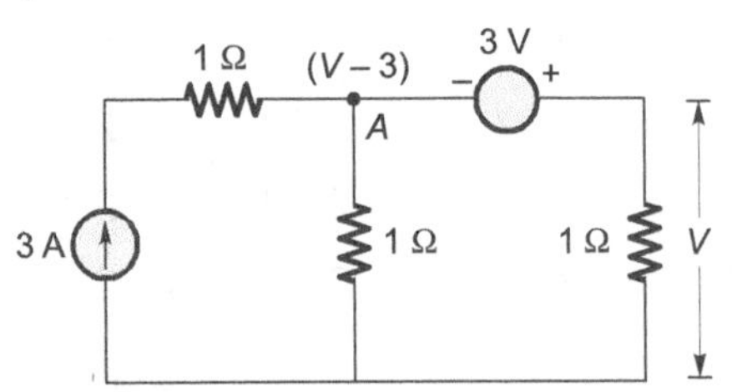

Voltage of node $A = V - 3$

Applying KCL at node A

$$-3 + \frac{V-3}{1} + \frac{V}{1} = 0$$

$$V = 3 \text{ Volt}$$

2.6 (d)

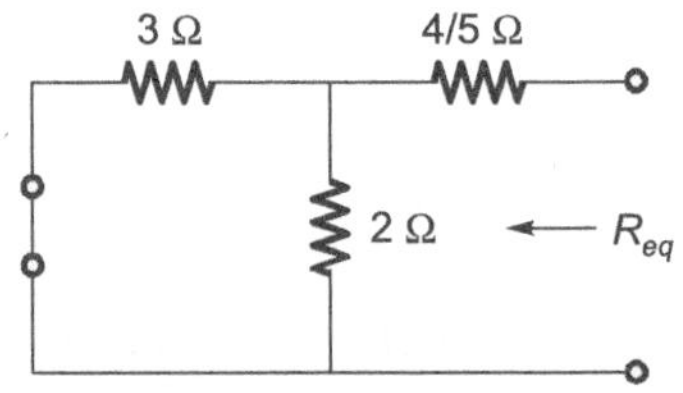

$$R_{eq} = (3 \parallel 2) + \frac{4}{5}$$

$$= \frac{3 \times 2}{3+2} + \frac{4}{5} = 2\ \Omega$$

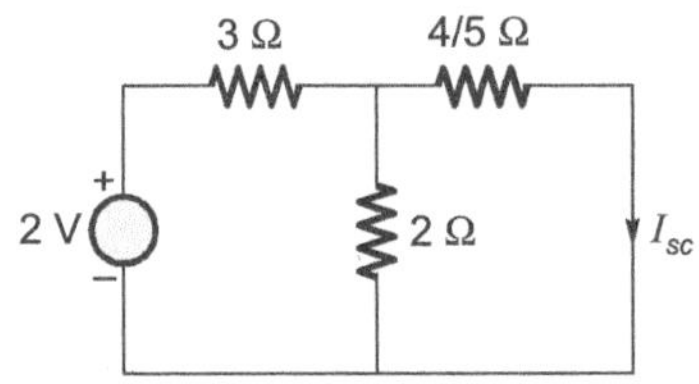

$$\therefore \quad I_{sc} = \left(\frac{2}{3 + \left(2 \parallel \frac{4}{5}\right)}\right) \times \left(\frac{2}{2 + \left(\frac{4}{5}\right)}\right) = \frac{2}{5} \text{A}$$

2.7 (c)

For Figure-I

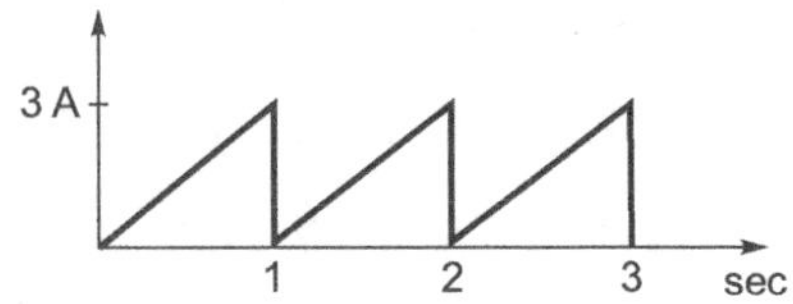

$$i(t) = \begin{cases} 3t & , \quad 0 < t < 1 \\ 3(t-1), & 1 < t < 2 \end{cases}$$

Heat produced in resistor of 1 Ω

$$H_1 = \int_0^2 i_1^2(t) \times 1 \times dt$$

$$H_1 = \int_0^2 9t^2 dt + \int_1^2 9(t-1)^2 dt$$

$$H_1 = \left[3t^3\right]_0^1 + \left[3(t-1)^3\right]_1^2$$

$$= 3 + 3 = 6 \text{ Joule}$$

For Figure-II

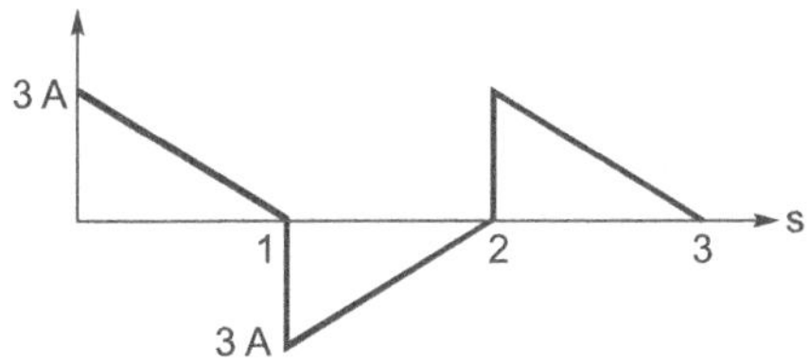

$$i(t) = \begin{cases} 3(1-t), & 0 < t < 1 \\ -3(2-t), & 1 < t < 2 \end{cases}$$

$$H_2 = \int_0^2 i_2^2(t) \times 1 \times dt$$

$$= \int_0^1 9(1-t)^2 dt + \int_1^2 9(2-t)^2 dt$$

$$= \left[-3(1-t)^3\right]_0^1 + \left[-3(2-t)^3\right]_1^2$$

$$= 3 + 3 = 6 \text{ Joule}$$

$\therefore \; H_1 : H_2 = 1 : 1$

Note: As heating depends on square of rms value of current, it is independent from the direction of low of current.

2.8 (d)

v-i characteristic is $i = \frac{1}{5}v - 4$

$$\therefore \qquad R_{AB} = \frac{dv}{di} = 5\,\Omega$$

2.9 (b)

For maximum power transfer, load impedance is complex conjugate of source impedance.

2.10 (c)

Superposition theorem is not applicable for nonlinear parameters.

$$\because \qquad P = I^2R \text{ or } \frac{V^2}{R}$$

$\therefore$ Not applicable for power calculation.

2.11 (a)

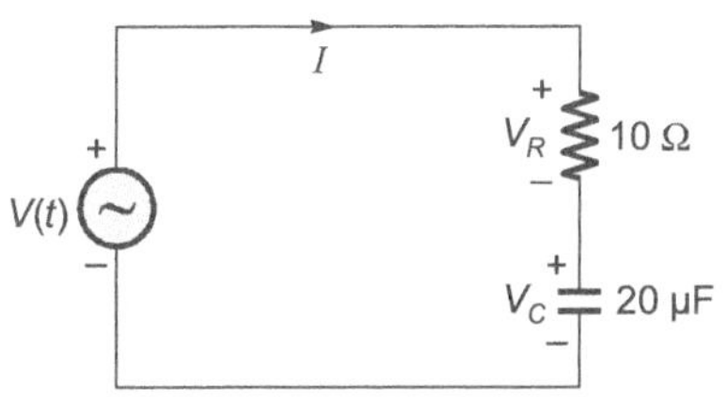

$$I = C\frac{dV_c}{dt}$$

$$2\cos 5000t = 20 \times 10^{-6} \times \frac{dV_c}{dt}$$

$$\Rightarrow \qquad V_c = \int \frac{2\cos 5000t\, dt}{20 \times 10^{-6}}$$

and $\quad V_R = -20 \sin 5000t\,\text{V}$

and $\quad V_R = IR = 20\cos 5000t\,\text{V}$

$$V(t) = V_R + V_C$$

$$= 20 \times \sqrt{2}\left[\frac{1}{\sqrt{2}}\cos 5000t - \frac{1}{\sqrt{2}}\sin 5000t\right]$$

$$= 28.28\cos(5000t - 45°)\,\text{V}$$

2.12 (d)

$$I_R = \sqrt{5^2 - 4^2} = 3 \text{ Amp}$$

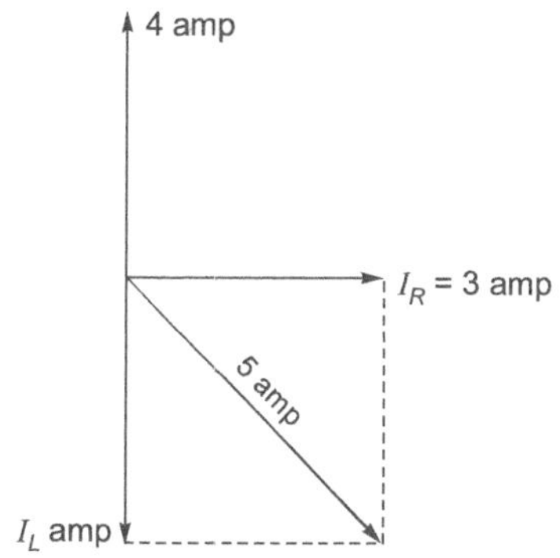

$$(I_L - 4)^2 + 3^2 = 5^2$$

$$I_L = 8 \text{ amp}$$

2.13 (c)

According to KCL at any node, sum of all the incoming or outgoing currents is zero

$\therefore \; i_1 + i_2 + i_3 = 0$

$$\Rightarrow \qquad i_3 = -(i_1 + i_2)$$

$$(i_1 + i_2) = 10\sin(400t + 60°) + 10\sin(400t - 60°)$$

$$= 10[2\sin 400t \times \cos 60°]$$

$$= 10 \sin 400t$$

$$\therefore \qquad i_3 = -(i_1 + i_2) = -10\sin 400t \text{ A}$$

2.14 (c)

$$I = \frac{V_{Th}}{R_{Th}} = 3 \text{ for } R = 0 \qquad ...(i)$$

$$I = \frac{V_{Th}}{R_{Th} + 2} = \frac{3}{2} \text{ for } R = 2 \qquad ...(ii)$$

From (*i*) and (*ii*) $V_{Th} = 6$ V, $R_{Th} = 2\,\Omega$

$\therefore I$ for $(R = 1) = \frac{6}{2+1} = 2$ A

2.15 (c)

When $\quad I_1 = 8$ A, $I_2 = 12$ A

$V = 80$ V

and when $I_1 = -8$ A, $I_2 = 4$ A

$$V = 0\text{ V}$$

By using superposition theorem

When I_1 is present, $I_2 = 0$

$$\Rightarrow \quad V' = I_1R_1$$

and when I_2 is present, $I_1 = 0$

$$\Rightarrow \quad V'' = I_2R_2$$

when I_1 and I_2 both are present

$$\Rightarrow \quad V = V' + V''$$

$$\therefore \quad V = I_1R_1 + I_2R_2$$

$$80 = 8R_1 + 12R_2 \quad ...(i)$$

and

$$0 = -8R_1 + 4R_2 \quad ...(ii)$$

By solving these two equation,

$$R_2 = 5\ \Omega \text{ and } R_1 = 5/2\ \Omega$$

$\therefore$ when $I_1 = I_2 = 10$ A

$$V = 10 \times \frac{5}{2} + 10 \times 5 = 75 \text{ volt}$$

2.16 (b)

$$Z(j\omega) = [(3+j4) \parallel (3-j4)] + \frac{17}{6}$$

$$= \left[\frac{(3+j4)(3-j4)}{3+j4+3-j4}\right] + \frac{17}{6}$$

$$= \left[\frac{9+16}{6}\right] + \frac{17}{6} = (7 + j0)\ \Omega$$

2.17 (a)

Only resister dissipates the power, neither inductor nor capacitor.

2.18 (a)

When, $V_s = 3$ V

Voltage of the node a

$$V_1 = I \times 3 = 4 \times 3 = 12 \text{ V}$$

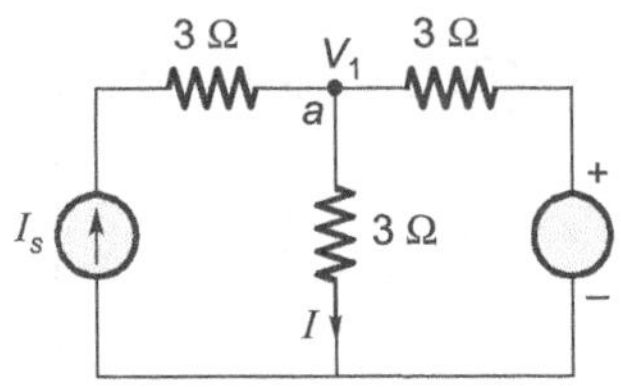

By applying KCL at node a

$$I_s = I + \frac{V_1 - V_s}{3} = 4 + \frac{12-3}{3} = 7\text{ A}$$

When, $V_s = 12$ V

Again applying KCL at node a,

$$I_s = \frac{V_1}{3} + \frac{V_1 - V_s}{3}$$

$$\Rightarrow \quad 7 = \frac{V_1}{3} + \frac{V_1 - 12}{3} \Rightarrow V_1 = 16.5 \text{ V}$$

$$I = \frac{V_1}{3} = \frac{16.5}{3} = 5.5\text{ A}$$

Alternate solution:

By superposition,

when $V_s = 3$ V, $I_s = 0 \Rightarrow I = 0.5$ A

when $V_s = 0$ and $I_s = I_s \Rightarrow I = 2 + 3.5 = 5.5$ A

when $V_s = 12$ V, $I_s = 0 \Rightarrow I = 2$ A

so, when $V_s = 12$ V, $I_s = I_s \Rightarrow I_s = 2 + 3.5 = 5.5$ A

2.19 (a)

Applying superposition theorem

Current due to current source

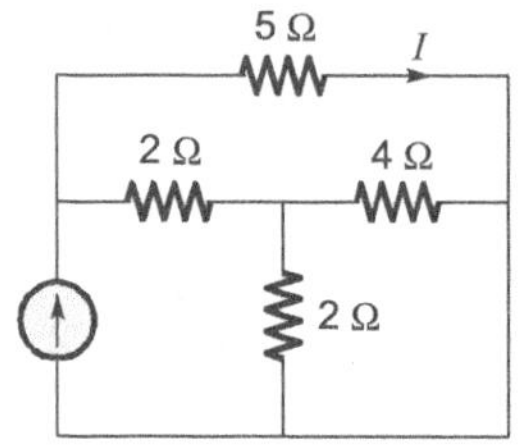

$$I = \frac{1}{5+(2+4 \parallel 2)} \times (2 + 4 \parallel 2) = 0.4 \text{ A}$$

Current due to voltage source

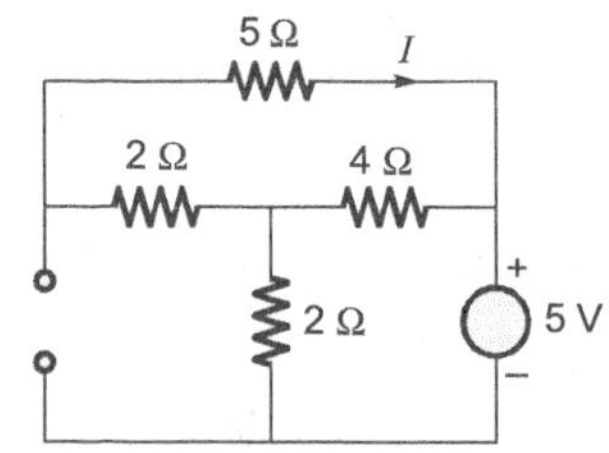

$$-I = \frac{5}{2+\{(5+2) \parallel 4\}} \times \left\{\frac{4}{4+(5+2)}\right\} = -0.4 \text{ A}$$

Current due to both sources = 0.4 + (–0.4) = 0

2.20 (c)

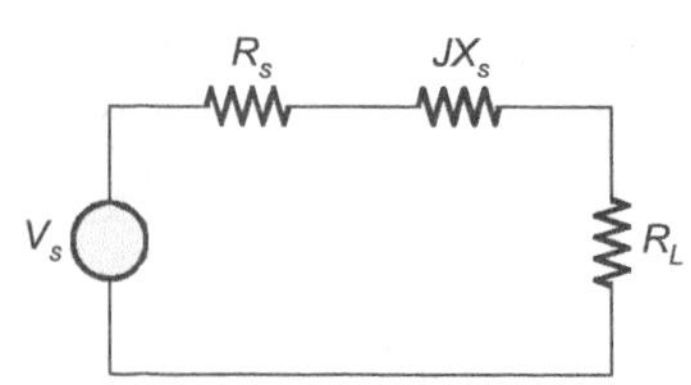

For maximum power transfer,

$$R_L = \sqrt{R_s^2 + X_s^2} = \sqrt{8^2 + 6^2} = 10\ \Omega$$

2.21 (b)

We calculate R_{Th} by short circuiting the voltage source

$$\therefore \quad R_{Th} = 40 \| 60$$

$$R_{Th} = 24\ \Omega$$

and $$I_{sc} = \frac{V_{OC}}{R_{Th}}$$

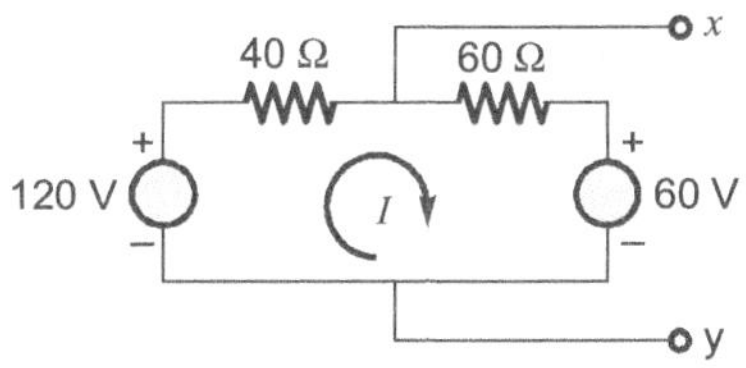

Let I current flows through loop

$$V_{OC} = 60I + 60$$

and $$120 - 40I - 60I - 60 = 0$$

$$\boxed{I = \frac{3}{5}}$$

$$\therefore \quad V_{OC} = 60 \times \frac{3}{5} + 60 = 96\ \text{V}$$

$$\boxed{I_{SC} = \frac{V_{OC}}{R_{Th}} = 4\ \text{A}}$$

2.22 (a)

For maximum power transfer from source,

$$R = R_{eq}$$

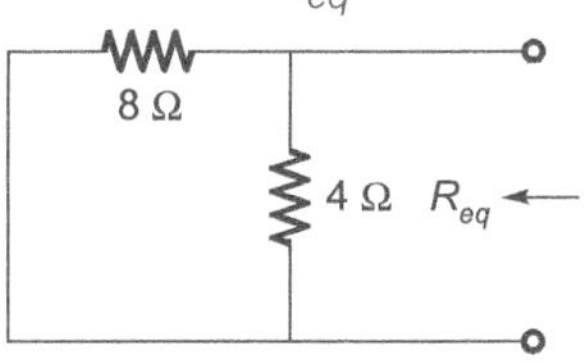

$$R_{eq} = 8 \| 4 = \frac{32}{12} = \frac{8}{3}\ \Omega$$

$$R = R_{eq} = \frac{8}{3}\ \Omega$$

2.23 (c)

Norton's equivalent is dual of Thevenin's equivalent not reciprocal.

2.24 (c)

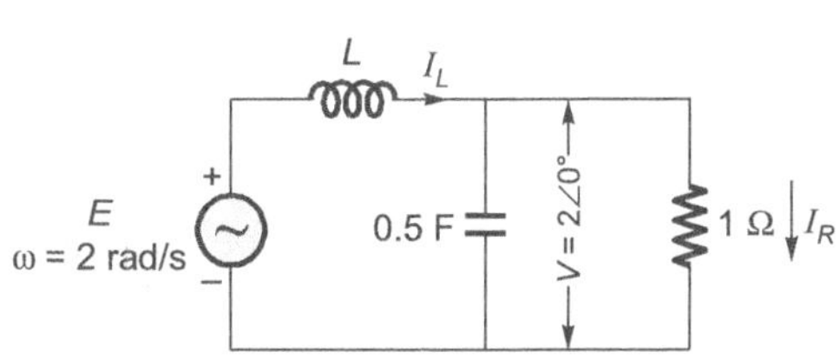

Let current through inductor, $I_L \angle \theta$

$$\therefore \quad I_L \angle \theta \cdot \frac{R}{\frac{1}{j\omega C} + R} = I_C \qquad \text{...(i)}$$

and $$I_C = \frac{2\angle 0^\circ}{\frac{1}{j\omega C}} \qquad \text{...(ii)}$$

From equation (*i*) and (*ii*),

$$I_L \angle \theta \cdot \frac{R}{1 + j\omega CR} = 2\angle 0^\circ$$

$$\Rightarrow \quad \theta - \tan^{-1}(\omega CR) = 0$$

$$\Rightarrow \quad \theta = \tan^{-1}(2 \times 0.5 \times 1) = 45^\circ$$

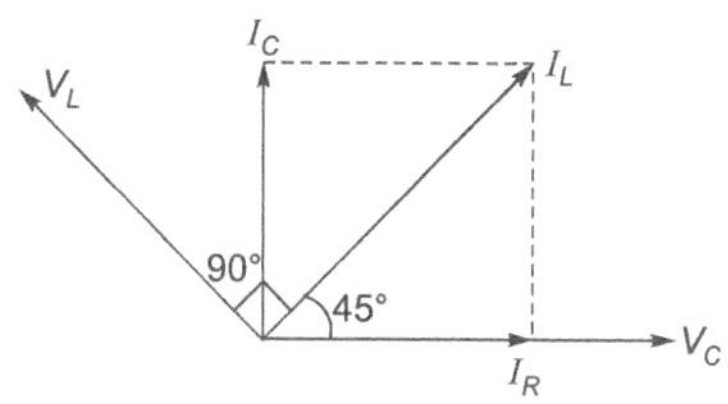

As we know, in inductor current lags the voltage by 90° so inductor voltage phase angle = 90° + 45° = 135° i.e. inductor voltage leads the capacitor voltage by 135°.

2.25 (b)

Superposition theorem is valid for linear network. Tellegen's theorem is valid irrespective of type of network, types of elements contained in the network, value of each element contained in the network, type of voltage and current sources present in the network (dependent/independent).

2.26 (d)

When, $I_s = 0$

$$I = 1 = \frac{V}{2} \Rightarrow V = 2\ \text{V}$$

When, $I_s = 2$ A

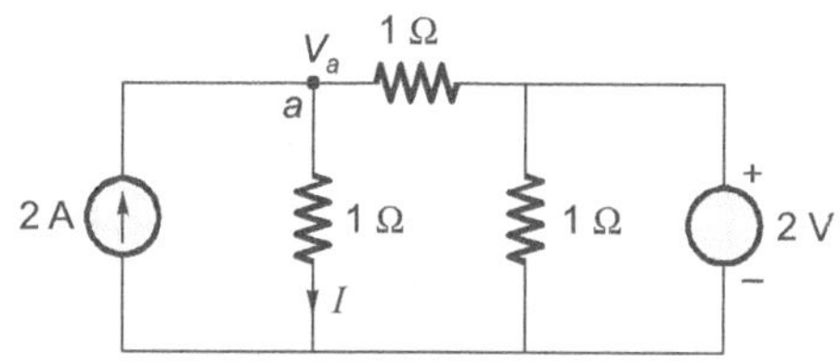

By applying KCL at node a

$$-2 + \frac{V_a}{1} + \frac{V_a - 2}{1} = 0$$

$$\Rightarrow \quad -2 + 2V_a - 2 = 0$$

$$V_a = 2\text{V}$$

$$I = \frac{V_a}{1} = \frac{2}{1} = 2\ \text{A}$$

2.27 (b)

Thevenin's equivalent of black box

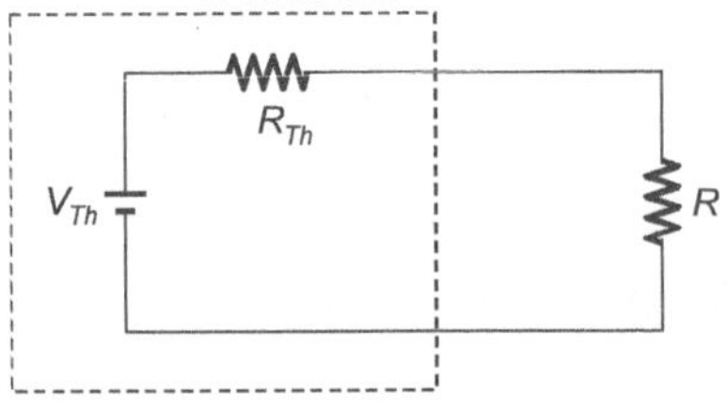

When $I = 3$ A and $R = 0$

$$3 = \frac{V_{Th}}{R_{Th} + 0} \quad ...(i)$$

When $I = \frac{3}{2}$ A and $R = 2\ \Omega$

$$\frac{3}{2} = \frac{V_{Th}}{R_{Th} + 2} \quad ...(ii)$$

From equation (*i*) and (*ii*),

$V_{Th} = 6$ V and $R_{Th} = 2\ \Omega$

when $R = 1\ \Omega$

$$I = \frac{6}{2+1} = 2 \text{ A}$$

2.28 (b)

$$Y = \frac{1}{60 + j60} + \frac{1}{-j120} = 8.33 \times 10^{-3} \angle 0°$$

$$I = VY = 120\angle 0° \times 8.33 \times 10^{-3} \angle 0°$$
$$= 1 + j0$$

2.29 (a)

$$\omega = 2$$

$$\Rightarrow \quad Z(j2) = \frac{(1+j) \times \left(\frac{1}{j2C}\right)}{1 + j + \frac{1}{j2C}}$$

$$= \frac{1+j}{j2C + 1 - 2C}$$

$$Z(j2) = \frac{(1+j) \times \{(1-2C) - j2C\}}{(1-2C)^2 + (2C)^2}$$

$$= \frac{(1-2C) - j2C + j(1-2C) + 2C}{(1-2C)^2 + (2C)^2}$$

Putting imaginary term to be zero.

$-2C + 1 - 2C = 0$

$$C = \frac{1}{4} \text{F}$$

2.30 (a)

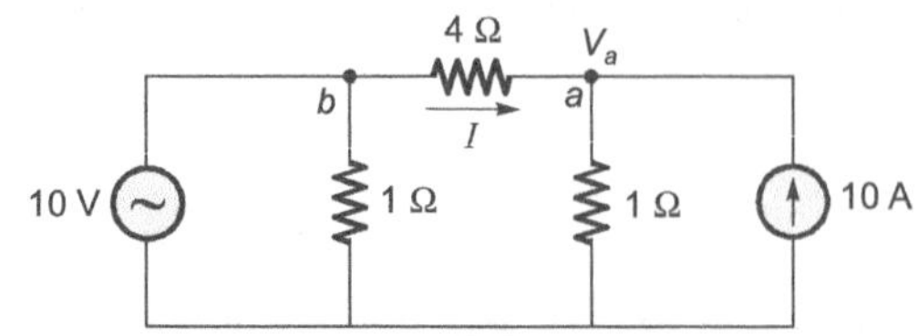

By applying KCL at node a

$$-10 + \frac{V_a}{1} + \frac{V_a - 10}{4} = 0 \Rightarrow V_a = 10 \text{ V}$$

$$I = \frac{10 - V_a}{4} = \frac{10-10}{4} = 0 \text{ A}$$

2.31 (c)

Thevenin's equivalent Network

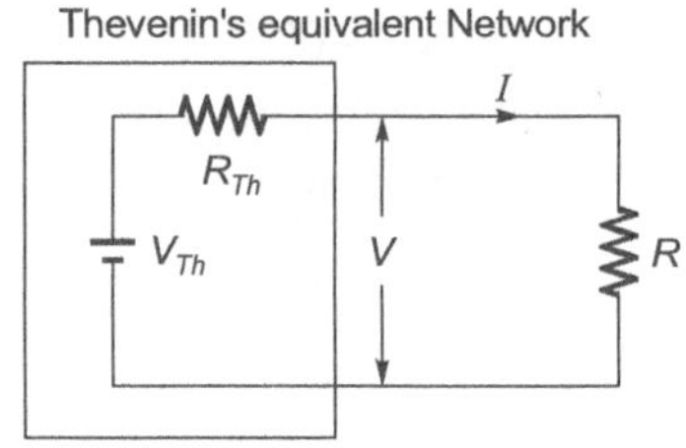

When $I = 0$, $V = 20$ V

$\Rightarrow \quad V_{Th} = 20$ V

When $R = 0$, $I = 10$ A

$$\Rightarrow \quad R_{Th} = \frac{V_{Th}}{I} = \frac{20}{10} = 2\ \Omega$$

If $\quad R = 3\ \Omega$

$$I = \frac{V_{Th}}{R_{Th} + R} = \frac{20}{2+3} = 4 \text{ A}$$

2.32 (b)

$$2 = \left(\frac{I_N}{R_N + R}\right) \times R_N$$

$$\Rightarrow \quad 2 = \frac{4}{2+R} \times 2$$

$$\Rightarrow \quad R = 2\ \Omega$$

2.33 (a)

Let voltage across the 1 Ω grounded resistor = V_1

$$\therefore \quad \frac{e_1(t) - V_1}{1} = \frac{V_1}{1} + \frac{V_1 - e_2(t)}{1}$$

$$\therefore \quad \frac{e_1(t) + e_2(t)}{3} = V_1$$

$$V_1 = \frac{\sqrt{3}\cos(\omega t + 30°) + \sqrt{3}\sin(\omega t + 60°)}{3}$$

After simplifying, $\boxed{V_1 = [\cos \omega t] V}$

2.34 (d)

Circuit for current due to voltage source only.

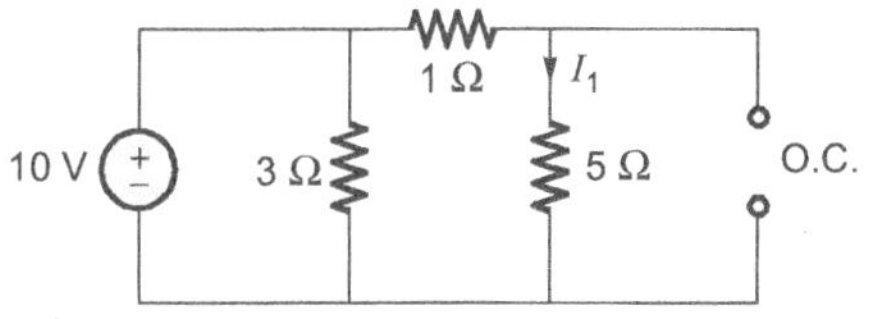

$$I_1 = \frac{10}{(5+1)}$$

Circuit for current due to current source only

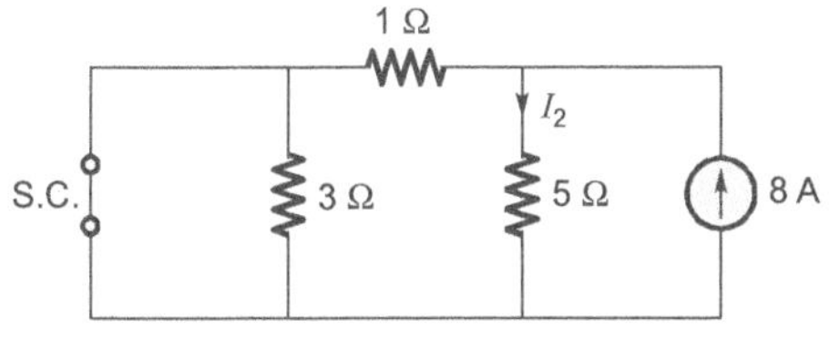

$$I_2 = \frac{1}{5+1}\times 8 \quad \text{(current division rule)}$$

So, total current

$$I_1 + I_2 = \frac{10}{6}+\frac{8}{6} = 3 \text{ Amp.}$$

2.35 (b)

Any four terminal passive network can be converted to equivalent T or π-network. Thevenin Theorem or Norton can also be used to get a transfer function.

But R is not correct explanation of A.

2.36 (b)

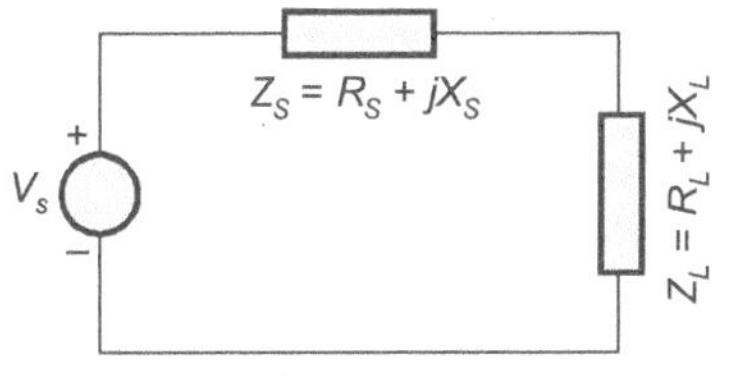

For maximum power transfer,

$$Z_L = Z_S^*$$

i.e. $\quad R_L = R_S$ and $X_L = -X_S$

2.37 (b)

$$\frac{V}{X} = 3 \text{ A} \quad (\text{I}^{\text{st}} \text{ condition})$$

$$\frac{V}{X+2} = 1.5 \text{ A} \quad (\text{II}^{\text{nd}} \text{ condition})$$

From equation (i) and (ii),

$V = 6$ V and $X = 2\ \Omega$

$$\therefore \quad \frac{6}{2+1} = I \qquad \therefore I = 2 \text{ A}$$

2.38 (c)

$$I_{SC} = \frac{V}{Z} = \frac{100\angle 30^\circ}{4+j3} = 20\angle{-6.86^\circ} \text{ A}$$

2.39 (c)

From maximum power transfer theorem

$$2 \,||\, 4 \,||\, R = 1$$
$$R = 4\ \Omega$$

2.40 (c)

Angle between voltage across R and inductor will be 30° as given

So, $\tan 30^\circ = \dfrac{X_L}{R}$; $R = \dfrac{X_L}{\tan 30^\circ} = \sqrt{3}\cdot X_L$

as $\quad X_L = 1\ \Omega$

$$R = \sqrt{3}\ \Omega$$

2.41 (b)

Total real power = Power for fundamental + Power for harmonics

$$= \frac{50\times 10}{2}\cos 60^\circ + \frac{50\times 8}{2}\cos 45^\circ$$

$$= 125 + 100\sqrt{2}$$

Total reactive power

$$= \frac{50\times 10}{2}\sin 60^\circ + \frac{50\times 8}{2}\sin 45^\circ$$

$$= 125\sqrt{3} + 100\sqrt{2}$$

Hence power factor $= \dfrac{\text{Real power}}{\text{Apparent power}}$

$$= \frac{125+100\sqrt{2}}{\sqrt{\left(125+100\sqrt{2}\right)^2 + \left(125\sqrt{3}+100\sqrt{2}\right)^2}}$$

or we can make two triangles,

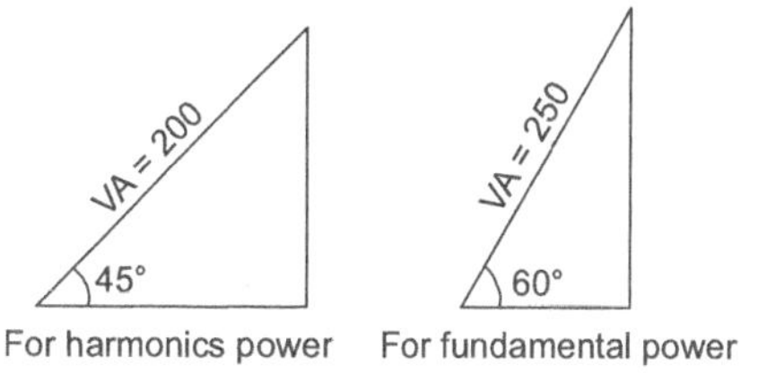

For harmonics power   For fundamental power

Hence resultant angle will be in between 45° and 60°. So PF will be 0.7 > PF > 0.5.

From option it is 0.6.

Shortcut Method:

Observe $V(t)$ and $i(t)$

p.f. angle will be between 60° and 45°.

⇒ p.f. will be between 0.5 and 0.7 viz. option (b)

2.42 (b)

$I = \frac{V}{X_C}$; V_3 for 3rd harmonics = $V/3$ as given

$$X_{C_3} = \frac{X_{C_1}}{3}$$

Hence,
$$I_{C_3} = \frac{V/3}{\frac{X_4}{3}} = I_{C_1}$$

2.43 (b)

$$I = \frac{V}{|Z|} = \frac{100}{5}$$

$$\text{p.f. of circuit} = \frac{3}{\sqrt{(3)^2+(4)^2}} = \frac{3}{5} = \cos\phi$$

$$\text{and, } \sin\phi = \frac{4}{5}$$

$$\text{So, real power} = 100\times\frac{100}{5}\times\frac{3}{5} = 1200\text{ W}$$

$$\text{Reactive Power} = 100\times\frac{100}{5}\times\frac{4}{5} = 1600\text{ VAR}$$

2.44 (a)

Assuming internal resistance of cell is R

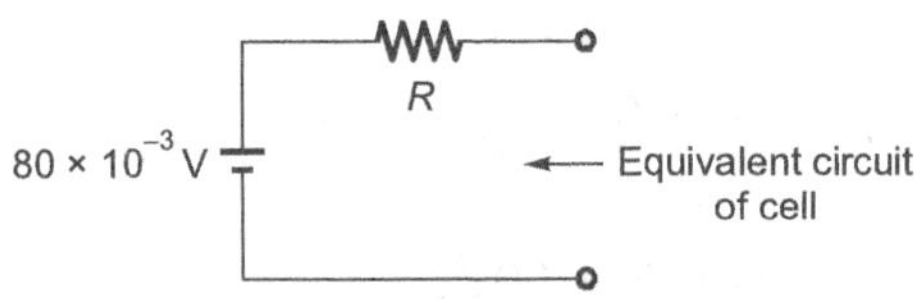

For $R_L = 6\times10^6\ \Omega$

$$I = \frac{80\times10^{-3}}{R+6\times10^6} = 5\times10^{-9}$$

$$R = 10\times10^6\ \Omega$$

For maximum power,

$$R_L = R = 10\times10^6\ \Omega$$

$$P_{max} = \left(\frac{V}{2R_L}\right)^2\times R_L = \frac{V^2}{4R_L}$$

$$= \frac{(80\times10^{-3})^2}{4\times10\times10^6}$$

$$= 1.6\times10^{-10}\text{ W} = 0.16\text{ nW}$$

2.45 (a)

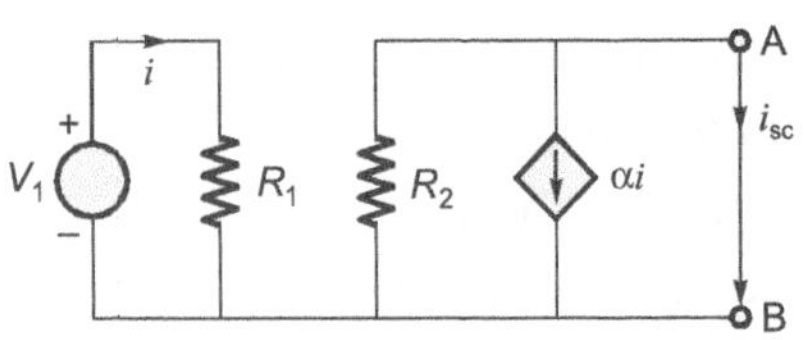

Norton's equivalent current = $i_{sc} = -\alpha i$

and
$$i = \frac{V_1}{R_1}$$

$$\therefore\quad i_{sc} = \frac{-\alpha V_1}{R_1}$$

$$\text{Conductance} = \frac{1}{R_{Th}} \text{ and } R_{Th} = R_2$$

$$\therefore\quad \text{Conductance} = \frac{1}{R_2}$$

2.46 (a)

According to reciprocity theorem, in a linear bilateral single source network, the ratio of excitation to response is constant when the position of excitation and response are interchanged

$$\frac{V_1}{I_1} = \frac{V_2}{I_2} \Rightarrow I_2 = \frac{V_2}{V_1}I_1$$

$$= 4\angle{-45^\circ}\times\frac{15\angle 90^\circ}{20\angle 30^\circ} = 3\angle 15^\circ\text{ A}$$

2.47 (a)

From the given data,

$$V_{OC} = 3\text{ V and } R_{Th} = 1\ \Omega$$

when, $E = 1$ V and I is replaced by a resistor of 2 ohm, then

$$V = 3\times\frac{2}{2+1} = 2\text{ V}$$

2.48 (c)

$$Y_{eq} = 1 + j1 - j1$$

$$V_{eq} = \frac{\Sigma EY}{\Sigma Y} = \frac{1+1+1}{1} = 3\text{ V}$$

$$\therefore\quad \boxed{Z_{eq} = \frac{1}{Y_{eq}} = 1}$$

2.49 (b)

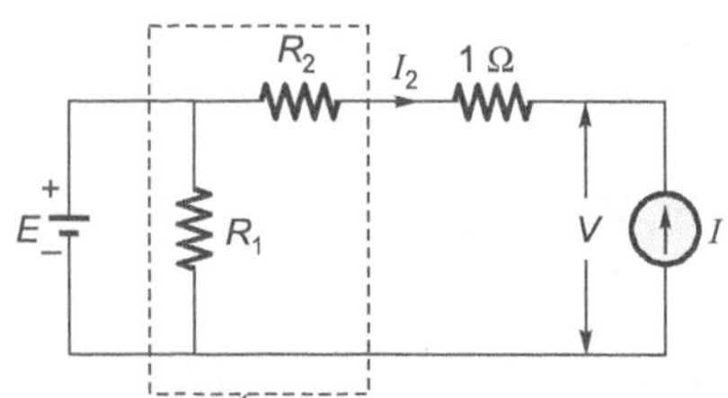

General Resistive circuit

When I is removed $I_2 = 0$

$$\therefore\quad V = E = E_1 = 5\text{ V}$$

$$\therefore\quad E_1 = 5\text{ V}$$

When $E = 0$ and $I = 1$ A

$\therefore \quad V = I(R_2 + 1)$

$5 = 1(R_2 + 1)$

$\therefore \quad R_2 = 4\ \Omega$

When I is replaced by 5 Ω

$E_1 = (R_2 + 1 + 5)\, I_2$

$5 = (4 + 1 + 5)\, I_2$

$\therefore \quad I_2 = \frac{1}{2}$ A and $V = \frac{1}{2} \times 5 = 2.5$ V

2.50 (b)

$$|V|_{rms} = 250\text{ V}$$

$$V_R = \sqrt{V^2 - V_L^2}$$

$$= \sqrt{250^2 - 150^2} = 200\text{ V}$$

$$|I|_{rms} = \frac{V_R}{R} = \frac{200}{100} = 2\text{ A}$$

$$X_L = \frac{V_L}{|I|_{rms}} = \frac{150}{2} = 75\ \Omega$$

$$\omega L = 75$$

$$\Rightarrow \quad L = \frac{75}{300} = 0.25\text{ H}$$

2.51 (d)

$$I_R = \frac{V_2}{R} = \frac{2\angle 0^\circ}{1} = 2\angle 0^\circ\text{A}$$

$$I_C = \frac{V_2}{X_C} = \frac{2\angle 0^\circ}{\left(\frac{1}{j\omega C}\right)}$$

$$= \frac{2\angle 0^\circ}{\left(\frac{1}{j2 \times 0.5}\right)} = 2\angle 90^\circ\text{A}$$

$$I_L = I_R + I_C$$

$$= 2\angle 0^\circ + 2\angle 90^\circ = 2\sqrt{2}\angle 45^\circ\text{A}$$

$$V_L = I_L \times X_L$$

$$= 2\sqrt{2}\angle 45^\circ \times (j\omega L)$$

$$= 4\sqrt{2}\angle 135^\circ\text{ V}$$

2.52 (a)

$$Z(j\omega) = R + j\omega L + \frac{1}{j\omega C}$$

$$Z(j\omega) = R + j\left(\omega L - \frac{1}{\omega C}\right)$$

$$\tan\phi = \frac{\omega L - \frac{1}{\omega C}}{R}$$

$$\tan\phi = \frac{4 \times 1 - \frac{1}{4 \times 0.05}}{R}$$

$$\tan\phi = -\frac{1}{R}$$

$$\phi = -\tan^{-1}\left(\frac{1}{R}\right)$$

$\because$ Current leads voltage by $\tan^{-1} 2$.

$\Rightarrow \quad R = 0.5\ \Omega$

2.53 (d)

$$V_{rms} = \frac{V_m}{\sqrt{2}} = \frac{150\sqrt{2}}{\sqrt{2}} = 150\text{ V}$$

$$V_L = \sqrt{V_{rms}^2 - V_R^2}$$

$$= \sqrt{150^2 - 120^2} = 90\text{ V}$$

$$|I| = \frac{V_R}{R} = \frac{120}{1 \times 10^3} = 0.12\text{A}$$

$$|V_L| = |I|\omega L$$

$$90 = 0.12 \times 500 \times L$$

$$L = 1.5\text{ H}$$

2.54 (a)

Reciprocity theorem is applicable for linear bilateral network.

2.55 (b)

$$R_n = 10\ \Omega$$

$$V_{Th} = 60\text{ V}$$

$$P_{max} = \frac{V_{Th}^2}{4R_{Th}} = \frac{60 \times 60}{4 \times 10} = 90\text{ W}$$

2.56 (a)

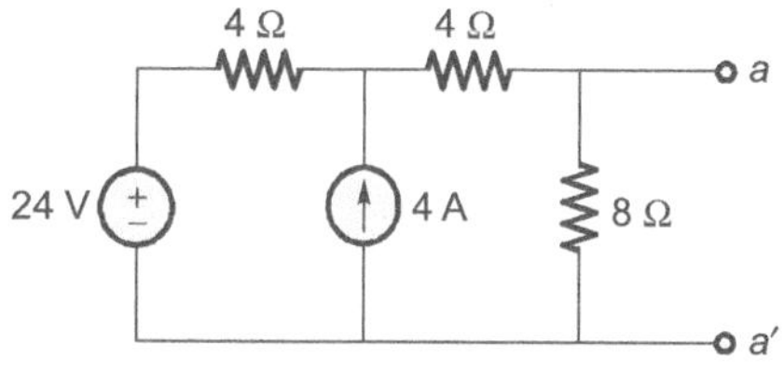

Finding out R_n,

By S.C. the voltage source and O.C. the current source.

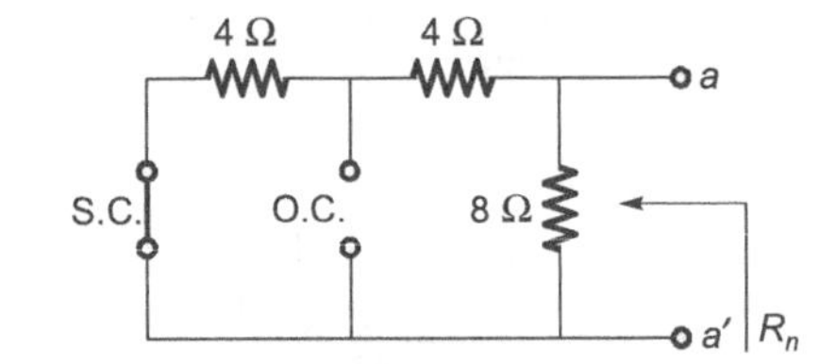

$R_n = 4\ \Omega$

Finding out I_N,

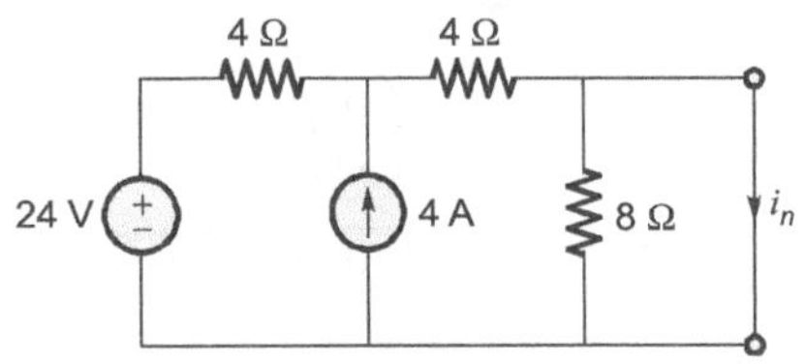

⇓

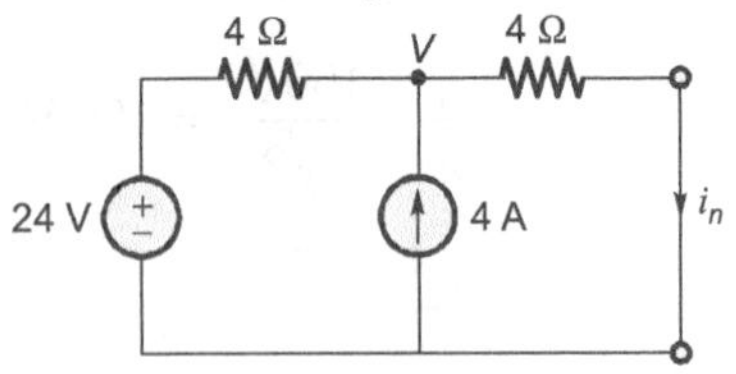

By applying KCL at node 'V'

$$\frac{V-24}{4}+\frac{V}{4} = 4$$

$$\frac{V}{2} = 10\text{ V}$$

$$V = 20\text{ V}$$

$$I_n = \frac{20}{4} = 5\text{ A}$$

2.57 (b)

$$Z_L = Z_{in}^*$$

$$Z_L = 10 - j50 = R_L + jX_L$$

Current through circuit,

$$I_{max} = \frac{50}{10+j50+10-j50}$$

$$= \frac{50}{20}\text{ A}$$

$$P_{max} = I_{max}^2 R_L = \left(\frac{50}{20}\right)^2 \times 10$$

$$= 62.5\text{ W}$$

2.58 (b)

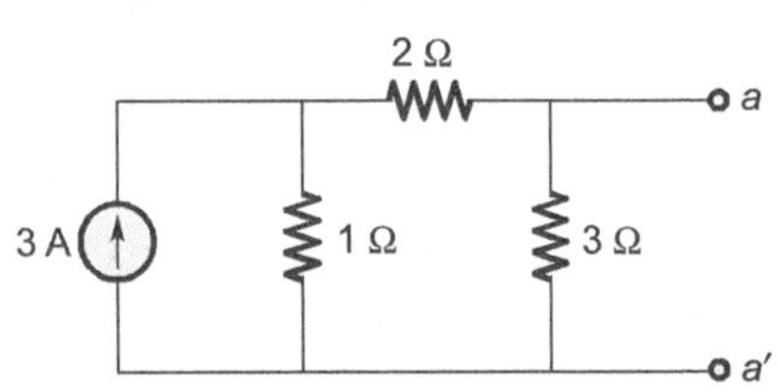

Finding of R_{Th}

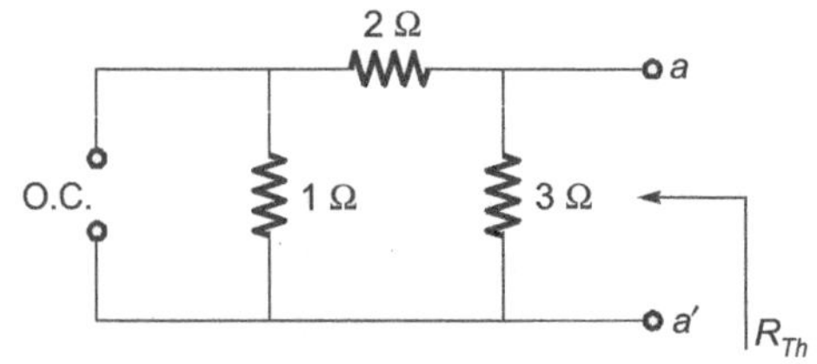

$R_{Th} = 1.5\ \Omega$

Finding of V_{Th}

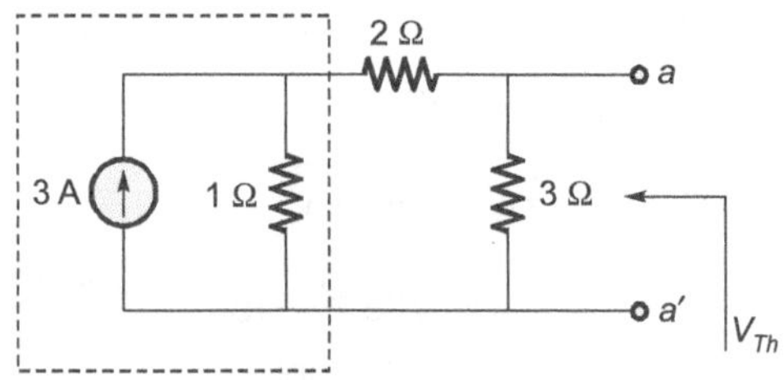

By applying source transformation,

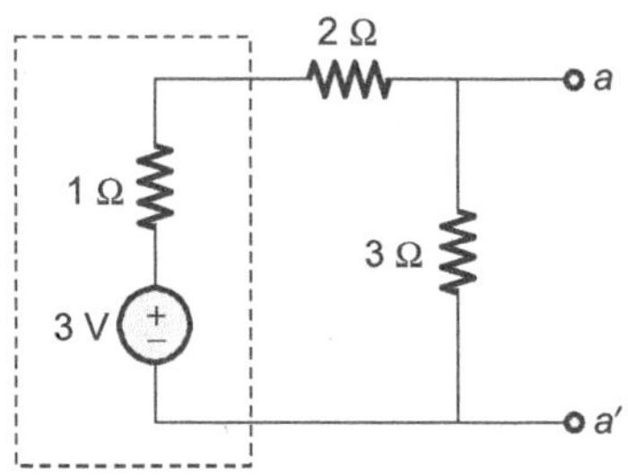

$$V_{Th} = \frac{3\times 3}{6} = 1.5\text{ V}$$

2.59 (b)

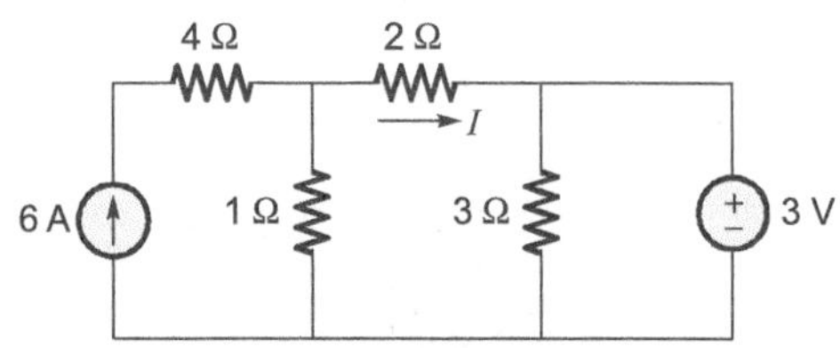

By offing the 6 A current source only,

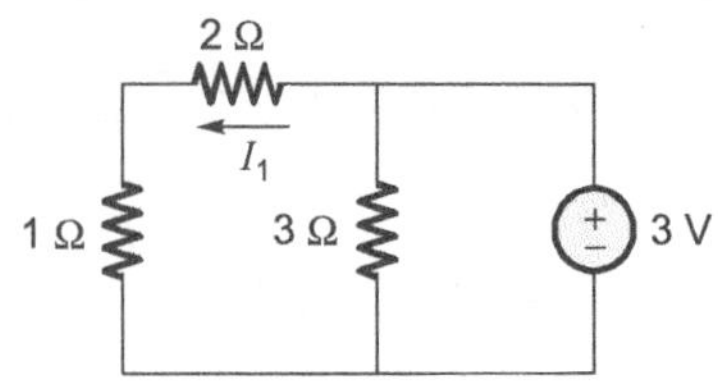

$I_1 = 1\text{ A}$

By offing 3 V voltage source only,

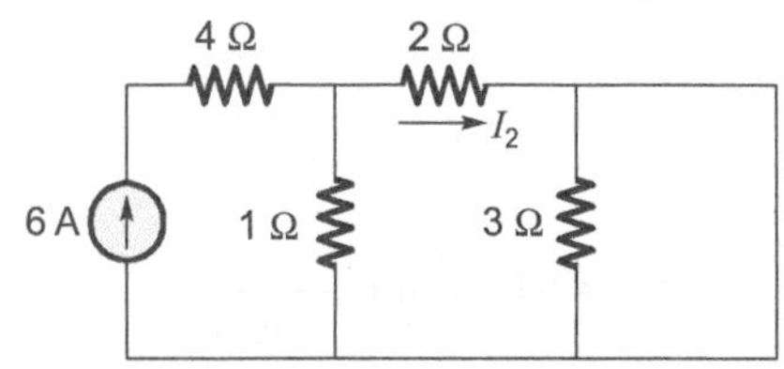

$$I_2 = \frac{6 \times 1}{3} = 2\text{ A}$$

By superposition theorem

$$I = I_2 - I_1$$
$$I = 2 - 1 = 1\text{ A}$$

2.60 (b)

$$V(t) = 160\sin(\omega t + 10°)$$
$$i(t) = 5\sin(\omega t - 200°)$$
$$Z = 32\angle 210°$$
$$Z = -27.7 - 16j$$
$$P_{reactive} = \frac{i_{max}^2 X}{2} = \frac{1}{2} \times 25 \times 16 = 200\text{ VAR}$$

2.61 (b)

Tellegen's theorem is applicable to any network made up of lumped two terminal elements. The circuit and/or circuit elements may be linear/non-linear, passive/active time invariant/time varying and the circuit may contain dependent/independent sources. Hence, option (b) is correct.

2.62 (d)

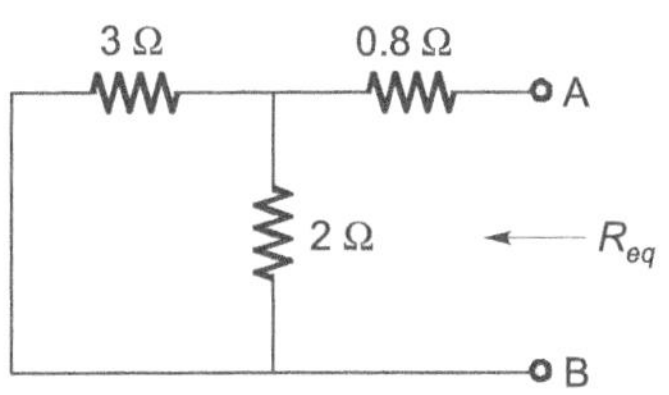

$$R_{eq.} = 3||2 + 0.8 = 2\ \Omega$$

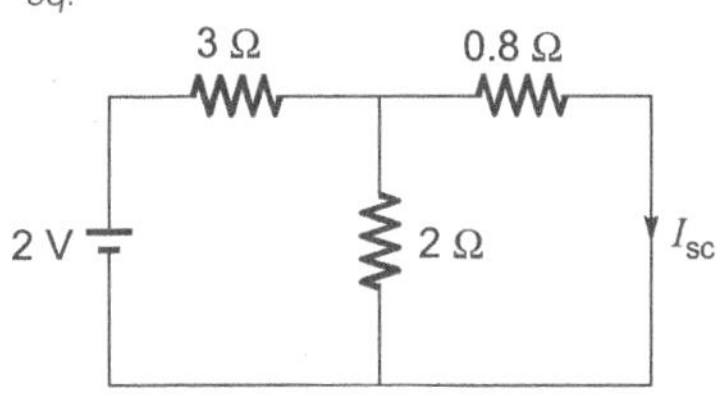

$$I_{sL} = \left(\frac{2}{3 + 2||0.8}\right) \times \left(\frac{2}{2 + 0.8}\right)$$
$$= 0.4\text{ A}$$

Hence, option (d) is correct.

2.63 (d)

For finding R_{Th},

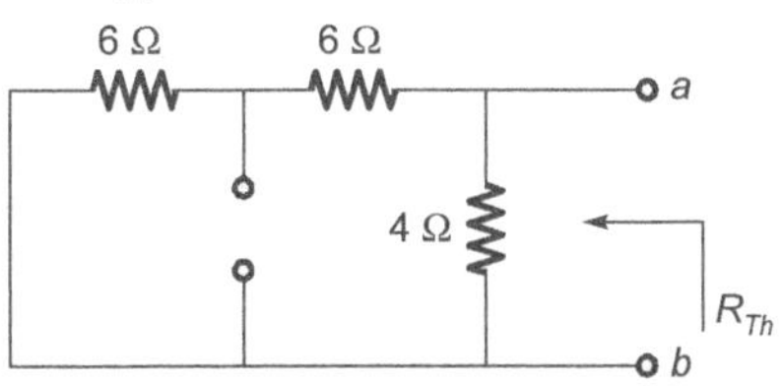

$$R_{Th} = \frac{4 \times (6 \times 6)}{(6+6)+4} = 3\ \Omega$$

For finding V_{Th},

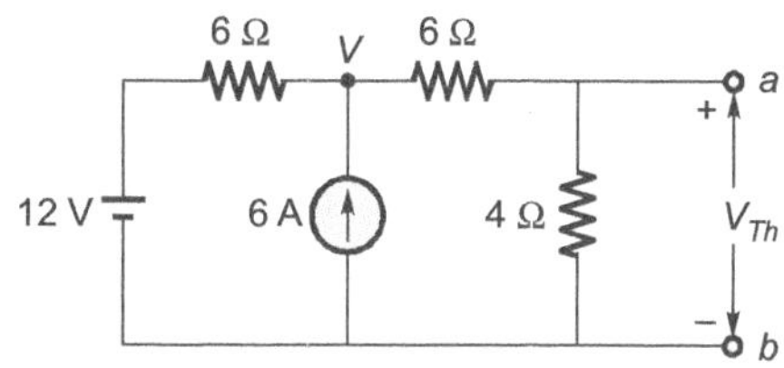

Applying KVL,

$$\frac{V-12}{6} + \frac{V}{10} - 6 = 0$$
$$V = 30\text{ V}$$
$$V_{Th} = \frac{V}{6+4} \times 4 = 12\text{ V}$$

Hence, option (d) is correct.

2.64 (b)

According to maximum power transfer theorem, the power transferred from voltage source to a load will be maximum when source impedance is equal to (complex conjugate of that of) the load impedance.

2.65 (c)

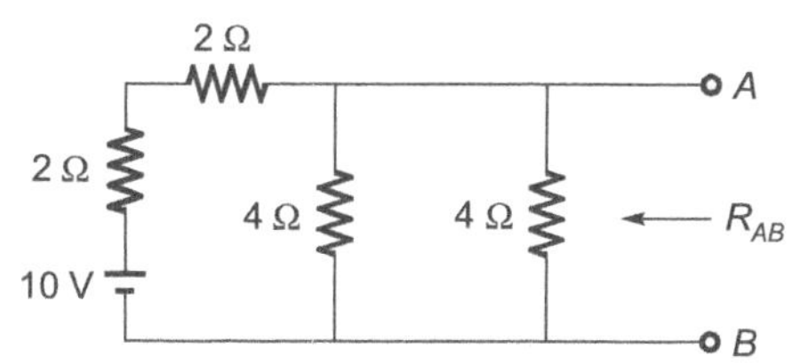

$$R_{AB} = (4)\ ||\ (4)\ ||\ (4)$$
$$= \frac{2 \times 4}{6} = \frac{4}{3}\ \Omega$$

2.66 (c)

$$V_{OC} = 5\text{ V}$$
$$I_{SC} = 2.5\text{ A}$$
$$\Rightarrow \quad R_{Th} = \frac{V_{OC}}{I_{SC}} = \frac{5}{2.5} = 2\ \Omega$$

∴ Overall network can be drawn as:

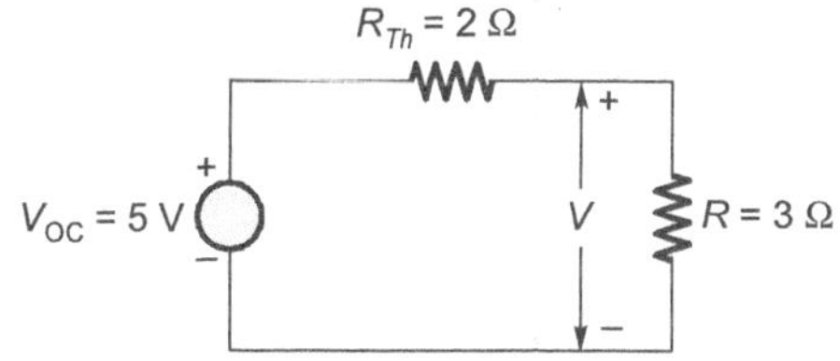

∴ Using voltage division rule,

$$V = \left(\frac{R}{R + R_{Th}}\right) V_{OC} = \left(\frac{3}{3+2}\right) \times 5 = 3\text{ V}$$

2.67 (b)

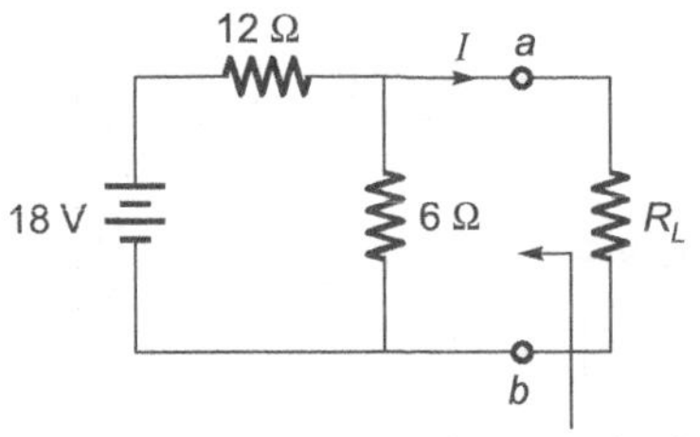

Resistance seen across the terminal *ab* is

$$= \frac{12\times 6}{12+6} = 4\ \Omega$$

For maximum power transfer,

$$R_L = 4\ \Omega$$

Now, current passing through R_L,

$$I = \frac{18}{(6\|4+12)} \times \frac{6}{6+4}$$

$$= \frac{18}{2.4+12} \times \frac{6}{10} = 0.75\text{ A}$$

Now maximum power absorbed by R_L is

$$= I^2R_L = (0.75)^2 \times 4 = 2.25\text{ W}$$

2.68 (d)

$$V_{OC} = 25\text{ V}$$

$$I_{SC} = -5\text{ A (in opposite direction)}$$

$$\therefore \quad R_{Th} = \frac{25}{5} = 5\ \Omega$$

(Resistance can't be negative)

2.69 (d)

$$Z_L = Z_S^*$$

2.70 (d)

For any value of *C*, I_1 will always lead, the V_s by 90° and I_2 will always lag, the V_s by 45°

⇒ I_1 leads I_2 by 90° + 45° = 135°.

2.71 (a)

$$R = \frac{V^2}{P} = \frac{(230)^2}{1000} = 52.9\ \Omega$$

2.72 (a)

Maximum power absorbed by load

$$= \frac{V_{Th}^2}{4R_{Th}}$$

Maximum power delivered by source

$$= \frac{V_{Th}^2}{4R_{Th}} \times 2$$

$$x = \frac{V_{Th}^2/4R_{Th}}{V_{Th}^2/4R_{Th} \times 2} = 50$$

$$x = 50\%$$

2.73 (b)

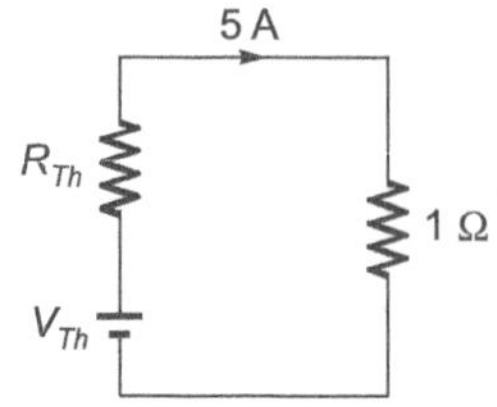

$$I = \frac{V_{Th}}{R_{Th}+1} \Rightarrow 5 = \frac{V_{Th}}{R_{Th}+1}$$

$$\Rightarrow \quad 5R_{Th} + 5 = V_{Th}$$

$$\Rightarrow \quad \boxed{V_{Th} - 5R_{Th} = 5}$$

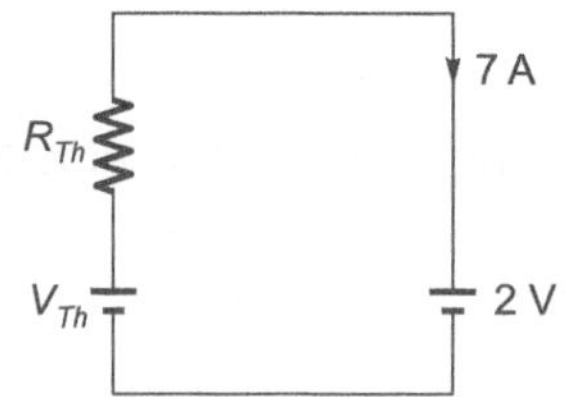

$$7 = \left(\frac{V_{Th}-2}{R_{Th}}\right)$$

$$7R_{Th} = V_{Th} - 2$$

$$\boxed{V_{Th} - 7R_{Th} = 2}$$

$$V_{Th} - 5R_{Th} = 5$$
$$V_{Th} - 7R_{Th} = 2$$
$$- \quad + \quad -$$
$$2R_{Th} = 3$$

$$\boxed{R_{Th} = 1.5\ \Omega}$$

$$V_{Th} - (1.5) = 5$$

$$V_{Th} - 7.5 = 5$$

$$\boxed{V_{Th} = 12.5\text{ V}}$$

2.74 (b)

$$V = i$$

$$V + i = 100$$

$$\Rightarrow \quad 2i = 100$$

$$\Rightarrow \quad i = 50\text{ A}$$

2.75 (d)

$$200 \times 2.5 \times \cos\phi = 300$$

$$\cos\phi = \frac{300}{200 \times 2.5} = 0.6$$

$$\sin\phi = 0.8$$

$$\text{Reactive power} = 200 \times 2.5 \times \sin\phi$$

$$= 200 \times 2.5 \times 0.8$$

$$= 400\text{ VAR}$$

2.76 (d)

For maximum power,

$$|X_C| = R$$
$$\phi = 45^\circ$$
$$\cos\phi = \cos 45^\circ = 0.707 \text{ lead}$$

2.77 (b)

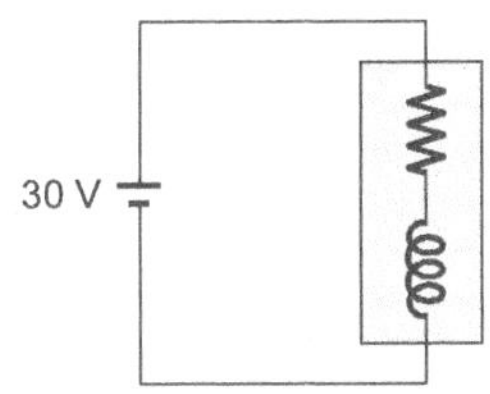

$$I_2 R = 300$$

$$\Rightarrow \quad \frac{V^2}{R} = 300 \Rightarrow \frac{30^2}{R} = 300$$

$$\Rightarrow \quad \frac{900}{R} = 300 \Rightarrow R = 3\ \Omega$$

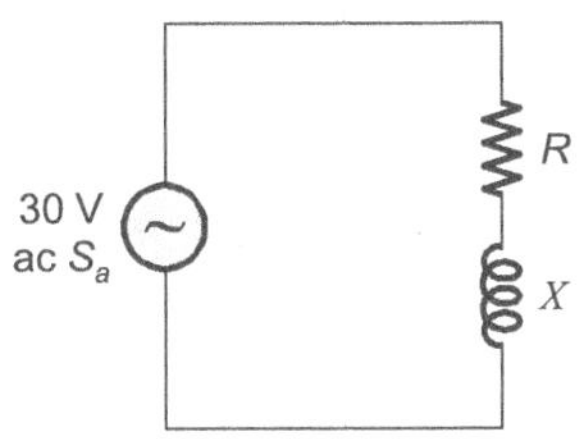

$$I = \frac{V}{\sqrt{R^2 + X^2}} = \frac{30}{\sqrt{R_2 + X^2}}$$

$$\left(\frac{30}{\sqrt{3^2 + X^2}}\right)^2 \times 3 = 108$$

$$5 = \sqrt{3^2 + X^2} \quad \Rightarrow \quad X = 4\ \Omega$$

2.78 (c)

Transforming the current to voltage source the new circuit can be represented as below.

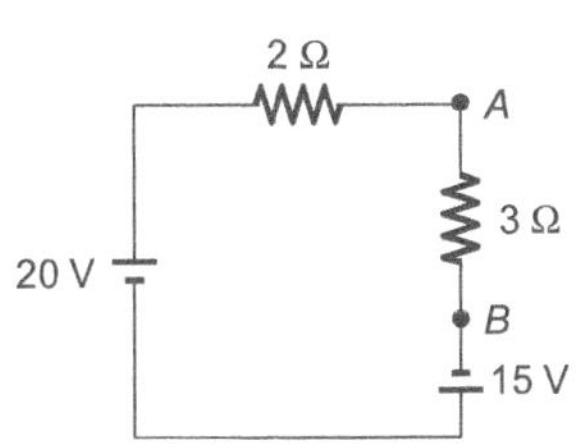

$\Rightarrow$ Hence Thevenin's equivalent across AB.

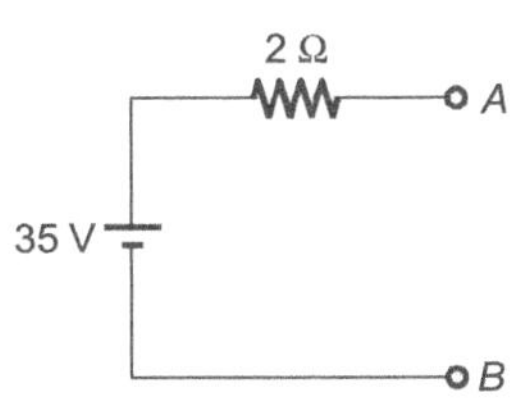

$$V_{Th} = 35 \text{ V}$$
$$R_{TH} = 2\ \Omega$$

2.79 (b)

$$Z = 2.38 - j0.667$$

$\therefore \quad Z = R - jX_C$

or $\quad Z = R - j(X_C - X_L)$ but $|X_C| > |X_L|$

$\therefore$ Thevenin impedance can be realised by using either R and C or by using R, L and C but minimal realisation will be with R and C.

2.81 (d)

$$e = L\frac{dI}{dt} = j\omega LI \qquad \text{(for ac source)}$$

$$e \propto \omega$$

2.82 (c)

$$Z_{coil} = R + j\omega L$$
$$V = 240 \text{ V},$$
$$I = 10 \text{ A}$$

$$Z_{coil} = \frac{240}{10} = 24\ \Omega,$$

$$X_L = \sqrt{24^2 - 5^2} = 23.47\Omega$$

$$\cos\theta = \frac{R}{Z} = \frac{5}{24} = 0.208 \text{ lag}$$

$$L = \frac{23.47}{100\pi} = 74.7 \text{ mH}$$

2.83 (c)

$$V = 100 \text{ volts},$$
$$z = (3 + j4)\ \Omega$$

$$\because \quad |S| = \frac{V^2}{Z} = \frac{100^2}{5} = 2000 \text{ VA}$$

$$\text{Real power } P = S\cos\left[\tan^{-1}\left(\frac{4}{3}\right)\right]$$

$$= 2000 \times \frac{3}{5} = 1200 \text{ W}$$

Reactive power,

$$Q = S\sin\left[\tan^{-1}\left(\frac{4}{3}\right)\right]$$

$$= 2000 \times \frac{4}{5} = 1600 \text{ VAR}$$

2.84 (a)

$$V = 100\angle 15^\circ \text{ volts},$$

$$I = 20\angle{-45^\circ} \text{A}$$

Total power, $S = VI^*$

$= 100\angle 15 \times 20\angle 45° = 2000\angle 60°$

Real power, $P = S\cos 60° = 1000$ W

Reactive power,

$Q = S\sin 60° = 1732$ VAR

2.85 (a)

$V = 200$ volts r.m.s

$P = 1200$ W,

$Q = 1600$ VAR

$$S = \sqrt{1200^2 + 1600^2} = 2000 \text{ VA}$$

$$\cos\theta = \frac{P}{S} = \frac{1200}{2000} = 0.6$$

$$I = \frac{S}{V} = \frac{2000}{200} = 10 \text{ A}$$

2.86 (*)

$V_A = 100\angle 30°$ V,

$V_B = 100\angle 0°$ V

$X_L = j5\ \Omega$

Real power transferred from node A to B

$$= \frac{V_A \cdot V_B}{|X_L|}\sin\delta$$

$$= \frac{100 \times 100}{5}\sin 30° = 1000 \text{ W}$$

2.88 (b)

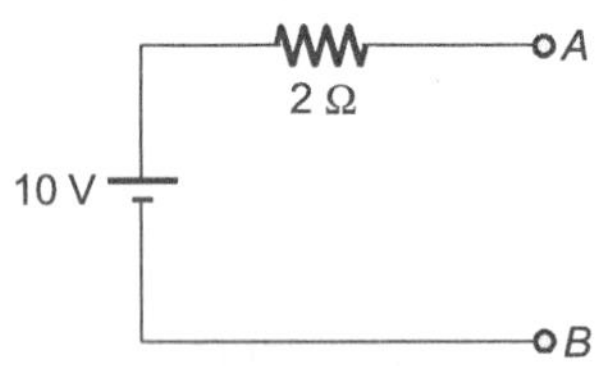

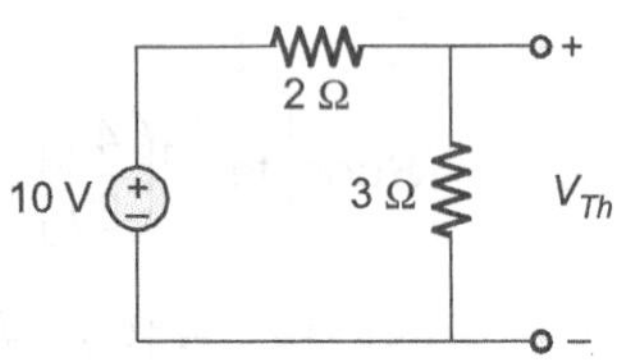

$$V_{TH} = \frac{10}{3+2} \times 3 = 6 \text{ V}$$

R_{TH}: Deactivate independent source

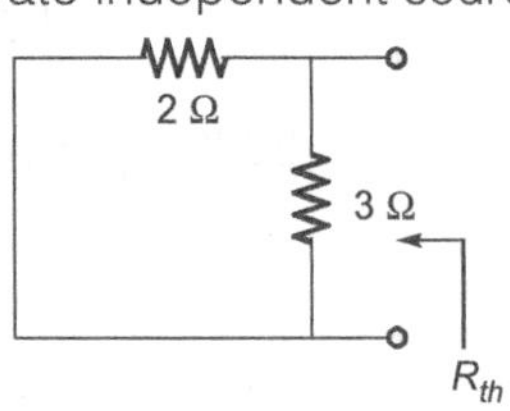

$$R_{TH} = \frac{3 \times 2}{3+2} = \frac{6}{5} = 1.2\ \Omega$$

2.89 (c)

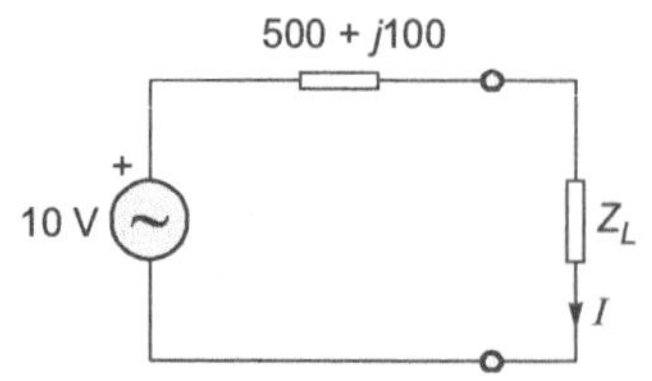

For maximum power transfer;

$Z_L = Z^*_{TH}$

$Z_L = 500 - j100$

$$I_{rms} = \frac{V_{rms}}{|z|}$$

$Z = Z_L + Z_{TH}$

$\Rightarrow \quad Z = 500 + j100 + 500 - j100$

$= 1000$

$$\therefore \quad I_{rms} = \frac{V_{rms}}{Z} = \frac{10}{1000} = \frac{1}{100} \text{ A}$$

$$P_{max} = I_{rms}^2 \cdot R_L$$

$$= \left(\frac{1}{100}\right)^2 \times 500 = 0.05 \text{ W}$$

2.90 (a)

Reciprocity theorem is applicable in case of linear and bilateral networks containing only one independent source.

2.91 (d)

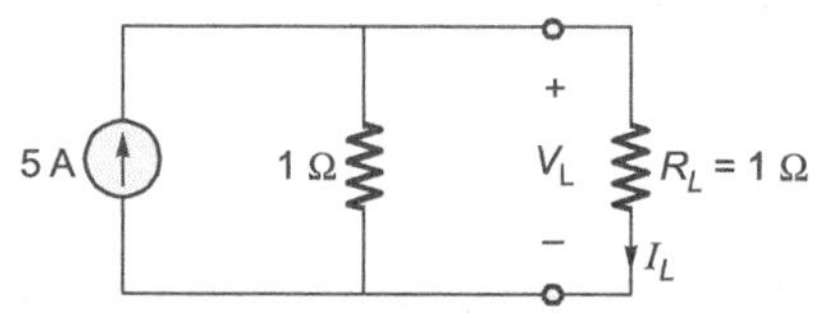

For maximum power transfer $R_L = 1\ \Omega$

$$I_L = \frac{5}{1+1} \times 1 = 2.5 \text{A}$$

$\therefore \quad V_L = 2.5 \times 1 = 2.5$ V

$$P_{max} = I_L^2 \cdot R = (2.5)^2 \times 1$$

$P_{max} = 6.25$ W

2.92 (d)

In an ideal transformer,

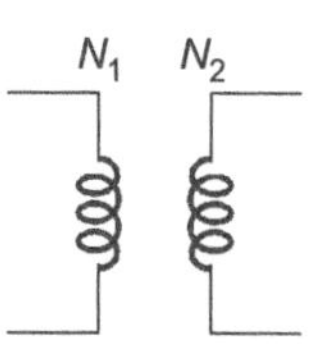

$$Z_i = Z_L\left(\frac{N_1}{N_2}\right)^2 \quad ...(1)$$

As, $\dfrac{V_1}{V_2} = \dfrac{N_1}{N_2}$

$$\Rightarrow \quad \frac{220}{110} = \frac{N_1}{N_2} = 2$$

From equation (1),

$$Z_i = Z_L(2)^2$$

$$\Rightarrow \quad Z_i = 4Z_L$$

And for maximum power transfer,

$$Z_i = 2\ \Omega$$

$$\therefore \quad 4Z_L = 2$$

$$\Rightarrow \quad Z_L = \frac{1}{2} = 0.5\ \Omega$$

2.93 (a)

$$i(t) = \sqrt{2}\sin(\omega t - 45°)$$

$$\omega L = 1\ \Omega$$

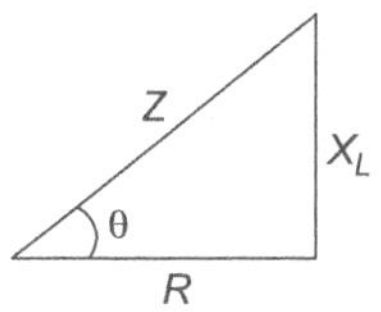

$$\tan\theta = \frac{X_L}{R}$$

$$\Rightarrow \quad R = \frac{X_L}{\tan\theta}$$

$$R = \frac{1}{\tan 45°} \Rightarrow R = 1\ \Omega$$

2.94 (*)

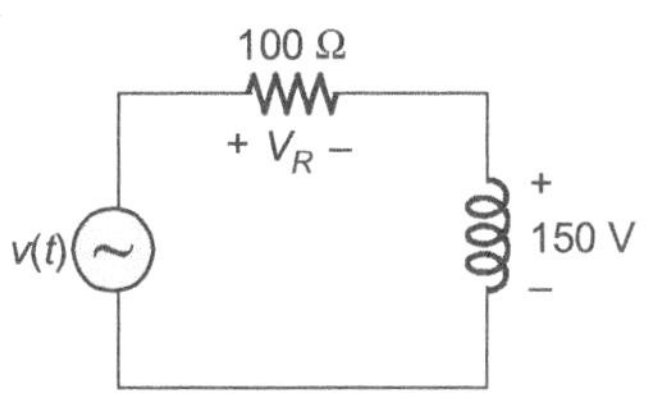

$$V^2 = V_R^2 + V_L^2$$

$$V_{rms} = \frac{250\sqrt{2}}{\sqrt{2}} = 250$$

$$(250)^2 = V_R^2 + (150)^2$$

$$\Rightarrow \quad V_R = 200\ \text{V}$$

$$I = \frac{V_R}{100} = \frac{200}{100} \Rightarrow I = 2\ \text{A}$$

$$V_L = IX_L \Rightarrow X_L = \frac{150}{2} = 75$$

$$\omega L = 75 \Rightarrow L = \frac{75}{600} = 0.125\ \text{H}$$

$$L = 0.125\ \text{H}$$

2.95 (d)

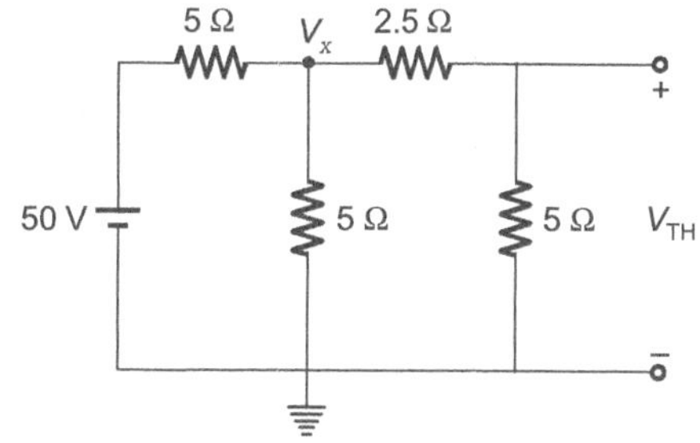

Case-1: For V_{TH}

$$\frac{V_x - 50}{5} + \frac{V_x}{5} + \frac{V_x}{7.5} = 0$$

$$3V_x - 150 + 3V_x + 2V_x = 0$$

$$8V_x = 150$$

$$V_x = \frac{150}{8}\ \text{V}$$

$$\therefore \quad V_{TH} = \frac{V_x}{2.5 + 5} \times 5$$

$$= \frac{150/8}{7.5} \times 5$$

$$= 12.5\ \text{V}$$

Case-2: For R_{TH}

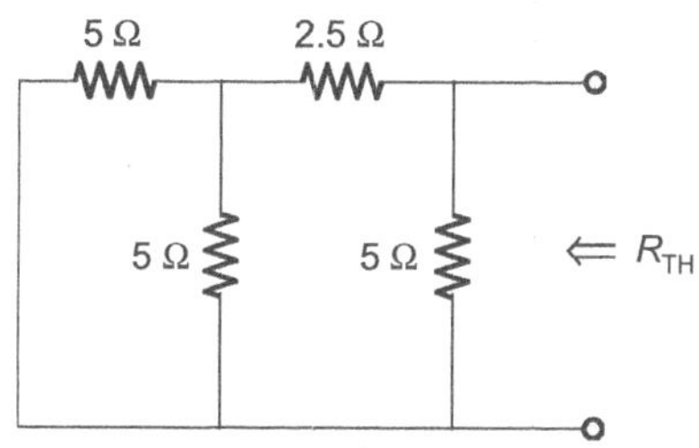

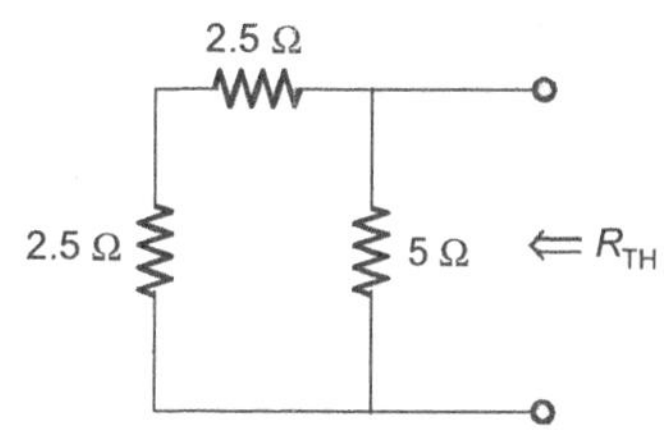

$$R_{TH} = [(5 \,||\, 5) + 2.5] \,||\, 5$$

$$= 5 \,||\, 5 = 2.5\ \Omega$$

2.96 (a)

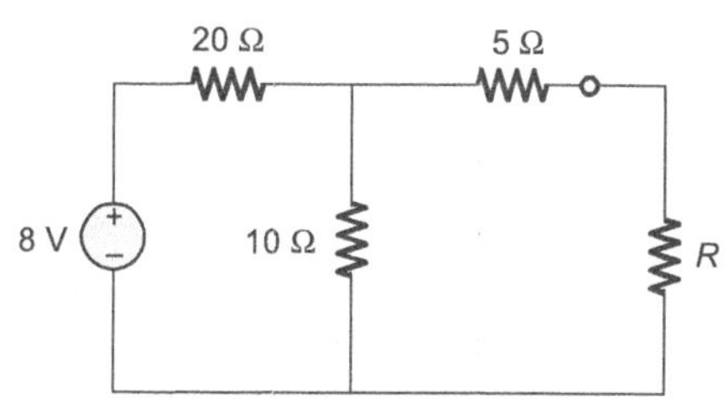

For maximum power transfer to load,

$$R = R_{TH}$$

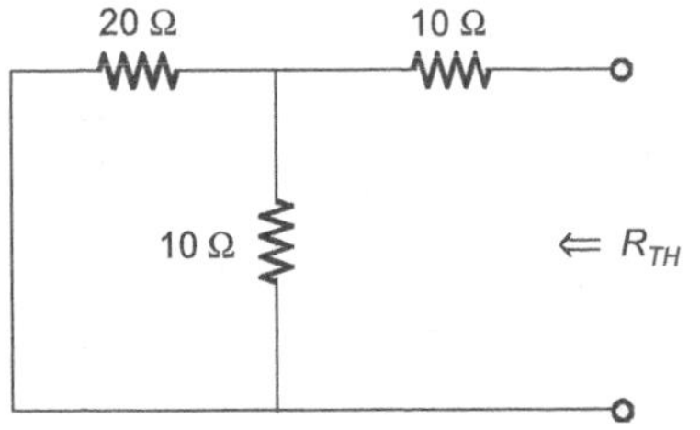

$$R_{TH} = (20 \parallel 10) + 5$$

$$= \frac{200}{30} + 5 = 11.66\ \Omega$$

$$\therefore \quad R = 11.66\ \Omega$$

2.97 (d)

First two statements are wrong as network theorems are derivable from Kirchhoff's laws and Norton current is found after short-circuiting the terminals across which the Norton equivalent is asked where as third statement is correct only.

2.98 (d)

Case-1: For V_{TH}

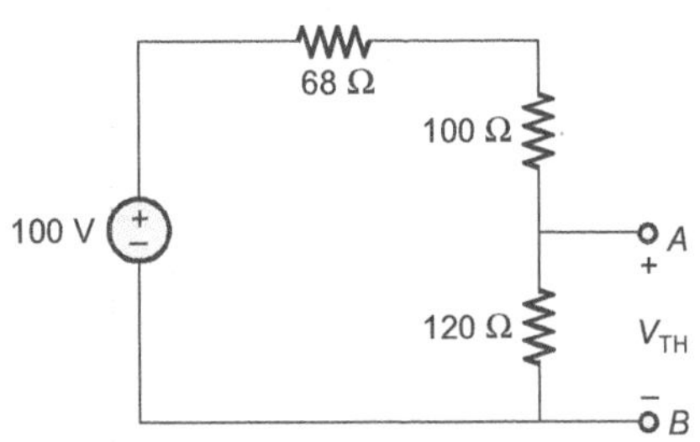

$$V_{TH} = \frac{100}{168 + 120} \times 120$$

$$= 41.67\text{ V}$$

Case-2: For R_{TH}

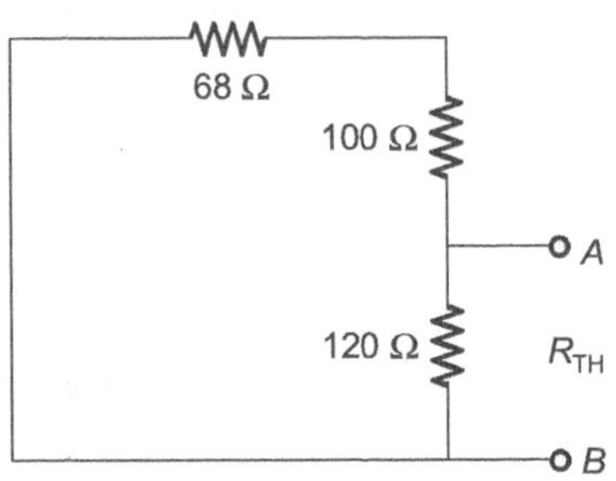

$$R_{TH} = \frac{168 \times 120}{168 + 120} = 70\ \Omega$$

2.99 (c)

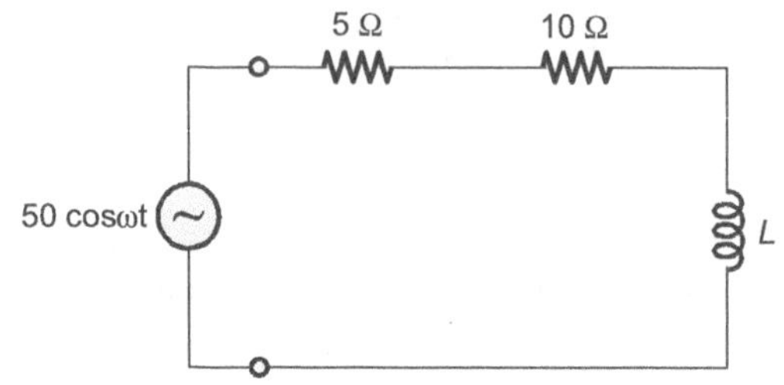

Power factor,

$$\cos\phi = \frac{V_R}{V_{rms}}$$

and power dissipated,

$$P = I_{rms}^2 \cdot R$$

$$\Rightarrow \quad 10 = I_{rms}^2 \cdot 5$$

$$\Rightarrow \quad I_{rms} = \sqrt{2}\text{ A}$$

$$\therefore \quad V_R = I_{rms}[5 + 10] = 15\sqrt{2}\text{ V}$$

$$V_{rms} = \frac{50}{\sqrt{2}}\text{ V}$$

$$\cos\phi = \frac{15\sqrt{2}}{50/\sqrt{2}} = \frac{30}{50} = 0.6$$

2.100 (b)

At source, $e = 200\sqrt{2}\sin(314t + 20)$

$$= 200\sqrt{2}\angle 20^\circ\text{ V}$$

Current, $i = 10\sqrt{2}\cos(314t - 25^\circ)$ A

$$= 10\sqrt{2}\sin(314t - 25^\circ + 90^\circ)$$

$$= 10\sqrt{2}\sin(314t + 65^\circ)$$

$$= 10\sqrt{2}\angle 65^\circ\text{ A}$$

Impedance,

$$Z = \frac{e}{i} = \frac{200\sqrt{2}\angle 20^\circ}{10\sqrt{2}\angle 65^\circ} = 20\angle -45^\circ\ \Omega$$

$$= 20[\cos 45^\circ - j\sin 45^\circ]$$

$$= 20\left[\frac{1}{\sqrt{2}} - j\frac{1}{\sqrt{2}}\right]$$

$$Z = 14.14 - j14.14$$

$$\therefore \quad R = 14.14\ \Omega \text{ and } X_C = 14.14\ \Omega$$

At, $X_C = 14.14\ \Omega = \frac{1}{\omega C} = 14.14$

$$\Rightarrow \quad C = \frac{1}{314 \times 14.14} = 225\ \mu\text{F}$$

2.102 (a)

$$Y_{TH} = 1+\frac{1}{j1}+\frac{R}{R^2+X_C^2}+j\frac{X_C}{R^2+X_C^2}$$

$$Y_{TH} = 1+\frac{1}{j1}+\frac{\frac{1}{2}}{\frac{1}{4}+\frac{9}{4}}+j\frac{\frac{3}{2}}{\frac{1}{4}+\frac{9}{4}}$$

$$= 1-j1+\frac{\frac{1}{2}}{\frac{10}{4}}+j\frac{\frac{3}{2}}{\frac{10}{4}}$$

$$= 1-j1+\frac{1}{5}+j\frac{3}{5} = \frac{6}{5}-j\frac{2}{5}$$

$$Y_{TH} = \frac{6-j2}{5}$$

$$Z_{TH} = \frac{5}{6-j2}\times\frac{6+j2}{6+j2}$$

$$= \frac{30+j10}{36+4} = \frac{30+j10}{40}$$

$$Z_{TH} = \frac{3+j}{4}$$

For maximum power transfer,

$$Z_L = Z_{TH}^* = \frac{3-j}{4}\,\Omega$$

2.104 (b)

Thevenin equivalent,

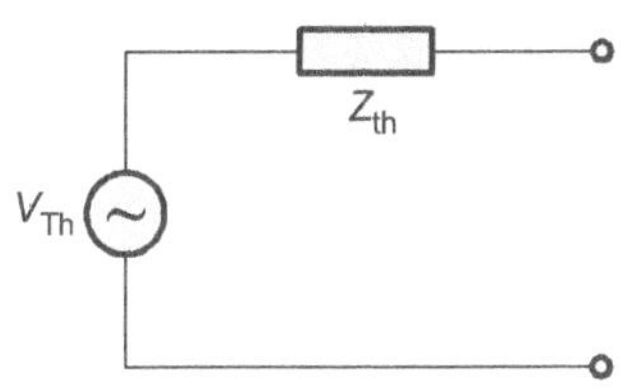

V_{Th} = Voltage source
Z_{Th} = Impedance (Ω)

2.105 (a)

Superposition theorem

2.106 (b)

Equal to zero.

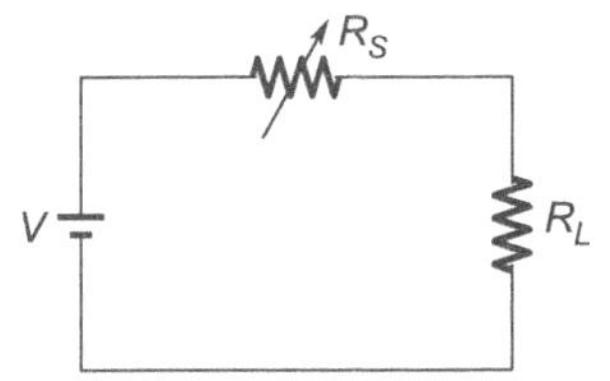

$$I = \frac{V}{R_S+R_L}$$

When $R_S = 0$. The current is maximum and power delivered to load is maximum.

2.107 (b)

From maximum power Transfer Theorem, when load resistance is equal to Thevenin equivalent resistance then the maximum power is delivered to the load.

2.108 (a)

$$I = 10\angle{-30^\circ};\quad V = 100\text{ V}$$

By Ohm's law,

$$Z = \frac{V}{I} = \frac{100}{10\angle{-30^\circ}} = 10\angle 30^\circ$$

$$Z = 10\cos 30^\circ + j10\sin 30^\circ$$

$$Z = (8.66+j5)\,\Omega$$

2.109 (b)

- The interchange of an ideal current source and an ideal voltmeter in any passive linear bilateral circuit will NOT change the voltmeter reading. Hence, the statement 3 is incorrect according to the reciprocity theorem.
- The statement 1 is the correct definition of reciprocity theorem.
- The statement 2 is the application of reciprocity theorem.

2.110 (b)

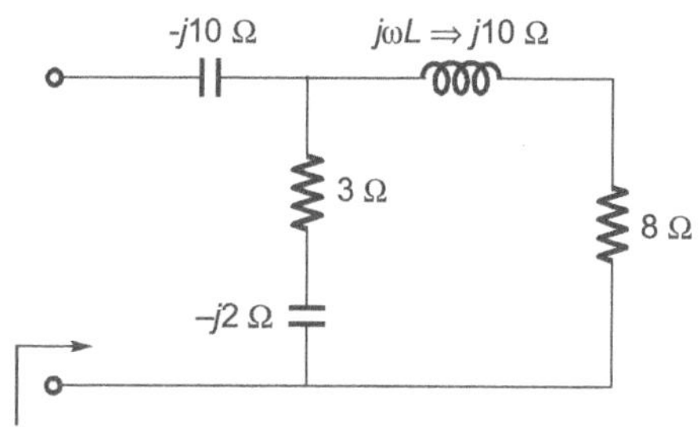

$$Z_{in} = \frac{(8+j10)(3-j2)}{8+j10+3-j2}-j10 = (3.22-j11.07)\,\Omega$$

2.111 (a)

Case-I : V_{th}

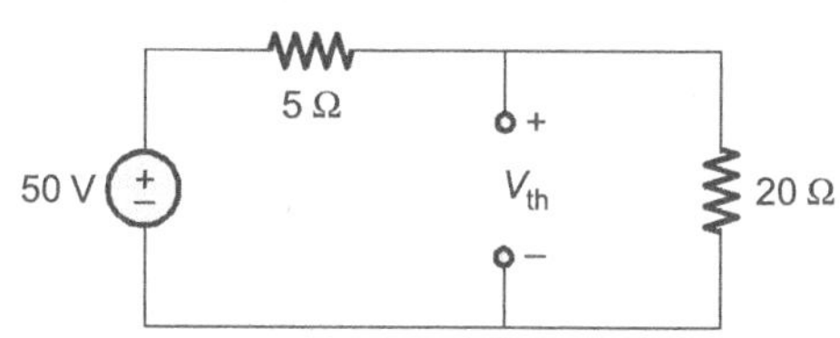

$$V_{th} = 50\times\frac{20}{20+5} = 40\text{ V}$$

Case-II : R_{th}

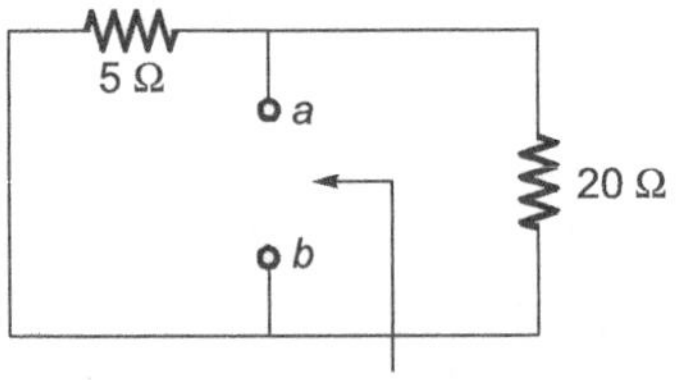

$$R_{th} = \frac{5\times 20}{5+20} = 4\ \Omega$$

2.112 (c)

$$i_{av} = \frac{\text{Area}}{\text{Time}}$$

$$= \frac{(1.5\times 2)+((-2)\times 2)+(1\times 1)}{6} = 0\ \text{A}$$

2.113 (c)

To calculate Thevenin equivalent resistance, deactivating independent sources and applying a test current source I_s = 1 A :

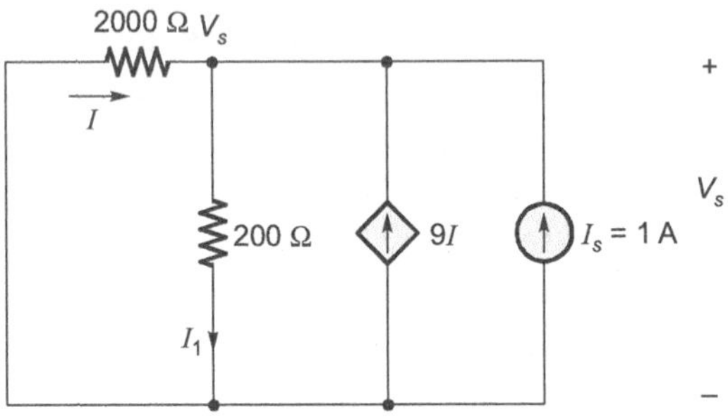

$$V_s = 200(I_1) = 200(I + 9I + 1)$$

$$V_s = 200(10I_1 + 1) \quad ...(1)$$

Also, $V_s = -2000I \quad ...(2)$

$$200(10I + 1) = -2000I$$

$$I = -0.05\ \text{Amp}$$

So, $R_{th} = \dfrac{V_s}{I_s} = \dfrac{-2000(-0.05)}{1} = 100\ \Omega$

2.114 (d)

Given : $R_L = 10\ \Omega$

So, $I_L = \dfrac{V_S}{R_S+R_L} = \dfrac{10}{30+10} = \dfrac{1}{4}$ Amp

$$P = (I_L)^2R_L = \left(\frac{1}{4}\right)^2 \times 10 = 625\ \text{mW}$$

2.115 (c)

$$Z_L = Z_S^*$$

$\because \quad Z_S = R + jXL$

$$= 500 + j100 \times 30 \times 10^{-3}$$

$$= 500 + j3$$

$\therefore \quad Z_L = (500 - j3)\ \Omega$

2.116 (c)

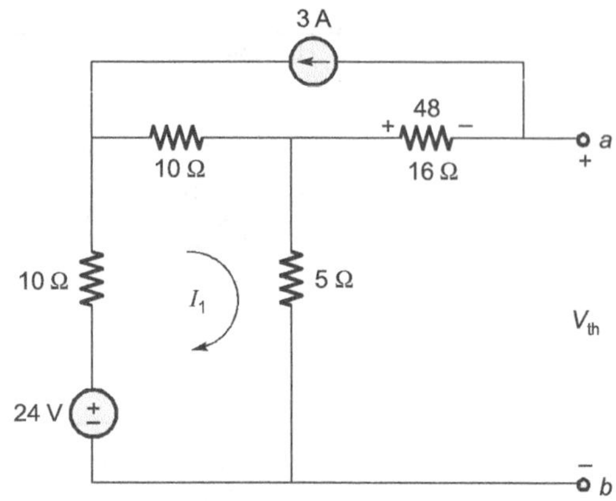

Applying KVL in loop 1

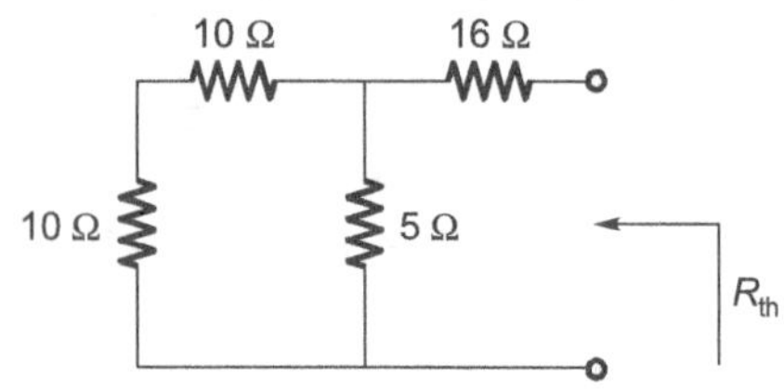

$$24 = 25I_1 + 30$$

$$I_1 = \frac{24-30}{25} = -\frac{6}{25}$$

$$V_{th} = 5(I_1) - 48$$

$$= 5\times\frac{-6}{25} - 48$$

$$= \frac{-6}{5} - 48$$

$$V_{th} = -49.2\ \text{V}$$

$$R_{th} = [(10 + 10) \parallel 5] + 16 = 20\ \Omega$$

2.117 (d)

Condition for maximum power transfer is

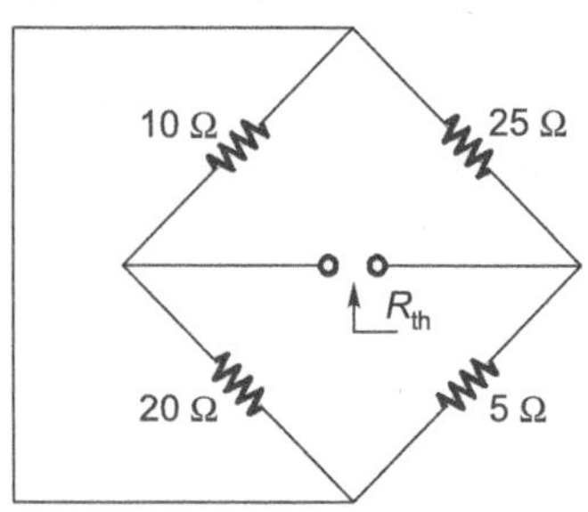

$$R_L = R_{th}$$

For R_{th} calculation :

$$R_{th} = (10 \parallel 20) + (25 \parallel 5)$$

$$R_{th} = \frac{10\times 20}{30} + \frac{25\times 5}{30}$$

$$R_{th} = \frac{325}{30}$$

For V_{th} calculation :

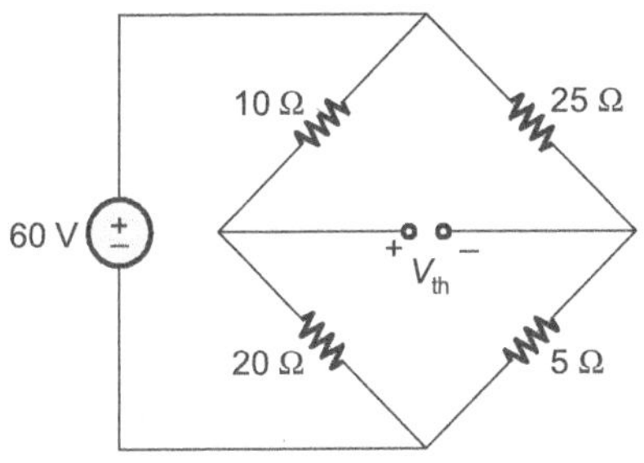

$$V_{th} = 60\times\frac{20}{30} - 60\times\frac{5}{30}$$

$$V_{th} = 30 \text{ V}$$

$$P_{max} = \frac{V_{th}^2}{4R_{th}} = \frac{(30)^2\times 30}{4\times\frac{325}{30}} = 20.7 \text{ W}$$

2.118 (c)

Applying KCL

$$\frac{V_o}{j6} + \frac{V_o}{-j3} = 4\angle 90^\circ$$

$$V_o\left[\frac{1}{j6} + \frac{1}{-j3}\right] = j4$$

$$V_o = 24 \text{ V}$$

■■■■

3 Transient Response of DC & AC Networks

3.1 **Assertion (A):** The current in a constant inductive system does not change instantaneously.

Reason (R): In constant inductive system the flux linkage is conserved initially.

(a) Both A and R are true and R is the correct explanation of A
(b) Both A and R are true but R is NOT the correct explanation of A
(c) A is true but R is false
(d) A is false but R is true

[ESE-2001]

3.2 **Assertion (A):** A unit current impulse applied to a capacitor of 'C' farads instantly inserts 1/2 C Joules of energy in it.

Reason (R): A unit current impulse has infinite current for zero duration and encloses a charge of one coulomb in it.

(a) Both A and R are true and R is the correct explanation of A
(b) Both A and R are true but R is NOT the correct explanation of A
(c) A is true but R is false
(d) A is false but R is true

[ESE-2001]

3.3 A unit step voltage $u(t-5)$ is applied to the RL network

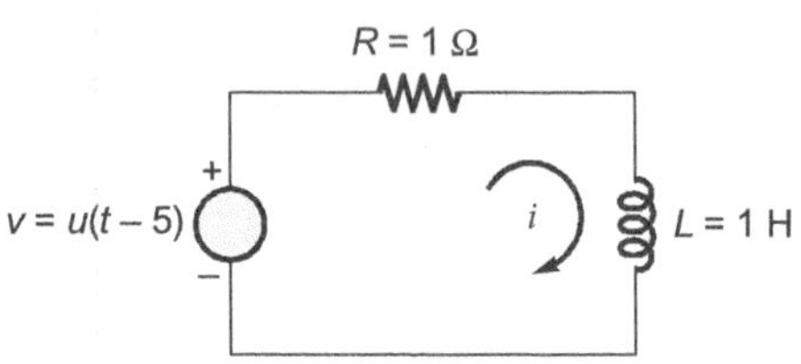

The current i is given by
(a) $1 - e^{-t}$
(b) $[1 - e^{-(t-5)}]\,u(t-5)$
(c) $(1 - e^{-t})\,u(t-5)$
(d) $1 - e^{-(t-5)}$

[ESE-2001]

3.4 The response of a network is $i(t) = Kt\,e^{-\alpha t}$ for $t \geq 0$ where α is real positive. The value of 't' at which the $i(t)$ will become maximum, is

(a) α (b) 2α
(c) $1/\alpha$ (d) α^2

[ESE-2001]

3.5 If $i = -10\,e^{-2t}$, the voltage of the source of the given circuit, V_s is given by

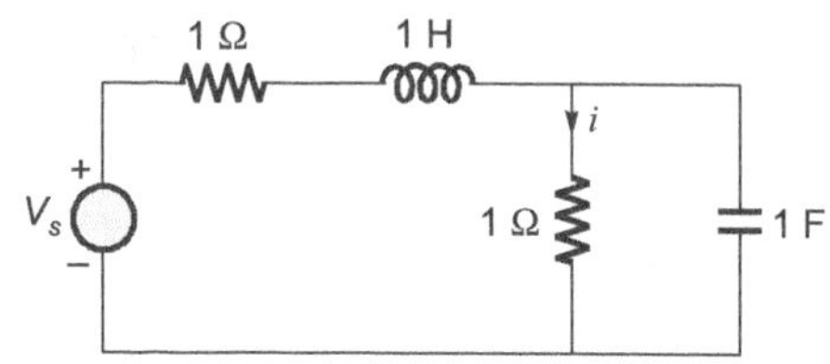

(a) $-10\,e^{-2t}$ (b) $-20\,e^{-2t}$
(c) $20\,e^{-2t}$ (d) $-30\,e^{-2t}$

[ESE-2001]

3.6 The steady state in the circuit, shown in the given figure is reached with S open. S is closed at $t = 0$. The current I at $t = 0^+$ is

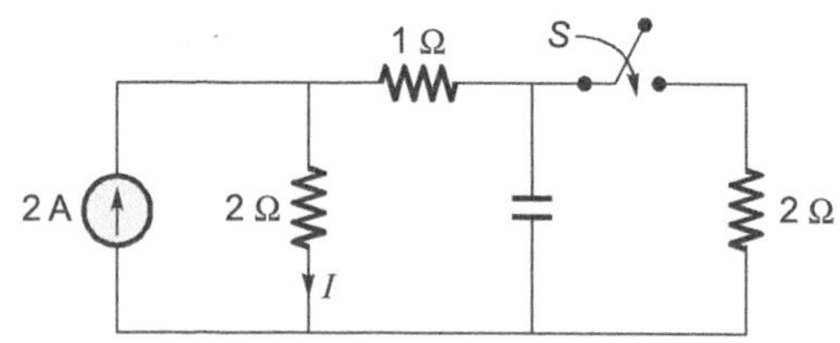

(a) 1 A (b) 2 A
(c) 3 A (d) 4 A

[ESE-2001]

3.7 For the circuit shown in the given figure below, if $C = 20\ \mu F$, $v(0^-) = -50$ V and $dv(0^-)/dt = 500$ V/s, then R is

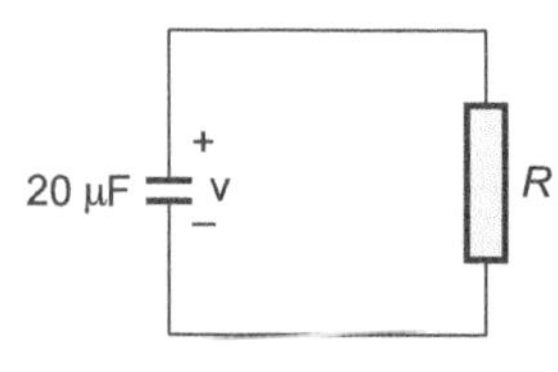

(a) 2 kΩ (b) 3 kΩ
(c) 5 kΩ (d) 10 kΩ

[ESE-2002]

3.8 In the circuit shown in the given figure, the switch is closed at $t = 0$.

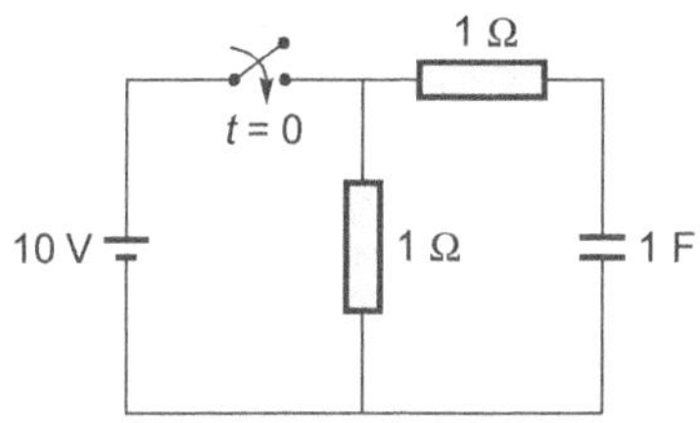

The current through the capacitor will decrease exponentially with a time constant:

(a) 0.5 s (b) 1 s
(c) 2 s (d) 10 s

[ESE-2002]

3.9 In the circuit shown in the given figure, the switch is moved from position A to B at time $t = 0$. The current i through the inductor satisfies the following conditions:

$$i(0) = -8\text{ A},\ \left.\frac{di}{dt}\right|_{t=0} = 3\text{ A/s},\ i(\infty) = 4\text{ A}.$$

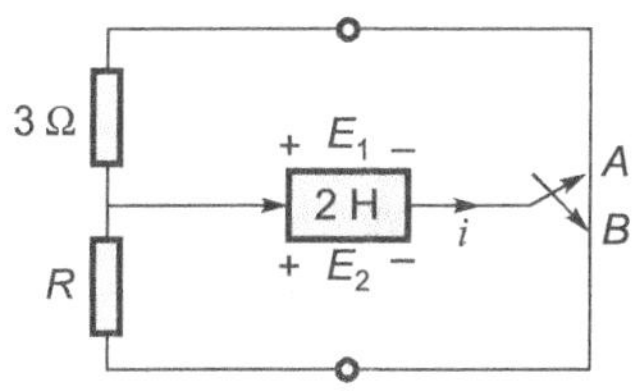

The value of R is

(a) 0.5 ohm (b) 2 ohm
(c) 4 ohm (d) 12 ohm

[ESE-2002]

3.10 In the circuit shown below, the initial voltage across the capacitor is 2 V and I is a unit step current source. Then the voltage across the capacitor for $t > 0$ is

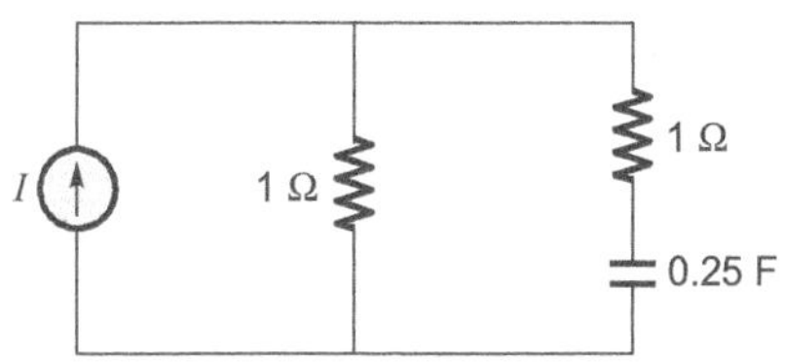

(a) $(2 - e^{-2t})\, u(t)$ (b) $(2 + e^{2t})\, u(t)$
(c) $(1 - e^{2t})\, u(t)$ (d) $(1 + e^{-2t})\, u(t)$

[ESE-2003]

3.11 The capacitor in the circuit as shown below is initially charged to 12 V with S_1 and S_2 open S_1 is closed at $t = 0$ while S_2 is closed at $t = 3$s. The waveform of the capacitor is represented by

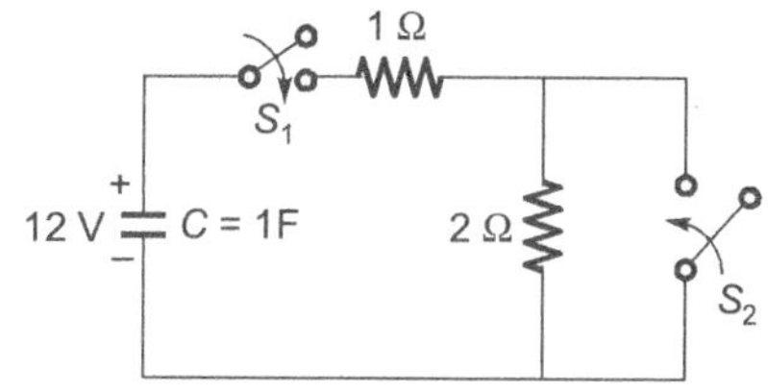

(a)

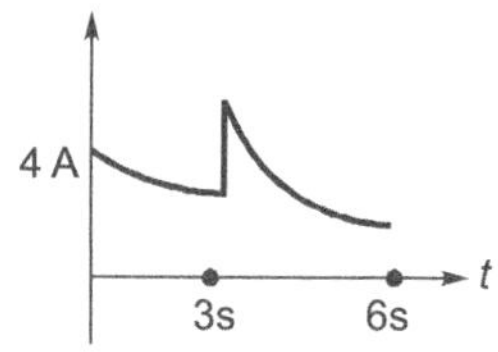

(b)

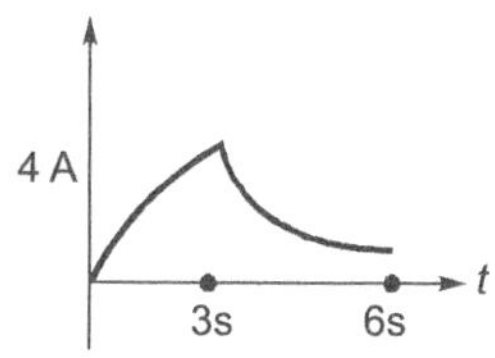

(c)

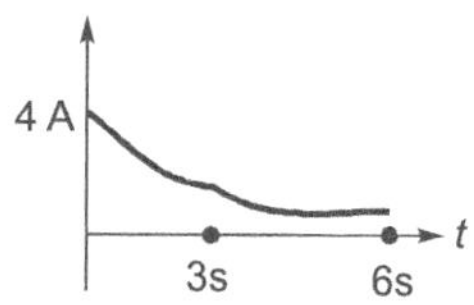

(d)

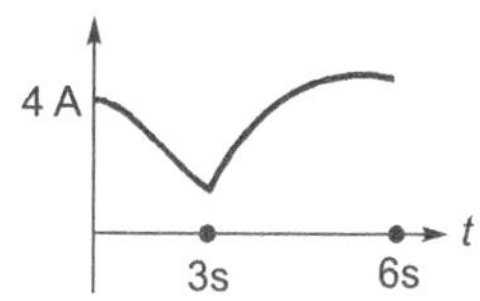

[ESE-2003]

3.12 Consider the following circuit:

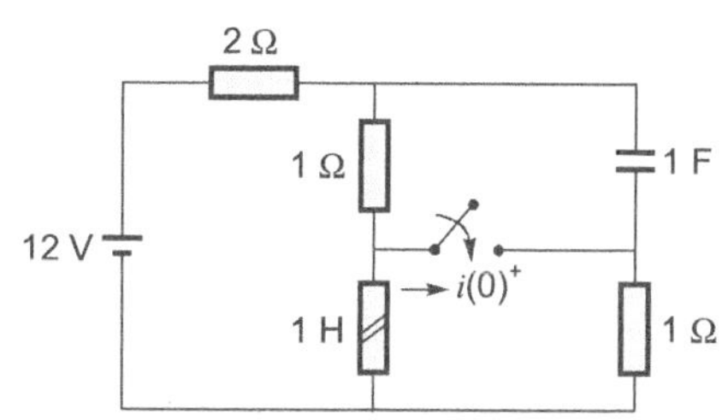

The circuit shown above is in steady state before closing the switch. What is the current $i(0^+)$ through the switch if the circuit is closed at $t = 0$?

(a) – 4 A (b) 0 A
(c) 4 A (d) 12 A

[ESE-2004]

3.13 Consider the following circuit:

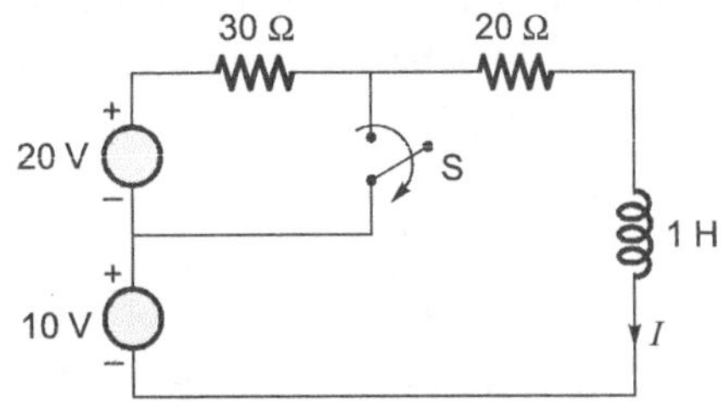

The circuit shown above attains a steady state with the switch S closed. What is the value of $i(t)$ for $t \geq 0$, if the switch S is opened at $t = 0$?

(a) $0.5 + 0.5\ e^{-50t}$ (b) $0.6 - 0.1\ e^{-50t}$
(c) $0.5 - 0.5\ e^{-50t}$ (d) $0.6 - 0.6\ e^{-20t}$

[ESE-2004]

3.14 In the circuit given below, the steady state is attained with S open. S is closed at $t = 0$. What is the value of current I at $t = 0^+$?

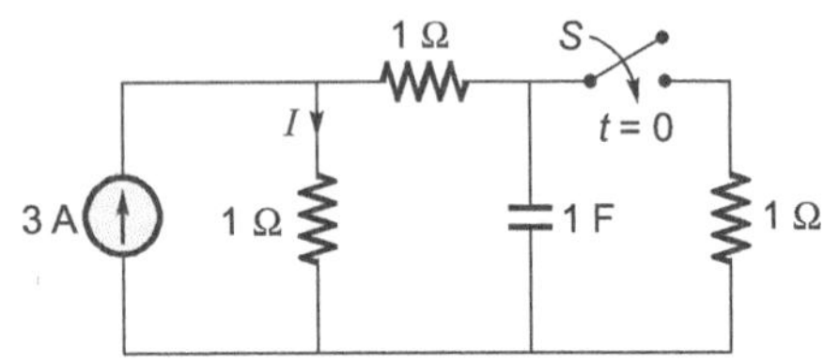

(a) 2 A (b) 2.25 A
(c) 3 A (d) 4 A

[ESE-2005]

3.15 A unit impulse voltage is applied at $t = 0$ to the R-L circuit shown below while $i(0^-) = 1$ A. What is the expression for $i(t)$?

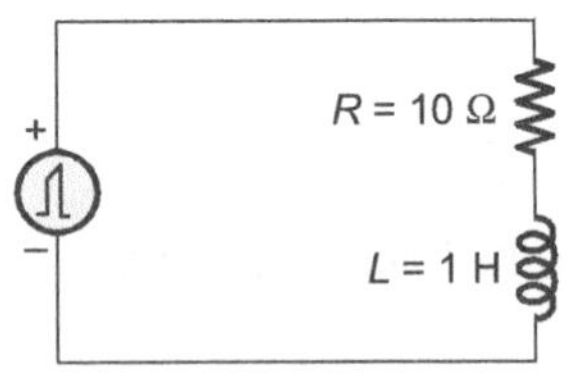

(a) e^{-10t} (b) $1.5\ e^{-10t}$
(c) $2\ e^{-10t}$ (d) $10\ e^{-10t}$

[ESE-2005]

3.16 In the circuit shown in the figure below, if $i_s = u(t)$ A, then what are the initial and steady-state voltages across the capacitor?

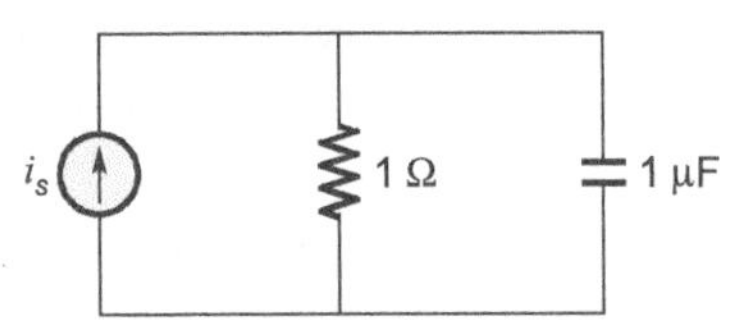

(a) 1 V and 1 V, respectively
(b) 1 V and 0, respectively
(c) 0 and 1 V, respectively
(d) 0 and 0, respectively [ESE-2006]

3.17 In the circuit shown below, the constant current source of value I is switched on at $t = 0$. What are the values of currents i_1 and i_2 at $t = 0$, with zero initial conditions?

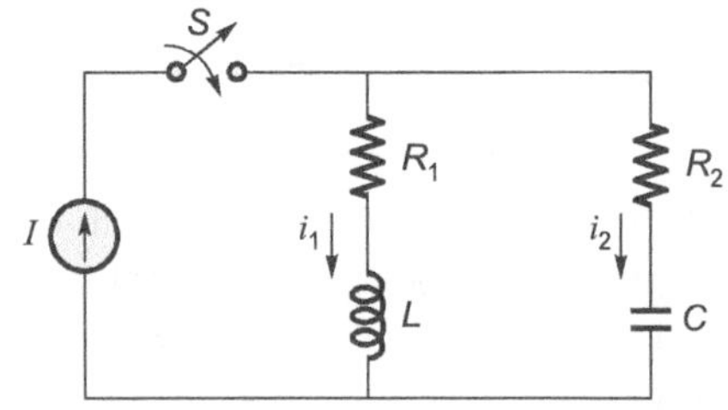

	i_1	i_2
(a)	I	0
(b)	0	I
(c)	$\dfrac{R_2}{R_1+R_2}I$	$\dfrac{R_1}{R_1+R_2}I$
(d)	0	0

[ESE-2006]

3.18 In the circuit shown in the figure given below, the switch is opened at $t = 0$ after having been closed for a long time. What is the current through 50 Ω resistor?

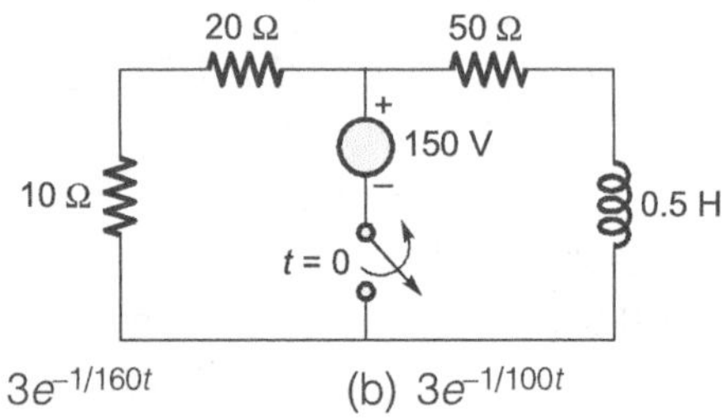

(a) $3e^{-1/160t}$ (b) $3e^{-1/100t}$
(c) $3e^{-100t}$ (d) $3e^{-160t}$

[ESE-2006]

3.19 At $t = 0$, the switch S is thrown from b to a of the circuit as shown below. What are the values of $v(0^+)$ and $i(0^+)$?

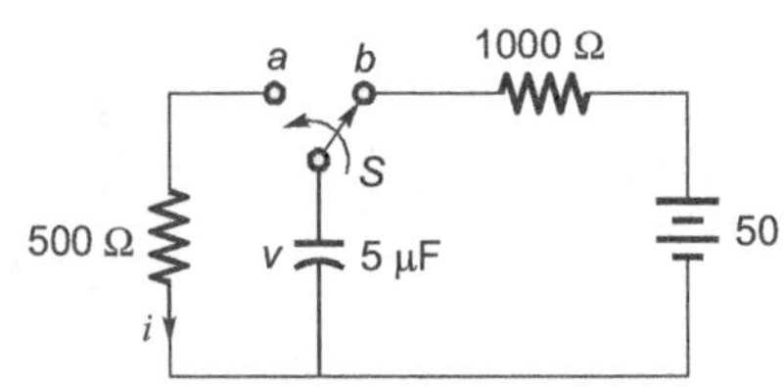

(a) 50 V, 90 mA (b) 50 V, 100 mA
(c) 50 V, 110 mA (d) 50 V, 120 mA

[ESE-2006]

3.20 An R-C series circuit, initially at rest has a step voltage signal. The response $v(t)$ across C is $v(t) = 1 - e^{-3t}$. If now there is an initial voltage at C of 3 volts, what is $v(t)$ for the same step signal?

(a) $1 + 3e^{-3t}$ (b) $1 + 2e^{-3t}$
(c) $3e^{-3t}$ (d) None of these

[ESE-2006]

3.21 Steady state is reached with S open. S is closed at $t = 0$. What is the value of the current I at $t = 0^+$?

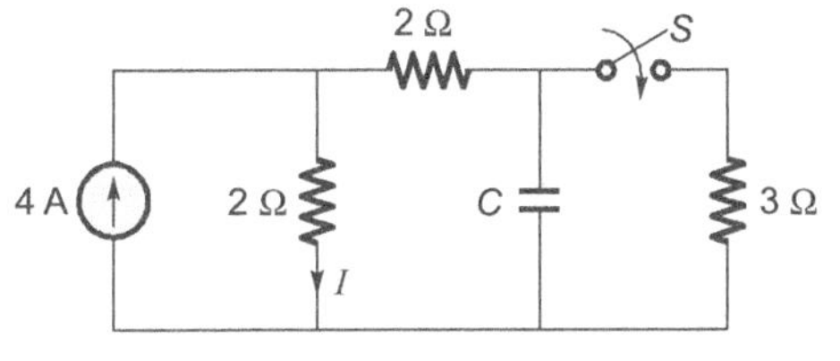

(a) 1 A (b) 2 A
(c) 3 A (d) 4 A

[ESE-2007]

3.22 The circuit shown in the figure given below is in steady state with switch S open. The switch is closed at $t = 0$. What are the values of $V_c(0^+)$ and $V_c(\infty)$, respectively?

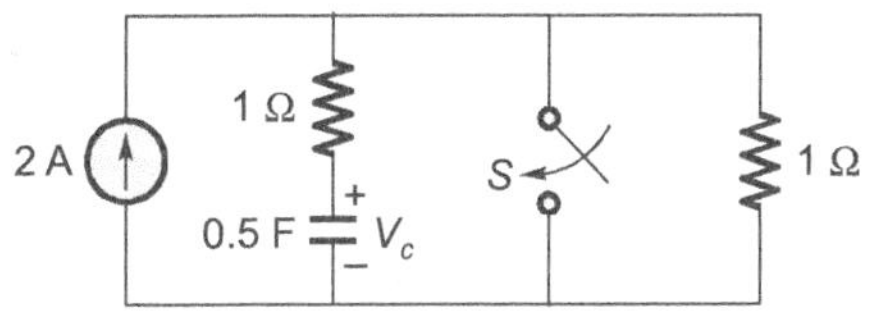

(a) 0 V, 0 V (b) 0 V, 2 V
(c) 2 V, 0 V (d) 2 V, 2 V

[ESE-2007]

3.23 In the below network, S is closed for a long time till steady state is attained. S is opened at $t = 0$. What are the values of voltage V at $t = 0^+$ and $t = \infty$?

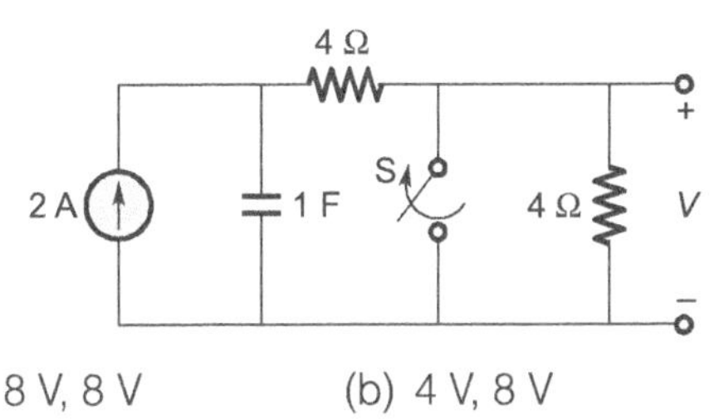

(a) 8 V, 8 V (b) 4 V, 8 V
(c) 8 V, 4 V (d) 4 V, 4 V

[ESE-2007]

3.24 What is the time constant of the circuit below?

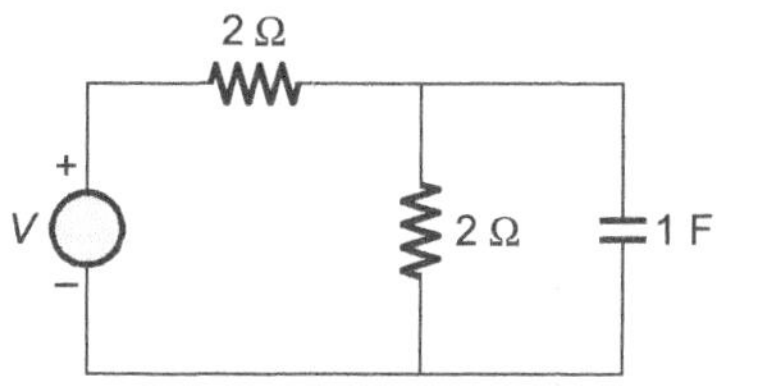

(a) 0.5s (b) 1s
(c) 2s (d) 4s

[ESE-2008]

3.25 In the network shown below, it is given that $v = 1$ V and $dv/dt = -10$ V/s at a time t, where t is the time after the switch S is closed. What is the value of C?

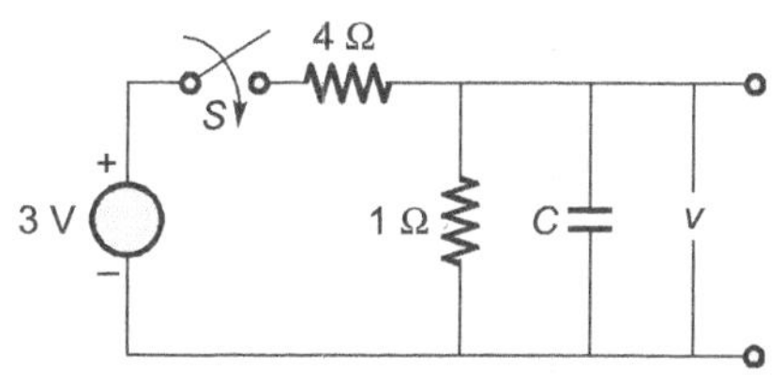

(a) 0.05 F (b) 0.1 F
(c) 0.15 F (d) 0.2 F

[ESE-2008]

3.26 A series R-L circuit is to be connected to an a.c. source $v(t) = V_m \sin(\omega t + \phi)$ volt. Which one of the following is correct?

The transient current will be absent if the source is connected at a time t_0 such that

(a) $\omega t_0 = 0$

(b) $\omega t_0 = \frac{\pi}{2}$

(c) $\omega t_0 = \tan^{-1}\frac{\omega L}{R}$

(d) ωt_0 has any arbitrary value

[ESE-2008]

3.27 A lossy capacitor is represented by an ideal capacitor C with a high resistance R in parallel. What is the Q of the circuit at frequency ω?

(a) ωCR (b) $1/(\omega CR)$
(c) $\omega C/R$ (d) $R/(\omega C)$

[ESE-2008]

3.28 The network shown below is initially at rest. What is the initial current I when the switch S is closed at $t = 0$?

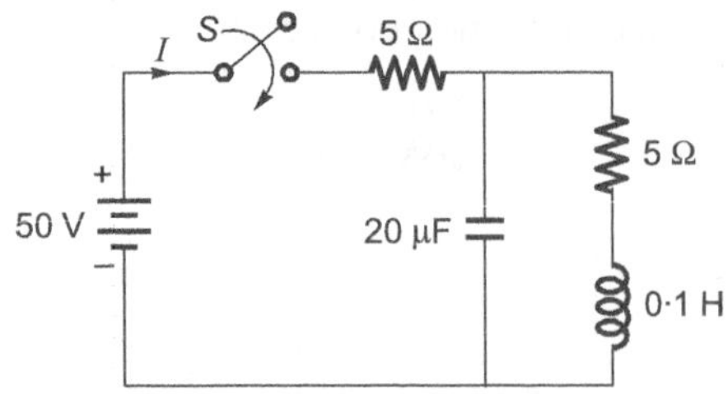

(a) 0 A (b) 5 A
(c) 10 A (d) 20 A [ESE-2009]

3.29 The circuit as shown below is in the steady state. The switch S is closed at $t = 0$. What are the values of V and $\frac{dV}{dt}$ at $t = 0^+$?

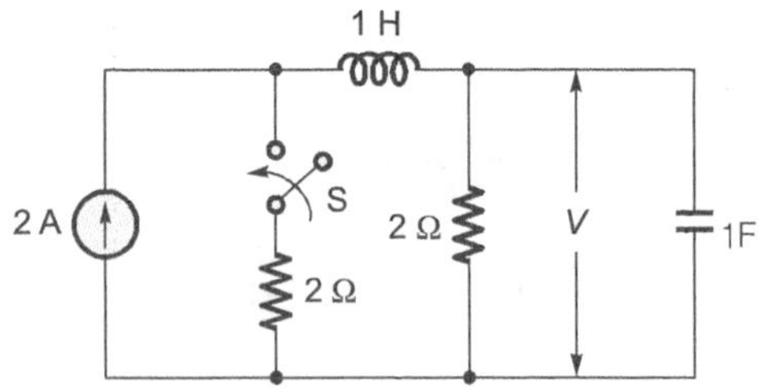

(a) 0 and 4 (b) 4 and 0
(c) 2 and 0 (d) 0 and 2 [ESE-2009]

3.30 The switch of below circuit was open for long, and at $t = 0$ it is closed. What is the final steady state voltage across the capacitor and the time-constant of the circuit?

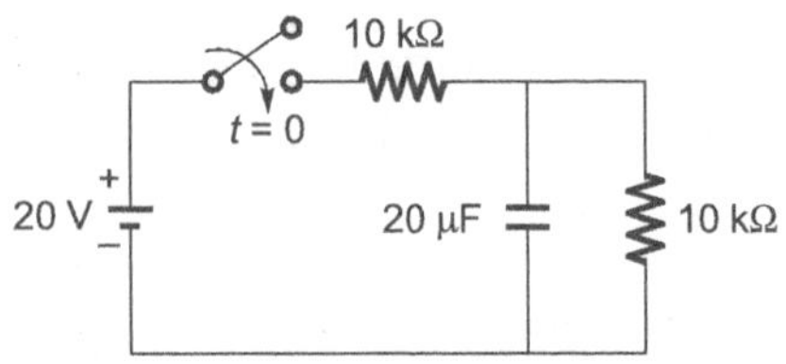

(a) 0 V and 0.1 sec. (b) 20 V and 0.2 sec.
(c) 10 V and 0.2 sec. (d) 10 V and 0.1 sec.
[ESE-2009]

3.31 In the below shown circuit, under what conditions the output V_0 will be proportional to dv_i/dt (with T = time period of input)

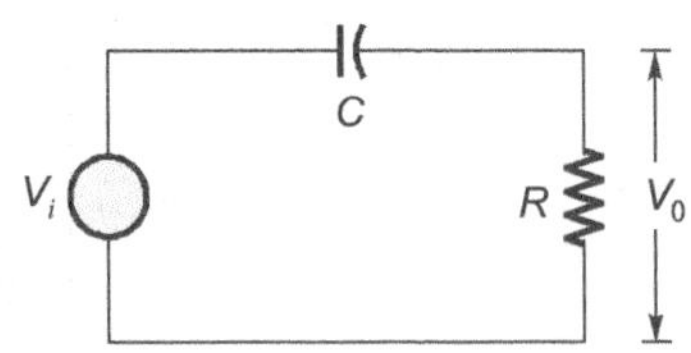

(a) $RC = T$ (b) $RC >> T$
(c) $RC << T$ (d) Both R and C large
[ESE-2009]

3.32 If the initial voltage across the capacitor of 2 F is $V(0) = 1$, the voltage and charge on the capacitor at $t = 3$ sec after connecting a current source $I_S = 2$ A at $t = 0$ are respectively

(a) 2 V, 4 coulomb (b) –2 V, 4 coulomb
(c) 4 V, 8 coulomb (d) 8 V, 4 coulomb
[ESE-2010]

3.33 For the circuit shown in figure below; the switch has been in position A for a long time. At $t = 0$, the switch is moved to B. Then, the capacitor voltage $V_C(t)$ for $t > 0$ is

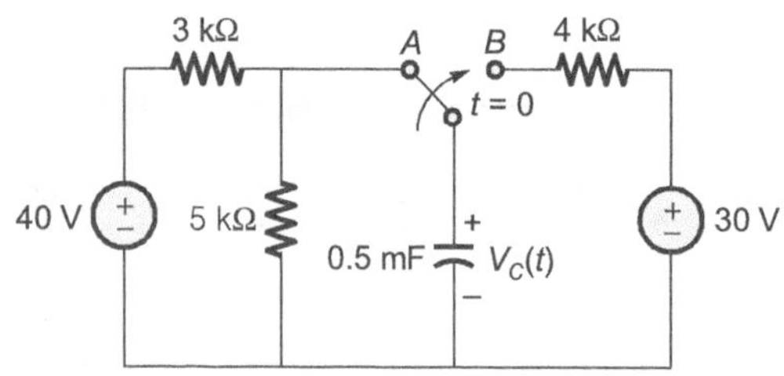

(a) $V_C(t) = (24 - 6\,e^{-2t})$ V
(b) $V_C(t) = (30 - 15\,e^{-0.5t})$ V
(c) $V_C(t) = (6 - 6\,e^{-2t})$ V
(d) $V_C(t) = (30 - 5\,e^{-0.5t})$ V [ESE-2010]

3.34 For the given circuit shown in the figure below, the initial inductor current and the voltage across the capacitor are zero and 2, respectively. When the switch S is closed at $t = 0$, the values of v and dv/dt are, respectively

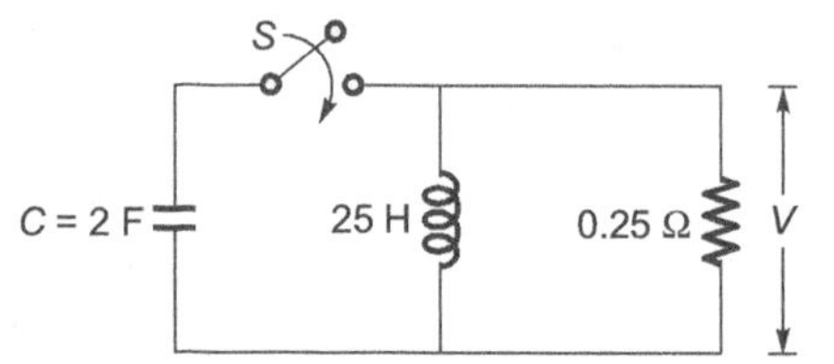

(a) 2, –4 (b) 0, 0.25
(c) 0, –0.5 (d) 2, 0
[ESE-2010]

3.35 In the circuit, if $V_C(0) = 25$ V, the expression for $V_C(t)$ for $t > 0$ is

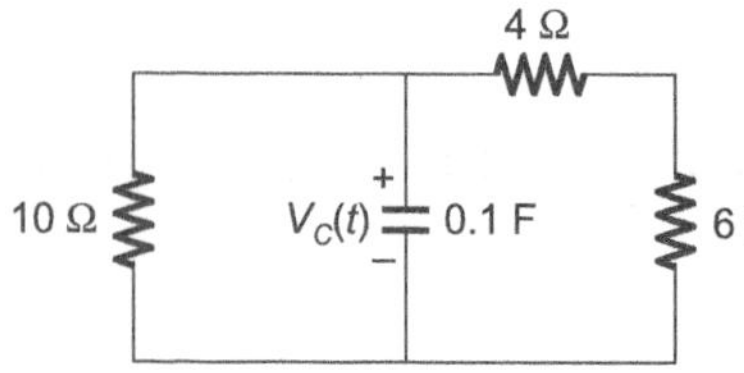

(a) $V_C(t) = 20\,e^{-0.4t}$ V (b) $V_C(t) = 25\,e^{-0.4t}$ V
(c) $V_C(t) = 20\,e^{-2.5t}$ V (d) $V_C(t) = 25\,e^{-2t}$ V
[ESE-2011]

3.36 A 0.2 H inductor with an initial current of 4 A is in parallel with a resistor of 100 Ω. The current at 0.8 ms is

(a) $4\,e^{-0.4}$ A
(b) $4\,e^{-16\times10^{-6}}$ A
(c) $4\,e^{-0.4\times10^{-3}}$ A
(d) $4\,e^{-16\times10^{-3}}$ A

[ESE-2011]

3.37 In the below network, the switch *K* is opened at $t = 0$. Then $\frac{dV}{dt}$ at $t = 0^+$ is

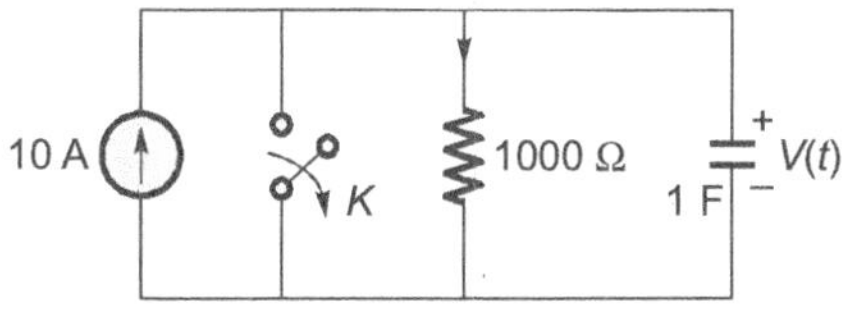

(a) 1000 V/sec
(b) 100 V/sec
(c) 10 V/sec
(d) 1 V/sec

[ESE-2011]

3.38 The nature of current response $i(t)$ for $t \geq 0$ for the network shown is

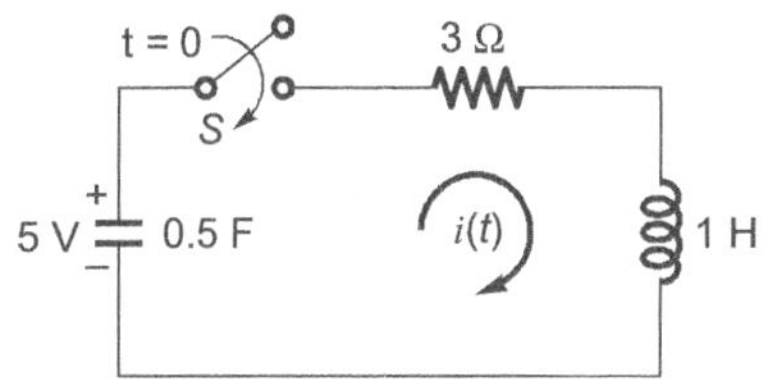

(a)

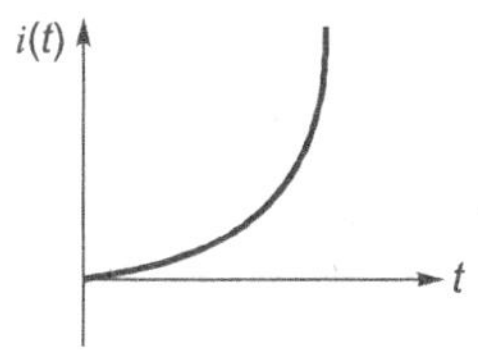

(b)

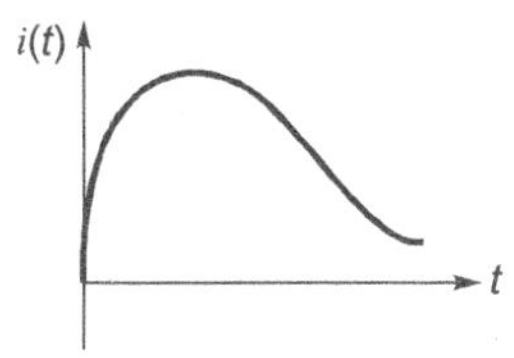

(c)

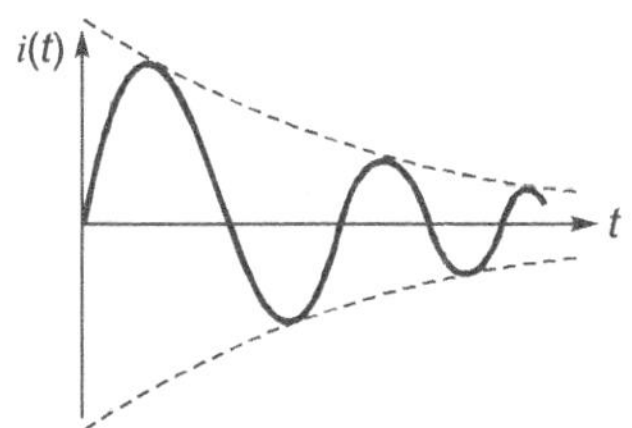

(d)

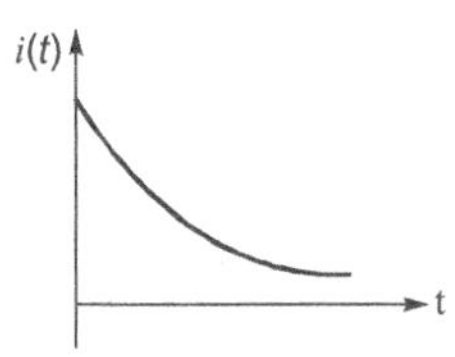

[ESE-2011]

3.39 A coil of resistance 20 Ω and inductance 0.8 H is connected to a 200 V DC supply. The rate of change of current at $t = 0^+$ is

(a) 16 A/s
(b) 160 A/s
(c) 250 A/s
(d) 4000 A/s

[ESE-2012]

3.40 The circuit given below is in steady state for a long time with switch *S* open. The switch is closed at $t = 0$. The current through *R* at $t = 0$ will be

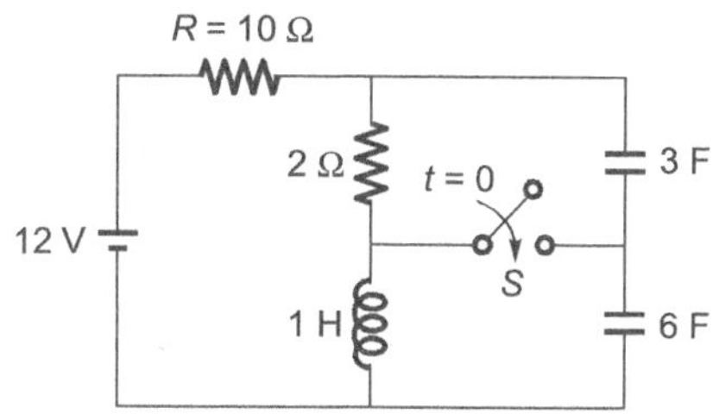

(a) $\frac{1}{3}$A
(b) $\frac{2}{3}$A
(c) 1 A
(d) 2 A

[ESE-2012]

3.41 Statement (I): Under steady-state conditions, a pure inductance acts as a short circuit for direct current.

Statement (II): The potential drop across an inductance is proportional to the rate of change of current.

(a) Both Statement (I) and Statement (II) are individually true and Statement (II) is the correct explanation of Statement (I).
(b) Both Statement (I) and Statement (II) are individually true but Statement (II) is not the correct explanation of Statement (I).
(c) Statement (I) is true but Statement (II) is false.
(d) Statement (I) is false but Statement (II) is true.

[ESE-2012]

3.42 Statement (I): Under steady-state condition, a pure capacitor behaves as an open circuit for direct voltage.

Statement (II): The current through a capacitor is proportional to the rate of change of voltage.

(a) Both Statement (I) and Statement (II) are individually true and Statement (II) is the correct explanation of Statement (I).

(b) Both Statement (I) and Statement (II) are individually true but Statement (II) is not the correct explanation of Statement (I).
(c) Statement (I) is true but Statement (II) is false.
(d) Statement (I) is false but Statement (II) is true.

[ESE-2014]

3.43 In any network the current will be seen to be consisting of a forced current and a natural current. A forced current is
(a) a steady-state current with external source but a natural current is a transient current in a closed circuit with no external source.
(b) a transient current with external source but a natural current is a steady-state current in a closed circuit with no external source.
(c) a steady-state current in a closed circuit without external source, while a natural current is a transient current with an external source.
(d) a transient current in a closed circuit without external source, while a natural current is a steady-state current with an external source.

[ESE-2014]

3.44 A coil of resistance 10 Ω and inductance 0.8 H is connected to 200 V dc supply. The initial rate of change of current is
(a) 16 A/s (b) 160 A/s
(c) 250 A/s (d) 4000 A/s

[ESE-2014]

3.45 There are no transients in pure resistance circuits because they
(a) offer high resistance
(b) obey ohm's law
(c) have no stored energy
(d) are linear circuits

[ESE-2014]

3.46 The initial and final values of

$$f(t) = 15 - 10t - 10\, e^{-20\,t}$$

are respectively
(a) 5 and ∞ (b) 5 and −∞
(c) 15 and ∞ (d) 15 and 10

[ESE-2014]

3.47 An ideal current source is connected to the disconnected circuit shown in the figure $t = 0$. The time constant of the circuit is

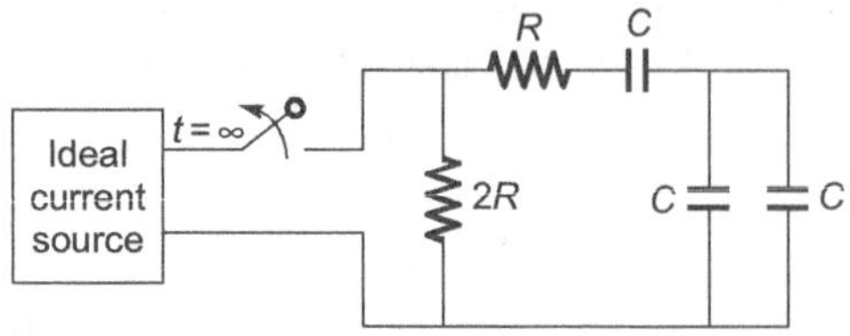

(a) $\frac{RC}{2}$ (b) RC
(c) $2RC$ (d) $\frac{9RC}{2}$

[ESE-2015]

3.48 Transients are caused because
1. the load is suddenly connected to or disconnected from the supply
2. of the sudden change in applied voltage from one finite value to the other
3. of the change in stored energy in inductors and capacitors

Which of the above statements are correct?
(a) 1 and 2 only (b) 1 and 3 only
(c) 2 and 3 only (d) 1, 2 and 3

[ESE-2015]

3.49 The response of a series *R-C* circuit is given by

$$I(s) = \frac{\frac{2V}{\pi} - \frac{q_0}{C}}{R\left(s + \frac{1}{RC}\right)}$$

where q_0 is the initial charge on the capacitor. What is the final value of the current?
(a) $\frac{1}{R}\left(\frac{2V}{\pi} - \frac{q_0}{C}\right)$ (b) $\frac{e^{t/RC}}{R}\left(\frac{2V}{\pi} - \frac{q_0}{C}\right)$
(c) Infinity (d) Zero

[ESE-2016]

3.50 What should be done to find the initial values of the circuit variables in a first-order *R-C* circuit excited by only initial conditions?
(a) To replace the capacitor by a short circuit
(b) To replace the capacitor by an open circuit
(c) To replace the capacitor by a voltage source
(d) To replace the capacitor by a current source

[ESE-2016]

3.51 Which of the following is true for the complete response of any network voltage or current variables for a step excitation to a first-order circuit?

(a) It has the form k_1e^{-at}
(b) It has the form k
(c) It may have either the form (a) or the form of (a) plus (b)
(d) It has the form e^{+at} [ESE-2016]

3.52 **Statement (I):** One series RC circuit and the other series RL circuit are connected in parallel across an ac supply. The circuit exhibits two resonances when L is variable.

Statement (II): The circuit has two values of L for which the imaginary part of the input admittance of the circuit is zero.

(a) Both Statement (I) and Statement (II) are individually true and Statement (II) is the correct explanation of Statement (I).
(b) Both Statement (I) and Statement (II) are individually true but Statement (II) is not the correct explanation of Statement (I).
(c) Statement (I) is true but Statement (II) is false.
(d) Statement (I) is false but Statement (II) is true.

[ESE-2017]

3.53 Consider the following statements:
If a high Q parallel resonant circuit is loaded with a resistance.
1. The circuit impedance reduces.
2. The resonant frequency remains the same.
3. The bandwidth reduces.

Which of the above statements is/are correct?
(a) 3 only (b) 1 and 2 only
(c) 1 only (d) 1, 2 and 3 [ESE-2017]

3.54 For the circuit as shown, consider that switch S_1 has been in position B for a very long time and switch S_2 has been open all the time. At time $t = 0$, the switch S_1 moves to position A and switch S_2 closes instantaneously. What is the value of V_{out} at $t = 0^+$, assuming initial charge on $C = 0$?

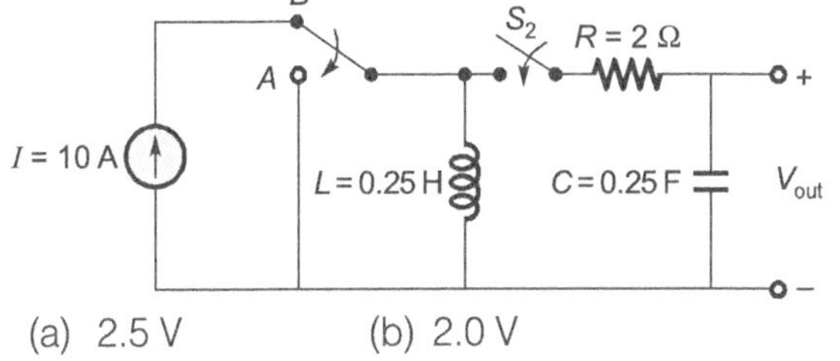

(a) 2.5 V (b) 2.0 V
(c) 1.5 V (d) 0 V [ESE-2018]

3.55 In the circuit as shown in figure, the switch is closed at $t = 0$. The current through the capacitor will decrease exponentially with a time constant of magnitude

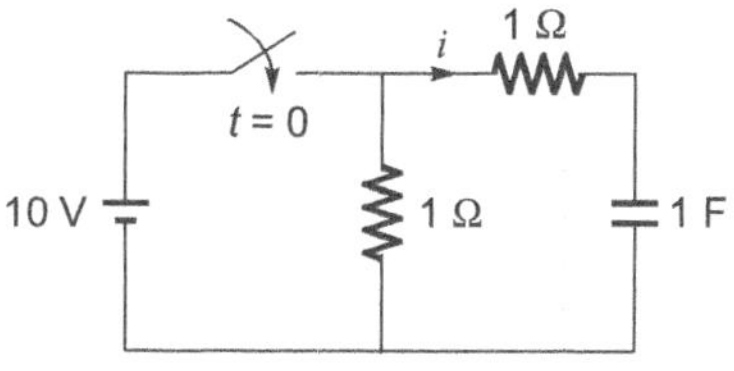

(a) 0.5s (b) 1s
(c) 2s (d) 4s

[ESE-2018]

3.56 A coil resistance 30 Ω and inductance 0.6 H is switched on to a 240 V supply. What are the rate of change of current at the instant of closing the switch at $t = 0$ and the magnitude of the final steady state current respectively?
(a) 80 A/sec and 80 A
(b) 400 A/sec and 8 A
(c) 8 A/sec and 80 A
(d) 400 A/sec and 80 A

[ESE-2021]

3.57 When angular frequency for d.c. sources is zero, capacitor and inductor will act like respectively
(a) short circuited, open circuited
(b) open circuited, short circuited
(c) open circuited, open circuited
(d) short circuited, short circuited

[ESE-2022]

3.58 The time constant of R-L series circuit may be defined as the time at which the current through the R-L series circuit rises to
(a) 36.8% of steady-state value
(b) 63.2% of steady-state value
(c) 23.2% of initial value
(d) 46.8% of initial value

[ESE-2022]

3.59 Transient disturbance is produced in a circuit whenever
(a) currents in electrical circuit are associated with resistors
(b) circuit is suddenly connected to or disconnected from the supply
(c) the source is overdamped
(d). the source is underdamped

[ESE-2022]

3.60 In the circuit of the figure shown below, the source voltage is 100 V, the resistance is 10 kΩ, and the capacitance is 0.005 μF. In how much time can the capacitor voltage be discharged to 5 V after the switch is turned to position 3?

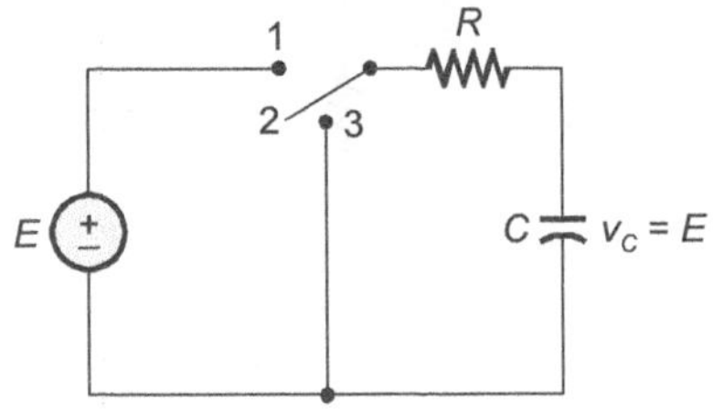

(a) 50 μs (b) 50 ms
(c) 150 ms (d) 150 μs

[ESE-2023]

3.61 Consider the following statements regarding the initial conditions for inductor and capacitor :

1. If there is no current flowing through the inductor at $t = 0^-$, the inductor will act as an open circuit at $t = 0^+$.
2. If there is no voltage across the capacitor at $t = 0^-$, the capacitor will act as an open circuit at $t = 0^+$.
3. If a current of value I_0 flows through the inductor at $t = 0^-$, the inductor can be regarded as a current source of I_0 ampere at $t = 0^+$.

Which of the above statements is/are correct?

(a) 1 only (b) 2 only
(c) 1 and 3 only (d) 1, 2 and 3

[ESE-2024]

Answers Transient Response of DC & AC Networks

3.1 (a)	3.2 (a)	3.3 (b)	3.4 (c)	3.5 (b)	3.6 (b)	3.7 (c)	3.8 (b)	3.9 (a)
3.10 (d)	3.11 (a)	3.12 (b)	3.13 (b)	3.14 (c)	3.15 (c)	3.16 (c)	3.17 (b)	3.18 (d)
3.19 (b)	3.20 (b)	3.21 (d)	3.22 (c)	3.23 (b)	3.24 (b)	3.25 (a)	3.26 (c)	3.27 (a)
3.28 (c)	3.29 (b)	3.30 (d)	3.31 (c)	3.32 (b, c)	3.33 (d)	3.34 (a)	3.35 (d)	3.36 (a)
3.37 (c)	3.38 (b)	3.39 (c)	3.40 (c)	3.41 (a)	3.42 (a)	3.43 (a)	3.44 (c)	3.45 (c)
3.46 (b)	3.47 (c)	3.48 (d)	3.49 (d)	3.50 (c)	3.51 (c)	3.52 (a)	3.53 (b)	3.54 (d)
3.55 (b)	3.56 (b)	3.57 (b)	3.58 (b)	3.59 (b)	3.60 (d)	3.61 (c)		

Explanations Transient Response of DC & AC Networks

3.1 (a)

Inductor does not allow sudden changes of voltage because flux linkages are constant.

3.2 (a)

$$\because \quad \int_{-\infty}^{+\infty} I\,dt = Q$$

For unit impulse current

$$\int_{-\infty}^{+\infty} I\,dt = \int_{-0}^{+0} I\,dt = 1$$

$$\Rightarrow \quad Q = 1\,C$$

$$\Rightarrow \quad W_E = \frac{Q^2}{2C} = \frac{1}{2C} \cdots \text{Joules}$$

3.3 (b)

By applying KVL in loop

$$u(t-5) = Ri + L\frac{di}{dt}$$

By taking Laplace transform

$$\frac{e^{-5s}}{s} = RI(s) + LsI(s)$$

$$\Rightarrow \quad \frac{e^{-5s}}{s} = I(s)\,(1+s)$$

$$\Rightarrow \quad I(s) = \frac{e^{-5s}}{s(1+s)} = e^{-5s}\left[\frac{1}{s} - \frac{1}{s+1}\right]$$

$$\therefore \quad i(t) = \mathcal{L}^{-1}[I(s)]$$

$$= u(t-5) - e^{-(t-5)}\, u(t-5)$$
$$i(t) = u(t-5)\,[1 - e^{-(t-5)}]$$

3.4 (c)

For $i(t)$ to be maximum

$$\frac{di(t)}{dt} = 0 \quad \text{and} \quad \frac{d^2i(t)}{dt^2} < 0$$

given, $i(t) = Kte^{-\alpha t}$

$$\frac{di(t)}{dt} = K[t \cdot e^{-\alpha t}(-\alpha) + e^{-\alpha t}]$$
$$= Ke^{-\alpha t}[1 - \alpha t]$$

For $\frac{di(t)}{dt} = 0$

$$\Rightarrow \quad 1 = \alpha t \Rightarrow t = \frac{1}{\alpha}$$

3.5 (b)

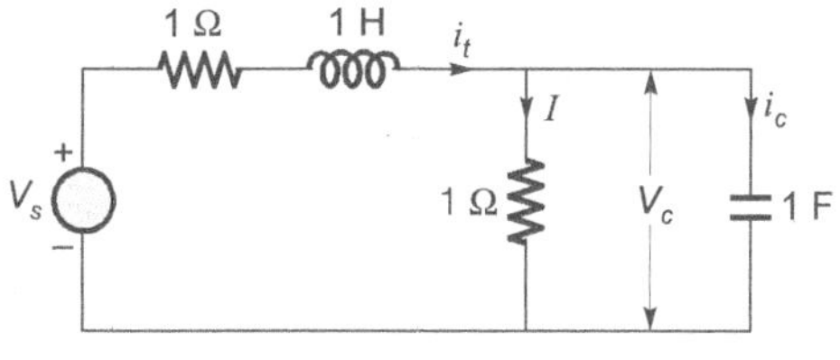

Voltage across capacitor

$$V_c = i \times 1 = -10e^{-2t}$$

Current through capacitor

$$i_c = C\frac{dV_c}{dt} = 20e^{-2t}$$

Total current,

$$i_t = i + i_c$$
$$= -10\,e^{-2t} + 20\,e^{-2t} = 10e^{-2t}$$

$$V_s = V_c + 1 \times i_t + L\frac{di_t}{dt}$$
$$= 10e^{-2t} - 10e^{-2t} - 20e^{-2t}$$
$$= -20e^{-2t}$$

3.6 (b)

With S open, capacitor get charged to

$$V_c(0^-) = 2 \times 2 = 4\text{ V}$$

The circuit at $t = 0^+$

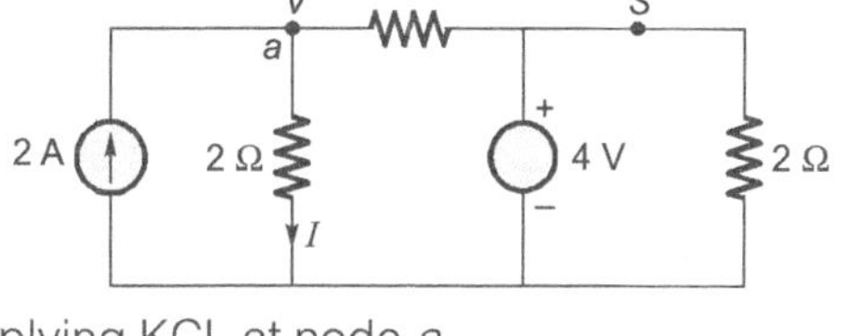

Applying KCL at node a,

$$-2 + \frac{V}{2} + \frac{V-4}{1} = 0 \Rightarrow V = 4\text{ Volt}$$

$$I = \frac{4}{2} = 2\text{ Amp}$$

Note: Charged capacitor would not let the current flow through the 1 Ω resistor as potential difference across it would be zero.

3.7 (c)

$$q = C \cdot V$$

$$\therefore \quad \frac{dq}{dt} = i = C \cdot \frac{dv}{dt}$$

(dv/dt would be negative as current would decrease).

Also, $v = iR = C \cdot \frac{dv}{dt} \cdot R$

$$\therefore \quad -50 = 20 \times 10^{-6} \times (-500) \times R$$
$$\Rightarrow \quad R = 5\text{ k}\Omega$$

3.8 (b)

$$v = 10 = i + \frac{1}{1}\int i\,dt$$

taking Laplace transform

$$\frac{10}{s} = I(s) + \frac{I(s)}{s}$$

$$\therefore \quad I(s) = \frac{10}{(s+1)}$$

$$\Rightarrow \quad i(t) = 10e^{-t} \Rightarrow \tau = 1\,s$$

OR

Time constant $\tau = R_{eq}C_{eq}$

To find time constant open all the current source and close all the voltage source in the given network,

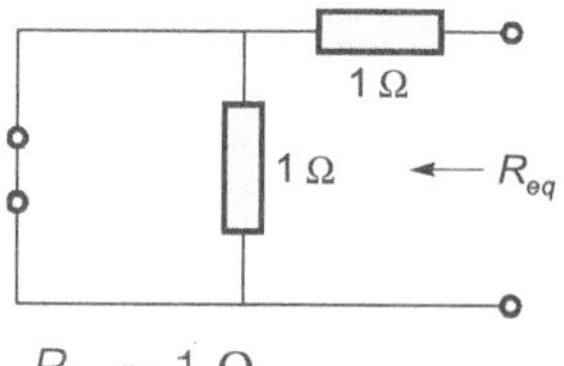

$$\therefore \quad R_{eq} = 1\ \Omega$$
$$\text{and} \quad C_{eq} = 1\text{ F}$$
$$\therefore \quad \tau = 1 \times 1 = 1\text{ sec}$$

3.9 (a)

$$E_2 = iR + L\frac{di}{dt}$$

at $t = 0$, $E_2 = -8R + 2 \times 3$ ...(*i*)

at $t = \infty$, $E_2 = 4 \times R$ ...(*ii*)

From equation (*i*) and (*ii*) $R = 0.5\ \Omega$

3.10 (d)

$$V_c(t) = V_c(\infty) - [V_c(\infty) - V_c(0^+)]\,e^{-t/\tau}$$

$V_c(0^+) = 2\,\text{V}$

At steady state, capacitor becomes open circuit.

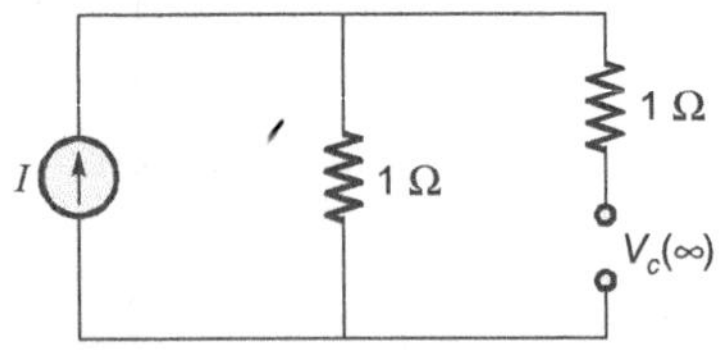

$V_c(\infty) = 1 \times 1 = 1\,\text{V}$

τ = time constant $= R_{eq}C_{eq}$

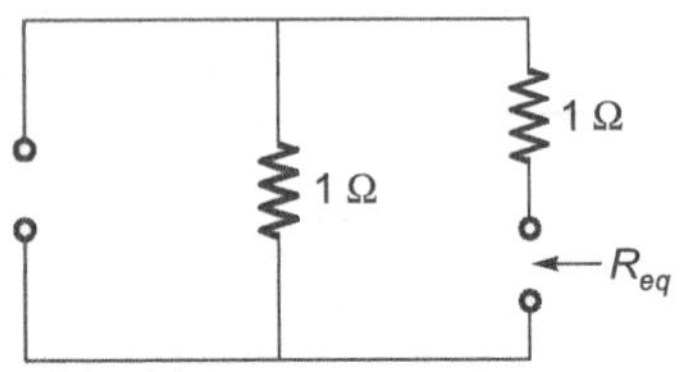

$R_{eq} = 2\,\Omega$

$\tau = 2 \times 0.25 = 0.5$ sec

$V_c(t) = V_c(\infty) - [V_c(\infty) - V_c(0^+)]e^{-t/\tau}$

$= 1 - [1 - 2]e^{-t/0.5} = (1 + e^{-2t})\,u(t)$

3.11 (a)

At $t = 3$ s, since $2\,\Omega$ is shorted, I_C jumps to a new value and then decreases with reduced time constant.

3.12 (b)

Circuit at $t = 0^-$

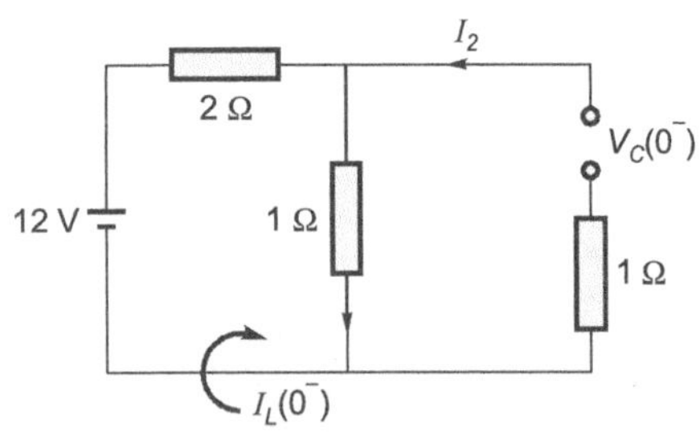

$$i_L(0^-) = \frac{12}{2+1} = 4\,\text{A}$$

$V_c(0^-) = 1 \times i_L(0^-) = 4\,\text{V}$

Circuit at $t = 0^+$

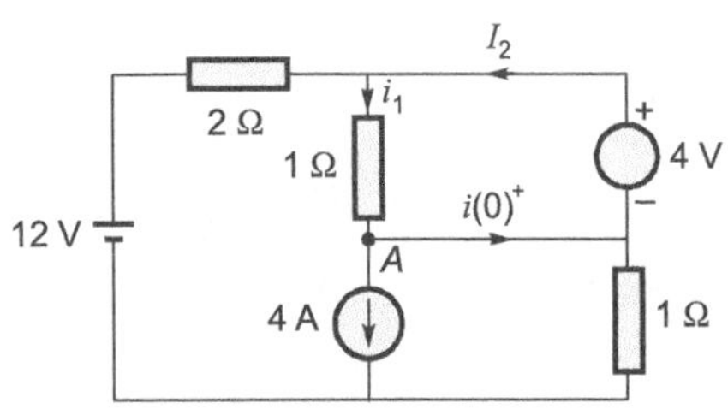

Current, $i_1 = \frac{4}{1} = 4\,\text{A}$

Applying KCL at node A

$i(0^+) + 4 - i_1 = 0 \Rightarrow i(0^+) + 4 = 0$

$i(0^+) = 0\,\text{A}$

3.13 (b)

At $t = 0^-$

$$i(0^-) = \frac{10}{20} = 0.5\,\text{A}$$

At $t = \infty$

$$i(\infty) = \frac{30}{50} = 0.6\,\text{A}$$

$$\tau = \frac{L}{R} = \frac{1}{30+20} = \frac{1}{50}\,\text{s}$$

and $i(t) = i(\infty) + [i(0^-) - i(\infty)]\,e^{-1/\tau}$

$= 0.6 + (0.5 - 0.6)\,e^{-50t}$

$i(t) = 0.6 - 0.1\,e^{-50t}$

3.14 (c)

At $t = 0^-$

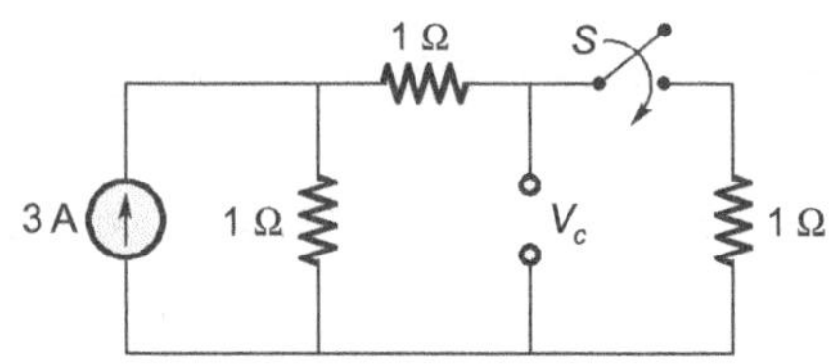

$V_c(0^-) = 3 \times 1 = 3\,\text{V}$

At $t = 0^+$

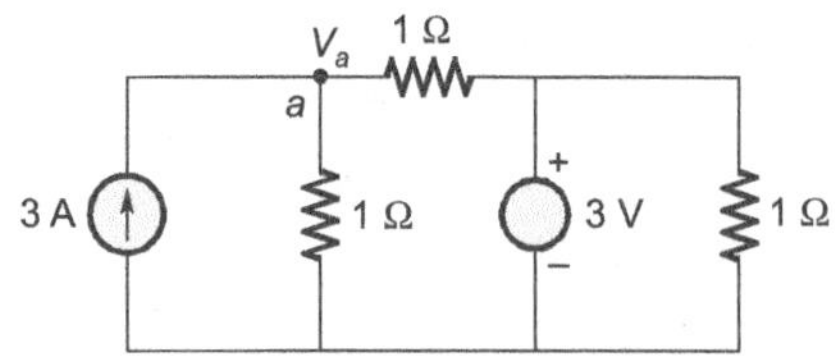

By applying KCL at node a

$$-3 + \frac{V_a}{1} + \frac{V_a - 3}{1} = 0$$

$V_a = 3\,\text{V}$

$$I = \frac{V_a}{1} = \frac{3}{1} = 3\,\text{A}$$

3.15 (c)

Taking Laplace transform of the circuit

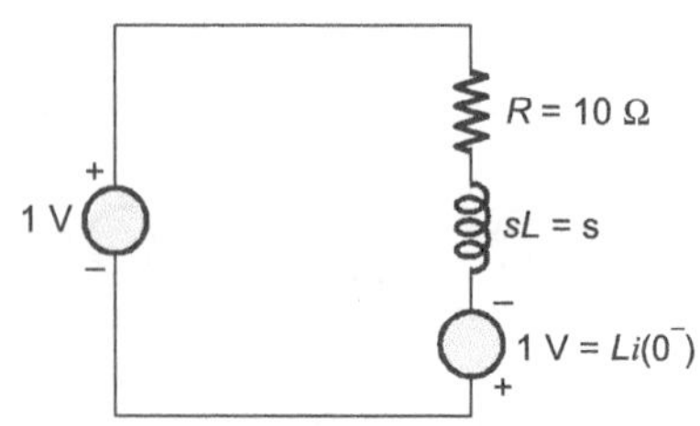

$$\Rightarrow \quad I(s) = \frac{(1+1)}{10+s} \Rightarrow i(t) = 2e^{-10t}$$

3.16 (c)

At steady state, capacitor acts as OC and at initial state it acts as SC to dc current.

3.17 (b)

At $t = 0$

Inductor acts as open circuit, so $i_1 = 0$

Capacitor acts as short circuit, so $i_2 = I$

3.18 (d)

$$i(0^+) = i(0^-) = \frac{150}{50} = 3\text{ A}$$

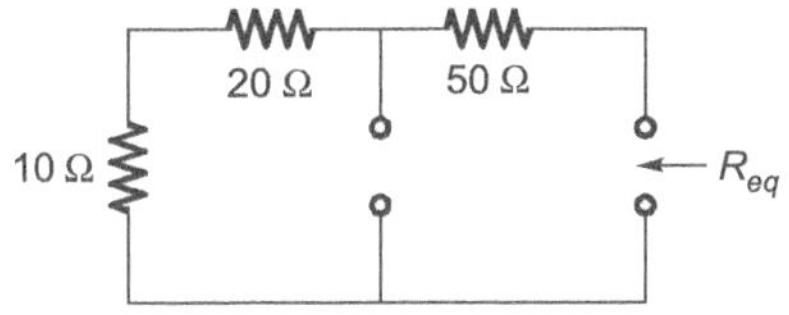

$$R_{eq} = 80\ \Omega$$

$$\tau = \frac{L}{R} = \frac{0.5}{80} = \frac{1}{160}$$

$$i(t) = i(0^-) \times e^{-t/\tau} = 3 \times e^{-160t}$$

3.19 (b)

$v(0^+) = 50$ V equal to battery emf at steady state when switch S is at b.

When switch S is at a. The voltage across capacitor does not change instantaneously.

$$\therefore \quad i(0^+) = \frac{50}{500} = 0.1\text{ A} = 100\text{ mA}$$

3.20 (b)

$$V(\infty) = 1\text{ V and } V(0^+) = 3\text{ V}$$

$$v(t) = V(\infty) + [V(0^+) - V(\infty)]e^{-t/\tau}$$

$$= 1 + [3 - 1]e^{-3t} = 1 + 2e^{-3t}$$

3.21 (d)

Equivalent circuit at $t < 0$

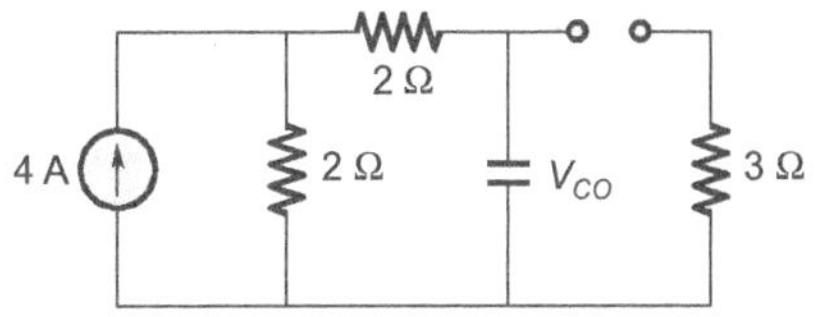

So, $V_{CO} = 4 \times 2$ volt

Capacitor will behave as open circuit at steady state. Equivalent circuit at $t > 0$,

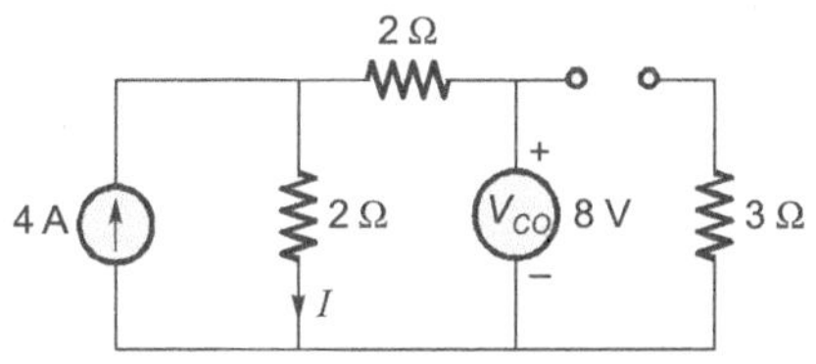

Using superposition

$$I = \frac{4\times 2}{2+2} + \frac{8}{(2+2)} = 4\text{ A}$$

3.22 (c)

Equivalent circuit for $t < 0$.

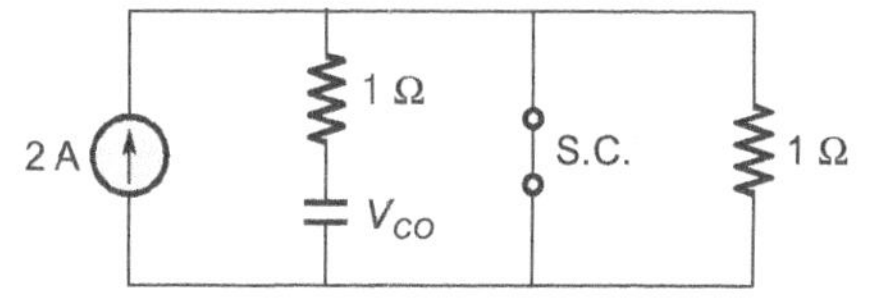

As capacitor will be open in steady state

$\Rightarrow \quad V_{CO} = 2 \times 1$ volt $= 2$ volt

equivalent circuit $t > 0$

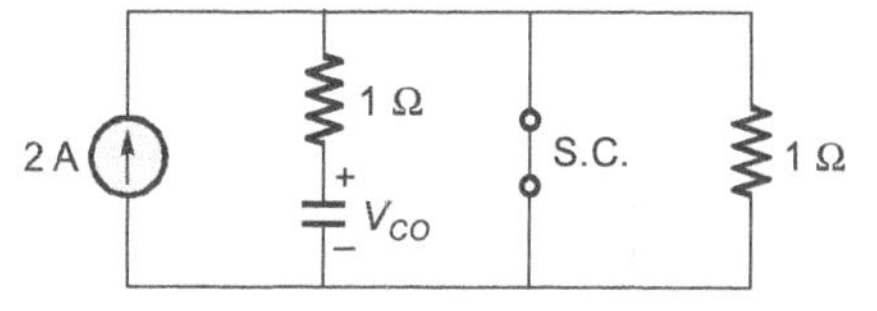

At $t = \infty$ capacitor will be again open. So, $V_{C(\infty)} = 0$ as capacitor is across a short circuit.

3.23 (b)

At $t = 0^-$

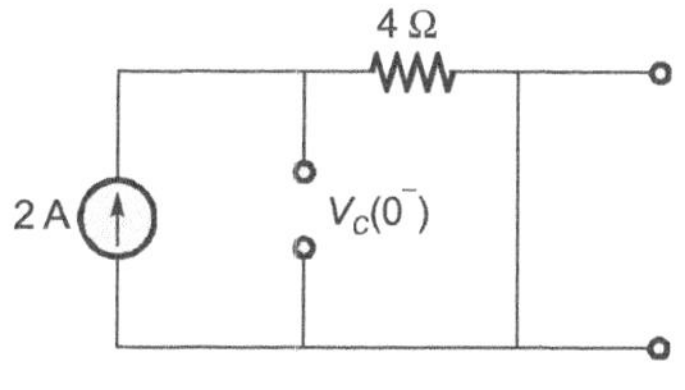

$$V_C(0^-) = 8\text{ V}$$

At $t = 0^+$

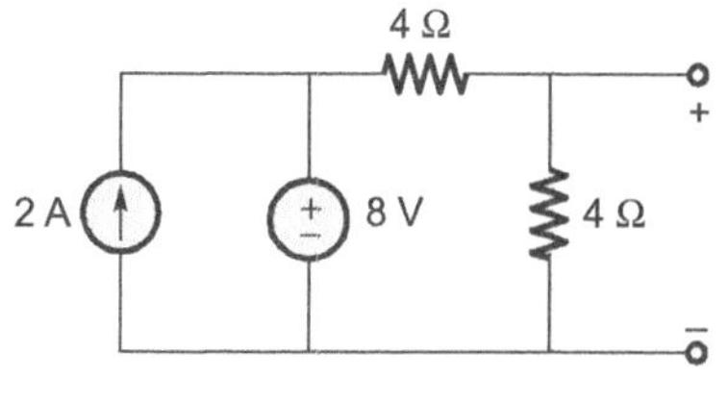

$$V(0^+) = \frac{8\times 4}{8}$$

$$= 4\text{ V}$$

At $t = \infty$

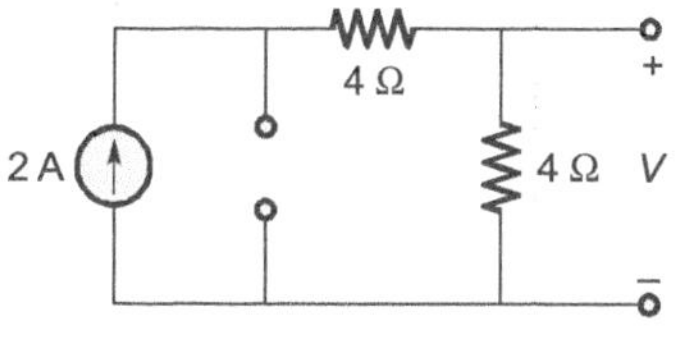

$$V(\infty) = 8\text{ V}$$

3.24 (b)

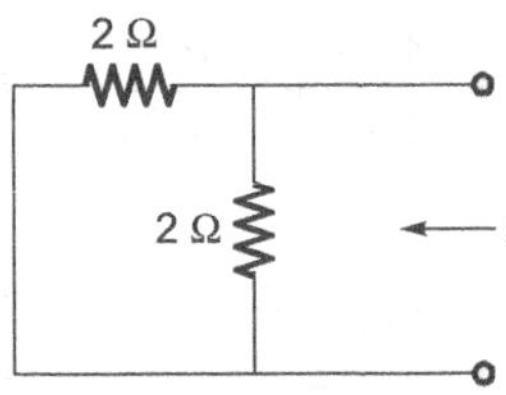

$R_{eq} = (2 \parallel 2) = 1\ \Omega$

Time constant $= R_{eq} \times C$

$= 1 \times 1 = 1$ s

3.25 (a)

Applying nodal analysis

$$\frac{3-V}{4} = \frac{V}{1} + \frac{CdV}{dt}$$

$$\frac{3-1}{4} = 1 + C(-10)$$

$$\frac{1}{2} - 1 = -10\,C$$

$$\boxed{C = \frac{1}{20} = 0.05\,\text{F}}$$

3.26 (c)

To have zero transient instantaneous angle of source should be equal to impedance angle.

3.27 (a)

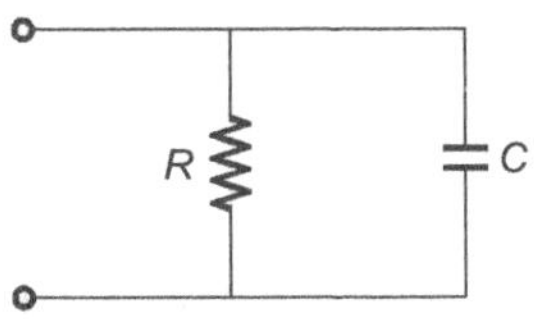

$$Q = \frac{2\pi \times \text{Max. Energy stored}}{\text{Energy dissipated per cycle}}$$

$$= \frac{2\pi \times \frac{1}{2}CV^2}{\left(\frac{V}{R\sqrt{2}}\right)^2 \times R \times \frac{1}{f}} = 2\pi fCR = \omega CR$$

Here, $V_{max.} = V$ So, $V_{rms} = \frac{V}{\sqrt{2}}$

As for capacitor we require V_{max} and for resistance V_{rms}.

3.28 (c)

When system is initially at rest: Inductor will behave as open circuit and capacitor act as short circuit

$$\therefore \quad I = \frac{50}{5} = 10\text{ A}$$

3.29 (b)

Before closing the switch, $t = 0^-$

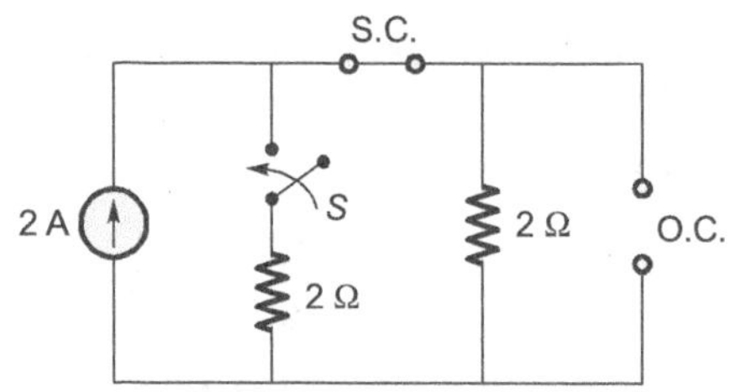

Inductor behaves as short circuit and capacitor behaves as open circuit.

So, $i_L(0^-) = 2$ A and $V(0^-) = 4$ V

$i_L(0^+) = i_L(0^-) = 2$ A

and $V(0^+) = V(0^-) = 4$ V

after closing the switch, $t = 0^+$

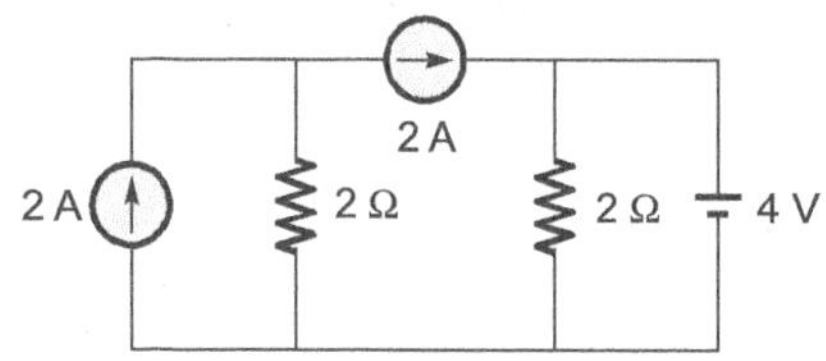

Applying KCL,

$$-2 + \frac{4}{2} + C\frac{dV}{dt} = 0$$

$$\frac{dV}{dt} = 0$$

3.30 (d)

In steady state, circuit looks like

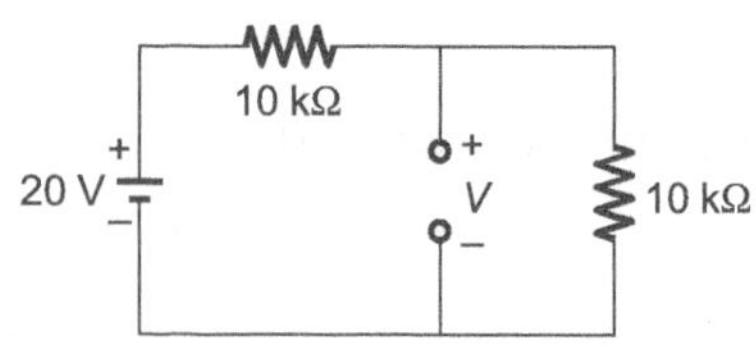

$$V = \left(\frac{10}{10+10}\right) \times 20 = 10\text{ V}$$

Time constant $= R_{eq} \times C$

$= (10\,k \parallel 10\,k) \times 20 \times 10^{-6}$

$= 5 \times 10^3 \times 20 \times 10^{-6}$

$= 0.1$ sec

3.31 (c)

For $V_0 \propto \frac{dv_i}{dt} \Rightarrow V_0(s) = CsV_i(s)$

where C is any constant

from the circuit $V_0(s) = \frac{sRC}{1+sRC} V_i(s)$

for the required condition $RC << T$

3.32 (b, c)

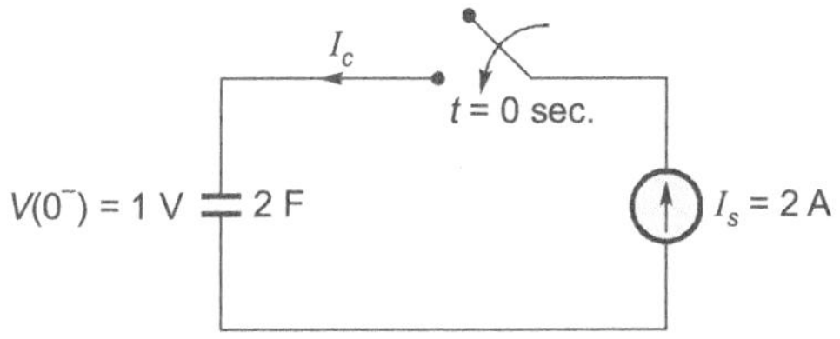

$$I_c = I_s$$

$$\because \quad C\frac{dV_c}{dt} = I_s \Rightarrow \frac{dV_c}{dt} = 1$$

i.e. capacitor will charge at 1 V per sec.

$\Rightarrow$ after 3 sec.

$$V_{c(t=3\text{ sec})} = V_{c(t=0)} + 1 \times 3 = 4 \text{ V}$$

$$\Rightarrow \quad Q = CV = 8\ C$$

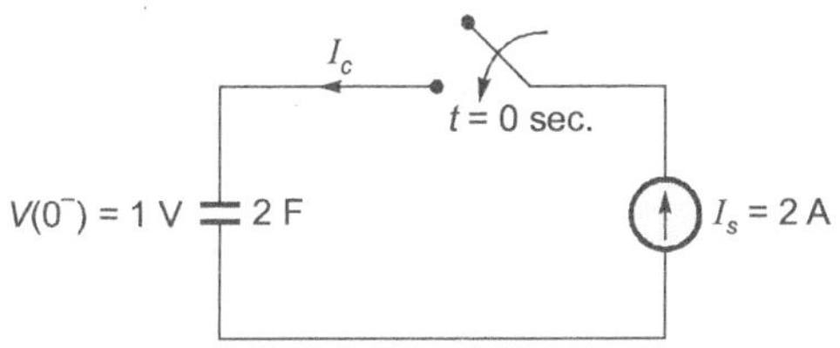

$$I_c = -I_s$$

$$\frac{CdV_c}{dt} = -I_s \Rightarrow \frac{dV_c}{dt} = -1$$

i.e. capacitor will discharge at 1 V per sec.

$\Rightarrow$ after 3 sec.

$$V_{c(t=3\text{ sec})} = V_{c(t=0)} + (-1) \times 3 = -2 \text{ V}$$

$$\Rightarrow \quad Q = CV = -4C$$

$\because$ Negative sign indicate only polarity of charging.

$\Rightarrow$ Magnitude of charge accumulated = $4C$.

3.33 (d)

At $t = 0^-$

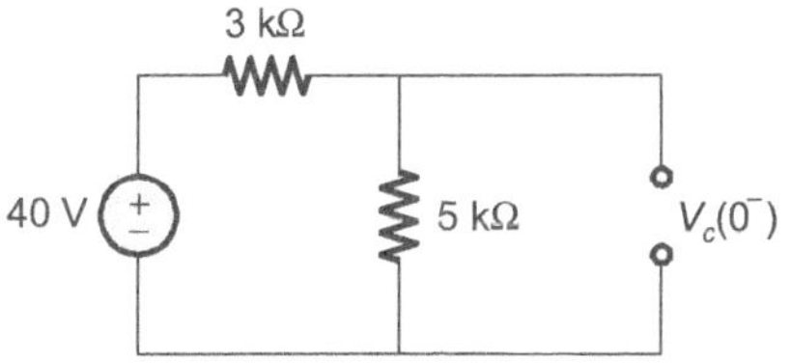

$$V_C(0^-) = \frac{5 \times 40}{8} = 25 \text{ V}$$

At $t = \infty$

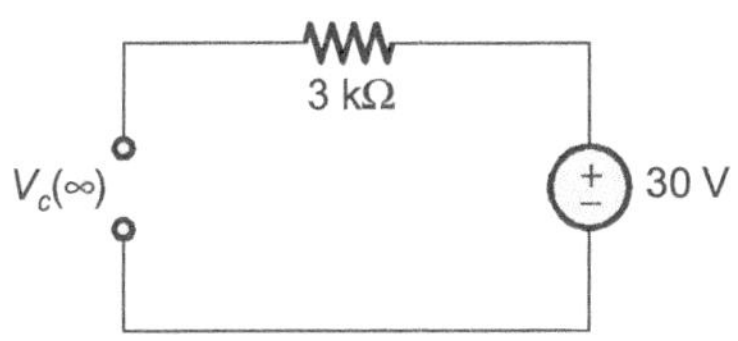

$$\tau = RC = 2 \text{ sec}$$

$$V_C(\infty) = 30 \text{ V}$$

So,
$$V_C(t) = V(\infty) - \{V(\infty) - V(0^-)\}e^{-t/\tau}$$
$$= 30 - (30 - 25)\ e^{-5t}$$
$$= 30 - 5e^{-0.5t}$$

3.34 (a)

At $t = 0^+$

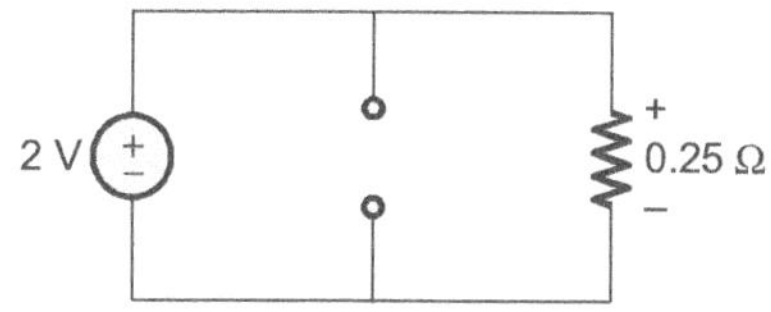

$$V = 2 \text{ V}$$

$$I = -\frac{cdV}{dt} = \frac{2}{0.25}$$

$$\Rightarrow \quad \frac{dV}{dt} = -4 \text{ V/sec}$$

3.35 (d)

$$V_c = V_c(0)e^{-t/R_{eq}C}$$

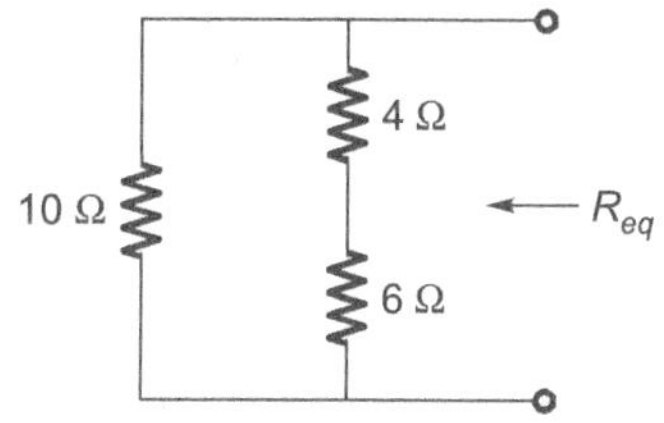

$$R_{eq} = \frac{10 \times 10}{10 + 10} = 5\ \Omega$$

$$V_c(t) = 25\ e^{-t/0.5} = 25\ e^{-2t} \text{ V}$$

Hence, option (d) is correct.

3.36 (a)

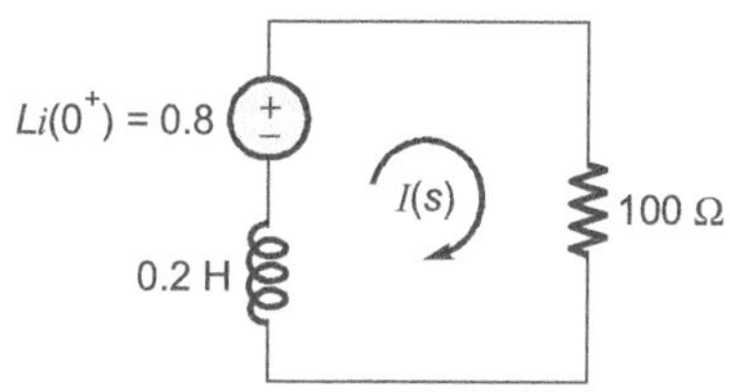

$$I(s) = \frac{0.8}{100 + 0.2s} = \frac{4}{8 + 500}$$

$$I(t) = \mathcal{L}^{-1}[I(s)] = \mathcal{L}^{-1}\left[\frac{4}{s+200}\right] = 4\ e^{-500t}$$

At $t = 0.8$ ms

$$i(t)\big|_{t=0.8\text{ms}} = 4\ e^{-500 \times 0.8 \times 10^{-3}} = 4\ e^{-0.4} \text{ A}$$

Hence, option (c) is correct.

3.37 (c)

$$V(0^-) = V(0^+) = 0$$

Applying KCL,

$$-10+\frac{V}{1000}+C\frac{dV}{dt}=0$$

$$-10+\frac{V(0^+)}{1000}+C\left.\frac{dV}{dt}\right|_{t=0^+}=0$$

$$C\left.\frac{dV}{dt}\right|_{t=0^+}=10$$

$$\left.\frac{dV}{dt}\right|_{t=0^+}=10 \text{ V/sec.}$$

Hence, option (c) is correct.

3.38 (b)

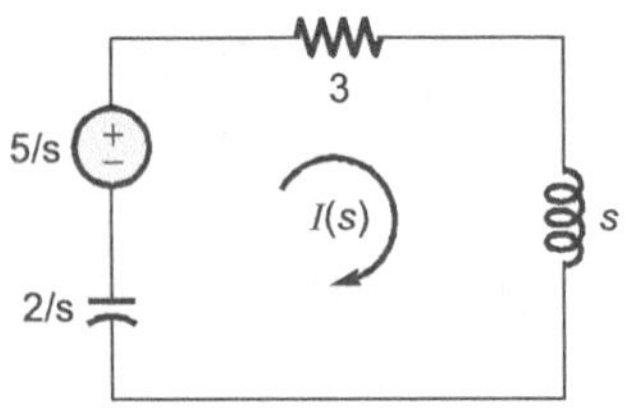

$$I(s) = \frac{5/s}{s+3+2/s}$$

$$= \frac{5}{s^2+3s+2}$$

$$s^2+2\xi\omega_n s+\omega_n^2 \equiv s^2+3s+2$$

$$\omega_n = \sqrt{2}$$

$$2\xi\omega_n = 3$$

$$\xi = \frac{3}{2\sqrt{2}} = 1.06$$

$\because$ $\xi > 1 \Rightarrow$ overdamped response.
Hence, option (b) is correct.

3.39 (c)

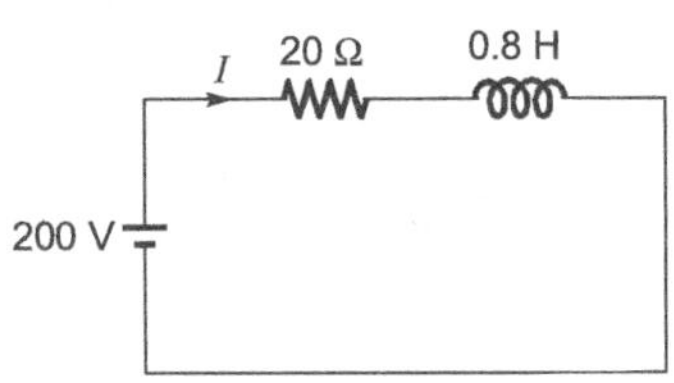

$$200 = 20i + 0.8\frac{di}{dt}$$

$$0.8\frac{di}{dt} = 200 - 20i$$

$$\left.\frac{di}{dt}\right|_{t=0^+} = \frac{1}{0.8}[200-0] = 250 \text{ A/sec}$$

3.40 (c)

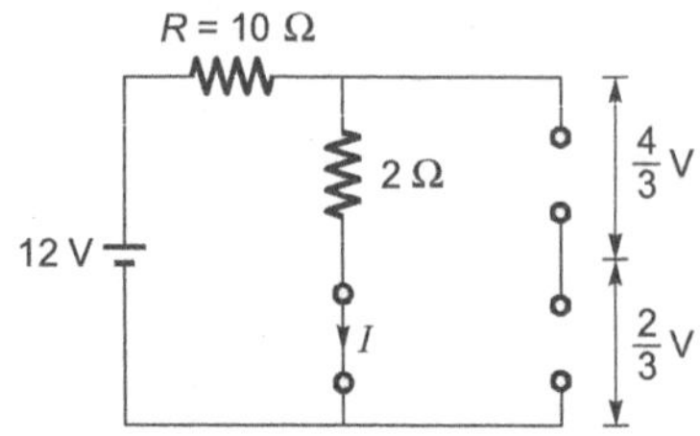

$$\text{Current } I = \frac{12}{10+2} = 1 \text{ A}$$

$\therefore$ $I_L(0^-) = I_L(0^+) = 1$ A

Voltage across 3 F capacitor (at $t = 0^-$)

$$= \frac{4}{3}\text{V}$$

Voltage across 6F capacitor (at $t = 0^-$)

$$= \frac{2}{3}\text{V}$$

By applying KVL in outer loop

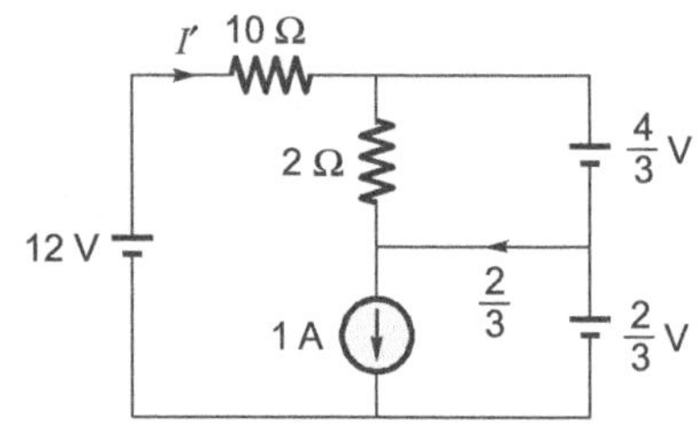

$$12 = 10I' + 2$$

$$I' = 1 \text{ A}$$

3.41 (a)

$$V_L = L\frac{di}{dt}$$

Under steady-state condition current through inductor does not change with time.

$\therefore$ $V_L = 0$

and inductor acts as short circuit.

3.42 (a)

Since, $I_C = c\dfrac{dV}{dt}$

under steady-state condition voltage through capacitor does not change with time thus,

$$I_C = 0$$

and capacitor acts as open circuit.

3.44 (c)

$$0.8\frac{di}{dt} = 200 \Rightarrow \frac{di}{dt} = 250 \text{ A/sec.}$$

3.45 (c)

When the network is having only resistive elements no transients are present in the network since.

(*i*) Resistor allows sudden change of current and voltage.

(*ii*) Resistor does not store any energy.

Transients are present when the circuit is having any energy storage element.

3.46 (b)

$$f(t) = 15 - 10t - 10e^{-20t}$$

Initial value,

$$\lim_{t\to 0} f(t) = 15 - 10(0) - 10e^{-20(0)} = 15 - 10 = 5$$

Final value,

$$\lim_{t\to\infty} f(t) = 15 - 10(\infty) - 10e^{-20(\infty)} = -\infty$$

3.47 (c)

For given circuit diagram,

$$R_{eq} = R + 2R = 3R$$

$$C_{eq} = \frac{C\cdot 2C}{C+2C} = \frac{2}{3}C$$

Time constant

$$\Rightarrow \quad \tau = R_{eq}C_{eq} = 2RC$$

3.49 (d)

$$\lim_{t\to\infty} i(t) = \lim_{s\to 0} sI(s)$$

$$= \lim_{s\to 0} s\cdot\frac{\left(\frac{2V}{\pi} - \frac{q_0}{C}\right)}{R\left(s+\frac{1}{RC}\right)} = 0$$

3.50 (c)

As capacitor does not allow the sudden change in voltage

∴ for initial values we replace capacitor by a voltage source.

3.51 (c)

Total Response = Forced Response + Natural Response

$$i(t) = i(\infty) + [i(0^+) - i(\infty)]\, e^{-t/\tau}$$

$$v(t) = v(\infty) + [v(0^+) - v(\infty)\, e^{-t/\tau}$$

So it can be either in the form

$K_1 \cdot e^{-at}$
↓
if final value is zero

or $K + K_1\, e^{-at}$

3.52 (a)

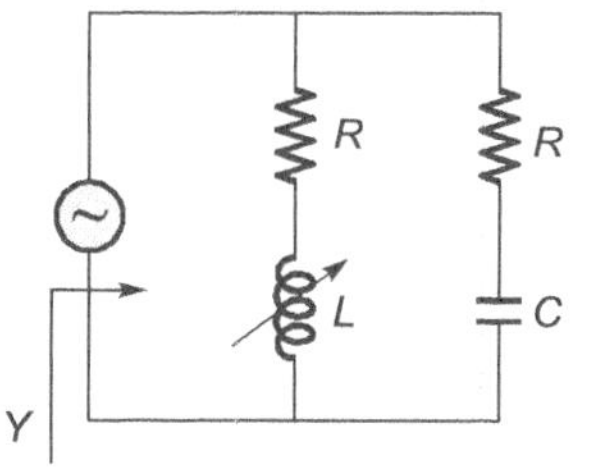

The admittance of the circuit is,

$$Y = \frac{1}{R+jX_L}\times\frac{R-jX_L}{R-jX_L} + \frac{1}{R-jX_C}\times\frac{R+jX_C}{R+jX_C}$$

$$Y = \frac{R-jX_L}{R^2+X_L^2} + \frac{R+jX_C}{R^2+X_C^2}$$

For the resonance imaginary part of admittance is equal to zero.

$$\frac{-X_L}{R^2+X_L^2} + \frac{X_C}{R^2+X_C^2} = 0$$

$$\frac{X_L}{R^2+X_L^2} = \frac{X_C}{R^2+X_C^2}$$

$$X_L^2 + R^2 = \frac{1}{X_C}[X_L R^2 + X_L X_C^2]$$

$$X_L^2 - \frac{X_L}{X_C}[R^2 + X_C^2] + R^2 = 0$$

The equation is quadratic, for the two resonances, it requires two difference values of inductor.

3.53 (b)

In parallel resonant circuit,

$$Q = \frac{\omega_0}{\text{Bandwidth}}$$

$$\omega_0 = \frac{1}{\sqrt{LC}}$$

$$\text{Bandwidth} = \frac{1}{RC}$$

$$\Rightarrow \quad Q = R\sqrt{\frac{C}{L}}$$

Since Q of circuit is high so R will be high, when we connect a resistance in parallel then total resistance of circuit will reduce thus, Q will reduce hence bandwidth will increase and resonant frequency remain same. Thus, only statement-2 is correct.

3.54 (d)

As capacitor does not allow the sudden change in voltage therefore it will hold its previous value.

$$\therefore \quad V_{out} = 0 \text{ V}$$

3.55 (b)

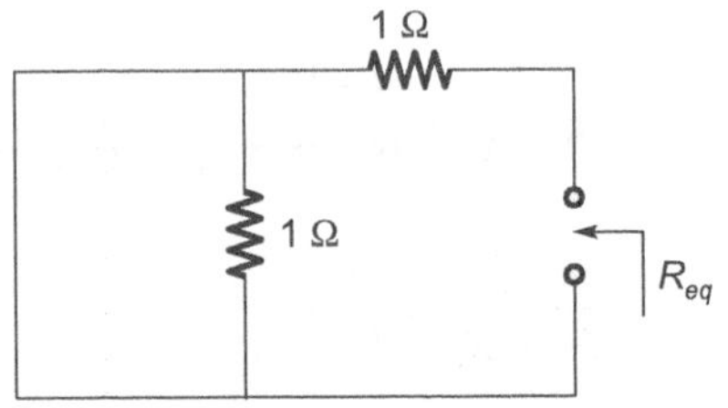

$$\tau = R_{eq} \cdot C$$

R_{eq} :

$$R_{eq} = 1\ \Omega$$
$$\tau = (1)(1)$$
$$\tau = 1 \text{ sec.}$$

3.56 (b)

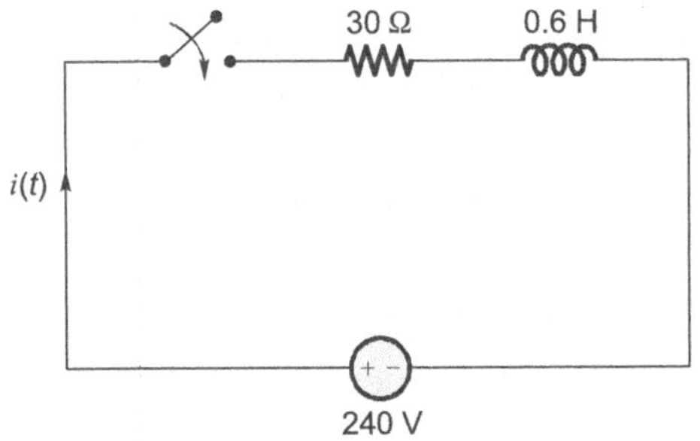

By applying KVL in loop :

$$240 = 30i(t) + 0.6\frac{di(t)}{dt}$$

$t = 0^+$, $i(0^+) = 0$ A (because inductor cannot change its current suddenly)

$$240 = 30i(0^+) + 0.6\frac{di(0^+)}{dt}$$

$$\frac{di(0^+)}{dt} = \frac{240}{0.6} \Rightarrow 400 \text{ A/s}$$

At steady state inductor acts as a short circuit.

$$i(\infty) = \frac{V}{R} = \frac{240}{30} = 8 \text{ A}$$

3.57 (b)

When angular frequency is zero,

$$\omega = 0$$

$\therefore \quad X_L = \omega L = 0$ or inductor behaves as short circuit

Similarly, $X_C = \dfrac{1}{\omega C} = \dfrac{1}{0} = \infty$ capacitor behaves as open circuit.

Option (b) is correct

3.58 (b)

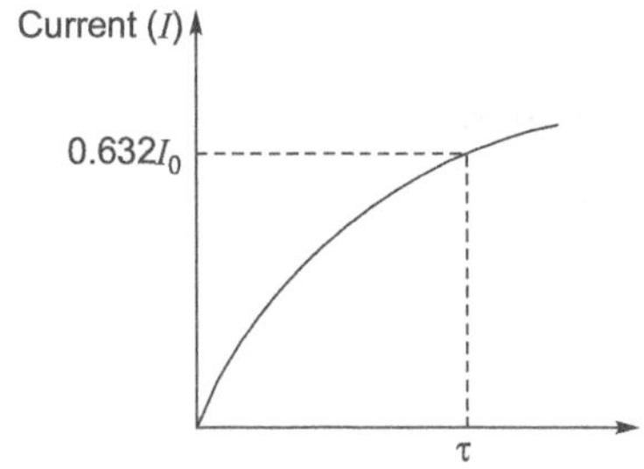

where, τ = time constant

I_0 is the steady state value or final value.

$\therefore$ Time constant of R-L series circuit, τ is the time when current through R-L series circuit rises to 63.2% of steady state value.

$\therefore$ (b) option is correct.

3.59 (b)

Transient disturbance is produced in circuit whenever circuit is suddenly connected to or disconnected from the supply.

$\therefore$ (b) option is correct.

3.60 (d)

The initial voltage,

$$V_c = 100 \text{ V}$$

Time constant,

$$\tau = RC$$
$$= 10 \times 10^3 \times 0.005 \times 10^{-6}$$
$$\tau = 50\ \mu\text{sec}$$

The final voltage,

$$V_c(t_o) = 5 \text{ V}$$

The % change in voltage

$$= \frac{100-5}{100} = 95\%$$

So, $\quad t_o = 3\tau = 3 \times 50 = 150\ \mu$sec

4 Resonance

4.1 The value of the capacitance 'C' in the given ac circuit to make it a constant resistance circuit OR for the supply current to be independent of its frequency is

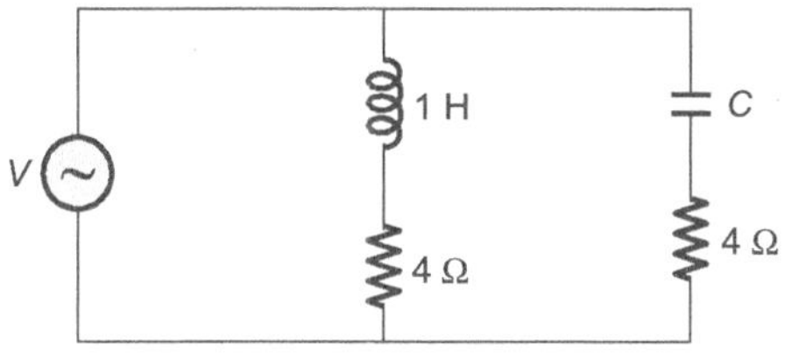

(a) 1/16 F (b) 1/12 F
(c) 1/8 F (d) 1/4 F

[ESE-2001]

4.2 For the circuit shown in the given figure, the current through L and the voltage across C_2 are respectively

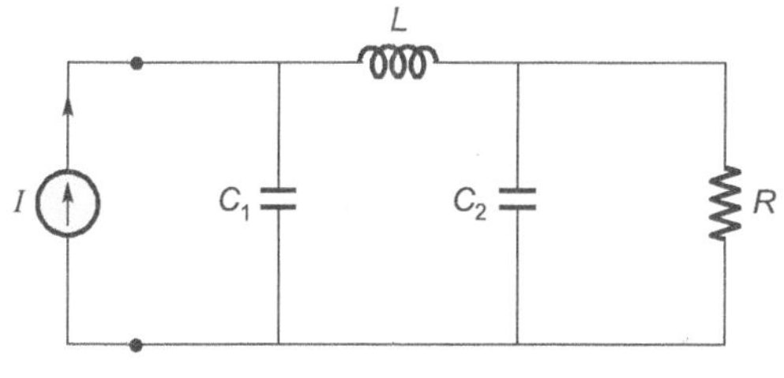

(a) Zero and RI (b) I and zero
(c) zero and zero (d) I and RI

[ESE-2001]

4.3 A parallel circuit consists of two branches. One branch has R_L and L connected in series and the other branch has R_C and C connected in series. Consider the following statements:

1. The two branch currents will be in quadrature if $R_L R_C = L/C$.
2. The impedance of the whole circuit is independent of frequency, if $R_L = R_C$ and $\omega = 1/\sqrt{LC}$.
3. The circuit is in resonance for all the frequencies if $R_L = R_C$.
4. The two branch currents will be in phase at $\omega = 1/\sqrt{LC}$.

Which of the above statements are correct?

(a) 1 and 2 (b) 2 and 3
(c) 1 and 3 (d) 3 and 4 [ESE-2002]

4.4 At resonant frequency a R-L-C series circuit draws maximum current due to the reason that,

(a) the difference between capacitive reactance and inductive reactance is zero.
(b) The impedance is more than the resistance.
(c) the voltage across the capacitor equals the applied voltage.
(d) the power factor is less than unity.

[ESE-2002]

4.5 In the given circuit, at resonance, I_R in Amperes is equal to

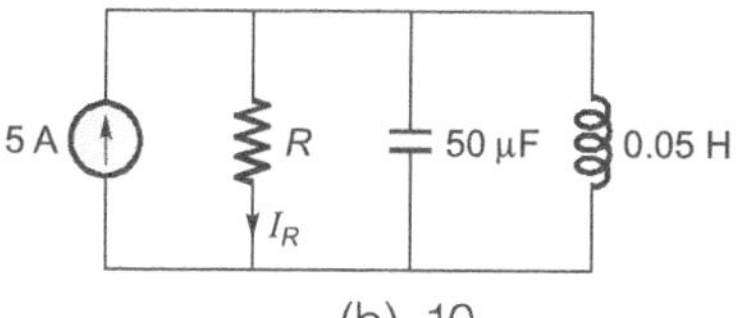

(a) 0 (b) 10
(c) 5 (d) 0.5 [ESE-2002]

4.6 Consider the following statements about the quality factor of a R-L-C circuit:

1. For the critically damped circuit, the quality factor $Q = 1/2$.
2. Higher the value of quality factor higher will be the bandwidth of the circuit.
3. Higher the value of quality factor lower will be the bandwidth of the circuit.
4. For underdamped circuits the value of Q is greater than 1/2.

Which of these statements are correct?

(a) 1 and 2 (b) 1 and 3
(c) 2 and 4 (d) 1, 3 and 4

[ESE-2002]

4.7 **Assertion (A):** In a series R-L-C circuit, the current is minimum at resonant frequency

Reason (R): The maximum voltage across the capacitor occurs at a frequency lower than the resonant frequency.

(a) Both A and R are true and R is the correct explanation of A
(b) Both A and R are true but R is NOT the correct explanation of A
(c) A is true but R is false
(d) A is false but R is true

[ESE-2003]

4.8 For a parallel RLC resonant circuit, the damped frequency is $\sqrt{8}$ r/s and bandwidth is 2 r/s. What is its resonant frequency?

(a) 2 (b) $\sqrt{7}$
(c) $\sqrt{10}$ (d) 3

[ESE-2004]

4.9 Consider the following circuit:

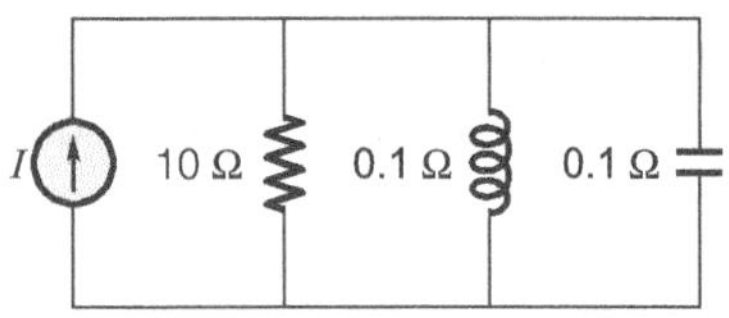

For the above circuit, the current source is sinusoidal with frequency equal to the resonant frequency of the circuit. What is the value of current through the resistor?

(a) 0 (b) $0.1I$
(c) I (d) $10I$

[ESE-2004]

4.10 In an a.c. series RLC circuit, the voltage across R and L is 20 V, voltage across L and C is 9 V and voltage across RLC is 15 V. What is the voltage across C?

(a) 7 V (b) 12 V
(c) 16 V (d) 21 V

[ESE-2004]

4.11 A circuit has two parallel branches. In one branch, R and L are connected in series while in the other, R and C are connected in series. If $R = \sqrt{L/C}$, which one of the following is not correct?

(a) The circuit is in resonance.
(b) The two branch currents are in quadrature.
(c) The circuit has an impedance independent of its frequency.
(d) The two branch currents are in phase.

[ESE-2005]

4.12 For a series RLC resonant circuit, what is the total reactance at the lower half power frequency?

(a) $\sqrt{2}R\angle 45°$ (b) $\sqrt{2}R\angle -45°$
(c) R (d) $-R$

[ESE-2005]

4.13 In a series R-L-C circuit, the maximum voltage across the capacitor occurs at a frequency

(a) double the resonant frequency
(b) equal to resonant frequency
(c) $\sqrt{2}$ times the resonant frequency
(d) below the resonant frequency

[ESE-2006]

4.14 A series R-L-C circuit, excited by a 100 V, variable frequency source has a resistance of 10 Ω and an inductive reactance of 50 Ω at 100 Hz. If the resonance frequency is 500 Hz, what is the voltage across the capacitor at resonance?

(a) 100 V (b) 500 V
(c) 2500 V (d) 5000 V

[ESE-2006]

4.15 Consider the following statements:

If a high-Q parallel resonant circuit is loaded, then

1. the circuit impedance reduces.
2. the resonant frequency reduces.
3. the bandwidth reduces.

Which of the statements given above is/are correct?

(a) 1 and 2 (b) 1 only
(c) 2 and 3 (d) 1, 2 and 3

[ESE-2007]

4.16 For a series RLC resonant circuit, which one of the following gives the impedances at the lower and upper half power frequencies, respectively?

(a) $\sqrt{2}R\angle 45°$; $\sqrt{2}R\angle -45°$
(b) $\left(\frac{1}{\sqrt{2}}\right)R\angle -45°$; $\left(\frac{1}{\sqrt{2}}\right)R\angle 45°$
(c) $R\angle 45°$; $R\angle -45°$
(d) $\sqrt{2}R\angle -45°$; $\sqrt{2}R\angle 45°$

[ESE-2007]

4.17 When is a series RLC circuit over-damped?

(a) $\frac{R^2}{4L^2} = \frac{1}{LC}$ (b) $\frac{R^2}{4L} < \frac{1}{C}$

(c) $\frac{R^2}{4L} > \frac{1}{C}$ (d) $\frac{R^2}{4C^2} = \frac{1}{LC}$

[ESE-2007]

4.18 In an RLC series resonant circuit, if the maximum stored energy is increased by 10% and at the same time the energy dissipated per cycle is reduced by 10%, it will result in which one of the following?

(a) An 11% decreases in quality factor
(b) An increase in the resonant frequency by 11%
(c) A 22% increases in quality factor
(d) A decrease in the resonant frequency by 22%

[ESE-2009]

4.19 For a parallel RLC circuit, if $R = 40\ \Omega$, $L = 2$ H and $C = 0.5$ F, the bandwidth and quality factor are respectively

(a) 20 rad/s, 0.05 (b) 10 rad/s, 20
(c) 20 rad/s, 10 (d) 0.05 rad/s, 20

[ESE-2010]

4.20 If a series RLC circuit resonates at 1.5 kHz and consumes 100 watts from a 100 volts AC source operating at resonant frequency with a bandwidth of 0.75 kHz; the values of R, L and Q-factor of the circuit are respectively

(a) $100\ \Omega$; $\frac{0.2}{3\pi}$ H; 2 (b) $50\ \Omega$; $\frac{0.1}{2\pi}$ H; 4

(c) $100\ \Omega$; $\frac{2}{3\pi}$ H; 1 (d) $50\ \Omega$; $\frac{0.3}{2\pi}$ H; 3

[ESE-2010]

4.21 The resonant frequency of an RLC series circuit is 1.5 MHz with the resonating capacitor of 150 pF. The bandwidth is 10 kHz. The effective value of the resistor is

(a) $16.3\ \Omega$ (b) $9.5\ \Omega$
(c) $7.4\ \Omega$ (d) $4.7\ \Omega$ [ESE-2010]

4.22 For a series RLC circuit energized with a sinusoidal voltage source of frequency 4 rad/s, the applied voltage lags the current by an angle of $\tan^{-1} 2$ degrees. Then the value of R for $L = 1$ H and $C = 0.05$ F is

(a) 4.0 ohm (b) 2.0 ohm
(c) 1.0 ohm (d) 0.5 ohm

[ESE-2010]

4.23 Consider the following statements:
A step voltage is applied to an under-damped series RLC circuit in which R is variable. If R is

1. increased, the steady-state voltage across C will be reduced.
2. increased, the frequency of transient oscillations across C will be reduced.
3. reduced, the transient oscillations will die down at a fast rate.
4. reduced to zero, the peak amplitude of the voltage across C will be double that of input voltage.

Which of these statements is/are correct?

(a) 2 only (b) 4 only
(c) 2 and 4 only (d) 1, 2, 3 and 4

[ESE-2010]

4.24 For a parallel resonant circuit, if the damped frequency is $\sqrt{8}$ rad/s and the bandwidth is 2 rad/s, the resonant frequency of the circuit is

(a) 10 rad/s (b) 7 rad/s
(c) 3 rad/s (d) 2 rad/s

[ESE-2010]

4.25 For the network function $T(s) = \frac{s}{s^2 + 2s + 100}$, the resonant frequency and bandwidth are respectively

(a) 10, 1 (b) 10, 2
(c) 100, 1 (d) 100, 2 [ESE-2010]

4.26 For the parallel RLC circuit shown, if $R = 10\ \Omega$, $L = 0.1$ H and $C = 0.1$ F, the current I is sinusoidal of frequency equal to the resonant frequency of the circuit, then the current through R is

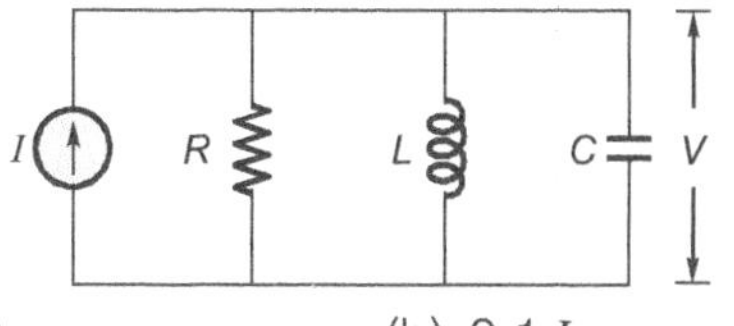

(a) 0 (b) $0.1I$
(c) I (d) $10I$

[ESE-2011]

4.27 **Assertion (A):** A network is said to be in resonance when the voltage and current at the network input terminals are in phase.

Reason (R): In a two-terminal network containing at least one inductor and one capacitor, the resonance is defined as the condition which exists when the input impedance of the network is purely resistive.

(a) Both A and R are individually true and R is the correct explanation of A
(b) Both A and R are individually true but R is not the correct explanation of A
(c) A is true but R is false
(d) A is false but R is true

[ESE-2011]

4.28 A series L-C-R circuit has a resonant frequency f_0, with $R = 1\ \Omega$, $L = 1$ H and $C = 1$ F. If the components' values are tripled, the new resonant frequency will be

(a) $3f_0$ (b) unaltered

(c) $\frac{f_0}{\sqrt{3}}$ (d) $\frac{f_0}{3}$

[ESE-2012]

4.29 In a series resonance circuit, at resonance, selectivity Q is equal to

(a) $\frac{1}{R\sqrt{LC}}$ (b) $\frac{1}{R}\sqrt{\frac{C}{L}}$

(c) $\frac{1}{R}\sqrt{\frac{L}{C}}$ (d) $\frac{1}{R}\sqrt{LC}$

[ESE-2012]

4.30 For an R-L-C series circuit in resonance, the following statement is not correct.

(a) The current is maximum.
(b) The voltage phasors across the capacitance and inductance are unequal.
(c) The voltage drop across the resistance is maximum.
(d) The voltage drops across the capacitance and inductance are unequal in magnitude.

[ESE-2012]

4.31 The circuit given below is constituted by an iron-cored coil and a capacitor. At resonance, the circuit behaves like

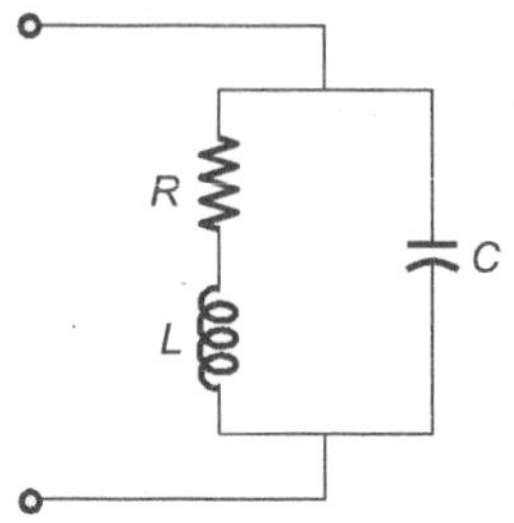

(a) an open circuit
(b) a short circuit
(c) a pure resistor of value R
(d) a pure resistor of value much higher than R

[ESE-2012]

4.32 The given network will act as an ideal current source with respect to terminals X and Y, if frequency f is

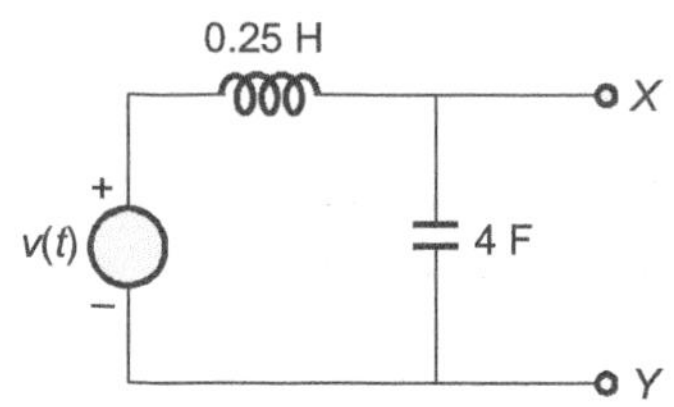

(a) $\frac{1}{\pi}$Hz (b) $\frac{1}{2\pi}$Hz

(c) $\frac{1}{4\pi}$Hz (d) $\frac{1}{3\pi}$Hz

[ESE-2012]

4.33 In a linear network, a 1 Ω resistor consumes a power of 4 W when voltage source of 4 V is applied to the entire circuit, and 16 W when the voltage source is replaced by an 8 V source. The power consumed by the 1 Ω resistor when 12 V is applied will be

(a) 0 W (b) 20 W
(c) 36 W (d) 144 W

[ESE-2012]

4.34 A series RLC circuit has a bandwidth of 300 rad/sec at a resonant frequency of 3000 rad/sec when exited by a voltage source of 100 V. The inductance of the coil is 0.1 H. The value of R and the voltage across C are, respectively

(a) 10 Ω and 100 V (b) 30 Ω and 100 V
(c) 30 Ω and 1000 V (d) 300 Ω and 1000 V

[ESE-2013]

4.35 The circuit comprises a coil of resistance R and inductance L, in parallel with an ideal capacitor C. At the resonant frequency, the impedance of the parallel combination is
(a) R (b) LC/R
(c) L/RC (d) ∞ [ESE-2013]

4.36 In RLC circuits, the current at resonance is
(a) maximum in series RLC and minimum in parallel RLC circuit
(b) maximum in parallel circuit and minimum in series circuit
(c) maximum in both circuits
(d) minimum in both circuits [ESE-2013]

4.37 Elements R, L and C are connected in parallel. The impedance of the parallel combination can be expressed as

$$Z(s) = \frac{10s}{s^2 + s + 400}$$

The value of the individual elements R, L and C are
(a) 10 Ω, 40 H and 0.1 F
(b) 4 Ω, 1 H and 0.1 F
(c) 10 Ω, $\frac{1}{40}$ H and 0.1 F
(d) 1 Ω, 40 H and 10 F [ESE-2013]

4.38 **Statement (I):** During resonance, an R-L-C series circuit behaves like a purely resistive circuit.
Statement (II): During resonance in an R-L-C series circuit, the voltages across the L and the C elements are in phase to each other.
Codes:
(a) Both Statement (I) and Statement (II) are individually true and Statement (II) is the correct explanation of Statement (I).
(b) Both Statement (I) and Statement (II) are individually true but Statement (II) is not the correct explanation of Statement (I).
(c) Statement (I) is true but Statement (II) is false.
(d) Statement (I) is false but Statement (II) is true.
[ESE-2014]

4.39 An RLC resonant circuit has a resonance frequency of 1.5 MHz and bandwidth of 10 kHz. If C = 150 pF, then effective resistance of the circuit will be
(a) 29.5 Ω (b) 14.75 Ω
(c) 9.5 Ω (d) 4.7 Ω [ESE-2014]

4.40 In a series resonant circuit, maximum voltage across L occurs at
(a) Resonant frequency
(b) Slightly below resonant frequency
(c) Slightly above resonant frequency
(d) At a frequency where I is maximum
[ESE-2014]

4.41 If the Q-factor of a coil at resonant frequency of 1·5 MHz is 150 for a series resonant circuit, then the corresponding bandwidth is
(a) 225 MHz (b) 1.06 MHz
(c) 50 kHz (d) 10 kHz [ESE-2015]

4.42 The low-frequency circuit impedance and the high-frequency circuit impedance for a series resonant circuit respectively are
(a) capacitive and inductive
(b) inductive and capacitive
(c) resistive and inductive
(d) capacitive and resistive [ESE-2015]

4.43 **Statement (I) :** In an R-L-C series circuit, excited from a variable frequency voltage source, the circuit behaves like a 'resistive' one at a particular frequency.
Statement (II) : The frequency at which an R-L-C series circuit becomes 'resistive' in character, a part of the input energy oscillates between the inductive and the capacitive elements of the circuit.
(a) Both Statement (I) and Statement (II) are individually true and Statement (II) is the correct explanation of Statement (I).
(b) Both Statement (I) and Statement (II) are individually true but Statement (II) is not the correct explanation of Statement (I).
(c) Statement (I) is true but Statement (II) is false.
(d) Statement (I) is false but Statement (II) is true.
[ESE-2015]

4.44 In a series R-L-C circuit supplied by a source of 125 V at a resonant frequency of 220 Hz, the magnitudes of the voltages across the capacitor and the inductor are found to be 4150 V. If the resistance of the circuit is 1 Ω, then the selectivity of the circuit is

(a) 33.20 (b) 3.32
(c) 0.0301 (d) 0.301 [ESE-2016]

4.45 A series *R-L-C* circuit is connected to a 25 V source of variable frequency. The circuit current is found to be a maximum of 0.5 A at a frequency of 400 Hz and the voltage across *C* is 150 V. Assuming ideal components, the values of *R* and *L* are respectively

(a) 50 Ω and 300 mH
(b) 12.5 Ω and 0.119 H
(c) 50 Ω and 0.119 H
(d) 12.5 Ω and 300 mH [ESE-2016]

4.46 The resonant frequency for the circuit

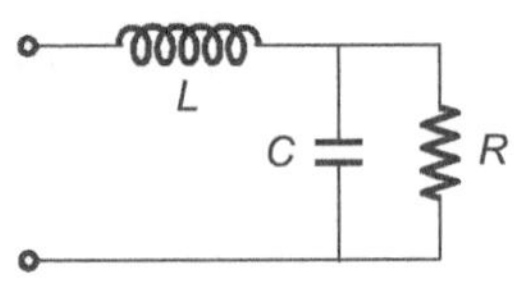

for $L = 0.2$ H, $R = 1$ Ω and $C = 1$ F, is

(a) 1 rad/s (b) 2 rad/s
(c) 3 rad/s (d) 4 rad/s [ESE-2016]

4.47 Two networks are connected in cascade in the figure. The equivalent *ABCD* constants are obtained for the combined network having $C = 0.1\angle 90°$.

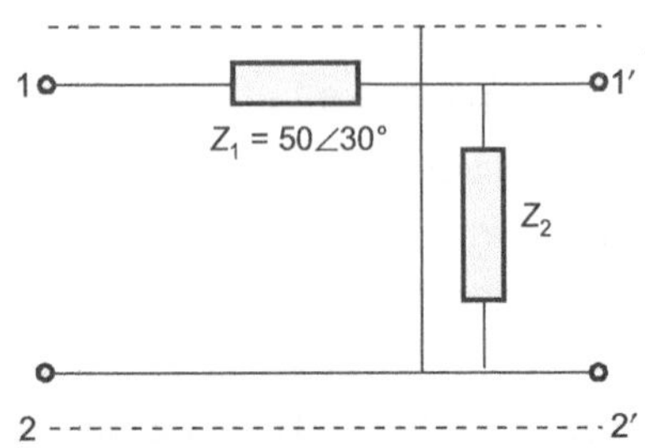

What is the value of Z_2?

(a) $500\angle -60°$ (b) $0.10j$
(c) $-10j$ (d) $50\angle -60°$ [ESE-2017]

4.48 A series RLC circuit has a resistance of 50 Ω, inductance of 0.4 H and a capacitor of 10 μF. The circuit is connected across a 100 V supply. The resonance frequency and the current through the resistance are

(a) 500 rad/s and 2 A
(b) 1000 rad/s and 2 A
(c) 500 rad/s and 0.5 A
(d) 1000 rad/s and 0.5 A [ESE-2018]

4.49 For the circuit as shown, what is the value of *C* that leads to maximum power transfer to the load, if the value of *L* is 0.5 H?

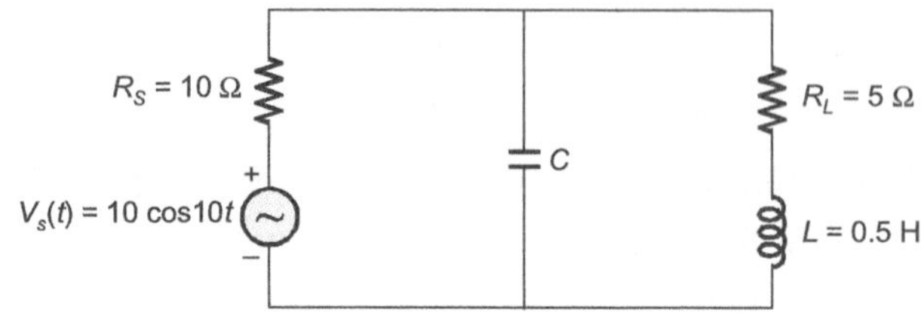

(a) 0.1 μF (b) 0.01 F
(c) 0.001 μF (d) 0.01 μF [ESE-2018]

4.50 In the network as shown, with $I_{S1} = 5$ A, $I_{S2} = 10$ A, $V_{AB} = 120$ V, and with $I_{S1} = 10$ A, $I_{S2} = 5$ A, $V_{AB} = 15$ V. What is the value of *k* to describe $I_{S1} = kI_{S2}$, such that $V_{AB} = 0$?

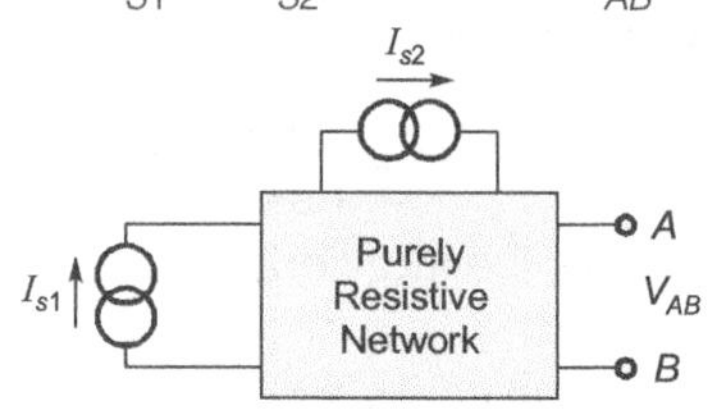

(a) 2.5 (b) 3.5
(c) 5.5 (d) 6.5 [ESE-2018]

4.51 **Statement I :** In resonant circuits, the capacitor voltage or inductor current could be much lower than the source voltage or source current, and a large input signal can produce a small output signal when resonance appears in a circuit.

Statement II : In resonant circuits, the quality factor for a coil is defined as the ratio of the inductive reactance and the winding resistance, which is a dimensionless parameter.

(a) Both Statement (I) and Statement (II) is the correct explanation of Statement (I).
(b) Both Statement (I) and Statement (II) are individually true, but Statement (II) is not the correct explanation of Statement (I).
(c) Statement (I) is true, but Statement (II) is false.
(d) Statement (I) is false, but Statement (II) is true.
[ESE-2021]

4.52 At off-resonance frequencies in parallel resonant circuit, the phase angle is greater than

(a) 0° (b) 30°
(c) 60° (d) 90°

[ESE-2022]

Answers Resonance

4.1 (a)	4.2 (d)	4.3 (a)	4.4 (a)	4.5 (c)	4.6 (d)	4.7 (d)	4.8 (d)	4.9 (c)
4.10 (a)	4.11 (d)	4.12 (b)	4.13 (d)	4.14 (c)	4.15 (b)	4.16 (d)	4.17 (c)	4.18 (c)
4.19 (d)	4.20 (a)	4.21 (d)	4.22 (d)	4.23 (c)	4.24 (c)	4.25 (b)	4.26 (c)	4.27 (a)
4.28 (d)	4.29 (c)	4.30 (d)	4.31 (d)	4.32 (b)	4.33 (c)	4.34 (c)	4.35 (c)	4.36 (c)
4.37 (c)	4.38 (c)	4.39 (d)	4.40 (c)	4.41 (d)	4.42 (a)	4.43 (b)	4.44 (a)	4.45 (c)
4.46 (b)	4.47 (c)	4.48 (a)	4.49 (b)	4.50 (a)	4.51 (d)	4.52 (a)		

Explanations Resonance

4.1 (a)

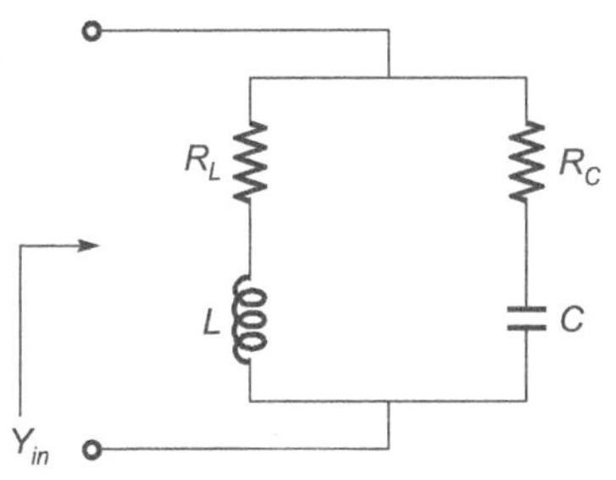

If $$R_L = R_C = \sqrt{\frac{L}{C}}$$

The input admittance of the network is purely real irrespective of the frequency of operation

$$\therefore \quad 4 = \sqrt{\frac{1}{C}} \Rightarrow C = \frac{1}{16}\text{ F}$$

4.2 (d)

At steady state, inductor behaves like short circuit and capacitor behaves like open circuit.

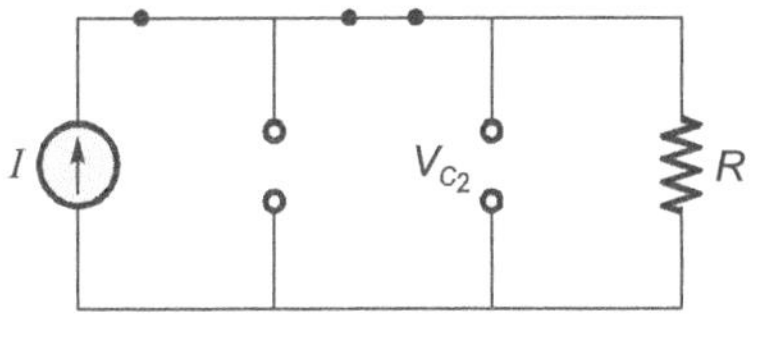

So, from circuit

Current through $L = I$

Voltage across C_2

$$V_{C2} = I \times R$$

4.3 (a)

Calculate $Y_{in}(j\omega)$. The impedance of the whole circuit is independent of frequency when imaginary $Y_{in}(j\omega) = 0$ which gives $R_L = R_C$ and $\omega = 1/\sqrt{LC}$. In other words, circuit is in resonance for $R_L = R_C$ and $\omega = 1/\sqrt{LC}$. i.e., not for all the frequencies.

Further, two branch currents will be in quadrature when

$$\tan^{-1}\left(\frac{1}{\omega C R_C}\right) + \tan^{-1}\left(\frac{\omega L}{R_L}\right) = 90°$$

which gives $R_L R_C = \frac{L}{C}$.

4.4 (a)

$$Z(j\omega) = R + j\left(\omega L - \frac{1}{\omega C}\right)$$

At resonance, imaginary term of $Z(j\omega)$ becomes zero i.e. capacitive reactance is equal to inductive reactance and $Z(j\omega)$ has minimum value which is equal to R.

4.5 (c)

Parallel R-L-C resonance has $Z_{in} = R$ at resonance so $I_R = 5$ A.

4.6 (d)

$$\because \quad Q = \frac{\omega_0}{\Delta\omega}$$

$$Q \propto \left(\frac{1}{\text{Bandwidth}}\right)$$

$$Q = \frac{1}{2\xi} \text{ or } \xi = \frac{1}{2Q}$$

For $\xi = 1$, critically damped, $Q = 1/2$

For $\xi < 1$, under damped, $Q > 1/2$

4.8 (d)

$$\omega_d = \sqrt{\omega_0^2 - \alpha^2}$$

where, $$\alpha = \frac{\text{B.W.}}{2} = \xi\omega_o$$

$$\therefore \quad \omega_0^2 = \omega_d^2 + \alpha^2 = (\sqrt{8})^2 + \left(\frac{2}{2}\right)^2 = 9$$

$$\Rightarrow \quad \omega_0 = 3 \text{ rad/s}$$

4.9 (c)

At resonance in parallel R-L-C circuit input impedance is equal to R.

4.10 (a)

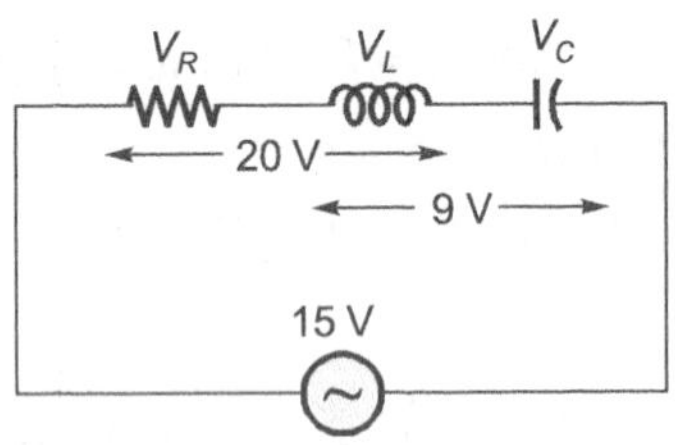

$$V_s = \sqrt{V_R^2 + (V_L - V_C)^2}$$

$$V_L - V_C = 9 \text{ V}$$

and $$V_S = 15 \text{ V}$$

$$V_R = \sqrt{V_S^2 - (V_L - V_C)^2}$$

$$= \sqrt{15^2 - 9^2} = 12 \text{ V}$$

$$\sqrt{V_R^2 + V_L^2} = 20 \text{ V}$$

$$\Rightarrow \quad V_L = \sqrt{20^2 - V_R^2}$$

$$= \sqrt{20^2 - 12^2} = 16 \text{ V}$$

$$V_L - V_C = 9 \text{ V}$$

$$\Rightarrow \quad V_C = V_L - 9 = 16 - 9 = 7 \text{ V}$$

4.11 (d)

If $R = \sqrt{\frac{L}{C}}$, the two branch currents are in quadrature not in phase.

4.12 (b)

In series RLC circuit Z is minimum at resonance frequency viz $Z = R$, otherwise $Z > R$ at all other frequencies.

At lower cut-off frequency $Z > R$ and Z will be capacitive in nature.

$\Rightarrow$ Option (b) is correct.

Similarly, at upper cut-off frequency $Z > R$ and Z will be inductive in nature.

$\Rightarrow$ option (b) is correct.

4.13 (d)

Series R-L-C circuit is capacitive below the resonant frequency. For more details refer to explanation of **Q.(34)** variation of capacitor voltage w.r.t. frequency.

4.14 (c)

Inductive reactance at 500 Hz

$$= \frac{50}{100} \times 500 = 250\ \Omega$$

$$Q = \frac{\omega L}{R}$$

$$\Rightarrow \quad Q = \frac{250}{10} = 25$$

$$V_C = Q \times V$$

$$= 25 \times 100 = 2500 \text{ V}$$

4.15 (b)

In parallel resonant circuit resonant frequency depends on L and C so will not reduce on loading hence point 2 is not correct.

4.16 (d)

At lower half frequency current in series RLC is $1/\sqrt{2}$ times and leading in nature where as at upper half frequency current is lagging in nature.

4.17 (c)

Characteristic equation for series RLC is

$\left(s^2 + \frac{R}{L}s + \frac{1}{LC} = 0\right)$ for over damped case $\xi > 1$

and here, $2\xi\omega_n = \frac{R}{L}$, $\omega_n = \frac{1}{\sqrt{LC}}$.

So, $\xi = \frac{R}{2}\sqrt{\frac{C}{L}} > 1$.

4.18 (c)

$$Q = \frac{\text{Energy stored per cycle}}{\text{Energy dissipated per cycle}}$$

$$Q_{\text{new}} = \frac{1.1 \times \text{Energy stored}}{0.9 \times \text{Energy dissipated}} = 1.22\ Q_{\text{old}}$$

Resonance frequency depends on L and C;

$$\omega_0 = \frac{1}{2\pi\sqrt{LC}}$$

$\Rightarrow$ Option (c) is correct.

4.19 (d)

Characteristic equation for parallel resonance

$$s^2 + \frac{1}{RC}s + \frac{1}{LC} = 0$$

$$\omega_0^2 = \frac{1}{LC} \qquad \text{B.W.} = \frac{1}{RC}$$

$$\omega_0 = \frac{1}{\sqrt{LC}} = 1 \text{ rad/sec}$$

$$\text{B.W.} = \frac{1}{20} \text{ rad/sec}$$

$$Q\text{-factor} = \frac{\omega_0}{\text{B.W.}} = \frac{1}{1/20} = 20$$

4.20 (a)

Characteristic equation series circuit

$$s^2 + \frac{R}{L}s + \frac{1}{LC} = 0$$

$$\text{B.W.} = 0.75 \text{ kHz}$$

$$f_0 = 1.5 \text{ kHz}$$

$$Q\text{-factor} = \frac{f_0}{\text{B.W.}} = \frac{1.5}{0.75} = 2$$

$\Rightarrow$ Option (a) is correct.

4.21 (d)

$$f_0 = 1.5 \text{ MHz}$$

$$C = 150 \text{ pF}$$

$$\text{B.W.} = 10 \text{ kHz}$$

$$Q = \frac{f_0}{\text{B.W.}} = \frac{1}{\omega RC}$$

$$\frac{1.5\times 10^6}{10\times 10^3} = \frac{1}{2\pi \times 1.5\times 10^6 \times R \times 150\times 10^{-9}}$$

So, $R = 4.7\ \Omega$

4.22 (d)

$$Z = R + j(X_L - X_C)$$

$$= \sqrt{R^2 + (X_L - X_C)^2}\angle \tan^{-1}\left(\frac{X_L - X_C}{R}\right)$$

$$\theta = \tan^{-1}\left(\frac{\omega L - \frac{1}{\omega C}}{R}\right)$$

$$\theta = \tan^{-1}\left(\frac{4-5}{R}\right)$$

So, voltage lags the current by $\tan^{-1}(1/R)$

So, $\frac{1}{R} = 2 = 0.5\ \Omega$

4.23 (c)

- Steady state voltage across C will be same as source voltage. If will be independent from value of R and L.

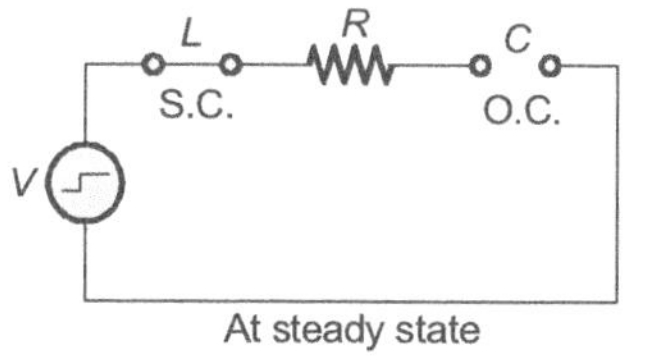

At steady state

- Frequency of transient oscillations

$$\omega_d = \omega_o\sqrt{1-\xi^2}$$

as R increases $\Rightarrow \xi$ increases
$\Rightarrow \omega_d$ decreases

- If R decreases $\Rightarrow \xi$ decreases
$\Rightarrow$ damping decreases
- When $R \to 0$
During the undamped oscillations following case will also be possible.

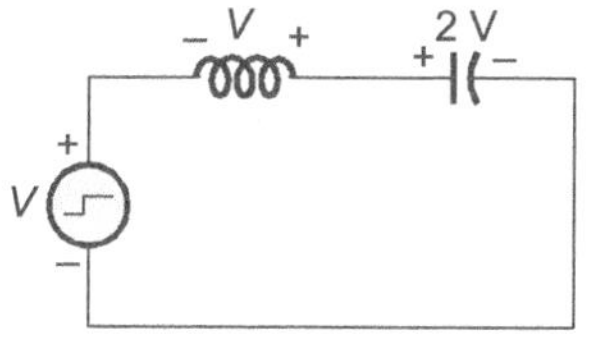

4.24 (c)

Damped frequency

$$\omega_d = \sqrt{8} \text{ rad/s}$$

$$\text{BW} = 2 \text{ rad/s}$$

$$\omega_d = \sqrt{\omega_n^2 - \alpha^2}$$

where, $\alpha = \frac{\text{BW}}{2} = 1$

$$\sqrt{8} = \sqrt{\omega_n^2 - 1}$$

$$\omega_n = 3 \text{ rad/sec}$$

4.25 (b)

Characteristic equation

$$s^2 + 2s + 100 = 0$$

Comparing with

$$s^2 + 2\xi\omega_n + \omega_n^2 = 0$$

Resonant frequency = $\sqrt{100}$ = 10 rad/sec

$\text{BW} = 2\,\xi\omega_n = 2$ rad/sec

4.26 (c)

For parallel RLC resonance, the current through L and C are in equal in magnitude but phase shifted by 180°. Therefore the current through L and C combination is equal to zero. Hence, option (c) is correct.

4.27 (a)

At resonance, the voltage across the inductor and across the capacitor are equal in magnitude but are phase shifted by 180°. So that, the net voltage across inductor and capacitor combination is zero. Hence, option (a) is correct.

4.28 (d)

Resonant frequency,

$$f_0 = \frac{1}{\sqrt{LC}}$$

If components are tripled,

$$f' = \frac{1}{\sqrt{3L \cdot 3C}} = \frac{1}{3} f_0$$

4.29 (c)

In a series resonance circuit,

$$Q = \frac{1}{R}\sqrt{\frac{L}{C}}$$

4.30 (d)

In a series RLC circuit, at resonance the voltage drops across the capacitance and inductance are equal in magnitude.

4.31 (d)

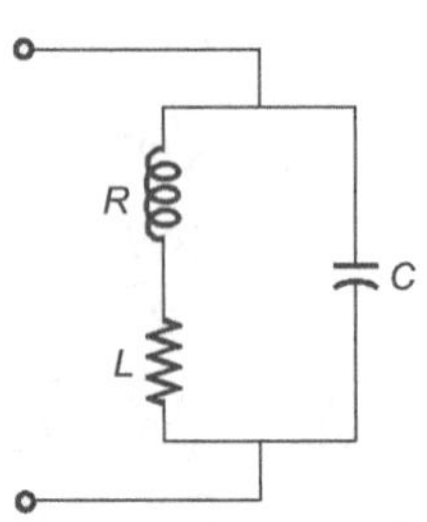

Resonant frequency

$$\omega_0 = \frac{1}{\sqrt{LC}}\sqrt{1-\frac{R^2C}{L}}$$

$$R < \sqrt{\frac{L}{C}}$$

Input impedance

$$Z_{in} = \frac{R^2+\omega_0^2 L^2}{R}$$

$$= R + \frac{1}{RLC}\left(1-\frac{R^2C}{L}\right)L^2$$

$$= R + \frac{1}{RC}\left[L - R^2C\right]$$

$$= R + \frac{L}{RC} - R$$

At resonance,

$$Z_{in} = \frac{L}{RC}\,\Omega$$

as $\quad R^2 < \frac{L}{C} \Rightarrow R < \frac{L}{RC}$

Circuit behaves like a pure resistor of value much higher than R.

4.32 (b)

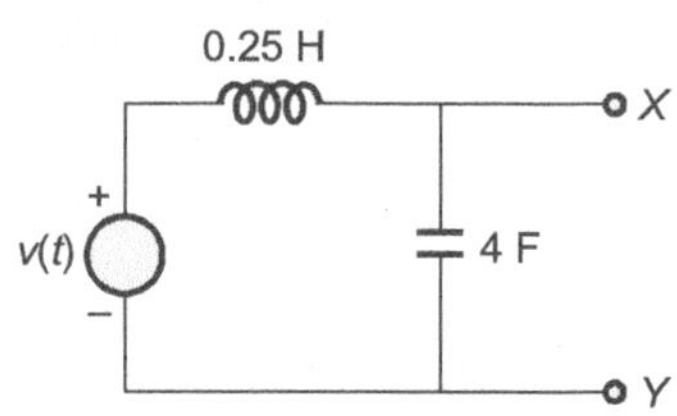

For given network to be act as ideal current source the input impedance should be infinite.

$$Y_{xy} = \frac{j\omega}{4} + \frac{1}{4 \cdot j\omega} = 0$$

$$\Rightarrow \quad \frac{j\omega}{4} = \frac{j}{4\omega} \Rightarrow \omega = 1 = 2\pi f$$

$$\Rightarrow \quad f = \frac{1}{2\pi}\text{ Hz}$$

4.33 (c)

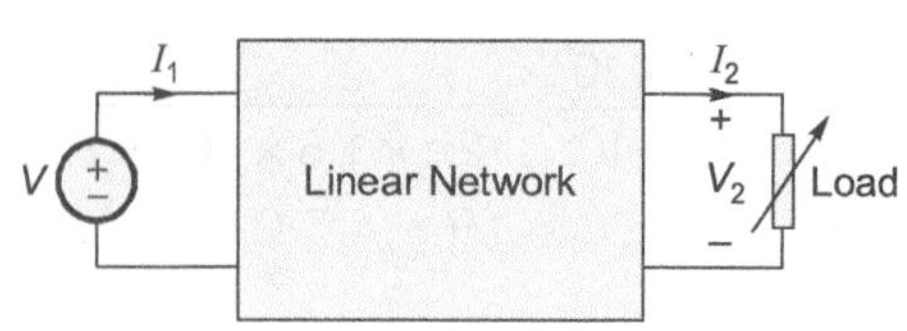

$$V_1 = AV_2 + BI_2 \quad \text{...(i)}$$

$$I_1 = CV_2 + DI_2 \quad \text{...(ii)}$$

When $\quad V = 4$ V and $P = 4$ W

$\because \quad P = I_2^2 R \Rightarrow I_2 = \sqrt{\frac{P}{R}}$

and $\quad P = \frac{V_2^2}{R} \Rightarrow V_2 = \sqrt{PR}$

$$4 = A\sqrt{PR} + B\sqrt{\frac{P}{R}}$$

As $R = 1\ \Omega$

$\therefore$ $4 = \sqrt{P}A + \sqrt{P}B$

$4 = 2A + 2B$ ...(iii)

When $V = 8$ V and $P = 16$ W

$8 = 4A + 4B$ ...(iv)

Adding equation (iii) and (iv)

$12 = 6A + 6B$

$\therefore$ $V_2 = 6$ V and $I_2 = 6$ A

Therefore power consumed by 1 Ω resistor

$$= I_2^2 R = 36 \text{ W}$$

4.34 (c)

In series RLC circuit,

$$\text{Bandwidth} = \frac{R}{L} = 300$$

$\Rightarrow$ $R = 300 \times 0.1 = 30\ \Omega$

$$Q = \frac{\omega_r}{\Delta\omega} = \frac{\text{Resonant frequency}}{\text{Bandwidth}}$$

$$= \frac{3000}{300} = 10$$

At resonance,

$$|V_L| = |V_C| = QV_s$$

$$= 10 \times 100 = 1000 \text{ V}$$

4.36 (c)

During resonance, an *RLC* series circuit behaves like a purely resistive circuit but voltage across *L* and *C*-elements are in opposite phase.

4.37 (c)

For a parallel RLC circuit,

$$Z(s) = \frac{10s}{s^2 + s + 400}$$

$$Y(s) = \frac{s^2 + s + 400}{10s}$$

$$= \frac{s}{10} + \frac{1}{10} + \frac{40}{s}$$

$$= \left(\frac{1}{10}\right)s + \frac{1}{10} + \frac{1}{\left(\frac{1}{40}\right)s}$$

$$= sC + \frac{1}{R} + \frac{1}{Ls}$$

$$\therefore \quad C = \frac{1}{10}\text{F},\ R = 10\ \Omega,\ L = \frac{1}{40}\text{H}$$

4.39 (d)

$$Q = \frac{1}{2\pi f_0 RC} = 150$$

$$= \frac{1}{2\pi(1.5 \times 10^6) \times R \times 150 \times 10^{-12}}$$

$$R = \frac{1}{1.5 \times 2\pi \times 1.5 \times 1.5 \times 10^{-12+2+6+2}}$$

$$= \frac{100}{(1.5)^3 \times 2\pi} = 4.7\ \Omega$$

4.40 (c)

Variation of voltage across capacitor and voltage across inductor with respect to frequency is

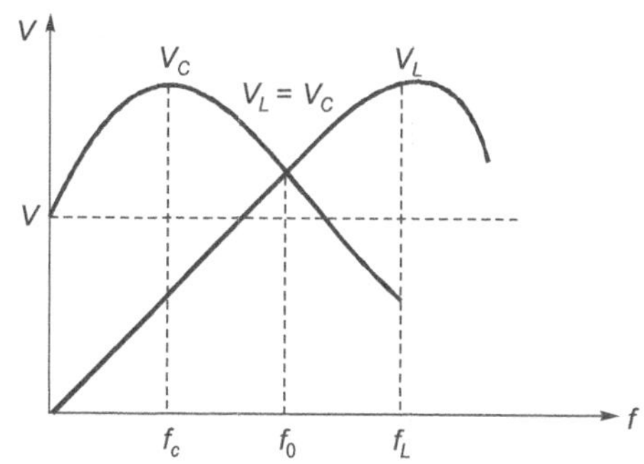

so, in series resonant circuit, maximum voltage across *L* occurs at slightly above resonant frequency.

4.41 (d)

$$Q = \frac{\omega_0}{\Delta\omega} = \frac{f_0}{\Delta f} = Q\text{-factor}$$

$$\Delta f = \text{B.W.} = \frac{1.5 \times 10^6}{150} = 10 \text{ kHz}$$

4.42 (a)

For a series RLC circuit,

$$X_{eq} = |X_L - X_C| = \omega L - \frac{1}{\omega C}$$

at low frequency, the circuit impedance.

$$X_{eq} = -\frac{1}{\omega C}$$

at high frequency, the circuit impedance

$$X_{eq} = \omega L$$

4.43 (b)

For series RLC circuit, at resonant frequency the circuit impedance, $Z = R$, (purely resistive) and at this frequency the energy stored in inductor and capacitor oscillates between them also.

4.44 (a)

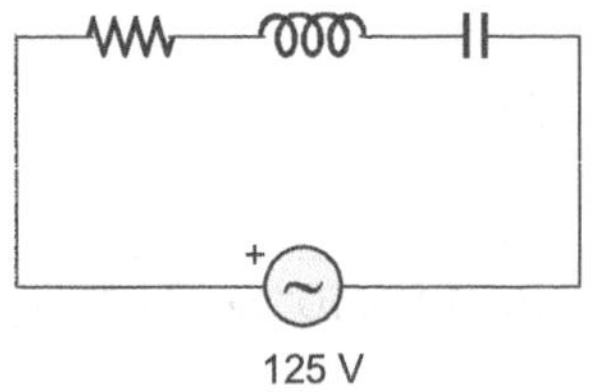

$f_0 = 220$ Hz

$V_C = V_L = 4150$ volts

$R = 1\ \Omega$

$$Q = \frac{V_L \text{ or } V_C}{V} = \frac{4150}{125} = 33.2$$

Selectivity = Q = 33.2

4.45 (c)

$I_{max} = 0.5$ A ; $f_r = 400$ Hz, $V_C = 150$ V

As the circuit is in resonance,

$\therefore \quad |V_L| = |V_C|$ and $V_R = V_S$

$\Rightarrow \quad V_R = 25$ V

$V_R = I \cdot R$

$$\Rightarrow \quad R = \frac{25}{0.5} \Rightarrow R = 50\ \Omega$$

$$V_L = I X_L \Rightarrow X_L = \frac{150}{0.5} = 300$$

$X_L = \omega L$

$$\Rightarrow \quad L = \frac{300}{2\pi \times 400} = 0.119 \text{ H}$$

4.46 (b)

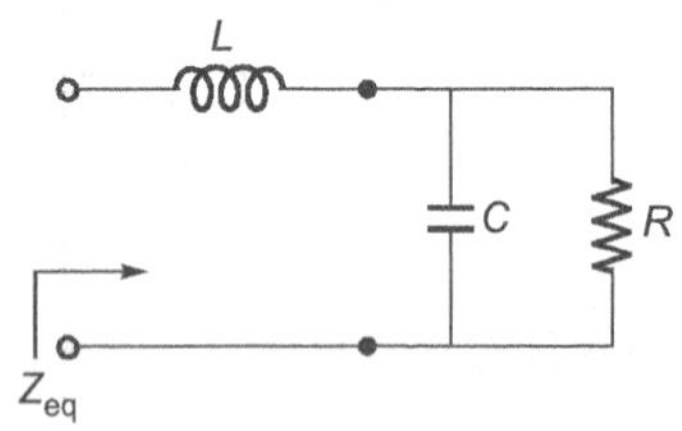

$\therefore \quad Z_{eq} = j\omega L + Z$

$$Z = \frac{R \times \frac{1}{j\omega C}}{R + \frac{1}{j\omega C}} = \frac{R}{1 + j\omega RC} \times \frac{1 - j\omega RC}{1 - j\omega RC}$$

$$Z = \frac{R - j\omega R^2 C}{1 + \omega^2 R^2 C^2}$$

$$Z_{eq} = j\omega L + \frac{R - j\omega R^2 C}{1 + \omega^2 R^2 C^2}$$

$$= \frac{j\omega L(1 + \omega^2 R^2 C^2) + R - j\omega R^2 C}{1 + \omega^2 R^2 C^2}$$

$$= \frac{R}{1 + \omega^2 R^2 C^2} + j\frac{[\omega L + \omega^3 R^2 C^2 L - \omega R^2 C]}{1 + \omega^2 R^2 C^2}$$

At resonance impedance is purely resistive in nature.

$\therefore$ Equating Imaginary part to zero.

$\omega L + \omega^3 R^2 C^2 L - \omega R^2 C = 0$

$\omega^3 R^2 C^2 L = \omega[R^2 C - L]$

$$\omega^2 = \frac{R^2 C}{R^2 C^2 L} - \frac{1}{R^2 C^2 L}$$

$$\omega^2 = \frac{1}{LC} - \frac{1}{R^2 C^2} = \frac{1}{0.2 \times 1} - \frac{1}{1}$$

$\omega^2 = 4$

$\Rightarrow \quad \omega = 2$ rad/sec.

4.47 (c)

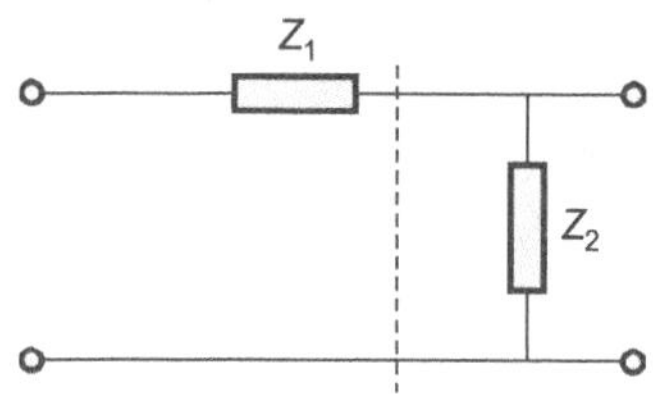

Where can divide the network in two networks as named A and B.

Network A;

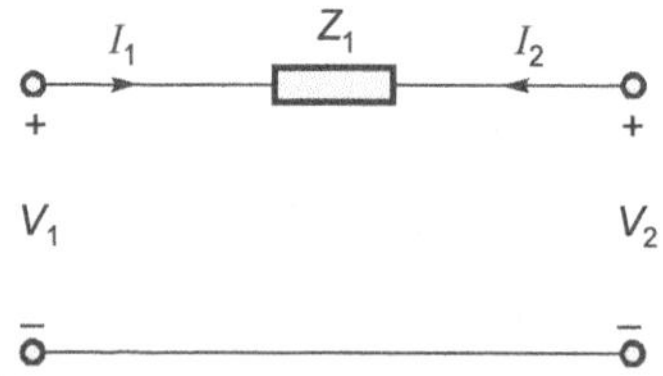

From network A;

$V_1 = V_2 - Z_1 I_2$

$I_1 = -I_2$

ABCD parameters;

$$[T_A] = \begin{bmatrix} 1 & Z_1 \\ 0 & 1 \end{bmatrix}$$

Network B;

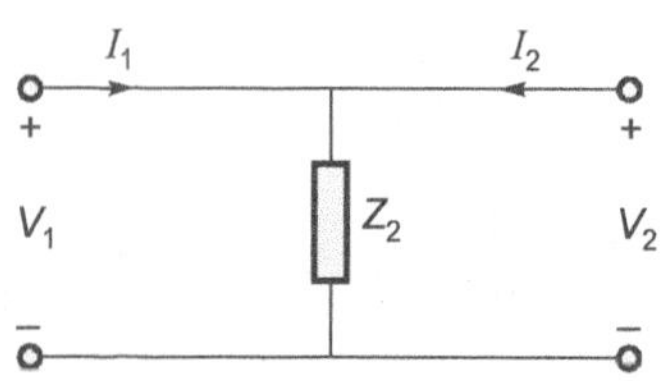

From network B;

$$V_1 = V_2$$

$$I_1 = \frac{V_2}{Z_2} - I_2$$

$$[T_B] = \begin{bmatrix} 1 & 0 \\ \frac{1}{Z_2} & 1 \end{bmatrix}$$

So, the given network consists of 2 networks A and B connected in cascade.

$$\therefore \quad [T] = [T_A] \times [T_B]$$

$$= \begin{bmatrix} 1 & Z_1 \\ 0 & 1 \end{bmatrix} \times \begin{bmatrix} 1 & 0 \\ \frac{1}{Z_2} & 1 \end{bmatrix}$$

$$= \begin{bmatrix} 1+\frac{Z_1}{Z_2} & Z_1 \\ \frac{1}{Z_2} & 1 \end{bmatrix}$$

$$\therefore \quad \frac{1}{Z_2} = C = 0.1\angle 90° = -j10 \ \mho$$

Alternative method:

For end condenser network:

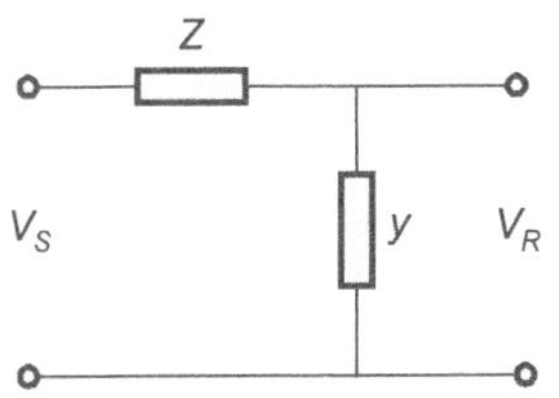

$$\begin{bmatrix} A & B \\ C & D \end{bmatrix} = \begin{bmatrix} (1+YZ) & Z \\ Y & 1 \end{bmatrix}$$

$$Y = C$$

$$Z_2 = \frac{1}{Y} = \frac{1}{C} = \frac{1}{0.1\angle 90°} = -10j$$

4.48 (a)

$$\omega_n = \frac{1}{\sqrt{LC}} = \frac{1}{\sqrt{4 \times 10^{-1} \times 10 \times 10^{-6}}}$$

$$= \frac{1000}{2} = 500 \text{ rad/sec.}$$

$$X_L = \omega L = (500)\,(0.4) = 200 \ \Omega$$

$$X_C = \frac{1}{\omega C}$$

$$= \frac{1}{500 \times 10 \times 10^{-6}} = 200 \ \Omega$$

$$I = \frac{100}{50 + j(200-200)} = \frac{100}{50} = 2 \text{ A}$$

4.49 (b)

Load impedance Z_L:

$$Z_L = \frac{(-jX_C)\,(R_L + jX_L)}{R_L + jX_L - jX_C}$$

$$Z_L = \frac{(-jX_C)\,(R_L + jX_L)}{R_L + j(X_L - X_C)} \times \frac{R_L - j(X_L - X_C)}{R_L - j(X_L - X_C)}$$

Separating Z_L into real and imaginary terms and equating the imaginary terms to zero for maximum power transfer.

$$\therefore \quad R_L^2 X_C + X_L X_C (X_L - X_C) = 0$$

$$R_L^2 + X_L^2 - X_L X_C = 0$$

$$X_C = \frac{R_L^2 + X_L^2}{X_L} = \frac{25+25}{5} = \frac{50}{5}$$

$$X_C = 10$$

$$\Rightarrow \quad \frac{1}{\omega C} = 10$$

$$\Rightarrow \quad C = \frac{1}{(10)\,(10)} = 0.01\text{F}$$

4.50 (a)

$$V_{AB} = \alpha I_{S_1} + \beta I_{S_2}$$

$$120 = \alpha(5) + \beta(10)$$

$$\alpha + 2\beta = 24 \quad \text{...(1)}$$

$$15 = \alpha(10) + \beta(5)$$

$$3 = 2\alpha + \beta \quad \text{...(2)}$$

Equation (1) – 2 × equation (2):

$$-3\alpha = 18 \Rightarrow \alpha = -6 \quad \text{...(3)}$$

Putting (3) in (2),

$$\beta = 15$$

$$-6I_{S_1} + 15I_{S_2} = 0$$

$$-6KI_{S_2} + 15I_{S_2} = 0$$

$$-6\,K + 15 = 0$$

$$K = \frac{15}{6} = \frac{5}{2} = 2.5$$

4.52 (a)

At resonance condition the current $\vec{I}$ and $\vec{V}$ are in phase with each other

$\vec{I}$ $\vec{V}$ angle is 0°

∴ At off resonance condition means not in resonance means angle between voltage and current is greater than 0°.

(a) option is correct.

■■■■

5 Two Ports and Network Functions

5.1 For the circuit shown in the below figure h_{11}, h_{12}, h_{21} and h_{22} are respectively

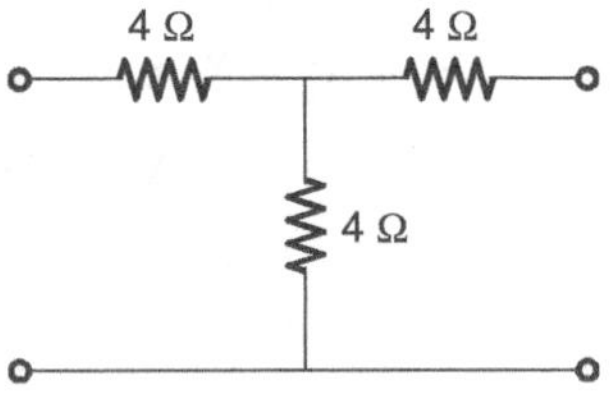

(a) –0.5, 0.5, 0.125 and 6
(b) 6, 0.5, –0.5, and 0.125
(c) 0.5, –0.5, 6 and 0.125
(d) 0.125, 6, 0.5 and –0.5 **[ESE-2001]**

5.2 The h parameters h_{11} and h_{22} are related to z and y parameters as

(a) $h_{11} = z_{11}$ and $h_{22} = 1/z_{22}$
(b) $h_{11} = z_{11}$ and $h_{22} = y_{22}$
(c) $h_{11} = 1/y_{11}$ and $h_{22} = 1/z_{22}$
(d) $h_{11} = 1/y_{11}$ and $h_{22} = y_{22}$ **[ESE-2001]**

5.3 The response i of a network is expressed by the differential equation $d^2i/dt^2 + i = v$. If $v = A\,e^{2t}$, the dominant solution of i for $t > 0$ is of the nature

(a) $K_1\,e^{t}$ (b) $K_1\,e^{-t}$
(c) $K_1\,e^{2t}$ (d) $K_2 \cos t + K_3 \sin t$
[ESE-2001]

5.4 The [y] parameters of the network given below, are given by

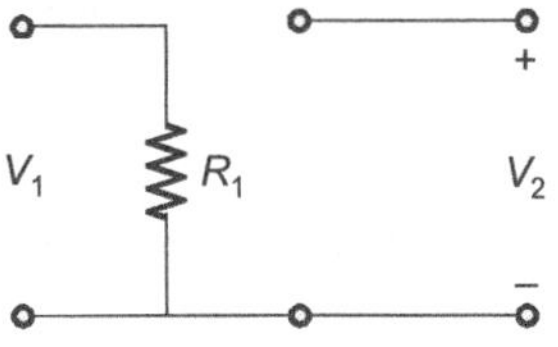

(a) $\begin{bmatrix} R_1^{-1} & R_1^{-1} \\ R_1^{-1} & R_1^{-1} \end{bmatrix}$ (b) $\begin{bmatrix} 0 & R_1^{-1} \\ R_1^{-1} & R_1^{-1} \end{bmatrix}$

(c) $\begin{bmatrix} R_1^{-1} & 0 \\ 0 & 0 \end{bmatrix}$ (d) $\begin{bmatrix} R_1^{-1} & 0 \\ R_1^{-1} & 0 \end{bmatrix}$

[ESE-2002]

5.5 For the two-port network as shown below, Y_{12} is equal to

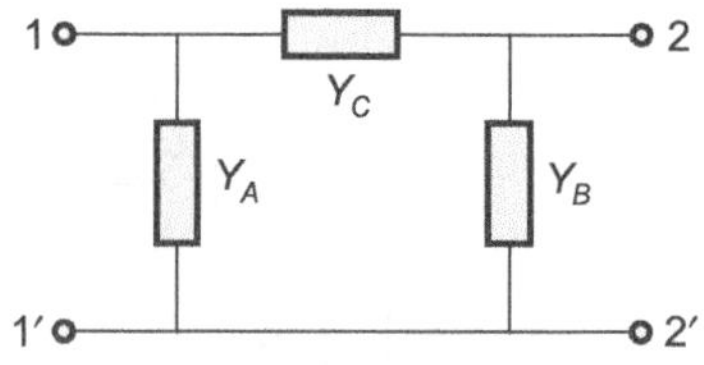

(a) $Y_A + Y_B$ (b) $Y_C + \left(\dfrac{Y_A Y_B}{Y_A + Y_B}\right)$

(c) $-Y_C$ (d) Y_C

[ESE-2003]

5.6 The lattice circuit has the following impedances $Z_A = 3 + j4$, $Z_B = 3 - j4$. Then the Z-parameters would be

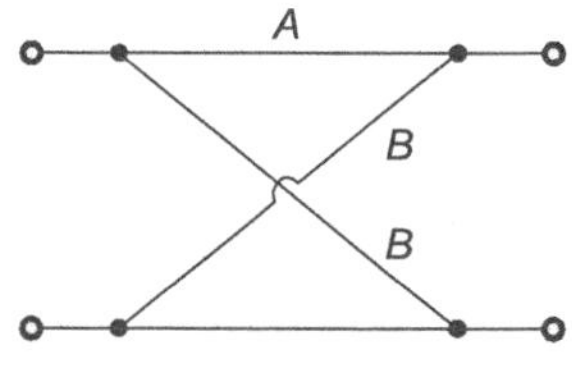

(a) $\begin{bmatrix} 3+j4 & 0 \\ 0 & 3-j4 \end{bmatrix}$ (b) $\begin{bmatrix} 3 & -j4 \\ -j4 & 3 \end{bmatrix}$

(c) $\begin{bmatrix} 3+j4 & 3 \\ 3 & 3+j4 \end{bmatrix}$ (d) $\begin{bmatrix} -j4 & 3 \\ 3 & +j4 \end{bmatrix}$

[ESE-2003]

5.7 A circuit is modelled by the following differential equation:

$$\frac{d^2 i(t)}{dt^2} + \frac{6\,di(t)}{dt} + 9\,i(t) = 0$$

The response $i(t)$ is of the form (with symbols having their standard meanings)

(a) $K_1 e^{-t} + K_2 e^{-9t}$ (b) $Ke^{-3t} \sin(t + \theta)$
(c) $(K_1 + K_2 t)\, e^{-3t}$ (d) $Ke^{-t} \cdot \sin(9t + \theta)$
[ESE-2003]

5.8 Consider the following circuit:

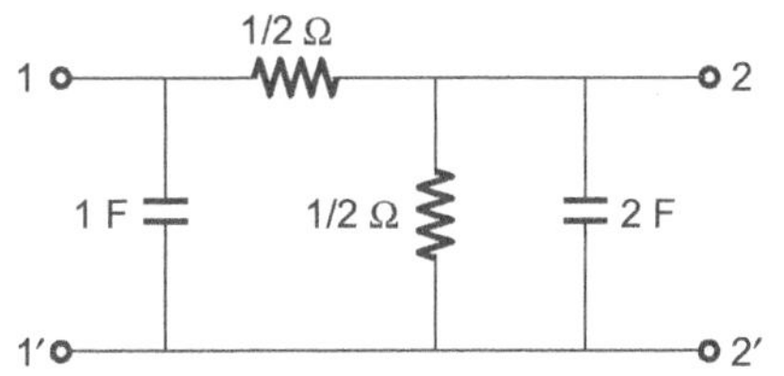

What is the short-circuit natural frequency at port 1 of the above circuit ?

(a) $s + 2 = 0$ (b) $s + 1 = 0$
(c) $2s + 1 = 0$ (d) $2s + 3 = 0$

[ESE-2004]

5.9 Two identical T-sections, such as one shown below are connected in series.

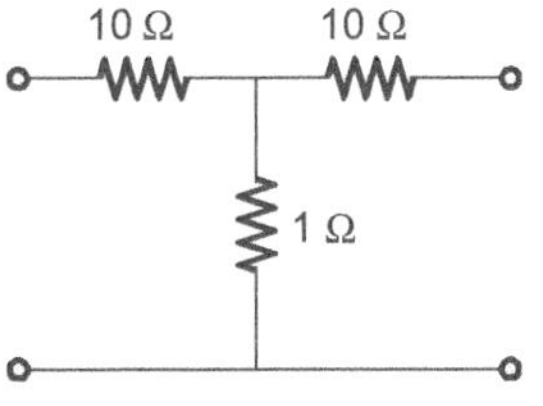

What is the y_{11} of the combination?

(a) 11/240 ℧ (b) 22/240 ℧
(c) 11/360 ℧ (d) 22/120 ℧

[ESE-2004]

5.10 For the 2-port network shown in the figure given below, what is the value of the parameter h_{21}?

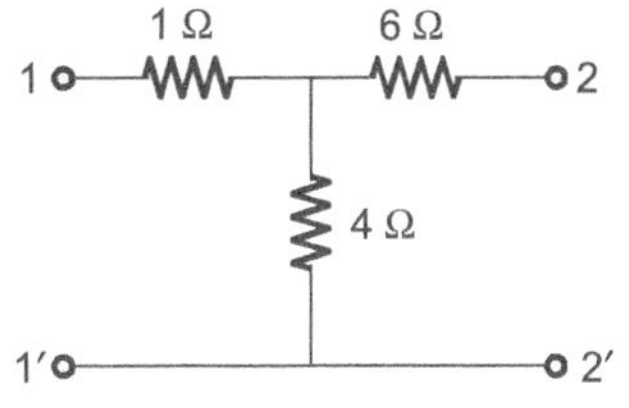

(a) 1.5 (b) –0.4
(c) 0.6 (d) –0.5

[ESE-2005]

5.11 Which one of the following gives the correct short circuit parameter matrix Y for the network shown below?

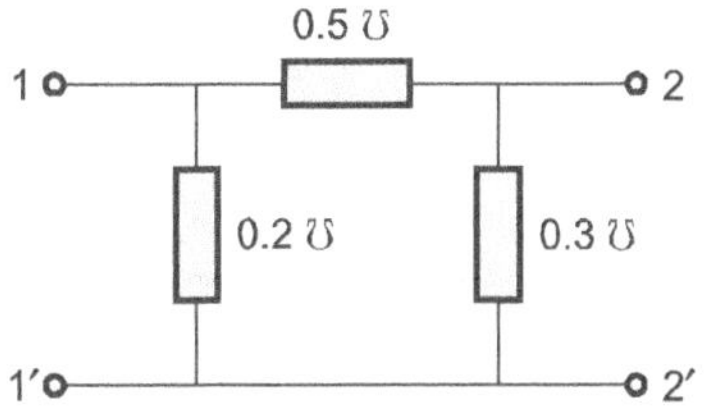

(a) $\begin{bmatrix} 0.7 & -0.5 \\ -0.5 & 0.8 \end{bmatrix}$ (b) $\begin{bmatrix} 0.7 & 0.5 \\ -0.5 & 0.8 \end{bmatrix}$

(c) $\begin{bmatrix} 0.8 & -0.5 \\ -0.5 & 0.7 \end{bmatrix}$ (d) $\begin{bmatrix} 0.7 & -0.5 \\ 0.5 & 0.8 \end{bmatrix}$

[ESE-2005]

5.12 Match **List-I** (Term) with **List-II** (Concept) and select the correct answer using the code given the below the lists:

List-I

A. Norton equivalent of one port
B. Open-circuit output admittance
C. Reciprocal network
D. Transmission parameters

List-II

1. Network where loop and node equations have a symmetric coefficient matrix
2. Hybrid parameter h_{22}
3. Parameters where V_1 and I_1 are expressed as functions of V_2 and $-I_2$
4. Current source in parallel with Thevenin impedance

Codes:

	A	B	C	D
(a)	1	3	4	2
(b)	4	2	1	3
(c)	1	2	4	3
(d)	4	3	1	2

[ESE-2006]

5.13 A, B, C and D represent the transmission parameters of a two-port network. When is the network reciprocal?

(a) $AB - CD = 1$ (b) $AD - BC = 1$
(c) $AB - CD = 0$ (d) $AD - BC = 0$

[ESE-2008]

5.14 Match **List-I** (Network parameter) with **List-II** (Dimension) and select the correct answer using the code given below the lists:

List-I	**List-II**
A. h_{11}	**1.** Impedance
B. h_{12}	**2.** Admittance
C. h_{22}	**3.** Current ratio
D. h_{21}	**4.** Voltage ratio

Codes:

	A	B	C	D
(a)	2	4	1	3
(b)	2	3	1	4
(c)	1	3	2	4
(d)	1	4	2	3

[ESE-2008]

5.15 A series R-L-C circuit is switched ON to a step voltage V at $t = 0$. What are the initial and final values of the current in the circuit, respectively?

(a) V/R, V/R (b) Zero, Infinity
(c) Zero, Zero (d) Zero, V/R

[ESE-2008]

5.16 For the 2-port network as shown below, what is the value of Y_{21} parameter?

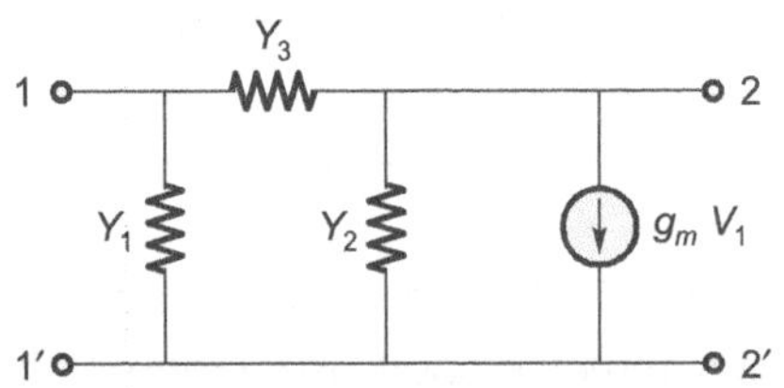

(a) $Y_1 + Y_3$ (b) $g_m - Y_2$
(c) $g_m - Y_3$ (d) $Y_1 + Y_2 + g_m$

[ESE-2009]

5.17 If the Z-parameters for the T-network as shown below are Z_{11} = 40 Ω, Z_{22} = 50 Ω and $Z_{12} = Z_{21}$ = 30 Ω, then what are the values of Z_1, Z_2 and Z_3?

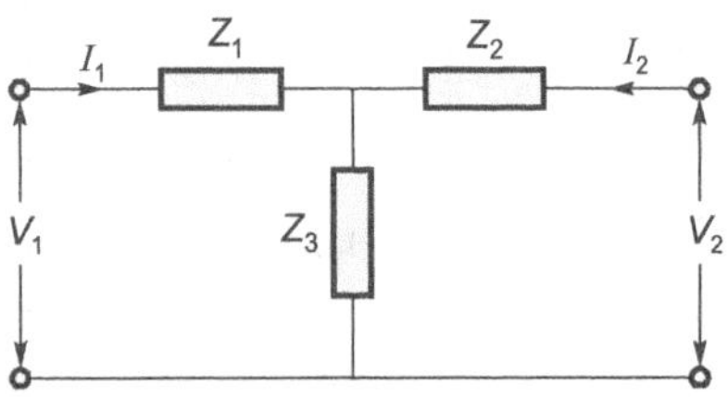

(a) 10 Ω, 20 Ω and 30 Ω
(b) 20 Ω, 30 Ω and 20 Ω
(c) 30 Ω, 40 Ω and 10 Ω
(d) 40 Ω, 50 Ω and 10 Ω

[ESE-2009]

5.18 A two-port network is defined by the relation:

$I_1 = 5\,V_1 + 3\,V_2$
$I_2 = 2\,V_1 - 7\,V_2$

The value of Z_{12} is

(a) 3 Ω (b) −3 Ω
(c) 3/41 Ω (d) 2/31 Ω

[ESE-2010]

5.19 The Y-parameters of the network shown below are

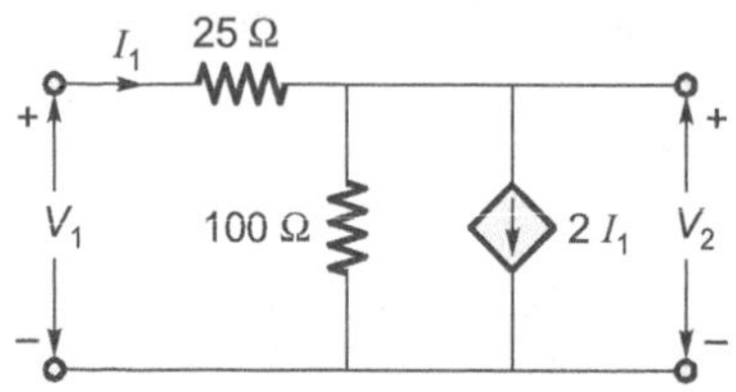

(a) $\begin{bmatrix} -0.04 & 0.04 \\ -0.04 & 0.03 \end{bmatrix}$ (b) $\begin{bmatrix} 0.04 & -0.04 \\ 0.04 & -0.03 \end{bmatrix}$

(c) $\begin{bmatrix} 0.04 & -0.03 \\ -0.04 & 0.03 \end{bmatrix}$ (d) $\begin{bmatrix} -0.04 & 0.04 \\ 0.04 & 0.03 \end{bmatrix}$

[ESE-2010]

5.20 A two-port network is described by the following equations:

$V_1 = 50I_1 + 20I_2$
$V_2 = 30I_1 + 10I_2$

Then, which one of the following is not correct?

(a) Z_{12} = 20 Ω (b) $Y_{12} = 0.2$ ℧
(c) $h_{12} = 2.0$ (d) $A = 25$

[ESE-2010]

5.21 Consider the two-port network as shown. The hybrid parameter h_{12} is

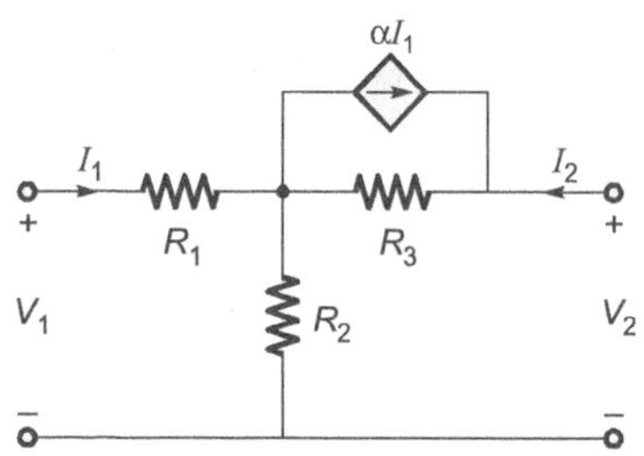

(a) $-\dfrac{(\alpha R_3 + R_2)}{R_2 + R_3}$ (b) $\dfrac{(1-\alpha)R_3}{R_2 + R_3}$

(c) $\dfrac{(1-\alpha)R_2}{R_2 + R_3}$ (d) $\dfrac{R_2}{R_2 + R_3}$

[ESE-2011]

5.22 A 2-port network is defined by the relation,

$$V_1 = \frac{3}{4}I_1 - \frac{1}{4}I_2,\ V_2 = -\frac{1}{2}I_1 + \frac{1}{2}I_2$$

Then y_{12} is

(a) $\frac{1}{2}$ ℧ (b) $-\frac{1}{2}$ ℧

(c) 1 ℧ (d) −1 ℧

[ESE-2011]

5.23 For the two-port network, the impedance parameter matrix $[Z]=\begin{bmatrix} Z_{11} & Z_{12} \\ Z_{21} & Z_{22} \end{bmatrix}$ is

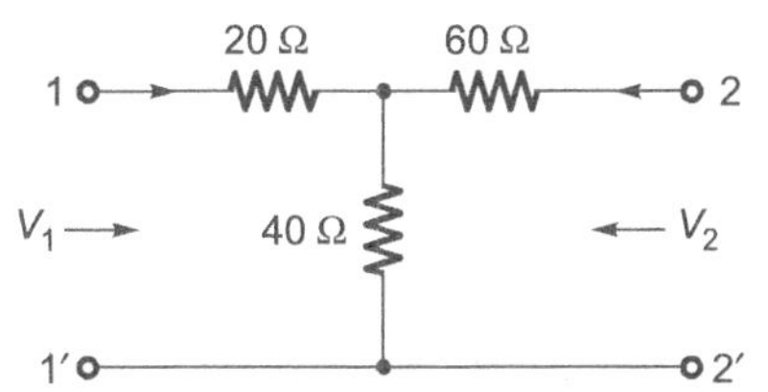

(a) $\begin{bmatrix} 60\Omega & 100\Omega \\ 40\Omega & 60\Omega \end{bmatrix}$ (b) $\begin{bmatrix} 40\Omega & 100\Omega \\ 60\Omega & 40\Omega \end{bmatrix}$

(c) $\begin{bmatrix} 60\Omega & 40\Omega \\ 40\Omega & 100\Omega \end{bmatrix}$ (d) $\begin{bmatrix} 40\Omega & 100\Omega \\ 100\Omega & 60\Omega \end{bmatrix}$

[ESE-2011]

5.24 When port-1 of a two-port network is short circuited, $I_1 = 4I_2$ and $V_2 = 0.5I_2$, then which of the following is true?

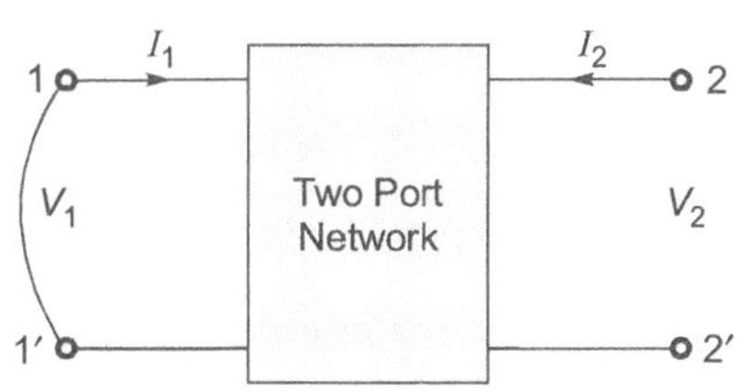

(a) $Y_{11} = 4$ mho (b) $Y_{12} = 8$ mho
(c) $Y_{21} = 16$ mho (d) $Y_{22} = 0.25$ mho

[ESE-2011]

5.25 A 100 Ω resistor has an effective inductance of 0.1 μH and a distributed capacitance of 10 pF. Its time constant at medium frequency is

(a) 0 ns (b) 1 ns
(c) 2 ns (d) 3 ns

[ESE-2011]

5.26 The condition for reciprocity for a two-port transmission network is expressed by

(a) $\begin{vmatrix} A & B \\ C & D \end{vmatrix} = 0$ (b) $\begin{vmatrix} A & D \\ B & C \end{vmatrix} = 1$

(c) $\begin{vmatrix} A & C \\ B & D \end{vmatrix} = 0$ (d) $\begin{vmatrix} A & B \\ C & D \end{vmatrix} = 1$

[ESE-2012]

5.27 Consider an L-R circuit in which a current $i = 5e^{-2t}$ A is flowing. The voltage across the R-L circuit is given by

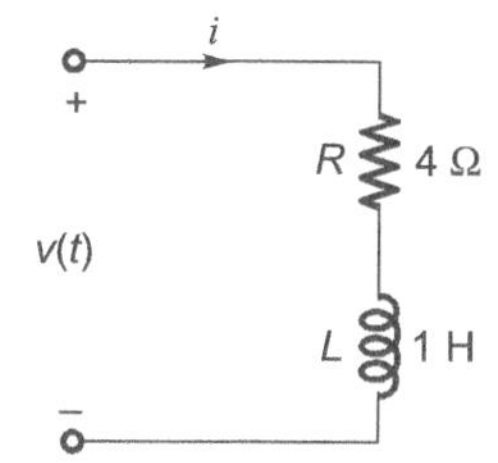

(a) $20e^{-2t}$ V (b) $-10e^{-2t}$ V
(c) $10e^{-2t}$ V (d) $5e^{-2t}$ V

[ESE-2012]

5.28 The current is given by

$$I(s) = \frac{(s+2)(s+4)}{s(s+1)(s+\alpha)}$$

If the steady-state current at $t = \infty$ is 12 A, then the value of α and initial value of current will be

(a) 1.5 and 1 A (b) 0.66 and 1 A
(c) 0.33 and 0.5 A (d) 0.25 and 0.5 A

[ESE-2012]

5.29 When a unit impulse voltage is applied to an inductor of 1 H, the energy supplied by the source is

(a) 2 J (b) 1 J
(c) 1/2 J (d) 1/4 J

[ESE-2012]

5.30 For the two port network as shown below, the parameters h_{11} and h_{21} are

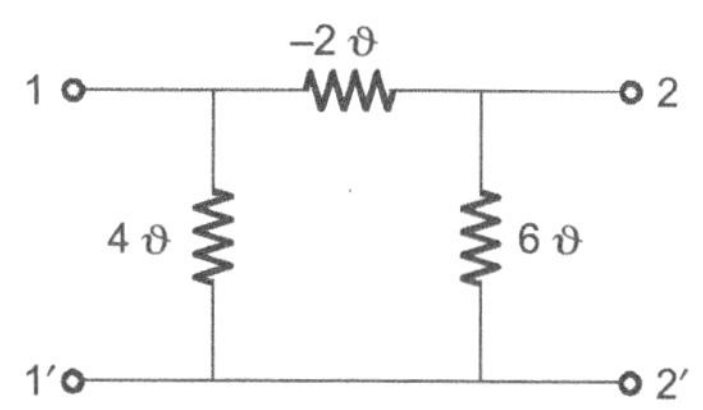

(a) 1 Ω and 2 Ω (b) 2 Ω and 1

(c) 1 and $\frac{1}{2}$ Ω (d) $\frac{1}{2}$ Ω and 1

[ESE-2013]

5.31 A 2-port network is represented by the following equations:

$$V_1 = 60I_1 + 20I_2$$
$$V_2 = 20I_1 + 40I_2$$

The *ABCD* parameters of the above network would be

(a) $\begin{bmatrix} 2 & \frac{1}{20} \\ 3 & 100 \end{bmatrix}$ (b) $\begin{bmatrix} 100 & 3 \\ 2 & \frac{1}{20} \end{bmatrix}$

(c) $\begin{bmatrix} 100 & 20 \\ 6 & 3 \end{bmatrix}$ (d) $\begin{bmatrix} 3 & 100 \\ \frac{1}{20} & 2 \end{bmatrix}$

[ESE-2013]

5.32 The terminal voltage and currents of a two-port network are indicated on the below figure. If the two-port is reciprocal, then

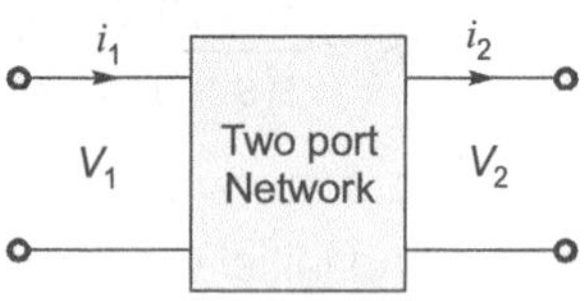

(a) $\frac{Z_{12}}{Y_{12}} = Z_{12}^2 - Z_{11} \cdot Z_{22}$

(b) $Z_{12} = \frac{1}{Y_{22}}$

(c) $h_{12} = -h_{21}$

(d) $AD - BC = 0$ [ESE-2013]

5.33 The Z-parameters of the 2-port network as shown below are

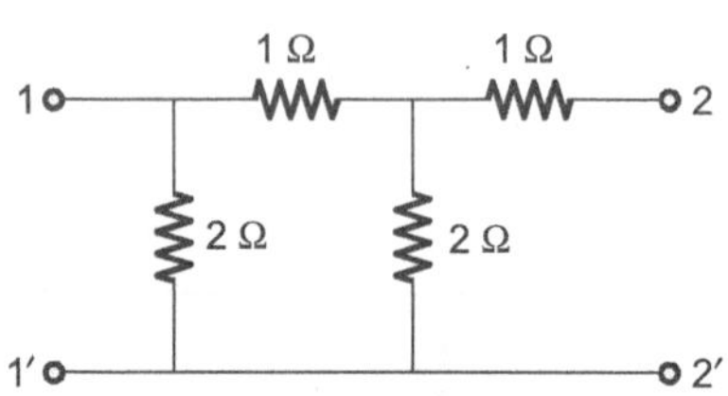

(a) $\begin{bmatrix} \frac{11}{5} & \frac{4}{5} \\ \frac{4}{5} & \frac{6}{5} \end{bmatrix}$ (b) $\begin{bmatrix} \frac{6}{5} & \frac{4}{5} \\ \frac{4}{5} & \frac{11}{5} \end{bmatrix}$

(c) $\begin{bmatrix} \frac{4}{5} & \frac{6}{5} \\ \frac{11}{5} & \frac{4}{5} \end{bmatrix}$ (d) $\begin{bmatrix} \frac{4}{5} & \frac{4}{5} \\ \frac{11}{5} & \frac{6}{5} \end{bmatrix}$

[ESE-2013]

5.34 The Z-parameter matrix of the two-port network as shown below is

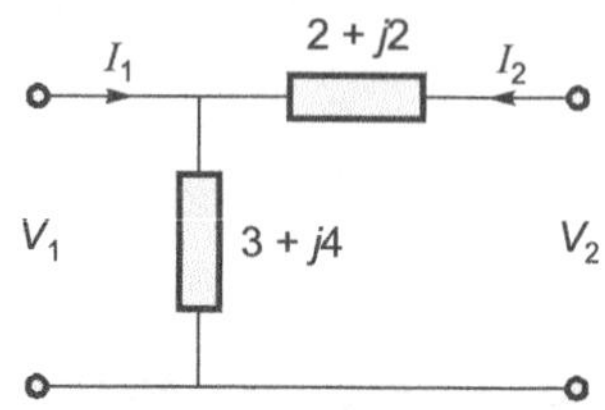

(a) $\begin{bmatrix} 3+j4 & 2+j2 \\ 2+j2 & 5+j6 \end{bmatrix}$ (b) $\begin{bmatrix} 3+j4 & 3+j4 \\ 3+j4 & 5+j6 \end{bmatrix}$

(c) $\begin{bmatrix} 2+j2 & 3+j4 \\ 2+j2 & 5+j6 \end{bmatrix}$ (d) $\begin{bmatrix} 3+j4 & 2+j2 \\ 1+j2 & 3+j4 \end{bmatrix}$

[ESE-2013]

5.35 Consider the following standard symbols for two-port parameters:

1. h_{12} and h_{21} are dimensionless.
2. h_{11} and B have dimensions of ohms.
3. *BC* is dimensionless.
4. *C* is dimensionless.

Which of the above are correct?

(a) 1, 2 and 3 only (b) 1, 2 and 4 only
(c) 3 and 4 only (d) 1, 2, 3 and 4

[ESE-2015]

5.36 A one-port network consists of a capacitor of 2 F in parallel with a resistor of $\frac{1}{3}$ Ω. Then the input admittance is

(a) $2s + 3$ (b) $3s + 2$

(c) $\frac{2}{s} + \frac{1}{3}$ (d) $\frac{s}{2} + 3$

[ESE-2015]

5.37 For the two-port network shown in the figure

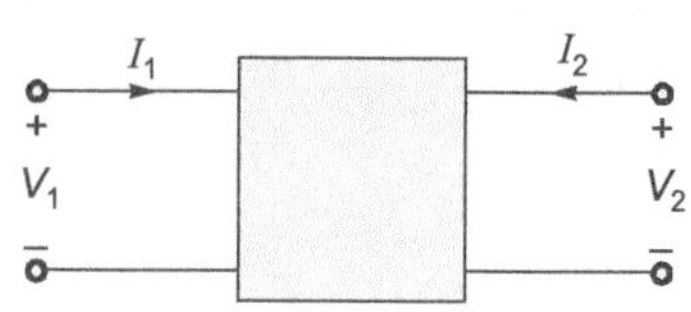

$V_1 = 60I_1 + 20I_2$ and $V_2 = 20I_1 + 40I_2$

Consider the following for the above network :

1. The network is both symmetrical and reciprocal.
2. The network is reciprocal.
3. $A = D$
4. $y_{11} = \frac{1}{50}$

Which of the above is/are correct?
(a) 2 only (b) 2 and 4
(c) 1 only (d) 1 and 3

[ESE-2016]

5.38 In terms of *ABCD* parameters of a 2-port network, the parameters Z_A, Z_B and Z_C of the equivalent T-network are, respectively

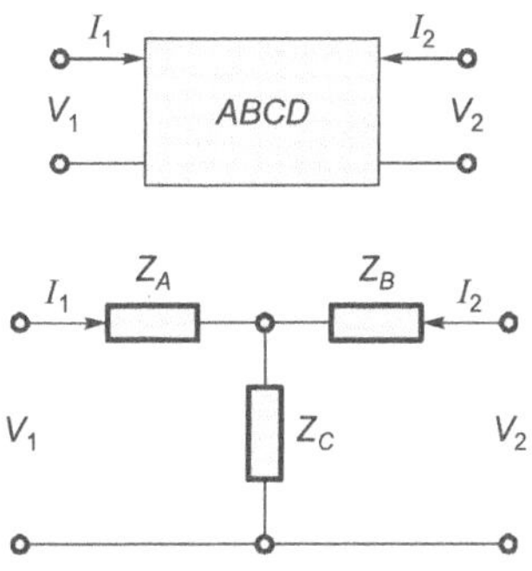

(a) $\frac{A-1}{C}, \frac{D-1}{C}$ and $\frac{1}{C}$

(b) $\frac{A}{C}, \frac{D-1}{C}$ and $\frac{1}{C}$

(c) $\frac{A-1}{C}, \frac{D}{C}$ and $\frac{1}{C}$

(d) $\frac{A}{C}, \frac{D}{C}$ and BC

[ESE-2018]

5.39 The *Z* parameters Z_{11}, Z_{12}, Z_{21} and Z_{22} for the circuit as shown in figure, respectively, are

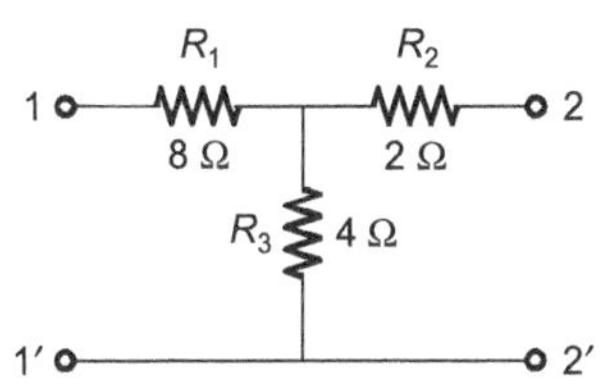

(a) 12 Ω, 4 Ω, 4 Ω and 6 Ω
(b) 8 Ω, 6 Ω, 4 Ω and 4 Ω
(c) 12 Ω, 6 Ω, 6 Ω and 4 Ω
(d) 8 Ω, 4 Ω, 6 Ω and 6 Ω [ESE-2018]

5.40 A series RLC circuit with $R = 2\ \Omega$, $L = \frac{1}{2}$ H, $C = \frac{1}{4}$ F is excited by a 100 V dc source. The circuit is initially in quiescent state. The expression for the current response $i(t)$ due to a dc source will be of the form (K, K_1, K_2 are constants)

(a) $Ke^{-4t}\sin\left(4t+\frac{\pi}{3}\right)$

(b) $Ke^{-2t}\sin\sqrt{8}t$

(c) $(K_1 + K_2t)e^{-2t}$

(d) $K_1e^{-2t} + K_2e^{-4t}$

[ESE-2018]

5.41 The defining equations for V_1 and V_2 analyzing a two-port network in terms of its impedance parameters are respectively
(a) $Z_{12}I_1 + Z_{12}I_2$ and $Z_{21}I_1 + Z_{21}I_2$
(b) $Z_{11}I_1 + Z_{12}I_2$ and $Z_{21}I_1 + Z_{22}I_2$
(c) $Z_{21}I_1 + Z_{21}I_2$ and $Z_{12}I_1 + Z_{12}I_2$
(d) $Z_{12}I_1 + Z_{12}I_2$ and $Z_{22}I_1 + Z_{22}I_2$

[ESE-2019]

5.42 When a number of two-port networks are cascaded then,
(a) *z*-parameters are added up
(b) *y*-parameters are added up
(c) *h*-parameters are multiplied
(d) ABCD-parameters are multiplied

[ESE-2019]

5.43 The *Z* parameters of a two-port network are $Z_{11} = 20\ \Omega$, $Z_{22} = 30\ \Omega$, $Z_{12} = Z_{21} = 10\ \Omega$. The corresponding values of *ABCD* parameters are

(a) $\begin{bmatrix} 2 & 0.1 \\ 50 & 3 \end{bmatrix}$ (b) $\begin{bmatrix} 3 & 50 \\ 0.1 & 2 \end{bmatrix}$

(c) $\begin{bmatrix} 2 & 40 \\ 0.1 & 3 \end{bmatrix}$ (d) $\begin{bmatrix} 2 & 50 \\ 0.1 & 3 \end{bmatrix}$

[ESE-2024]

■■■■

Answers Two Ports and Network Functions

5.1 (b)	5.2 (c)	5.3 (c)	5.4 (c)	5.5 (c)	5.6 (b)	5.7 (c)	5.8 (a)	5.9 (a)
5.10 (b)	5.11 (a)	5.12 (b)	5.13 (b)	5.14 (d)	5.15 (c)	5.16 (c)	5.17 (a)	5.18 (c)
5.19 (b)	5.20 (d)	5.21 (d)	5.22 (c)	5.23 (c)	5.24 (b)	5.25 (c)	5.26 (d)	5.27 (c)
5.28 (b)	5.29 (c)	5.30 (d)	5.31 (d)	5.32 (c)	5.33 (b)	5.34 (b)	5.35 (a)	5.36 (a)
5.37 (b)	5.38 (a)	5.39 (a)	5.40 (b)	5.41 (b)	5.42 (d)	5.43 (d)		

Explanations Two Ports and Network Functions

5.1 (b)

We know h-parameters are defined as

$$\begin{bmatrix} V_1 \\ I_2 \end{bmatrix} = \begin{bmatrix} h_{11} & h_{12} \\ h_{21} & h_{22} \end{bmatrix} \begin{bmatrix} I_1 \\ V_2 \end{bmatrix}$$

so, $V_1 = h_{11}I_1 + h_{12}V_2$ ...(i)

and $I_2 = h_{21}I_1 + h_{22}V_2$ ...(ii)

so, $h_{11} = \left.\dfrac{V_1}{I_1}\right|_{\text{with } V_2 = 0}$

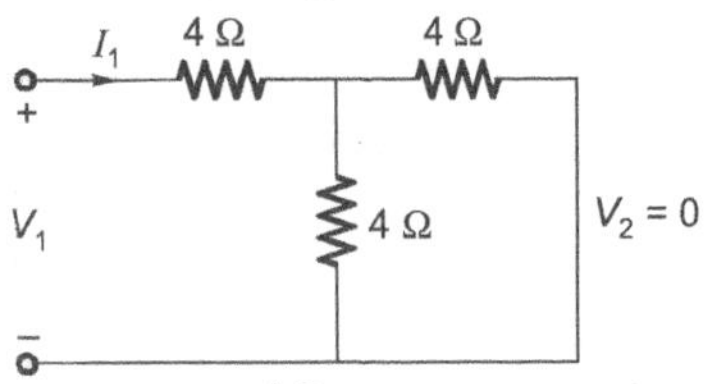

⇒

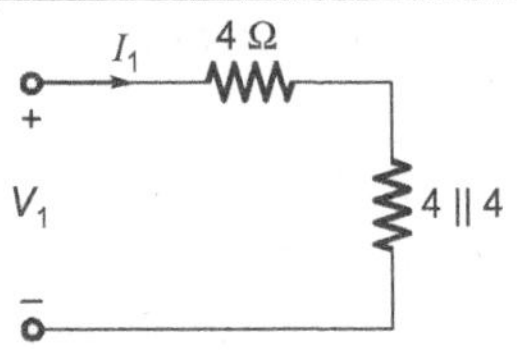

⇒ $h_{11} = \left.\dfrac{V_1}{I_1}\right|_{V_2 = 0} = 4 + 4 \,||\, 4 = 6\ \Omega$

$h_{12} = \left.\dfrac{V_1}{V_2}\right|_{I_1 = 0}$

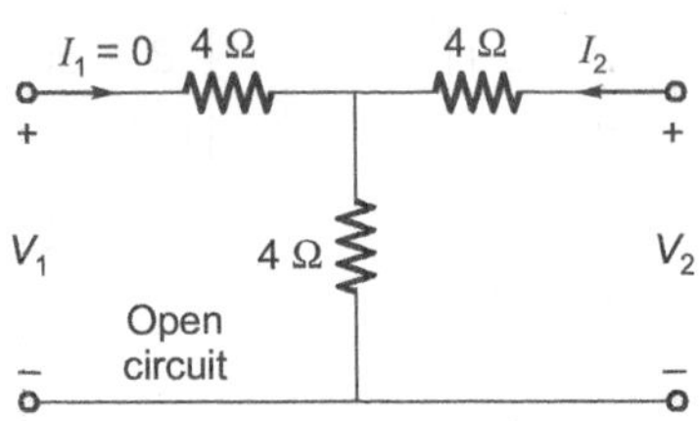

$I_2 = \dfrac{V_2}{8}$ A

∴ $V_1 = 4I_2$

⇒ $V_1 = 4 \times \dfrac{V_2}{8}$

$\dfrac{V_1}{V_2} = 0.5 = h_{12}$

$h_{21} = \left.\dfrac{I_2}{I_1}\right|_{V_2 = 0}$

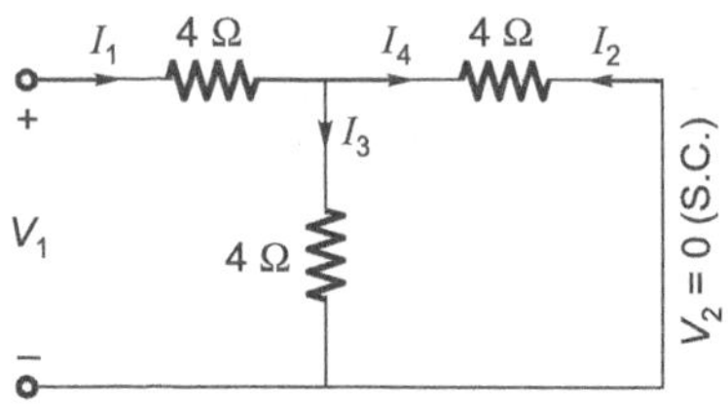

From the figure,

⇒ $I_2 = I_4$ ...(i)

and $I_1 = I_3 + I_4$

also since 4 Ω and 4 Ω are connected in parallel so I_3 and I_4 will be equal due to equal resistances.

$I_3 = I_4$

∴ $I_1 = I_4 + I_4 = 2I_4$

$I_1 = -2I_2$

$\dfrac{I_2}{I_1} = -\dfrac{1}{2} = h_{21} = -0.5$

$h_{22} = \left.\dfrac{I_2}{V_2}\right|_{I_1 = 0}$

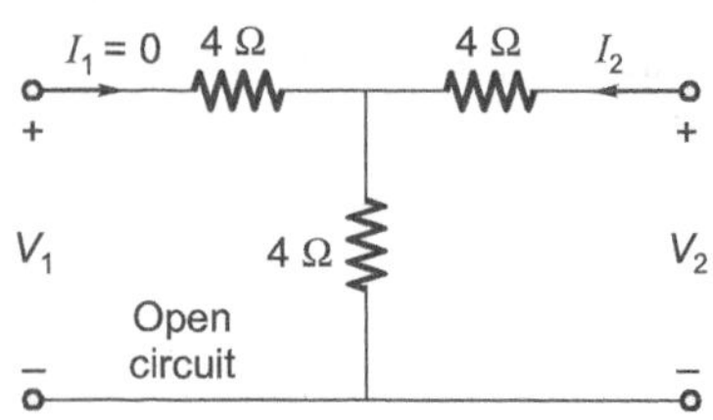

So, $V_2 = 8\,I_2$

$\dfrac{I_2}{V_2} = \dfrac{1}{8} = 0.125$

Alternate solution:

$V_1 = 8I_1 + 4I_2$ ...(i)

$$V_2 = 4(I_1 + I_2) + 4I_2$$
$$= 4I_1 + 8I_2$$
$$8I_2 = V_2 - 4I_1$$
$$I_2 = \frac{V_2}{8} - \frac{I_1}{2} = -\frac{I_1}{2} + \frac{V_2}{8} \quad ...(ii)$$

From equation (*i*),

$$V_1 = 8I_1 + 4\left(-\frac{I_1}{2} + \frac{V_2}{8}\right)$$
$$V_1 = 6I_1 + \frac{V_2}{2}$$

In terms of h-parameter,

$$\begin{bmatrix} V_1 \\ I_2 \end{bmatrix} = \begin{bmatrix} 6 & 0.5 \\ -0.5 & 0.125 \end{bmatrix} \begin{bmatrix} I_1 \\ V_2 \end{bmatrix}$$

5.2 (c)

$$\because \quad h_{11} = \left.\frac{V_1}{I_1}\right|_{V_2=0} \text{ and } h_{22} = \left.\frac{I_2}{V_2}\right|_{I_1=0}$$
$$z_{11} = \left.\frac{V_1}{I_1}\right|_{I_2=0} \text{ and } z_{22} = \left.\frac{V_2}{I_2}\right|_{I_1=0}$$
$$y_{11} = \left.\frac{I_1}{V_1}\right|_{V_2=0} \text{ and } y_{22} = \left.\frac{I_2}{V_2}\right|_{V_1=0}$$

By comparing,

$$h_{11} = \frac{1}{y_{11}} \text{ and } h_{22} = \frac{1}{z_{22}}$$

5.3 (c)

Differential equation,

$$\frac{d^2i}{dt^2} + i = v$$

taking Laplace transform on both side

$$s^2I(s) + I(s) = V(s)$$
$$I(s) = \frac{V(s)}{(s^2+1)} \quad ...(i)$$

Given, $v = Ae^{2t}$

By taking Laplace,

$$V(s) = \frac{A}{s-2}$$

From equation (*i*),

$$I(s) = \frac{A}{(s-2)(s^2+1)}$$

Which is an unstable function due to pole at $s = 2$.
The term $(s-2)$ is dominating one $(s^2 + 1)$.
Hence the dominant solution of i is k_1e^{2t}.

5.4 (c)

$$Y_{11} = \left.\frac{I_1}{V_1}\right|_{V_2=0} = \frac{1}{R_1}$$
$$Y_{21} = \left.\frac{I_2}{V_1}\right|_{V_2=0} = \frac{0}{V_1} = 0$$
$$Y_{12} = \left.\frac{I_1}{V_2}\right|_{V_1=0} = \frac{0}{V_2} = 0$$
$$Y_{22} = \left.\frac{I_2}{V_2}\right|_{V_1=0} = \frac{0}{V_2} = 0$$

5.5 (c)

$$Y_{12} = \left.\frac{I_1}{V_2}\right|_{V_1=0}$$

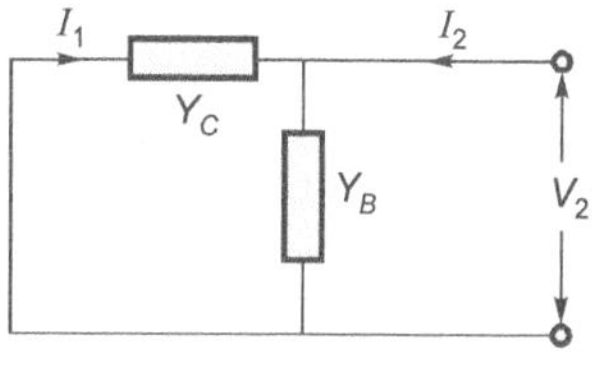

$$-I_1 = V_2Y_c$$
$$Y_{12} = \left.\frac{I_1}{V_2}\right|_{V_1=0} = -Y_c$$

5.6 (b)

For a Lattice circuit, Z-parameters

$$[z] = \begin{bmatrix} \frac{Z_A+Z_B}{2} & \frac{Z_B-Z_A}{2} \\ \frac{Z_B-Z_A}{2} & \frac{Z_A+Z_B}{2} \end{bmatrix}$$

$$= \begin{bmatrix} \frac{(3+j4)+(3-j4)}{2} & \frac{(3-j4)-(3+j4)}{2} \\ \frac{(3-j4)-(3+j4)}{2} & \frac{(3+j4)+(3-j4)}{2} \end{bmatrix}$$

$$\Rightarrow \quad [Z] = \begin{bmatrix} 3 & -j4 \\ -j4 & 3 \end{bmatrix}$$

5.7 (c)

$$I(s) = \frac{K}{s^2+6s+9} = \frac{K}{(s+3)^2}$$
$$= \frac{K_1}{(s+3)} + \frac{K_2}{(s+3)^2}$$
$$\therefore \quad I(t) = (K_1 + K_2t)\,e^{-3t}$$

5.8 (a)

$$Y_{11}(s) = \left.\frac{I_1}{V_1}\right|_{V_2=0} = (2+s)$$

5.9 (a)

In series,

$$[Z] = [Z_a] + [Z_b]$$

as $Z_{11} = Z_{22} = 11\ \Omega$

and $Z_{12} = Z_{21} = 1\ \Omega$

$\therefore$ $[Z]$ for combination

$$= \begin{bmatrix} 11+11 & 1+1 \\ 1+1 & 11+11 \end{bmatrix} = \begin{bmatrix} 22 & 2 \\ 2 & 22 \end{bmatrix}$$

$$\therefore\ [Y] = [Z]^{-1} = \begin{bmatrix} \frac{11}{240} & -\frac{1}{240} \\ -\frac{1}{240} & \frac{11}{240} \end{bmatrix}$$

$$\Rightarrow \quad Y_{11} = \frac{11}{240}\ \mho$$

5.10 (b)

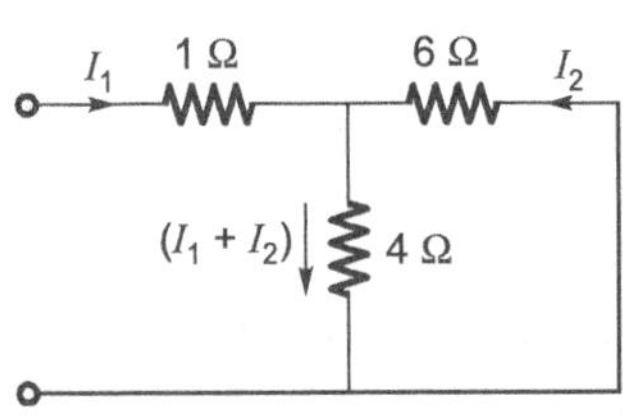

By applying KVL,

$$6I_2 + 4(I_1 + I_2) = 0$$

$$\Rightarrow \quad \frac{I_1 + I_2}{I_2} = -\frac{6}{4}$$

$$\frac{I_1}{I_2} = -\frac{10}{4}$$

$$\because \quad h_{21} = \left.\frac{I_2}{I_1}\right|_{V_2=0}$$

$$\therefore \quad h_{21} = -\frac{4}{10} = -0.4$$

5.11 (a)

$$Y_{11} = \left.\frac{I_1}{V_1}\right|_{V_2=0} = 0.5 + 0.2 = 0.7\ \mho$$

$$Y_{21} = \left.\frac{I_2}{V_1}\right|_{V_2=0} = -0.5\ \mho$$

$$Y_{12} = \left.\frac{I_1}{V_2}\right|_{V_1=0} = -0.5\ \mho$$

$$Y_{22} = \left.\frac{I_2}{V_2}\right|_{V_1=0} = 0.5 + 0.3 = 0.8\ \mho$$

5.12 (b)

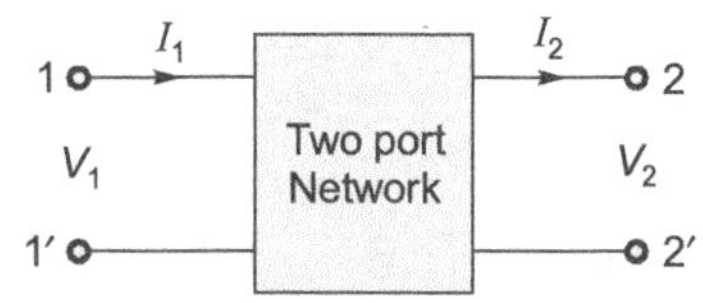

Transmission parameters A, B, C, D

$$V_1 = AV_2 - BI_2$$

$$I_1 = CV_2 - DI_2$$

$$h_{22} = \left.\frac{I_2}{V_2}\right|_{I_1=0}$$

= Open circuit output admittance

Norton equivalent is current source in parallel with Z_{Th}. So option (b) is correct.

5.13 (b)

Parameters	Condition for reciprocal network
z	$z_{12} = z_{21}$
y	$y_{12} = y_{21}$
$ABCD$	$AD - BC = 1$
h	$h_{12} = -h_{21}$
g	$g_{12} = -g_{21}$

5.14 (d)

$$h_{11} = \left.\frac{V_1}{I_1}\right|_{V_2=0} = \text{Impedance}$$

$$h_{12} = \left.\frac{V_1}{V_2}\right|_{I_1=0} = \text{Voltage ratio}$$

$$h_{22} = \left.\frac{I_2}{V_2}\right|_{I_1=0} = \text{Admittance}$$

$$h_{21} = \left.\frac{I_2}{I_1}\right|_{V_2=0} = \text{Current ratio}$$

5.15 (c)

At $t = 0$ inductor will behave as open and at $t = \infty$ Capacitor will behave as open

So, $I_0 = 0$ and $I_\infty = 0$

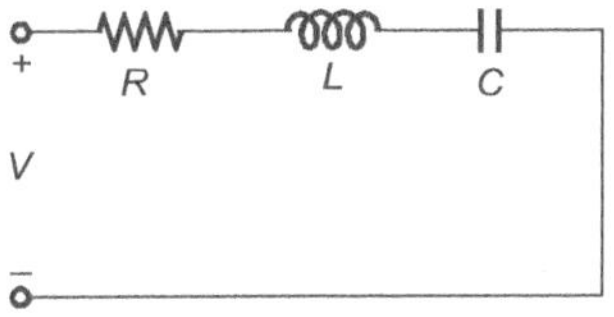

5.16 (c)

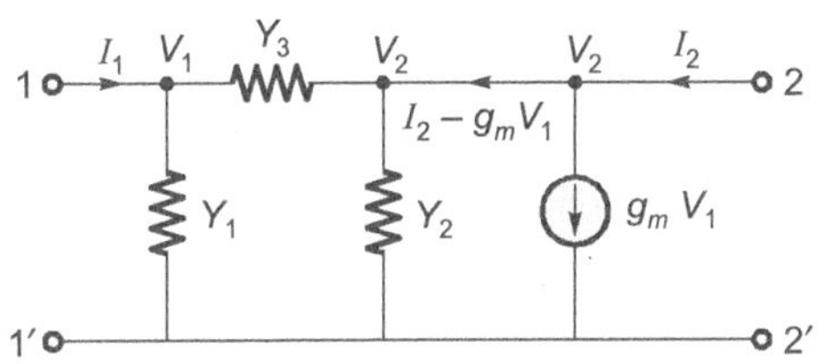

Applying Nodal Analysis

$(V_1 - V_2)\, Y_3 + I_2 - g_m V_1 = V_2 Y_2$

$$Y_{21} = \left.\frac{I_2}{V_1}\right|_{V_2=0}$$

Putting $V_2 = 0$ in above equation

$$V_1 Y_3 + I_2 - g_m V_1 = 0$$

$$\therefore \quad \boxed{\frac{I_2}{V_1} = g_m - Y_3 = Y_{21}}$$

5.17 (a)

$$V_1 = Z_{11} I_1 + Z_{12} I_2$$
$$V_2 = Z_{21} I_1 + Z_{22} I_2$$

From the given network

$$V_1 = Z_1 I_1 + Z_3 (I_1 + I_2)$$
$$V_1 = (Z_1 + Z_3) I_1 + Z_3 I_2 \quad \ldots(i)$$

and

$$V_2 = Z_2 I_2 + Z_3 (I_1 + I_2)$$
$$V_2 = Z_3 I_1 + (Z_3 + Z_2) I_2 \quad \ldots(ii)$$

Comparing equation (*i*) and (*ii*) with standard equations

$$Z_{11} = Z_1 + Z_3$$
$$40 = Z_1 + Z_3 \text{ and } Z_{12} = Z_3$$
$$\therefore \quad Z_3 = 30\ \Omega$$
$$\therefore \quad Z_1 = 10\ \Omega \text{ and } Z_3 + Z_2 = Z_{22}$$
$$Z_3 + Z_2 = 50$$
$$\therefore \quad Z_2 = 20\ \Omega$$

5.18 (c)

$$Z_{12} = \left.\frac{V_1}{I_2}\right|_{I_1=0}$$

$$I_1 = 5\,V_1 + 3\,V_2$$

when, $\quad I_1 = 0$

$$\Rightarrow \quad V_2 = -\frac{5}{3} V_1$$

$$I_2 = 2\,V_1 - 7\,V_2$$

$$2\,V_1 - 7\left(-\frac{5}{3} V_1\right) = \frac{41}{3} V_1$$

$$Z_{12} = \left.\frac{V_1}{I_2}\right|_{I_1=0} = \frac{3}{41}\ \Omega$$

5.19 (b)

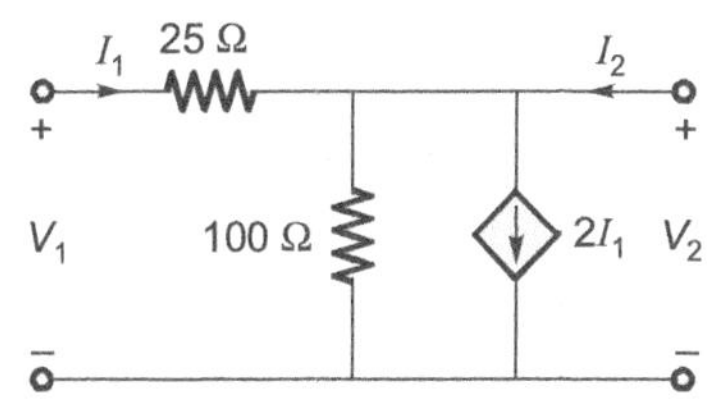

By applying source transformation

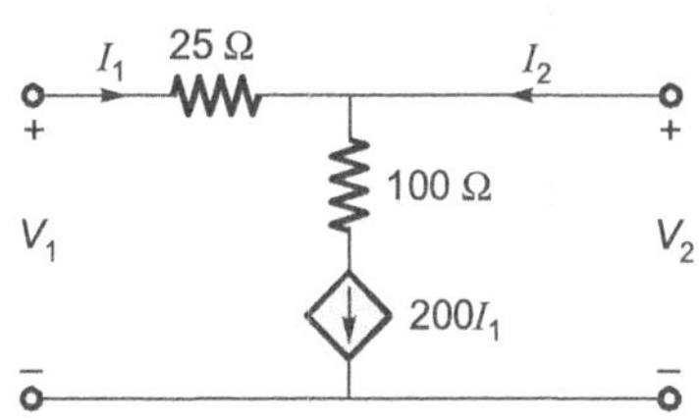

$$V_1 = 25 i_1 + 100(i_1 + i_2) - 200 i_1$$
$$V_2 = 100(i_1 + i_2) - 200 i_1$$

So, $\quad V_1 = -75 i_1 + 100 i_2 \quad \ldots(i)$

$\quad V_2 = -100 i_1 + 100 i_2 \quad \ldots(ii)$

$$Y_{12} = \left.\frac{i_1}{V_2}\right|_{V_1=0}$$

$$75 i_1 = 100 i_2$$

By equation (*i*),

$$V_2 = -100 i_1 + 75 i_1$$
$$V_2 = -25 i_1$$

$$Y_{12} = \frac{i_1}{V_2} = -0.04\ \mho$$

Alternate solution:

$$I_1 = \frac{V_1 - V_2}{25}$$

$$I_2 = 2I_1 + \frac{V_2}{10} + \frac{V_2 - V_1}{100}$$

solving the above two equation we get,

$$[Y] = \begin{bmatrix} 0.04 & -0.04 \\ 0.04 & -0.03 \end{bmatrix}$$

5.20 (d)

$$V_1 = 50i_1 + 20i_2 \quad ...(i)$$
$$V_2 = 30i_1 + 10i_2 \quad ...(ii)$$

So, $Z_{12} = 20$

$$Z = \begin{bmatrix} 50 & 20 \\ 30 & 10 \end{bmatrix}$$

$$Y = [Z]^{-1}$$

$$Y = \frac{-1}{100}\begin{bmatrix} 10 & -20 \\ -30 & 50 \end{bmatrix}$$

$$Y_{12} = 0.2\,℧$$

$$h_{12} = \left.\frac{V_1}{V_2}\right|_{i_1=0}$$

So by equation (*i*) and (*ii*),

$$V_1 = 20i_2$$
$$V_2 = 10i_2$$
$$h_{12} = \frac{V_1}{V_2} = 2$$

5.21 (d)

$$h_{12} = \left.\frac{V_1}{V_2}\right|_{I_1=0}$$

when, $I_1 = 0$

$$V_1 = R_2 I_2$$
$$V_2 = (R_2 + R_3)\,I_2$$
$$h_{12} = \left.\frac{V_1}{V_2}\right|_{I_1=0} = \frac{R_2}{R_2 + R_3}$$

Hence, option (d) is correct.

5.22 (c)

$$y_{12} = \left.\frac{I_1}{V_2}\right|_{V_1=0}$$

$$V_1 = \frac{3}{4}I_1 - \frac{1}{4}I_2$$

when, $V_1 = 0$

$$I_2 = 3I_1$$
$$V_2 = -\frac{I_1}{2} + \frac{I_2}{2} = -\frac{I_1}{2} + \frac{3I_1}{2} = I_1$$
$$y_{12} = \left.\frac{I_1}{V_2}\right|_{V_1=0} = 1\,℧$$

Hence, option (c) is correct.

5.23 (c)

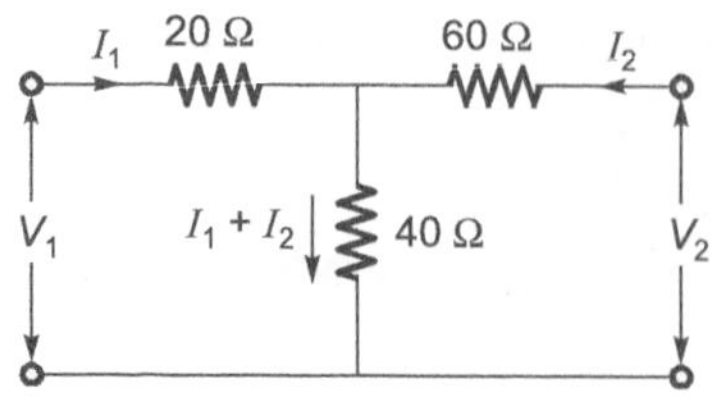

$$V_1 = 20I_1 + 40(I_1 + I_2)$$
$$= 60I_1 + 40I_2$$
$$V_2 = 60I_2 + 40(I_1 + I_2)$$
$$= 40I_1 + 100I_2$$
$$V_1 = Z_{11}I_1 + Z_{12}I_2$$
$$V_2 = Z_{21}I_1 + Z_{22}I_2$$
$$Z = \begin{bmatrix} Z_{11} & Z_{12} \\ Z_{21} & Z_{22} \end{bmatrix} = \begin{bmatrix} 60 & 40 \\ 40 & 100 \end{bmatrix}\Omega$$

Hence, option (c) is correct.

5.24 (b)

$$Y_{12} = \left.\frac{I_1}{V_2}\right|_{V_1=0} = \frac{4I_2}{0.5I_2} = 8 \text{ mho}$$

Hence, option (b) is correct.

5.25 (c)

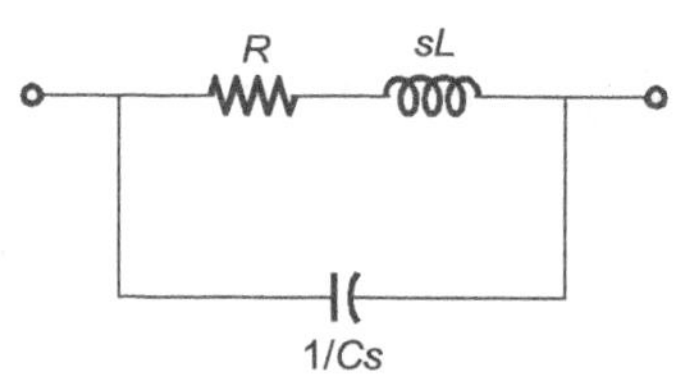

$$Z(s) = \frac{(R+sL)\times\left(\frac{1}{Cs}\right)}{(R+sL)+\frac{1}{Cs}}$$

$$Z(s) = \frac{s(R+sL)}{L\left(s^2 + \frac{R}{L}s + \frac{1}{LC}\right)}$$

Characteristic equation $= s^2 + \frac{R}{L}s + \frac{1}{LC} = 0$

Comparing with $s^2 + 2\xi\omega_n s + \omega_n^2 = 0$

$$\xi\omega_n = \frac{R}{2L}$$

$$\text{Time constant} = \frac{1}{\xi\omega_n} = \frac{2L}{R}$$

$$= \frac{2\times 0.1\times 10^{-6}}{100} = 2 \text{ ns}$$

Hence, option (c) is correct.

5.26 (d)

Condition for a network to be reciprocal.

$$AD - BC = 1 \quad \text{or} \quad \begin{vmatrix} A & B \\ C & D \end{vmatrix} = 1$$

5.27 (c)

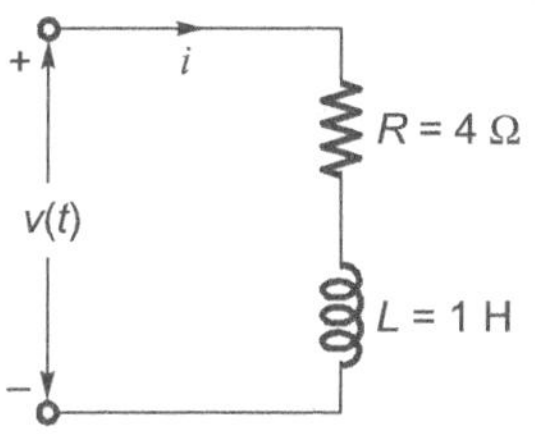

$$v(t) = 4i + L \cdot \frac{di}{dt}$$

$$= 4 \times 5e^{-2t} + 5 \times (-2)\, e^{-2t}$$

$$= 20\, e^{-2t} - 10\, e^{-2t} = 10\, e^{-2t}\ \text{V}$$

5.28 (b)

$$I(s) = \frac{(s+2)(s+4)}{s(s+1)(s+\alpha)}$$

$$\underset{t\to\infty}{Lt}\ I(t) = \underset{s\to 0}{Lt}\ s \cdot I(s)$$

$$= \underset{s\to 0}{Lt} \frac{(s+2)(s+4)}{(s+1)(s+\alpha)} = \frac{2\times 4}{1\times \alpha}$$

$$\Rightarrow \quad \frac{8}{\alpha} = 12$$

$$\alpha = \frac{2}{3} = 0.66$$

Initial value of current

$$= \underset{s\to\infty}{Lt}\ s \cdot \frac{(s+2)(s+4)}{s(s+1)(s+0.66)}$$

$$= \frac{1\times 1}{1\times 1} = 1\,\text{A}$$

5.29 (c)

$$v = L\frac{di}{dt}$$

$$\Rightarrow \quad \int_{-\infty}^{\infty} v\,dt = \int_0^i L\,di$$

$$\int_{-\infty}^{\infty} \delta(t)\,dt = \int_0^i L\,di$$

$$1 = Li \Rightarrow i = \frac{1}{L} = 1\,\text{A}$$

Energy supplied by source = energy and forced in the inductor

$$= \frac{1}{2}Li^2 = \frac{1}{2}\times 1\times (1)^2 = \frac{1}{2}\,\text{J}$$

5.30 (d)

$$h_{11} = \left.\frac{V_1}{I_1}\right|_{V_2=0} \quad \text{and} \quad h_{21} = \left.\frac{I_2}{I_1}\right|_{V_2=0}$$

$\therefore$ At $V_2 = 0$ circuit will seem like

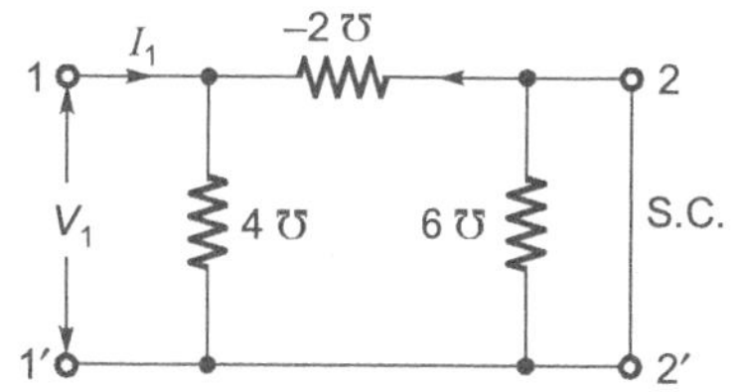

$$\therefore \quad h_{11} = \frac{V_1}{I_1} = \frac{1}{4-2} = \frac{1}{2}\,\Omega$$

$$h_{21} = \frac{I_2}{I_1} = -\left[\frac{-2}{-2+4}\right] = 1$$

5.31 (d)

$$V_1 = 60I_1 + 20I_2 \qquad \ldots(i)$$

$$V_2 = 20I_1 + 40I_2 \qquad \ldots(ii)$$

From equation (*ii*),

$$20I_1 = V_2 - 40I_2 \qquad \ldots(iii)$$

On putting it in equation (*i*),

$$V_1 = 60\,(0.05\,V_2 - 2I_2)$$

$$V_1 = 3\,V_2 - 100I_2 \qquad \ldots(iv)$$

From equation (*iii*) and (*iv*),

$$V_1 = 3\,V_2 - 100I_2 = AV_2 - BI_2$$

$$I_1 = 0.05\,V_2 - 2I_2 = CV_2 - DI_2$$

$$\therefore \quad \begin{bmatrix} A & B \\ C & D \end{bmatrix} = \begin{bmatrix} 3 & 100 \\ \frac{1}{20} & 2 \end{bmatrix}$$

5.32 (c)

For reciprocal networks,

$$Z_{12} = Z_{21}$$

$$Y_{12} = Y_{21}$$

$$\begin{vmatrix} A & B \\ C & D \end{vmatrix} = 1$$

$$h_{12} = -h_{21}$$

$$g_{12} = -g_{21}$$

5.33 (b)

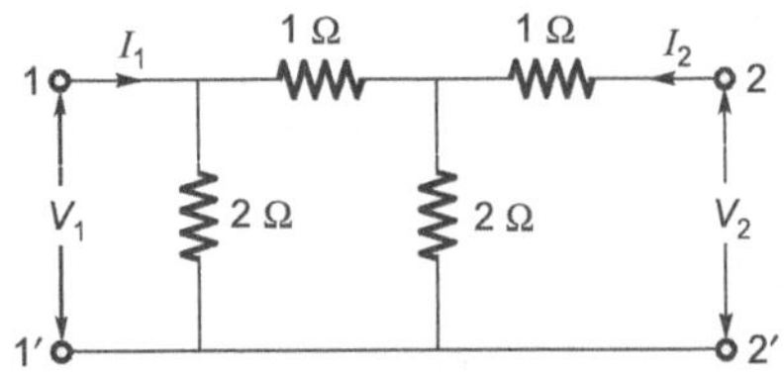

$$Z_{11} = \left.\frac{V_1}{I_1}\right|_{I_2=0} = 2\|(1+2) = \frac{6}{5}\,\Omega$$

$$Z_{12} = \left.\frac{V_1}{I_2}\right|_{I_1=0}$$

$$V_1 = 2\cdot\frac{2}{2+3}$$

$$\Rightarrow \quad Z_{12} = \frac{4}{5}\,\Omega$$

$$Z_{21} = \left.\frac{V_2}{I_1}\right|_{I_2=0}$$

$$V_2 = 2\left(I_1\cdot\frac{2}{2+3}\right)$$

$$\Rightarrow \quad Z_{21} = \frac{4}{5}\,\Omega$$

$$Z_{22} = \left.\frac{V_2}{I_2}\right|_{I_1=0} = \frac{3\times 2}{3+2}+1 = \frac{11}{5}\,\Omega$$

$$\therefore \quad \begin{bmatrix} Z_{11} & Z_{12} \\ Z_{21} & Z_{22} \end{bmatrix} = \begin{bmatrix} \frac{6}{5} & \frac{4}{5} \\ \frac{4}{5} & \frac{11}{5} \end{bmatrix}$$

5.34 (b)

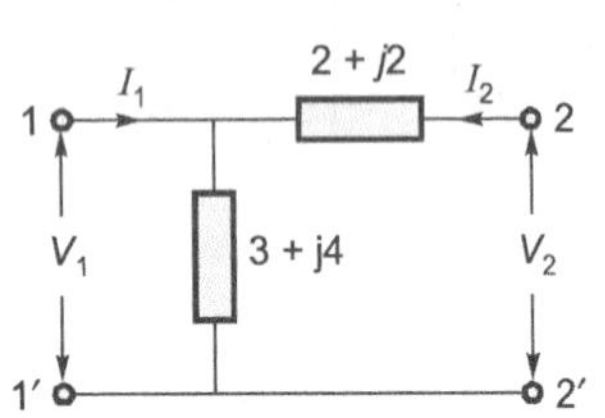

$$Z_{11} = \left.\frac{V_1}{I_1}\right|_{I_2=0} = 3+j4$$

$$Z_{12} = \left.\frac{V_1}{I_2}\right|_{I_1=0} = 3+j4$$

$$Z_{21} = \left.\frac{V_2}{I_1}\right|_{I_2=0} = 3+j4$$

$$Z_{22} = \left.\frac{V_2}{I_2}\right|_{I_1=0} = 3+j4+2+j2$$

$$= 5 + j6$$

$$\therefore \quad \begin{bmatrix} Z_{11} & Z_{12} \\ Z_{21} & Z_{22} \end{bmatrix} = \begin{bmatrix} 3+j4 & 3+j4 \\ 3+j4 & 5+j6 \end{bmatrix}$$

Alternate solution:

From KVL,

$$V_1 = (3 + j4)(I_1 + I_2)$$

$$V_2 = (2 + j2)I_2 + (3 + j4)(I_1 + I_2)]$$

$$\begin{bmatrix} Z_{11} & Z_{12} \\ Z_{21} & Z_{22} \end{bmatrix} = \begin{bmatrix} 3+j4 & 3+j4 \\ 3+j4 & 5+j6 \end{bmatrix}$$

5.35 (a)

$$V_1 = h_{11} I_1 + h_{12} V_2$$

$$I_2 = h_{21} I_1 + h_{22} V_2$$

$$\begin{bmatrix} V_1 \\ I_2 \end{bmatrix} = \begin{bmatrix} h_{11} & h_{12} \\ h_{21} & h_{22} \end{bmatrix}\begin{bmatrix} I_1 \\ V_2 \end{bmatrix} \quad \ldots(i)$$

and

$$\begin{bmatrix} V_1 \\ I_1 \end{bmatrix} = \begin{bmatrix} A & B \\ C & D \end{bmatrix}\begin{bmatrix} V_2 \\ -I_2 \end{bmatrix} \quad \ldots(ii)$$

Here we can conclude,

(a) h_{12} and h_{21} are dimensionless

(b) h_{11} and B have dimension of Ω.

(c) BC is dimensionless

5.36 (a)

Input admittance;

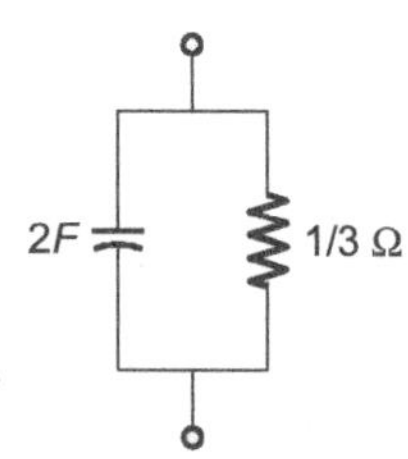

$$Y = \frac{1}{R} + sC$$

$$Y(s) = 3 + 2s$$

5.37 (b)

$$V_1 = 60I_1 + 20I_2 \quad \ldots(1)$$

$$V_2 = 20I_1 + 40I_2 \quad \ldots(2)$$

Comparing (1) and (2) with standard Z-parameters equations i.e.

$$V_1 = Z_{11}I_1 + Z_{12}I_2$$

$$V_2 = Z_{21}I_1 + Z_{22}I_2$$

$$Z_{11} = 60 \quad Z_{12} = 20$$

$$Z_{21} = 20 \quad Z_{22} = 40$$

- For symmetry

  $Z_{11} = Z_{22}$ but as in the given case

  $60 \neq 40$ ∴ Network is not symmetrical

- For reciprocity

$$Z_{12} = Z_{21}$$

and $20 = 20$

$\therefore$ Network is reciprocal.

From equation (2),

$$20I_1 = V_2 - 40I_2$$

$$\Rightarrow \quad I_1 = \frac{1}{20}V_2 - 2I_2 \quad \text{...(3)}$$

Substituting (3) in (1):

$$V_1 = 60\left[\frac{1}{20}V_2 - 2I_2\right] + 20I_2$$

$$V_1 = 3V_2 - 120\,I_2 + 20I_2$$

$$V_1 = 3V_2 - 100I_2 \quad \text{...(4)}$$

Comparing (3) and (4) with standard *ABCD* parameters equation i.e.

$$V_1 = AV_2 - BI_2$$

$$I_1 = CV_2 - DI_2$$

$$A = 3, \quad B = 100$$

$$C = \frac{1}{20}, \quad D = 2$$

$$\therefore \quad A \neq D$$

$$y_{11} = \frac{Z_{22}}{|Z|} = \frac{Z_{22}}{Z_{11}Z_{22} - Z_{12}Z_{21}}$$

$$= \frac{40}{60 \times 40 - 20 \times 20}$$

$$= \frac{40}{2000} = \frac{1}{50}$$

$$y_{11} = \frac{1}{50}$$

5.38 (a)

Easiest way is to check the options.

As, $Z_{11} = \frac{A}{C}$ and $Z_{22} = \frac{D}{C}$

and for T-network,

$$Z_{11} = Z_A + Z_C$$

$$Z_{22} = Z_C + Z_B$$

$$Z_{11} = \frac{A-1}{C} + \frac{1}{C} = \frac{A}{C}$$

$$Z_{22} = \frac{D-1}{C} + \frac{1}{C} = \frac{D}{C}$$

$\therefore$ Option (a) is the correct option.

5.39 (a)

$$Z_{11} = 8 + 4 = 12\ \Omega$$

$$Z_{12} = Z_{21} = 4\ \Omega$$

$$Z_{22} = 4 + 2 = 6\ \Omega$$

5.40 (b)

$$\xi = \frac{R}{2}\sqrt{\frac{C}{L}} = \frac{2}{2}\sqrt{\frac{1}{4}\cdot 2} = \frac{1}{\sqrt{2}}$$

$$= 0.707 \quad \text{(underdamped)}$$

$$\tau = \frac{2L}{R} = 2\cdot\frac{1}{2}\cdot\frac{1}{2} = \frac{1}{2}\text{ sec.}$$

5.41 (b)

$$V_1 = Z_{11}I_1 + Z_{12}I_2$$

$$V_2 = Z_{21}I_1 + Z_{22}I_2$$

5.43 (d)

$$V_1 = 20I_1 + 10I_2 \quad \text{...(1)}$$

$$V_2 = 10I_1 + 30I_2 \quad \text{...(2)}$$

ABCD parameters :

$$V_1 = AV_2 - BI_2$$

$$I_1 = CV_2 - DI_2$$

From equation (2)

$$I_1 = \frac{V_2}{10} - \frac{30}{10}I_2 \quad \text{...(3)}$$

Substitute equation (3) in equation (1)

$$V_1 = 20\left(\frac{V_2}{10} - 3I_2\right) + 10I_2$$

$$V_1 = 2V_2 - 50I_2 \quad \text{...(4)}$$

From eqn. (3) and eqn. (4)

Values of ABCD parameters are $\begin{bmatrix} 2 & 50 \\ 0.1 & 3 \end{bmatrix}$

■■■■

6 Magnetically Coupled Circuits

6.1 The resonant frequency of the given series circuit is

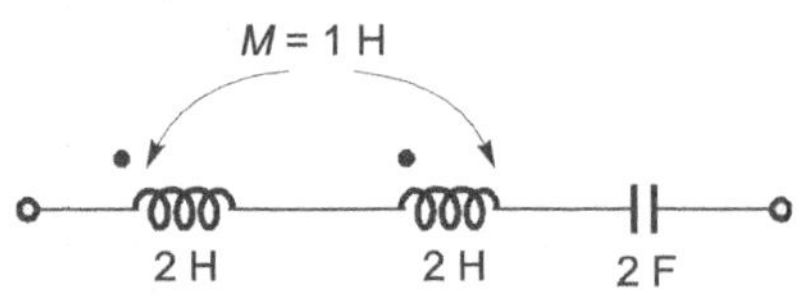

(a) $1/2\pi\sqrt{3}$ Hz (b) $1/4\pi\sqrt{3}$Hz
(c) $1/4\pi\sqrt{2}$Hz (d) $1/\pi\sqrt{2}$ Hz

[ESE-2001]

6.2 The inductance matrix of a system of two mutually coupled inductors shown in Figure-1 is given by

$$L = \begin{bmatrix} 5 & -4 \\ -4 & 7 \end{bmatrix}$$

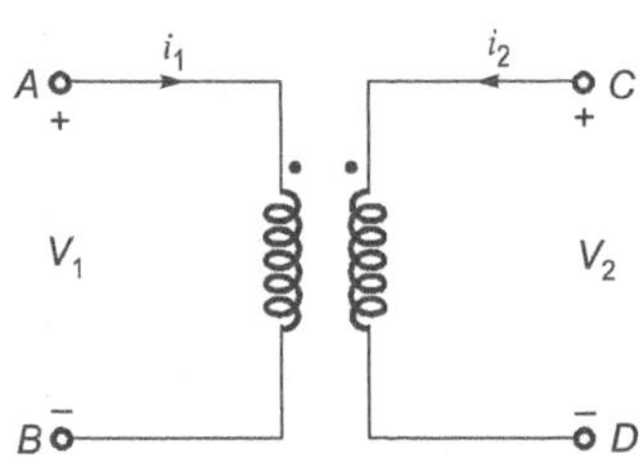

Figure-1

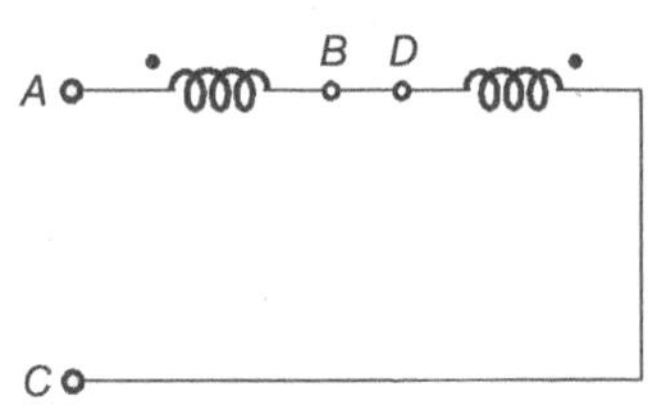

Figure-2

When the inductors are connected as shown in Figure-2, the equivalent inductance of the system is given by
(a) 20 H (b) 4 H
(c) 16 H (d) 8 H

[ESE-2002]

6.3 A pole of driving point admittance function implies
(a) zero current for a finite value of driving voltage
(b) zero voltage for a finite value of driving current
(c) an open circuit condition
(d) None of (a), (b) and (c) mentioned in the question

[ESE-2001]

6.4 The driving point function of the circuit shown in the given figure when $s \to 0$ and $s \to \infty$, (the elements are normalised) will respectively be

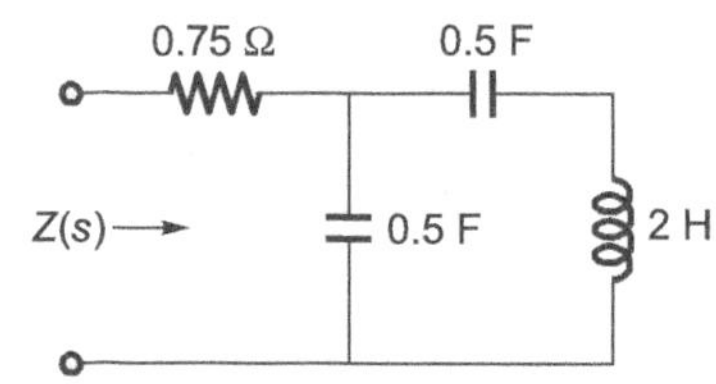

(a) 1/s and 2/s (b) 1/s and 0.75
(c) 0.75 and 2/s (d) 2/s and 0.75

[ESE-2001]

6.5 The driving point impedance $Z(s) = \frac{s+2}{s+3}$. The system is initially at rest. For a voltage signal of unit step, the current $i(t)$ through the impedance Z is given by

(a) $2 - e^{-t}$ (b) $\frac{3}{2} - \frac{1}{2}e^{-3t}$

(c) $\frac{3}{2} - \frac{1}{2}e^{-2t}$ (d) $3 - 2e^{-2t}$

[ESE-2001]

6.6 The system function $H(s) = \frac{1}{s+1}$. For an input signal cos t, the steady state response is

(a) $\frac{1}{\sqrt{2}}\cos\left(t - \frac{\pi}{4}\right)$ (b) $\cos t$

(c) $\cos\left(t - \frac{\pi}{4}\right)$ (d) $\frac{1}{\sqrt{2}\cos t}$

[ESE-2001]

6.7 The inductance matrix of a system of two mutually coupled inductors shown in Figure-1 is given by

$$L = \begin{bmatrix} 5 & -4 \\ -4 & 7 \end{bmatrix}$$

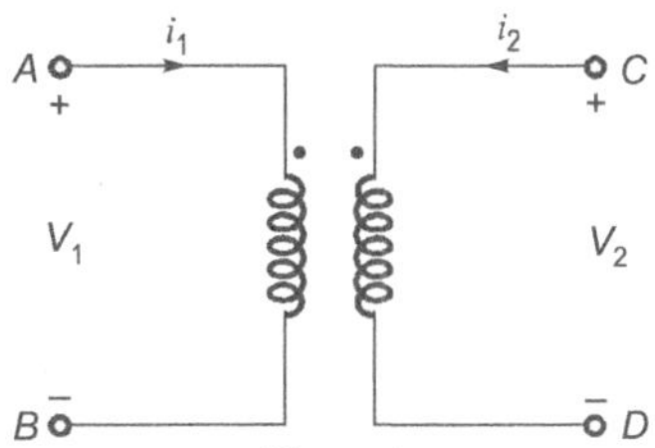

Figure-1

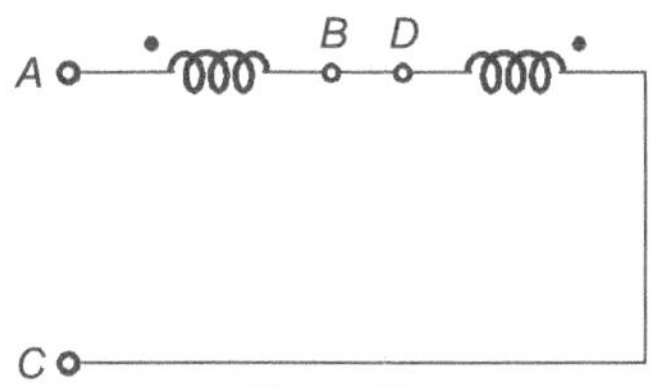

Figure-2

When the inductors are connected as shown in Figure-2, the equivalent inductance of the system is given by

(a) 20 H (b) 4 H
(c) 16 H (d) 8 H

[ESE-2002]

6.8 **Assertion (A):** Transfer impedance of network is always the reciprocal of its transfer admittance.
Reason (R): The impedance of a network is the reciprocal of its admittance.

(a) Both A and R are true and R is the correct explanation of A
(b) Both A and R are true but R is NOT the correct explanation of A
(c) A is true but R is false
(d) A is false but R is true

[ESE-2002]

6.9 The impedance of a parallel RLC network is $Z(s) = \dfrac{5s}{s^2 + 0.5s + 100}$. Then the value of R, L and C are, respectively

(a) $10\ \Omega, \frac{1}{20}\text{H}, \frac{1}{5}\text{F}$ (b) $1\ \Omega, \frac{1}{2}\text{H}, \frac{1}{5}\text{F}$
(c) $10\ \Omega, \frac{1}{20}\text{H}, \frac{1}{2}\text{F}$ (d) $2\ \Omega, \frac{1}{20}\text{H}, \frac{1}{5}\text{F}$

[ESE-2003]

6.10 For a driving point impedance function $Z(s) = \dfrac{s+\alpha}{s+\beta}$; the voltage will lead the current sinusoidal input, if

(a) α and β real positive and $\alpha > \beta$
(b) α is real positive and β is real negative and $\alpha' > \beta$
(c) α and β are real positive and $\beta > \alpha$
(d) α and β are real negative and $\beta > \alpha$

[ESE-2003]

6.11 The impedance $Z(s)$ in the below circuit is

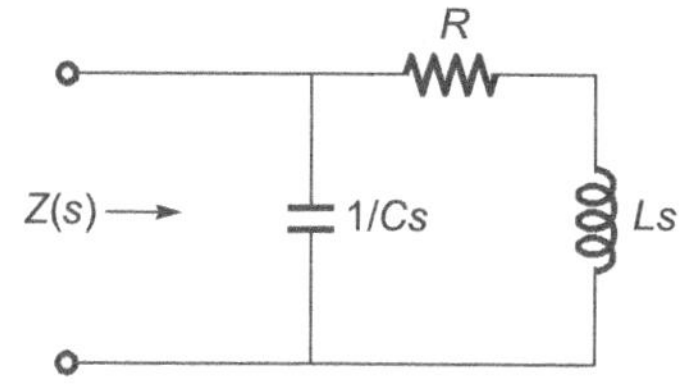

(a) $\dfrac{1}{C}\left(\dfrac{s+(R/L)}{s^2+(R/L)s+(1/LC)}\right)$

(b) $\dfrac{1}{L}\left(\dfrac{s+(1/RC)}{s^2+(1/RC)s+(1/LC)}\right)$

(c) $\dfrac{1}{L}\left(\dfrac{s+(R/L)}{s^2+(1/RC)s+(1/LC)}\right)$

(d) $\dfrac{1}{C}\left(\dfrac{s+(1/RC)}{s^2+(R/L)s+(1/LC)}\right)$

[ESE-2003]

6.12 A unit step current of 1 A is applied to a network whose driving point impedance is

$$Z(s) = \frac{V(s)}{I(s)} = \frac{s+3}{(s+2)^2}$$

Then the steady state and initial values of the voltage developed across the source are respectively

(a) $\left(\frac{3}{4}\text{V}, 1\text{V}\right)$ (b) $\left(\frac{1}{4}\text{V}, \frac{3}{4}\text{V}\right)$
(c) $\left(\frac{3}{4}\text{V}, 0\text{V}\right)$ (d) $\left(1\text{V}, \frac{3}{4}\text{V}\right)$

[ESE-2003]

6.13 The transfer function of a system

$Z(s) = \frac{V(s)}{I(s)} = \frac{s}{s+3}$. The system is at rest for $t < 0$. What will be the value of $v(t)$ for $t \geq 0$, if $i(t) = 3\,u(t)$, where $u(t)$ is a unit step?

(a) e^{-t}
(b) $2\,e^{-3t}$
(c) $3\,e^{-3t}$
(d) $4\,e^{-t}$

[ESE-2004]

6.14 Consider the following pole-zero diagram of a system function $Z(s) = V(s)/I(s)$.

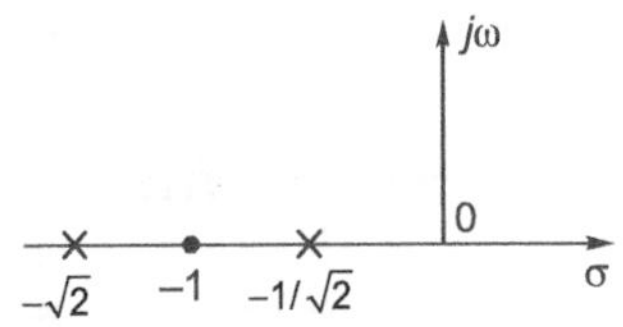

What will be the magnitude of the voltage phasor for $i(t) = \sin t$?

(a) $\frac{3}{2}$
(b) 1
(c) $\frac{2}{3}$
(d) $\frac{1}{3}$

[ESE-2004]

6.15 Consider the following circuit:

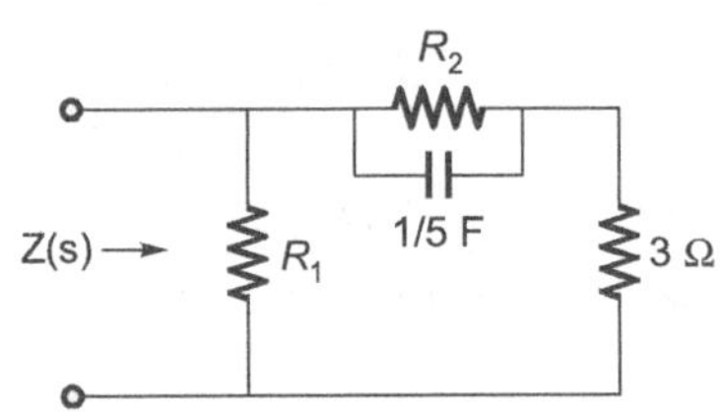

In the above circuit $Z(s) = 2$ as $s \to \infty$ and $Z(s) = 3$ as $s \to 0$. What are, respectively the values of R_1 and R_2?

(a) 6 Ω, 3 Ω
(b) 4 Ω, 2 Ω
(c) 3 Ω, 6 Ω
(d) 6 Ω, 2 Ω

[ESE-2004]

6.16 The realization of the driving point impedance function $Z(s) = (\alpha s^2 + 7s + 3)/(s^2 + 3s + \beta)$ is shown in the circuit given below:

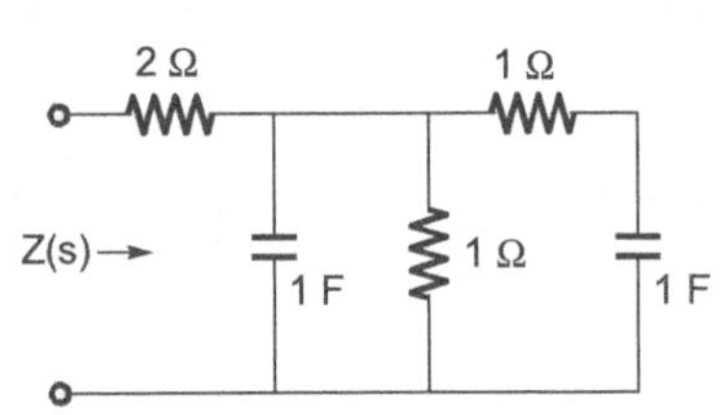

What are, respectively the values of α and β?

(a) 5, 2
(b) 2, 5
(c) 2, 3
(d) 2, 1

[ESE-2004]

6.17 A network has a zero at $s = -1$ and poles at $s = -1 \pm j1$, the multiplier being unity. If the input is unit-step function, then what is the steady state response ?

(a) $2\angle 0°$
(b) $1\angle -45°$
(c) $3\angle 90°$
(d) $0.5\angle 0°$

[ESE-2005]

6.18 What are the poles and zeroes of $Z(s)$ of the below network?

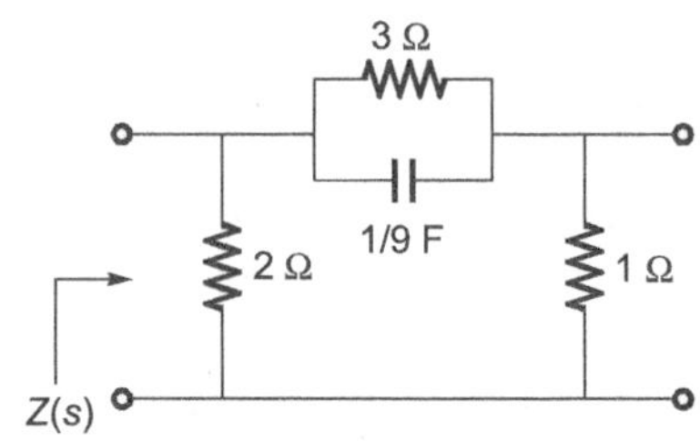

(a) $s = -12$, $s = -6$
(b) $s = -6$, $s = -12$
(c) $s = -3$, $s = -6$
(d) $s = -2$, $s = -4$

[ESE-2005]

6.19 For the two port network shown below, what is the voltage transfer function $V_2(s) / V_1(s)$?

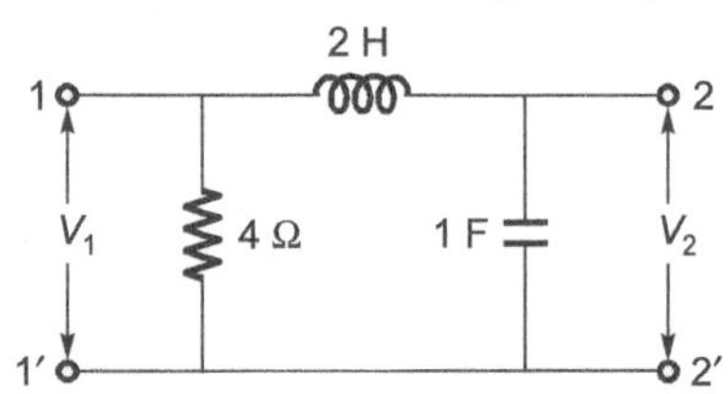

(a) $\frac{s}{(1+2s^2)}$
(b) $\frac{s}{(1+2s)}$
(c) $\frac{1}{(1+2s^2)}$
(d) $\frac{1}{(1+2s)}$

[ESE-2005]

6.20 The current behaviour in a circuit is expressed by: $i(t) = 2e^{-t} - e^{-5t}$ for $t \geq 0$. Which one of the following figures shows the pole-zero pattern of $I(s)$?

(a)

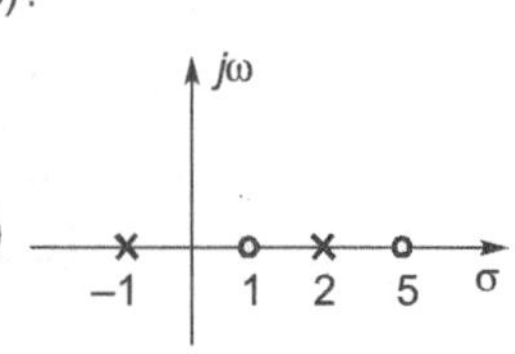

(b) 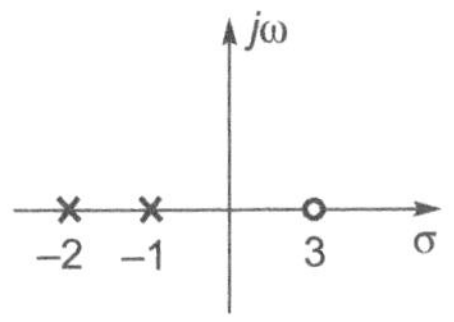

(c) 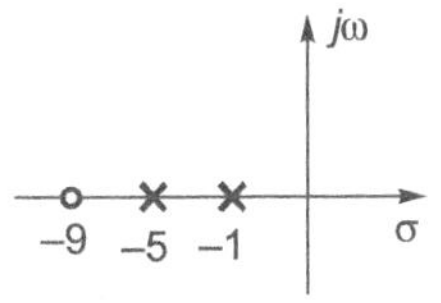

(d) 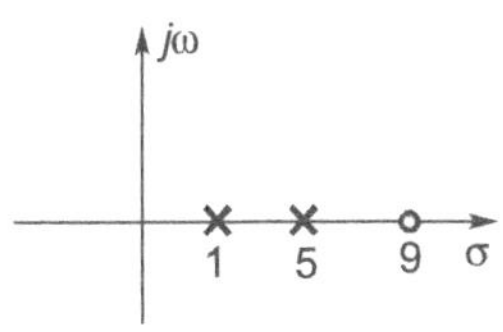

[ESE-2005]

6.21 For the circuit shown below, the poles of the driving-point impedance function are at which one of the following locations?

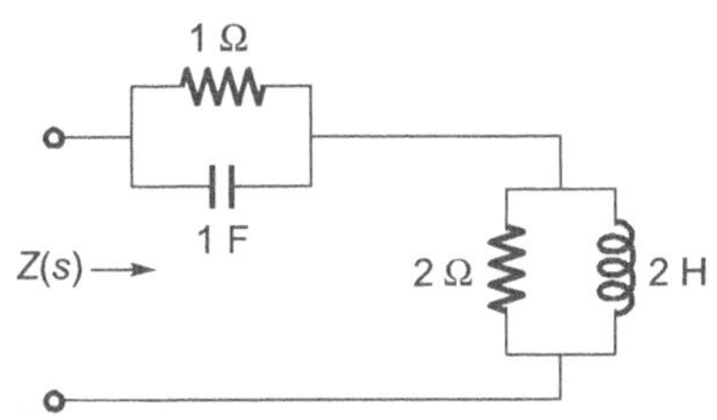

(a) A pole at $s = -2$
(b) A pole at $s = -1$
(c) A double pole at $s = -1$
(d) Poles at $s = -1$ and $s = -2$

[ESE-2005]

6.22 If the driving-point impedance Z_{11} of the network shown in the figure below is given by

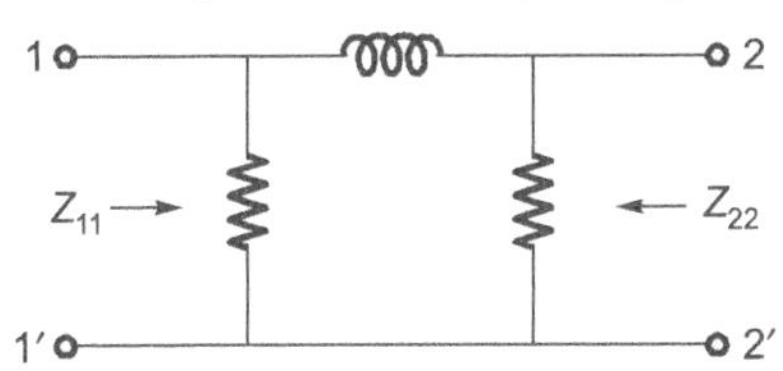

$Z_{11} = k_1\left(\frac{s+3}{s+8}\right)$ then what is the driving-point impedance Z_{22}?

(a) $k_2\left(\frac{s+5}{s+3}\right)$ (b) $k_2\left(\frac{s+5}{s+8}\right)$

(c) $k_2\left(\frac{s+3}{s+5}\right)$ (d) $k_2\left(\frac{s+8}{s+5}\right)$

(k_1 and k_2 are scale factors)

[ESE-2006]

6.23 What is the driving point operational impedance of the network given below?

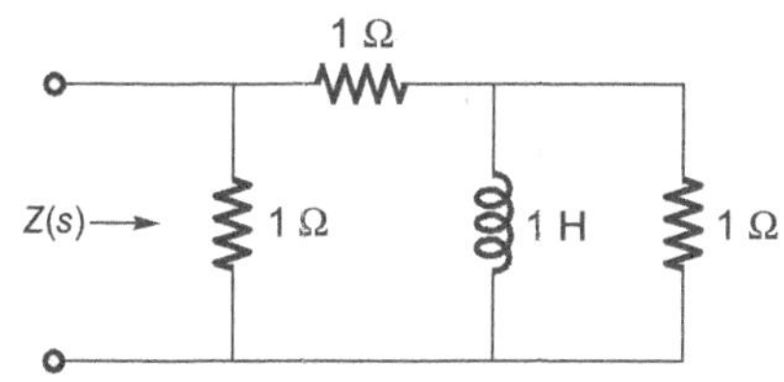

(a) $\frac{(2s+1)}{(3s+2)}$ (b) $\frac{(s+1)}{(2s+1)}$

(c) $\frac{s}{(3s+2)}$ (d) $\frac{(s+2)}{(s+3)}$

[ESE-2007]

6.24 If $Z(s) = \frac{s^2+8s+15}{s^2+4s}$ for the circuit shown below, then what is the value of R_1?

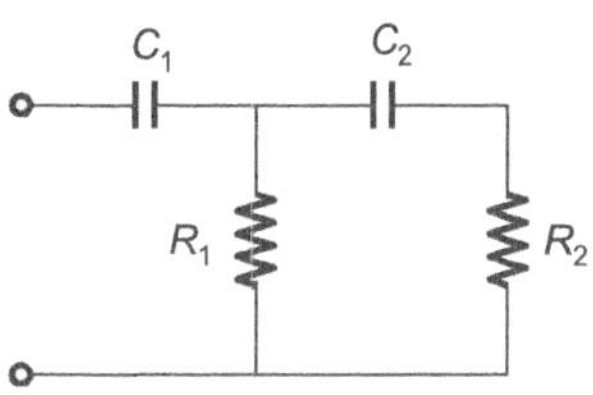

(a) $\frac{17}{16}\,\Omega$ (b) $\frac{9}{8}\,\Omega$

(c) $\frac{19}{16}\,\Omega$ (d) $\frac{21}{16}\,\Omega$

[ESE-2007]

6.25 For the network shown in the figure given below, $Z(0) = 3\ \Omega$ and $Z(\infty) = 2\ \Omega$. What are the values of R_1 and R_2, respectively?

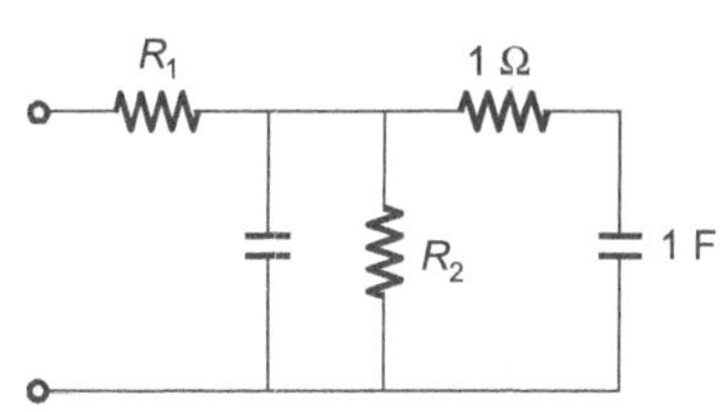

(a) 2 Ω, 1 Ω (b) 3 Ω, 2 Ω
(c) 1 Ω, 2 Ω (d) 2 Ω, 3 Ω

[ESE-2007]

6.26 An impedance function has poles and zeros as follows:
Poles −1, −3
Zeros −2, −4
and impedance has its value $Z(0) = 8$.

Then which one of the following expresses $Z(s)$?

(a) $\frac{(s^2+4s+3)}{(s^2+6s+8)}$ (b) $\frac{(3s^2+12s+9)}{(s^2+6s+8)}$

(c) $\frac{(s^2+6s+8)}{(s^2+4s+3)}$ (d) $\frac{(3s^2+18s+24)}{(s^2+4s+3)}$

[ESE-2007]

6.27 The pole zero diagram of an impedance $Z(s)$ is shown in the below figure. $K = 1$, $\times \Rightarrow$ pole, $O \Rightarrow$ zero

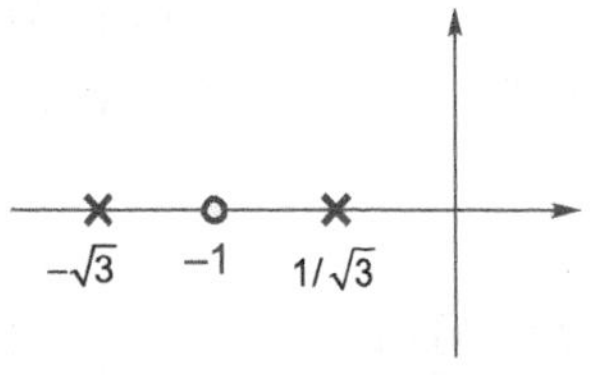

Poles (×) are at $\frac{-1}{\sqrt{3}}$ and $-\sqrt{3}$ and zero (O) is at $s = -1$.

For a signal $I = \cos t$, the steady state response across $Z(s)$ is $V = V_m \cos(t+\theta)$. What is the value of θ?

(a) –45° (b) 45°
(c) –90° (d) 90°

[ESE-2007]

6.28 Two coils are coupled in such a way that the mutual inductance between them is 16 mH. If the inductances of the coils are 20 mH and 80 mH respectively, the coefficient of coupling is

(a) 0.01 (b) 0.4
(c) 0.1 (d) 0.0025

[ESE-2008]

6.29 A network function $Z(s) = \frac{V(s)}{I(s)}$ has a single pole at $s = -\frac{1}{\sqrt{3}}$ and a single zero $s = -\sqrt{3}$. If the excitation $v(t) = \sin t$, then what is the angle of lead or lag of the current?

(a) Lead the voltage by 30°
(b) Lag the voltage by 30°
(c) Lead the voltage by 90°
(d) Lag the voltage by 90°

[ESE-2009]

6.30 The poles and zeroes of a driving point impedance function, $z(s)$ are as

Poles 0, – 2
Zeroes – 1, – 3
and $z(\infty) = 4$
then what is $z(s)$?

(a) $\frac{(s^2+2s)}{(2s^2+8s+6)}$ (b) $\frac{(2s^2+4s)}{(s^2+4s+3)}$

(c) $\frac{(s^2+4s+3)}{(s^2+2s)}$ (d) $\frac{(4s^2+16s+12)}{(s^2+2s)}$

[ESE-2009]

6.31 For the below shown network, the function $G(s) = \frac{V_0(s)}{V_1(s)}$ is $\frac{4s}{s^2+4s+20}$ when R is 2 Ω. What is the value of L and C?

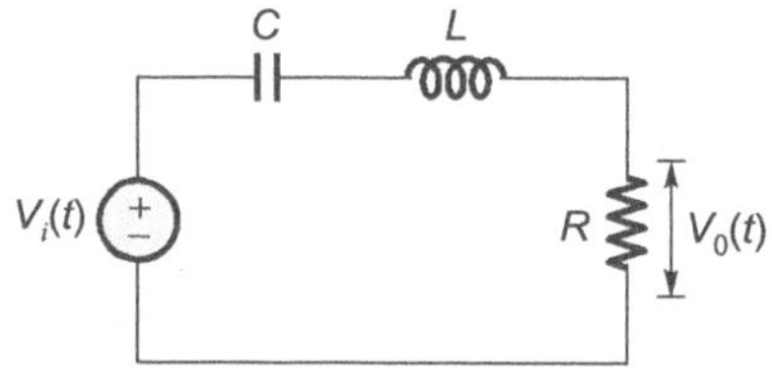

(a) 0.3 H and 1 F (b) 0.4 H and 0.5 F
(c) 0.5 H and 0.1 F (d) 0.5 H and 0.01 F

[ESE-2009]

6.32 For the driving point impedance function, $Z(s) = \frac{as^2+7s+3}{s^2+3s+b}$, the circuit realization is shown below. The values of 'a' and 'b' respectively are:

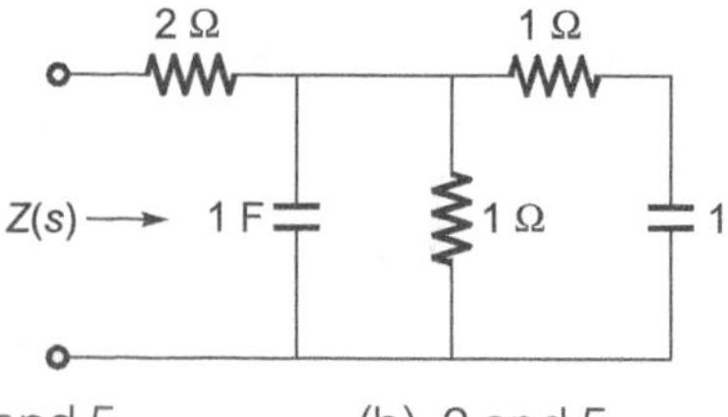

(a) 4 and 5 (b) 2 and 5
(c) 2 and 1 (d) 2 and 3

[ESE-2010]

6.33 For the circuit shown above, the natural frequencies at port 2 are given by $s + 2 = 0$ and $s + 5 = 0$, without knowing which refers to open-circuit and which to short-circuit. Then the impedances Z_{11} and Z_{22} are given respectively by

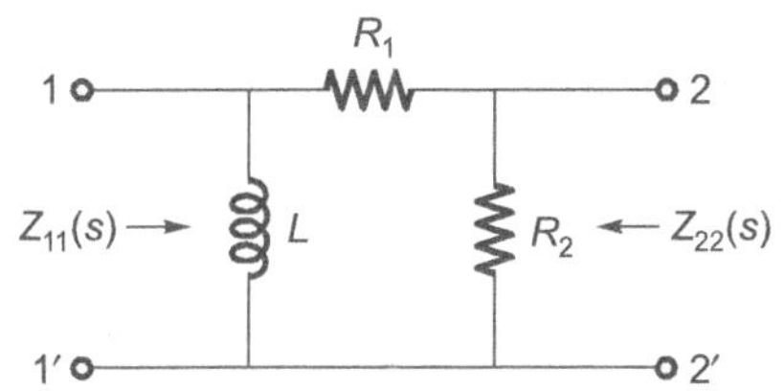

(a) $K_1\dfrac{s+5}{s+2}$, $K_2\dfrac{s+2}{s+5}$

(b) $K_1\dfrac{s+2}{s+5}$, $K_2\dfrac{s+5}{s+2}$

(c) $K_1\dfrac{s}{s+5}$, $K_2\dfrac{s+2}{s+5}$

(d) $K_1\dfrac{s+2}{s+5}$, $K_2\dfrac{s+2}{s+5}$ [ESE-2010]

6.34 For the circuit shown below, the two natural frequencies of the driving-point impedance $Z(s)$ are given by $s + 1 = 0$ and $s + 4 = 0$. It is not known, which is for open-circuit and which is for short-circuit. Then $Z(s)$ is given by

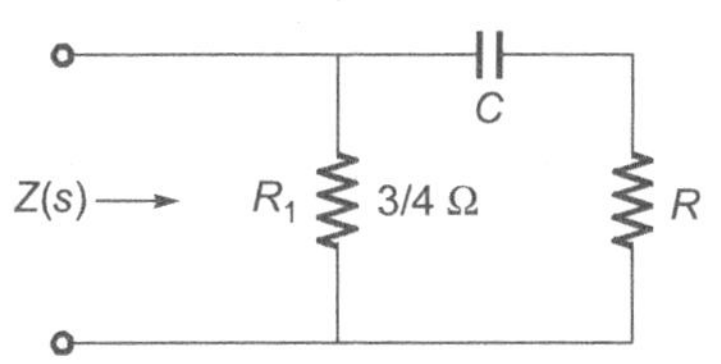

(a) $\dfrac{3}{16}\dfrac{(s+4)}{(s+1)}$ (b) $\dfrac{3}{4}\dfrac{(s+4)}{(s+1)}$

(c) $3\dfrac{(s+1)}{(s+4)}$ (d) $\dfrac{1}{3}\dfrac{(s+4)}{(s+1)}$

[ESE-2010]

6.35 For the circuit shown in figure below, the values of $Z(s)$ as $s \to 0$ and $s \to \infty$ are respectively given by

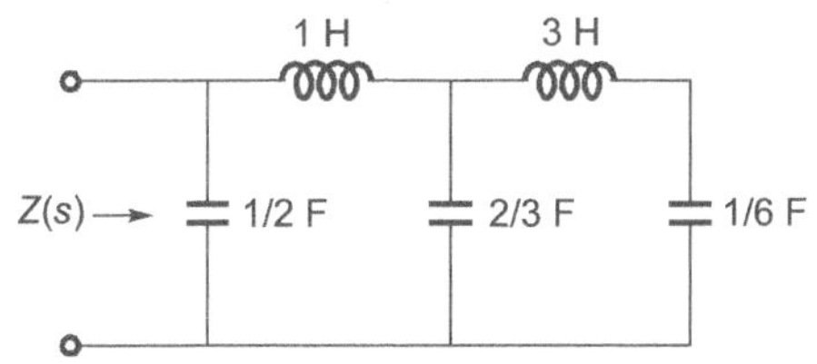

(a) $\dfrac{2}{s}, s$ (b) $\dfrac{6}{8s}, \dfrac{2}{s}$

(c) $\dfrac{8s}{6}$ (d) $\dfrac{s}{2}$

[ESE-2010]

6.36 For the below circuit, the current gain function $I_0(s)/I_i(s)$ has poles and zeros as

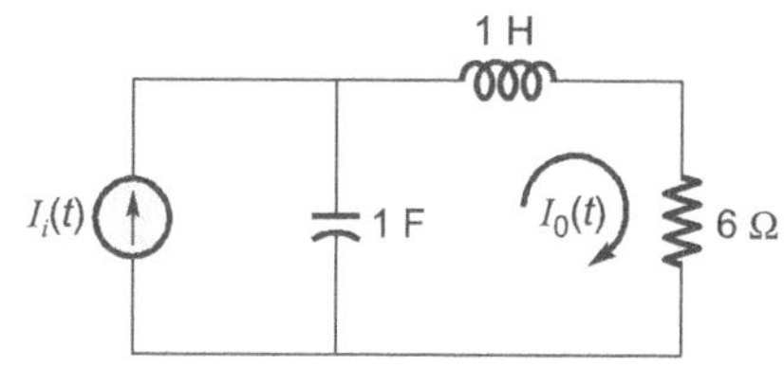

(a) No zeros, $P_1 = -3 + 2\sqrt{2}$, $P_2 = -3 - 2\sqrt{2}$

(b) $Z_1 = 0$, $Z_2 = 2$, $P_1 = -3 + 2\sqrt{2}$, $P_2 = +3 - 2\sqrt{2}$

(c) No zeros, $P_1 = 1$, $P_2 = 1$

(d) $Z_1 = 1$, $Z_2 = 1$, $P_1 = 3 + j2\sqrt{2}$, $P_2 = 3 - j2\sqrt{2}$

[ESE-2011]

6.37 For the network function $\dfrac{V(s)}{I(s)} = \dfrac{s+3}{2s+3}$, the $v(t)$ at $t = 0$ for the relaxed circuit with unit step $i(t)$, is

(a) 0.5 V (b) 1.0 V

(c) 1.5 V (d) 2.0 V

[ESE-2011]

6.38 The voltage transfer function of the network is

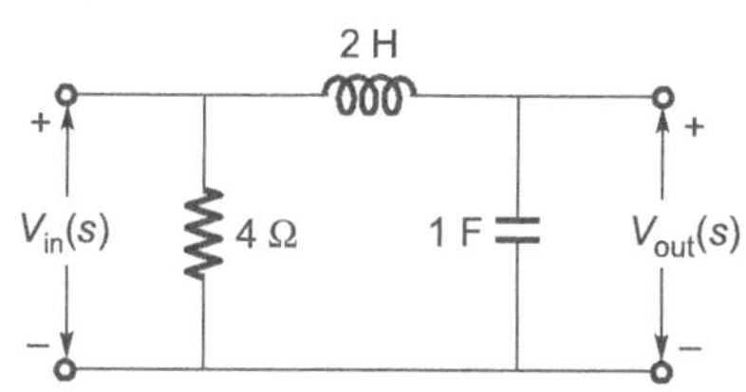

(a) $\dfrac{1}{1+2s}$ (b) $1 + 4s$

(c) $6 - s$ (d) $\dfrac{1}{1+2s^2}$

[ESE-2011]

6.39 The transfer function $\dfrac{V_0(s)}{V_i(s)}$ of the 2-port network is

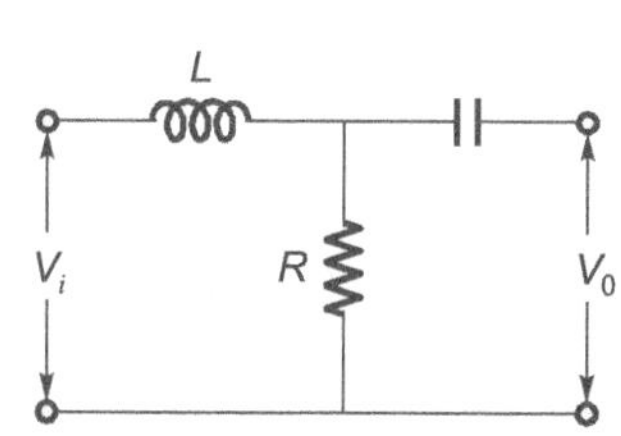

(a) $\dfrac{1}{1+s\dfrac{L}{R}}$ (b) $\dfrac{1+sCR}{sC(R+sL)}$

(c) $\dfrac{\dfrac{1}{sC}}{R+sL}$ (d) $\dfrac{s}{s+\dfrac{R}{L}}$

[ESE-2011]

6.40 The current in a coil changes from 5 A to 1 A in 0.4 second. The induced voltage is 40 V. The self-inductance in henry is

(a) 1 (b) 2
(c) 4 (d) 10 [ESE-2012]

6.41 A network has a transfer function

$$H(s) = \frac{V(s)}{I(s)} = \frac{2s+5}{s+2}$$

If the current $i(t)$ is a unity step function, the steady-state value of $v(t)$ is given by

(a) 0 (b) 2.5 A
(c) 2 A (d) infinity

[ESE-2012]

6.42 Two coupled coils with $L_1 = L_2 = 0.6$ H have a coupling coefficient of $K = 0.8$. The turns ratio N_1/N_2 is

(a) 4 (b) 2
(c) 1 (d) 0.5 [ESE-2013]

6.43 An air-cored solenoid of 250 turns has a cross-sectional area $A = 80\ \text{cm}^2$ and length $l = 100$ cm. The value of its inductance is

(a) 0.425 mH (b) 0.628 mH
(c) 0.751 mH (d) 0.904 mH

[ESE-2016]

6.44 The current in a coil changes uniformly from 10 A to 1 A in half a second. A voltmeter connected across the coil gives a reading of 36 V. The self-inductance of the coil is

(a) 0.5 H (b) 1 H
(c) 2 H (d) 4 H [ESE-2016]

6.45 In a mutually coupled circuit, the primary current is reduced from 4 A to zero in 10 μs. A voltage of 40000 V is observed across the secondary. The mutual inductance between the coils is

(a) 100 H (b) 10 H
(c) 0.1 H (d) 0.01 H

[ESE-2016]

6.46 For a two-port reciprocal network, the three transmission parameters are $A = 4$, $B = 7$ and $C = 5$. What is the value of D?

(a) 9.5 (b) 9.0
(c) 8.5 (d) 8.0 [ESE-2017]

6.47 Consider the following as representation of reciprocity in terms of z-parameters:

1. $z_{11} = z_{12}$
2. $z_{12} = z_{22}$
3. $z_{12} = z_{21}$

Which of the above representation is/are correct?

(a) 1 only (b) 2 only
(c) 3 only (d) 1, 2 and 3

[ESE-2017]

6.48 The y-parameters for the network shown in the figure can be represented by

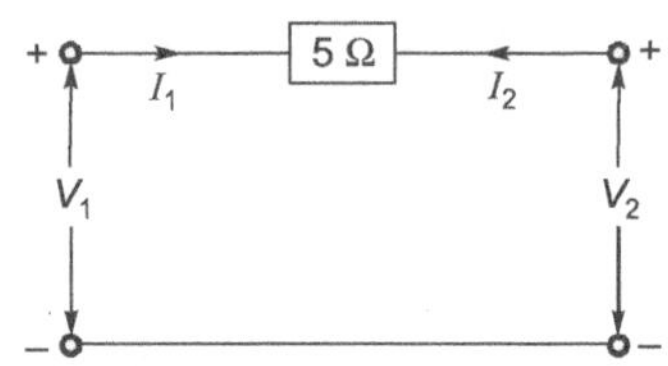

(a) $[y] = \begin{bmatrix} -\frac{1}{5} & \frac{1}{5} \\ \frac{1}{5} & -\frac{1}{5} \end{bmatrix} \mho$

(b) $[y] = \begin{bmatrix} \frac{1}{5} & -\frac{1}{5} \\ -\frac{1}{5} & \frac{1}{5} \end{bmatrix} \mho$

(c) $[y] = \begin{bmatrix} -5 & 5 \\ 5 & -5 \end{bmatrix} \mho$

(d) $[y] = \begin{bmatrix} 5 & -5 \\ -5 & 5 \end{bmatrix} \mho$

[ESE-2017]

6.49 In the two-port network shown, which of the following is correct?

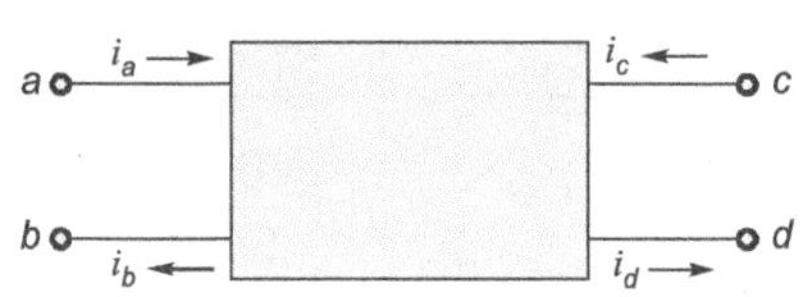

(a) $i_b = i_c$ (b) $i_a = i_d$
(c) $i_c = i_d$ (d) $i_a = i_b$

[ESE-2017]

6.50 Consider the following statements:

1. Mutual inductance describes the voltage induced at the ends of a coils due to the magnetic field generated by a second coil.
2. The dot convention allows a sign to be assigned to the voltage induced due to mutual inductance term.
3. The coupling coefficient is given by $k = \frac{M}{\sqrt{L_1 + L_2}}$.

Which of the above statements are correct?

(a) 1, 2 and 3 (b) 1 and 3 only
(c) 1 and 2 only (d) 2 and 3 only

[ESE-2019]

6.51 Two perfectly coupled coils each of 1 H self-inductance are connected in parallel so as to aid each other. What is the value of overall inductance?

(a) 0 H (b) 0.5 H
(c) 1 H (d) 2 H

[ESE-2024]

6.52 What is the value of equivalent inductance of the network shown in the figure?

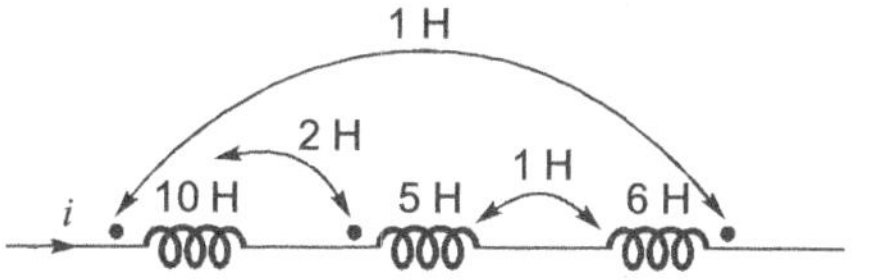

(a) 19 H (b) 21 H
(c) 23 H (d) 25 H

[ESE-2024]

Answers Magnetically Coupled Circuits

6.1 (b)	6.2 (a)	6.3 (b)	6.4 (b)	6.5 (c)	6.6 (a)	6.7 (a)	6.8 (d)	6.9 (a)
6.10 (c)	6.11 (a)	6.12 (c)	6.13 (c)	6.14 (c)	6.15 (a)	6.16 (d)	6.17 (d)	6.18 (b)
6.19 (c)	6.20 (c)	6.21 (b)	6.22 (b)	6.23 (a)	6.24 (a)	6.25 (a)	6.26 (d)	6.27 (a)
6.28 (b)	6.29 (a)	6.30 (d)	6.31 (c)	6.32 (c)	6.33 (c)	6.34 (a)	6.35 (b)	6.36 (a)
6.37 (a)	6.38 (d)	6.39 (a)	6.40 (c)	6.41 (b)	6.42 (c)	6.43 (b)	6.44 (c)	6.45 (c)
6.46 (b)	6.47 (c)	6.48 (b)	6.49 (c & d)		6.50 (c)	6.51 (c)	6.52 (b)	

Explanations Magnetically Coupled Circuits

6.1 (b)

$$L_{\text{equivalent}} = L_1 + L_2 + 2\text{ M} = 2 + 2 + 2 \times 1 = 6\text{ H}$$

Resonant frequency

$$= f_r = \frac{1}{2\pi\sqrt{LC}} = \frac{1}{2\pi\sqrt{6 \times 2}} = \frac{1}{4\pi\sqrt{3}}\text{ Hz}$$

6.2 (a)

Inductance matrix

$$\begin{bmatrix} V_1 \\ V_2 \end{bmatrix} = \begin{bmatrix} j\omega L_1 & +j\omega M \\ +j\omega M & j\omega L_2 \end{bmatrix}\begin{bmatrix} I_1 \\ I_2 \end{bmatrix}$$

By comparing given inductance matrix

$$M = -4$$
$$L_1 = 5, L_2 = 7$$

and for figure 2

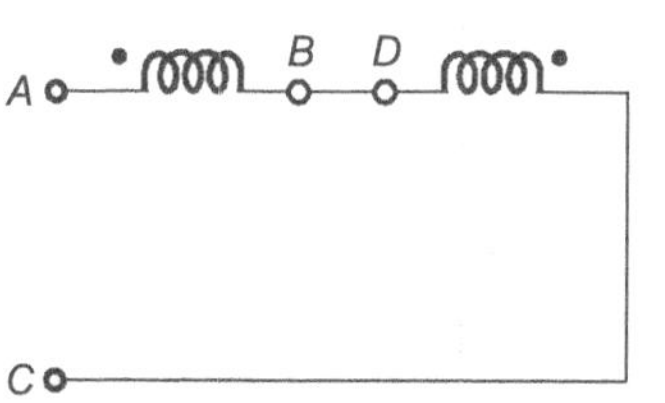

$$L_{eq} = L_1 + L_2 - 2\text{ M} = 5 + 7 - 2(-4) = 20\text{ H}$$

6.3 (b)

At pole, the driving point admittance function becomes infinite i.e.

$$Y_{11} = \infty$$

and we know,

$$Y_{11} = \frac{I_1}{V_1} = \infty$$

$\therefore$ for a finite value of driving current there is a zero voltage.

6.4 (b)

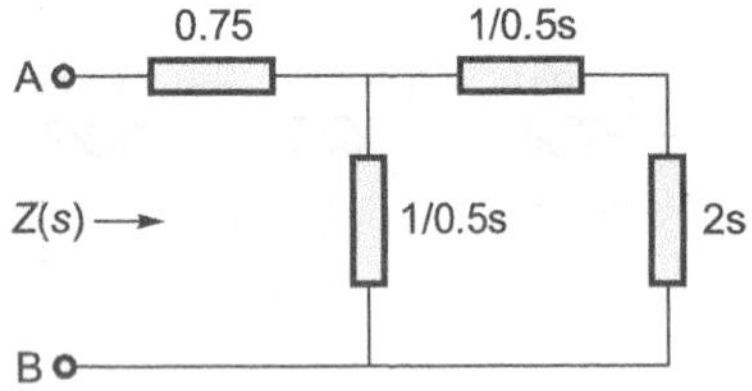

$$\therefore \quad Z(s) = \left[\left(\frac{2}{s} + 2s\right) \| \frac{2}{s}\right] + 0.75$$

$$\Rightarrow \quad Z(s) = \frac{2(1+s^2)}{s(s^2+2)} + \frac{3}{4}$$

$$\therefore \quad Z(s)|_{s \to 0} = \frac{8}{8s} = \frac{1}{s}$$

$$\text{also} \quad Z(s) = \frac{3 + \frac{8}{s} + \frac{6}{s^2} + \frac{8}{s^3}}{4\left(1 + \frac{2}{s^2}\right)}$$

$$\therefore \quad Z(s)|_{s \to \infty} = \frac{3}{4} = 0.75$$

6.5 (c)

$$Z(s) = \frac{V(s)}{I(s)} = \frac{s+2}{s+3}$$

$$I(s) = \left(\frac{s+3}{s+2}\right) V(s)$$

For unit step voltage signal

$$V(s) = \frac{1}{s}$$

$$\therefore \quad I(s) = \left(\frac{s+3}{s+2}\right)\frac{1}{s} = \frac{3}{2s} - \frac{1}{2(s+2)}$$

$$i(t) = \mathcal{L}^{-1}[I(s)] = \frac{3}{2} - \frac{1}{2}e^{-2t}$$

6.6 (a)

Given input signal is cos*t*

$$\therefore \quad \omega = 1$$

$$\text{and} \quad H(s) = \frac{1}{s+1}$$

$$\therefore \quad H(j\omega) = \frac{1}{j+1}$$

$$\text{and} \quad |H(j\omega)| = \frac{1}{\sqrt{1^2+1^2}} = \frac{1}{\sqrt{2}}$$

$$\angle H(j\omega) = -\tan^{-1}(1) = -\frac{\pi}{4}$$

$\therefore$ steady state response

$$= \frac{1}{\sqrt{2}}\cos\left(t - \frac{\pi}{4}\right)$$

6.7 (a)

Inductance matrix

$$\begin{bmatrix} V_1 \\ V_2 \end{bmatrix} = \begin{bmatrix} j\omega L_1 & +j\omega M \\ +j\omega M & j\omega L_2 \end{bmatrix}\begin{bmatrix} I_1 \\ I_2 \end{bmatrix}$$

By comparing given inductance matrix

$$M = -4$$

$$L_1 = 5, L_2 = 7$$

and for figure 2

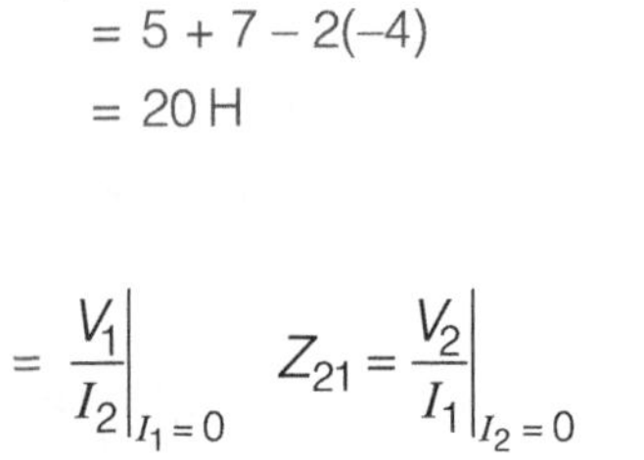

$$L_{eq} = L_1 + L_2 - 2M$$

$$= 5 + 7 - 2(-4)$$

$$= 20 \text{ H}$$

6.8 (d)

$$Z_{12} = \left.\frac{V_1}{I_2}\right|_{I_1=0} \quad Z_{21} = \left.\frac{V_2}{I_1}\right|_{I_2=0}$$

$$Y_{12} = \left.\frac{I_1}{V_2}\right|_{V_1=0} \quad Y_{21} = \left.\frac{V_2}{I_1}\right|_{I_2=0}$$

$\because$ These transfer functions are defined conditionally,

$$\Rightarrow \quad Z_{12} \neq \frac{1}{Y_{12}} \text{ and } Z_{21} \neq \frac{1}{Y_{21}}$$

6.9 (a)

For parallel RLC network, go for $Y(s)$ instead of $Z(s)$

$$Y(s) = \frac{s^2+0.5s+100}{5s}$$

$$= \frac{s}{5}+0.1+\frac{20}{s}$$

Comparing with

$$Y(s) = Cs+\frac{1}{R}+\frac{1}{Ls}$$

$$C = \frac{1}{5}F;\ R = \frac{1}{0.1} = 10\ \Omega$$

$$L = \frac{1}{20}\text{H}$$

6.10 (c)

$$V(s) = I(s)\left(\frac{s+\alpha}{s+\beta}\right)$$

$$\phi = \tan^{-1}\left(\frac{\omega}{\alpha}\right)-\tan^{-1}\left(\frac{\omega}{\beta}\right)$$

$$= \tan^{-1}\left(\frac{\frac{\omega}{\alpha}-\frac{\omega}{\beta}}{1+\left(\frac{\omega}{\alpha}\right)\left(\frac{\omega}{\beta}\right)}\right)$$

$$\Rightarrow \quad \phi = \tan^{-1}\left\{\frac{\omega(\beta-\alpha)}{\alpha\beta+\omega^2}\right\}$$

For voltage to be lead current

$\phi > 0$ or $\beta > \alpha$

6.11 (a)

$$Z(s) = (R+Ls)\parallel\left(\frac{1}{Cs}\right) = \frac{(R+Ls)\left(\frac{1}{Cs}\right)}{R+Ls+\frac{1}{Cs}}$$

$$= \frac{L\left(s+\frac{R}{L}\right)\left(\frac{1}{Cs}\right)}{L\left(\frac{R}{L}+s+\frac{1}{CLs}\right)}$$

$$Z(s) = \frac{(s+R/L)}{C\left(s^2+\frac{R}{L}s+\frac{1}{LC}\right)}$$

6.12 (c)

As $\quad I(s) = \frac{1}{s}$

$$\therefore \quad V(s) = Z(s)I(s) = \frac{1}{s}\frac{(s+3)}{(s+2)^2}$$

Steady state value of voltage

$$= \underset{s\to 0}{Lt}[sV(s)]$$

$$= \underset{s\to 0}{Lt}\left[s\left\{\frac{1}{s}\frac{(s+3)}{(s+2)^2}\right\}\right] = \frac{3}{4}\text{ V}$$

Initial value of voltage

$$= \underset{s\to\infty}{Lt}[sV(s)]$$

$$= \underset{s\to\infty}{Lt}\left[\frac{s\left(\frac{1+3}{s}\right)}{s^2\left(1+\frac{2}{s}\right)^2}\right] = 0\text{ V}$$

6.13 (c)

$$I(s) = \frac{3}{s}$$

$$V(s) = Z(s)I(s)$$

$$\therefore \quad V(s) = \left(\frac{s}{s+3}\right)\times\left(\frac{3}{s}\right) = \frac{3}{s+3}$$

$$v(t) = \mathcal{L}^{-1}[V(s)] = 3e^{-3t}$$

6.14 (c)

$$Z(s) = \frac{(s+1)}{(s+\sqrt{2})\left(s+\frac{1}{\sqrt{2}}\right)}$$

For $\quad i(t) = \sin t,\ \omega = 1$

Put $\quad s = j\omega = j$

$$Z(j) = \frac{(j+1)}{(1+\sqrt{2})\left(j+\frac{1}{\sqrt{2}}\right)} = \frac{2}{3}\angle -45^\circ$$

$$|V| = |Z||i(t)| = \frac{2}{3}\times 1 = \frac{2}{3}$$

6.15 (a)

$$Z_{\text{capacitor}} = \frac{5}{s}$$

For $s\to\infty$, $Z_{\text{capacitor}} = 0$, short-circuit

For $s\to 0$, $Z_{\text{capacitor}} = \infty$, open circuit

As $s \to \infty$, the circuit is

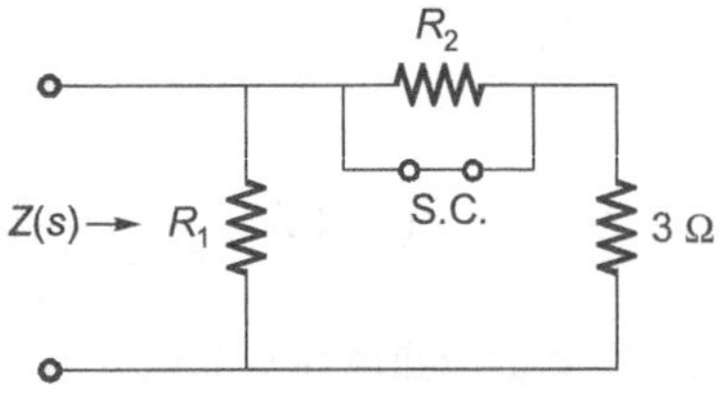

$$Z(s) = R_1 \parallel 3 = \frac{3R_1}{3+R_1} = 2$$

$$R_1 = 6\ \Omega$$

As $s \to 0$, the circuit is

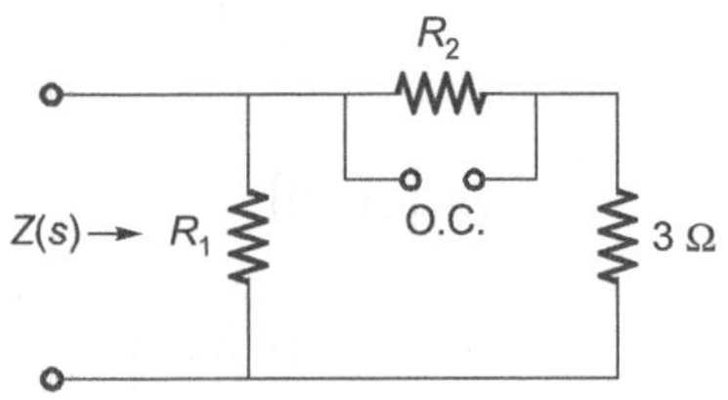

$$Z(s) = R_1 \parallel (3 + R_2)$$
$$= 6 \parallel (3 + R_2)$$

$$\Rightarrow \quad 3 = \frac{6\times(3+R_2)}{6+3+R_2} \Rightarrow R_2 = 3\ \Omega$$

6.16 (d)

$$Z(s) = 2+\left(\frac{1}{s}\right)\left|\left|1\right|\right|\left(1+\frac{1}{s}\right) = \frac{2s^2+7s+3}{s^2+3s+1}$$

Comparing with the given $Z(s)$, $\alpha = 2$ and $\beta = 1$

6.17 (d)

$$\frac{C(s)}{R(s)} = \frac{(s+1)}{(s+1-j)(s+1+j)}$$

For unit step function,

$$R(s) = \frac{1}{s}$$

$$\therefore \quad C(s) = \frac{(s+1)}{s(s+1-j)(s+1+j)}$$

Steady state response

$$= \underset{s\to 0}{Lt}\,[sC(s)]$$

$$= \frac{1}{(1-j)(1+j)} = 0.5\angle 0^\circ$$

6.18 (b)

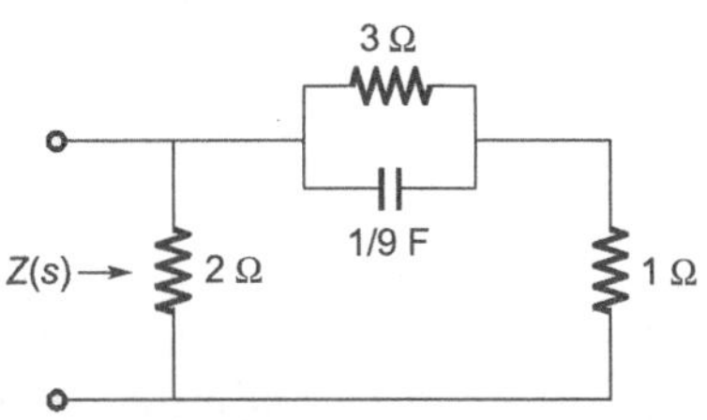

$$Z(s) = 2\left|\left|\left\{\left(3\left|\left|\frac{9}{s}\right.\right.\right)+1\right\}\right.\right. = \frac{2(s+12)}{3(s+6)}$$

6.19 (c)

$$V_2(s) = \frac{V_1(s)}{\left(2s+\frac{1}{s}\right)}\cdot\frac{1}{s}$$

$$\Rightarrow \quad \frac{V_2(s)}{V_1(s)} = \frac{1}{1+2s^2}$$

6.20 (c)

$$I(s) = \mathcal{L}[i(t)] = \mathcal{L}\left[2e^{-t} - e^{-5t}\right]$$

$$= \frac{2}{s+1} - \frac{1}{s+5} = \frac{s+9}{(s+1)(s+5)}$$

Zero = –9

Poles = –1 and –5

6.21 (b)

$$Z(s) = \left(1 \parallel \frac{1}{s}\right) + (2 \parallel 2s)$$

$$= \frac{1\times 1/s}{1+1/s} + \frac{2\times 2s}{2+2s} = \frac{1}{s+1} + \frac{2s}{s+1}$$

$$Z(s) = \frac{2s+1}{s+1}$$

Hence pole is at –1.

6.22 (b)

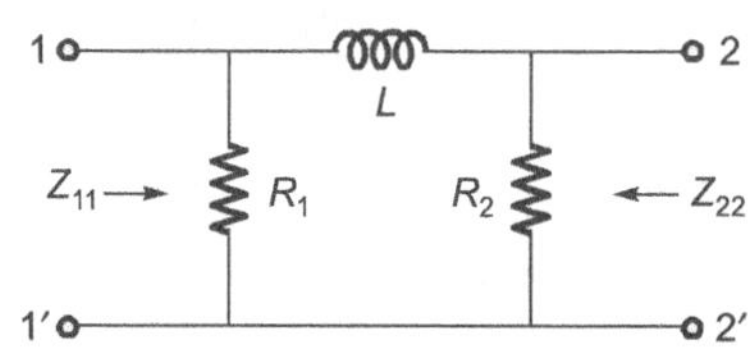

$$Z_{11} = R_1 \parallel (R_2 + Ls)$$

$$\Rightarrow \quad Z_{11} = \frac{R_1\times(R_2+Ls)}{R_1+R_2+Ls} = \frac{R_1\left(s+\frac{R_2}{L}\right)}{\left(s+\frac{R_1+R_2}{L}\right)}$$

Comparing with,

$$Z_{11} = k_1\left(\frac{s+3}{s+8}\right)$$

$$\frac{R_2}{L} = 3 \text{ and } \frac{R_1+R_2}{L} = 8$$

$$\Rightarrow \quad \frac{R_1}{L} = 5$$

Now, $Z_{22} = R_2 || (R_1 + Ls)$

$$= \frac{R_2 \times (R_1 + Ls)}{R_1 + R_2 + Ls}$$

$$= R_2\left(\frac{s + \frac{R_1}{L}}{s + \frac{R_1 + R_2}{L}}\right) = k_2\left(\frac{s+5}{s+8}\right)$$

Alternate solution:

Hint: Denominator in Z_{11} and Z_{22} is same.

6.23 (a)

$$Z = [(1||s) + 1]||(1)$$

$$\Rightarrow \quad Z = \frac{2s+1}{3s+2}$$

6.24 (a)

It is application of Cauer's-II realization because capacitor is series element. In Cauer's-II continued fraction after writing the functions in increasing order of powers of s.

$4s + s^2 \,)\, 15 + 8s + s^2 \,(\, \frac{15/4}{s} \rightarrow Z_1$ (Series)

$\quad 15 + \frac{15s}{4}$

$\frac{17s}{4} + s^2 \,)\, 4s + s^2 \,(\, \frac{16}{17} \rightarrow Y_2$ (Parallel)

$\quad 4s + \frac{16}{17}s^2$

$\frac{1}{17}s^2 \,)\, \frac{17}{4}s + s^2 \,(\, \frac{17 \times 17}{4s} \rightarrow Z_3$ (Series)

$\quad \frac{17}{4}s$

$s^2 \,)\, \frac{1}{17}s^2 \,(\, \frac{1}{17} \rightarrow Y_4$ (Parallel)

$\quad \frac{1}{17}s^2$

$\quad \times$

$$Z = Z_1 + \frac{1}{Y_2 + \frac{1}{Z_3} + \frac{1}{Y_4} + \ldots}$$

So, $Y_2 = \frac{16}{17}$ hence, $R_1 = \frac{17}{16}\,\Omega$

6.25 (a)

For $s = 0$ capacitor will be open circuits so equivalent circuit will be

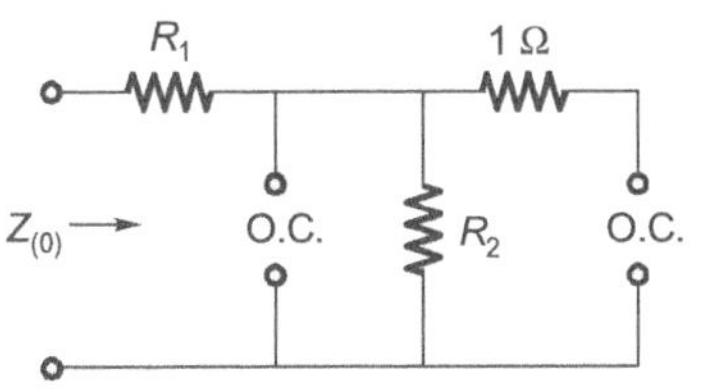

$\Rightarrow \quad Z_{(0)} = R_1 + R_2 = 3$ (as given)

And, $s = \infty$ capacitor will be short circuited so equivalent circuit will be

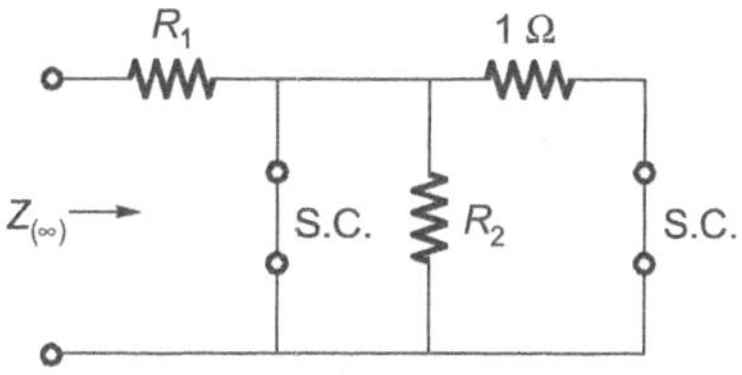

$\Rightarrow \quad Z_{(\infty)} = R_1 = 2$ (as given)

$R_2 = 3 - R_1 = 1$

6.26 (d)

$$Z(s) = \frac{K(s+z_1)(s+z_2)}{(s+P_1)(s+P_2)}$$

$$= \frac{K(s+2)(s+4)}{(s+1)(s+2)}$$

$$Z(0) = \frac{K(0+2)(0+4)}{(0+1)(s+3)} = 8$$

So, $K = 3$

Hence, $Z(s) = \frac{3(s+2)(s+4)}{(s+1)(s+3)}$

6.27 (a)

$$Z(s) = \frac{(s+1)}{(s+\sqrt{3})\left(s+\frac{1}{\sqrt{3}}\right)}$$

$\omega = 1$ rad/s

$$Z(j\omega) = 2(j) = \frac{(j+1)}{(j+\sqrt{3})\left(j+\frac{1}{\sqrt{3}}\right)}$$

$$= 0.613\angle{-45^\circ}\,\Omega$$

Angle of V = angle of I + angle of Z

$\therefore \quad \theta = 0 + (-45^\circ) = -45^\circ$

6.28 (b)

$$M = k\sqrt{L_1 L_2}$$

So, $k = \frac{16}{\sqrt{20 \times 80}} = 0.4$

6.29 (a)

$$\frac{V(s)}{I(s)} = \frac{s+\sqrt{3}}{s+\frac{1}{\sqrt{3}}}$$

$$\frac{V(j\omega)}{I(j\omega)} = \frac{j\omega+\sqrt{3}}{j\omega+\frac{1}{\sqrt{3}}}$$

and $\omega = 1$

$$\frac{V(j\omega)}{I(j\omega)} = \frac{j+\sqrt{3}}{j+\frac{1}{\sqrt{3}}}$$

$$\therefore \quad \angle\frac{V(j\omega)}{I(j\omega)} = \tan^{-1}\frac{1}{\sqrt{3}} - \tan^{-1}\sqrt{3} = -30°$$

$\therefore$ Current lead the voltage by 30°

6.30 (d)

$$Z(s) = \frac{k(s+1)(s+3)}{s(s+2)}$$

$$Z(\infty) = 4$$

$$4 = \frac{k\left(1+\frac{1}{\infty}\right)\left(1+\frac{3}{\infty}\right)}{\left(1+\frac{2}{\infty}\right)}$$

$$\therefore \quad k = 4$$

$$Z(s) = \frac{4\,(s^2+4s+3)}{s^2+2s}$$

$$Z(s) = \frac{4s^2+16s+12}{s^2+2s}$$

6.31 (c)

$$\frac{V_0(s)}{V_i(s)} = \frac{R}{R+sL+\frac{1}{sC}}$$

$$\frac{V_0(s)}{V_i(s)} = \frac{sRC}{sRC+1+s^2LC}$$

$$= \frac{(RC)s}{LC\left(s^2+s\cdot\frac{R}{L}+\frac{1}{LC}\right)}$$

$$\frac{V_0(s)}{V_i(s)} = \frac{\frac{R}{L}\times s}{s^2+s\cdot\frac{R}{L}+\frac{1}{LC}}$$

Comparing it with given equation

$$\frac{R}{L} = 4 \text{ and } R = 2\,\Omega \text{ (given)}$$

$$\therefore \quad L = 0.5\text{ H}$$

$$\frac{1}{LC} = 20 \Rightarrow C = 0.1\text{ F}$$

6.32 (c)

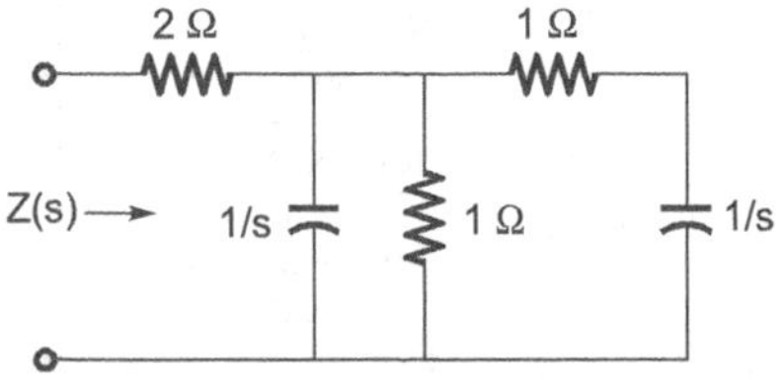

At $s = 0$

$$Z(0) = \frac{3}{b}$$

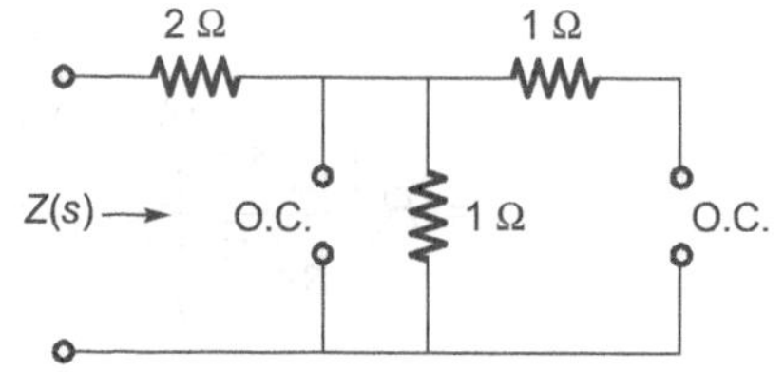

From circuit,

$$Z(0) = 3$$

$$\frac{3}{b} = 3 \Rightarrow b = 1$$

At $s = \infty$

$$Z(\infty) = a$$

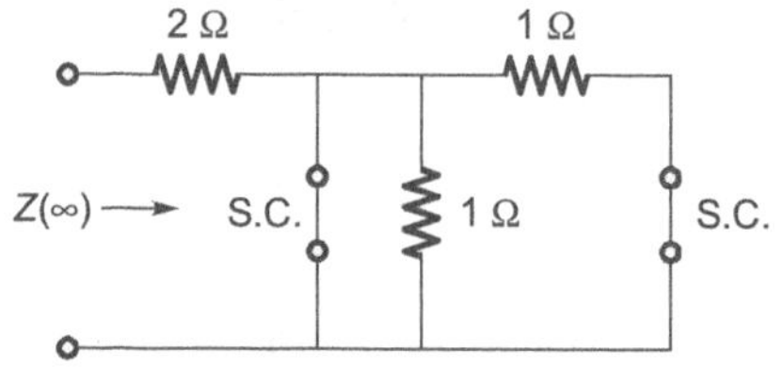

From circuit,

$$Z(\infty) = 2$$

So, $a = 2$

6.33 (c)

$$Z_{11} = (R_1+R_2)\,||\,Ls$$

$$= \frac{(R_1+R_2)Ls}{R_1+R_2+sL} = \frac{Ks}{R_1+R_2+sL}$$

Also, $$Z_{22} = \frac{(R_1+sL)R_2}{R_1+R_2+sL}$$

So option (c) is correct.

6.34 (a)

For RC impedance function, pole must be nearer to origin and zero must be nearer to ∞

$\Rightarrow$ Option (c) eliminated.

Further at steady state, capacitor → O.C.

$\Rightarrow Z(s)$ will be $\rightarrow \frac{3}{4}\,\Omega$ at $s \rightarrow 0$

$\Rightarrow$ Only option (a) is correct.

Alternate method:

$$Z(s) = \frac{3}{4}\,\Big|\Big|\left(\frac{1}{sC}+R\right)$$

$$= \frac{\frac{3}{4}\left(\frac{1}{sC}+R\right)}{\left(\frac{3}{4}\right)+\left(\frac{1}{sC}+R\right)}$$

$$= \left(\frac{\frac{3}{4}R}{R+\frac{3}{4}}\right)\frac{\left(s+\frac{1}{CR}\right)}{s+\frac{1}{C\left(R+\frac{3}{4}\right)}}$$

$$\because \quad \frac{1}{CR} > \frac{1}{C\left(R+\frac{3}{4}\right)}$$

$$\Rightarrow \quad \frac{1}{CR} = 4 \text{ and } \frac{1}{C\left(R+\frac{3}{4}\right)} = 1$$

$$\Rightarrow \quad \frac{1}{\frac{1}{4}+\frac{3}{4}C} = 1 \Rightarrow C = 1\text{ F}$$

$$\Rightarrow \quad R = \frac{1}{4}\Omega \Rightarrow \frac{\frac{3}{4}R}{R+\frac{3}{4}} = \frac{3}{16}$$

Hence, $Z(s) = \frac{3}{16}\left(\frac{s+4}{s+1}\right)$

6.35 (b)

On solving,

$$Z(s) = \frac{2(s^4+4s^2+3)}{s(s^4+6s^2+8)}$$

6.36 (a)

$$\frac{I_0(s)}{I_i(s)} = \frac{1/s}{6+s+\frac{1}{s}} = \frac{1}{s^2+6s+1}$$

No zeros,

Poles $P_1 = -3+2\sqrt{2}$

$P_2 = -3-2\sqrt{2}$

Hence, option (a) is correct.

6.37 (a)

$$V(s) = I(s)\left(\frac{s+3}{2s+3}\right) = \frac{1}{s}\left(\frac{s+3}{2s+3}\right)$$

$$V(0) = \underset{s\to\infty}{Lt}\,[sV(s)]$$

$$V(0) = \underset{s\to\infty}{Lt}\left\{\left(\frac{s+3}{2s+3}\right)\right\}$$

$$= \underset{s\to\infty}{Lt}\left(\frac{1+3/s}{2+3/s}\right) = 0.5\text{ V}$$

Hence, option (a) is correct.

6.38 (d)

$$V_{out}(s) = \frac{V_{in}(s)}{2s+\frac{1}{s}}\times\frac{1}{s}$$

$$\text{Voltage transfer function} = \frac{V_{out}(s)}{V_{in}(s)} = \frac{1}{1+2s^2}$$

Hence, option (a) is correct.

6.39 (a)

$$V_0(s) = \frac{V_{in}(s)}{R+Ls}\times R = \frac{V_{in}(s)}{1+\frac{L}{R}s}$$

$$\frac{V_0(s)}{V_{in}(s)} = \frac{1}{1+\frac{L}{R}}$$

Hence, option (a) is correct.

6.40 (c)

$$\text{Induced emf} = -L\frac{di}{dt}$$

$$40 = -L\times\frac{(1-5)}{(0.4-0)}$$

$$L = \frac{40\times0.4}{4} = 4\text{ Henry}$$

6.41 (b)

$$H(s) = \frac{V(s)}{I(s)} = \frac{2s+5}{s+2}$$

$$\Rightarrow \quad V(s) = \left(\frac{2s+5}{s+2}\right)I(s)$$

Given, $I(s) = \frac{1}{s}$

$$\therefore \quad V(s) = \frac{(2s+5)}{s+2}\times\frac{1}{s}$$

$$\underset{t\to\infty}{Lt}\,v(t) = \underset{s\to0}{Lt}\,sV(s)$$

$$= \underset{s\to0}{Lt}\left(\frac{2s+5}{s+2}\right) = \frac{5}{2}$$

6.42 (c)

$\because \quad L \propto N^2$

as $L_1 = L_2$

$\Rightarrow \quad N_1 = N_2$

$$\Rightarrow \quad \frac{N_1}{N_2} = \frac{1}{1}$$

6.43 (b)

$$L = \frac{N^2 \mu a}{l}$$

$$= \frac{250 \times 250 \times 4\pi \times 10^{-7} \times 80 \times 10^{-4}}{100 \times 10^{-2}}$$

$= 0.628$ mH

$\mu = \mu_0 \mu_r$ as air $\mu_r \simeq 1$

6.44 (c)

$$V = L\frac{di}{dt}$$

$$36 = L\frac{9}{1/2} \Rightarrow L = 2\,\text{H}$$

6.45 (c)

$$V = M\frac{di}{dt}$$

$$\Rightarrow \quad 4000 = M\frac{4}{10 \times 10^{-6}}$$

Mutual inductance,

$$M = 10^{-1}$$

$$\Rightarrow \quad M = 0.1\ \text{H}$$

6.46 (b)

As the 2-port network is reciprocal,

$\therefore \quad AD - BC = 1$

$4D - 7 \times 5 = 1$

$4D = 36$

$D = 9$

6.47 (c)

Reciprocity condition in terms of Z-parameters is

$$Z_{12} = Z_{21}$$

6.48 (b)

From the given network,

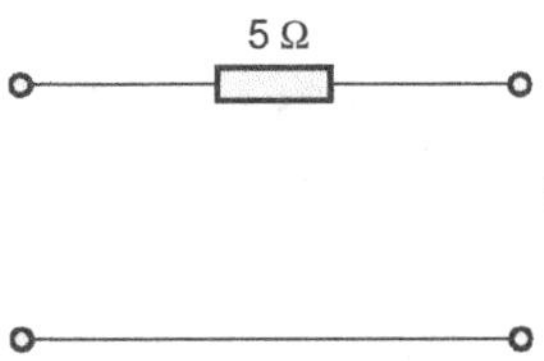

Comparing it with standard circuit

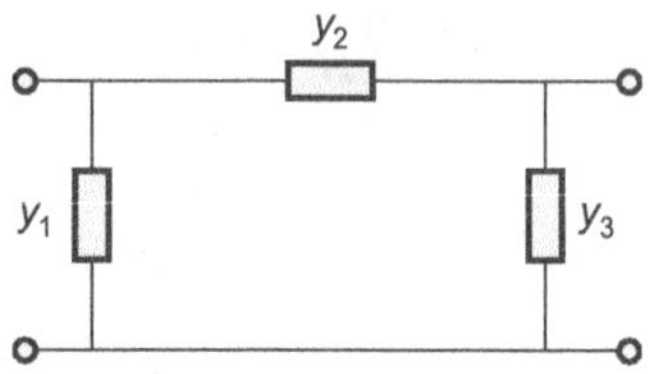

On comparing, we get

$$y_1 = 0, \quad y_2 = \frac{1}{5}℧, \; y_3 = 0$$

$$\therefore \quad y_{11} = y_1 + y_2 = \frac{1}{5}℧$$

$$y_{12} = y_{21} = -y_2 = -\frac{1}{5}℧$$

$$y_{22} = y_2 + y_3 = \frac{1}{5}℧$$

$$[Y] = \begin{bmatrix} \frac{1}{5} & -\frac{1}{5} \\ -\frac{1}{5} & \frac{1}{5} \end{bmatrix}$$

6.49 (c & d)

Since KCL has to be satisfied at input and output port so,

$$i_a = i_b$$

and $\quad i_c = i_d$4.51

(c & d)

Since KCL has to be satisfied at input and output port so,

$$i_a = i_b$$

and $\quad i_c = i_d$

6.50 (c)

The coupling coefficient is

$$k = \frac{M}{\sqrt{L_1 L_2}}$$

$\therefore$ 3rd statement is wrong.

6.51 (c)

$$M = K\sqrt{L_1 L_2}$$

$K = 1$ (for perfect coupling)

$$M = 1\sqrt{1 \times 1} = 1\ \text{H}$$

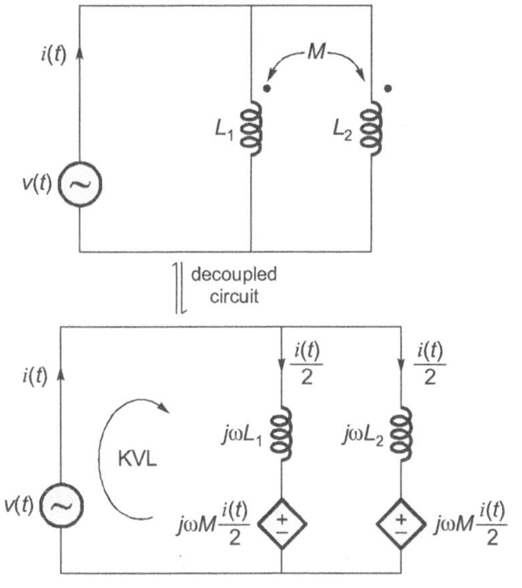

∵ Both inductors are identical.

∴ Current is divided equally.

By KVL, $v(t) = \dfrac{i(t)\times j\omega L_1}{2} + \dfrac{j\omega M i(t)}{2}$

$$\frac{v(t)}{i(t)} = j\omega\left(\frac{L_1}{2}+\frac{M}{2}\right) = j\omega\left(\frac{1}{2}+\frac{1}{2}\right)$$

$$\frac{v(t)}{i(t)} = j\omega(1) = j\omega L_{eff} \Rightarrow L_{eff} = 1\text{ H}$$

6.52 (b)

$$L_{eq} = L_1 + L_2 + L_3 + 2M_{12} - 2M_{23} - 2M_{13}$$
$$L_{eq} = 10 + 5 + 6 + 2(2) - 2(1) - 2(1)$$
$$L_{eq} = 21H$$

■■■■

7 Graph Theory & Duality

7.1 An electric circuit with 10 branches and 7 nodes will have
(a) 3 loop equations (b) 4 loop equations
(c) 7 loop equations (d) 10 loop equations
[ESE-2001]

7.2 The transfer function $\frac{V_2(s)}{V_1(s)} = \frac{10s}{s^2 + 10s + 100}$ is for an active
(a) low pass filter (b) band pass filter
(c) high pass filter (d) all pass filter
[ESE-2001]

7.3 The incidence matrix A of a connected graph is given below:

Nodes \ Branches	1	2	3	4	5	6
1	1	0	0	–1	0	–1
2	0	0	0	1	–1	0
3	0	1	1	0	1	1
4	–1	–1	–1	0	0	0

The following statements are associated with the above matrix:
1. Branches 2, 3 are in series.
2. Branches 4, 5 are in series.
3. Voltages of branches 4, 5 and 6 form an independent set.

Which of the statements given above is/are correct?
(a) 1 only (b) 2 only
(c) 1 and 3 (d) 1, 2 and 3
[ESE-2007]

7.4 The total number of branches in a network is equal to b. The graph of the network has n number of branches. The minimum number of line currents is
(a) $b + n$ (b) b
(c) $b - n$ (d) n [ESE-2012]

7.5 In an electric circuit, the number of independent meshes M is
(where B is number of branches and N is number of nodes).
(a) $2B - N + 1$ (b) $B - N + 1$
(c) $2B - N - 1$ (d) $B - N - 1$
[ESE-2015]

7.6 Two networks are said to be dual when
(a) their node equations are the same
(b) the loop equations of one network are analogous to the node equations of the other
(c) their loop equations are the same
(d) the voltage sources of one networks are the current sources of the other
[ESE-2016]

7.7 In a connected graph, the total number of branches is b and the total number of nodes is n. Then the number of links L of a co-tree is
(a) $b - n$ (b) $b - n - 1$
(c) $b + n - 1$ (d) $b - n + 1$
[ESE-2017]

7.8 A network graph with its tree shown by firm lines is given in the figure. The fundamental cut-set for the tree-branch number 2 is

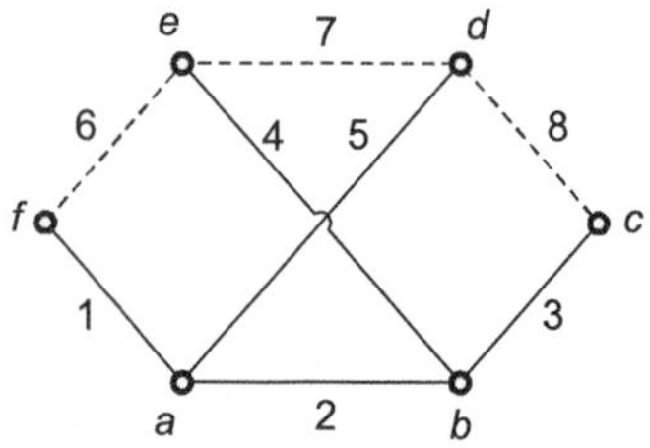

(a) 1, 2, 3, 4 and 5 (b) 1, 2 and 5
(c) 2, 6, 7 and 8 (d) 2, 3 and 4
[ESE-2018]

7.9 The reduced incidence matrix for a network is given as

$$A = \begin{matrix} & 1 & 2 & 3 & 4 & 5 & 6 \\ a & 1 & -1 & -1 & -1 & 0 & 0 \\ b & 0 & 1 & 0 & 0 & -1 & 1 \\ c & 0 & 0 & -1 & 0 & 1 & 0 \end{matrix}$$

Which of the following sets constitute a tree?

(a) 2, 3 and 5 (b) 1, 2 and 6
(c) 1, 2 and 4 (d) 1, 2 and 3

[ESE-2018]

7.10 Consider the following statements for a network graph, if B_f is its fundamental tie set matrix, and B_t and B_l are its sub-matrices corresponding to twigs and links, respectively:

1. B_t is a unit matrix.
2. B_l is a rectangular matrix.
3. Rank of B_f is $(b - n - 1)$.

where b is the number of branches n is the number of nodes.

Which of the above statements are correct?

(a) 1 and 2 only (b) 1 and 3 only
(c) 2 and 3 only (d) None of the above

[ESE-2018]

7.11 For a network graph having its fundamental loop matrix B_f and its sub-matrices B_t and B_l corresponding to twigs and links, which of the following statements are correct?

1. B_l is always an identity matrix.
2. B_t is an identity matrix.
3. B_f has a rank of $b - (n - 1)$, where b is the number of branches and n is the number of nodes of the graph.

(a) 1 and 2 only (b) 2 and 3 only
(c) 1 and 3 only (d) 1, 2 and 3

[ESE-2020]

7.12 A network has 8 branches and 4 independent loops. How many nodes are there in the network?

(a) 11 (b) 13
(c) 5 (d) 3

[ESE-2021]

■■■■

Answers Graph Theory & Duality

7.1 (b) 7.2 (b) 7.3 (b) 7.4 (d) 7.5 (b) 7.6 (b) 7.7 (d) **7.8 (c)** 7.9 (*)
7.10 (d) 7.11 (c) 7.12 (c)

Explanations Graph Theory & Duality

7.1 (b)

($b - n + 1$) links associated with fundamental loops. So, $b - n + 1 = 10 - 7 + 1 = 4$

7.2 (b)

$$\frac{V_2(s)}{V_1(s)} = \frac{10s}{s^2 + 10s + 100}$$

For $s \to 0$

$$\frac{V_2(s)}{V_1(s)} = 0$$

For $s \to \infty$

$$\left.\frac{V_2(s)}{V_1(s)}\right|_{S=\infty} = \frac{10s}{s^2\left(1 + \frac{10}{s} + \frac{100}{s^2}\right)} = 0$$

But for $0 < s < \infty$, $\frac{V_2(s)}{V_1(s)} \neq 0$

So, transfer function represents band pass filter.

7.3 (b)

The connected graph is

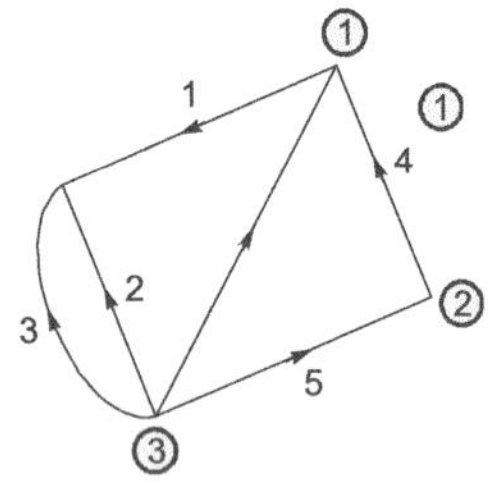

Branches 2 and 3 are not in series so, statement 1 is incorrect.

7.4 (d)

Minimum number of line current should be equal to the number of branches in the graph of a network.

7.6 (b)

Duality means mathematical representation of both the networks should be identical (KVL and KCL).

∴ Loop equations of one network are analogous to the node euqations of the other.

7.7 (d)

Number of links of a co-tree is given by,

$$L = b - (n - 1)$$
$$= b - n + 1$$

7.8 (c)

As fundamental cut set consistis of only one tree branch, therefore by looking at the options we can find that option which contains only one tree branch and rest links.

7.9 (*)

Question wrong because branch 3 has both entires (–1) which is not possible because a branch is connected between 2 nodes and it starts from one node and end at the other so at one node –1 and at the other +1.

7.12 (c)

No. of independent loops/fundamental loops/F-loops = No.of links (l) = 4

No. of links (l)= $b - n + 1$

$$l = b - n + 1$$
$$4 = 8 - n + 1$$
$$n = 5$$

■■■■

8 Miscellaneous (Three-Phase Circuit + Filters)

8.1 **Assertion (A) :** There are pulsations in power of a balanced three-phase system

Reason (R) : The three-phase generators produce sinusoidal voltage and current

(a) Both A and R are true and R is the correct explanation of A
(b) Both A and R are true but R is NOT the correct explanation of A
(c) A is true but R is false
(d) A is false but R is true

[ESE-2002]

8.2 A 3-phase star-connected balanced balanced load consumes P watts of power from a 400 V (line to line voltage) supply. If the same load is connected in delta across the same supply, what is the power consumption ?

(a) $P/3$ W (b) P W
(c) $\sqrt{3}P$ W (d) $3P$ W [ESE-2004]

8.3 A balanced delta connected load has an impedance of $9\angle 30°$ ohms per phase. What is the impedance per phase of its equivalent star?

(a) $27\angle 30°\ \Omega$ (b) $27\angle 90°\ \Omega$
(c) $3\angle 30°\ \Omega$ (b) $3\angle 20°\ \Omega$

[ESE-2004]

8.4 Two resistors of equal value are connected in series across the lines B and C of a symmetrical 3-phase 400 V system. What is the magnitude of the voltage between the line A and the junction of the resistors?

(a) $400/\sqrt{3}$ (b) $200\sqrt{3}$
(c) $400/3$ (d) $100\sqrt{3}$

[ESE-2004]

8.5 A 3-phase delta-connected symmetrical load consumes P watt of power from a balanced supply. If the same load is connected in star to the same supply, then what is the power consumption?

(a) $P/3$ (b) P
(c) $\sqrt{3}P$ (d) $3P$ [ESE-2005]

8.6 A 3-phase 3-wire supply feeds a star connected load consisting of 3 equal resistors. If one of the resistors is to be removed, then what is the reduction in power as compared to the original power?

(a) 25% of the original power
(b) 33.3% of the original power
(c) 50% of the original power
(d) 66.6% of the original power

[ESE-2007]

8.7 A balanced RYB-sequence, Y-connected (Star connected) source with V_{RN} = 100 volts is connected to a Δ-connected (Delta connected) balanced load of $(8 + j6)$ ohms per phase. Then the phase current and line current values respectively, are

(a) 10 A; 30 A (b) $10\sqrt{3}$ A; 30 A
(c) 10 A; 10 A (d) $10\sqrt{3}$ A ; $10\sqrt{3}$ A

[ESE-2010]

8.8 A 3-phase, 4-wire system supplies power to a balanced star-connected load. The current in each phase is 15 A. The current in the neutral wire will be

(a) 15 A (b) 45 A
(c) 8.66 A (d) 0 A [ESE-2012]

8.9 The output power of filter is 100 mW, when the signal frequency is 5 kHz. When the frequency is increased to 25 kHz, the output power falls to 50 mW. What is the dB change in power?

(a) –3 dB (b) –5 dB
(c) –7 dB (d) –2 dB

[ESE-2014]

8.10 Three equal impedance are first connected in delta across a 3-phase balanced supply. If the same impedances are connected in star across the same supply then,
(a) Phase current will be one-third
(b) Line current will be one-third
(c) Power consumed will be one-third
(d) Phase current will remain the same
[ESE-2014]

8.11 The pole-zero pattern of a certain filter is shown in figure. The filter must be

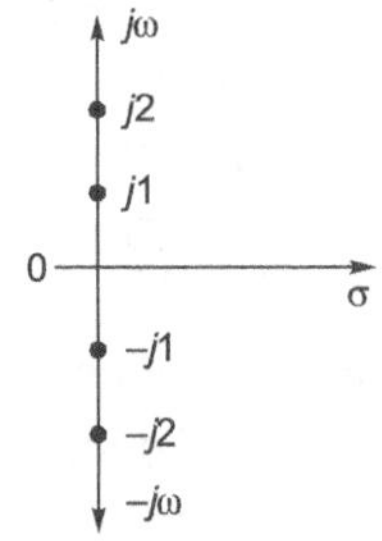

(a) Low-pass type
(b) High-pass type
(c) Band-pass type
(d) All-pass type
[ESE-2014]

8.12 A balanced delta-connected load $(16 + j12)\ \Omega$/phase is connected to a 3-phase 230 V balanced supply. The line current and the real power drawn respectively are
(a) 19.9 A and 3.17 kW
(b) 11.5 A and 7.34 kW
(c) 19.9 A and 6.34 kW
(d) 11.5 A and 3.17 kW [ESE-2015]

8.13 A balanced load of $5 + j4$ is connected in delta. What is the impedance per phase of the equivalent star connection?
(a) $5 + j4$ (b) $1.66 + j1.33$
(c) $15 + j12$ (d) $2.5 + j2$
[ESE-2015]

8.14 A three-phase star-connected load is operating at a power factor angle ϕ, with ϕ being the angle between
(a) line voltage and line current
(b) phase voltage and phase current
(c) line voltage and phase current
(d) phase voltage and line current
[ESE-2017]

8.15 A filter that allows high and low frequencies to pass but attenuates any signal with a frequency between two corner frequencies is a
(a) Notch filter (b) Band pass filter
(c) Band stop filter (d) Multiband filter
[ESE-2019]

8.16 A 3-phase star-connected 1000 volt alternator supplied power to a 500 kW delta-connected induction motor. If the motor power factor is 0.8 lagging and its efficiency 0.9, then the current in each alternator and motor phase respectively are nearly
(a) 321 A and 231.5 A
(b) 401 A and 231.5 A
(c) 321 A and 185.4 A
(d) 401 A and 185.4 A [ESE-2019]

8.17 A three-phase motor can be regarded as a balanced Y-load. A three-phase motor draws 5.6 kW, when the line voltage is 220 V and the line current is 18.2 A. What is the power factor of the motor?
(a) 0.8075 (b) 0.6134
(c) 0.9593 (d) 0.7947
[ESE-2021]

8.18 What is the minimum capacitance for the supply transformer rated 5 kVA with secondary voltage of 120 V RMS and switching frequency of 400 Hz?
(a) 0.92 μF (b) 0.76 μF
(c) 0.52 μF (d) 0.25 μF
[ESE-2021]

8.19 The value of the power consumed by the balanced star-connected load in terms of the balanced delta-connected load is
(a) $P_Y = \left[\frac{P_\Delta}{\sqrt{3}}\right]$ (b) $P_Y = \left[\sqrt{3}P_\Delta\right]$
(c) $P_Y = \left[\frac{P_\Delta}{3}\right]$ (d) $P_Y = [3P_\Delta]$
[ESE-2024]

Answers Miscellaneous (Three-Phase Circuit + Filters)

8.1 (d) 8.2 (d) 8.3 (c) 8.4 (b) 8.5 (a) 8.6 (c) 8.7 (b) 8.8 (d) 8.9 (a)
8.10 (c) 8.11 (a) 8.12 (c) 8.13 (b) 8.14 (b) 8.15 (c) 8.16 (b) 8.17 (a) 8.18 (*)
8.19 (c)

Explanations Miscellaneous (Three-Phase Circuit + Filters)

8.2 (d)

Take 1st case, circuit is in Y-connection

Current flowing in each resistor,

$$I = \frac{V}{\sqrt{3}R}$$

Total power,

$$P = 3I^2 R = 3\times\left(\frac{V}{\sqrt{3}R}\right)^2 \times R = \frac{V^2}{R}$$

Take 2nd case i.e. Δ-connection.

Current flowing in each resistor,

$$I = \frac{V}{R}$$

Total power,

$$P = 3\times I^2R = 3\times\left(\frac{V}{R}\right)^2 \times R = 3\frac{V^2}{R} = 3P$$

Hint: Power consumption in delta is 3 times than that of star.

8.3 (c)

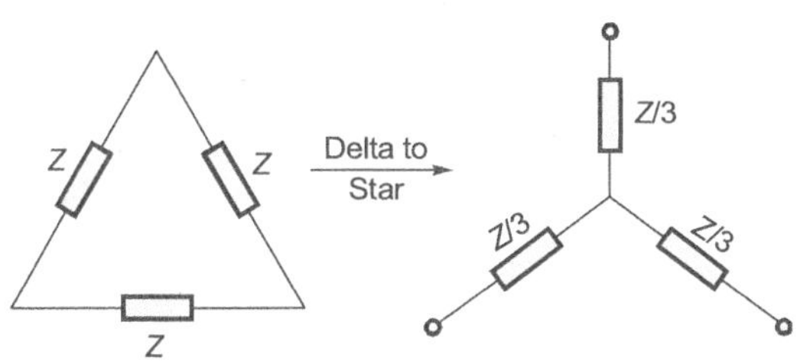

$$Z_{eq} = \frac{Z}{3} = \frac{9\angle 30^\circ}{3} = 3\angle 30^\circ\ \Omega$$

8.4 (b)

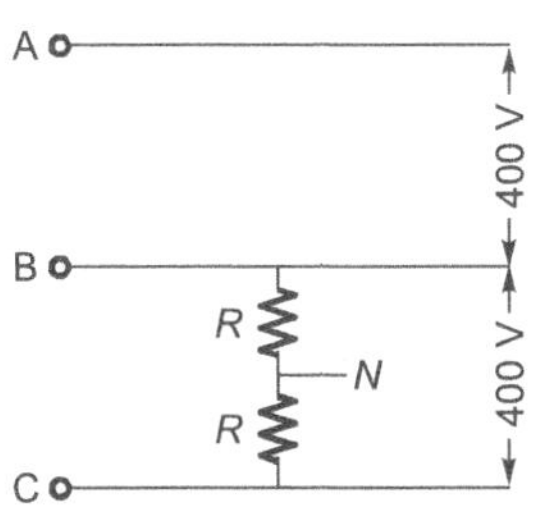

$$V_{AN} = V_{AB} + V_{BN}$$
$$= 400 + 200\angle -120^\circ$$
$$= (300 - j100\sqrt{3})$$
$$\therefore \quad |V_{AN}| = 200\sqrt{3}\ \text{V}$$

8.6 (c)

Power when one resistor is removed, star connected load,

$$'P_1' = \frac{V_L^2}{2R}$$

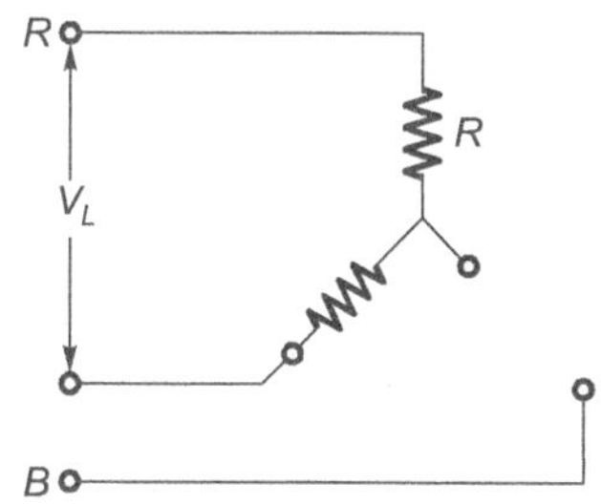

Power with all three resistors

$$P_2 = 3\left(\frac{V_L}{\sqrt{3}}\right)^2 \times \frac{1}{R} = \frac{V_L^2}{R}$$

8.8 (d)

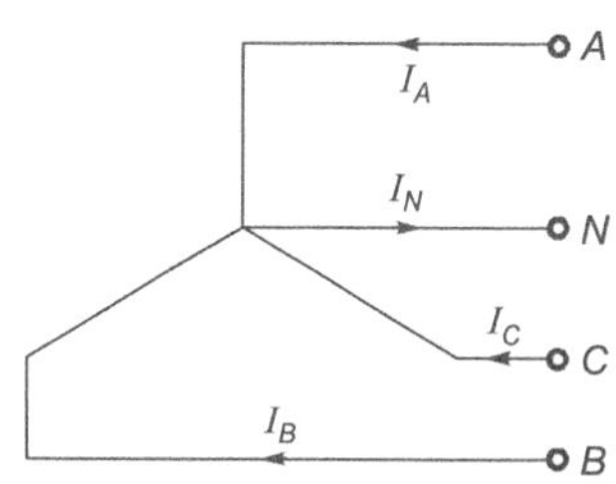

Current in neutral wire

$$I_N = I_A + I_B + I_C$$
$$= 15\angle 0^\circ + 15\angle 120^\circ + 15\angle 24^\circ = 0$$

8.9 (a)

dB value at 100 mW

$$= 20\log(100) = 40\ \text{dB}$$

dB value at 50 mW

$$= 20\log 50 = 34\ \text{dB}$$

dB change in power

$$= 40 - 34\ \text{dB}$$

⇒ dB/decade change in power

$$\frac{+6\ \text{dB}}{5-25} = -\frac{6}{20} = -3\ \text{dB/decade}$$

8.10 (c)

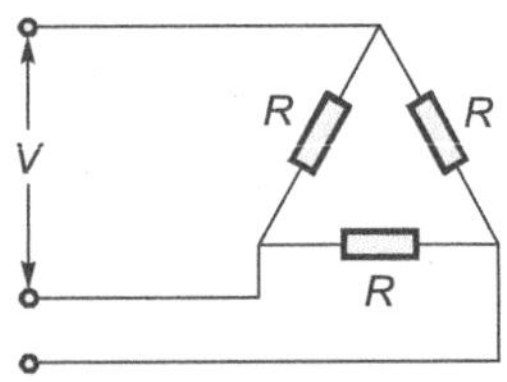

$$P_{\Delta} = \left(\frac{V^2}{R}\right) \times 3$$

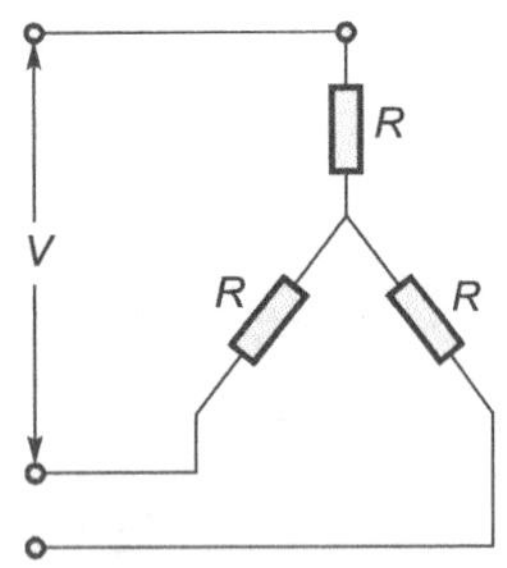

$$P_{\lambda} = \frac{(V/\sqrt{3})^2}{R} \times 3$$

$$= \frac{V^2}{R} \times \frac{1}{3} \times 3 = \frac{V^2}{R} = \frac{P_{\Delta}}{3}$$

8.12 (c)

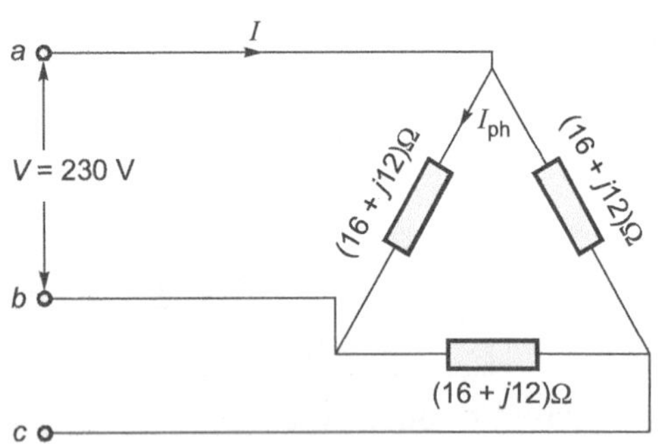

Phase current I_{ph}

$$I_{ph} = \frac{230}{\sqrt{16^2+12^2}} = \frac{230}{20} = 11.5 \text{ A}$$

line current = $\sqrt{3} \times I_{ph}$ = 19.9 A

and real power

$$P = \sqrt{3} V_L I_L \cos\theta$$

$$P = \sqrt{3} \times 230 \times 19.9 \times \cos\left\{\tan^{-1}\left(\frac{12}{16}\right)\right\}$$

$$P = 6.34 \text{ kW}$$

8.13 (b)

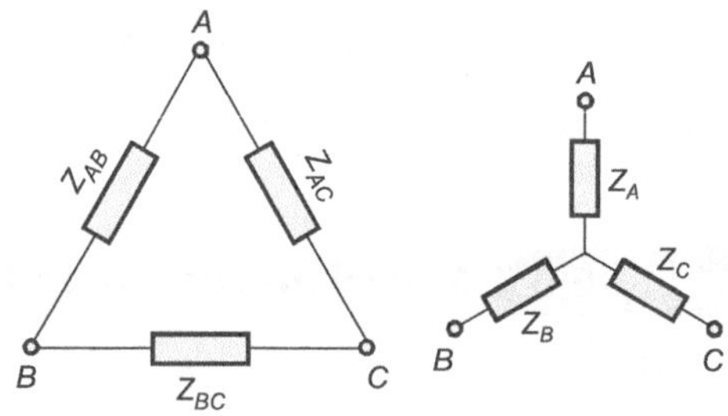

If $\quad Z_{AB} = Z_{AC} = Z_{BC} = Z$

$$Z_A = Z_B = Z_C = \frac{Z}{3}$$

$$Z_A = Z_B = Z_c = \left(\frac{5+j4}{3}\right)$$

$$Z = 1.66 + j\,1.33 \ \Omega$$

8.14 (b)

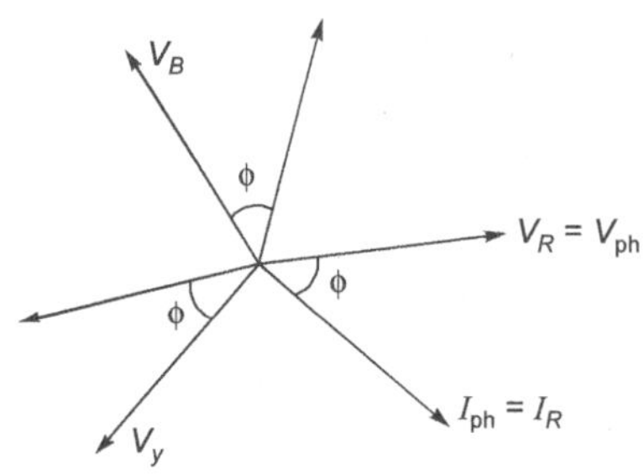

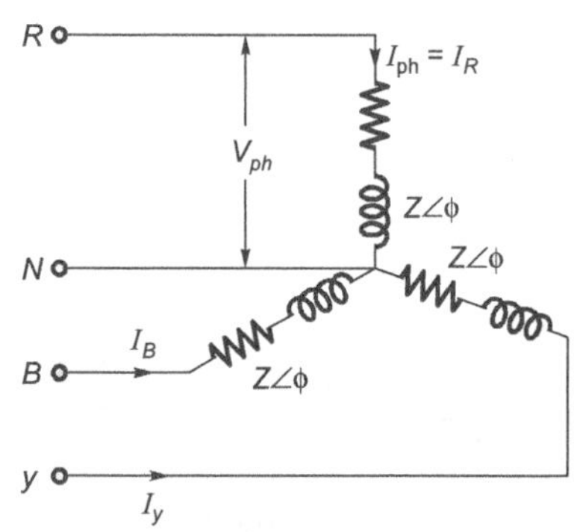

A load is connected between phase to neutral. The power factor (φ) of load is the angle between phase voltage and phase current.

8.15 (c)

Band stop filter attenuates any signal between two frequency components (lower cut-off frequency and upper cut-off frequency).

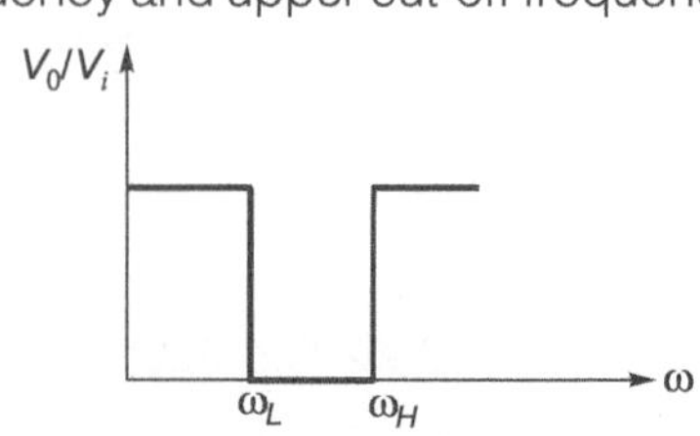

8.16 (b)

$$\text{Input of motor} = \frac{\text{Output}}{\eta} = \frac{500}{0.9}$$

$$= 555.55 \text{ kW}$$

For motor, $P_{3\phi} = \sqrt{3} V_L I_L \cos\theta$

$$555.55 \text{ kW} = \sqrt{3} \times 1k \times I_L \times 0.8$$

$$I_L = 401 \text{ A}$$

$$\Rightarrow \quad I_{Ph} = \frac{401}{\sqrt{3}} = 231.5 \text{ A}$$

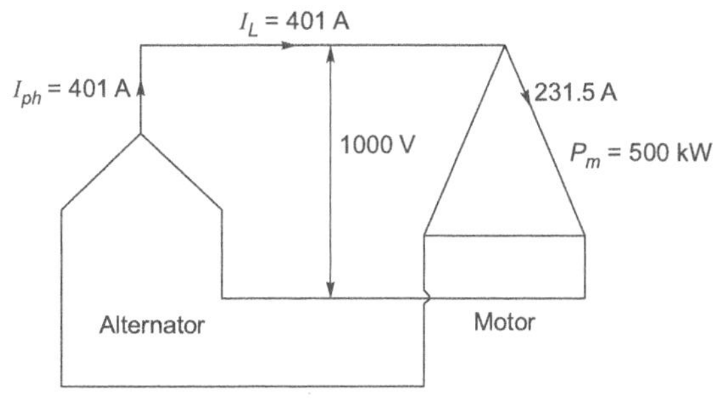

Correct answer is 401 A for alternator and 231.5 A for motor.

8.17 (a)

$$\text{Active power input} = \sqrt{3} V_L I_L \cos\phi = 5600$$

$$\cos\phi = \frac{5600}{\sqrt{3} \times 220 \times 18.2} = 0.8075$$

8.18 (*)

Data is insufficient

8.19 (c)

$$Z_\Delta \text{ (converted to equivalent } Y) = \frac{1}{3} Z_Y$$

$$\therefore \quad P_\Delta = 3P_Y$$

$$\therefore \quad P_Y = \frac{1}{3} P_\Delta$$

■■■■

9 Network Synthesis (Not in Syllabus)

9.1 Cauer and Foster forms of realizations are used only for
(a) driving point reactance functions
(b) transfer reactance functions
(c) driving point impedance functions
(d) transfer impedance functions
[ESE-2001]

9.2 For the given driving point impedance function

$Z(s) = \dfrac{4s^4 + 7s^2 + 1}{s(2s^2 + 3)}$, the circuit realization is shown in the given figure. Then the values of L and C are, respectively

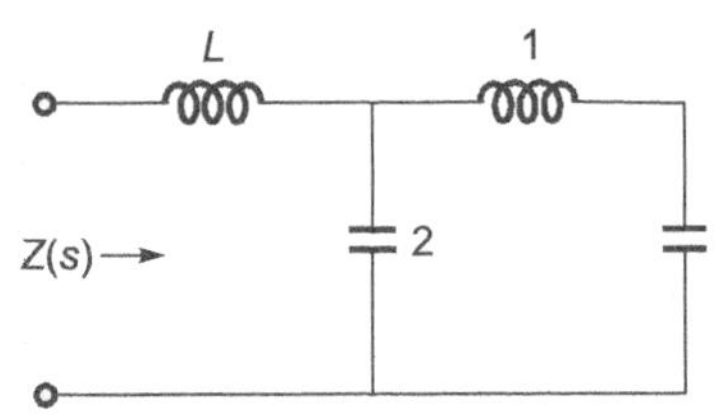

(a) 0.5 and 5 (b) 2 and 1
(c) 4 and 1 (d) 4 and 3
[ESE-2001]

9.3 The driving point impedance function

$Z(s) = \dfrac{s^2 + 2s + 2}{s^2 + s + 1}$ can be realized as a

(a) R-C network (b) R-L network
(c) L-C network (d) R-L-C network
[ESE-2003]

9.4 Impedance $Z(s) = [(s + 3)(s + 5)]/[(s + 4)(s + 6)]$ is realized in the Foster's second form. What will be the values, respectively of R and L in one of the series R-L branches?
(a) 2 Ω, 1/3 H (b) 2 Ω, 4/3 H
(c) 2 Ω, 5/3 H (d) 2 Ω, 2/3 H
[ESE-2004]

9.5 From the given list of driving point impedance functions, which one can be realised using R and C elements only

$$Z_1(s) = \frac{(s+8)}{(s+2)(s+6)}$$

$$Z_2(s) = \frac{(s+2)(s+6)}{(s+4)}$$

$$Z_3(s) = \frac{s(s+4)}{(s+2)}$$

$$Z_4(s) = \frac{(s+2)}{(s+1)(s+3)}$$

(a) $Z_1(s)$ (b) $Z_2(s)$
(c) $Z_3(s)$ (d) $Z_4(s)$
[ESE-2005]

9.6 Consider the following network functions:

1. $\dfrac{(s^2+1)(s^2+4)}{(s^2+3)(s^2+9)}$ 2. $\dfrac{(s^2+1)(s^2+9)}{s(s^2+4)}$

3. $\dfrac{s(s^2+9)}{(s^2+1)(s^2+4)}$ 4. $\dfrac{s(s^2+4)}{(s^2+1)}$

Which of the above functions can represent L-C driving-point immittances?
(a) Only 1 and 2 (b) Only 2 and 3
(c) Only 2 and 4 (d) Only 1 and 3
[ESE-2006]

9.7 Consider the following statements in connection with the properties of R-C impedance functions:
1. All the poles and zeros are simple.
2. Poles and zeros interlace.
3. Poles and zeros are located on the negative real axis of the s-plane

Which of the statements given above are correct?
(a) 1, 2 and 3 (b) Only 1 and 2
(c) Only 2 and 3 (d) Only 1 and 3
[ESE-2006]

9.8 $Z(s) = \dfrac{2(s^2+1)(s^2+9)}{s(s^2+4)}$ may be synthesized into Cauer first form as

(a) 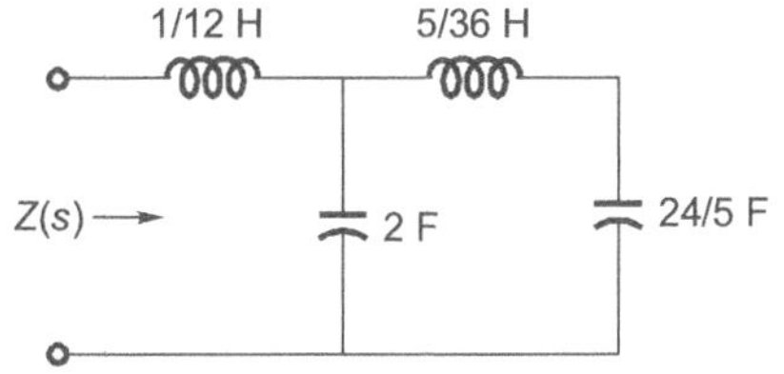

(b) 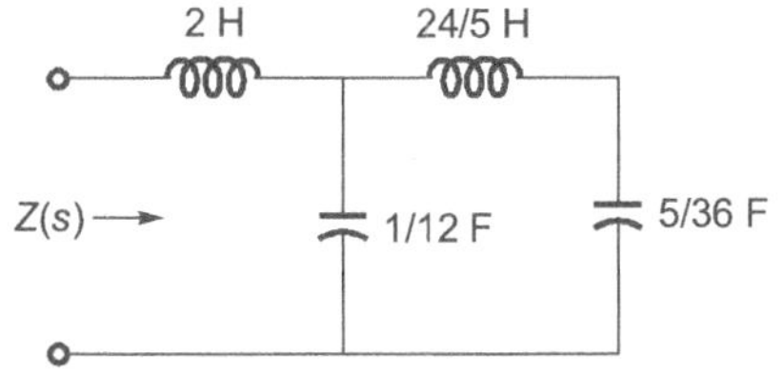

(c) 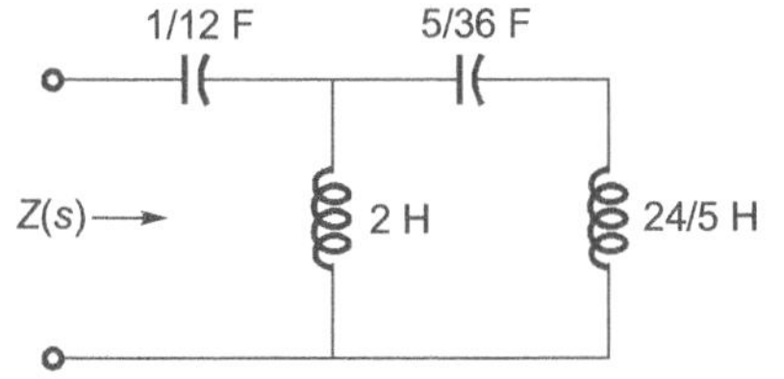

(d) 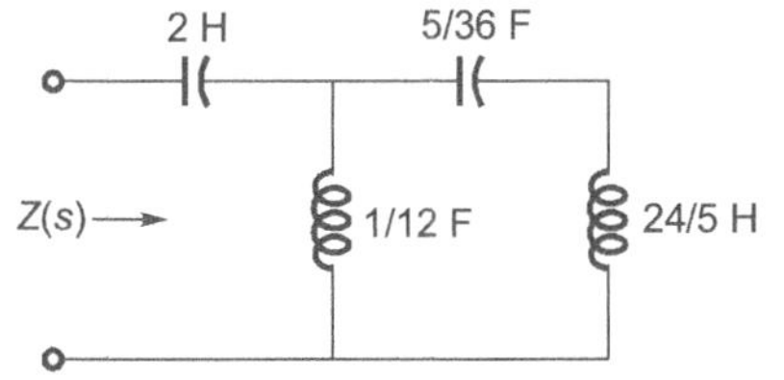

[ESE-2006]

9.9 The driving point impedance, $Z(s)$ of an RC network is given by

$$Z(s) = \left[\frac{s^2+4s+3}{s(s+2)}\right]$$

If it is realized in the first foster form, then what shall be the first and the last elements?

	First element	Last element
(a)	Resistance	Capacitance
(b)	Capacitance	Resistance
(c)	Resistance	Resistance
(d)	Capacitance	Capacitance

[ESE-2007]

9.10 Which one of the following is correct?
The impedance function

$Z(s) = \dfrac{(s^2+1)(s^2+3)}{s(s^2+2)}$ is an

(a) LC driving point function
(b) RL driving point function
(c) RC driving point function
(d) RLC driving point function

[ESE-2008]

9.11 The network function,

$$F(s) = \frac{(s+2)}{(s+1)(s+3)}$$

represents an
(a) RL impedance only
(b) RC impedance only
(c) RL admittance and RC impedance
(d) RC admittance and RL impedance

[ESE-2008]

9.12 Consider the following driving point immittance functions:

1. $z(s) = \dfrac{Ks(s^2+6)}{(s^2+2)(s^2+4)}$

2. $z(s) = \dfrac{(s^5+3s^3+5s)}{(3s^4+6s^2)}$

3. $z(s) = \dfrac{K(s^2+4)(s^2+9)}{(s^2+2)(s^2+6)}$

4. $z(s) = \dfrac{K(s^2+4)(s^2+9)}{s(s^2+6)}$

Which of these are LC immittance functions?
(a) 1 and 2 (b) 3 and 4
(c) 2 and 3 (d) 4 only

[ESE-2009]

9.13 Why is the network function,

$N(s) = \dfrac{s^3+3s}{s^4+4s^2+4}$ not positive real?

(a) The highest degree of numerator and denominator polynomials differ by one
(b) The terms of the lowest degree in the numerator and denominator polynomials differ in degree by one
(c) The poles and zeroes have zero real parts
(d) It has multiple poles on the imaginary axis

[ESE-2009]

9.14 The network realization of RC impedance function,

$$Z(s) = \frac{(\alpha s^2 + 7s + 3)}{(s^2 + 3s + \beta)}$$ is as shown below. What are the values of α and β?

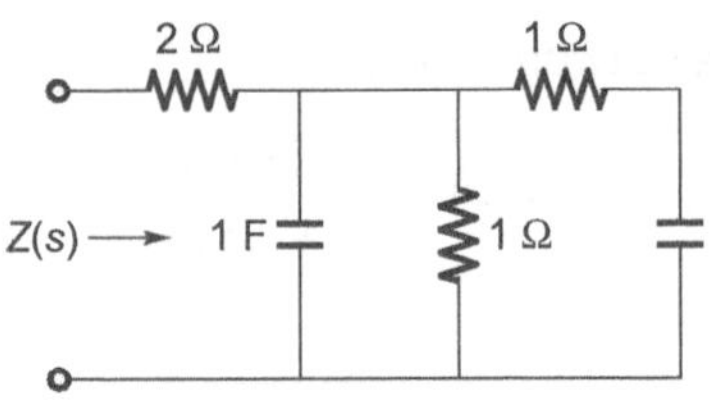

(a) 1 and 2 (b) 2 and 1
(c) 2 and 3 (d) 3 and 2

[ESE-2009]

9.15 Assertion (A): A capacitor has one pole at s = infinity and one zero at $s = 0$, where $s = j\omega$, ω is the angular frequency.

Reason (R): The driving point impedance of a capacitor is $1/sC$.

(a) Both A and R are true and R is the correct explanation of A
(b) Both A and R are true but R is not the correct explanation of A
(c) A is true but R is false
(d) A is false but R is true

[ESE-2009]

9.16 For the following driving point impedance functions, which of the following statements is true ?

$$Z_1(s) = \frac{s+2}{s^2 + 3s + 5}$$

$$Z_2(s) = \frac{s+2}{s^2 + 5}$$

$$Z_3(s) = \frac{s+3}{s^2 + 2s + 1}$$

$$Z_4(s) = \frac{(s+2)(s+4)}{(s+1)(s+3)}$$

(a) Z_1 is not positive real
(b) Z_2 is positive real
(c) Z_3 is positive real
(d) Z_4 is positive real

[ESE-2010]

9.17 A reactance network in the Foster's *I* form has poles at $\omega = 0$ (zero) and $\omega = \infty$ (infinity). The element in box-1 in the above network is

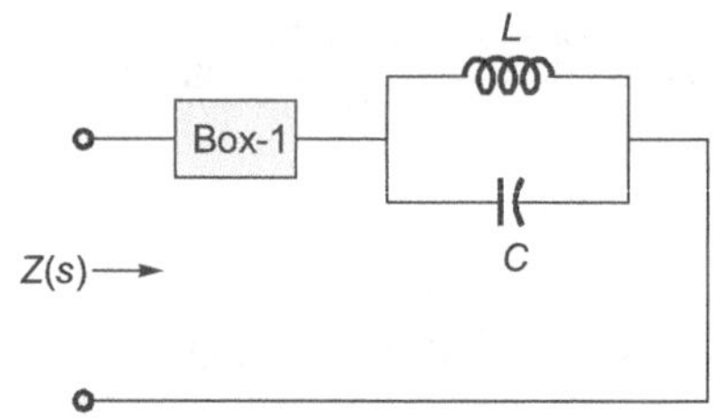

(a) a capacitor (b) an inductor
(c) a parallel LC circuit (d) a series LC circuit

[ESE-2010]

9.18 For the driving point impedance function of a circuit, $Z(s) = \frac{s+\alpha}{s+\beta}$, α and β real. Then voltage will lead the current if α and β are

(a) positive and $\alpha > \beta$
(b) positive and $\alpha < \beta$
(c) positive and real negative, respectively
(d) negative and real positive, respectively

[ESE-2011]

9.19 The driving point impedance function,

$$Z(s) = \frac{s^2 + 2s + 2}{s^2 + s + 1},$$ may be realized by

(a) R-C network
(b) R-L network
(c) L-C network
(d) None of the above networks

[ESE-2011]

9.20 If an RC driving point impedance function, $Z(s)$ has equal number of poles and zeros at finite locations, then

(a) $Z(0) \le Z(\infty)$ (b) $Z(0) \ge Z(\infty)$
(c) $Z(0) < Z(\infty)$ (d) $Z(0) > Z(\infty)$

[ESE-2011]

9.21 Which one of the following networks represents the Cauer's IInd form for the given driving point impedance function $Z(s) = \frac{12s^4 + 10s^2 + 1}{6s^3 + 4s}$?

(a) 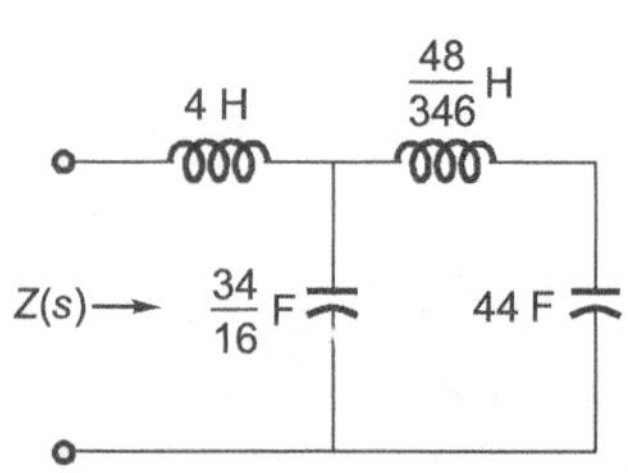

(b) 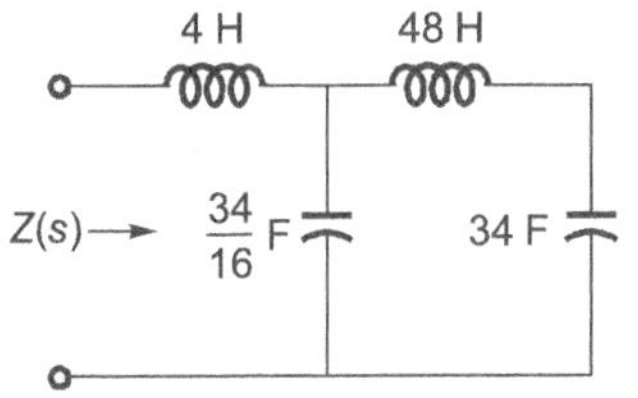

(c) 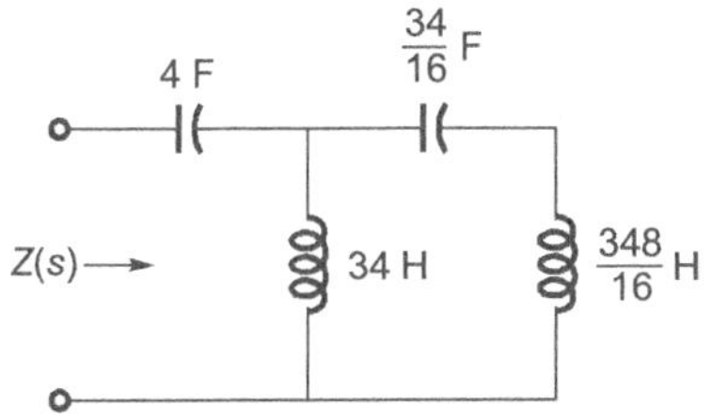

(d) 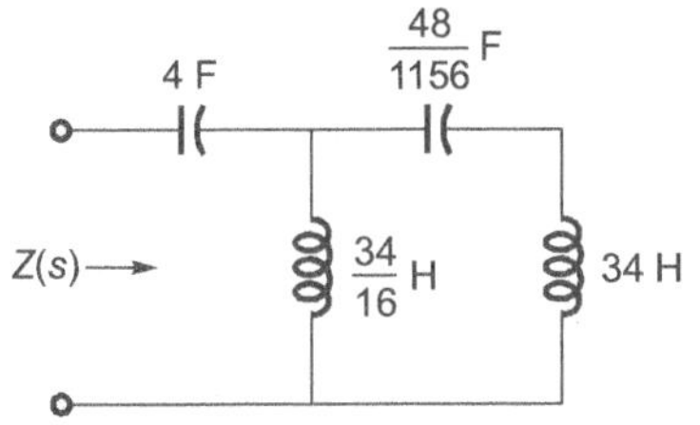

[ESE-2011]

9.22 The lowest and the highest critical frequencies of R-C driving-point impedance are respectively

(a) a zero and a pole (b) a pole and a pole
(c) a zero and a zero (d) a pole and a zero

[ESE-2012]

9.23 The driving-point impedance of an RC network is given by

$$Z(s) = \frac{(2s^2 + 7s + 3)}{(s^2 + 3s + 1)}$$

Its canonical realization will be

(a) 6 elements (b) 5 elements
(c) 4 elements (d) 3 elements

[ESE-2013]

9.24 A Hurwitz polynomial $D(s)$ must satisfy two conditions. One is the polynomial is real when s is real. What is the other condition?

(a) Roots of $D(s)$ have real parts which are positive and non-zero.
(b) Roots of $D(s)$ have imaginary parts which are negative.
(c) Roots of $D(s)$ have real parts which are either zero or negative.
(d) Roots of $D(s)$ have real parts which are positive or zero.

[ESE-2014]

9.25 Consider the following functions:

1. $\frac{(s^2+1)(s^2+3)}{s(s^2+2)}$ 2. $\frac{s(s^2+1)(s^2+3)}{(s^2+0.5)(s^2+2)}$

3. $\frac{(s^4+4s^2+3)}{s^2+2s}$ 4. $\frac{s^5+4s^3+3s}{s^4+2.5s+1}$

Which of the above functions are *LC* driving point impedance?

(a) 1, 2, 3 and 4 (b) 2 and 3 only
(c) 1 and 2 only (d) 3 and 4 only

[ESE-2014]

9.26 In the first Cauer network, with a pole at infinity, the first element must be,

(a) Series capacitor (b) Series inductor
(c) Shunt capacitor (d) Shunt inductor

[ESE-2017]

9.27 Consider the following driving point impedance functions:

$$Z_1(s) = \frac{(s+2)}{(s^2+3s+5)}$$

$$Z_2(s) = \frac{(s+2)}{(s^2+5)}$$

$$Z_3(s) = \frac{(s+2)}{(s^2+2s+1)}$$

$$Z_4(s) = \frac{(s+2)(s+4)}{(s+1)(s+3)}$$

Which one of the above is positive real?

(a) Z_1 (b) Z_2
(c) Z_3 (d) Z_4 [ESE-2017]

■■■■

Answers Network Synthesis (Not in Syllabus)

9.1 (c)	9.2 (b)	9.3 (d)	9.4 (d)	9.5 (d)	9.6 (c)	9.7 (a)	9.8 (b)	9.9 (a)
9.10 (a)	9.11 (c)	9.12 (d)	9.13 (d)	9.14 (b)	9.15 (d)	9.16 (d)	9.17 (d)	9.18 (b)
9.19 (d)	9.20 (d)	9.21 (d)	9.22 (d)	9.23 (b)	9.24 (c)	9.25 (c)	9.26 (b)	9.27 (d)

Explanations Network Synthesis (Not in Syllabus)

9.2 (b)

$$Z(s) = \frac{4s^4 + 7s^2 + 1}{2s^3 + 3s}$$

∵ Given figure is in Cauer first form.

⇒ Using continued fraction expansion:

$$\begin{array}{l}
2s^3+3s\,\big)\,4s^4+7s^2+1\,\big(\,2s \leftarrow Z_1 \text{ (series)}\\
\qquad\qquad\;\; 4s^4+6s^2\\
\qquad\qquad\qquad s^2+1\,\big)\,2s^3+3s\,\big(\,2s \leftarrow Y_2 \text{ (parallel)}\\
\qquad\qquad\qquad\qquad\quad\; 2s^3+2s\\
\qquad\qquad\qquad\qquad\qquad s\,\big)\,s^2+1\,\big(\,s \leftarrow Z_3 \text{ (series)}\\
\qquad\qquad\qquad\qquad\qquad\quad\;\; s^2\\
\qquad\qquad\qquad\qquad\qquad\qquad 1\,\big)\,s\,\big(\,s \leftarrow Y_4 \text{ (parallel)}\\
\qquad\qquad\qquad\qquad\qquad\qquad\quad\; s\\
\qquad\qquad\qquad\qquad\qquad\qquad\quad\; \times
\end{array}$$

$$\Rightarrow \quad L_1 = 2\text{ H and } C_4 = 1\text{ F}$$

9.3 (d)

$$Z(s) = \frac{s^2 + 2s + 2}{s^2 + s + 1}$$

Continued fraction expansion:

$$\begin{array}{l}
s^2+s+1\,\big)\,s^2+2s+2\,\big(\,1 \leftarrow Z_1 \text{ (series)}\\
\qquad\qquad\quad s^2+s+1\\
\qquad\qquad\qquad s+1\,\big)\,s^2+s+1\,\big(\,s \leftarrow Y_2 \text{ (parallel)}\\
\qquad\qquad\qquad\qquad\quad s^2+s\\
\qquad\qquad\qquad\qquad\qquad 1\,\big)\,s+1\,\big(\,s \leftarrow Z_3 \text{ (series)}\\
\qquad\qquad\qquad\qquad\qquad\quad\; s\\
\qquad\qquad\qquad\qquad\qquad\qquad 1\,\big)\,1\,\big(\,1 \leftarrow Y_4 \text{ (parallel)}\\
\qquad\qquad\qquad\qquad\qquad\qquad\quad 1\\
\qquad\qquad\qquad\qquad\qquad\qquad\quad \times
\end{array}$$

RLC network.

9.4 (d)

$$Y(s) = \frac{1}{Z(s)} = \frac{(s+4)(s+6)}{(s+3)(s+5)}$$

$$\Rightarrow Y(s) = 1 + \frac{2s+9}{(s+3)(s+5)} = 1 + \frac{3/2}{s+3} + \frac{1/2}{s+5}$$

$$\Rightarrow Y(s) = 1 + \frac{1}{2 + \frac{2s}{3}} + \frac{1}{2s+10}$$

9.5 (d)

For RC elements, the impedance function should have alternate pole and zero on negative real axis, and also power of denominator should be greater than numerator. Further highest critical frequency is a zero which may or may not lie at ∞. Also, lowest critical frequency is a pole which may or may not lie at origin.

9.6 (c)

The immittance function of the LC network is the ratio of even to odd or odd to even polynomial. All the poles & zeros lie on the jω axis and they are alternatively placed. There must be either a zero or a pole at the origin or ∞.

The difference between highest power of s of numerator & denominator polynomial must be unity. The lowest power of s of the numerator and denominator polynomial also differ by unity.

9.7 (a)

Properties of R-C impedance or R-L admittance function:

- All the poles and zeros lie on the negative real axis and they are alternatively placed.
- The singularity nearest to or at the origin must be a pole whereas singularity nearest to or at infinity must be a zero.

9.8 (b)

$$Z(s) = \frac{2(s^2+1)(s^2+9)}{s(s^2+4)}$$

$$= \frac{2s^4 + 20s^2 + 18}{s^3 + 4s}$$

Continued fraction expansion

$$\begin{array}{l}
s^3+4s\,\big)\,2s^4+20s^2+18\,\big(\,2s \leftarrow Z_1 \text{ (Series)}\\
\qquad\qquad\; 2s^4+8s^2\\
\qquad\qquad\qquad 12s^2+18\,\big)\,s^3+4s\,\big(\,s/12 \leftarrow Y_2 \text{ (Parallel)}\\
\qquad\qquad\qquad\qquad\qquad\quad s^3+(3/2)s\\
\qquad\qquad\qquad\qquad\qquad (5/2)s\,\big)\,12s^2+18\,\big(\,(24/5)s \leftarrow Z_3 \text{ (Series)}\\
\qquad\qquad\qquad\qquad\qquad\qquad\quad 12s^2\\
\qquad\qquad\qquad\qquad\qquad\qquad\qquad 18\,\big)\,(5/2)s\,\big(\,(5/36)s \leftarrow Y_4 \text{ (Parallel)}\\
\qquad\qquad\qquad\qquad\qquad\qquad\qquad\quad (5/2)s\\
\qquad\qquad\qquad\qquad\qquad\qquad\qquad\quad \times
\end{array}$$

$$Z_1 = L_1 s \Rightarrow L_1 = 2\text{H}$$

$$Y_2 = C_2 s = \frac{s}{12} \Rightarrow C_2 = \frac{1}{12}\text{F}$$

$$Z_3 = L_3 s = \frac{24}{5}s \Rightarrow L_3 = \frac{24}{5}\text{H}$$

$$Y_4 = \frac{5}{36}s = C_4 s \Rightarrow C_4 = \frac{5}{36}\text{F}$$

9.9 (a)

Question has some mistake as through foster we can't comment on first and last element. But if we consider it as Cauer's-I we can comment on first and last element being a ladder network.

$$Z(s) = \frac{s^2+4s+3}{s(s+2)} = \frac{(s+3)(s+1)}{s(s+2)}$$

Pole nearest to origin hence RC function.

So, ladder will be →

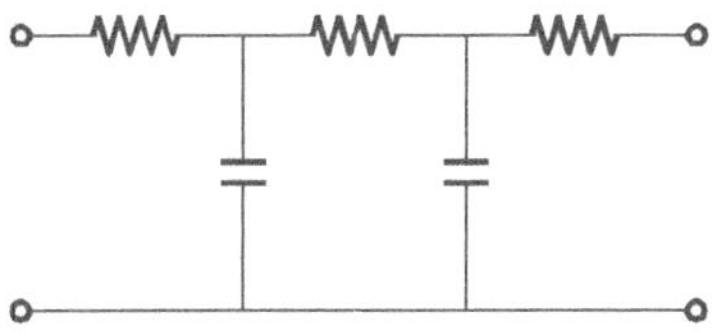

to check first element $s = \infty$ if gives 'O' then shunt element is first otherwise series element is first

$Z(\infty) = 1$ Hence series element is first so $Z(s)$ will be used for continued fraction.

$s^2 + 2s$) $s^2 + 4s + 3$ ($1 \rightarrow Z_1$ (Series)
$\quad s^2 + 2s$
$\quad 2s + 3$) $s^2 + 2s$ ($\frac{s}{2} \rightarrow Y_2$ (Parallel)
$\quad\quad s^2 + \frac{3s}{2}$
$\quad\quad \frac{s}{2}$) $2s + 3$ ($4 \rightarrow Z_3$ (Series)
$\quad\quad\quad 2s$
$\quad\quad\quad 3$) $\frac{s}{2}$ ($\frac{s}{3 \times 2} \rightarrow Y_4$ (Parallel)
$\quad\quad\quad\quad \frac{s}{2}$
$\quad\quad\quad\quad$ x

$$Z(s) = Z_1 + \cfrac{1}{Y_2 + \cfrac{1}{Z_3 + \cfrac{1}{Y_4}}}$$

Hence first element is resistance and last element is capacitor.

9.10 (a)

$$Y(s) = \frac{\frac{1}{2}s}{s^2+1} + \frac{\frac{1}{2}s}{s^2+3} = \frac{1}{2s+\frac{2}{s}} + \frac{1}{2s+\frac{6}{s}}$$

So, LC driving point function.

9.11 (c)

$$F(s) = \frac{(s+2)}{(s+1)(s+3)}$$

Since, the function has pole close to origin thus it is RL admittance or RC impedance.

9.12 (d)

Properties of LC function

- LC functions are ratio of either even to odd or odd to even polynomials
- Degree of denominator and numerator must differ by 1.
- Lowest power of 's' in N^r and D^r must differ by 1.
- Poles and zeros of LC function always lie on $j\omega$-axis and should be simple
- There is always either a zero or a pole at origin and infinity.
- The residues must be real and positive.

9.13 (d)

Properties of positive real function

- Both $P(s)$ and $Q(s)$ should be Hurwitz polynomial.
- All poles and zeros of function must lie in left half of s-plane.
- Imaginary poles and zeros should be simple.
- The highest and lowest power of s in numerator and denominator can differ atmost by 1.

9.14 (b)

$$Z(s) = 2 + \left[\frac{1}{s} \parallel \left\{1 \parallel \left(1+\frac{1}{s}\right)\right\}\right]$$

$$= 2 + \left[\frac{1}{s} \parallel \left\{\frac{s+1}{2s+1}\right\}\right] = 2 + \left[\frac{\frac{1}{s} \times \left(\frac{s+1}{2s+1}\right)}{\frac{1}{s} + \frac{s+1}{2s+1}}\right]$$

$$Z(s) = \frac{2s^2+7s+3}{s^2+3s+1}$$

Comparing with given equation

$\alpha = 2$ and $\beta = 1$

9.15 (d)

A capacitor have one pole at $s = 0$

and $\quad Z(s) = \frac{1}{sC}$

9.16 (d)

Condition for PRF:

1. $F(s)$ must have no poles in RHS.
2. $F(s)$ may have simple poles on $j\omega$-axis with real coefficients.

9.17 (d)

Note:

1. Power of numerator will be more than that of denominator power since pole at ∞ present. $\Rightarrow L$ element.
2. Pole at zero so C element. And both should be connected to fulfill above conditions.

9.18 (b)

As voltage leads current $Z(s)$ must be RL impedance function.

For RL impedance function, the singularity nearest to origin must be a zero where as the singularity nearest to infinity must be a pole. Poles and zeros must lie on the negative real axis.

Hence, option (b) is correct.

9.19 (d)

For LC Network:

These must be either a zero or a pole at the origin or ∞. All the poles and zeros must lie on the $j\omega$-axis and they alternate.

For RC or RL Network:

Residues of the poles must be real.

Hence, option (d) is correct.

9.20 (d)

For RC driving point impedance function, the singularity nearest to origin must be pole whereas singularity nearest to or at infinity must be a zero. Hence, option (d) is correct.

9.21 (d)

$$Z(s) = \frac{1+10s^2+12s^4}{4s+6s^3}$$

$4s + 6s^3 \overline{)\, 1 + 10s^2 + 12s^4 \,}\left(\frac{1}{4s} \leftarrow z_1\right.$ (Series)

$1 + \frac{6}{4}s^2$

$\frac{34}{4}s^2 + 12s^4 \overline{)\, 4s + 6s^3 \,}\left(\frac{16}{34s} \leftarrow y_2\right.$ (Parallel)

$4s + \frac{12 \times 16}{34}s^3$

$\frac{12}{34}s^3 \overline{)\, \frac{34}{4}s^2 + 12s^4 \,}\left(\frac{1156}{48s} \leftarrow z_3\right.$ (Series)

$\frac{34}{4}s^2$

$12s^4 \overline{)\, \frac{12}{34}s^3 \,}\left(\frac{1}{34s} \leftarrow y_4\right.$ (Parallel)

Hence, option (d) is correct.

9.22 (d)

In R-C driving point impedance function the singularity nearest to or at the origin must be a pole where as nearest to or at infinity must be a zero. Hence option (d) is correct.

9.23 (b)

$$Z(s) = \left(\frac{2s^2+7s+3}{s^2+3s+1}\right)$$

$s^2 + 3s + 1 \overline{)\, 2s^2 + 7s + 3 \,}\, 2 \longrightarrow$ 1 element

$-2s^2 + 6s + 2$

$s + 1 \overline{)\, s^2 + 3s + 1 \,}\, s \longrightarrow$ 1 element

$-s^2 + s$

$2s + 1 \overline{)\, s + 1 \,}\, 1/2 \longrightarrow$ 1 element

$-s + 1/2$

$1/2 \overline{)\, 2s + 1 \,}\, 4s + 2 \longrightarrow$ 2 elements

∴ Total 5 elements are there.

9.24 (c)

A Hurwitz polynomial $D(s)$ must satisfy two conditions:

1. Polynomial is real when s is lead
2. No root of the polynomial must lie in the right half of s-plane.

9.25 (c)

Poles and zeroes of an LC function must lie on $j\omega$-axis and also they must should interlace.

Hint: Always find residue to check LC driving point immitances.

9.26 (b)

For Cauer-I network; a pole at infinity indicates a finite zero present. That represents a series inductor.

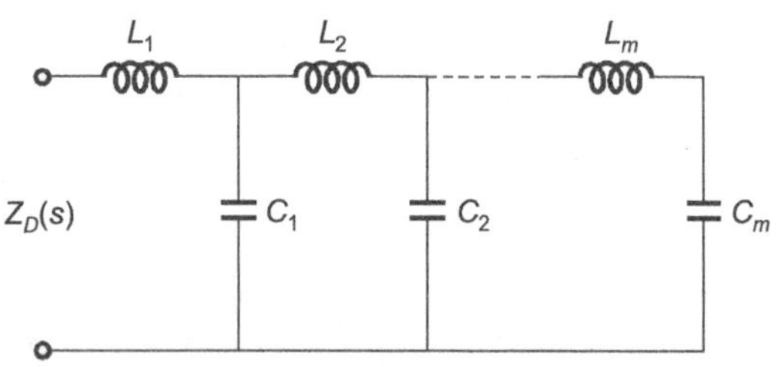

9.27 (d)

Among all driving point impedances only $Z_4(s)$ satisfy the criteria of positive real function.

■■■■

UNIT

III

Electric Fields

Syllabus

Gauss theorem, electric field and potential due to point, line, plane and spherical charge distributions, Ampere's and Biot-Savart's laws; inductance, dielectrics, capacitance; Maxwell's equations.

Contents

OOOO

1 Coordinate Systems and Vector Calculus

1.1 If $\bar{A} = \hat{a}_r + \hat{a}_\phi + \hat{a}_z$, the value of $\oint \bar{A} \cdot \overline{dl}$ around the closed circular quadrant shown in the given figure is

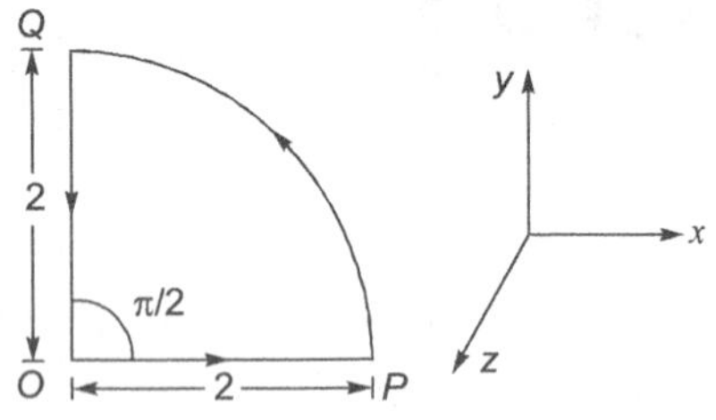

(a) π (b) $\frac{\pi}{2} + 4$

(c) $\pi + 4$ (d) $\frac{\pi}{2} + 2$

[ESE-2001]

1.2 Which of the following equations is correct?

(a) $\hat{a}_x \times \hat{a}_x = |a_x|^2$

(b) $(\hat{a}_x \times \hat{a}_y) + (\hat{a}_y \times \hat{a}_x) = 0$

(c) $\hat{a}_x \times (\hat{a}_y \times \hat{a}_z) = \hat{a}_x \times (\hat{a}_z \times \hat{a}_y)$

(d) $\hat{a}_r \cdot \hat{a}_\theta + \hat{a}_\theta \cdot \hat{a}_r = 0$

[ESE-2002]

1.3 Match **List-I** (Term) with **List-II** (Type) and select the correct answer:

List-I	**List-II**
A. Curl $(\bar{F}) = 0$	1. Laplace equation
B. Div $(\bar{F}) = 0$	2. Irrotational
C. Div Grad $(\phi) = 0$	3. Solenoidal
D. Div Div $(\phi) = 0$	4. Not defined

Codes:

	A	B	C	D
(a)	2	3	1	4
(b)	4	1	3	2
(c)	2	1	3	4
(d)	4	3	1	2

[ESE-2002]

1.4 An electromagnetic field is said to be conservative when

(a) $\nabla^2 E = \mu\epsilon(\partial^2 E/\partial t^2)$

(b) $\nabla^2 H = \mu\epsilon(\partial^2 H/\partial t^2)$

(c) Curl of the field is zero

(d) Divergence of the field is zero

[ESE-2003]

1.5 Given a vector field $\bar{A} = 2r\cos\phi T_r$ in cylindrical coordinates. For the contour as shown below, $\oint \bar{A} \cdot \overline{dl}$ is

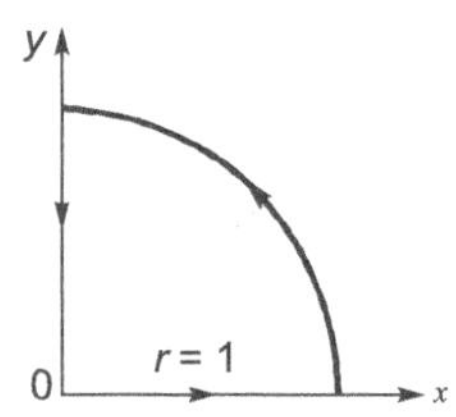

(a) 1 (b) $1 - (\pi/2)$

(c) $1 + (\pi/2)$ (d) –1

[ESE-2003]

1.6 Which one of the following pairs is **NOT** correctly matched ?

(a) Gauss Theorem : $\oint_s \vec{D} \cdot \vec{ds} = \oint_v \nabla \cdot \vec{D} dv$

(b) Gauss's Law : $\oint \vec{D} \cdot \vec{ds} = \int_v \rho dv$

(c) Coulomb's Law : $V = -\frac{d\phi_m}{dt}$

(d) Stoke's Theorem : $\oint_l \vec{E} \cdot \vec{dl} = \int_s (\nabla \times \vec{E}) \cdot \vec{ds}$

[ESE-2003]

1.7 What is the value of the integral $\int_c \vec{dl}$ along the curve c (c is the curve $ABCD$ in the direction of the arrow) ?

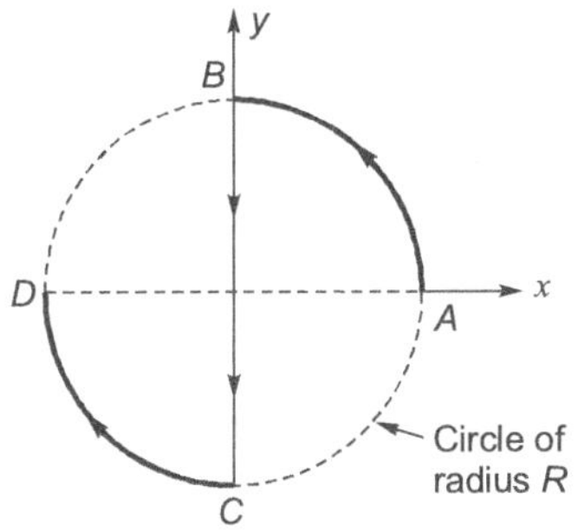

(a) $2R(\hat{a}_x + \hat{a}_y)/\sqrt{2}$

(b) $-2R(\hat{a}_x + \hat{a}_y)/\sqrt{2}$

(c) $2R\hat{a}_x$

(d) $-2R\hat{a}_y$

[ESE-2005]

1.8 The maximum space rate of change of the function which is in increasing direction of the function is known as

(a) curl of the vector function
(b) gradient of the scalar function
(c) divergence of the vector function
(d) Stokes theorem

[ESE-2015]

1.9 A phasor

1. may be a scalar or a vector
2. is a time-dependent quantity
3. is a complex quantity

Which of the above statements are correct?

(a) 1, 2 and 3
(b) 1 and 2 only
(c) 1 and 3 only
(d) 2 and 3 only

[ESE-2015]

Answers Coordinate Systems & Vector Calculus

1.1 (a) 1.2 (b, d) 1.3 (a) 1.4 (c) 1.5 (a) 1.6 (c) 1.7 (d) 1.8 (b) 1.9 (a)

Explanations Coordinate Systems & Vector Calculus

1.1 (a)

$\oint \vec{A}\cdot\vec{dL} = 3$ line integrals or paths

for path OP,

$$\oint \vec{A}\cdot\vec{dL} = \int_0^2 d\rho = 2$$

for path PQ,

$$\oint \vec{A}\cdot\vec{dL} = \int_0^{\pi/2} \rho\, d\phi = 2\times\frac{\pi}{2} = \pi$$

for path QO,

$$\oint \vec{A}\cdot\vec{dL} = \int_2^0 d\rho = -2$$

$$\therefore \oint \vec{A}\cdot\vec{dL} = 2+\pi-2 = \pi$$

1.2 (b, d)

$$(\hat{a}_x\times\hat{a}_x) = 0$$

Since cross product with same vector is zero because $\theta = 0$ so $\sin\theta = 0$

$$\hat{a}_x\times\hat{a}_y = \hat{a}_z$$

$$\hat{a}_y\times\hat{a}_x = -\hat{a}_z$$

$$(\hat{a}_x\times\hat{a}_y)+(\hat{a}_y\times\hat{a}_x) = \hat{a}_z+(-\hat{a}_z) = 0$$

1.3 (a)

curl $\vec{F} = 0$ that means vector $\vec{F}$ is irrotational

div $\vec{F} = 0$ that means vector $\vec{F}$ is solenoidal since flux coming out from a solenoid is zero.

div(grad ϕ) = $\nabla\cdot(\nabla\phi) = 0 \rightarrow$ Laplace equation

div(divϕ) = not defined because divϕ $\rightarrow$ scalar quantity and div of a scalar quantity is not defined.

1.4 (c)

$\nabla\times E = 0$ for conservative electric field.

1.5 (a)

$$\oint \vec{A}\cdot\vec{dl} = \int_0^1 2r\cos\phi\, T_r\cdot drT_r + \int_0^{\pi/2} 2r\cos\phi\cdot r\,d\phi(T_r\cdot T_\phi) + \int_{01}^0 2r\cos\phi\cdot T_r\cdot drT_r$$

as $T_r\cdot T_\phi$ and $\cos\frac{\pi}{2} = 0$

$$\therefore \oint \vec{A}\cdot\vec{dl} = 1+0+0 = 1$$

1.6 (c)

It is Faraday's Law $\left(V = -\frac{d\phi_m}{dt}\right)$

1.7 (d)

$$\int_{AB} \vec{dl} = \int_0^{\pi/2} R\cdot d\theta\,\hat{a}_\theta = \frac{\pi}{2}R\,\hat{a}_\theta$$

$$\int_{BC} \vec{dl} = \int_{+R}^{-R} dl(-\hat{a}_y) = -2R\hat{a}_y$$

$$\int_{CD} \vec{dl} = \int_{-\pi/2}^{-\pi} R\cdot d\theta\,\hat{a}_\theta = -\frac{\pi}{2}R\,\hat{a}_\theta$$

$$\therefore \int_C \vec{dl} = \int_{AB}\vec{dl} + \int_{BC}\vec{dl} + \int_{CD}\vec{dl} = -2R\,\hat{a}_y$$

1.8 (b)

Gradient of a scalar;

∇A = maximum rate of change of scalar A with respect to given coordinates system.

1.9 (a)

A voltage phase is a scalar and electric field phasor is a vector. Phasor is time or space dependent quantity. Phasor is complex with amplitude and phase variation.

■■■■

2 Electrostatic Fields

2.1 **Assertion (A) :** When there is no charge in the interior of a conductor the electric field intensity is infinite.

Reason (R): As per Gauss's law the total outward electric flux through any closed surface constituted inside the conductor must vanish.

(a) Both A and R are true and R is the correct explanation of A

(b) Both A and R are true but R is NOT the correct explanation of A

(c) A is true but R is false

(d) A is false but R is true

[ESE-2001]

2.2 A coaxial cable has two concentric dielectrics separated by a sheath as shown in the given figure. The distribution of electric field 'E' and potential 'ϕ' in the coaxial cable exist as

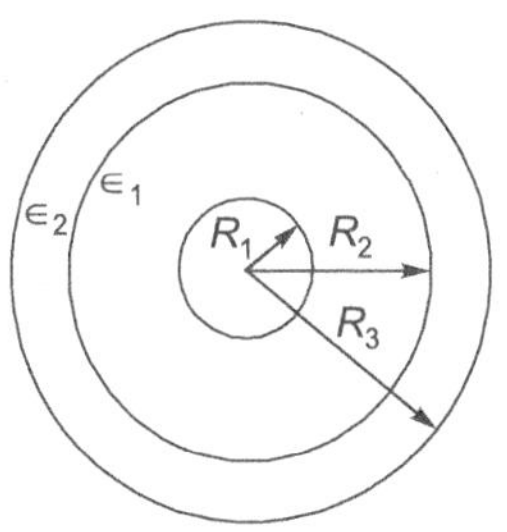

(a)
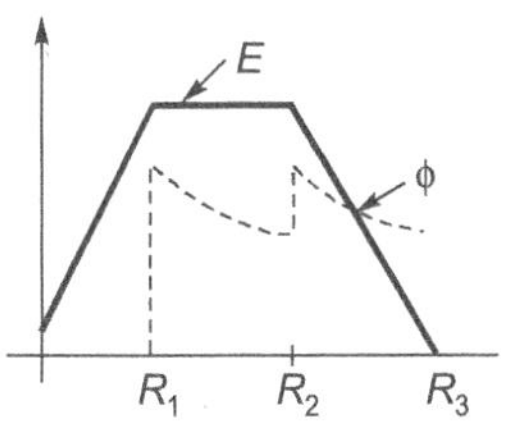

(b)
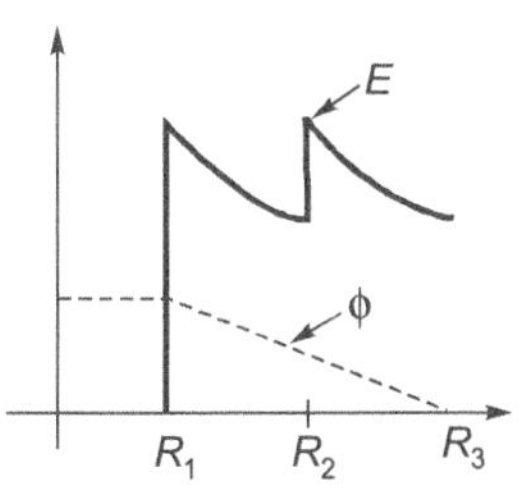

(c)
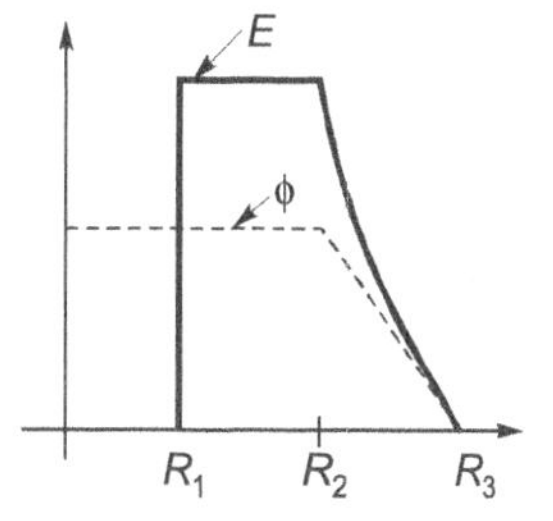

(d)
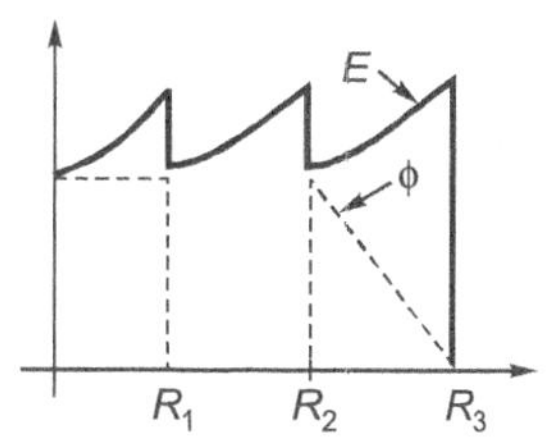

[ESE-2001]

2.3 A point charge $+Q$ is brought near a corner of two right angle conducting planes which are at zero potential as shown in the given Figure-I. Which one of the following configurations describes the total effect of the charges for calculating the actual field in the first quadrant?

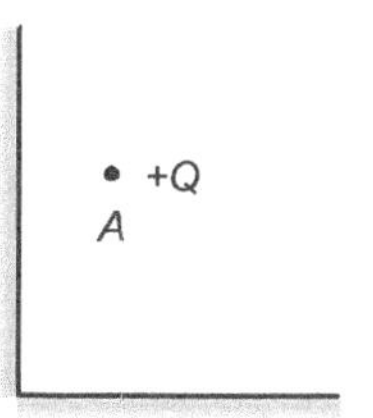

Figure-I

(a)
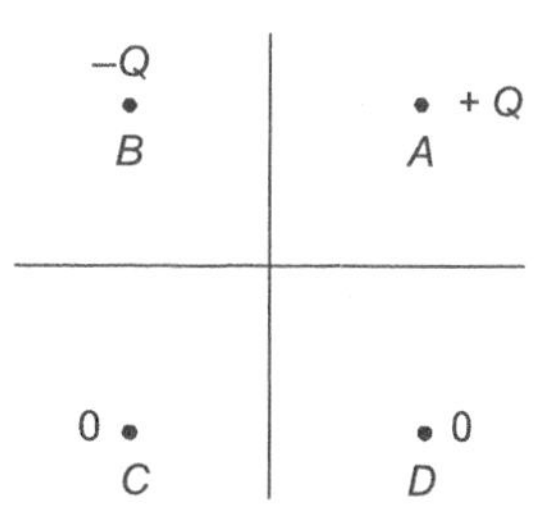

(b)

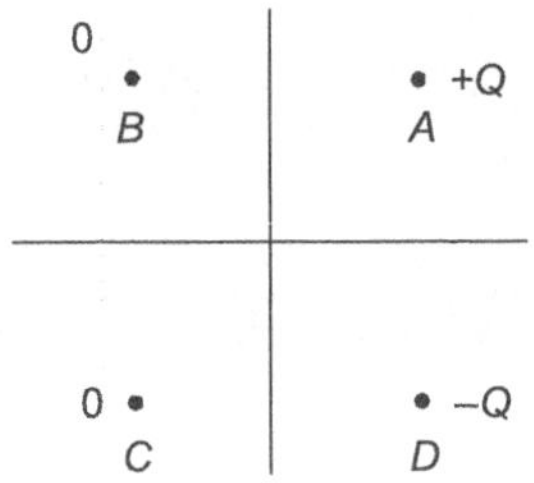

(c)

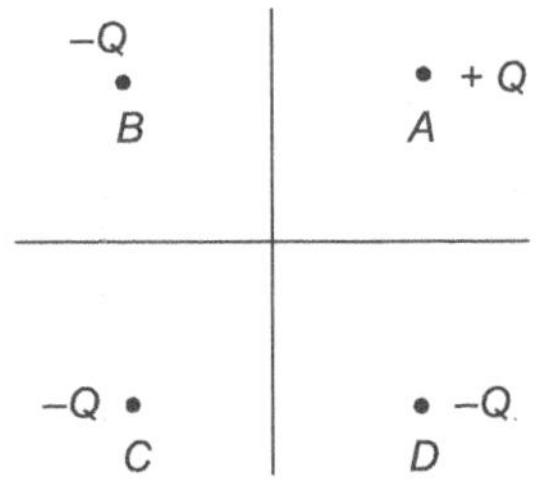

(d)

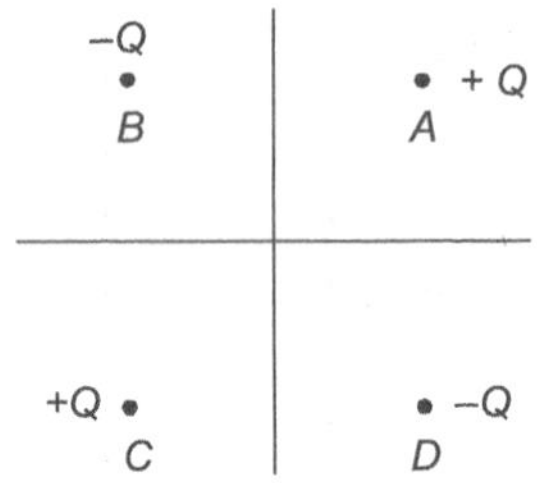

[ESE-2001]

2.4 The electric field across a dielectric-air interface is shown in the given figure. The surface charge density on the interface is

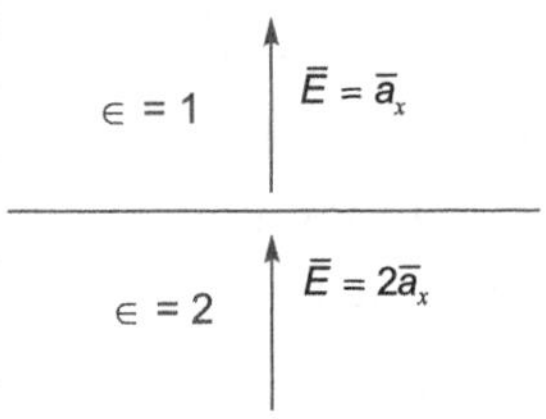

(a) $-4\epsilon_0$ (b) $-3\epsilon_0$
(c) $-2\epsilon_0$ (d) $-\epsilon_0$

[ESE-2001]

2.5 When air pocket is trapped inside a dielectric of relative permittivity '5', for a given applied voltage across the dielectric, the ratio of stress in the air pocket to that in the dielectric is equal to

(a) 1/5 (b) 5
(c) 1 + 5 (d) 5 – 1

[ESE-2001]

2.6 A potential field is given by $V = 3x^2y - yz$. Which of the following is NOT true?

(a) At the point (1, 0, –1), V and the electric field $\bar{E}$ vanish
(b) $x^2y = 1$ is an equipotential plane in the xy-plane
(c) The equipotential surface $V = -8$ passes through the point $P(2, -1, 4)$
(d) A unit normal to the equipotential surface $V = -8$ at P is $(-0.83\hat{x} + 0.55\hat{y} + 0.07\hat{z})$

[ESE-2002]

2.7 The relation between electric intensity E, voltage applied V and the distance d between the plates of a parallel plate condenser is

(a) $E = V/d$ (b) $E = V \times d$
(c) $E = V/(d)^2$ (d) $E = V \times (d)^2$

[ESE-2002]

2.8 Ohm's law in point form in field theory can be expressed as

(a) $V = RI$ (b) $\bar{J} = \bar{E}/\sigma$
(c) $\bar{J} = \sigma\bar{E}$ (d) $R = \rho l/A$

[ESE-2002]

2.9 Consider the arrangement of two equal and opposite charges of magnitude q separated by an infinitesimal distance l as shown in the figure given below. If $\bar{a}_r$ is the unit vector in the direction r and $\bar{a}_\phi$ is the unit vector in the direction θ, the electric field at the point P is

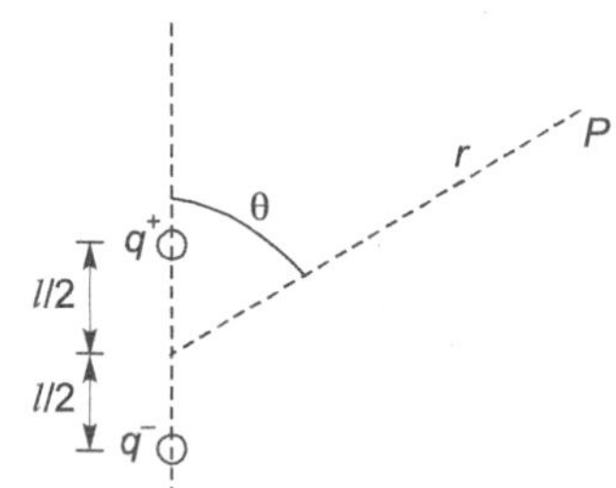

(a) $\dfrac{-2ql\cos\theta}{4\pi\varepsilon r^3}\bar{a}_r - \dfrac{ql\sin\theta}{4\pi\varepsilon r^3}\bar{a}_\theta$

(b) $\dfrac{-2ql\cos\theta}{4\pi\varepsilon r^3}\bar{a}_r + \dfrac{ql\sin\theta}{4\pi\varepsilon r^3}\bar{a}_\theta$

(c) $\dfrac{2ql\cos\theta}{4\pi\varepsilon r^3}\bar{a}_r + \dfrac{ql\sin\theta}{4\pi\varepsilon r^3}\bar{a}_\theta$

(d) $\dfrac{2ql\cos\theta}{4\pi\varepsilon r^3}\bar{a}_r - \dfrac{ql\sin\theta}{4\pi\varepsilon r^3}\bar{a}_\theta$

[ESE-2002]

2.10 **Assertion (A):** The test charge may have any value while defining electric field intensity.
Reason (R): The test charge should not disturb the field being measured.
(a) Both A and R are true and R is the correct explanation of A
(b) Both A and R are true but R is NOT the correct explanation of A
(c) A is true but R is false
(d) A is false but R is true
[ESE-2003]

2.11 **Assertion (A):** For steady current in an arbitrary conductor, the current density is solenoidal.
Reason (R): The reciprocal of the resistance is the conductivity.
(a) Both A and R are true and R is the correct explanation of A
(b) Both A and R are true but R is NOT the correct explanation of A
(c) A is true but R is false
(d) A is false but R is true
[ESE-2003]

2.12 A charge is uniformly distributed throughout the sphere of radius a. Taking the potential at infinity as zero, the potential at $r = b < a$ is

(a) $-\int_{\infty}^{b} \frac{Qr}{4\pi\varepsilon_0 a^3} dr$

(b) $-\int_{\infty}^{b} \frac{Q}{4\pi\varepsilon_0 r^2} dr$

(c) $-\int_{\infty}^{a} \frac{Q}{4\pi\varepsilon_0 r^2} dr - \int_{a}^{b} \frac{Qr}{4\pi\varepsilon_0 a^3} dr$

(d) $-\int_{\infty}^{a} \frac{Q}{4\pi\varepsilon_0 r^2} dr$

[ESE-2003]

2.13 A plane slab of dielectric having dielectric constant 5, placed normal to a uniform field with a flux density of 2 C/m^2, is uniformly polarized. The polarization of the slab is
(a) 0.4 C/m^2 (b) 1.6 C/m^2
(c) 2.0 C/m^2 (d) 6.4 C/m^2
[ESE-2003]

2.14 Consider the following statements relating to Laplace's equation:
1. Solution of Laplace's equation with two different approved methods lead to different answers.
2. Every physical problem satisfying Laplace's equation must contain at least two conducting boundaries.
3. Every field (if $\rho_v = 0$) satisfies Laplace's equation.
4. Every conceivable configuration of electrodes or conductors produces a field for which $\nabla^2 V = 0$.

Which of these statements are correct?
(a) 1, 3 and 4 (b) 3 and 4
(c) 1 and 2 (d) 2, 3 and 4
[ESE-2003]

2.15 Plane $z = 10$ m carries surface charge density 20 nC/m^2. What is the electric field at the origin?
(a) $-10\hat{a}_z$ V/m (b) $-18\pi\hat{a}_z$ V/m
(c) $72\pi\hat{a}_z$ V/m (d) $-360\pi\hat{a}_z$ V/m
[ESE-2004]

2.16 Consider the following diagram :

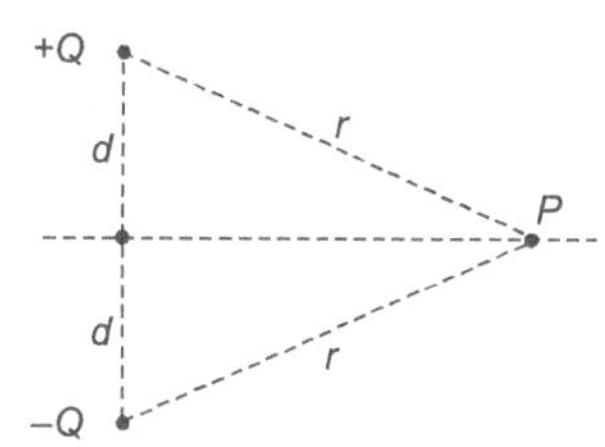

Which of the following statements is correct ?
The electric field E at a point P due to the presence of dipole as shown in the above diagram (considering distance $r >>$ distance d) is proportional to
(a) $1/r$ (b) $1/r^2$
(c) $1/r^3$ (d) $1/r^4$ [ESE-2004]

2.17 What is the value of total electric flux coming out of a closed surface?
(a) Zero
(b) Equal to volume charge density
(c) Equal to the total charge enclosed by the surface
(d) Equal to the surface charge density
[ESE-2004]

2.18 Two extensive homogenous isotropic dielectrics meet on a plane $z = 0$. For $z \geq 0$, $\epsilon_{r1} = 4$ and for $z \leq 0$, $\epsilon_{r2} = 2$. A uniform electric field exists at $z \geq 0$ as $\bar{E}_1 = 5\hat{a}_x - 2\hat{a}_y + 3\hat{a}_z$ kW/m. What is the value of $\bar{E}_{Z2}$ in the region $z \leq 0$?

(a) $3\hat{a}_z$ (b) $5\hat{a}_x - 2\hat{a}_y$

(c) $6\hat{a}_z$ (d) $\hat{a}_x - \hat{a}_y$

[ESE-2004]

2.19 A flat slab of dielectric, $\epsilon_r = 5$ is placed normal to a uniform field with a flux density $D = 1$ C/m^2. The slab is uniformly polarized. What is the polarization P of the slab in Coulomb/m^2?

(a) 0.8 (b) 1.2

(c) 4 (d) 6

[ESE-2004]

2.20 Which one of the following gives the approximate value of the capacitance between two spheres, whose separation is very much larger than their radii R?

(a) $2\pi/\varepsilon_0 R$ (b) $2\pi\epsilon_0 R$

(c) $2\pi\epsilon_0/R$ (d) $4\pi\epsilon_0/R$

[ESE-2004]

2.21 Which one of the following is the Poisson's equation for a linear and isotropic but inhomogeneous medium?

(a) $\nabla^2 E = -\rho/\epsilon$ (b) $\bar{\nabla}\cdot(\epsilon\bar{\nabla}V) = -\rho$

(c) $\bar{\nabla}\cdot\bar{\nabla}(\epsilon V) = -\rho$ (d) $\nabla^2 V = -\rho/\epsilon$

[ESE-2005]

2.22 For electrostatic fields in charge free atmosphere, which one of the following is correct?

(a) $\nabla\times\vec{E} = 0$ and $\nabla\cdot\vec{E} = 0$

(b) $\nabla\times\vec{E} \neq 0$ and $\nabla\cdot\vec{E} = 0$

(c) $\nabla\times\vec{E} = 0$ and $\nabla\cdot\vec{E} \neq 0$

(d) $\nabla\times\vec{E} \neq 0$ and $\nabla\cdot\vec{E} \neq 0$

[ESE-2005]

2.23 If the electric field established by three point charges Q, $2Q$ and $3Q$ exerts a force $3\vec{F}$ on $3Q$ and $2\vec{F}$ on $2Q$, then what is the force exerted on the point charge Q?

(a) $\vec{F}$ (b) $-\vec{F}$

(c) $5\vec{F}$ (d) $-5\vec{F}$

2.24 Which one of the following statements ***does not*** state that electrostatic field is conservative?

(a) The curl of $\vec{E}$ is identically zero

(b) The potential difference between two points is zero

(c) The electrostatic field is a gradient of a scalar potential

(d) The work done in a closed path inside the field is zero

[ESE-2006]

2.25 Where is the Laplace's equation valid?

(a) Only in free space

(b) Only in conductors

(c) Only in charge free dielectric regions

(d) Only in cavities bounded on all sides by conducting walls

[ESE-2006]

2.26 Sphere of radius a with a uniform charge density ρ_v C/m^3 shall have electric flux density at $r = a$, equal to

(a) $\frac{a}{3}\rho_v\hat{i}_r$ C/m^2 (b) $\frac{1}{3}\rho_v\hat{i}_r$ C/m^2

(c) $a\rho_v\hat{i}_r$ C/m^2 (d) $\frac{a}{4}\rho_v\hat{i}_r$ C/m^2

[ESE-2006]

2.27 Image theory is applicable to problems involving

(a) electrostatic field only

(b) magnetostatic field only

(c) both electrostatic and magnetostatic fields

(d) neither electrostatic nor magnetostatic field

[ESE-2006]

2.28 Six capacitors of different capacitances C_1, C_2, C_3, C_4, C_5 and C_6 are connected in series. $C_1 > C_2 > C_3 > C_4 > C_5 > C_6$. What is the total capacitance almost equal to?

(a) C_1 (b) C_3

(c) C_4 (d) C_6

[ESE-2006]

2.29 Equipotential surfaces about a pair of equal and opposite linear charges exist in what form?

(a) Concentric spheres
(b) Concentric cylinders
(c) Non-concentric cylinders
(d) Planes

[ESE-2006]

2.30 In free space, if $\rho = 0$, the Poisson's equation becomes

(a) Maxwell's divergence equation $\nabla \cdot \vec{B} = 0$
(b) Laplacian equation $\nabla^2 V = 0$
(c) Kirchhoff's voltage equation $\Sigma V = 0$
(d) None of the above

[ESE-2006]

2.31 Point charges of –10 nC and 10 nC are located in free space at (–1, 0, 0) m and (1, 0, 0) m respectively. What is the energy stored in the field?

(a) Zero (b) 450 nJ
(c) – 450 nJ (d) –900 nJ

[ESE-2007]

2.32 A parallel plate air capacitor carries a charge Q at its maximum withstand voltage V. If the capacitor is half filled with an insulating slab of dielectric constant 4 as shown in the figure given below, what are the maximum withstand voltage and the charge on the capacitor at this voltage, respectively?

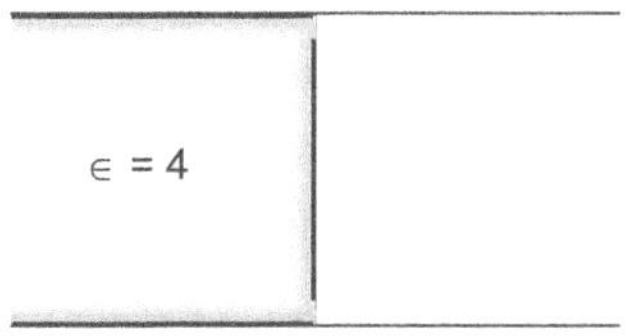

(a) 2.5 V, Q (b) 4 V, 2.5 Q
(c) V, 2.5 Q (d) V/4, Q

[ESE-2007]

2.33 A spherical balloon of radius a is charged. The energy density in the electric field at a point P shown in the figure given below is w. If the balloon is inflated to a radius b without altering its charge, what is the energy density at P?

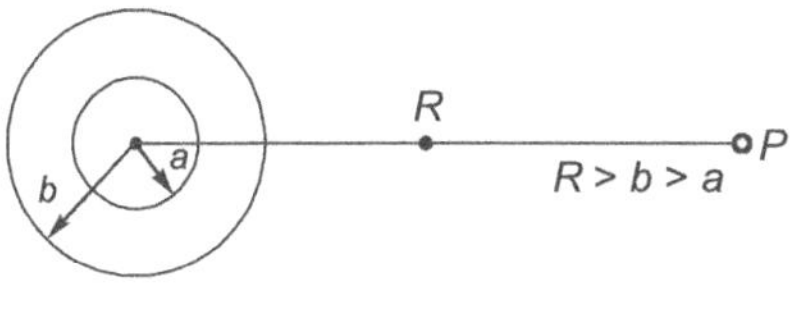

(a) $w\left(\frac{b}{a}\right)^3$ (b) $w\left(\frac{b}{a}\right)^2$
(c) $w\left(\frac{b}{a}\right)$ (d) w

[ESE-2007]

2.34 When an infinite charged conducting plate is placed between two infinite conducting grounded surfaces as shown in the figure given below, what would be the ratio of the surface charge densities ρ_1 and ρ_2 on the two sides of the plate?

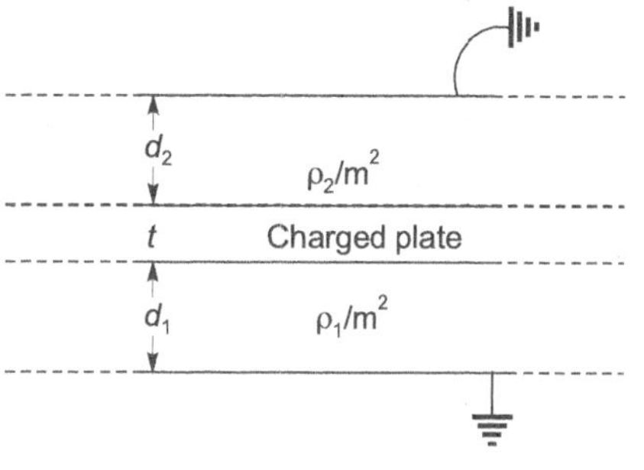

(a) $\frac{(d_1 + t)}{(d_2 + t)}$ (b) $\frac{(d_2 + t)}{(d_1 + t)}$
(c) $\frac{d_1}{d_2}$ (d) $\frac{d_2}{d_1}$

[ESE-2007]

2.35 The polarization in a solid dielectric is related to the electric field $\vec{E}$ and the electric flux density $\vec{D}$ according to which one of the following equations?

(a) $\vec{E} = \epsilon_0 \vec{D} + \vec{P}$ (b) $\vec{D} = \epsilon_0 (\vec{E} + \vec{P})$
(c) $\vec{D} = \epsilon_0 \vec{E} + \vec{P}$ (d) $\vec{E} = \vec{D} + \epsilon_0 \vec{P}$

[ESE-2007]

2.36 Consider the following statements for an electrostatic field:

1. Its curl is zero.
2. It is negative of gradient of scalar potential ϕ.
3. It is conservative.

Which of these statements are correct?

(a) 1 and 2 (b) 2 and 3
(c) 1 and 3 (d) 1, 2 and 3

[ESE-2008]

2.37 Which one of the following is the correct statement?

Equipotential lines and field lines
(a) are parallel
(b) are anti-parallel
(c) are orthogonal
(d) bear no definite relationship [ESE-2008]

2.38 The capacitance of an insulated conducting sphere of radius R in vacuum is
(a) $2\pi \varepsilon_0 R$ (b) $4\pi \varepsilon_0 R$
(c) $4\pi \varepsilon_0 R^2$ (d) $4\pi \varepsilon_0/R$
[ESE-2008]

2.39 Consider the following statements in connection with boundary relations of electric field:
1. In a single medium electric field is continuous.
2. The tangential components are the same on both sides of a boundary between two dielectrics.
3. The tangential electric field at the boundary of a dielectric and a current carrying conductor with finite conductivity is zero.
4. Normal component of the flux density is continuous across the charge-free boundary between two dielectrics.

Which of these statements is/are correct?
(a) 1 only (b) 1, 2 and 3
(c) 1, 2 and 4 (d) 3 and 4 only
[ESE-2009]

2.40 Point charges of Q_1 = 2 nC and Q_2 = 3 nC are located at a distance apart. With regard to this situation, which one of the following statements is **NOT** correct?
(a) The force on the 3 nC charge is repulsive.
(b) A charge of –5 nC placed midway between Q_1 and Q_2 will experience no force.
(c) The forces Q_1 and Q_2 are same in magnitude.
(d) The forces on Q_1 and Q_2 will depend on the medium in which they are placed
[ESE-2010]

2.41 The normal components of electric flux density across a dielectric-dielectric boundary
(a) are discontinuous
(b) are continuous
(c) depend on the magnitude of the surface charge density
(d) depend on electric field intensity
[ESE-2011]

2.42 **Assertion (A):** The electric field around a positive charge is outward.
Reason (R): Gauss law states that the differential of the normal component of the outward electric flux density over a closed surface yields the positive charge enclosed.
(a) Both A and R are true and R is the correct explanation of A
(b) Both A and R are true but R is NOT the correct explanation of A
(c) A is true but R is false
(d) A is false but R is true
[ESE-2011]

2.43 The electric field strength at any point a distance 'r' from the point charge 'q' located in homogeneous isotropic medium with dielectric constant ε_r is given by
(a) $E = \dfrac{q \in^{-1}}{4\pi r^2}\hat{r}$ (b) $E = \oint D\, da \cos\theta$
(c) $E = \dfrac{q \in}{4\pi r^2}\hat{r}$ (d) $E = \dfrac{q^2}{4\pi \in r^2}\hat{r}$
[ESE-2012]

2.44 The electric field in the vicinity of two oppositely charged parallel conductors is
(a) radial uniformly.
(b) in parallel lines between the two imaginary parallel planes passing through the centres of the two conductors.
(c) not uniform and its direction changes from point to point.
(d) in parallel circular paths between the two conductors, with the centre of the circles located at the mid-point of a line joining the two centres of the two conductors.
[ESE-2012]

2.45 Two charges are placed at a distance apart. Now, if a glass slab is inserted between them, then the force between the charges will
(a) reduce to zero (b) increase
(c) decrease (d) not change
[ESE-2012]

2.46 The following point charges are located in air:
+0.008 μC at (0, 0) m
+0.05 μC at (3, 0) m
–0.009 μC at (0, 4) m
The total electric flux over a sphere of 5 m radius with centre (0, 0) is
(a) 0.058 μC (b) 0.049 μC
(c) 0.029 μC (d) 0.016 μC
[ESE-2012]

2.47 Electric flux through a surface area is the integral of the
(a) normal component of the electric field over the area.
(b) parallel component of the electric field over the area.
(c) normal component of the magnetic field over the area.
(d) parallel component of the magnetic field over the area.
[ESE-2012]

2.48 The vector statement of Gauss's law is
(a) $\oint_V D \cdot da = \int_S \sigma dv$
(b) $\int_V D \cdot da = \oint_S \rho dv$
(c) $\iint_S D \cdot da = \int_V \rho^2 dv$
(d) $\oint_S D \cdot da = \int_V \rho dv$ [ESE-2012]

2.49 The equation of continuity defines the relation between
(a) electric field and magnetic field
(b) electric field and charge density
(c) flux density and charge density
(d) current density and charge density
[ESE-2012]

2.50 If $E = 0$ at all points on a closed surface,
1. The electric flux through the surface is zero.
2. The total charge enclosed by the surface is zero.
3. Charge resides on the surface.
(a) 1 and 2 only (b) 1 and 3 only
(c) 2 and 3 only (d) 1, 2 and 3
[ESE-2013]

2.51 The electrostatic force of repulsion between two α-particles of charges 4.0×10^{-19} C each, and separated by a distance of 10^{-10} cm is
(Given $\epsilon_0 = 8.854 \times 10^{-12}$ Nm²/Coul²)
(a) 57.6×10^{-4} N (b) 28.8×10^{-4} N
(c) 14.4×10^{-4} N (d) 3.6×10^{-4} N
[ESE-2013]

2.52 A parallel plate capacitor of area A cm² and separating distance 'a' cm is dipped in ethyl alcohol up to a depth of $a/2$. Given the dielectric constant ϵ_r of ethyl alcohol to be 25, the ratio of capacitance after dipping to that before dipping would be
(a) $\frac{26}{50}$ (b) $\frac{45}{50}$
(c) $\frac{50}{26}$ (d) $\frac{3}{1}$
[ESE-2013]

2.53 The electric field lines and equipotential lines
(a) are parallel to each other
(b) are one and the same
(c) cut each other orthogonally
(d) can be inclined to each other at any angle
[ESE-2014]

2.54 According to Gauss's Law, the surface integral of the normal component of electric flux density D over a closed surface containing charge Q is
(a) $\frac{Q}{\varepsilon_0}$ (b) $\varepsilon_0 Q$
(c) Q (d) $\frac{Q^2}{\varepsilon_0}$
[ESE-2014]

2.55 Consider the following statements associated with the basic electrostatic properties of ideal conductors:
1. The resultant field inside is zero.
2. The net charge density in the interior is zero.
3. Any net charges reside on the surface.
4. The surface is always equipotential.
5. The field just outside is zero.
Which of the above statements are correct?
(a) 1, 2, 3 and 4 (b) 3, 4 and 5 only
(c) 1, 2 and 3 only (d) 2 and 3 only
[ESE-2014]

2.56 In practice, Earth is chosen as a place of zero electric potential because it
(a) is non-conducting
(b) is easily available reference
(c) keeps losing and gaining electric charge every day
(d) has almost constant potential
[ESE-2014]

2.57 The capacitance of a concentric spherical capacitor of shell radii x and y $(x \times y)$ is

(a) $\frac{1}{4\pi\varepsilon_0} ln\frac{x}{y}$ (b) $\frac{4\pi\varepsilon_0 xy}{x-y}$

(c) $4\pi\varepsilon_0 \, ln\frac{y}{x}$ (d) $\frac{1}{4\pi\varepsilon_0}\left[\frac{1}{y}-\frac{1}{x}\right]$

[ESE-2014]

2.58 The field strength at a point of finite distance from an infinitely long straight uniformly charged conductor is obtained by considering the radial (R) component and the longitudinal (L) component of the forces acting on a unit charge at the point, by the charges on the elemental length of the conductor. The resultant field strength is
(a) the sum of R-components, when the sum of L-components is zero
(b) the sum of L-components, when the sum of R-components is zero
(c) the sum of both R- and L-components
(d) average of the sums of R- and L- components
[ESE-2015]

2.59 A parallel-plate air capacitor as shown below has a total charge Q and a breakdown voltage V. A slab of dielectric constant 6 is inserted as shown. The maximum breakdown voltage and the charge at this voltage respectively would be

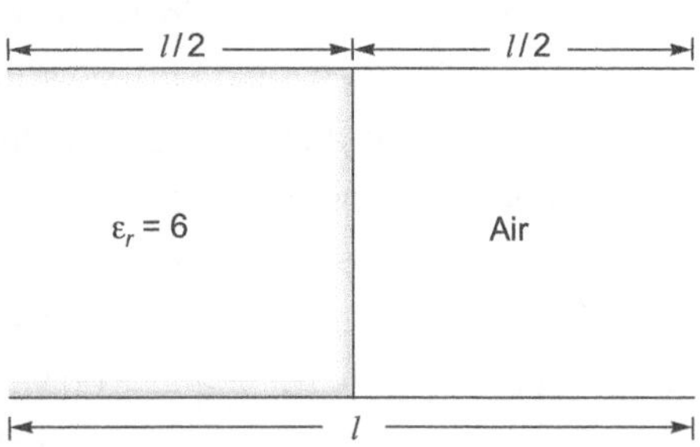

(a) V and 3.5 Q (b) 3.5 V and Q
(c) V/6 and 3.5 Q (d) 6V and 3 Q
[ESE-2015]

2.60 The capacitance of a conducting sphere of radius r with a total charge of q uniformly distributed on its surface is
(a) proportional to qr
(b) independent of r
(c) proportional to q/r
(d) independent of q
[ESE-2016]

2.61 Consider the following types of transmission lines:
1. Open-wire line
2. Twin-lead wire
3. Coaxial cable

The capacitance per metre will be least in which of the above transmission lines?
(a) 1 only (b) 2 only
(c) 3 only (d) 1, 2 and 3
[ESE-2016]

2.62 Three equal point charges are located at the vertices of an equilateral triangle on the circumference of a circle of radius r. The total electric field intensity at the centre of the circle would be

(a) zero (b) $\frac{3q}{4\pi\varepsilon_0 r^2}$

(c) $\frac{q}{12\varepsilon_0 r^2}$ (d) $\frac{q}{3\pi\varepsilon_0 r^2}$

[ESE-2016]

2.63 The dielectric strength of rubber is 40000 V/mm at frequency of 50 Hz. What is the thickness of insulation required on an electrical conductor at 33 kV to sustain the breakdown?
(a) 0.83 mm (b) 8.3 mm
(c) 8.3 cm (d) 0.083 mm
[ESE-2016]

2.64 **Statement (I):** Voltage is the energy per unit charge created by charge separation.
Statement (II): Power is energy per unit of time.
[ESE-2016]

2.65 Gauss's theorem states that total electric flux-ϕ emanating from a closed surface is equal to

(a) total current density on the surface
(b) total charge enclosed by that surface
(c) total current on the surface
(d) total charge density within the surface

[ESE-2017]

2.66 "Electrical flux enclosed by a surface surrounding a charge is equal to the amount of charge enclosed." This is the statement of
(a) Faraday's law
(b) Lenz's law
(c) Modified Ampere's law
(d) Gauss's law

[ESE-2017]

2.67 If a positively charge body is placed inside a spherical hollow conductor, what will be the polarity of charge inside and outside the hollow conductor?
(a) Inside positive, outside negative
(b) Inside negative, outside positive
(c) Both negative
(d) Both positive

[ESE-2017]

2.68 Which of the following equations represent Gauss's law adapted to a homogenous isotropic medium?

1. $\oint_s \vec{D}.d\vec{s} = \oint_v \rho dv$ 2. $\nabla \times \vec{H} = \vec{D}$

3. $\nabla \cdot \vec{J} + \rho = 0$ 4. $\nabla \cdot \vec{E} = \frac{\rho}{\varepsilon}$

5. $\nabla^2 \Phi = 0$

(a) 1 and 4 only (b) 2 and 3 only
(c) 3 and 5 only (d) 1, 2, 4 and 5 only

[ESE-2017]

2.69 The maximum potential-gradient that can be imposed in air at atmospheric pressure without breakdown is 30 kV/cm. The corresponding energy density is nearly
(a) 30 J/m^3 (b) 35 J/m^3
(c) 40 J/m^3 (d) 45 J/m^3

[ESE-2018]

2.70 A hollow metallic sphere of radius R is charged to a surface density of σ. The strength of the electric field inside the sphere at a radius $r(<R)$ is

(a) $\frac{\sigma}{\pi r^2}$ (b) $\frac{\sigma}{2\pi r^2}$

(c) $\frac{\sigma}{4\pi r^2}$ (d) zero

[ESE-2018]

2.71 A parallel-plate capacitor with air between the plates has a capacitance of 10 pF. If the distance between the parallel plates is halved and the space between the plates is filled with a material of dielectric constant 5, the newly formed capacitor will have a capacitance of
(a) 10 pF (b) 50 pF
(c) 100 pF (d) 150 pF

[ESE-2018]

2.72 A point charge of 10^{-9} C is placed at a point A in the free space. The potential difference between the two points 20 cm and 10 cm away from the charge at A will be
(a) 40 V (b) 45 V
(c) 50 V (d) 55 V

[ESE-2020]

2.73 According to Gauss's theorem, the surface integral of the normal component of electric flux density D over a closed surface, containing free charge is

(a) Q (b) $\frac{Q}{\varepsilon_0}$

(c) $\varepsilon_0 Q$ (d) $\frac{Q^2}{\varepsilon_0}$

[ESE-2020]

2.74 A capacitor of 10 pF is connected to a voltage source of 100 V. If the distance between the capacitor plates is reduced to 50%, while it remains connected to the 100 V supply, the value of potential gradient in the second case will be
(a) Half of earlier value
(b) Same as earlier value
(c) Twice of earlier value
(d) One-fourth of earlier value

[ESE-2020]

2.75 A point charge of 10^{-9} C is placed at a point A in free space. What is the intensity of electrostatic field on the surface of a sphere of radius 5 cm at centre A?

(Take $\frac{1}{4\pi\varepsilon_0} = 9 \times 10^9$ in SI units)

(a) 1800 V/m (b) 3600 V/m
(c) 5360 V/m (d) 2820 V/m

[ESE-2021]

2.76 **Statement I :** The pattern of the electric field lines strongly depends on the geometric arrangement of charge carriers and the field lines always enter or exit the charge carrier vertically.

Statement II : From the field pattern, a small distance between adjacent field lines (high field line density) indicates low field strength.

(a) Both Statement (I) and Statement (II) is the correct explanation of Statement (I).
(b) Both Statement (I) and Statement (II) are individually true, but Statement (II) is not the correct explanation of Statement (I).
(c) Statement (I) is true, but Statement (II) is false.
(d) Statement (I) is false, but Statement (II) is true.

[ESE-2021]

2.77 Which one of the following statements is not correct regarding potential due to a point charge?

(a) It is directly proportional to the magnitude of the charge.
(b) It is inversely proportional to the distance from the charge.
(c) It is inversely proportional to the relative permittivity of the medium in which the charge is placed.
(d) It is directly proportional to the electric field intensity.

[ESE-2022]

2.78 The relation between electric flux density (D) and field intensity (E) with absolute and relative permittivity is expressed as

(a) $E = \frac{\varepsilon_0 \varepsilon_r}{D}$ (b) $E = \frac{D}{\varepsilon_0 \varepsilon_r}$

(c) $E = \frac{D\varepsilon_0}{\varepsilon_r}$ (d) $E = \frac{D\varepsilon_r}{\varepsilon_0}$

[ESE-2022]

2.79 A cross-sectional view of a Printed Circuit Board (PCB) consisting of two conducting lands on the surface of a dielectric board above a reference ground plane is shown in the figure. The three-conductor system characterized by a partial capacitance scheme is also shown. For the multiple conductor system, what is the capacitance matrix [C]?

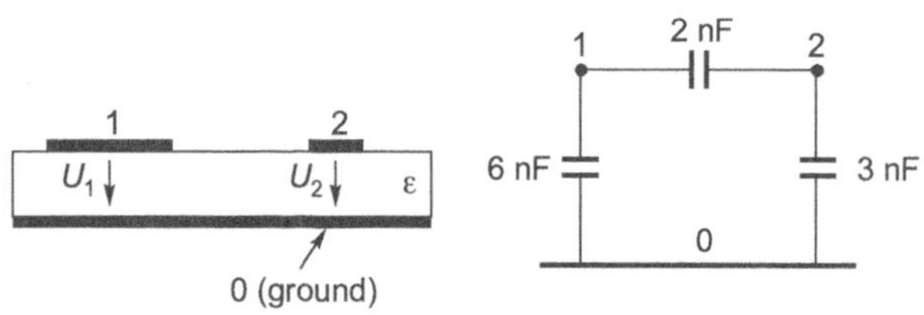

(a) $[C] = \begin{bmatrix} -2 & 8 \\ 5 & -2 \end{bmatrix}$ nF

(b) $[C] = \begin{bmatrix} -2 & 5 \\ 8 & -2 \end{bmatrix}$ nF

(c) $[C] = \begin{bmatrix} 2 & -8 \\ -5 & 2 \end{bmatrix}$ nF

(d) $[C] = \begin{bmatrix} 8 & -2 \\ -2 & 5 \end{bmatrix}$ nF

[ESE-2023]

■■■■

Answers Electrostatic Fields

2.1 (d)	2.2 (b)	2.3 (d)	2.4 (b)	2.5 (b)	2.6 (a)	2.7 (a)	2.8 (c)	2.9 (c)
2.10 (d)	2.11 (c)	2.12 (d)	2.13 (b)	2.14 (b)	2.15 (d)	2.16 (c)	2.17 (c)	2.18 (c)
2.19 (a)	2.20 (b)	2.21 (b)	2.22 (a)	2.23 (d)	2.24 (b)	2.25 (c)	2.26 (a)	2.27 (a)
2.28 (d)	2.29 (d)	2.30 (b)	2.31 (c)	2.32 (c)	2.33 (d)	2.34 (c)	2.35 (c)	2.36 (d)
2.37 (c)	2.38 (b)	2.39 (c)	2.40 (b)	2.41 (c)	2.42 (c)	2.43 (a)	2.44 (c)	2.45 (c)
2.46 (b)	2.47 (a)	2.48 (d)	2.49 (d)	2.50 (a)	2.51 (c)	2.52 (c)	2.53 (c)	2.54 (c)
2.55 (a)	2.56 (d)	2.57 (b)	2.58 (a)	2.59 (a)	2.60 (d)	2.61 (a)	2.62 (a)	2.63 (a)
2.64 (b)	2.65 (b)	2.66 (d)	2.67 (b)	2.68 (a)	2.69 (c)	2.70 (d)	2.71 (c)	2.72 (b)
2.73 (a)	2.74 (c)	2.75 (b)	2.76 (c)	2.77 (d)	2.78 (b)	2.79 (d)		

Explanations Electrostatic Fields

2.1 (d)

When there is no charge in the interior of a conductor the electric field intensity is zero not infinite as per Gauss's Law.

2.2 (b)

Inside a conductor electric field is zero and potential is constant everywhere i.e.

for $0 \le R < R_1$; $E = 0$, V = constant

and for $R > R_1$; $E \propto 1/R$, decreases with R these two condition satisfied (b) option only.

2.3 (d)

We know that on the surface of conductor tangential component of electric field is zero only normal component exist in order to find this we use the concept of method of images i.e. by replacing the surface by placing a charge of a magnitude and polarity so that the electric field component existing remains normal to surface and for this we replace surfaces by placing equal and opposite charges at a distance same as existing charge's distance from surface on the other side of surface so from the options, option D is correct.

2.4 (b)

According to boundary condition for electric field.

$$(D_{n_1} - D_{n_2}) = -\rho_s \quad \text{...(1)}$$

$\rho_s \rightarrow$ surface charge density on the interface.

D_{n_1} normal component of electric flux density from medium 1 to 2 (i.e. $\hat{a}_{12}$)

So, $D_{n_1} = \in \vec{E} \cdot \hat{a}_x$

$$= 2 \in_0 2\hat{a}_x \cdot \hat{a}_x = 4\in_0$$

Similarly, $D_{n_2} = \in \vec{E} \cdot \hat{a}_x = \in_0 \hat{a}_x \cdot \hat{a}_x = \in_0$

From equation (1),

$$(4\in - \in_0) = -\rho_s$$

$$\rho_s = -3\in_0$$

2.5 (b)

$$D = \in_0 \in_r E \Rightarrow E \propto \frac{1}{\in_r \in_0}$$

$$\therefore \quad \frac{E_{\text{air}}}{E_{\text{dielectric}}} = \frac{\in_r \in_0}{\in_0} = 5$$

2.6 (a)

$$V = 3x^2y - yz$$

$$\bar{E} = -\nabla V = -\left[\frac{\partial V}{\partial x}\hat{a}_x + \frac{\partial V}{\partial y}\hat{a}_y + \frac{\partial V}{\partial z}\hat{a}_z\right]$$

$$\bar{E} = -\left[6xy\,\hat{a}_x + (3x^2 - z)\hat{a}_y - y\hat{a}_z\right]$$

So, $\bar{E}$ at $(+1, 0, -1) = -\left[0 + 3\hat{a}_y - 0\right]$

$$= -3\hat{a}_y \neq 0$$

2.7 (a)

For a parallel plate capacitor electric field intensity between the plates

$$= \frac{\text{Potential difference between the plates}}{\text{Distance between the plates}}$$

$$E = \frac{V}{d}$$

2.8 (c)

$\vec{J} = \sigma\vec{E}$ is point form of ohm's law.

2.9 (c)

Electric field $\vec{E}$ at a distance r from centre of a dipole is given by

$$\vec{E} = \frac{p}{4\pi \in r^3}\left[2\cos\theta\cdot\hat{a}_r + \sin\theta\cdot\hat{a}_\theta\right]$$

where, $p = q \times l = ql$

2.10 (d)

Test charge can not have any value. It should be too small to disturb the field being measured.

2.11 (c)

For steady state current $\vec{J}_c$ will be a constant vector as magnitude of current is constant and area of cross-section also so $\nabla\cdot\vec{J}_c = 0$ divergence of a constant vector is zero.

so $\vec{J}_c$ is a solenoidal vector.

A → is correct.

R → is wrong (reciprocal of R is conductance not the conductivity).

2.12 (d)

Charge is uniformly distributed on sphere.

So electric field $\vec{E} = \frac{Q}{4\pi \in r^2}\hat{a}_r$ for $r \geq a$

$= 0 \quad r < a$ inside

∴ Potential at a point $r = b < a$

$$V = -\int_\infty^b \vec{E}\cdot d\vec{r}$$

$$V = -\int_\infty^a \vec{E}\cdot d\vec{r} - \int_a^b \vec{E}\cdot d\vec{r}$$

$$V = -\int_\infty^a \frac{Q}{4\pi \epsilon_0 r^2}dr + 0$$

$(\because E = 0 \text{ for } r < a)$

$$V = -\int_\infty^a \frac{Q}{4\pi \epsilon_0 r^2}dr$$

2.13 (b)

Polarisation,

$$\vec{P} = \vec{D}\left(1 - \frac{1}{\epsilon_r}\right) = 2\left(1 - \frac{1}{5}\right) = 1.6 \text{ C/m}^2$$

2.14 (b)

- Every physical problem must contain at least one conducting boundary and usually contains two or more.
- Solution of Laplace's equation with two different approved methods lead to same answer.
- Laplace's equation is all-embracing for applying as it does wherever volume charge density (ρ_v) is zero, it states that every conceivable configuration of electrodes or conductors produces a field for which $\nabla^2 V = 0$.

2.15 (d)

$$E = \frac{\rho_s}{2\epsilon_0}(-\hat{a}_z)$$

$$= \frac{20\times10^{-9}\times4\pi\times9\times10^9}{2\times1}(-\hat{a}_z)$$

$$= -360\pi\,\hat{a}_z \text{ V/m}$$

2.16 (c)

$\vec{E}$ (at a point P)

$$= \frac{2Qd}{4\pi\epsilon_0 r^3}(2\cos\theta\,\hat{a}_r + \sin\theta\,\hat{a}_\theta)$$

$$= \frac{2Qd}{4\pi\epsilon_0 r^3} \text{ as } \theta = 90°$$

2.17 (c)

Net flux coming out from a closed surface i.e.

$\oint \vec{D}\cdot\vec{ds} = Q_{enclosed}$ total charge enclosed within the surface this is Gauss's law for electric field.

2.18 (c)

From the boundary condition

$$D_{n_1} = D_{n_2} \text{ (Charge-free)}$$

$$\therefore \quad \epsilon_0\epsilon_{r_1}E_{z_1} = \epsilon_0\epsilon_{r_2}E_{z_2}$$

$$\therefore \quad 4\times3 = 2\times E_{z_2}$$

$$\therefore \quad E_{z_2} = 6\hat{a}_z$$

2.19 (a)

$$\vec{P} = \left(\frac{\epsilon_r - 1}{\epsilon_r}\right)\vec{D} = \left(\frac{5-1}{5}\right)\cdot1 = 0.8 \text{ C/m}^2$$

2.20 (b)

Potential of the sphere with a zero reference at infinity is,

$$V = \frac{Q}{4\pi\epsilon_0 R}$$

So, $$C = \frac{Q}{V} = 4\pi\epsilon_0 R$$

Capacitance between two spheres is given by

$$\frac{1}{C} = \frac{1}{C_1} + \frac{1}{C_2} \quad \text{(series combination)}$$

$$\therefore \quad C = \frac{C_1 C_2}{C_1 + C_2} = 2\pi\epsilon_0 R$$

Trick: Only option (b) has dimensions of capacitor.

2.21 (b)

We known that,

$$\nabla \cdot \vec{D} = \rho_v$$

But, $$D = \epsilon \vec{E}$$

$$\therefore \quad \nabla \cdot (\epsilon \vec{E}) = \rho_v$$

But, $$\vec{E} = -\nabla V$$

V-potential (Scalar), $\nabla \cdot (-\epsilon \nabla V) = \rho_v$

$$\boxed{\nabla \cdot (\epsilon \nabla V) = -\rho_v}$$

Medium is inhomogeneous i.e. ϵ is function of dimensions x, y, z so we can't take it outside.

2.22 (a)

For electrostatic fields $\nabla \times \vec{E} = 0$ always since electrostatic fields are conservative and since region is charge free therefore $\vec{E}$ will be constant therefore $\nabla \cdot \vec{E} = 0$.

2.23 (d)

In the equilibrium of whole system (total force on all the charges) $\Sigma \vec{F} = 0 \Rightarrow$ Force exerted on the point charge $Q = -5\ F$.

2.24 (b)

Electrostatic field is conservative

i.e. $$\nabla \times \vec{E} = 0$$

$$\Rightarrow \quad \oint_l \vec{E} \cdot dl = 0$$

i.e. integration of $\vec{E}$ over a closed path is zero

$$\because \quad \int_l \vec{E} \cdot \vec{dl} = V \text{ (Potential difference)}$$

So on close path net potential difference is zero.

$$\vec{E} = -\nabla V$$

Since potential difference along a closed path is $V = 0$. Therefore work done in moving a charge in closed path

$$W = qV = 0$$

2.25 (c)

Laplace's equation is all-embracing, for applying as it does where volume charge density is zero, it states that every conceivable configuration of electrodes or conductors produces a field for which $\nabla^2 V = 0$.

2.26 (a)

Charge density = ρ_v C/m^3

We know that from Gauss's law,

$$\int_s \vec{D} \cdot \vec{ds} = Q_{enclosed}$$

Since, D is radially outward and constant through out the spherical surface.

$$\therefore \quad D \cdot 4\pi r^2 = Q_{enclosed}$$

At $r = a$,

$$D \cdot 4\pi a^2 = Q_{enclosed}$$

$$D \cdot 4\pi a^2 = \text{Density} \times \text{volume}$$

$$D \cdot 4\pi a^2 = \rho_v \times \frac{4}{3}\pi a^2$$

$$D = \frac{a}{3}\rho_v \text{ C/m}^2$$

$$\vec{D} = D\hat{i}_r \text{ C/m}^2 = \frac{a}{3}\rho_v \hat{i}_r \text{ C/m}^2$$

2.27 (a)

Image theory is applicable to problem involving electrostatic fields only normally to find the field on a conductor surface (Normal component).

2.28 (d)

Net capacitance C is given by, for series connection

$$\frac{1}{C} = \frac{1}{C_1} + \frac{1}{C_2} + \frac{1}{C_3} + \frac{1}{C_4} + \frac{1}{C_5} + \frac{1}{C_6}$$

$$\because \quad C_1 > C_2 > C_3 > C_4 > C_5 > C_6$$

$$\therefore \quad \frac{1}{C_1} < \frac{1}{C_2} < \frac{1}{C_3} < \frac{1}{C_4} < \frac{1}{C_5} < \frac{1}{C_6}$$

$$\therefore \quad \frac{1}{C} \simeq \frac{1}{C_6}$$

$$\therefore \quad C \cong C_6$$

2.29 (d)

\- - - - - - - - - - - - -

\+ + + + + + + + + + + + +

On either side of these linear charges potential will be same in a plane parallel to these linear charges because of uniformity.

2.30 (b)

For a homogeneous medium Poisson's equation is

$$\nabla^2 V = -\frac{\rho_v}{\epsilon}$$

if $\rho_v = 0$ (charge free region)

$\nabla^2 V = 0$ this is Laplace equation

2.31 (c)

$$\text{Energy of the system} = \frac{q_1 q_2}{4\pi\epsilon_0 r}$$

$$= \frac{-10\times10^{-9}\times10\times10^{-9}}{4\times\pi\times8.85\times10^{-12}\times2}$$

$$= -449.59 \text{ nJ} \simeq -450 \text{ nJ}$$

2.32 (c)

Two capacitors are connected is parallel so voltage rating will be same.

But net capacitance will be now,

$$C_{eq} = C_1 + C_2$$

$$= \frac{\epsilon_0\epsilon_r A'}{d} + \frac{\epsilon_0 A'}{d}$$

$$= \frac{4\epsilon_0 A/2}{d} + \frac{\epsilon_0 A/2}{d}$$

$$= \frac{2\epsilon_0 A}{d} + \frac{\epsilon_0 A}{2d} = 2\cdot C + \frac{C}{2}$$

$$\left(\because \frac{\epsilon_0 A}{d} = C\right) \text{ initial capacitance}$$

$$C_{eq} = 2.5\,C$$

Now new charge storage at withstand voltage V

$$Q' = C_{eq}V$$

$$Q' = 2.5\,CV$$

$Q' = 2.5\,Q$ (as $Q = CV$ initially)

2.33 (d)

$$\text{Energy stored, } W_e = \frac{1}{2}\epsilon E^2$$

$E \propto \frac{1}{R^2}$ where R is distance from center to point 'P'. Due to inflation of balloon R is not changing.

2.34 (c)

Potential at both sides of conducting sheet will be same.

2.35 (c)

$$\vec{D} = \epsilon_0\vec{E} + \vec{P}$$

$\epsilon_0\vec{E} \rightarrow$ Flux density due to free charges

$\vec{P} \rightarrow$ Flux density due to bound charges, also called polarization

and $\vec{P} = X_e\epsilon_0\vec{E}$

$X_e \rightarrow$ Electrical susceptibility

$\vec{D} \rightarrow$ Total flux density

2.36 (d)

$\nabla\times\vec{E} = -\frac{d\vec{B}}{dt}$ for static field $\frac{d\vec{B}}{dt} = 0$

$\vec{E} = -\nabla V$ for V as scalar potential

$\oint \vec{E}\cdot\vec{dl} = 0$ so conservative

2.37 (c)

Equipotential lines: Suppose scalar potential V is function of Z so for a particular Z, a plane parallel to xy plane will have equipotential lines, we know that $\vec{E} = -\nabla V$, gradient of scalar potential V will give a line $\perp$ to plane on which V lie i.e. $\perp$ to xy plane so equipotential lines and field lines are orthogonal to each other.

2.38 (b)

Suppose there is charge Q is placed on conducting sphere. Due to conductivity of sphere this charge will be uniformly distributed on the sphere. So potential at the surface of such sphere is

$$V = \frac{Q}{4\pi\epsilon_0 R} \quad \text{for } r \le R$$

$$\therefore \quad C = \frac{Q}{V}$$

$$\boxed{C = 4\pi\epsilon_0 R}$$

2.39 (c)

According to boundary condition for electric field

$$E_{t_1} = E_{t_2}$$

i.e. tangential component is continuous.

$$D_{n_1} - D_{n_2} = -\rho_s$$

if region is charge free $\rho_s = 0$

$$D_{n_1} = D_{n_2}$$

i.e. normal component of flux density is continuous across the charge free boundary.

In case of perfect conductor $E_{t_2} = 0$ since electric field does not exist inside a perfect conductor but in case of finite conductivity it can't be zero so statement 3 is wrong.

2.40 (b)

$$Q_1 = 2\,\text{nC}$$
$$Q_2 = 3\,\text{nC}$$
$$Q_3 = -5\,\text{nC}$$

Force on Q_3 by Q_1

$$F_{13} = \frac{1}{4\pi \epsilon_0}\frac{Q_1Q_3}{r^2}$$
$$= \frac{1}{4\pi \epsilon_0}\frac{-10\times10^{-9}\times10^{-9}}{d^2/4}$$
$$= \frac{1}{4\pi \epsilon_0}\frac{-40\times10^{-9}\times10^{-9}}{d^2}$$

Force on Q_3 by Q_2

$$F_{23} = \frac{1}{4\pi \epsilon_0}\frac{Q_2Q_3}{r^2}$$
$$= \frac{1}{4\pi \epsilon_0}\frac{-15\times10^{-9}\times10^{-9}}{d^2/4}$$
$$= \frac{1}{4\pi \epsilon_0}\frac{-60\times10^{-9}\times10^{-9}}{d^2}$$

Total force $F = \vec{F}_{13} + \vec{F}_{23} \neq 0$

So option (b) is wrong.

2.41 (c)

For dielectric boundary

$D_{n_1} - D_{n_2} = \rho_s$, where D_{n_1} and D_{n_2} are normal component of electric flux density on the two side of boundary. So whether electric flux density (normal component) will be continuous or not depend on ρ_s, if $\rho_s = 0$, it will be continuous and if $\rho_s \neq 0$ it will be discontinuous.

2.42 (c)

Electric field lines are outward due to +ve charge so field will be outward. So A is correct. R is wrong because Gauss law states that net flux coming out from a closed surface is equal to net charge enclosed in surface.

2.43 (a)

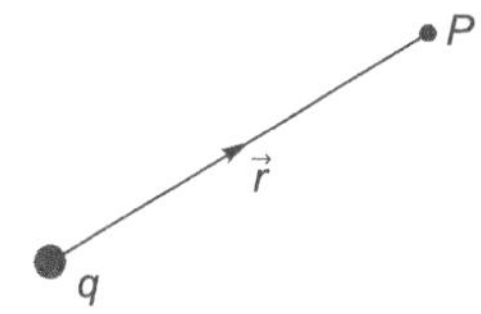

Electric field strength at point P at a distance $\vec{r}$ from charge q is given by

$$\vec{E} = \frac{1}{4\pi\epsilon_o\epsilon_r}\frac{q}{r^3}\vec{r} = \frac{q}{4\pi\epsilon r^2}\cdot\hat{r}$$

as $$\hat{r} = \frac{\hat{r}}{|\vec{r}|} = \frac{\hat{r}}{r}$$

$\therefore$ $$\vec{E} = \frac{q\epsilon^{-1}}{4\pi r^2}\hat{r} \qquad (\text{as } \epsilon = \epsilon_o\epsilon_r)$$

2.45 (c)

Force between charges

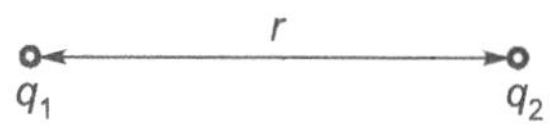

$$F = \frac{1}{4\pi\epsilon_o}\frac{q_1q_2}{r^2} \quad \text{(in free space)}$$

If glass slab is inserted between charges then

$$F' = \frac{1}{4\pi\epsilon_o\epsilon_r}\frac{q_1q_2}{r^2}$$
$$F' = \frac{F}{\epsilon_r}$$

As $\epsilon_r \geq 1$

$\therefore$ force decreases

2.46 (b)

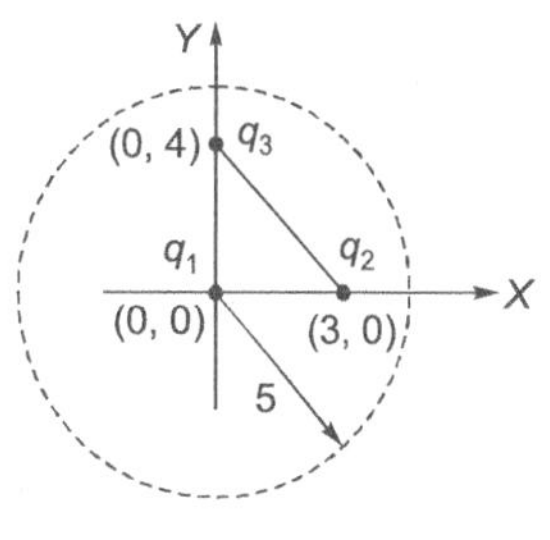

Total flux over the sphere

= total charge enclosed by the sphere

$= (q_1 + q_2 + q_3)$

$= (0.008 + 0.05 - 0.009) = 0.049\ \mu\text{C}$

2.47 (a)

$$\Psi = \int_s \vec{D}\cdot d\vec{s} = \epsilon_o\int_s \vec{E}\cdot d\vec{s}$$

Therefore, electric flux through a surface area is the integral of the normal component of field over the area.

2.48 (d)

Gauss's law states that the total electric flux through any closed surface is equal to the total charge enclosed by that surface

$$Q = \oint_s \vec{D}\cdot d\vec{s} = \int_v \rho_v \, dv$$

2.49 (d)

Continuity equation, $\nabla\cdot\vec{J} = -\dfrac{\partial \rho_v}{\partial t}$

The above equation relates current density and charge density.

2.51 (c)

$$F = \frac{1}{4\pi\epsilon_0}\frac{Q^2}{r^2}$$

$$= 9\times 10^9 \times \frac{(4\times 10^{-19})^2}{(10^{-10}\times 10^{-2})^2}$$

$$= 9\times 16\times 10^{-5} = 14.4\times 10^{-4}\ \text{N}$$

2.52 (c)

Overall capacitance before dipping $= \dfrac{\epsilon_0 A}{a}$

After dipping in ethyl alcohol upto a depth of $a/2$ it will work as 2 capacitors connected in series connection, so the overall capacitance in this case will be

$$= \frac{\dfrac{\epsilon_o A}{\frac{a}{2}} \times \dfrac{\epsilon_o \epsilon_r A}{\frac{a}{2}}}{\dfrac{\epsilon_o A}{\frac{a}{2}} + \dfrac{\epsilon_o \epsilon_r A}{\frac{a}{2}}}$$

$$= \frac{2\times 25\times 2}{2+25\times 2}\frac{\epsilon_0 A}{a} = \frac{100}{52}\frac{\epsilon_0 A}{a}$$

$$= \frac{50}{26}\frac{\epsilon_o A}{a}$$

2.53 (c)

The electric field lines and equipotential lines cut each other orthogonally.

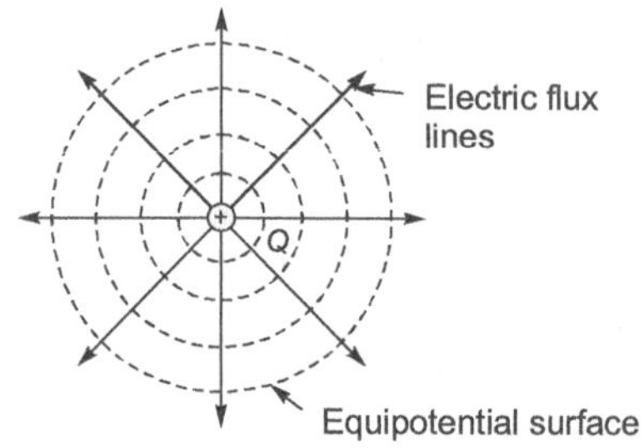

2.54 (c)

Gauss's law states that the total electric flux Ψ through any closed surface is equal to the total charge enclosed by that surface

$$\Psi = Q_{enclosed}$$

$$Q = \oint_s D\cdot ds$$

2.55 (a)

The basic electrostatic properties of ideal conductors are:

1. the resultant field inside is zero.
2. the net charge density in the interior is zero.
3. any net charges reside on the surface.
4. the surface is always equipotential.

The field just outside the conductor is not zero.

2.56 (d)

In practice, Earth is chosen as a place of zero electric potential because it has almost constant potential.

2.57 (b)

$$V = \frac{Q}{4\pi\cos}\left[\frac{1}{y}-\frac{1}{x}\right]$$

$$Q = CV$$

$$C = \frac{Q}{V} = \frac{Q}{\dfrac{Q}{4\pi}\left[\dfrac{1}{y}-\dfrac{1}{x}\right]} = \frac{4\pi\varepsilon_0 xy}{x-y}$$

2.58 (a)

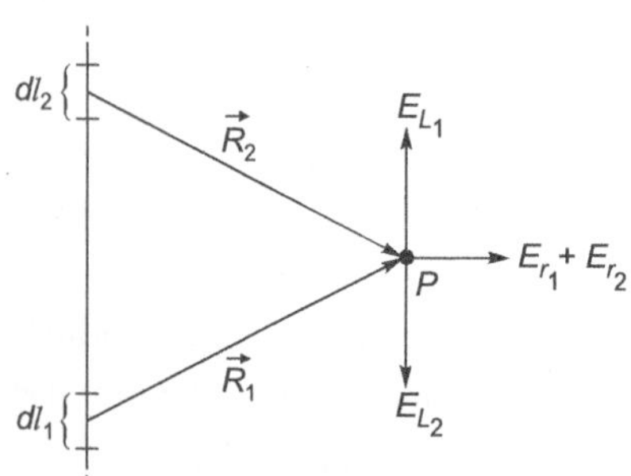

According to coulomb's law the force or field strength at point '*P*' will act only in radial (*R*) direction while longitudinal components (*L*) will get cancelled out.

2.59 (a)

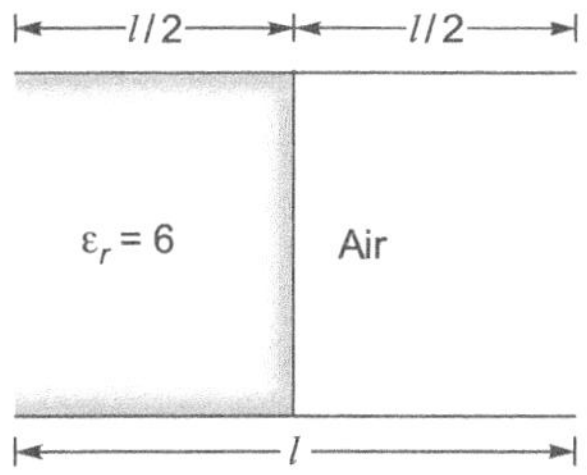

$$C_{old} = \frac{\varepsilon_0 lw}{d} = \frac{Q}{V} \qquad ...(i)$$

Let *w* is width of parallel plates and *d* is distance between them

C_{new} = capacitance after inclusion of dielectric

$$C_{new} = \frac{\varepsilon_r \varepsilon_0 \times (l/2) \times w}{d} + \frac{\varepsilon_0 \times (l/2) \times w}{d}$$

$$= \frac{\varepsilon_0 \times l \times w}{2d}[\varepsilon_r + 1]$$

$$= \frac{7}{2}\frac{\varepsilon_0 lw}{d} = \frac{7}{2}C_{old}$$

$$C_{new} = \frac{3.5Q}{V};$$

i.e., $V_{new} = V,$

$Q_{new} = 3.5\ Q$

2.63 (a)

Dielectric strength of rubber is = 40000 V/mm

Applied voltage = 33 kV

Let us assume thickness is '*t*' then

$$\frac{40000\text{ V}}{10^{-3}\text{m}} = \frac{33 \times 10^3\text{V}}{t}$$

$$\Rightarrow \qquad t = \frac{33 \times 10^3 \times 10^{-3}}{40000}$$

$= 8.25 \times 10^{-4}$ m

$= 0.825 \times 10^{-3}$ m

= 0.83 mm

2.64 (b)

As we know that, voltage is the work done per unit charge

i.e., $\frac{dW}{dQ}$

and power is the energy per unit of time

i.e. $\frac{dW}{dt}$

∴ Both are correct.

2.65 (b)

Gauss theorem states that the total electric flux-ϕ emanating from a closed surface is equal to the total charge enclosed by that closed surface.

2.66 (d)

Given statement is related Gauss law.

2.67 (b)

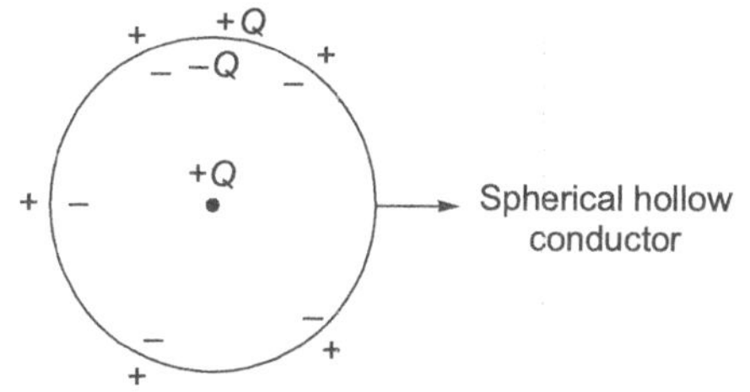

2.68 (a)

According to gauss law,

$$\oint \vec{D}.d\vec{S} = Q_{enclosed} \text{ (or)}$$

$$\oint \vec{D}.d\vec{S} = \iiint \rho_v dv \text{ (or)}$$

in differential form

$$\vec{\nabla} \cdot \vec{D} = \rho_v$$

$$\vec{\nabla} \cdot \in \vec{E} = \rho_v$$

For homogeneous medium ∈ is constant.

$$\in (\vec{\nabla} \cdot \vec{E}) = \rho_v$$

$$\vec{\nabla} \cdot \vec{E} = \frac{\rho_v}{\in}$$

2.69 (c)

E = 10 kV/cm

$$\frac{dW_E}{dV} = \frac{1}{2}\epsilon_0 |E^2|$$

$$= \frac{1}{2} \times 8.854 \times 10^{-12}\left(30 \times \frac{10^3}{10^{-2}}\right)^2$$

$= 39.843 \times 10^{-12} \times 10^{12}$

= 40 J/m^3

2.71 (c)

$$C = \frac{\in A}{d}$$

$$C \propto \frac{\epsilon}{d}$$

$$\frac{C_2}{C_1} = \frac{\epsilon_2 / d_2}{\epsilon_1 / d_1}$$

$$\Rightarrow \quad \frac{C_2}{C_1} = \frac{\epsilon_2 d_1}{\epsilon_1 d_2}$$

$$\Rightarrow \quad \frac{C_2}{10\text{ pF}} = \frac{5\epsilon_1 d}{\epsilon_1 \frac{d}{2}}$$

$$C_2 = 100\text{ pF}$$

2.72. (b)

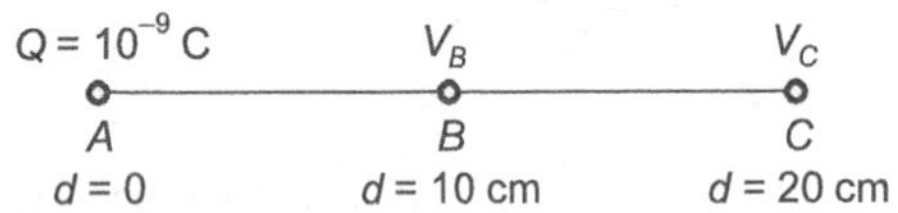

$$V = \frac{Q}{4\pi \epsilon_0 d}$$

$$V_{BC} = V_B - V_C$$

$$= \frac{Q}{4\pi\epsilon_0}\left[\frac{1}{r_B} - \frac{1}{r_C}\right]$$

$$= (10^{-9})(9\times10^{9})\left(\frac{100}{10} - \frac{100}{20}\right)$$

$$= 9\,(10 - 5) = 45\text{ Volt}$$

2.73. (a)

Gauss law say,

$$\oiint \vec{D}\cdot\vec{ds} = Q_{enc}$$

$$= Q$$

2.74. (c)

$$E = \frac{V}{d};$$

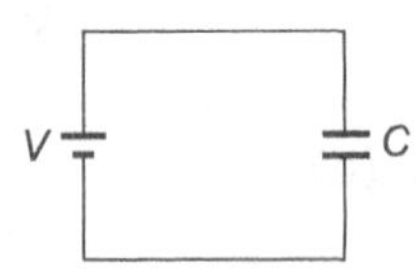

Where V is constant.

$$E \propto \frac{1}{d}$$

$$\Rightarrow \quad \frac{E_2}{E_1} = \frac{d_1}{d_2}$$

$$\Rightarrow \quad E_2 = \frac{d_1}{\left(\frac{1}{2}d_1\right)} E_1 = 2E_1$$

2.75 (b)

$\because$ Sphere of 5 cm centered at A is a gaussian surface for point charge 10^{-9} C at A.

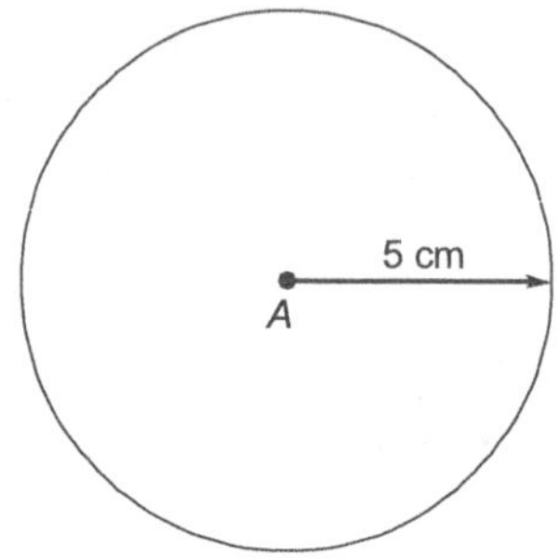

$\therefore$ By Gauss theorem,

$$\oiint \vec{D}\cdot\vec{ds} = Q_{enclosed} = 10^{-9}\text{ C}$$

$$D_r \times 4\pi r^2 = Q$$

$$D_r = \frac{Q}{4\pi r^2}$$

$$\because \quad \vec{E} = \frac{\vec{D}}{\varepsilon_0}$$

$$\therefore \quad \vec{E} = \frac{D_r}{\varepsilon_0}\hat{a}_r = \frac{Q}{4\pi\varepsilon_0 r^2}\hat{a}_r$$

$$E = \frac{Q}{4\pi\varepsilon_0 r^2} = \frac{10^{-9}\times 9\times 10^{9}}{(5\times10^{-2})^2}$$

$$= 3600\text{ V/m}$$

2.77 (d)

Potential due to point charge,

$$V_Q = \frac{KQ}{r}\ ; \quad K = \frac{1}{4\pi\epsilon}$$

$$\Rightarrow \quad V_Q = \frac{Q}{4\pi\epsilon r}$$

Here, $V \propto Q$

$$V \propto \frac{1}{r}$$

$$V \propto \frac{1}{\epsilon}$$

2.78 (b)

As we know,

$$D = \varepsilon E$$

$$\Rightarrow \quad D = \varepsilon_0 \varepsilon_r E$$

$$\Rightarrow \quad E = \frac{D}{\varepsilon_0 \varepsilon_r}$$

2.79 (d)

From the circuit,

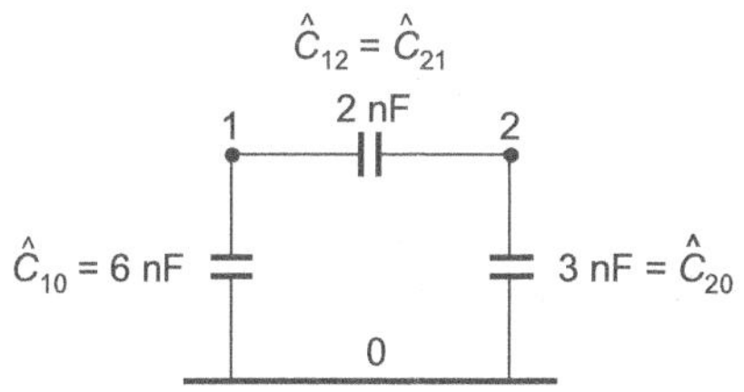

$$Q_1 = \hat{C}_{10}V_1 + \hat{C}_{12}(V_1 - V_2)$$

$$\Rightarrow \quad Q_1 = \hat{C}_{10}V_1 + \hat{C}_{12}V_1 - \hat{C}_{12}V_2$$

$$\Rightarrow \quad Q_1 = (\hat{C}_{10} + \hat{C}_{12})V_1 - \hat{C}_{12}V_2$$

$$\Rightarrow \quad Q_1 = C_{11}V_1 + C_{12}V_2$$

$$\therefore \quad C_{11} = \hat{C}_{10} + \hat{C}_{12} = 8 \text{ nF}$$

$$C_{12} = -\hat{C}_{12} = -2 \text{ nF}$$

Similarly,

$$Q_2 = \hat{C}_{20}V_2 + \hat{C}_{21}(V_2 - V_1)$$

$$\Rightarrow \quad Q_2 = \hat{C}_{20}V_2 + \hat{C}_{21}V_2 - \hat{C}_{21}V_1$$

$$\Rightarrow \quad Q_2 = (\hat{C}_{20} + \hat{C}_{21})V_2 - \hat{C}_{21}V_1$$

$$\Rightarrow \quad Q_2 = C_{22}V_2 + C_{21}V_1$$

$$\therefore \quad C_{22} = \hat{C}_{20} + \hat{C}_{21} = 5 \text{ nF}$$

$$C_{21} = -\hat{C}_{21} = -2 \text{ nF}$$

$$\therefore \quad [C] = \begin{bmatrix} 8 & -2 \\ -2 & 5 \end{bmatrix} \text{nF}$$

3 Magnetostatic Fields

3.1 **Assertion (A):** The relationship between Magnetic Vector potential $\vec{A}$ and the current density J in free space is

$$\nabla \times (\nabla \times \vec{A}) = \mu_0 \vec{J}$$

For a magnetic field in free space due to a dc or slowly varying current is $\nabla^2 \vec{A} = -\mu_0 \vec{J}$.

Reason (R): For magnetic field due to dc or slowly varying current $\nabla \cdot \vec{A} = 0$

(a) Both A and R are true and R is the correct explanation of A
(b) Both A and R are true but R is NOT the correct explanation of A
(c) A is true but R is false
(d) A is false but R is true

[ESE-2001]

3.2 Plane defined by $z = 0$ carry surface current density $2\hat{a}_x$ A/m. The magnetic intensity 'H_y' in the two regions $-\infty < z < 0$ and $0 < z < \infty$ are respectively

(a) $\hat{a}_y$ and $-\hat{a}_y$ (b) $-\hat{a}_y$ and $\hat{a}_y$
(c) $\hat{a}_x$ and $-\hat{a}_x$ (d) $-\hat{a}_x$ and $\hat{a}_x$

[ESE-2001]

3.3 A solid cylindrical conductor of radius 'R' carrying a current 'I' has a uniform current density. The magnetic field intensity 'H' inside the conductor at the radial distance 'r' ($r < R$) is

(a) Zero (b) $\dfrac{I}{2\pi r}$
(c) $\dfrac{Ir}{2\pi R^2}$ (d) $\dfrac{IR^2}{2\pi r^3}$

[ESE-2001]

3.4 The magnetic flux through each turn of a 100 turn coil is $(t^3 - 2t)$ milli-Webers where t is in seconds. The induced e.m.f. at $t = 2s$ is

(a) 1 V (b) – 1 V
(c) 0.4 V (d) – 0.4 V

[ESE-2002]

3.5 If a vector field $\overline{B}$ is solenoidal, which of these is true ?

(a) $\oint_L \vec{B} \cdot \vec{dl} = 0$ (b) $\oint_S \vec{B} \cdot \vec{ds} = 0$
(c) $\nabla \times \vec{B} = 0$ (d) $\nabla \cdot \vec{B} \neq 0$

[ESE-2002]

3.6 **Assertion (A):** $\int_s \vec{B} \cdot \vec{ds} = 0$ where, $\vec{B}$ is magnetic flux density, $\vec{ds}$ = vector with direction normal to surface element ds.

Reason (R): Tubes of magnetic flux have no sources or sinks.

(a) Both A and R are true and R is the correct explanation of A
(b) Both A and R are true but R is NOT the correct explanation of A
(c) A is true but R is false
(d) A is false but R is true

[ESE-2002]

3.7 If $\overline{A}$ and $\overline{J}$ are the vector potential and current density vectors associated with a coil, then $\int \overline{A} \cdot \overline{J} dv$ has the units of

(a) flux-linkage (b) power
(c) energy (d) inductance

[ESE-2002]

3.8 **Assertion (A):** The scalar magnetic potential is evidently the quantity whose equipotential surfaces will form curvilinear squares with the stream lines of H.

Reason (R): The scalar magnetic potential satisfies Laplace's equation where $J = 0$.

(a) Both A and R are true and R is the correct explanation of A
(b) Both A and R are true but R is **NOT** the correct explanation of A
(c) A is true but R is false
(d) A is false but R is true [ESE-2003]

3.9 Which one of the following formulae is **NOT** correct for the boundary between two magnetic materials?

(a) $B_{n_1} = B_{n_2}$
(b) $B_2 = \sqrt{B_{n_2} + B_{t_2}}$
(c) $H_1 = H_{n_1} + H_{t_1}$
(d) $\hat{a}_{n_{21}} \times (\vec{H}_1 - \vec{H}_2) = \vec{K}$ where $\hat{a}_{n_{21}}$ is a unit vector normal to the interface and directed from region 2 to region 1.

[ESE-2003]

3.10 Interface of two regions of two magnetic materials is current-free. The region 1, for which relative permeability $\mu_{r1} = 2$ is defined by $z < 0$, and region 2, $z > 0$ has $\mu_{r2} = 1$.
If $\vec{B}_1 = 1.2\hat{a}_x + 1.6\,\hat{a}_y + 0.4\,\hat{a}_z T$; then $\vec{H}_2$ is

(a) $\frac{1}{\mu_0}[0.6\,\hat{a}_x + 0.8\,\hat{a}_y + 0.4\,\hat{a}_z]$ A/m
(b) $\frac{1}{\mu_0}[1.2\,\hat{a}_x + 0.8\,\hat{a}_y + 0.8\,\hat{a}_z]$ A/m
(c) $\frac{1}{\mu_0}[1.2\,\hat{a}_x + 0.4\,\hat{a}_y + 0.4\,\hat{a}_z]$ A/m
(d) $\frac{1}{\mu_0}[0.6\,\hat{a}_x + 0.4\,\hat{a}_y + 0.8\,\hat{a}_z]$ A/m

[ESE-2003]

3.11 What is the force experienced per unit length by a conductor carrying 5 A current in positive Z-direction and placed in a magnetic field $\bar{B} = (3\vec{a}_x + 4\vec{a}_y)$?

(a) $15\vec{a}_x + 20\vec{a}_y$ N/m
(b) $-20\vec{a}_x + 15\,\vec{a}_y$ N/m
(c) $20\vec{a}_x - 15\,\vec{a}_y$ N/m
(d) $-20\vec{a}_x - 20\,\vec{a}_y$ N/m [ESE-2004]

3.12 Which one of the following is not the valid expression for magnetostatic field vector $\vec{B}$?

(a) $\vec{B} = \nabla \cdot \vec{A}$ (b) $\vec{B} = \nabla \times \vec{A}$
(c) $\nabla \cdot \vec{B} = 0$ (d) $\nabla \times \vec{B} = \mu_0 \vec{J}$

[ESE-2005]

3.13 Equation $\nabla \cdot \vec{B} = 0$ is based on

(a) Gauss's Law (b) Lenz's Law
(c) Ampere's Law (d) Continuity Equation

[ESE-2006]

3.14 Which of the following is zero as applied to electromagnetic fields?

(a) grad div $\vec{A}$ (b) div grad V
(c) div curl $\vec{A}$ (d) curl curl $\vec{A}$

[ESE-2006]

3.15 What is the force on a unit charge moving with velocity $\vec{v}$ in presence of electric field $\vec{E}$ and magnetic field $\vec{B}$?

(a) $\vec{E} - \vec{v} \cdot \vec{B}$ (b) $\vec{E} + \vec{v} \cdot \vec{B}$
(c) $\vec{E} + \vec{B} \times \vec{v}$ (d) $\vec{E} + \vec{v} \times \vec{B}$

[ESE-2006]

3.16 Plane $y = 0$ carries a uniform current density $30\hat{k}$ mA/m. At (1, 20, –2) m, what is the magnetic field intensity?

(a) $-15\hat{i}$ mA/m (b) $15\hat{i}$ mA/m
(c) $18.85\hat{j}$ mA/m (d) $25\hat{i}$ mA/m

[ESE-2006]

3.17 What is the magnetic field due to an infinite linear current carrying conductor?

(a) $H = \frac{\mu I}{2\pi r}$ A/m (b) $H = \frac{I}{2\pi r}$ A/m
(c) $H = \frac{\mu I}{2r}$ A/m (d) $H = \frac{I}{r}$ A/m

[ESE-2006]

3.18 If a very flexible wire is laid out in the shape of a hairpin with its two ends secured, what shape will the wire tend to assume if a current is passed through it?

(a) Parabolic (b) Straight line
(c) Circle (d) Ellipse

[ESE-2007]

3.19 Consider the following:

Lorentz force $\vec{F} = e(\vec{v} \times \vec{B})$

where e, $\vec{v}$ and $\vec{B}$ are respectively the charge of the particle, velocity of the particle and flux density of uniform magnetic field.

Which one of the following statements is **NOT** correct?

(a) Acceleration is normal to the plane containing the particle path and $\vec{B}$.
(b) If the direction of the particle path is normal to $\vec{B}$, the acceleration is maximum.
(c) If the particle is at rest, the field will deflect the particle.
(d) If the particle path is in the same direction of $\vec{B}$, there will be no acceleration.

[ESE-2007]

3.20 The force on a charge moving with velocity $\vec{v}$ under the influence of electric and magnetic fields is given by which one of the following?

(a) $q(\vec{E} + \vec{B} \times \vec{v})$ (b) $q(\vec{E} + \vec{v} \times \vec{H})$
(c) $q(\vec{H} + \vec{v} \times \vec{E})$ (d) $q(\vec{E} + \vec{v} \times \vec{B})$

[ESE-2008]

3.21 The magnetic vector potential $\vec{A}$ obeys which equations?

1. $\vec{B} = \nabla \times \vec{A}$
2. $\nabla^2 \cdot \vec{A} = -\mu_0 \vec{J}$
3. $\vec{A} = \int \frac{\mu_0 I d\vec{l}}{4\pi R}$

Select the correct answer using the code given below:

(a) 1 and 2 (b) 2 and 3
(c) 1 and 3 (d) 1, 2 and 3

[ESE-2008]

3.22 A long straight wire carries a current $I = 10$ A. At what distance is the magnetic field $H = 1$ Am^{-1}?

(a) 1.19 m (b) 1.39 m
(c) 1.59 m (d) 1.79 m

[ESE-2008]

3.23 A long straight wire carries a current $I = 1$ A. At what distance is the magnetic field 1 Am^{-1}?

(a) 1.59 m (b) 0.159 m
(c) 0.0159 m (d) 0.00159

[ESE-2009]

3.24 Magnetic field intensity is

$$\vec{H} = 3\hat{a}_x + 7y\hat{a}_y + 2x\hat{a}_z \text{ A/m}$$

What is the current density $\vec{J}$ A/m^2?

(a) $-2\hat{a}_y$ (b) $-7\hat{a}_z$
(c) $3\hat{a}_x$ (d) $12\hat{a}_y$

[ESE-2009]

3.25 How much current must flow in a loop of radius 1 m to produce a magnetic field 1 mAm^{-1}?

(a) 1.0 mA (b) 1.5 mA
(c) 2.0 mA (d) 2.5 mA

[ESE-2009]

3.26 What does the expression $\frac{1}{2}\vec{J} \cdot \vec{A}$ represent?

(a) Power density
(b) Radiation resistance
(c) Magnetic energy density
(d) Electric energy density

[ESE-2009]

3.27 **Assertion (A):** Knowing magnetic vector potential $\vec{A}$ at a point, the flux density $\vec{B}$ at that point can be obtained.

Reason (R): $\vec{\nabla} \cdot \vec{A} = 0$.

(a) Both A and R are true and R is the correct explanation of A
(b) Both A and R are true but R is not the correct explanation of A
(c) A is true but R is false
(d) A is false but R is true

[ESE-2009]

3.28 A conductor 2 metre long lies along the Z-axis with a current of 10 A in $\hat{a}_Z$ direction. If the magnetic field is $\bar{B} = 0.05\, \hat{a}_X T$, the force on the conductor is

(a) $4.0\hat{a}_Y N$ (b) $1.0\hat{a}_Z N$
(c) $1.0\hat{a}_Y N$ (d) $3.0\hat{a}_Z N$

[ESE-2010]

3.29 Match **List-I** with **List-II** and select the correct answer using the code given below the lists:

List-I

A. Work
B. Electric field strength
C. Magnetic flux
D. Magnetic field strength

List-II

1. Ampere per meter
2. Weber
3. Volt per meter
4. Joule

Codes:

	A	B	C	D
(a)	4	3	2	1
(b)	1	3	2	4
(c)	4	2	3	1
(d)	1	2	3	4

[ESE-2011]

3.30 The following equation is not valid for magnetostatic field in inhomogeneous magnetic materials:

(a) $\nabla \cdot \vec{B} = 0$

(b) $\nabla \cdot \vec{H} = 0$

(c) $\nabla \times \vec{A} = \vec{B}$

(A is magnetic vector potential)

(d) $\vec{\nabla} \times \vec{H} = \vec{J}$ [ESE-2011]

3.31 The unit of magnetic flux density is

(a) gauss (b) tesla
(c) bohr (d) weber/sec

[ESE-2012]

3.32 The magnetic flux density created by an infinitely long conductor carrying a current I at a radial distance R is

(a) $\frac{\mu_0 I}{2\pi R}$ (b) $\frac{I}{2\pi R}$

(c) $\frac{\mu_0 I}{2\pi R^3}$ (d) $\frac{4\pi R^2 I}{3}$

[ESE-2012]

3.33 A long straight wire carries a current $I = 10$ A, the magnetic field at a distance of 1.59 m is

(a) 0.1 Am^{-1} (b) 1 Am^{-1}
(c) 10 Am^{-1} (d) 100 Am^{-1}

[ESE-2013]

3.34 Consider the following statements:

A current I flows through a circular coil of one turn of radius R in the counter-clockwise direction.

1. The magnetic field at the center is inversely proportional to R.
2. The magnetic moment of the coil is directly proportional to R^2.
3. The magnetic field at its center is directly proportional to R^2.

Which of the above statements is/are correct?

(a) 1 and 2 (b) 1 only
(c) 2 and 3 (d) 3 only [ESE-2014]

3.35 A magnetic ring has a mean circumference of 20 cm and a cross-section of 20 cm^2 and has 800 numbers of turns of wire. When the exciting current is 5 A, the flux is 2 mWb. The relative permeability of iron is nearly

(a) 3.98 (b) 398
(c) 0.398 (d) 39.8 [ESE-2014]

3.36 A conductor having a cross-sectional area a sq m carrying current $I\vec{j}$ A, lies in a magnetic field

$$\vec{B} = B_0(\vec{i} + \vec{j}) \text{ Wb/m}^2$$

The force density on the conductor is

where $\vec{i}, \vec{j}$ and $\vec{k}$ are orthogonal unit vectors.

(a) $\frac{B_0}{a} I\vec{k}$ (b) $-\frac{B_0}{a} I\vec{i}$

(c) $-\frac{B_0}{a} I\vec{k}$ (d) $\frac{B_0}{a} I\vec{j}$

[ESE-2015]

3.37 If the current density inside a straight conductor is uniform over its cross-section, the flux density variation inside the conductor at different distances from its centre is

(a) linear
(b) square of the distance
(c) inverse of the distance
(d) exponential [ESE-2015]

3.38 The law which states that the line integral of the magnetic field around a closed curve is equal to the free current through a surface, is

(a) Gauss' law (b) Tellegen's theorem
(c) Coulomb's law (d) Ampere's law

[ESE-2015]

3.39 Two conductors of a transmission line carry equal current I in opposite directions. The force on each conductor is proportional to

(a) I
(b) I^2
(c) the distance between the conductors
(d) I^3 **[ESE-2015]**

3.40 The magnitude of magnetic field strength H is independent of

(a) current only
(b) distance only
(c) permeability of the medium only
(d) both current and distance **[ESE-2016]**

3.41 The variation of $|B|$ with distance r from a very long straight conductor carrying a current I is correctly represented by

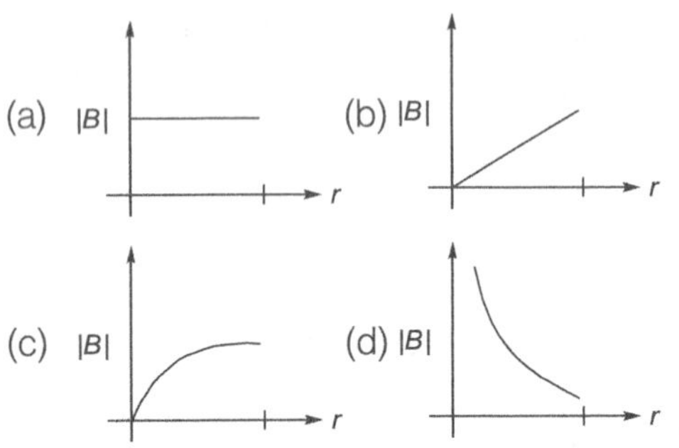

[ESE-2016]

3.42 The magnetic susceptibility of aluminium is 2.1×10^{-5}. The permeability and relative permeability are, respectively

(a) 12.6×10^{-7} and 1.0021
(b) 1.26×10^{-7} and 1.0021
(c) 12.6×10^{-7} and 1.000021
(d) 1.26×10^{-7} and 1.000021 **[ESE-2018]**

3.43 An iron rod of 10^{-3} m^3 volume and relative permeability of 1150 is placed inside a long solenoid wound with 5 turns/cm. If a current of 0.5 A is allowed to pass through the solenoid, the magnetic moment of the rod is

(a) 2.87×10^4 A-m^2
(b) 28.7×10^3 A-m^2
(c) 2.87×10^2 A-m^2
(d) 28.7×10^2 A-m^2 **[ESE-2018]**

3.44 The flux-density at a distance of 0.1 m from a long straight wire, carrying a current of 200 A is

(a) 5×10^{-4} Wb/m^2
(b) 4×10^{-4} Wb/m^2
(c) 3×10^{-4} Wb/m^2
(d) 2×10^{-4} Wb/m^2 **[ESE-2018]**

3.45 A steady flow of 10 A is maintained in a thin wire placed along the X-axis from (0, 0, 0) to (2, 0, 0) to find the value of the magnetic field intensity H at (0, 0, 5). When end effects are ignored, H is

(a) $-59.1\,\hat{a}_y$ mA/m
(b) $59.1\,\hat{a}_y$ mA/m
(c) $-118.2\,\hat{a}_y$ mA/m
(d) $118.2\,\hat{a}_y$ mA/m **[ESE-2018]**

3.46 A unit magnetic pole may be defined as that pole which when placed in vacuum at a distance of one metre from a similar and equal pole repels it with a force of

(a) $\frac{1}{4\pi}$ Newtons (b) $\frac{\mu_0}{4\pi}$ Newtons
(c) $\frac{\pi}{4\mu_0}$ Newtons (d) $\frac{1}{4\pi\mu_0}$ Newtons

[ESE-2020]

3.47 An analogous of magnetic circuit 'permeability' in electrical circuit is

(a) Reluctivity (b) Conductance
(c) Conductivity (d) Resistivity

[ESE-2020]

3.48 The magnetizing force at the centre of a circular coil varies

1. Directly as the number of its turns.
2. Directly as the current.
3. Directly as its radius.
4. Inversely as its radius.

Which of the above statements are correct?

(a) 1, 2 and 3 only (b) 1 and 4 only
(c) 1, 2 and 4 only (d) 2 and 3 only

[ESE-2020]

3.49 If L is self-inductance, I is current, λ is flux linkage of a magnetic field, then the energy stored in the magnetic field is

(a) $\frac{0.5\lambda^2}{L}$ (b) $\frac{2\lambda^2}{L}$
(c) $\frac{0.5\lambda}{L^2}$ (d) $\frac{2\lambda}{L^2}$

[ESE-2021]

3.50 What is the value of magnetic field on the axis that is perpendicular to the plane containing a circular loop of current shown in the figure using the Biot-Savart law?

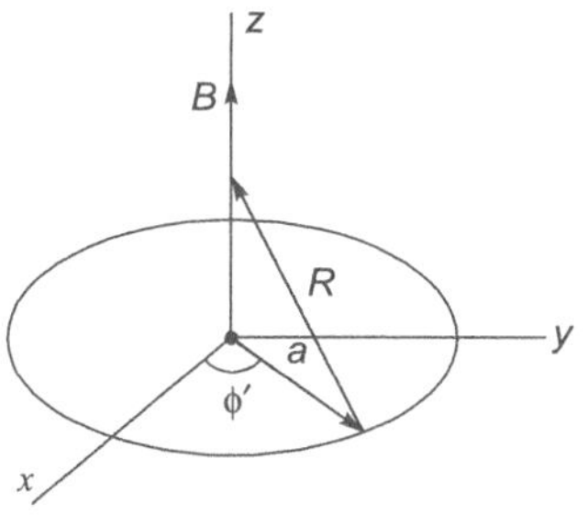

(a) $B(z) = \left[\frac{\mu_0 m}{2\pi R^3}\right]$ (b) $B(z) = \left[\frac{\mu_0 m}{2\pi R^2}\right]$

(c) $B(z) = \left[\frac{\mu_0 m}{2\pi R^4}\right]$ (d) $B(z) = \left[\frac{\mu_0 m}{2\pi R}\right]$

[ESE-2024]

Answers Magnetostatic Fields

3.1 (a)	3.2 (a)	3.3 (c)	3.4 (b)	3.5 (b)	3.6 (a)	3.7 (c)	3.8 (b)	3.9 (b)
3.10 (a)	3.11 (b)	3.12 (a)	3.13 (a)	3.14 (c)	3.15 (d)	3.16 (a)	3.17 (b)	3.18 (b)
3.19 (c)	3.20 (d)	3.21 (d)	3.22 (c)	3.23 (b)	3.24 (a)	3.25 (c)	3.26 (c)	3.27 (b)
3.28 (c)	3.29 (a)	3.30 (b)	3.31 (b)	3.32 (a)	3.33 (b)	3.34 (a)	3.35 (d)	3.36 (c)
3.37 (a)	3.38 (d)	3.39 (b)	3.40 (c)	3.41 (d)	3.42 (c)	3.43 (c)	3.44 (b)	3.45 (a)
3.46 (b)	3.47 (c)	3.48 (c)	3.49 (a)	3.50 (a)				

Explanations Magnetostatic Fields

3.1 (a)

$\nabla\times\nabla\times\vec{A} = \nabla(\nabla\cdot\vec{A}) - \nabla^2\vec{A}$; for steady magnetic field $\nabla\cdot\vec{A} = 0$; $\nabla^2\vec{A} = -\mu_o\vec{J}$ (Poisson's equation) so that $\nabla\times\nabla\times\vec{A} = \mu_o\vec{J}$

3.2 (a)

for $-\infty < z < 0$ $\quad \vec{H}_y = \frac{1}{2}(2)\,\hat{a}_y = \hat{a}_y$

for $0 < z < \infty$ $\quad \vec{H}_y = \frac{1}{2}(2)\,(-\hat{a}_y) = -\hat{a}_y$

3.3 (c)

In general $\vec{H} = \frac{I}{2\pi r}\hat{a}_\phi$. Current enclosed at distance,

$$r = \frac{I}{\pi R^2}\times\pi r^2 = \frac{Ir^2}{R^2}$$

$$\therefore \quad \vec{H}(r) = \frac{Ir^2}{2\pi r\cdot R^2} = \frac{Ir}{2\pi R^2}\hat{a}_\phi$$

3.4 (b)

Total flux linkage $\lambda = N\phi$

$$V = -\frac{d\lambda}{dt} = -\frac{d}{dt}(100\times(t^3-2t)\times10^{-3})$$

$$\therefore \quad V = -0.1\,(3t^2-2)$$

$$\therefore \quad V(\text{at } t = 2s) = -0.1\,(12-2) = -1\text{ V}$$

3.5 (b)

According to gauss law for magnetic field

$\nabla\cdot\vec{B} = 0$ i.e. no existence of monopoles

$$\oint_s \vec{B}\cdot\vec{ds} = 0 \qquad (\because \text{Divergence theorem})$$

$\therefore$ divergence of $\vec{B} = 0$ i.e. $\vec{B}$ is solenoidal

3.6 (a)

$$\int_s \vec{B}\cdot\vec{ds} = 0$$

Maxwell's 2nd equation → i.e. single pole does not exist because there is no source or sink separately so both are correct and R is correct explanation for A.

3.7 (c)

Compare $W_M = \int_v \vec{A}\cdot\vec{J}\,dv$ with electrical energy stored

$$W_E = \frac{1}{2}\int_v \rho_v \cdot V\,dv$$

3.8 (b)

Magnetic field intensity $\vec{H} = -\nabla V_m$ ...(1)
If we can define this then V_m will be scalar magnetic potential.
Now from equation (1),

$$\nabla\times\vec{H} = -\nabla\times(\nabla V_m)$$

$$\vec{J}_c+\vec{J}_d = -\nabla\times(\nabla V_m)$$

$$\vec{J}_c = -\nabla\times\nabla V_m$$

$(\because \vec{J}_d = 0$ for static fields)

So we can define scalar magnetic potential

if $\nabla\times\nabla V_m = 0$ i.e. $\vec{J}_c = 0$

It A is also true but R is not the reason for A.

3.9 (b)

From the boundary condition for the magnetic field.

$$\vec{B}_{n_1} = \vec{B}_{n_2}$$

$$H_{t_1} - H_{t_2} = J_s$$

where J_s is surface current density

or $\vec{a}_{21}\times\left[\vec{H}_1-\vec{H}_2\right] = \vec{J}_s$

$$\vec{H}_1 = \vec{H}_{t_1}+\vec{H}_{n_1}$$

So, $B_2 = \sqrt{B_{n_2}+B_{t_2}}$ is incorrect

as $B_2 = \overrightarrow{B_{n_2}}+\overrightarrow{B_{t_2}}$

3.10 (a)

$$\vec{H}_1 = \frac{\vec{B}_1}{\mu_0\mu_{r1}} = \frac{1}{\mu_0}(0.6\hat{a}_x+0.8\hat{a}_y+0.2\hat{a}_z)T$$

$$H_{t_1} = H_{t_2}$$

and $B_{n_1} = B_{n_2}$

$\therefore$ $H_{x_1} = H_{x_2}, H_{y_1} = H_{y_2}$

$$B_{z_1} = B_{z_2} \Rightarrow \frac{0.4}{\mu_0\mu_{r2}} = H_{z_2}$$

$$\therefore\ \vec{H}_2 = \frac{1}{\mu_0}\left(0.6\hat{a}_x+0.8\hat{a}_y+0.4\hat{a}_z\right)\text{A/m}$$

3.11 (b)

Force on current carrying conductor due to magnetic field

$$\vec{B} = 3\hat{a}_x+4\hat{a}_y$$

$$\vec{F} = I(l\,\vec{a}_z\times\vec{B})$$

Force per unit length,

$$\vec{F}_1 = I(\hat{a}_z\times\vec{B})$$

$$= 5\left[\hat{a}_z\times(3\hat{a}_x+4\hat{a}_y)\right]$$

$$= 5[3\hat{a}_y-4\hat{a}_x]$$

$$\vec{F}_1 = -20\hat{a}_x+15\hat{a}_y \text{ N/m}$$

3.12 (a)

$\nabla\cdot\vec{B} = 0$ Maxwell's second equation gauss law for magnetic field.

div curl $\vec{F} = 0$

$\therefore$ $\vec{B} = \nabla\times\vec{A}$

Such that,

$$\nabla\cdot\vec{B} = \nabla\cdot(\nabla\times\vec{A}) = 0$$

$$\nabla\times\vec{H} = \vec{J}_C+\vec{J}_D$$

(from the fourth maxwell equation)

$$\nabla\times\vec{H} = \vec{J}$$

$$\therefore\ \left(\nabla\times\frac{\vec{B}}{\mu_0}\right) = \vec{J} \qquad (\vec{J} = \vec{J}_C+\vec{J}_D)$$

$$\boxed{\nabla\times\vec{B} = \mu_0\vec{J}}$$

Trick: $\vec{B}$ is a vector, but $\nabla\cdot\vec{A}$ is a scaler.
Option (a) is correct answer.

3.13 (a)

Magnetic monopoles are not physically possible as per Gauss's Law i.e. $\nabla\cdot\vec{B} = 0$.

3.14 (c)

$\nabla \cdot (\nabla \times \vec{A})$ i.e. Div curl $\vec{A}$ is always zero.

and $\nabla \times (\nabla \phi)$ i.e. curl gradϕ is always zero.

3.15 (d)

Force due to electric field on unit charge

$$\vec{F}_1 = q\vec{E}$$

$$\vec{F}_1 = \vec{E} \qquad (\because q = 1\ c)$$

Now force on a moving charge due to magnetic field.

$$\vec{F}_2 = q(\vec{v} \times \vec{B})$$

where $\vec{v}$ is velocity of charge particle.

$$\vec{F}_2 = \vec{v} \times \vec{B} \quad (\because q = 1\ c)$$

$\therefore$ Net force on the moving unit charge

$$\vec{F} = \vec{F}_1 + \vec{F}_2$$

$$\vec{F} = q\vec{E} + \vec{v} \times \vec{B}$$

$$\vec{F} = \vec{E} + \vec{v} \times \vec{B}$$

3.16 (a)

$$\vec{H} = \frac{1}{2} \times 30\hat{k} \times \hat{a}_n$$

$$(\hat{a}_n \perp \text{vector to plane} = \hat{j})$$

$$H = \frac{1}{2} \times 30\hat{k} \times \hat{j} = -15\hat{i}\ \text{mA/m}$$

3.17 (b)

r is $\perp$ distance of point A from infinitely long current carrying wire then magnetic field $\vec{H}$ at point B is given by

$$\vec{H} = \frac{I}{2\pi r}\hat{a}_\phi$$

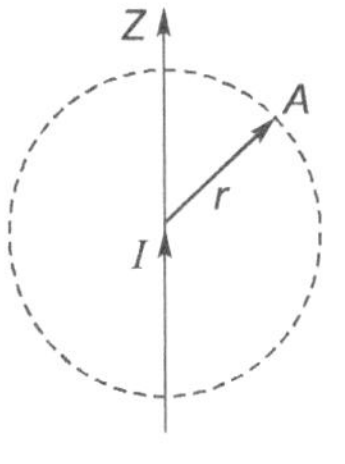

Trick: Using dimensional analysis, eliminate options (a) and (c).

3.18 (b)

Direction of current is opposite so repulsion force will be present till the wire become straight.

3.19 (c)

$$\vec{F} = e(\vec{v} \times \vec{B})$$

$\vec{a}$ acceleration will be in the direction of force.

and $$\vec{a} = \frac{e}{m}(\vec{v} \times \vec{B})$$

$$= \frac{e}{m}|\vec{v}|\,|\vec{B}|\sin\theta\,\hat{a}_n$$

$\theta \rightarrow$ angle between $\vec{v}$ and $\vec{B}$, $\hat{a}_n$ normal vector $\perp$ to both $\vec{v}$ and $\vec{B}$.

So, a_{max} when $\theta = 90°$ as $\sin\theta = 1$ maximum

when $\theta = 0 \Rightarrow a = 0$ i.e. minimum

when $|\vec{v}| = 0 \Rightarrow a = 0$

so no deflection to the charge particle.

3.20 (d)

Force on charge due to electric field $\vec{E}$

$$\vec{F}_1 = q\vec{E} \qquad ...(1)$$

force on a moving charge moving with velocity $\vec{v}$ due to magnetic field $\vec{B}$ i.e. Lorentz force

$$\vec{F}_2 = q(\vec{v} \times \vec{B}) \qquad ...(2)$$

so net force on the moving charge q is

$$\vec{F} = \vec{F}_1 + \vec{F}_2$$

$$= q\vec{E} + q(\vec{v} \times \vec{B})$$

$$= q[\vec{E} + \vec{v} \times \vec{B}]$$

3.21 (d)

We know that $\nabla \cdot \vec{B} = 0$ Guass's law for Magnetic field.

We known that div curl $\vec{F} = 0$

$$\therefore \qquad \vec{B} = \nabla \times \vec{A}$$

here $\vec{A}$ is known as magnetic vector potential

we know $\nabla \times \vec{B} = \mu_0[\vec{J}_C + \vec{J}_D]$ 4th equation of Maxwell

So, $$\nabla \times \vec{B} = \mu_0\vec{J}$$

$$\vec{B} = \nabla \times \vec{A}$$

$$\therefore \nabla \times (\nabla \times \vec{A}) = \mu_0\vec{J}$$

$$\nabla(\nabla \cdot \vec{A}) - \nabla^2\vec{A} = \mu_0\vec{J}$$

$$-\nabla^2\vec{A} = \mu_0\vec{J} \qquad (\because \nabla \cdot \vec{A} = 0)$$

$$\nabla^2\vec{A} = -\mu_0\vec{J}$$

Also, $\vec{A} = \int \frac{\mu_0}{4\pi}\frac{I\,\vec{dl}}{R}$ for current element

3.22 (c)

Due to long straight wire field at a distance r(r being ⊥ distance from wire) at a point A

$$\vec{H} = \frac{I}{2\pi r}\hat{a}_\phi$$

$$|\vec{H}| = \frac{I}{2\pi r}$$

$$I = 10 \text{ A}, |\vec{H}| = 1 \text{ A/m}$$

$$\therefore \quad 1 = \frac{10}{2\pi r}$$

$$r = \frac{10}{2\pi} = 1.59 \text{ m}$$

3.23 (b)

Field (magnetic) at a distance r due to a long current carrying wire carrying a current I.

Where r is ⊥ distance of point from wire.

$$\vec{H} = \frac{I}{2\pi r}\hat{a}_\phi$$

$$|\vec{H}| = \frac{I}{2\pi r}$$

$$1 = \frac{1}{2\pi r} \quad [\because I = 1 \text{ A}, |\vec{H}| = 1 \text{ A/m}]$$

$$r = \frac{1}{2\pi} = 0.159 \text{ m}$$

3.24 (a)

$$J = \nabla \times \vec{H} = \begin{bmatrix} \hat{a}_x & \hat{a}_y & \hat{a}_z \\ \frac{\partial}{\partial x} & \frac{\partial}{\partial y} & \frac{\partial}{\partial z} \\ 3 & 7y & 2x \end{bmatrix}$$

$$\vec{J} = -2\vec{a}_y$$

3.25 (c)

Magnetic field at the center of the loop in which current-I is flowing

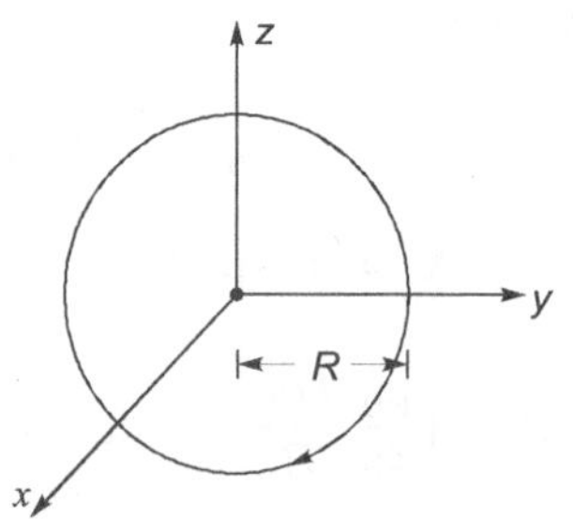

$$\vec{H} = \frac{I}{2r}\hat{a}_z$$

$$|\vec{H}| = \frac{I}{2r}$$

given that $|\vec{H}| = 1 \text{ mA/m} = 1 \times 10^{-3} \text{ A/m}$

$$r = 1 \text{ m}$$

$$\therefore 1 \times 10^{-3} \text{ A/m} = \frac{I}{2 \times 1}$$

$$I = 2 \times 10^{-3} \text{A}$$

$$\boxed{I = 2 \text{ mA}}$$

3.26 (c)

$$\frac{1}{2}\vec{J}\cdot\vec{A} = \frac{1}{2}\frac{A}{m}\cdot\frac{wb}{m^2} \text{ (Linear current density)}$$

$$\frac{1}{2}\vec{J}\cdot\vec{A} = \frac{1}{2}\vec{H}\cdot\mu\vec{H} = \frac{1}{2}\mu H^2$$

Magnetic current density as

A/m → $\vec{H}$

$\vec{B}$ → Wb/m^2 → $\mu\vec{H}$

3.27 (b)

Since magnetic vector potential is defined as

$$\nabla \times \vec{A} = \vec{B}$$

So knowledge of $\vec{A}$ will be useful in determining $\vec{B}$ so A is correct.

$$\nabla \cdot \vec{A} = 0$$

it is also true but not the reason for A.

3.28 (c)

$$l = 2 \text{ m}$$

$$I = 10\,\hat{a}_z \text{ A}$$

$$B = 0.05\,\hat{a}_x T$$

Force on conductor

$$\vec{F} = l\cdot(\vec{I} \times \vec{B})$$

$$= 2(10\hat{a}_z \times 0.05\,\hat{a}_x)$$

$$\vec{F} = \hat{a}_y N$$

3.29 (a)

Work → Joule, $A \rightarrow 4$

Electric field strength $\rightarrow$ V/m, $B \rightarrow 3$
Magnetic flux $\rightarrow$ Weber, $C \rightarrow 2$
Magnetic field strength $\rightarrow$ A/m, $D \rightarrow 1$

3.30 (b)

$\nabla \cdot \vec{B} = 0$ is Maxwell's second equation true for any general medium but $\nabla \cdot H$ will be zero or not we can not say as for inhomogeneous medium

$$\nabla \cdot \vec{B}/\mu \neq \frac{1}{\mu} \nabla \cdot \vec{B} \qquad \left(\because \vec{H} = \frac{\vec{B}}{\mu}\right)$$

3.31 (b)

Unit of magnetic flux density is weber/m^2 or tesla.

3.32 (a)

Using ampere's law

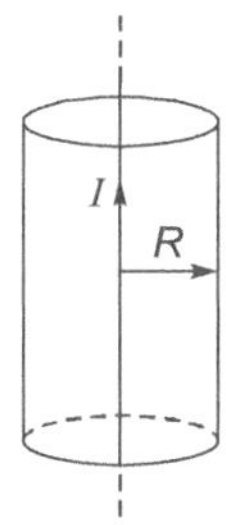

$$\oint_L \vec{H} \cdot \vec{dl} = I_{enc}.$$

$$H \times 2\pi R = I$$

$$H = \frac{I}{2\pi R}$$

Magnetic flux density $B = \mu H$

$$\therefore \quad B = \frac{\mu_o I}{2\pi R}$$

3.33 (b)

$$H = \frac{I}{2\pi R} = \frac{10}{2\pi \times 1.59} = 1 \text{ A/m}$$

3.34 (a)

Magnetic field at the center $= \dfrac{I}{2\pi R}$

Magnetic moment of the coil $= IA = I(\pi R)^2$

3.35 (d)

Reluctance $(R) = \dfrac{l}{\mu_0 \mu_r A}$

$$\Rightarrow \quad MMF = \text{Reluctance} \times \text{flux}$$

$$\Rightarrow \quad NI = R \times \phi$$

$$\Rightarrow \quad = \left(\frac{l}{\mu_0 \mu_r A}\right)\phi$$

$$\Rightarrow \quad 800 \times 5 = \left(\frac{20 \times 10^{-2}}{4\pi \times 10^{-7} \times \mu_r \times 20 \times 10^{-4}}\right) \times 2 \times 10^{-3}$$

$$\Rightarrow \quad \mu_r = \left(\frac{20 \times 10^{-2} \times 2 \times 10^{-3}}{4\pi \times 10^{-7} \times 20 \times 10^{-4} \times 800 \times 5}\right)$$

$$\Rightarrow \quad \mu_r = 39.8$$

3.36 (c)

$$\vec{I} = I\hat{j}A$$

$$\vec{B} = B_0(\hat{i} + \hat{j}) \text{ Wb/m}^2,$$

$$\vec{F} = l(\overline{I} \times \overline{B})$$

$$= l\{I\hat{j} \times B_0(i + \hat{j})\}$$

Force density,

$$\frac{F}{\text{Vol.}} = -\frac{B_0}{a} I \hat{k}$$

3.37 (a)

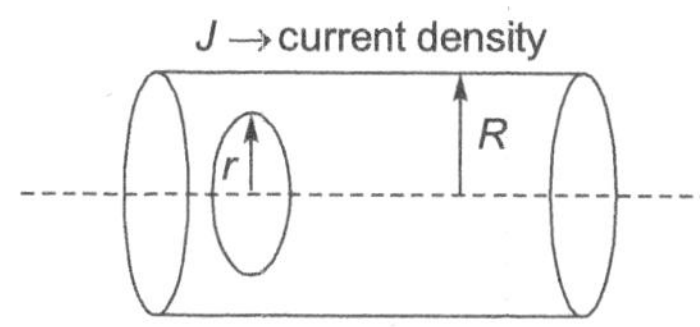

For $0 < r < R$:

$$\oint H \cdot dl = \int J \cdot ds$$

$$H \cdot 2\pi r = J \cdot \pi r^2$$

$$H = \frac{J \cdot r}{2}$$

$$B = \frac{\mu J \cdot r}{2}$$

3.38 (d)

Ampere's law:

It states that the line integral of the magnetic field strength around a closed curve is equal to the net current enclosed by that surface area.

Integral form

$$\oint_c H.dl = I_{enclosed}$$

Point form, $\nabla \times H = J$

3.39 (b)

Force between two conductors carrying current I in opposite direction,

$$F = \frac{\mu_0 I_1 I_2}{4\pi r^2}$$

or $\quad F \propto I^2$

3.42 (c)

Magnetic susceptibility,

$$\chi_m = 2.1 \times 10^{-5}$$

Relative permeability,

$$\mu_r = \chi_m + 1$$
$$= 1.000021$$

Permeability, $\mu = \mu_o \times \mu_r$
$$= 4\pi \times 10^{-7} \times 1.000021$$
$$= 12.6 \times 10^{-7} \text{ H/m}$$

3.43 (c)

Magnetic moment of the rod
= Magnetisation × Volume

Magnetization,
$$M = \chi_m H$$
$$M = (\mu_r - 1) H$$
$$= (\mu_r - 1) \times Ni$$
$$= \frac{(1150-1)\times 5\times 0.5}{10^{-2}}$$

Magnetic moment,
$$m = M \times V$$
$$= \frac{1149\times 5\times 0.5\times 10^{-3}}{10^{-2}}$$
$$= 2.87 \times 10^2 \text{ A-m}^2$$

3.44 (b)

$$r = 0.1 \text{ m}$$
$$I = 200 \text{ A}$$
$$B = \mu H = \frac{\mu I}{2\pi\rho}$$
$$B = \mu H = \frac{\mu_0 I}{2\pi\rho}$$
$$= \frac{4\pi\times 10^{-7}(200)}{2\pi(0.1)}$$
$$= 2 \times 200 \times 10^{-6}$$
$$= 4 \times 10^{-4} \text{ Wb/m}^2$$

3.45 (a)

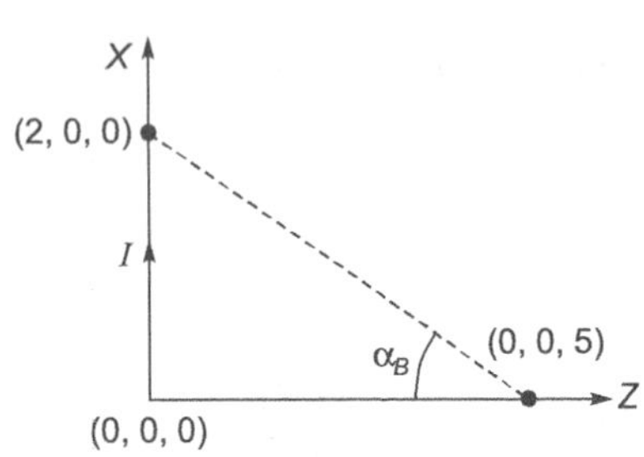

$$\vec{H} = \frac{I}{4\pi\rho}[\sin\alpha_A + \sin\alpha_B]\,\hat{a}_H$$
$$= \hat{a}_x \times \hat{a}_z = -\hat{a}_y$$
$$\sin\alpha_A = \frac{0-0}{5} \Rightarrow \alpha_A = 0$$
$$\sin\alpha_B = \frac{2}{\sqrt{2^2+5^2}} = \frac{2}{\sqrt{29}}$$
$$\vec{H} = \frac{10}{4\pi(5)}\left[0+\left(\frac{2}{\sqrt{29}}\right)\right](-\hat{a}_y)$$
$$= \frac{2}{4\pi}\left(\frac{2}{\sqrt{29}}\right)(-\hat{a}_y)$$
$$= 0.0591(-\hat{a}_y)$$

3.46 (b)

Force between two point magnetic poles present in free space is
$$F = \frac{\mu_0 m_1 m_2}{4\pi r^2} \text{ (Newton)}$$

Where, m_1, m_2 are strengths of two magnetic poles in terms of unit magnet poles a unit magnet pole is defined as one that exerts a force of one dyne (10^{-7} Newton) on another unit magnet pole when poles one in free space and separated by 1 meter distance in MKS system.

Unit magnet pole = 10^{-7} Newton
$$= \frac{\mu_0}{4\pi} \text{ Newton}$$

3.47 (c)

Permeability, μ = (H/m)

$\Rightarrow$ $\vec{B} = \mu\vec{H}$

Conductivity, σ = (℧/m)

$\Rightarrow$ $\vec{J} = \sigma\vec{E}$

3.48 (c)

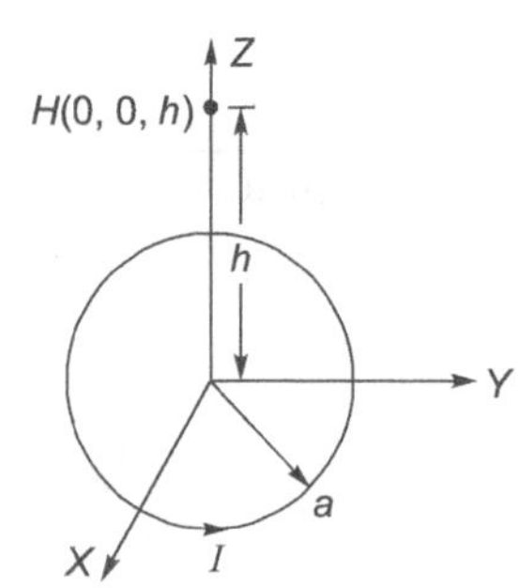

$$H(0,0,h) = \frac{Ia^2}{2(a^2+h^2)^{3/2}}$$

At centre,
$$H(0,0,0) = \frac{Ia^2}{2(a^2)^{3/2}} = \frac{I}{2a}$$

If N-turn circular coil then replace I with NI

$$H = \frac{NI}{2a}$$

N is number of turns. I is current in coil. a is radius of circular coil.

3.49 (a)

As we know that,

$$\lambda \propto I$$

$$\Rightarrow \quad \lambda = LI$$

$$\Rightarrow \quad I = \frac{\lambda}{L}$$

∴ Magnetic energy,

$$W_m = \frac{1}{2}LI^2$$

$$W_m = \frac{1}{2}\cdot L \cdot \frac{\lambda^2}{L^2}$$

$$W_m = \frac{\lambda^2}{2L} = \frac{0.5\lambda^2}{L}$$

3.50 (a)

The magnetic field due to circular loop on the axis is given by

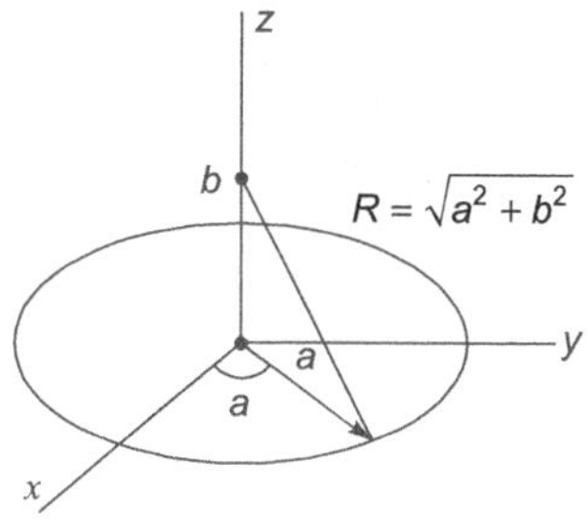

$$B(z) = \frac{ia^2}{2(a^2 + b^2)^{3/2}}$$

where, a = radius of loop

b = height along z-axis

$$B(z) = \frac{\mu_0 ia^2}{2(a^2 + b^2)^{3/2}}$$

From the figure,

$$\sqrt{a^2 + b^2} = R$$

$$\therefore \quad B(z) = \frac{\mu_0 ia^2}{2(R)^3} = \frac{\mu_0 m}{2\pi R^3}$$

where, m = magnetic dipole moment = $i\pi a^2$

Hence, $B(z) = \dfrac{\mu_0 m}{2\pi R^3}$

■■■■

4 Time Varying Electromagnetic Fields

4.1 Given that: $\nabla \times \vec{H} = \vec{J} + \dfrac{\partial \vec{D}}{\partial t}$

Assertion (A): In the equation, the additional term $\dfrac{\partial \vec{D}}{\partial t}$ is necessary.

Reason (R): The equation will be consistent with the principle of conservation of charge.

(a) Both A and R are true and R is the correct explanation of A
(b) Both A and R are true but R is NOT the correct explanation of A
(c) A is true but R is false
(d) A is false but R is true

[ESE-2001]

4.2 Match **List-I** (Law/quantity)with **List-II** (Mathematical expression) and select the correct answer :

List-I	List-II
A. Gauss's law	1. $\nabla \cdot \vec{D} = \rho$
B. Ampere's law	2. $\nabla \times \vec{E} = -\dfrac{\partial \vec{B}}{\partial t}$
C. Faraday's law	3. $S = \vec{E} \times \vec{H}$
D. Poynting vector	4. $\vec{F} = q(\vec{E} + \vec{v} \times \vec{B})$
	5. $\nabla \times \vec{H} = \vec{J}_c + \dfrac{\partial \vec{D}}{\partial t}$

Codes:

	A	B	C	D
(a)	1	2	4	3
(b)	3	5	2	1
(c)	1	5	2	3
(d)	3	2	4	1

[ESE-2001]

4.3 Consider coils C_1, C_2, C_3 and C_4 (shown in the given figures) which are placed in the time-varying electric field $\vec{E}(t)$ and electric field produced by the coils C_2', C_3' and C_4' carrying time varying current $I(t)$ respectively:

1. 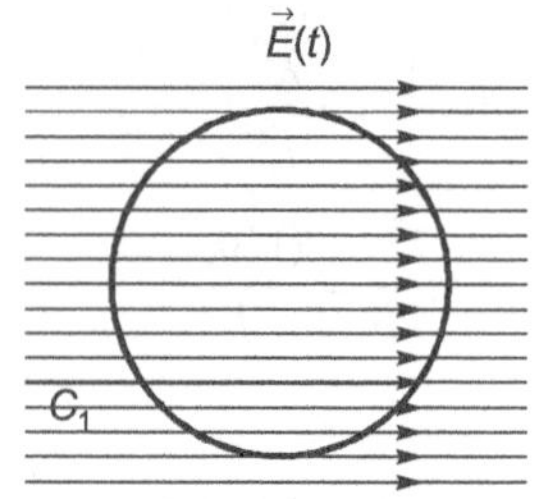

Time varying electric field $\vec{E}(t)$ parallel to the plane of coil C_1

2. 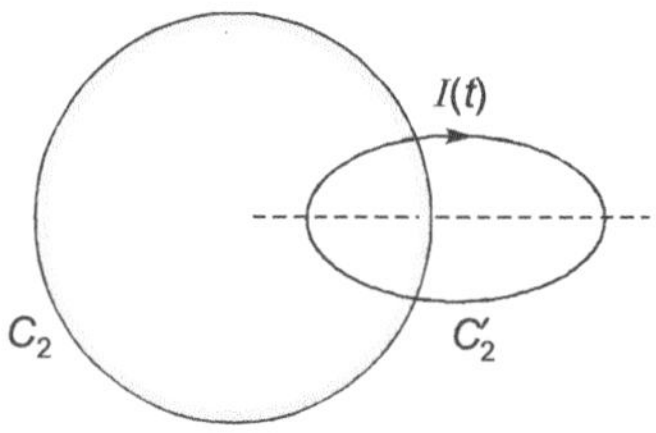

Coil planes are orthogonal

3. 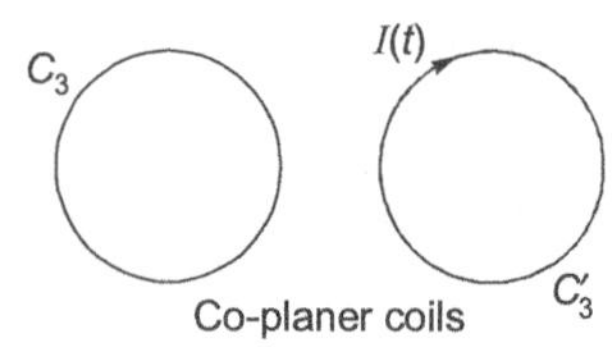

Co-planer coils

4. 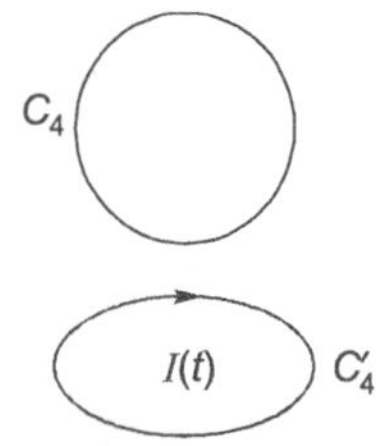

Coil-planes are orthogonal

The electric field will induce an emf in the coils

(a) C_1 and C_2
(b) C_2 and C_3
(c) C_1 and C_3
(d) C_2 and C_4

[ESE-2001]

4.4 A circular loop is rotating about the *y*-axis as a diameter in a magnetic field

$$\bar{B} = B_0 \sin\omega t \, \hat{a}_x \text{ Wb/m}^2$$

The induced emf in the loop is

(a) due to transformer emf only
(b) due to motional emf only
(c) due to a combination of transformer and motional emf
(d) zero

[ESE-2001]

4.5 Which of the following equations is/are not Maxwell's equation(s)?

1. $\nabla \cdot \vec{J} = -\dfrac{\partial \rho_v}{\partial t}$

2. $\nabla \cdot \vec{D} = \rho_v$

3. $\nabla \cdot \vec{E} = -\dfrac{\partial \vec{B}}{\partial t}$

4. $\oint \vec{H} \cdot \vec{dl} = \int_s \left(\sigma \vec{E} + \varepsilon \dfrac{\partial \vec{E}}{\partial t} \right) \cdot \vec{ds}$

Select the correct answer using the codes given below:

(a) 2 and 4 (b) 1 alone
(c) 1 and 3 (d) 1 and 4

[ESE-2002]

4.6 A medium behaves like dielectric when the

(a) displacement current is just equal to the conduction current
(b) displacement current is less than the conduction current
(c) displacement current is much greater than the conduction current
(d) displacement current is almost negligible

[ESE-2002]

4.7 A copper wire carries a conduction current of 1.0 A at 50 Hz. For copper wire $\varepsilon = \varepsilon_0$, $\mu = \mu_0$, $\sigma = 5.8 \times 10$ mho/m. What is the displacement current in the wire ?

(a) 2.8×10 A
(b) 4.8×10^{-11} A
(c) 1 A
(d) It cannot be calculated with the given data

[ESE-2002]

4.8 Two conducting coils 1 and 2 (identical except that 2 is split) are placed in a uniform magnetic field which decreases at a constant rate as in the figure. If the planes of the coils are perpendicular to the field lines, the following statements are made:

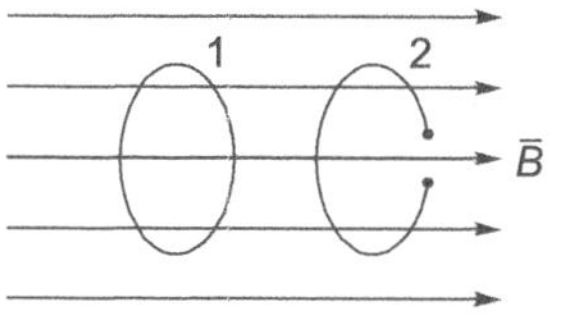

1. an e.m.f. is induced in the split coil 2
2. e.m.f. *s* are induced in both coils
3. equal Joule heating occurs in both coils
4. Joule heating does not occur in any coil

Which of the above statements is/are true?

(a) 1 and 4 (b) 2 and 4
(c) 3 only (d) 2 only

[ESE-2002]

4.9 Maxwell equation $\vec{\nabla} \times \vec{E} = -(\partial \vec{B}/\partial t)$ is represented in integral form as

(a) $\oint \vec{E} \cdot \vec{dl} = -\dfrac{\partial}{\partial t} \oint \vec{B} \cdot \vec{dl}$

(b) $\oint \vec{E} \cdot \vec{dl} = -\dfrac{\partial}{\partial t} \int_s \vec{B} \cdot \vec{ds}$

(c) $\oint \vec{E} \times \vec{dl} = -\dfrac{\partial}{\partial t} \oint \vec{B} \cdot \vec{dl}$

(d) $\oint \vec{E} \times \vec{dl} = -\dfrac{\partial}{\partial t} \int_s \vec{B} \cdot \vec{dl}$

[ESE-2003]

4.10 **Assertion (A):** Displacement current can have only a.c components.

Reason (R): It is generated by a change in electric flux.

(a) Both A and R are true and R is the correct explanation of A
(b) Both A and R are true but R is NOT the correct explanation of A
(c) A is true but R is false
(d) A is false but R is true

[ESE-2003]

4.11 Match **List-I** with **List-II** and select the correct answer using the codes given below:

List-I
A. Continuity equation
B. Ampere's law
C. Displacement current
D. Faraday's law

List-II
1. $\nabla \times \vec{H} = \vec{J} + \frac{\partial \vec{D}}{\partial t}$
2. $\vec{J} = \frac{\partial \vec{D}}{\partial t}$
3. $\nabla \times \vec{E} = -\frac{\partial \vec{B}}{\partial t}$
4. $\nabla \times \vec{J} = -\frac{\partial \rho_v}{\partial t}$

Codes:

	A	B	C	D
(a)	4	3	2	1
(b)	4	1	2	3
(c)	2	3	4	1
(d)	2	1	4	3

[ESE-2004]

4.12 Match **List-I** (Type of field denoted by $\vec{A}$) with **List-II** (Behaviour) and select the correct answer using the codes given below:

List-I	List-II
A. A static electric field in a charge free region	1. $\nabla \cdot \vec{A} = 0$, $\nabla \times \vec{A} \neq 0$
B. A static electric field in a charged region	2. $\nabla \cdot \vec{A} \neq 0$, $\nabla \times \vec{A} = 0$
C. A steady magnetic field in a current carrying conductor	3. $\nabla \cdot \vec{A} \neq 0$, $\nabla \times \vec{A} \neq 0$
D. A time-varying electric field in a charged medium with time-varying magnetic field	4. $\nabla \cdot \vec{A} = 0$, $\nabla \times \vec{A} = 0$

Codes:

	A	B	C	D
(a)	4	2	3	1
(b)	4	2	1	3
(c)	2	4	3	1
(d)	2	4	1	3

[ESE-2004]

4.13 Which one of the following equations is not Maxwell's equation for a static electro-magnetic field in a linear homogeneous medium?

(a) $\nabla \cdot \vec{B} = 0$ (b) $\nabla \times \vec{D} = \vec{0}$

(c) $\oint_c \vec{B} \cdot \vec{dl} = \mu_0 I$ (d) $\nabla^2 \vec{A} = \mu_0 \vec{J}$

[ESE-2007]

4.14 A straight current carrying conductor and two conducting loops *A* and *B* are shown in the figure given below. What are the induced currents in the two loops?

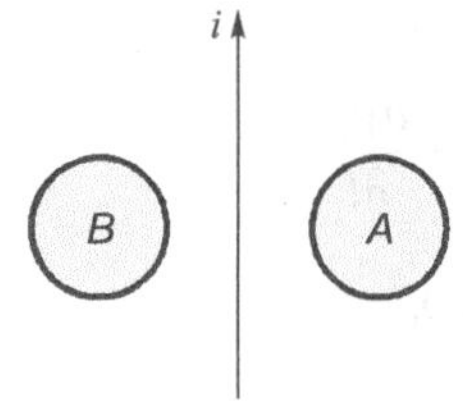

(a) Anticlockwise in A and clockwise in B
(b) Clockwise in A and anticlockwise in B
(c) Clockwise both in A and B
(d) Anticlockwise both in A and B

[ESE-2007]

4.15 Consider the following three equations:

1. $\nabla \times \vec{E} = -\frac{\partial \vec{B}}{\partial t}$
2. $\nabla \times \vec{H} = \vec{J} + \frac{\partial \vec{D}}{\partial t} 0$
3. $\nabla \cdot \vec{B} = 0$

Which of the above appears in Maxwell's equations?

(a) 1, 2 and 3 (b) 1 and 2
(c) 2 and 3 (d) 1 and 3

[ESE-2008]

4.16 What is the generalized Maxwell's equation

$\nabla \times \vec{H} = \vec{J}_e + \frac{\partial \vec{D}}{\partial t}$ for free space?

(a) $\nabla \times \vec{H} = 0$ (b) $\nabla \times \vec{H} = \vec{J}_e$

(c) $\nabla \times \vec{H} = \frac{\partial \vec{D}}{\partial t}$ (d) $\nabla \times \vec{H} = \vec{D}$

[ESE-2009]

4.17 Which one of the following is not a Maxwell's equation?

(a) $\nabla \times \vec{H} = (\sigma + j\omega \in) \vec{E}$

(b) $\vec{F} = Q(\vec{E} + \vec{v} \times \vec{B})$

(c) $\oint_C \vec{H} \cdot \vec{ds} = \int_s \vec{J} \cdot \vec{ds} + \int_s \frac{\partial \vec{D}}{\partial t} \cdot \vec{ds}$

(d) $\oint_s \vec{B} \cdot \vec{ds} = 0$

[ESE-2009]

4.18 A circular loop placed perpendicular to a uniform sinusoidal magnetic field of frequency ω_1 is revolved about an axis through its diameter at an angular velocity ω_2 rad/sec ($\omega_2 < \omega_1$) as shown in the figure below. What are the frequencies for the e.m.f. induced in the loop?

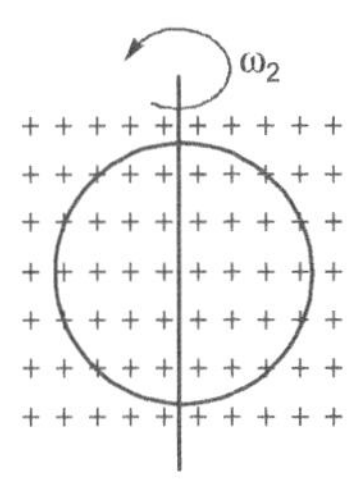

(a) ω_1 and ω_2
(b) ω_1, $\omega_1 + \omega_2$ and ω_2
(c) ω_2, $\omega_1 - \omega_2$ and ω_2
(d) $\omega_1 - \omega_2$ and $\omega_1 + \omega_2$ [ESE-2009]

4.19 The inconsistency of continuity equation for time varying fields was corrected by Maxwell and the correction applied was

(a) Ampere's law, $\frac{\partial \vec{D}}{\partial t}$

(b) Gauss's law, $\vec{J}$

(c) Faraday's law, $\frac{\partial \vec{B}}{\partial t}$

(d) Ampere's law, $\frac{\partial \vec{P}}{\partial t}$ [ESE-2013]

4.20 A quantitative relation between induced emf and rate of change of flux linkage is known as
(a) Maxwell's law (b) Stoke's law
(c) Lenz's law (d) Faraday's law
[ESE-2013]

4.21 Two identical coaxial circular loops carry the same current circulating in the same direction. If the loops approached each other, then the current in
(a) each one of them will increase
(b) both of them will remain the same
(c) each one of them will decrease
(d) one will increase while in the other the current will decrease
[ESE-2013]

4.22 If the magnetic flux through each turn of the coil consisting of 200 turns is ($t^2 - 3t$) milli-Webers, where t is in seconds, then the induced emf in the coil at $t = 4$ sec is
(a) –1 V (b) 1 V
(c) –0.1 V (d) 0.1 V [ESE-2013]

4.23 ($\nabla \times H$) = J is differential form of
(a) Gauss's law
(b) Ampere's circuital law
(c) Poisson's equation
(d) Laplace's equation [ESE-2015]

4.24 Maxwell equations
1. are extension of the works of Gauss, Faraday and Ampere.
2. help studying the application of electrostatic fields only.
3. can be written in integral form and point form.
4. need not be modified depending upon the media involved in the problem.

Which of the above statements are correct?
(a) 1 and 3 (b) 1 and 4
(c) 2 and 3 (d) 3 and 4
[ESE-2015]]

4.25 A conductor of length 100 cm moves at right angle to a uniform field flux density of 1.5 Wb/m with a velocity of 50 m/s. The e.m.f. induced in the conductor will be
(a) 150 V (b) 75 V
(c) 50 V (d) 37.5 V
[ESE-2015]

4.26 $\nabla \times \bar{H} = \sigma E + \varepsilon\left(\frac{\partial E}{\partial t}\right)$ is
(a) Modified Faraday's law
(b) Gauss's law
(c) Biot-Savart law
(d) Modified Ampere's law
[ESE-2017]

■■■■

Answers Time Varying Electromagnetic Fields

4.1 (a)	4.2 (c)	4.3 (c)	4.4 (c)	4.5 (c)	4.6 (c)	4.7 (b)	4.8 (d)	4.9 (b)
4.10 (a)	4.11 (b)	4.12 (b)	4.13 (b)	4.14 (a)	4.15 (a)	4.16 (c)	4.17 (b)	4.18 (d)
4.19 (a)	4.20 (d)	4.21 (a)	4.22 (a)	4.23 (b)	4.24 (a)	4.25 (b)	4.26 (d)	

Explanations Time Varying Electromagnetic Fields

4.1 (a)

$\frac{\partial \vec{D}}{\partial t}$ refers to displacement current through the dielectric medium. Presence of this term in the Faraday's law conforms to the principle of conservation of charge.

4.2 (c)

$\nabla \cdot \vec{D} = \rho_v$ → Gauss's low for electric field

$\nabla \times \vec{H} = \vec{J_c} + \dot{\vec{D}}$

→ Amperes law (generalized form)

$\nabla \times \vec{E} = -\frac{\partial \vec{B}}{\partial t}$

→ Faraday law of electromagnetic induction

$\vec{E} \times \vec{H}$

→ Poynting vector – representing direction of power flow.

4.3 (c)

Refer time-varying fields. To induce an emf, magnetic field produced by one coil must link with the other coil.

In C_1 and C_2 coils magnetic field is perpendicular, so emf is induced.

4.4 (c)

Magnetic field $\bar{B}$ is varying through the circular loop with respect to space (motional emf) as well as time (transformer emf).

4.5 (c)

$\nabla \cdot \vec{J} = \frac{-\partial \rho_v}{\partial t}$ is not a Maxwell's equation

as $\nabla \cdot \vec{D} = \rho_v$ is a Maxwell's first equation

$\nabla \times \vec{E} = \frac{-\partial \vec{B}}{\partial t}$ is Maxwell's third equation

Faraday's law

So, $\nabla \cdot \vec{E} = \frac{-\partial \vec{B}}{\partial t}$ is not a Maxwell's equation

4.6 (c)

In case of dielectric medium (good dielectric) $\epsilon_r >> 1$ and $\sigma \to 0$ so displacement current density.

$|J_D| >> |J_C|$ conduction current density

$\omega\epsilon >> \sigma$ for good dielectric $\Rightarrow \frac{\sigma}{\omega\epsilon} << 1$

4.7 (b)

$$\frac{J_C}{J_D} = \frac{\sigma}{\omega\epsilon} = \frac{I_C.A}{I_D.A}$$

$$\Rightarrow \quad I_D = I_C \cdot \frac{\omega\epsilon}{\sigma}$$

$$= 1 \times \frac{2 \times \pi \times 50 \times 10^{-9}}{5.8 \times 10 \times 36\pi}$$

$$= 4.8 \times 10^{-11} \text{ A}$$

4.8 (d)

emf will be induced in both the coil due to rate of change of magnetic field. Since coil 2 is split so no current will flow in it so no Joule heating but in coil 1 Joule heating will occur as current flows through it.

4.9 (b)

$$\nabla \times \vec{E} = -\frac{\partial \vec{B}}{\partial t}$$

taking integration on both the sides on a surface $\vec{ds}$.

$$\int_s \nabla \times \vec{E} \cdot \vec{ds} = -\frac{\partial}{\partial t}\int_s \vec{B} \cdot \vec{ds} \quad ...(1)$$

But according to stokes theorem

$$\oint \vec{E} \cdot \vec{dl} = \int_s curl\,\vec{E} \cdot \vec{ds} = \int_s \nabla \times \vec{E} \cdot \vec{ds} \quad ...(2)$$

So from equation (1) and (2) we have

$$\oint \vec{E}\cdot\vec{dl} = -\frac{\partial}{\partial t}\int_s \vec{B}\cdot\vec{ds}$$

4.10 (a)

The term displacement current comes into picture only with time varying (ac) fields. Displacement current flows through the dielectric medium and is generated by a change in electric flux as it conforms to principle of conservation of charge.

4.11 (b)

Continuity equation → Derived from charge conservation law is

$$\nabla\times\vec{J} = -\frac{\partial\rho_v}{\partial t}$$

ρ_v → Volume charge density.

Ampere's generalized law is given by Maxwell's 4th equation

$$\nabla\times\vec{H} = \vec{J}_c + \vec{J}_d$$

where, $\vec{J}_C$ → Conduction current density

$\vec{J}_d$ → Displacement current density

Displacement current density

$$\vec{J}_D = \frac{\partial\vec{D}}{\partial t}$$

Faraday's law → also Maxwell's 3rd equation

$$\nabla\times\vec{E} = -\frac{\partial\vec{B}}{\partial t}$$

emf is produced due to variation (rate of change of flux) through a surface (coil).

4.12 (b)

- Curl of static electric field is always zero i.e. it is conservative.
- Div of steady magnetic field is always zero. No monopole can exist.
- Curl of time-varying electric field is never zero.

4.13 (b)

$\nabla\cdot\vec{B} = 0$ Maxwell's 2nd equation

$\nabla\times\vec{D} = 0$ is not a Maxwell's equation

$\nabla\cdot\vec{D} = \rho_v$ is Maxwell's 1st equation

$\nabla\times\vec{E} = -\frac{\partial\vec{B}}{\partial t}$ is Maxwell's 3rd equation

$\nabla\times\vec{H} = \vec{J}_C + \frac{\partial\vec{D}}{\partial t}$ is Maxwell's 4th equation

4.14 (a)

Use right hand thumb rule.

Direction of current induced will be such that it will oppose its very cause.

4.15 (a)

$$\nabla\times\vec{E} = -\frac{\partial\vec{B}}{\partial t}$$

Maxwell's 3rd equation is derived from Faraday's law of electromagnetic induction.

$$\nabla\times\vec{H} = \vec{J}_C + \frac{\partial\vec{D}}{\partial t}$$

Maxwell's 4th equation is generalized form of Ampere's law.

$$\nabla\cdot\vec{B} = 0$$

Maxwell's 2nd equation is also Gauss law for Magnetic fields.

4.16 (c)

$$\nabla\times\vec{H} = \vec{J}_e + \frac{\partial\vec{D}}{\partial t}$$

since free space is a perfect dielectric medium.

$\therefore \quad \sigma = 0$

$\therefore \quad \vec{J}_e = \sigma\vec{E} = 0$

$\therefore \quad \boxed{\nabla\times\vec{H} = \frac{\partial\vec{D}}{\partial t}}$ for free space

4.17 (b)

$$\vec{F} = Q(\vec{E} + \vec{v}\times\vec{B})$$

is equation of force on moving charge in presence of both electric field and magnetic and is not a Maxwell's equation.

4.18 (d)

Since magnetic field has frequency ω_1 and induced emf due to varying field assuming loop is at rest i.e. transformer emf = $\frac{-\partial B}{\partial t}$ has frequency ω_1 because differentiation doesn't change frequency.

And due to motion of loop assuming B to be constant we have induced emf having frequency $= \omega_2$ (frequency of rotation) so overall induced emf has frequency component at $\omega_1 + \omega_2$ and $\omega_1 - \omega_2$.

4.19 (a)

Modified ampere's law:

$$\nabla \times \vec{H} = \vec{J} + \frac{\partial \vec{D}}{\partial t}$$

4.20 (d)

Faraday's law of electromagnetic induction says that magnitude of induced emf is directly proportional to the rate of change of flux.

4.21 (a)

Suppose, 2 identical circular loops carry the same current circulating in clockwise direction, when seen from up side. In this case the lower front of upper loop will behave like north pole and upper front of lower loop will behave like south pole i.e. there will be attraction between the loops and hence these will support the current flowing in each other. So, the current in each of them will increase.

4.22 (a)

$$e = -N\frac{d\phi}{dt}$$

$$= -200 \times \left[\frac{d}{dt}(t^2 - 3t)\right]_{\text{at } t = 4 \text{ sec.}}$$

$$= -200 \times [2t - 3]_{\text{at } t = 4 \text{ sec.}}$$

$$= -1000 \text{ mV}$$

$$= -1 \text{ V}$$

4.24 (a)

Maxwell's equations comprise of Gauss's, Faraday's and Ampere's equations which help studying the application of electromagnetic field to any medium. These equations can be represented in point as well as integral form.

4.25 (b)

Induced emf in the conductor

$$= V = v \times B$$

$$|V| = 50 \times 1.5$$

$$= 75 \text{ Volts}$$

4.26 (d)

From ampere's circuital law,

$$\vec{\nabla} \times \vec{H} = \vec{J}$$

$$\vec{\nabla} \times \vec{H} = \sigma \vec{E}$$

Maxwell modified ampere's circuital law by adding $\frac{\partial \vec{D}}{\partial t}$.

$$\vec{\nabla} \times \vec{H} = \sigma E + \frac{\partial \vec{D}}{\partial t} = \sigma \vec{E} + \frac{\partial (\in \vec{E})}{\partial t}$$

■■■■

5 EM Waves and Transmission Lines

5.1 The open circuit impedance of a certain length of a loss-less line is 100 Ω. The short circuit impedance of the same line is also 100 Ω. The characteristic impedance of the line is

(a) $100\sqrt{2}\ \Omega$ (b) $50\ \Omega$

(c) $100/\sqrt{2}\ \Omega$ (d) $100\ \Omega$

[ESE-2001]

5.2 **Assertion (A):** The solution to the wave equation at the critical diffracting condition in a crystal yields standing waves.

Reason (R): Standing waves have periodic variation both in amplitude as well as in the electron probability density in the crystal.

(a) Both A and R are true and R is the correct explanation of A
(b) Both A and R are true but R is NOT the correct explanation of A
(c) A is true but R is false
(d) A is false but R is true

[ESE-2001]

5.3 In the relation $S = \frac{1+|\Gamma|}{1-|\Gamma|}$; the values of S and Γ (where S stands for wave ratio and Γ is reflection coefficient), respectively, vary as

(a) 0 to 1 and – 1 to 0
(b) 1 to ∞ and –1 to +1
(c) –1 to +1 and 1 to ∞
(d) –1 to 0 and 0 to 1

[ESE-2001]

5.4 In the source free wave equation

$$\nabla^2 \vec{E} - \mu_0 \in_0 \mu \in \frac{\partial^2 \vec{E}}{\partial t^2} - \mu_0 \mu\sigma \frac{\partial \vec{E}}{\partial t} = 0$$

The term responsible for the attenuation of the wave is

(a) $\mu_0 \mu \sigma \frac{\partial \vec{E}}{\partial t}$

(b) $\mu_0 \in_0 \mu \in \frac{\partial^2 \vec{E}}{\partial t^2}$

(c) $\nabla^2 \vec{E}$

(d) $\mu_0 \mu \sigma \frac{\partial \vec{E}}{\partial t}$ and $\mu_0 \in_0 \mu \in \frac{\partial^2 \vec{E}}{\partial t^2}$

[ESE-2001]

5.5 Three media are characterised by

1. $\in_r = 8, \mu_r = 2, \sigma = 0$
2. $\in_r = 1, \mu_r = 9, \sigma = 0$
3. $\in_r = 4, \mu_r = 4, \sigma = 0$

$\in_r$ is relative permittivity, μ_r is relative permeability and σ is conductivity.

The value of the intrinsic impedances of the media 1, 2 and 3 respectively are

(a) 188 Ω, 377 Ω and 1131 Ω
(b) 377 Ω, 1131 Ω and 188 Ω
(c) 188 Ω, 1131 Ω and 377 Ω
(d) 1131 Ω, 188 Ω and 377 Ω

[ESE-2001]

5.6 A plane em wave $(\vec{E_i}, \vec{H_i})$ travelling in a perfect dielectric medium of surge impedance 'Z' strikes normally on an infinite perfect dielectric medium of surge impedance 2Z. If the refracted em wave is $(\vec{E_r}, \vec{H_r})$, the ratios of E_i/E_r and H_i/H_r are respectively

(a) 3 and -3 (b) 3/2 and 1/3
(c) 3/4 and 3/2 (d) 3/4 and 2/3

[ESE-2001]

5.7 For a perfect conductor, the field strength at a distance equal to the skin depth is X% of the field strength at its surface. The value 'X%' is

(a) Zero (b) 50%
(c) 36% (d) 26% [ESE-2001]

5.8 For linear isotropic materials, both $\vec{E}$ and $\vec{H}$ have the time dependence $e^{j\omega t}$ and regions of interest are free of charge. The value of $\vec{\nabla} \times \vec{H}$ is given by

(a) $\sigma\vec{E}$ (b) $j\omega \in \vec{E}$
(c) $\sigma\vec{E} + j\omega \in \vec{E}$ (d) $\sigma\vec{E} - j\omega \in \vec{E}$

[ESE-2002]

5.9 In free space

$\vec{H}(z, t) = 0.10 \cos(4 \times 10^7 t - \beta z)\ \hat{a}_x$ A/m.

The expression for $\vec{E}(z, t)$ is

(a) $\vec{E}(z,t) = 37.7 \cos(4 \times 10^7 t - \beta z)\hat{a}_y$
(b) $\vec{E}(z,t) = 2.65 \times 10 \cos(4 \times 10^7 t - \beta z)\hat{a}_z$
(c) $\vec{E}(z,t) = 37.7 \cos(4 \times 10^7 t - \beta z)\ \hat{a}_x$
(d) $\vec{E}(z,t) = -37.7 \cos(4 \times 10^7 t - \beta z)\ \hat{a}_y$

[ESE-2002]

5.10 A plane wave whose electric field is given by $\vec{E} = 100\cos(\omega t - 6\pi x)\hat{z}$ passes normally from a material '*A*' having $\epsilon_r = 4$, $\mu_r = 1$ and $\sigma = 0$ to a material '*B*' having $\epsilon_r = 9$, $\mu_r = 4$ and $\sigma = 0$. Match items in **List-I** with **List-II** and select the correct answer:

List-I	List-II
A. Intrinsic impedance of medium '*B*'	1. 6π
B. Reflection coefficient	2. 80π
C. Transmission Coefficient	3. 1/7
D. Phase shift constant of medium '*A*'	4. 8/7

Codes:

	A	B	C	D
(a)	4	1	2	3
(b)	2	3	4	1
(c)	4	3	2	1
(d)	2	1	4	3

[ESE-2002]

5.11 In free space $\vec{E}(z,t) = 50\cos(\omega t - \beta z)\ \hat{a}_x$ V/m and $\vec{H}(z,t) = 5/12\pi \cos(\omega t - \beta z)\ \hat{a}_y$ A/m. The average power crossing a circular area of radius $\sqrt{24}$ m in plane z = constant is

(a) 200 W (b) 250 W
(c) 300 W (d) 350 W

[ESE-2002]

5.12 Consider a plane electromagnetic wave incident normally on the surface of a good conductor. The wave has an electric field of amplitude 1(V/m) and the skin depth for the conductor is 10 cm.

Assertion (A) : The amplitude of electric field is $(1/e^2)$ (V/m) after the wave has travelled a distance of 20 cm in the conductor.

Reason (R) : Skin depth is the distance in which the wave amplitude decays to $(1/e)$ of its value at the surface.

(a) Both A and R are true and R is the correct explanation of A
(b) Both A and R are true but R is NOT the correct explanation of A
(c) A is true but R is false
(d) A is false but R is true [ESE-2002]

5.13 **Assertion (A):** When an alternating field is applied to a dielectric, the relative permittivity is a complex quantity.

Reason (R): The imaginary part of the relative permittivity is responsible for dielectric loss.

(a) Both A and R are true and R is the correct explanation of A
(b) Both A and R are true but R is NOT the correct explanation of A
(c) A is true but R is false
(d) A is false but R is true [ESE-2002]

5.14 **Assertion (A):** A uniform plane wave is a transverse electromagnetic wave.

Reason (R): A uniform plane wave can physically exist and represent finite energy.

(a) Both A and R are true and R is the correct explanation of A
(b) Both A and R are true but R is NOT the correct explanation of A
(c) A is true but R is false
(d) A is false but R is true [ESE-2003]

5.15 Given that :

$\vec{H} = 0.5 \exp[-0.1x] \sin(10^6 t - 2x)\ \hat{a}_z$ (A/m), which one of the following statements is NOT correct?

(a) Wave is linearly polarized along $\hat{a}_z$
(b) The velocity of the wave is 5×10^5 m/s
(c) The complex propagation constant is $(0.1 + 2j)$
(d) The wave is travelling along $\hat{a}_x$

[ESE-2003]

5.16 For a conducting medium with conductivity σ, permeability μ, and permittivity ϵ, the skin depth for an electromagnetic signal at an angular frequency ω is proportional to
(a) σ (b) $1/\omega$
(c) $1/\sqrt{\sigma}$ (d) $1/\mu$

[ESE-2003]

5.17 The electric field of a uniform plane wave is given by:

$\vec{E} = 10\sin(\omega t - \pi z)\,\hat{a}_x + 10\cos(\omega t - \pi z)\hat{a}_y$ (V/m).

The polarization of the wave is
(a) Circular (b) Elliptical
(c) Linear (d) Undefined

[ESE-2003]

5.18 Consider the following statements:
Characteristic impedance of a transmission line is given by

1. $\sqrt{\dfrac{R + j\omega L}{G + j\omega C}}$, ($R$, L, G and C are line constants)
2. $\sqrt{Z_{oc} Z_{sc}}$, (Z_{oc} and Z_{sc} are the open and short circuit impedances of the line)
3. V'/I', (V' and I' are the voltage and current of the wave travelling in the positive $\hat{y}$ direction)

Which of these are correct?
(a) 1, 2 and 3 (b) 1 and 2
(c) 2 and 3 (d) 1 and 3

[ESE-2003]

5.19 A loss-less transmission line of characteristic impedance Z_0 and $l < \lambda/4$ is terminated at the load end by a short circuit. Its input impedance Z_s is
(a) $Z_s = -jZ_0 \tan \beta l$ (b) $Z_s = jZ_0 \cot \beta l$
(c) $Z_s = jZ_0 \tan \beta l$ (d) $Z_s = -jZ_0 \cot \beta l$

[ESE-2003]

5.20 A loss-less transmission line with characteristic impedance of 600 ohms is terminated in a purely resistive load of 900 ohms. The reflection coefficient is
(a) 0.2 (b) 0.5
(c) 0.667 (d) 1.5

[ESE-2003]

5.21 What will be the reflected wave for an elliptically polarized wave incident on the interface of a dielectric at the Brewster angle?
(a) Elliptically polarized
(b) Linearly polarized
(c) Right circularly polarized
(d) Left circularly polarized

[ESE-2004]

5.22 Which one of the following is the correct electromagnetic wave equation in terms of vector potential $\vec{A}$?

(a) $\nabla^2 \vec{A} - \dfrac{\partial^2 \vec{A}}{\partial t^2} = -\vec{J}$

(b) $\nabla^2 \vec{A} - \dfrac{\mu}{\varepsilon}\dfrac{\partial^2 \vec{A}}{\partial t^2} = -\mu\vec{J}$

(c) $\nabla^2 \vec{A} - \dfrac{\partial^2 \vec{A}}{\partial t^2} = -\mu\vec{J}$

(d) $\nabla^2 \vec{A} - \mu\varepsilon\dfrac{\partial^2 \vec{A}}{\partial t^2} = -\mu\vec{J}$ [ESE-2004]

5.23 Which one of the following statements is correct?
The wavelength of a wave propagating in a wave guide is
(a) smaller than the free space wavelength
(b) greater than the free space wavelength
(c) directly proportional to the group velocity
(d) inversely proportional to the phase velocity

[ESE-2004]

5.24 Which one of the following statements is correct?
For a lossless dielectric medium, the phase constant for a travelling wave, β is proportional to
(a) ϵ_r (b) $\sqrt{\epsilon_r}$
(c) $1/\epsilon_r$ (d) $1/\sqrt{\epsilon_r}$

[ESE-2004]

5.25 In a lossless medium the intrinsic impedance $\eta = 60\,\pi$ and $\mu_r = 1$. What is the value of the dielectric constant ϵ_r?
(a) 2 (b) 1
(c) 4 (d) 8
[ESE-2004]

5.26 Which one of the following statements is correct? A wave guide can be considered to be analogous to a
(a) low pass filter (b) high pass filter
(c) band pass filter (d) band stop filter
[ESE-2004]

5.27 Consider the following statements about the effective length of a half wave dipole (Elevation angle θ is measured from the dipole axis):
1. Effective length is a function of θ.
2. Effective length is maximum for $\theta = \pi/2$
3. Maximum effective length is larger than physical length
4. Effective length is the same for the antenna in transmitting and receiving modes.

Which of the statements given above are correct?
(a) 1, 2 and 4 (b) 2, 3 and 4
(c) 1, 2 and 3 (d) 1, 3 and 4
[ESE-2004]

5.28 A lossless transmission line of length 50 cm with $L = 10$ μH/m, $C = 40$ pF/m is operated at 30 MHz. What is its electric length (βl)?
(a) 20λ (b) 0.2λ
(c) 108° (d) 40π
[ESE-2004]

5.29 **Assertion (A) :** In a lossless transmission line the voltage and current distributions along the line are always constant.
Reason (R) : The voltage and current distributions in an open line are such that at a distance $\lambda/4$ from the load end, the line looks like a series resonant circuit.
(a) Both A and R are true and R is the correct explanation of A
(b) Both A and R are true but R is NOT the correct explanation of A
(c) A is true but R is false
(d) A is false but R is true
[ESE-2004]

5.30 A uniform plane wave has a wavelength of 2 cm in free space and 1 cm in a perfect dielectric. What is the relative permittivity of the dielectric ?
(a) 2.0 (b) 0.5
(c) 4.0 (d) 0.25
[ESE-2005]

5.31 With the increase in frequency of an electromagnetic wave in free space, how do the velocity V_C and characteristic impedance Z_C change ?
(a) V_C increases and Z_C decreases
(b) V_C decreases and Z_C increases
(c) Both V_C and Z_C increase
(d) Both V_C and Z_C remain unchanged
[ESE-2005]

5.32 The E field of a plane electromagnetic wave travelling in a non-magnetic non-conducting medium is given by $\vec{E} = \hat{a}_x\, 5\cos(10^9 t + 30Z)$. What is the dielectric constant of the medium?
(a) 30 (b) 10
(c) 9 (d) 81 [ESE-2005]

5.33 In the wave equation $\nabla^2\vec{E} = \mu\epsilon\dfrac{\partial^2\vec{E}}{\partial t^2} + \mu\sigma\dfrac{\partial\vec{E}}{\partial t}$ which term is responsible for attenuation of the wave?
(a) $\nabla^2\vec{E}$ (b) $\mu\epsilon\dfrac{\partial^2\vec{E}}{\partial t^2}$
(c) $\mu\sigma\dfrac{\partial\vec{E}}{\partial t}$ (d) All of these three
[ESE-2005]

5.34 Consider the following statements:
1. Poisson's equation finds application in vacuum tube and gaseous discharge problems.
2. Gauss's law is useful for determining field and potential distribution about bodies having unsymmetrical geometry.
3. For the propagation of electro-magnetic waves, the time varying electric fields must support time varying magnetic fields.
4. The unit of Poynting's vector is W/m^2

Which of these statements are correct ?
(a) 1, 2 and 3 (b) 1, 3 and 4
(c) 2, 3 and 4 (d) 1, 2 and 4
[ESE-2005]

5.35 What is the value of standing wave ratio (SWR) in free space for reflection coefficient $\Gamma = -1/3$?
(a) 2/3 (b) 0.5
(c) 4.0 (d) 2.0 [ESE-2005]

5.36 What is the phase velocity of plane wave in a good conductor?
(a) $\sqrt{\pi f \mu \sigma}$ (b) $\sqrt{\frac{\pi f \sigma}{(\mu \sigma)}}$
(c) $\sqrt{\frac{\pi f}{(\mu \sigma)}}$ (d) $2\sqrt{\frac{\pi f}{(\mu \sigma)}}$
[ESE-2005]

5.37 The open circuit and short circuit impedances of a line are 100 Ω each. What is the characteristic impedance of the line?
(a) $100\sqrt{2}\ \Omega$ (b) $100\ \Omega$
(c) $100/\sqrt{2}\ \Omega$ (d) $50\ \Omega$
[ESE-2005]

5.38 A load impedance of (75 – *j*50) is connected to a transmission line of characteristic impedance Z_0 = 75 Ω. The best method of matching comprises
(a) A short circuit stub at load
(b) A short circuit stub at some specific distance from load
(c) An open stub at load
(d) Two short circuited stubs at specific distances from load
[ESE-2005]

5.39 When a lossless transmission line is terminated by a resistance equal to surge impedance, then what is value of the reflection coefficient ?
(a) 1 (b) –1
(c) 0 (d) 0.5 [ESE-2005]

5.40 The instantaneous electric field of a plane wave propagating in z-direction is

$$\vec{E}(t) = \left[\hat{a}_x E_1 \cos\omega t - \hat{a}_y E_2 \sin\omega t\right] e^{-jkz}$$

This wave is
(a) Linearly polarised
(b) Elliptically polarised
(c) Right hand circularly polarised
(d) Left hand circularly polarised
[ESE-2005]

5.41 **Assertion (A) :** For a lossy transmission line, the voltage standing wave ratio (VSWR) is higher near the load end than that near the source end.
Reason (R) : In the presence of losses, the propagation constant of the line becomes a complex quantity.
(a) Both A and R are true and R is the correct explanation of A
(b) Both A and R are true but R is NOT the correct explanation of A
(c) A is true but R is false
(d) A is false but R is true
[ESE-2005]

5.42 **Assertion (A) :** Skin depth is the depth by which electromagnetic wave has been increased to 37% of its original value.
Reason (R) : The depth of penetration of wave in a lossy dielectric increases with increasing wavelength.
(a) Both A and R are true and R is the correct explanation of A
(b) Both A and R are true but R is NOT the correct explanation of A
(c) A is true but R is false
(d) A is false but R is true [ESE-2005]

5.43 For a line of characteristic impedance Z_O terminated in a load of Z_R such that $Z_R = Z_O/3$, what is the reflection coefficient Γ_L?
(a) 1/3 (b) 2/3
(c) –1/3 (d) –1/2
[ESE-2006]

5.44 Impedance inversion may be obtained with
(a) a short-circuited stub
(b) an open-circuited stub
(c) a quarter-wave line
(d) a half-wave line
[ESE-2006]

5.45 What is the Poynting's vector on the surface of a long straight conductor of radius *b* and conductivity σ which carries current *I* in the z-direction?
(a) $-\frac{I^2}{2\sigma\pi^2 b^3}\hat{i}_r$ (b) $\frac{I^2}{2\sigma\pi^2 b^2}\hat{i}$
(c) $\frac{I^2}{\sigma\pi b^2}\hat{i}_z$ (d) $\frac{I}{2\pi b}\hat{i}_\phi$
[ESE-2006]

5.46 Consider the following statements regarding EM wave:

1. An EM wave incident on a perfect dielectric is partially transmitted and partially reflected.
2. An EM wave incident on a perfect conductor is fully reflected.
3. When an EM wave is incident from a more dense medium to less dense medium at an angle equal to or exceeding the critical angle, the wave suffers total internal reflection.

Which of the statements given above are correct?

(a) Only 1 and 2 (b) Only 2 and 3
(c) Only 1 and 3 (d) 1, 2 and 3

[ESE-2006]

5.47 **Assertion (A):** The velocity of electro-magnetic waves is same as velocity of light.
Reason (R): Electrons also travel with the same velocity as photons.

(a) Both A and R are true and R is the correct explanation of A
(b) Both A and R are true but R is not the correct explanation of A
(c) A is true but R is false
(d) A is false but R is true

[ESE-2007]

5.48 How is the attenuation factor in parallel plate guides represented?

(a) α = Power lost/power transmitted
(b) $\alpha = 2 \times$ Power lost/power transmitted
(c) α = Power lost per units length/ ($2 \times$ power transmitted)
(d) α = Power lost/(power lost + power transmitted)

[ESE-2007]

5.49 If '*a*' is the width of the rectangular wave guide and λ is the wavelength, then which one of the following is correct?

(a) $a > \lambda/4$ but $< \lambda/2$ (b) $a > \lambda/2$ but $< 3\lambda/2$
(c) $a > 3\lambda/2$ (d) $a < \lambda/4$

[ESE-2007]

5.50 What does the standing wave ratio (SWR) of unity imply?

(a) Transmission line is open circuited
(b) Transmission line is short circuited
(c) Transmission line's characteristic impedance is equal to load impedance
(d) Transmission line's characteristic impedance is not equal to load impedance

[ESE-2007]

5.51 Fields are said to be circularly polarized if their magnitudes are

(a) Equal and they are in phase
(b) Equal and they differ in phase by $\pm 90°$
(c) Unequal and they differ in phase by $\pm 90°$
(d) Unequal and they are in phase

[ESE-2007]

5.52 For a transmission line load matching over a range of frequencies, it is best to use a

(a) balun transformer
(b) single stub of adjustable position
(c) double stub
(d) broad band directional coupler

[ESE-2008]

5.53 Which one of the following modes has the highest cut off wavelength in a rectangular wave guide?

(a) TE_{10} (b) TE_{01}
(c) TM_{01} (d) TM_{11}

[ESE-2008]

5.54 A plane wave travelling in air is incident on a conducting medium. Which one of the following is correct?
The magnetic field intensity

(a) becomes approximately half
(b) gets approximately doubled
(c) remains unchanged
(d) cannot be determined

[ESE-2008]

5.55 What causes electromagnetic wave polarization?

(a) Refraction
(b) Reflection
(c) Longitudinal nature of electro-magnetic wave
(d) Transverse nature of electromagnetic wave

[ESE-2008]

5.56 Which of the following is a vector quantity?

(a) Standing wave ratio only
(b) Reflection coefficient only
(c) Gain
(d) Standing wave ratio and reflection coefficient

[ESE-2008]

5.57 Consider the following statements for transmission lines:

1. When a transmission line is terminated by its characteristic impedance the line will not have any reflected wave.
2. For a finite line terminated by its characteristic impedance the voltage and current at all points on the line are exactly same.
3. For a lossless half wave transmission line the input impedance is not equal to load impedance.

Which of these statements are correct?

(a) 1 and 2 (b) 2 and 3
(c) 1 and 3 (d) 1, 2 and 3

[ESE-2008]

5.58 What is the radiation resistance of a dipole antenna $\lambda/20$ long approximately equal to?

(a) 2 Ω (b) 40 Ω
(c) 0.6 Ω (d) 20 Ω

[ESE-2008]

5.59 In free space

$$\vec{E}(Z,t) = 120\pi \cos(\omega t - \beta Z)\,\hat{a}_x\ V_m^{-1}$$

What is the average power in Wm^{-2}?

(a) $30\pi\,\hat{a}_z$ (b) $60\pi\,\hat{a}_z$
(c) $90\pi\,\hat{a}_z$ (d) $120\pi\,\hat{a}_z$

[ESE-2009]

5.60 Consider the following statements in connection with cylindrical waveguides:

1. At low frequency the propagation constant is real and wave does not propagate.
2. At intermediate frequency the propagation constant is zero and wave cuts off.
3. At high frequency the propagation constant is imaginary and wave propagates.
4. At transition condition the cut-off frequency is inversely proportional to the eigen values of the Bessel function for the respective TE_{nr} mode.

Which of these statements is/are correct?

(a) 1, 2 and 3 (b) 2 only
(c) 2 and 3 only (d) 2, 3 and 4

[ESE-2009]

5.61 The electric field of a uniform plane wave is given by:

$$\vec{E} = 10\sin(3\pi\times10^8\,t - \pi Z)\,\hat{a}_X + 10\cos(3\pi\times10^8 t - \pi Z)\,\hat{a}_Y\ Vm^{-1}$$

What is the corresponding magnetic field $\vec{H}$?

(a) $\frac{10}{377}\sin(3\pi\times10^8 t - \pi Z)\hat{a}_Y + \frac{10}{377}\cos(3\pi\times10^8 t - \pi Z)(-\hat{a}_X)\ Am^{-1}$

(b) $\frac{10}{377}\sin(3\pi\times10^8\,t - \pi Z)(-\hat{a}_Y) + \frac{10}{377}\cos(3\pi\times10^8\,t - \pi Z))(-\hat{a}_X)\ Am^{-1}$

(c) $\frac{10}{377}\sin(3\pi\times10^8 t - \pi Z)\,\hat{a}_Y + \frac{10}{377}\cos(3\pi\times10^8 t - \pi Z)(\hat{a}_X)\ Am^{-1}$

(d) $\frac{10}{377}\sin(3\pi\times10^8 t - \pi Z)(-\hat{a}_Y) + \frac{10}{377}\sin(3\pi\times10^8 t - \pi Z)(-\hat{a}_X)Am^{-1}$

[ESE-2009]

5.62 Consider the following statements in connection with electromagnetic waves:

1. Conducting medium behaves like an open circuit to the electromagnetic field.
2. At radio and microwave frequencies the relaxation time is much less than the period
3. In loss-less dielectric the relaxation time is infinite.
4. Intrinsic impedance of a perfect dielectric medium is a pure resistance.

Which of these statements is/are correct?

(a) 1 only (b) 1 and 2 only
(c) 2 and 3 only (d) 2, 3 and 4

[ESE-2009]

5.63 A transmission line of characteristic impedance of 50 ohm is terminated by a load impedance of (15 – j20) ohm. What is the normalized load impedance?

(a) 0.6 – j0.8 (b) 0.3 – j0.6
(c) 0.3 – j0.4 (d) 0.3 + j0.4

[ESE-2009]

5.64 Two loss-less resistive transmission lines each of characteristic impedance Z are connected as shown in the circuits below. If the maximum voltage on the two lines is the same and the power transmitted by line A is w_1, then what is the power transmitted by the line B?

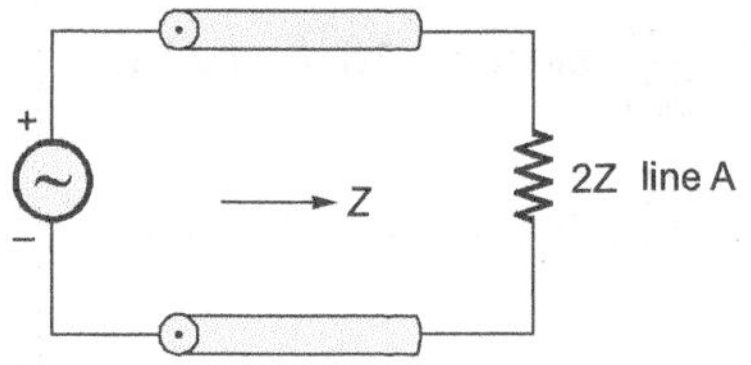

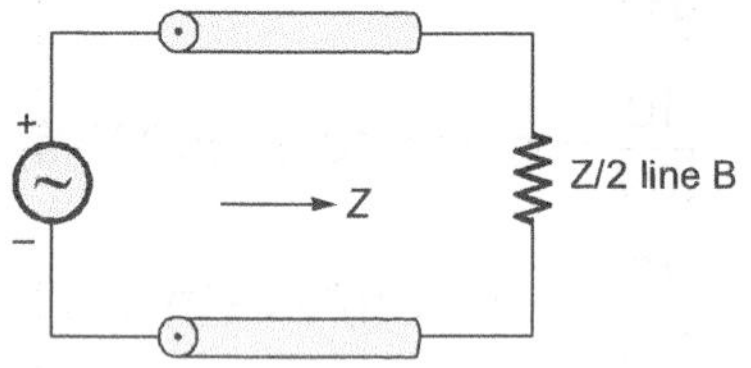

(a) 4 W_1 (b) 3 W_1
(c) 2 W_1 (d) 1 W_1

[ESE-2009]

5.65 A transmission line section shows an input impedance of 36 Ω and 64 Ω respectively, when short circuited and open circuited. What is the characteristic impedance of the transmission line?

(a) 100 Ω (b) 50 Ω
(c) 45 Ω (d) 48 Ω

[ESE-2009]

5.66 If reflection coefficient for voltage be 0.6, the voltage standing wave ratio (VSWR) is

(a) 0.66 (b) 4
(c) 1.5 (d) 2 [ESE-2010]

5.67 If the load impedance is 100 ohm and input impedance is 25 ohm, then the characteristic impedance of the transmission line is

(a) 70 ohm (b) 60 ohm
(c) 50 ohm (d) 40 ohm

[ESE-2010]

5.68 A signal of 10 V is applied to a 50 ohm coaxial transmission line, terminated in a 100 ohm load. The voltage reflection coefficient is

(a) 1/4 (b) 1/3
(c) 1/2 (d) 1

[ESE-2010]

5.69 A uniform plane wave is propagating in a material for which $\in = 4\in_0$, $\mu = 7\mu_0$ and $\sigma = 0$. The skin depth for the material is

(a) zero (b) infinity
(c) 28 m (d) 14 m

[ESE-2010]

5.70 Consider the following statements:

1. In conducting medium the field attenuates exponentially with increasing depth.
2. Conducting medium behaves like an open circuit to the electromagnetic field.
3. In lossless dielectric relaxation time is infinite.
4. In charge-free region, the Poisson's equation becomes Laplace's equation.

Which of the above statements are correct?

(a) 1, 2 and 3 only (b) 1, 3 and 4 only
(c) 2, 3 and 4 only (d) 1, 2, 3 and 4

[ESE-2010]

5.71 In free space $\vec{E}(Z, t) = 60\pi \cos(\omega t - \beta z)\,\hat{a}_x$ V/m.

The average power crossing a circular area of π square metres in the plane Z = constant is

(a) 16π watt/m^2 (b) 15π watt/m^2
(c) 14π watt/m^2 (d) 13π watt/m^2

[ESE-2010]

5.72 **Assertion (A)** : A thin sheet of conducting material can act as a low-pass filter for electromagnetic waves.

Reason (R) : The depth of penetration is inversely proportional to the square root of the frequency.

(a) Both A and R are true and R is the correct explanation of A
(b) Both A and R are true but R is not the correct explanation of A
(c) A is true but R is false
(d) A is false but R is true [ESE-2010]

5.73 An aerial having a directivity of 90 is operating at a wavelength of 2 m. The maximum effective aperture of the aerial is

(a) 7.16 m^2 (b) 14.32 m^2
(c) 28.64 m^2 (d) 57.28 m^2

[ESE-2011]

5.74 The electric field intensity phasor of an EM wave in free space is $\vec{E} = 10e^{-j4y}\hat{a}_x$ V/m. The angular frequency ω, in radian per second, is

(a) $4 \times 3 \times 10^8$ (b) $4y \times 3 \times 10^8$
(c) $t \times 3 \times 10^8$ (d) $10 \times 3 \times 10^8$

[ESE-2011]

5.75 In free space $\vec{H}$ field is given as

$$\vec{H}(z, t) = -\frac{1}{6\pi}\cos(\omega t + \beta z)\hat{a}_y \cdot \vec{E}(z, t) \text{ is}$$

(a) $20\cos(\omega t + \beta z)\hat{a}_x$
(b) $20\cos(\omega t + \beta z)\hat{a}_z$
(c) $20\sin(\omega t + \beta z)\hat{a}_y$
(d) $20\sin(\omega t + \beta z)\hat{a}_x$

[ESE-2011]

5.76 Screw projecting into the waveguide is

(a) capacitive discontinuity
(b) inductive discontinuity
(c) may be capacitive or inductive depending upon the position inside the guide
(d) none of the above

[ESE-2011]

5.77 Depth of penetration δ is equal to λ/2π for

(a) Good insulator (b) Good conductor
(c) Lossy medium (d) Low values of λ

[ESE-2011]

5.78 When a plane wave propagates in a dielectric medium?

(a) The average electric energy and the average magnetic energy densities are not equal.
(b) The average electric and the average magnetic energy densities are equal.
(c) The net average energy density is finite.
(d) The average electric energy density is not dependent on the average magnetic energy density.

[ESE-2011]

5.79 If the maximum and minimum voltages on a transmission line are 4 V and 2 V, respectively, for a typical load, VSWR is

(a) 1.0 (b) 0.5
(c) 2.0 (d) 8.0

[ESE-2011]

5.80 A higher directivity is specified by

(a) High gain high bandwidth
(b) Low gain high bandwidth
(c) High gain low bandwidth
(d) Low gain low bandwidth

[ESE-2011]

5.81 A lossless transmission line of characteristic impedance 300 Ω and length λ/2 is shorted at one end and is terminated in its characteristic impedance at the other. The input impedance measured at the mid section of the line is

(a) 0 Ω (b) 100 Ω
(c) 300 Ω (d) 150 Ω

[ESE-2011]

5.82 **Assertion (A):** Electromagnetic waves propagate being guided by parallel plate perfect conductor surface.

Reason (R): Tangential component of electric field intensity and normal component of magnetic field intensity are zero on a perfect conductor surface.

(a) Both A and R are true and R is the correct explanation of A
(b) Both A and R are true but R is NOT the correct explanation of A
(c) A is true but R is false
(d) A is false but R is true

[ESE-2011]

5.83 **Assertion (A):** A thin sheet of conducting material can act as a low-pass filter for electromagnetic waves.

Reason (R): The penetration depth is inversely proportional to the square root of the frequency.

(a) Both A and R are true and R is the correct explanation of A
(b) Both A and R are true but R is NOT the correct explanation of A
(c) A is true but R is false
(d) A is false but R is true

[ESE-2011]

5.84 A signal of 10 V is applied to a 50 Ω coaxial transmission line terminated in 200 Ω. The magnitude of the reflected voltage will be

(a) 6 V (b) 10 V
(c) 0 V (d) 1 V

[ESE-2012]

5.85 The vector magnetic potential of a particular wave travelling in free space is given by

$$\vec{A} = \hat{a}_x A_x \sin(\omega t - \beta z)$$

where A_x is a constant. The expression for the electric field will be

(a) $-\hat{a}_x \beta A_x \sin(\omega t - \beta z)$

(b) $-\hat{a}_y A_x \sin(\omega t - \beta z)$

(c) $-\hat{a}_y \omega A_x \cos(\omega t - \beta z)$

(d) $-\hat{a}_x \omega A_x \cos(\omega t - \beta z)$

[ESE-2012]

5.86 As a result of reflections from a plane conducting wall, electromagnetic waves acquire an apparent velocity greater than the velocity of light in space. This is called

(a) velocity propagation
(b) normal velocity
(c) group velocity
(d) phase velocity

[ESE-2012]

5.87 A 75 Ω transmission line is first short-terminated and the minima locations are noted. When the short is replaced by a resistive load R_L, the minima locations are not altered and the VSWR is measured to be 3. The value of R_L is

(a) 25 Ω (b) 50 Ω
(c) 225 Ω (d) 250 Ω

[ESE-2012]

5.88 The depth of penetration of a wave in a lossy dielectric increases with increasing

(a) conductivity (b) permeability
(c) wavelength (d) permittivity

[ESE-2012]

5.89 Loss-tangent in plane waves in lossy dielectrics will be

(a) proportional to the Y component of the magnetic field intensity (H_Y)
(b) inversely proportional to the Y component of the magnetic field intensity (H_Y)
(c) inversely proportional to the X component of the magnetic field intensity (H_X)
(d) proportional to X component of the magnetic field intensity (H_X)

[ESE-2013]

5.90 Transverse Electromagnetic waves are characterized by

(a) During wave propagation in Z-direction the components of H and E are transverse 60° to the direction of propagation of the waves.
(b) During wave propagation in Z-direction, the components of H and E are transverse to the direction of propagation of the waves.
(c) During wave propagation in Z-direction, the components of H and E are transverse 120° to the direction of propagation of the waves.
(d) None of the above

[ESE-2013]

5.91 Two media are characterized as:

1. $\epsilon_r = 1, \mu_r = 4$ and $\sigma = 0$
2. $\epsilon_r = 4, \mu_r = 4$ and $\sigma = 0$

where, ϵ_r = relative permittivity
μ_r = relative permeability
σ = conductivity

The ratio of the intrinsic impedance of the media 2 to media 1 is

(a) 2 : 1 (b) 1 : 2
(c) 1 : 1 (d) 2 : 2

[ESE-2013]

5.92 A transmission line of characteristic impedance 50 Ω is terminated at one end by $+j50$ Ω. The VSWR produced by the line is

(a) +1 (b) 0
(c) ∞ (d) $+j$

[ESE-2013]

5.93 A loss-less transmission line having characteristic impedance Z_0 is terminated in a load of Z_R. If the value of Z_R is exactly half of Z_0 then reflection coefficient G_L is

(a) $\frac{1}{3}$ (b) $\frac{2}{3}$
(c) $-\frac{1}{3}$ (d) $-\frac{2}{3}$

[ESE-2013]

5.94 The depth of penetration of a wave in a lossy dielectric medium increases with

(a) increasing wavelength
(b) increasing conductivity
(c) decreasing wavelength
(d) increasing permittivity

[ESE-2014]

5.95 The frequency of the power wave associated with an electromagnetic wave having an E field as

$$E = e^{-z/\delta}\cos\left(\omega t - \frac{z}{\delta}\right), \text{ is given by}$$

(a) $\frac{\omega}{8\pi}$ (b) $\frac{\omega}{4\pi}$

(c) $\frac{\omega}{2\pi}$ (d) $\frac{\omega}{\pi}$ [ESE-2014]

5.96 When the wave travels in 's' conducting medium, the ratio of attenuation is decided by
(a) Attenuation constant
(b) Phase constant
(c) Both attenuation constant and phase constant
(d) Neither attenuation constant nor phase constant

[ESE-2014]

5.97 Uniform plane wave is
(a) longitudinal in nature
(b) transverse in nature
(c) neither longitudinal nor transverse in nature
(d) x-directed

[ESE-2014]

5.98 The impedance Z offered by transmission line for a travelling wave which damps out the low frequency oscillation rapidly is also called
(a) surge impedance
(b) natural impedance
(c) both surge and natural impedances
(d) neither surge nor natural impedance

[ESE-2014]

5.99 In a coaxial transmission line, the useful power flows through
(a) the interface of the two conductors
(b) both inner and outer conductors
(c) inner conductor
(d) outer conductor

[ESE-2014]

5.100 **Statement (I):** At 50 Hz the depth of penetration is 8.5 mm. At 30 GHz the depth of penetration is 0.00038 mm.

Statement (II): A high frequency field attenuates as it penetrates conduction in a shorter distance than a low frequency field.

Codes:
(a) Both Statement (I) and Statement (II) are individually true and Statement (II) is the correct explanation of Statement (I).
(b) Both Statement (I) and Statement (II) are individually true but Statement (II) is not the correct explanation of Statement (I).
(c) Statement (I) is true but Statement (II) is false.
(d) Statement (I) is false but Statement (II) is true.

[ESE-2014]

5.101 A 10 GHz plane wave travelling in free space has amplitude of 15 V/m. The propagation coefficient β is
(a) 209.4 rad/m (b) 173.6 rad/m
(c) 543.5 rad/m (d) 3.97×10^{-2} rad/m

[ESE-2015]

5.102 Brewster angle is the angle when a wave is incident on the surface of a perfect dielectric at which there is no reflected wave and the incident wave is
(a) parallelly polarized
(b) perpendicularly polarized
(c) normally polarized
(d) none of the above

[ESE-2015]

5.103 A lossless transmission line has a characteristic impedance of z_0 and capacitance per unit length of C. The velocity of propagation of the travelling wave on the line is

(a) z_0C (b) $\frac{1}{z_0C}$

(c) $\frac{z_0}{C}$ (d) $\frac{C}{z_0}$

[ESE-2015]

5.104 The value of characteristic impedance in free space is equal to

(a) $\sqrt{\frac{\mu_0}{\varepsilon_0}}$ (b) $\sqrt{\mu_0\varepsilon_0}$

(c) $\sqrt{\frac{1}{\mu_0\varepsilon_0}}$ (d) $\sqrt{\frac{\varepsilon_0}{\mu_0}}$

[ESE-2016]

5.105 The Poynting vector on the surface of a long straight conductor of radius a and conductivity σ_0, which carries current I in the z-direction, is

(a) $\frac{I^2}{\sigma_0 \pi b^3}\hat{a}_r$ (b) $\frac{-I^2}{2\sigma_0 \pi^2 a^2}\hat{a}_r$

(c) $\frac{I^2}{\sigma_0 \pi^2 a^3}\hat{a}_r$ (d) $\frac{-I^2}{2\sigma_0 \pi^2 a^3}\hat{a}_r$

[ESE-2016]

5.106 Consider the following applications in respect of a square corner reflector:

1. Radio astronomy
2. Point-to-point communication
3. TV broadcast

Which of the above applications is/are correct?

(a) 1 only (b) 1 and 2 only
(c) 2 and 3 only (d) 1, 2 and 3

[ESE-2016]

5.107 Which of the following statements are correct regarding uniform plane waves?

1. Uniform plane waves are transverse.
2. The relation between E and H is $\frac{E}{H} = \sqrt{\frac{\epsilon}{\mu}}$.
3. $E \times H$ gives the direction of the wave travel.
4. For a uniform plane wave travelling in x-direction, $E_x = 0$.

(a) 1, 2 and 3 only (b) 1, 3 and 4 only
(c) 1, 2 and 4 only (d) 2, 3 and 4 only

[ESE-2018]

Answers EM Waves and Transmission Lines

5.1 (d)	5.2 (b)	5.3 (b)	5.4 (a)	5.5 (c)	5.6 (a)	5.7 (a)	5.8 (c)	5.9 (d)
5.10 (b)	5.11 (b)	5.12 (a)	5.13 (a)	5.14 (c)	5.15 (a)	5.16 (c)	5.17 (a)	5.18 (b)
5.19 (c)	5.20 (a)	5.21 (b)	5.22 (d)	5.23 (b)	5.24 (b)	5.25 (c)	5.26 (b)	5.27 (d)
5.28 (c)	5.29 (d)	5.30 (c)	5.31 (d)	5.32 (d)	5.33 (c)	5.34 (b)	5.35 (d)	5.36 (d)
5.37 (b)	5.38 (d)	5.39 (c)	5.40 (d)	5.41 (b)	5.42 (d)	5.43 (d)	5.44 (c)	5.45 (a)
5.46 (d)	5.47 (a)	5.48 (b)	5.49 (b)	5.50 (c)	5.51 (b)	5.52 (c)	5.53 (a)	5.54 (d)
5.55 (d)	5.56 (c)	5.57 (a)	5.58 (a)	5.59 (b)	5.60 (a)	5.61 (a)	5.62 (d)	5.63 (c)
5.64 (a)	5.65 (d)	5.66 (b)	5.67 (c)	5.68 (b)	5.69 (b)	5.70 (d)	5.71 (b)	5.72 (a)
5.73 (c)	5.74 (a)	5.75 (a)	5.76 (c)	5.77 (b)	5.78 (b)	5.79 (c)	5.80 (c)	5.81 (c)
5.82 (a)	5.83 (a)	5.84 (a)	5.85 (d)	5.86 (d)	5.87 (c)	5.88 (c)	5.89 (b)	5.90 (b)
5.91 (b)	5.92 (c)	5.93 (c)	5.94 (a)	5.95 (c)	5.96 (a)	5.97 (b)	5.98 (*)	5.99 (c)
5.100 (a)	5.101 (a)	5.102 (a)	5.103 (b)	5.104 (a)	5.105 (d)	5.106 (d)	5.107 (b)	

Explanations EM Waves and Transmission Lines

5.1 (d)

$$Z_0^2 = Z_{OC} \cdot Z_{SC}$$

5.2 (b)

Both are correct but R is not the explanation, i.e. in case of internal reflection reflected and incident waves meet in phase or out of phase or in between to form standing wave.

5.3 (b)

$$S = \frac{1+|\Gamma|}{1-|\Gamma|}$$

as reflection coefficients magnitude can be maximum = 1 and minimum = –1

i.e. $-1 \le |\Gamma| \le 1$

so standing wave ratio

$S = 1$ For $|\Gamma| = 0$ (Minimum value)

$S \to \infty$ For $|\Gamma| = 1$ (Maximum value)

$$1 \le S < \infty$$

5.4 (a)

The term $\mu_0\mu\sigma \dfrac{\partial \bar{E}}{\partial t}$ involves the parameter σ i.e., conductivity of the lossy medium that correspond to losses in EM wave propagation.

5.5 (c)

For all the mediums $\sigma = 0$ (lossless) and intrinsic impedance for lossless medium

$$\eta = \sqrt{\frac{\mu_r \mu_0}{\epsilon_0 \epsilon_r}} = 120\pi\sqrt{\frac{\mu_r}{\epsilon_r}}$$

For $\epsilon_r = 8, \mu_r = 2$

$$\eta = 120\pi\sqrt{\frac{2}{8}}$$

$$\eta = 60\,\pi = 188.49\,\Omega$$

For $\epsilon_0 = 1, \mu_r = 9$

$$\eta = 120\pi\sqrt{\frac{9}{1}} = 360\,\pi$$

$$\eta = 1130.97\,\Omega \simeq 1131\Omega$$

For $\epsilon_r = 4, \mu_r = 4$

$$\eta = 120\pi\sqrt{\frac{4}{4}} = 120\pi = 377\,\Omega$$

5.6 (a)

$$\because \quad \frac{E_r}{E_i} = -\frac{H_r}{H_i} = \Gamma = \frac{Z_2 - Z_1}{(Z_2 + Z_1)}$$

$$\frac{E_i}{E_r} = -\frac{H_i}{H_r} = \frac{Z_2 + Z_1}{Z_2 - Z_1}$$

5.7 (a)

Inside a perfect conductor electric and magnetic fields are zero at any distance from the surface.

5.8 (c)

We known that from Maxwell's 4th equation

$$\nabla \times \vec{H} = \vec{J}_C + \dot{D} = \vec{J}_C + \frac{d}{dt}\vec{D}$$

$$\vec{J}_C = \sigma\vec{E}, \quad \frac{d}{dt}\vec{D} = j\omega\vec{D}$$

$$\frac{d}{dt}\vec{D} = j\omega \in \vec{E}$$

as $\dfrac{d}{dt} = j\omega, \quad \vec{D} = \in \vec{E}$

so, $\nabla \times \vec{H} = \vec{J}_C + j\omega \in \vec{E}$

5.9 (d)

Examination of the phase, $(\omega t - \beta z)$ shows that the direction of propagation is $+z$. Since $E \times H$ must also be in the $+z$-direction, $\bar{E}$ must have the direction $-\hat{a}_y$.

$$\frac{|E_y|}{|H_x|} = 377\,\Omega$$

$$\Rightarrow \quad \bar{E}(z,t) = -37.7\cos(4\times 10^7 t - \beta z)\hat{a}_y$$

5.10 (b)

$A \to \epsilon_r = 4, \mu_r = 1$

$B \to \epsilon_r = 9, \mu_r = 4$

So, intrinsic impedances of material,

$$\eta_1 = \sqrt{\frac{\mu}{\in}} = \sqrt{\frac{\mu_0\mu_r}{\epsilon_0\epsilon_r}}$$

$$= 120\pi\sqrt{\frac{\mu_r}{\epsilon_r}} = 120\pi\sqrt{\frac{1}{4}} = 60\pi$$

$$\eta_2 = \sqrt{\frac{\mu}{\epsilon}} = \sqrt{\frac{\mu_0\mu_r}{\epsilon_0\epsilon_r}}$$

$$= \sqrt{\frac{\mu_0}{\epsilon_0}\times\frac{4}{9}} = \frac{2}{3}\times 120\pi = 80\pi$$

∴ Reflection coefficient,

$$G = \frac{\eta_2-\eta_1}{\eta_2+\eta_1} = \frac{80\pi-60\pi}{80\pi+60\pi}$$

$$= \frac{20\pi}{140\pi} = \frac{1}{7}$$

Transmission coefficient,

$$= \frac{2\eta_2}{\eta_1+\eta_2} = \frac{2\times 80\pi}{140\pi} = \frac{8}{7}$$

5.11 (b)

$$P_{avg} = \frac{1}{2}R_e(\vec{E}\times\vec{H}^*)\times\pi(\sqrt{24})^2\cdot\hat{a}_z \text{ W}$$

$$= \frac{1}{2}\times 50\times\frac{5}{12\pi}\times\pi\times 24 = 250 \text{ W}$$

5.12 (a)

Skin depth is the depth (distance) after travelling which electric field intensity fall to $1/e$ of it's initial value so after travelling 10 cm it becomes $\frac{1}{e}\cdot 1 = \frac{1}{e}$ V/m.

After travelling another 10 cm (i.e. total 20 cm) it becomes $\frac{1}{e}\times\left(\frac{1}{e}\right) = \frac{1}{e^2}$ V/m. So both are correct and R is also the explanation.

5.13 (a)

When dielectric material is subjected to alternating electric field it absorbs some of electric energy known as dielectric losses and relative permittivity will become a complex quantity where complex part represent the losses due to absorption.

So both A and R are correct and R is also the reason.

5.14 (c)

A uniform plane wave can not physically exist for it extends to infinity in two dimensions at least and represents an infinite amount of energy.

5.15 (a)

$$\vec{H} = 0.5e^{-0.1x}\sin(10^6t-2x)\ \hat{a}_z \text{ A/m.}$$

Comparing it with standard equation

$$\vec{H} = Ae^{-\alpha x}\sin(\omega t-\beta x)\ \hat{a}_z$$

$$\alpha = 0.1,\ \beta = 2,\ \omega = 10^6$$

$$\therefore \quad \gamma = \alpha + j\beta = 0.1 + 2j$$

velocity of wave

$$V = \omega/\beta = 5\times 10^5 \text{ m/sec}$$

and wave is travelling in the x-direction so option (a) is incorrect.

5.16 (c)

Skin depth $\delta = \frac{1}{\sqrt{\pi f\mu\sigma}}$

5.17 (a)

$|\vec{E}_x| = |\vec{E}_y|$ and angle between $\vec{E}_x$ and $\vec{E}_y$ is 90°

which means polarisation is circular.

5.18 (b)

Intrinsic impedance of transmission line is given by

$$\eta = \sqrt{\frac{R+j\omega L}{G+j\omega C}}$$

for lossless line if

$Z_{OC} \to$ open circuit input impedance

$Z_{SC} \to$ short circuit input impedance

then, characteristic impedance of line

$$Z_0 = \sqrt{Z_{OC}\cdot Z_{SC}}$$

V/I at any point on the line does not give intrinsic (characteristic) impedance of the line as

$$V = V_+e^{-\gamma z} + V_-e^{+\gamma z}$$

and $$I = \frac{1}{Z_0}[V_-e^{+\gamma z} - V_+e^{-\gamma z}]$$

so $$\frac{V}{I} \neq Z_0$$

5.19 (c)

$$Z_s = Z_{01}\frac{Z_{02}+jZ_{01}\tan\beta l}{Z_{01}+jZ_{02}\tan\beta l}$$

$$Z_s = jZ_0\tan\beta l$$

as $$Z_{02} = 0 \text{ (S.C.) and } Z_{01} = Z_0$$

5.20 (a)

$$T_R = \frac{Z_L-Z_0}{Z_L+Z_0} = \frac{900-600}{900+600} = 0.2$$

5.21 (b)

When elliptically polarized light falls on the interface of dielectric at the Brewster angle reflected wave is linearly polarized and transmitted wave is elliptically polarized.

5.22 (d)

Refer Retarded potentials (time-varying potentials)

5.23 (b)

Guide wave length of a wave propagating in a wave guide is given by

$$\lambda_g = \frac{\lambda}{\sqrt{1-\left(\frac{f_c}{f}\right)^2}}$$

where $\lambda \to$ free space wave length

$f_c \to$ cutoff frequency for a particular mode

$f \to$ frequency of operation

$$\because \quad \sqrt{1-\left(\frac{f_c}{f}\right)^2} < 1$$

$$\therefore \quad \lambda_g > \lambda$$

5.24 (b)

$$\because \quad \beta = \frac{\omega}{v}$$

$$\beta = \omega\sqrt{\mu\epsilon} \qquad \beta\text{-phase constant}$$

$$\beta = \omega\sqrt{\mu_0\mu_r\,\epsilon_0\epsilon_r}$$

$$\beta \propto \sqrt{\epsilon_r}$$

5.25 (c)

$$\eta = \eta_0\sqrt{\frac{\mu_r}{\epsilon_r}}$$

$$\Rightarrow \quad 60\pi = 120\pi\sqrt{\frac{1}{\epsilon_r}} \Rightarrow \epsilon_r = 4$$

5.26 (b)

A waveguide allows the waves (EM waves) to travel only if it's frequency is greater than a cutoff frequency defined for that mode of propagation so this is analogues to high pass filter where frequencies having value greater than certain cutoff value f_e will present at the output only.

5.27 (d)

- Effective length is a function of θ as well as wavelength (λ), so it is not maximum for $\theta = \pi/2$.
- Effective length refers to the fact that antennas behave as though (electrically) they were taller than their physical length.

5.28 (c)

$$L = 10\ \mu\text{H/m} = 10^{-5}\ \text{H/m}$$

$$C = 40\ \text{pF/m} = 4 \times 10^{-11}\ \text{F/m}$$

$$\therefore \quad \beta = \omega\sqrt{LC}$$

$$= 2 \times \pi \times 10^6 \times 30\sqrt{4 \times 10^{-16}}$$

$$= 2 \times \pi \times 10^6 \times 30 \times 2 \times 10^{-8}$$

$$\beta = 1.2\pi\ \text{rad/m}$$

$$\beta l = 1.2\pi \times 0.5 \quad (\because l = 50\ \text{cm} = 0.5\ \text{m})$$

$$\beta l = 0.6\pi\ \text{rad}$$

$$\beta l = 0.6 \times 180° = 108°$$

5.29 (d)

A is not true as voltage and current on lossless transmission line is time harmonically varying so not constant until a standing wave is formed which is not always the case.

$R \to$ at a distance $\lambda/4$ from the load end input impedance

$$Z_{in} = \frac{Z_0^2}{Z_L}$$

given that, $Z_L = \infty$ (as O.C.)

so $Z_{in} = 0$

minimum impedance as in case of series resonant circuit so true statement.

5.30 (c)

We know that is any medium,

$$f \times \lambda = \frac{1}{\sqrt{\mu\epsilon}}$$

so in free space,

$$\mu_r = \epsilon_r = 1$$

$$\therefore \quad f \times \lambda_1 = \frac{1}{\sqrt{\mu_0\epsilon_0}}$$

Now in dielectric,

$$f \times \lambda_2 = \frac{1}{\sqrt{\mu_0\epsilon_0\epsilon_r \cdot \mu_r}}$$

$$f \times \lambda_2 = \frac{1}{\sqrt{\mu_0 \epsilon_0 \epsilon_r}}$$

as $\mu_r = 1$ for perfect dielectric assuming non-magnetic f will be same as independent of medium.

$$\therefore \quad \frac{\lambda_1}{\lambda_2} = \sqrt{\epsilon_r}$$

$$\frac{2}{1} = \sqrt{\epsilon_r}$$

$$\epsilon_r = 4$$

5.31 (d)

Characteristic impedance of free space

$$\eta = \sqrt{\frac{\mu_0}{\epsilon_0}} = 120\pi$$

Independent of f (frequency)

Velocity of EM wave in free space

$$= \frac{1}{\sqrt{\mu_0 \epsilon_0}} = 3 \times 10^8 \text{ m/sec}$$

constant and independent of f(frequency).

5.32 (d)

$\vec{E}$ field is given as

$$\vec{E} = 5\cos(10^9 t + 30Z)\,\hat{a}_x$$

Comparing standard equation of EM wave travelling in –Z-direction.

$$\vec{E} = A\cos[\omega t + \beta Z)\,\hat{a}_x$$

$$\omega = 10^9 \text{ rad/sec}$$

$$\beta = 30$$

$$\therefore \quad v = \omega/\beta = \frac{10^9}{30} = \frac{10^8}{3} \quad \ldots(1)$$

but also, $$v = \frac{1}{\sqrt{\mu \epsilon}} = \frac{1}{\sqrt{\mu_0 \epsilon_0 \mu_r \epsilon_r}}$$

$$= \frac{3 \times 10^8}{\sqrt{\mu_r \epsilon_r}}$$

$$v = \frac{3 \times 10^8}{\sqrt{\mu_r \epsilon_r}} = \frac{3 \times 10^8}{\sqrt{\epsilon_r}} \quad \ldots(2)$$

($\because$ for nonmagnetic medium $\mu_r = 1$)

From equation (1) and (2) we have,

$$\frac{10^8}{3} = \frac{3 \times 10^8}{\sqrt{\epsilon_r}}$$

$$\sqrt{\epsilon_r} = 9$$

$$\epsilon_r = 81$$

5.33 (c)

$\mu\sigma \frac{\partial \bar{E}}{\partial t}$ involves σ i.e. conductivity of a lossy medium, which is responsible for attenuation of the wave.

5.34 (b)

Application of Gauss's Law depends on symmetry.

5.35 (d)

SWR is defined as

$$S = \frac{1 + |\Gamma|}{1 - |\Gamma|}$$

Given that, $$\Gamma = -\frac{1}{3} = \frac{1}{3}\angle 180^\circ$$

$$\therefore \quad |\Gamma| = \frac{1}{3}$$

$$\therefore \quad S = \frac{1 + 1/3}{1 - 1/3} = 2$$

5.36 (d)

For a good conducting medium

$$\alpha = \beta = \sqrt{\frac{\omega\mu\sigma}{2}} = \sqrt{\pi f \mu \sigma}$$

$\therefore$ Phase velocity,

$$v_p = \left(\frac{\omega}{\beta}\right) = \frac{\omega}{\sqrt{\frac{\omega\mu\sigma}{2}}}$$

$$= \sqrt{\frac{2\omega}{\mu\sigma}} = \sqrt{\frac{4\pi f}{\mu\sigma}} = 2\sqrt{\frac{\pi f}{\mu\sigma}}$$

5.37 (b)

$$Z_0 = \sqrt{Z_{SC} Z_{OC}}$$

5.38 (d)

Best matching will be short circuit stub (because it has better matching option and there is no radiation losses as in open circuit stub).

Best method will be two short circuit stub at specific distances from load because matching options will be two one from each stub so easily match to characteristic impedance of T_x line.

5.39 (c)

Since load impedance is equal to surge impedance

i.e. $Z_L = Z_0$

$\therefore$ Reflection coefficient

$$\Gamma = \frac{Z_L - Z_0}{Z_L + Z_0} = \frac{Z_0 - Z_0}{Z_0 + Z_0} = 0$$

5.40 (d)

Given that,

$$\vec{E} = \left[E_1 \cos\omega t\,\hat{a}_x - E_2 \sin\omega t\,\hat{a}_y\right]e^{-JKZ}$$

wave is travelling in Z direction.

considering the wave at $Z = 0$

$$\vec{E}(0, t) = E_1 \cos\omega t\ \hat{a}_x - E_2 \sin\omega t\ \hat{a}_y$$

phase difference $\phi = 90°$ so this will be elliptically polarized as

$$E_1 \neq E_2$$

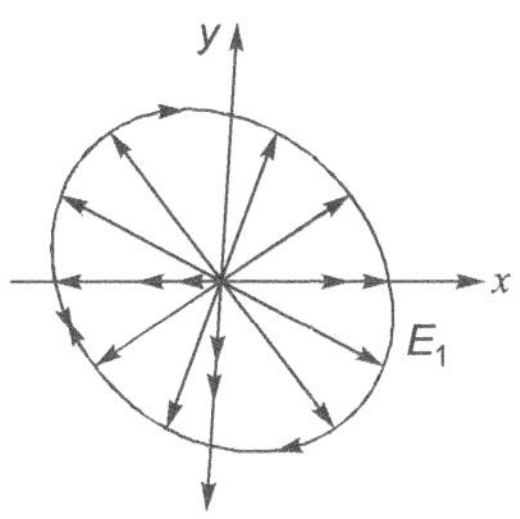

Left handed elliptically polarized.

5.41 (b)

For a lossy transmission line as we move away from the load end the reflected wave goes on decreasing so values of maximas decreasing the VSWR thus A is correct. In case of losses, attenuation constant α will be present and $\gamma = \alpha + j\beta$, propagation constant become a complex quantity. R is correct but not the explanation for R as there is no link between the two statements.

5.42 (d)

At skin depth δ, electric field strength is $1/e$ times of the electric field strength at the surface. i.e. decreased to 37% of its original value.

5.43 (d)

$$|\Gamma| = \frac{Z_L - Z_0}{Z_L + Z_0} = \frac{\frac{Z_0}{3} - Z_0}{\frac{Z_0}{3} + Z_0} = -\frac{1}{2}$$

5.44 (c)

For quarter wave line

input impedance $Z_{in} = \dfrac{Z_0^2}{Z_L}$

where Z_L is load impedance

so if $Z_L = \infty$ (OC)

$Z_{in} = 0$ (SC)

and if $Z_L = 0$ (SC)

$Z_{in} \to \infty$ (OC)

so impedance inversion takes place.

5.45 (a)

$$I^2R = I^2 \cdot \frac{1}{\sigma} \cdot \frac{l}{\pi b^2}$$

$\therefore$ Poynting vector (W/m^2) on the surface

$$= \frac{I^2}{\sigma\pi b^2} \cdot \frac{l}{2\pi bl}(\hat{k} \times \hat{j})$$

$$= -\frac{I^2}{2\sigma\pi^2 b^3} \cdot \hat{i}_r (\text{W/m}^2)$$

5.46 (d)

When EM wave falls on dielectric it is partially reflected and partially transmitted.

When EM wave falls on perfect conductor it is totally reflected.

Therefore Electric field and Magnetic field inside a perfect conductor is zero.

When EM wave is incident from a more dense medium to less dense medium at an angle greater than the angle (critical angle)

$$\theta > \theta_c$$

$$\theta > \sin^{-1}\sqrt{\frac{\epsilon_2}{\epsilon_1}}$$

$$\theta_c = \sin^{-1}\sqrt{\frac{\epsilon_2}{\epsilon_1}}$$

Here, $\epsilon_1 > \epsilon_2$

Then total internal reflection takes place.

5.47 (a)

The velocity of EM wave is same as velocity of light in a particular medium A is correct.

Electron and photon moves (travel) with the same velocity 3×10^8 R $\to$ correct. R is also the reason for A as EM corresponds to movement of e^{-s} while light wave correspond to movement of photon.

5.48 (b)

In case of parallel plate attenuation constant is defined as

$$\alpha = \frac{2 \times \text{Power lost}}{\text{Power transmitted}}$$

5.49 (b)

In case of dominant mode TE_{10} (for $a > b$)
cutoff wave length $\lambda_C = 2a$
therefore wavelength of operation must be less than this values i.e. frequency must be greater than cut-off frequency f_c.

$$\therefore \quad \lambda < \lambda_C$$
$$\lambda < 2a$$

$\boxed{a > \lambda/2}$ lower limit

but there is also a upper limit

i.e. $\boxed{a < 3\lambda/2}$

5.50 (c)

$$\text{SWR} = 1$$

$$\therefore |\Gamma| \text{ (reflection coefficient)} = \frac{\text{VSWR}-1}{\text{VSWR}+1}$$

$$= \frac{1-1}{1+1} = 0$$

i.e. there is no reflection from load that is load is perfectly matched with characteristic impedance of line i.e. $Z_L = Z_0$.

5.51 (b)

Polarization is orientation of electric field in the EM region.

If $E_y = 0$ and $E_x \neq 0$
then horizontally polarized
if $E_y \neq 0$ and $E_x = 0$
Vertically polarized
if E_x, $E_y \neq 0$ and $\phi = 0$ or $180°$
Linearly polarized
if $E_x = E_y$ and $\phi = \pm 90$
Circularly polarized
if $E_x \neq E_y$ and $\phi = \pm 90$
elliptically polarized and in all other cases also it is elliptically polarized.

5.52 (c)

Balun transformer is used for balance and unbalance line. Double stub (short circuit stub with parallel connection) is best suited for impedance matching because we have 4 matching options. Single stub is also used but having two matching options, so best is double stub.

5.53 (a)

If $a > b$ then TE_{10} has lowest cutoff frequency and

$$\because \quad \lambda \propto \frac{1}{f}$$

Therefore TE_{10} mode has highest cut-off wave length.

5.54 (d)

It is not given in question at which angle it falls on the interface, so we can't determine the magnetic field, as from the boundary condition

$$B_{n_1} = B_{n_2} \text{ and } \vec{H}_{t_1} - \vec{H}_{t_2} = \hat{a}_{n12} \times \vec{J}_s$$

5.55 (d)

Polarization means the orientation of electric field in EM region which is always ⊥ to direction of movement of EM wave so it is transverse nature of EM wave.

5.56 (c)

$$\text{VSWR} = \frac{1+|\Gamma|}{1-|\Gamma|} \text{ is a scalar quantity}$$

$$\text{reflection coefficient} = \frac{Z_L - Z_0}{Z_L + Z_0}$$

is a scalar quantity as Z_L and Z_0 are scalar quantities and gain is a vector quantity as depends on direction.

5.57 (a)

$$Z_{in} = Z_0\left[\frac{Z_L + jZ_0 \tan\beta l}{Z_0 + jZ_L \tan\beta l}\right]$$

$$\text{Here,} \quad l = \frac{\lambda}{2}, \ \beta = \frac{2\pi}{\lambda}$$

$Z_{in} = Z_L$ hence point 3 is wrong.

5.58 (a)

$$dl = \frac{\lambda}{20}$$

since length is very small compare to wave length so we can treat it as Hertzian dipole therefore radiation resistance for Hertzian dipole.

$$R_r = 80\pi^2\left(\frac{dl}{\lambda}\right)^2$$

$$R_r = 80\pi^2 \times \frac{1}{400} = 1.973\ \Omega \simeq 2\ \Omega$$

5.59 (b)

$$\vec{P} = \frac{1}{2}\times\frac{E_m^2}{\eta} = \frac{1}{2}\times\frac{(120\pi)^2}{120\pi}\hat{a}_z = 60\pi\,\hat{a}_z$$

5.60 (a)

Propagation constant for cylindrical waveguide is

$$\gamma = \sqrt{\left(\frac{K'_{np}}{a}\right)^2 - \omega^2\mu_0\epsilon_0} \quad \text{for TE mode}$$

So at frequency,

$$\omega < \frac{K'_{np}C}{a}$$

$$\left(\frac{K'_{np}}{a}\right)^2 > \omega^2\mu_0\epsilon_0$$

where, $C = \dfrac{1}{\sqrt{\mu_0\epsilon_0}}$

$\omega \rightarrow$ frequency of operation, K_{np} – p^{th} root of derivative of nth order Bessel function.

$\therefore \quad \gamma > 0$

so $\quad \gamma = \alpha + j\beta$

α = some constant

$\beta = 0$

So wave will attenuate as travel in wave guide so no propagation.

when, $\omega > \dfrac{K'_{np}C}{a}$

then $\left(\dfrac{K'_{np}C}{a}\right)^2 > \omega^2\mu_0\epsilon_0$

and $\gamma \rightarrow$ imaginary quantity

$\therefore \quad \alpha = 0$ and β = some constant

so wave propagation takes place

so 1, 3 $\rightarrow$ correct

at $\quad \omega = \dfrac{K'_{np}C}{a} \Rightarrow \gamma = 0$

and will oscillate between the walls of wave guide and it will not propagate

so 1, 2, 3 $\rightarrow$ correct

here cutoff frequency

$$\omega_c = \frac{K'_{np}C}{a}$$

i.e. directly proportional to eigen values of Bessel functions derivative.

5.61 (a)

$$\because \quad \frac{E_x}{H_y} = \frac{E_y}{H_x} = \eta$$

wave is travelling in free space

as $\quad v = \dfrac{\omega}{\beta} = \dfrac{3\pi\times10^8}{\pi} = 3\times10^8$ m/sec

Speed of wave in free space

($\omega = 3\pi\times10^8$ rad/sec, $\beta = \pi$ rad/m)

$$\therefore \quad H_y = \frac{E_x}{\eta} = \frac{10}{377}$$

$$H_x = \frac{E_y}{\eta} = \frac{10}{377}$$

$$\therefore \quad \vec{H} = H_x\cos(3\pi\times10^8t-\pi Z)(-\hat{a}_x) + H_y\sin(3\pi\times10^8t-\pi Z)\hat{a}_y$$

$$= -\frac{10}{377}\cos(3\pi\times10^8t-\pi Z)\,\hat{a}_x + \frac{10}{377}\sin(3\pi\times10^8t-\pi Z)\hat{a}_y$$

($-\hat{a}_x$, because wave is travelling + Z-direction

so, $\quad \vec{E}\times\vec{H} \Rightarrow \hat{a}_z$

for corresponding $\hat{a}_y\,(\vec{E}_y)$

$\vec{H}_x \rightarrow$ will be $\rightarrow -\hat{a}_x$

such that, $\quad \vec{E}_y\times\vec{H}_x \Rightarrow \hat{a}_z$

$$\hat{a}_y\times(-\hat{a}_x) = +\hat{a}_z$$

5.62 (d)

Relaxation time, $T_r = \dfrac{\epsilon}{\sigma}$

for loss less dielectric $\sigma = 0$

$\therefore \quad T_r = \infty$

Intrinsic impedance of a general medium

$$\eta = \sqrt{\frac{j\omega\mu}{\sigma + j\omega\epsilon}}$$

for perfect dielectric $\sigma = 0$

$$\therefore \quad \eta = \sqrt{\frac{\mu}{\epsilon}}$$

real quantity so pure resistance so is correct so option (d).

5.63 (c)

$$\vec{Z}_L = \frac{Z_L}{Z_0} = \frac{15-j20}{50} = 0.3 - j0.4$$

5.64 (a)

Since maximum voltage on the two lines is same so voltage available across the load will be same suppose V so power available at load end in case of A

$$P_1 = \frac{V^2}{2Z}$$

in case B, $P_2 = \frac{V^2}{Z/2} = \frac{2V^2}{Z}$

Since power available at load will be in proportion to power transmitted

$$\therefore \quad W_1 \propto P_1$$
$$W_2 \propto P_2$$
$$\frac{W_1}{W_2} = \frac{P_1}{P_2}$$
$$\frac{W_1}{W_2} = \frac{1}{4}$$
$$\boxed{W_2 = 4W_1}$$

5.65 (d)

$$Z_0 = \sqrt{Z_{o.c.} \times Z_{s.c}} = \sqrt{36 \times 64} = 48\ \Omega$$

5.66 (b)

Reflection coefficient $\rho = 0.6$

$$S = \frac{1+\rho}{1-\rho} = \frac{1.6}{0.4} = 4$$

5.67 (c)

$$Z_L = 100\ \Omega$$
$$Z_{in} = 25\ \Omega$$

For quarter wave ine,

$$Z_0 = \sqrt{Z_{in} Z_L} = \sqrt{25 \times 100}$$
$$Z_0 = 50\ \Omega$$

5.68 (b)

$$Z_0 = 50$$
$$Z_L = 100$$
$$\rho = \frac{Z_L - Z_0}{Z_L + Z_0} = \frac{50}{150} = \frac{1}{3}$$

5.69 (b)

$\sigma = 0 \rightarrow$ perfect dielectric

- Skin depth is ∞ for perfect dielectric.

$\because$ Skin depth $\propto \frac{1}{\sqrt{\sigma}}$

5.73 (c)

Directivity, $D = 90$

$l = 2$ m

Ae is aperture area

$$\because \quad D = \frac{4\pi Ae}{\lambda^2}$$

$$\therefore \quad A_e = \frac{D \times \lambda^2}{4\pi} = \frac{90 \times 4}{4 \times \pi} = 28.64\ \text{m}^2$$

5.74 (a)

Given, $E = 10 \times e^{-j4y}\ \hat{a}_x$ V/m in free space phase velocity $V_p = C = 3 \times 10^8$ m/sec. Comparing above equation with standard phasor equation

$$= E = E_0\, e^{-j\beta y}\ \hat{a}_x,\ \beta = 4 \text{ rad/m}$$

Now, $V_p = \omega / \beta$

$$3 \times 10^8 = \omega / 4$$
$$\omega = 12 \times 10^8 \text{ rad/sec.}$$

5.75 (a)

Given that, $\vec{H}(z,t) = -\frac{1}{6\pi}\cos(\omega t + \beta z)\,\hat{a}_y$

Now, wave is travelling in –z direction.

So, electric field component will be in x-direction as

$$\hat{a}_x \times (-\hat{a}_y) = -\hat{a}_z$$

Now, $\frac{E_{x0}}{H_{y0}} = \eta_0$

$$E_{x0} = H_{y_0} \times \eta_0$$
$$= \frac{1}{6\pi} \times 120\pi = 20$$

$$\therefore \quad \vec{E}(z,t) = 20\cos(\omega t + \beta z)\,\hat{a}_x$$

5.76 (c)

Screw projecting into the waveguide may capacitive or inductive depending upon the position inside the guide.

5.77 (b)

Skin depth $\delta = \frac{1}{\alpha}$ and for good conductor

$$\alpha = \beta = \sqrt{\frac{\omega \mu \sigma}{2}}$$

Also, $\alpha = \beta = \frac{2\pi}{\lambda}$

as $\beta = \frac{2\pi}{\lambda}$ always

$\therefore$ $\delta = \frac{1}{\alpha} = \frac{1}{\beta} = \frac{\lambda}{2\pi}$

5.78 (b)

When plane wave propagate in dielectric (losses we can assume to zero) so average electric energy density and magnetic energy density will be equal.

5.79 (c)

$$V_{max} = 4$$
$$V_{min} = 2$$
$$V_{SWR} = \frac{V_{max}}{V_{min}} = \frac{4}{2} = 2$$

5.80 (c)

A high directivity means high gain in a particular direction

$\because$ gain × bandwidth is a constant for a particular system.

$\therefore$ for high gain, bandwidth will be low.

5.81 (c)

Given, $\eta = 300\ \Omega$

$$l = \frac{\lambda}{2}$$

at mind section length

$$l_2' = \frac{\lambda}{4}$$

and for $\frac{\lambda}{4}$ line $Z_{in} = \frac{Z_0^2}{Z_L}$

$\because$ $Z_L = 0$ or Z_0

For $Z_L = 0 \Rightarrow Z_{in} = \infty$

and for $Z_L = 300\ \Omega \Rightarrow Z_{in} = 300\ \Omega$

5.82 (a)

Electromagnetic waves propagate being guided by parallel plate perfect conductor surface. So (a) is correct. R is also correct and also the reason for A as conductor (perfect) guide the wave by reflection of EM waves.

5.83 (a)

For conducting material, skin depth

$$d = \frac{1}{\alpha} = \frac{1}{\sqrt{\pi f \mu \sigma}}$$

$$\delta \propto \frac{1}{\sqrt{f}}$$

so as $f\uparrow$ $d\downarrow$ so EM wave can not be pass through conducting material if frequency is above certain frequency so act as low pass filter and R is also correct as $d \propto \frac{1}{\sqrt{f}}$ and also the reason for A.

5.84 (a)

The reflection coefficient

$$\Gamma = \frac{Z_L - Z_0}{Z_L + Z_0} = \frac{200 - 50}{200 + 50} = \frac{150}{250} = \frac{3}{5}$$

Magnitude of reflected voltage

$$= \frac{3}{5} \times 10 = 6\text{ V}$$

5.85 (d)

$$\vec{B} = \nabla \times \vec{A}$$
$$= \left(\frac{\partial A_z}{\partial y} - \frac{\partial A_y}{\partial z}\right)\hat{a}_x + \left(\frac{\partial A_x}{\partial z} - \frac{\partial A_z}{\partial x}\right)\hat{a}_y + \left(\frac{\partial A_y}{\partial x} - \frac{\partial A_x}{\partial y}\right)\hat{a}_z$$
$$= (0-0)\hat{a}_x + \left(\frac{\partial A_x}{\partial z} - 0\right)\hat{a}_y + (0-0)\hat{a}_z$$
$$= \frac{\partial A_x}{\partial z}\hat{a}_y$$
$$= A_x \cos(\omega t - \beta z) \times -\beta\,\hat{a}_y$$
$$= -A_x \beta \cos(\omega t - \beta z)\hat{a}_y$$
$$\frac{\partial \vec{B}}{\partial t} = -A_x\beta(-\sin(\omega t - \beta z)) \times \omega\,\hat{a}_y$$
$$= A_x \beta\omega \sin(\omega t - \beta z)\hat{a}_y$$

We know,

$$\nabla \times \vec{E} = -\frac{\partial \vec{B}}{\partial t}$$
$$= -A_x \beta\omega \sin(\omega t - \beta z)\hat{a}_y$$
$$\frac{\partial E_x}{\partial z}\hat{a}_y = -A_x \beta\omega \sin(\omega t - \beta z)\hat{a}_y$$

After integrating, option (d) is satisfying the above equation.

5.86 (d)

Phase velocity,

$$u_p = \frac{u'}{\sqrt{1 - \left(\frac{f_c}{f}\right)^2}}$$

Where u' is the medium velocity.
if $u' = c$, then phase velocity is greater than the speed of light in vacuum.

5.87 (c)

$$\text{VSWR} = \frac{1+|\Gamma_L|}{1-|\Gamma_L|} = 3$$

$$\Rightarrow \quad |\Gamma_L| = \frac{2}{4} = \frac{1}{2}$$

$$\frac{Z_L - Z_o}{Z_L + Z_o} = \frac{1}{2}$$

$$Z_L = \frac{3}{1}Z_o = 3\times 75 = 225\ \Omega$$

5.88 (c)
Depth of penetration

$$\delta = \frac{1}{\alpha} = \frac{1}{\omega\sqrt{\frac{\mu\in}{2}\left[\sqrt{1+\frac{\sigma^2}{\omega\in}}-1\right]}}$$

$$\therefore \quad \delta \propto \frac{1}{\sqrt{\omega\mu\sigma}}$$

δ will increase with increase in σ.
$\Rightarrow$ Option (c) is correct.

5.90 (b)

$\vec{H}, \vec{E}$ and direction of propagation of electromagnetic wave, these all 3 are perpendicular to each other.

5.91 (b)
For $\sigma = 0$

$$\eta \propto \sqrt{\frac{\mu_r}{\epsilon_r}}$$

$$\Rightarrow \quad \frac{\eta_2}{\eta_1} = \sqrt{\frac{\mu_{r_2}}{\epsilon_{r_2}}\times\frac{\epsilon_{r_1}}{\mu_{r_1}}} = \sqrt{\frac{4\times 1}{4\times 4}} = \frac{1}{2}$$

5.92 (c)

$$Z_C = 50$$
$$Z_L = j50$$

$$|\rho| = \left|\frac{50-j50}{50+j50}\right| = 1$$

$$\text{VSWR} = \frac{1+|\rho|}{1-|\rho|} = \frac{1+1}{1-1} = \infty$$

5.93 (c)
Reflection coefficient,

$$\Gamma_L = \frac{Z_R - Z_o}{Z_R + Z_o} = \frac{\frac{Z_o}{2} - Z_o}{\frac{Z_o}{2} + Z_o} = -\frac{1}{3}$$

5.94 (a)

Depth of penetration $\propto \frac{1}{\sqrt{f}}$ and $f \propto \frac{1}{\lambda}$

5.97 (b)
Uniform plane wave is transverse in nature.

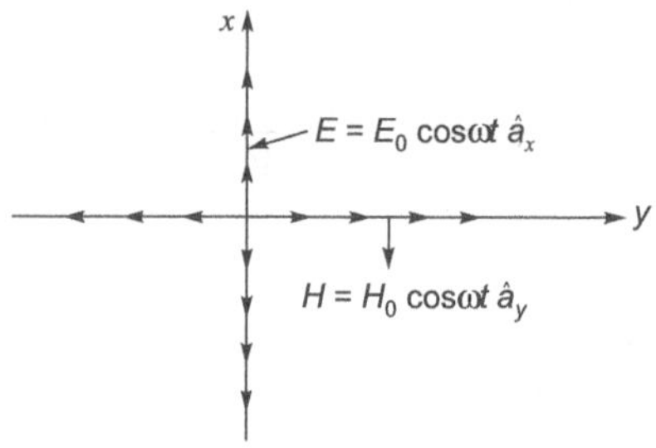

5.101 (a)

Depth of penetration $(\delta) = \frac{1}{\sqrt{\pi f \mu\sigma}}$

As the frequency increases the depth of penetration decreases.

5.102 (a)

$$f = 10\text{ GHz} = 10\times 10^9\text{ Hz}$$

$$\beta = \frac{2\pi}{\lambda} = \frac{2\pi}{v/f}$$

$$= \frac{2\pi}{\frac{3\times 10^8}{10^{10}}}\ [\text{assuming } v = c]$$

$$= \frac{200\pi}{3} = 209.4\text{ rad/m}$$

5.103 (b)
Velocity of propagation

$$v = \frac{1}{\sqrt{LC}}$$

Characteristic impedance

$$Z_0 = \sqrt{\frac{L}{C}}$$

Hence, $$v = \frac{1}{Z_0 C}$$

■■■■

UNIT

IV

Electrical & Electronic Measurements

Syllabus

Principles of measurement, accuracy, precision and standards; Bridges and potentiometers; moving coil, moving iron, dynamometer and induction type instruments, measurement of voltage, current, power, energy and power factor, instrument transformers, digital voltmeters and multi-meters, phase, time and frequency measurement, Q-meters, oscilloscopes, potentiometric recorders, error analysis, Basics of sensors, Transducers, basics of data acquisition systems.

Contents

OOOO

1 Characteristics of Instruments and Measurement Systems

1.1 Match **List-I** with **List-II** and select the correct answer:

List-I

A. Precision
B. Accuracy
C. Resolution
D. Static

List-II

1. The smallest change inthe input quantity which can be detected with its certainty
2. Closeness of the reading with its true value
3. Measure of reproductibility of the measurements
4. Ratio of infinitesimal change sensitivity in output to infinitesimal change in input

Codes:

	A	B	C	D
(a)	2	3	1	4
(b)	3	2	4	1
(c)	3	2	1	4
(d)	2	3	4	1

[ESE-2001]

1.2 Match **List-I** (Error parameters) with **List-II** (Values) and select the correct answer : (σ is the standard deviation of Gaussian error):

List-I	List-II
A. Precision index	**1.** $0.67\,\sigma$
B. Probable error	**2.** 3σ
C. Error limit	**3.** $0.39/\sigma$
D. Peak probability density of error	**4.** $0.71/\sigma$

Codes:

	A	B	C	D
(a)	4	2	1	3
(b)	4	1	2	3
(c)	3	1	2	4
(d)	3	2	1	4

[ESE-2001]

1.3 Which of the following are the characteristics of a thermocouple type of indicating instrument?

1. Its accuracy is very high, as high as about 1 percent.
2. It has linear scale because a d'Arsonval movement is used for measuring the output.
3. It is an RF instrument and can be used for frequency up to about 50 MHz.
4. It cannot be damaged by overloads.

(a) 1 and 2 (b) 2 and 3
(c) 3 and 4 (d) 1 and 3

[ESE-2001]

1.4 An ac voltmeter using full-wave rectification and having a sinusoidal input has an ac sensitivity equal to

(a) 1.414 times dc sensitivity
(b) dc sensitivity
(c) 0.90 times dc sensitivity
(d) 0.707 times dc sensitivity

[ESE-2001]

1.5 Consider the following statements about the D'Arsonval Movement :

1. It is best suited for d.c. current measurement.
2. It responds to the average value of current.
3. It measures the r.m.s. value of a.c. currents.
4. It could be used for power measurements.

Which of these statements is/are correct?

(a) Only 1 (b) 1 and 2
(c) 2 and 3 (d) 1, 2 3 and 4

[ESE-2002]

1.6 Two meters *X* and *Y* require 40 mA and 50 mA, respectively, to give full-scale deflection, then

(a) sensitivity can not be judged with given information.
(b) both are equally sensitive.
(c) *X* is more sensitive.
(d) *Y* is more sensitive.

[ESE-2002]

1.7 What is the correct sequence of the following types of ammeters and voltmeters with increasing accuracy?
1. Moving iron
2. Moving-coil permanent magnet
3. Induction
Select the correct answer using the codes given below:
(a) 1, 3, 2 (b) 1, 2, 3
(c) 3, 1, 2 (d) 2, 1, 3
[ESE-2002]

1.8 The total current $I = I_1 + I_2$ in a circuit is measured as I_1 = 150 ± 1 A, I_2 = 250 ± 2 A, where the limits of error are given as standard deviations. I is measured as
(a) (400 ± 3) A (b) (400 ± 2.24) A
(c) (400 ± 1/5) A (d) (400 ± 1) A
[ESE-2002]

1.9 Torque/Weight ratio of an instrument indicates
(a) Selectivity (b) Accuracy
(c) Fidelity (d) Sensitivity
[ESE-2003]

1.10 A first order instrument is characterized by
(a) Time constant only
(b) Static sensitivity and time constant
(c) Static sensitivity and damping coefficient
(d) Static sensitivity, damping coefficient and natural frequency of oscillations
[ESE-2003]

1.11 Decibel scale is useful while measuring voltages covering
(a) Wide frequency ratio
(b) Wide voltage ratio
(c) Narrow frequency range
(d) Narrow voltage range
[ESE-2003]

1.12 For time and frequency, the working standard is
(a) Microwave oscillator
(b) Crystal controlled oscillator
(c) Laser
(d) ARF oscillator
[ESE-2003]

1.13 The difference between the indicated value and the true value of a quantity is
(a) Gross error (b) Absolute error
(c) Dynamic error (d) Relative error
[ESE-2003]

1.14 Integrating principle in the digital measurement is the conversion of
(a) Voltage to time
(b) Voltage to frequency
(c) Voltage to current
(d) Current to voltage [ESE-2003]

1.15 Match **List-I** (Accuracy) with **List-II** (Type of the standard) and select the correct answer:

List-I	List-II
A. Least accurate	1. Primary
B. More accurate	2. Secondary
C. Much more accurate	3. Working
D. Highest possible accurate	4. International

Codes:

	A	B	C	D
(a)	3	4	1	2
(b)	1	4	3	2
(c)	3	2	1	4
(d)	1	2	3	4

[ESE-2004]

1.16 Which one of the following statements is correct? The application of the instrument in wrong manner in the procedure of measurement results in a/an
(a) systematic error (b) random error
(c) gross error (d) instrument error
[ESE-2004]

1.17 Which one of the following statements correctly represents the systematic errors?
(a) These errors can be calculated from the details of the instruments.
(b) These are the residual errors.
(c) These errors may occur under controlled conditions.
(d) These are the errors committed by the experiments.
[ESE-2004]

1.18 Which one of the following is basically a current sensitive instrument ?
(a) Permanent magnet moving coil instrument
(b) Cathode ray oscilloscope
(c) Electrostatic instrument
(d) FET input electronic voltmeter
[ESE-2004]

1.19 Match **List-I** (Instruments) with **List-II** (Application) and select the correct answer using the codes given below:

List-I
A. Dynamometer instrument
B. Thermocouple based instrument
C. Ramp generator
D. Weston Standard Cell

List-II
1. True r.m.s. value meter
2. Transfer instrument between a.c. and d.c.
3. Time base of CRO
4. Standard of Electromotive force (Emf)

Codes:

	A	B	C	D
(a)	4	1	3	2
(b)	4	3	1	2
(c)	2	1	3	4
(d)	2	3	1	4

[ESE-2004]

1.20 Which one of the following decides the time of response of an indicating instrument?
(a) Deflecting system
(b) Controlling system
(c) Damping system
(d) Pivot and Jewel bearing

[ESE-2004]

1.21 A 0 to 200 V voltmeter has a guaranteed accuracy of 1% of full scale reading. The voltage measured by this instrument is 50 V. What is the limiting error?
(a) 4% (b) 2%
(c) 1% (d) 0.25%

[ESE-2004]

1.22 A set of independent current measurements taken by four observers was recorded as: 117.02 mA, 117.11 mA, 117.08 mA and 117.03 mA. What is the range of error?
(a) ± 0.045 (b) ± 0.054
(c) ± 0.065 (d) ± 0.056

[ESE-2005]

1.23 Match **List-I** (Material used in Instruments) with **List-II** (Purpose) and select the correct answer using the code given below the lists:

List-I	List-II
A. Phosphor-bronze	1. Coils
B. Manganin	2. Discs
C. Aluminium	3. Control Springs
D. Eureka	4. Shunts
	5. Multipliers

Codes:

	A	B	C	D
(a)	3	5	2	4
(b)	2	4	1	5
(c)	3	4	2	5
(d)	2	5	1	4

[ESE-2005]

1.24 Match **List-I** (Type of Electronic Voltmeter) with **List-II** (Major Characteristic) and select the correct answer using the code given below the lists:

List-I
A. Amplifier-rectifier dynamic range
B. Rectifier-amplifier
C. True R.M.S.
D. Logarithmic

List-II
1. Wide input-signal
2. High sensitivity, limited bandwidth
3. Limited sensitivity, large bandwidth
4. Capability to read non-sinusoidal ac

Codes:

	A	B	C	D
(a)	4	1	2	3
(b)	2	3	4	1
(c)	4	3	2	1
(d)	2	1	4	3

[ESE-2005]

1.25 To measure 5 volts, if one selects a (0-100) volt range voltmeter which is accurate within ±1%, then the error in this measurement may be up to
(a) ± 1.5% (b) ± 2.5%
(c) ± 7.5% (d) ± 20%

[ESE-2005]

Precision is not a guarantee for accuracy.

1.26 **Assertion (A):** It is always desirable to take the reading of an indicating instrument very close to the full-scale reading.

Reason (R): Accuracy of an indicating instrument is maximum at the full-scale deflection and error increases as reading comes closer to the beginning of the scale.

(a) Both A and R are true and R is the correct explanation of A
(b) Both A and R are true but R is NOT the correct explanation of A
(c) A is true but R is false
(d) A is false but R is true

[ESE-2006]

1.27 A rectifier instrument is used to measure an alternating square wave of amplitude 100 V. What is the meter reading?

(a) 100 V (b) 70.7 V
(c) 111 V (d) None of these

[ESE-2006]

1.28 When reading is taken at half scale in the instrument, the error is

(a) exactly equal to half of full-scale error
(b) equal to full-scale error
(c) less than full-scale error
(d) more than full-scale error

[ESE-2006]

1.29 Match **List-I** (Instrument) with **List-II** (Error) and select the correct answer using the code given below the lists:

List-I
A. PMMC voltmeter
B. AC ammeter
C. Current transformer
D. Energy meter

List-II
1. Eddy current error
2. Phase angle error
3. Braking system error
4. Temperature error

Codes:

	A	B	C	D
(a)	2	3	4	1
(b)	4	1	2	3
(c)	2	1	4	3
(d)	4	3	2	1

[ESE-2006]

1.30 For defining the standard meter, wavelength of which material is considered?

(a) Neon (b) Krypton
(c) Helium (d) Xenon

[ESE-2006]

1.31 **Assertion (A):** The precision instruments are always accurate.

Reason (R): Precision indicates the degree of agreement within a set of measurements of the same quantity by the instrument, and accuracy refers to the degree of closeness to the true value of the quantity to be measured.

(a) Both A and R are true and R is the correct explanation of A
(b) Both A and R are true but R is NOT the correct explanation of A
(c) A is true but R is false
(d) A is false but R is true

[ESE-2007]

1.32 Match **List-I** (Name of Instrument) with **List-II** (Classification) and select the correct answer using the code given below the lists:

List I	List II
A. Ohmmeter	**1.** Absolute
B. Watt hour meter	**2.** Indicating
C. Null balance	**3.** Recording recorders
D. Rayleigh current	**4.** Integrating balance

Codes:

	A	B	C	D
(a)	1	2	3	4
(b)	2	4	3	1
(c)	2	4	1	3
(d)	1	2	4	3

[ESE-2007]

1.33 Match **List-I** (Property) with **List-II** (Type of the Instrument) and select the correct answer using the code given below the lists:

List-I
A. Linear scale
B. True r.m.s. up to RF range
C. r.m.s. only for sinusoidal input
D. Reads r.m.s. value using square law scale

List-II
1. Thermocouple type
2. Rectifier type
3. Moving iron type
4. Permanent magnet moving coil type

Codes:

	A	B	C	D
(a)	1	4	3	2
(b)	4	1	2	3
(c)	4	1	3	2
(d)	1	4	2	3

[ESE-2007]

1.34 **Assertion (A):** A precision instrument is always accurate.

Reason (R): A precision instrument is one where the degree of reproducibility of the measurements is very good.

(a) Both A and R are true and R is the correct explanation of A

(b) Both A and R are true but R is NOT the correct explanation of A

(c) A is true but R is false

(d) A is false but R is true [ESE-2008]

1.35 What is the prefix tera equivalent to?

(a) 10^3 (b) 10^6

(c) 10^9 (d) 10^{12}

[ESE-2008]

1.36 The errors introduced by an instrument fall in which category?

(a) Systematic errors

(b) Random errors

(c) Gross errors

(d) Environmental errors [ESE-2008]

1.37 A waveform shown in the figure below, is fed to a a.c. ammeter. What is the reading shown by the meter?

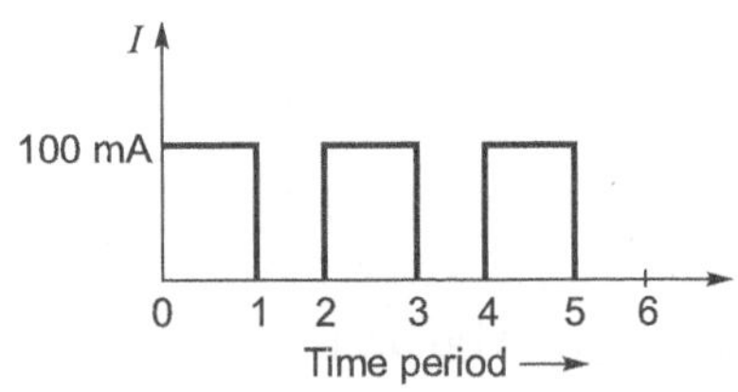

(a) zero (b) 50 mA

(c) 75 mA (d) 100 mA

[ESE-2008]

1.38 The instrumentation amplifiers are used principally to amplify signals from which of the following?

(a) Transducers (b) Active filters

(c) Choppers (d) D/A converters

[ESE-2008]

1.39 A 0 to 300 V voltmeter has an error of ± 2% of fsd. What is the range of readings if true voltage is 30 V?

(a) 24 V – 36 V (b) 20 V – 40 V

(c) 29.4 V – 30.6 V (d) 20 V – 30 V

[ESE-2009]

1.40 Consider the following:

1. Human errors
2. Improper application of instruments
3. Error due to worn parts of an instrument
4. Errors due to effects of environment

Which of the above come under the type of systematic errors?

(a) 1 and 2 (b) 2 and 3

(c) 3 and 4 (d) 1 and 4

[ESE-2009]

1.41 **Assertion (A):** Random errors can be minimized by statistical methods.

Reason (R): These are caused by arithmetic error while taking readings.

(a) Both A and R are true and R is the correct explanation of A

(b) Both A and R are true but R is NOT the correct explanation of A

(c) A is true but R is false

(d) A is false but R is true

[ESE-2009]

1.42 Match **List-I** (Type of Instrument) with **List-II** (Example) and select the correct answer using the code given below the lists:

List-I	List-II
A. Indicating	1. Wattmeter
B. Absolute	2. Tangent galvanometer
C. Recording	3. Aneroid barometer
D. Integrating	4. Energy meter

Codes:

	A	B	C	D
(a)	1	2	3	4
(b)	4	2	3	1
(c)	1	3	2	4
(d)	4	3	2	1

[ESE-2009]

1.43 Match **List-I (Meter)** with **List-II (Type)** and answer using the select the correct code given below the lists:

List-I	List-II
A. Reed frequency meter	1. Moving iron
B. Weston frequency meter	2. Vibrating
C. Weston Synchroscope	3. Moving coil
D. Ohm meter	4. Electro-dynamic

Codes:

	A	B	C	D
(a)	2	1	4	3
(b)	3	1	4	2
(c)	2	3	4	1
(d)	3	4	1	2

[ESE-2010]

1.44 Consider the following statements about hot-wire instruments:

1. They read equally well on dc and/or ac circuits.
2. They are simple and robust in construction and power consumption is low.
3. They are quite suitable for measurement of currents at very high frequencies.

Which of these statements are correct?

(a) 1 and 2 only (b) 2 and 3 only
(c) 1 and 3 only (d) 1, 2, 3 and 4

[ESE-2010]

1.45 A resistance of 105 ohms is specified using significant figures as indicated below:

1. 105 ohms
2. 105.0 ohms
3. 0.000105 MΩ

Among these

(a) 1 represents greater precision than 2 and 3.
(b) 2 and 3 represent greater precision than 1.
(c) 1, 2 and 3 represent same precision.
(d) 2 represents greater precision but 1 and 3 represent same precision.

[ESE-2010]

1.46 **Assertion (A)** : Comparison methods of direct measurements are most widely used in electrical engineering practice.

Reason (R) : Comparison methods of direct measurements give high accuracy.

(a) Both A and R are true and R is the correct explanation of A
(b) Both A and R are true but R is NOT the correct explanation of A
(c) A is true but R is false
(d) A is false but R is true

[ESE-2010]

1.47 A (0 – 250) V voltmeter has a guaranteed accuracy of 2 percent of full scale reading. The voltage measured by this voltmeter is 150 volts. The limiting error in percentage for this instrument is

(a) 2.5% (b) 0.05%
(c) 3.33% (d) 5.0%

[ESE-2010]

1.48 The pressure in a tank varies from 20 psi to 100 psi. Further pressure in the tank is desired to be kept at 50 psi. The full scale error, when pressure inside the tank is 30 psi, is given by

(a) –62.5% (b) 25%
(c) 80% (d) –2.5%

[ESE-2010]

1.49 What are the causes of gross error in the instruments?

1. Misreading of instruments.
2. Incorrect adjustment of instruments.
3. Errors due to defective instrument.
4. Errors due to effect of environment on the instrument.

(a) 1 and 2 (b) 2 and 3
(c) 3 and 1 (d) 4 and 1

[ESE-2011]

1.50 The measured value of a capacitor is 205.5 μF; whereas its true value is 202.4 μF. The relative error is

(a) 1.87% (b) 1.94%
(c) 1.53% (d) 1.73%

[ESE-2011]

1.51 Which of the following types of errors come under systematic errors?

1. Irregular spring tension.
2. Improper readings of an instrument.
3. Loading effects.
4. Error due to the presence of electric field or magnetic field.

(a) 1 and 2 (b) 2 and 3
(c) 3 and 1 (d) 4 and 1

[ESE-2011]

1.52 The value of a quantity and its uncertainty are given as 26455 ± 3754 without rounding off. Only two significant digits are relevant for error. Value of error rounded off to two significant figures is

(a) 26500 ± 3800 (b) 26400 ± 3800
(c) 26460 ± 3750 (d) 26400 ± 3700

[ESE-2011]

1.53 Match **List-I** (Excitation) with **List-II** (Two-port parameters) and select the correct answer using the code given below the lists:

List-I

A. Moving iron and hot wire type
B. Galvanometer
C. PMMC type
D. Electrostatic type

List-II

1. Air friction
2. Electromagnetic
3. Fluid friction
4. Eddy current

Codes:

	A	B	C	D
(a)	1	4	2	3
(b)	3	4	2	1
(c)	1	2	4	3
(d)	3	1	2	4

[ESE-2011]

1.54 Match **List-I** with **List-II** and select the correct answer using the code given below the lists:

List-I

A. Average value of current
B. Frequency of a waveform
C. Strain-gauge resistance

List-II

1. Self-balancing bridge
2. Wein's bridge
3. PMMC ammeter

Codes:

	A	B	C
(a)	2	1	3
(b)	3	1	2
(c)	1	2	3
(d)	3	2	1

[ESE-2011]

1.55 **Assertion (A):** It is always desirable to take measurements as close to the full-scale as possible.

Reason (R): The magnitude of the limiting error is a fixed quantity based on the full-scale reading of the meter and error increases as reading decreases.

(a) Both A and R are true and R is the correct explanation of A
(b) Both A and R are true but R is NOT the correct explanation of A
(c) A is true but R is false
(d) A is false but R is true

[ESE-2011]

1.56 A 0-100 V voltmeter has an accuracy of 1 percent at full-scale reading. What will be the error if it reads 50 V?

(a) 1 percent (b) 2 percent
(c) 0.5 percent (d) 4 percent

[ESE-2011]

1.57 If a high frequency AC signal, whose rms value is $\sqrt{2}$ V, is applied to a PMMC instrument, then the reading of the instrument will be

(a) 2 V (b) $\sqrt{2}$ V
(c) 1 V (d) zero

[ESE-2012]

1.58 The following is not essential for the working of an indicating instrument

(a) deflecting torque (b) braking torque
(c) damping torque (d) controlling torques

[ESE-2012]

1.59 Systematic error of an instrument for measurement can be minimized by

(a) selecting a proper measuring device for the particular application.
(b) calibrating the measuring device against a standard device.
(c) applying correction factors for change of ambient conditions.
(d) carrying out all of the above.

[ESE-2012]

1.60 Two resistances $R_1 = 100 \pm 10\%\ \Omega$ and $R_2 = 300 \pm 5\%\ \Omega$ are connected in series. The resulting limiting error of the series combination is

(a) 5 Ω (b) 15 Ω
(c) 25 Ω (d) 30 Ω

[ESE-2012]

1.61 **Statement (I):** A hot-wire instrument gives the rms value of the current measured.

Statement (II): The heat generated is dependent on the average value of the current.

(a) Both Statement (I) and Statement (II) are individually true and Statement (II) is the correct explanation of Statement (I).
(b) Both Statement (I) and Statement (II) are individually true but Statement (II) is not the correct explanation of Statement (I).
(c) Statement (I) is true but Statement (II) is false.
(d) Statement (I) is false but Statement (II) is true.

[ESE-2012]

1.62 Consider the following statements:
The causes of error in the measurement of temperature using a thermistor are

1. Self heating
2. Poor sensitivity
3. Non-linear characteristics

Which of these statements are correct?
(a) 1, 2 and 3 (b) 1 and 2 only
(c) 2 and 3 only (d) 1 and 3 only

[ESE-2013]

1.63 Five observers have taken a set of independent voltage measurements and recorded as 110.10 V, 110.20 V, 110.15 V, 110.30 V and 110.25 V. Under the situation mentioned above, the range of error is
(a) ±0.3 (b) ±0.1
(c) ±0.2 (d) ±1.0 [ESE-2013]

1.64 The technique used to check quantitatively whether the given data distribution is close to Gaussian distribution is
(a) curve fitting
(b) method of least squares
(c) Chi-square test
(d) standard deviation of mean [ESE-2013]

1.65 The unknown resistance R_4 measured in a Wheatstone bridge by the formula $R_4 = \frac{(R_2 R_3)}{R_1}$

with
$R_1 = 100 \pm 0.5\%\ \Omega$,
$R_2 = 1000 \pm 0.5\%\ \Omega$,
$R_3 = 842 \pm 0.5\%\ \Omega$
resulting in R_4
(a) $8420 \pm 0.5\%\ \Omega$ (b) $8420 \pm 1.0\%\ \Omega$
(c) $8420 \pm 1.5\%\ \Omega$ (d) $8420 \pm 0.125\%\ \Omega$

[ESE-2013]

1.66 A current $i = 5 + 14.14 \sin(314t + 45°)$ is passed through a centre-zero PMMC, hot-wire, and moving iron instrument, the respective readings are
(a) -5, 15 and $\sqrt{125}$
(b) 5, $\sqrt{125}$ and $\sqrt{125}$
(c) -5, $\sqrt{125}$ and 19.14
(d) 5, 10 and 10 [ESE-2013]

1.67 A frequency counter needs to measure a frequency of 15 Hz. Its signal geting time is 2s. What is the percentage accuracy of the counter, taking into account the getting error?
(a) 3.33% (b) 13.33%
(c) 98.67% (d) 96.67%

[ESE-2013]

1.68 Which of the following instrument will be used to measure a small current of very high frequency?
(a) Electrodynamic ammeter
(b) Moving coil galvanometer
(c) Thermocouple type instrument
(d) Induction type instrument

[ESE-2013]

1.69 **Statement (I):** For random error with normal distribution, probable error = ±0.6745 σ, where σ is the standard deviation.
Statement (II): Probable error ξ_p is that error value where there is a 50% chance that any observation has a random error no greater than $\pm\xi_p$.
(a) Both Statement (I) and Statement (II) are individually true and Statement (II) is the correct explanation of Statement (I).
(b) Both Statement (I) and Statement (II) are individually true but Statement (II) is not the correct explanation of Statement (I).
(c) Statement (I) is true but Statement (II) is false.
(d) Statement (I) is false but Statement (II) is true.

[ESE-2013]

1.70 What is the number of turns of wire needed to provide a potentiometer with a resolution of 0.05 percent?
(a) 200 turns (b) 2000 turns
(c) 20 turns (d) 20000 turns

[ESE-2014]

1.71 Four ammeters M_1, M_2, M_3 and M_4 with the following specifications are available. (Full scale, accuracy value as percentage of *FS*)
$M_1 = 20 \pm 0.10$; $M_2 = 10 \pm 0.20$; $M_3 = 5 \pm 0.50$; and $M_4 = 1 \pm 1.00$.
A current of 1 A is to be measured. To obtain minimum error in the reading one should select meter
(a) M_1 (b) M_2
(c) M_3 (d) M_4 [ESE-2014]

1.72 The reliability of an instrument refers to
(a) the measurement of changes due to temperature variation.
(b) the degree to which repeatability continues to remain within specified limits.
(c) the life of an instrument.
(d) the extent to which the characteristics remain linear. [ESE-2015]

1.73 The degree to which an instrument indicates the changes in measured variable without dynamic error is
(a) repeatability (b) hysteresis
(c) precision (d) fidelity [ESE-2016]

1.74 Loading by the measuring instruments introduces an error in the measured parameter. Which of the following devices gives the most accurate result?
(a) PMMC (b) Hot-wire
(c) CRO (d) Electrodynamic [ESE-2016]

1.75 Which of the following statements are correct?
1. Accuracy is the closeness with which an instrument approaches the true value of the quantity being measured.
2. Precision is a measure of the reproducibility of the measurement.
3. Precision of an instrument can be improved upon by calibration.
4. Accuracy may be specified in terms of limits of errors.

(a) 1, 2 and 3 only (b) 1, 3 and 4 only
(c) 1, 2 and 4 only (d) 2, 3 and 4 only [ESE-2020]

1.76 A 0-150 V voltmeter has a guaranteed accuracy of 1% of full scale reading. The voltage measured by this instrument is 75 V. What is the percentage of limiting error?
(a) 1% (b) 2%
(c) 3% (d) 4% [ESE-2021]

1.77 Consider a measuring system consisting of a sensor, an amplifier and an oscilloscope. The sensitivity of each equipment is as follows : Sensor sensitivity : 0.4 mV/°C, Amplifier gain : 5.0 V/mV, and Oscilloscope sensitivity : 10 mV/V. What is the sensitivity of complete measurement system?
(a) 5.0 V/mV (b) 10 mV/V
(c) 15.4 mV/°C (d) 20 mV/°C [ESE-2021]

1.78 In estimating the cost of a pile of bricks measured as 2 m × 15 m × 1.2 m, the tape is stretched 1% beyond the standard length. If the count is 450 bricks to 1 cu.m. and bricks cost Rs. 530 per 1000, what is the approximate error in the cost?
(a) Rs. 257.58 (b) Rs. 152.65
(c) Rs. 345.41 (d) Rs. 329.49 [ESE-2021]

1.79 Consider a parallel circuit having three branches the current in first branch is 50 ± 2 A, in the second branch is 100 ± 3 A and in the third branch is 200 ± 5 A. What is the value of the total current, assuming the errors as standard deviations?
(a) 350 ± 10 A (b) 350 ± 3.16 A
(c) 350 ± 6.16 A (d) 350 ± 5 A [ESE-2022]

1.80 A set of independent current measurements is recorded as 10.03 A, 10.10 A, 10.11 A and 10.08 A. What are the values of average current and range of error respectively?
(a) 10.08 A and ±0.03 A
(b) 10.08 A and ±0.04 A
(c) 10.09 A and ±0.04 A
(d) 10.09 A and ±0.03 A [ESE-2024]

1.81 During the measurement of low resistance using a potentiometer, the following readings were obtained :
The voltage drop across the low resistance under test = 0.4221 V.
The voltage drop across a 0.1 Ω standard resistance = 1.0235 V.
What are the values of unknown resistance and current respectively?
(a) 0.041208 Ω and ±10.235 A
(b) 0.031208 Ω and ±10.235 A
(c) 0.021208 Ω and ±10.235 A
(d) 0.041208 Ω and ±11.235 A [ESE-2024]

■■■■

Answers Characteristics of Instruments and Measurement Systems

1.1 (c)	1.2 (b)	1.3 (d)	1.4 (c)	1.5 (b)	1.6 (c)	1.7 (c)	1.8 (b)	1.9 (d)
1.10 (b)	1.11 (a)	1.12 (b)	1.13 (b)	1.14 (b)	1.15 (c)	1.16 (d)	1.17 (a)	1.18 (a)
1.19 (c)	1.20 (c)	1.21 (a)	1.22 (a)	1.23 (c)	1.24 (b)	1.25 (d)	1.26 (a)	1.27 (a)
1.28 (d)	1.29 (b)	1.30 (b)	1.31 (d)	1.32 (b)	1.33 (b)	1.34 (d)	1.35 (d)	1.36 (a)
1.37 (c)	1.38 (a)	1.39 (a)	1.40 (b, c)	1.41 (c)	1.42 (a)	1.43 (a)	1.44 (c)	1.45 (d)
1.46 (a)	1.47 (c)	1.48 (b)	1.49 (a)	1.50 (c)	1.51 (c)	1.52 (a)	1.53 (c)	1.54 (d)
1.55 (a)	1.56 (b)	1.57 (d)	1.58 (b)	1.59 (d)	1.60 (c)	1.61 (c)	1.62 (d)	1.63 (b)
1.64 (d)	1.65 (c)	1.66 (b)	1.67 (d)	1.68 (c)	1.69 (a)	1.70 (b)	1.71 (d)	1.72 (b)
1.73 (d)	1.74 (c)	1.75 (c)	1.76 (b)	1.77 (d)	1.78 (a)	1.79 (c)	1.80 (b)	1.81 (a)

Explanations Characteristics of Instruments and Measurement Systems

1.3 (d)

- Thermo-couple instrument has a "square-law response" so it does not have a linear scale.
- Their accuracy can be as high as 1% for frequencies upto 50 MHz and can be classified as RF instruments. Above this frequency the effective resistance of heater wire is increased on account of skin effect, thereby reducing the accuracy. Tubular designs are used for heater wire above 3 A to reduce errors on account of skin effect produced by high frequencies. For small currents upto 3 A, the heater wire is solid and very thin.
- At normal rated current, the heater attains a temperature of 300° C. If we pass twice the rated current, the heater would give a temperature of nearly 4 times the normal temperature i.e., 1200° C. It is obvious that square law rate will bring the heater to nearly its burn out temperature. Thus it is imperative that the heater be protected against damaging over loads. The fuses do not provide any protection as due to overload the heater wire may burn out before fuse blows out. Thus the overload capacity of thermo-couple instruments is small as compared with other instruments and is about 150 percent of the full scale current.

1.4 (c)

For full-wave rectifier type of instruments

$V_{avg} = 0.9\ V$ where $V = V_m/\sqrt{2}$.

1.5 (b)

D'Arsonval ammeter is PMMC Ammeter which measures average value.

1.6 (c)

$$S = \frac{\theta}{I_{FS}}$$

1.7 (c)

Induction principle is more generally used for watt-hour meters than for ammeters and voltmeters owing to their comparatively high cost, and inaccuracy of induction instruments of the later types and PMMC is most accurate.

1.8 (b)

$$\sigma_I = \sqrt{\left(\frac{\partial I}{\partial I_1}\right)^2 \sigma_{I_1}^2 + \left(\frac{\partial I}{\partial I_2}\right)^2 \sigma_{I_2}^2}$$

$$\frac{\partial I}{\partial I_1} = \frac{\partial I}{\partial I_2} = 1$$

$$\therefore \quad \sigma_I = \sqrt{(1)^2(1)^2 + (1)^2(2)^2}$$

$$= 2.24\ A$$

$$\therefore \quad I = (400 \pm 2.24)\ A$$

1.9 (d)
Torque/weight ratio of instrument simply represents sensitivity.

1.10 (b)
First order,

$$\text{T.F.} = \frac{K}{1+sT}$$

K = Static sensitivity
T = Time constant

1.21 (a)
Magnitude of limiting error of instrument is

$$= \frac{1}{100} \times 200$$
$$= 2\text{ V}$$

$$\therefore \text{ relative limiting error} = \frac{2}{50} \times 100 = 4\%$$

1.22 (a)
Average current

$$= \frac{117.02 + 117.08 + 117.11 + 117.03}{4}$$
$$= 117.06\text{ mA}$$

as $I_{max} = 117.11$ mA
and $I_{min} = 117.02$ mA

∴ Range of error

$$= \pm \frac{(I_{max} - I_{av}) + (I_{av} - I_{min})}{2}$$
$$= \pm \frac{0.05 + 0.04}{2}$$
$$= \pm 0.045\text{ mA}$$

Alternate solution:

$$\text{Range of error} = \frac{\text{Max. value} - \text{Min. value}}{2}$$
$$= \pm \frac{117.11 - 117.02}{2}$$
$$= \pm 0.045$$

1.25 (d)

$$\text{Magnitude of error} = \frac{1}{100} \times 100 = 1\text{ V}$$

$$\%\text{ of error} = \frac{1}{5} \times 100 = \pm 20\%$$

1.27 (a)

$$\text{Meter reading} = 1 \times \text{average value}$$
$$= 1 \times 100$$
$$= 100\text{ V}$$

1.28 (d)
Accuracy is defined as 'percentage of true value'.

1.29 (b)
Current transformer – Phase angle error
Energy meter – Braking system error

1.32 (b)
Watt Hour meter is integrating type instrument. Rayleigh Current balance is absolute type instrument as it measures current in terms of physical quantities.

1.33 (b)
PMMC has linear scale, Rectified type instruments are calibrated for sinusoidal type input. MI type instrument has square law scale.

1.34 (d)
Precision is not guarantee for accuracy.

1.37 (c)
Reading of ac ammeter = rms valve

$$\text{rms value} = I_{max}\sqrt{\alpha}$$

where, α = duty cycle

$$\alpha = \frac{1}{2}$$

$$\text{rms valve} = 100\text{ mA}\sqrt{\frac{1}{2}} \cong 71\text{ mA}$$

1.39 (a)

$$\text{Error} = \frac{2}{100} \times \text{FSD}$$
$$= \frac{2}{100} \times 300 = 6\text{ V}$$

∴ Range for 30 V = 30 ± 6 = 24 V – 36 V

1.40 (b and c)
2, 3, 4 comes under systematic error
While first one comes under gross error.

1.41 (c)
Random errors or uncertainties are computed using statistical methods. Random errors indicates precision of instrument.

1.44 (c)
The hot wire instrument have many disadvantages like high power consumption, instability due to stretching of wire. That is why these instruments are obsolete now.

1.47 (c)

$$\text{error} = \frac{250 \times 2}{100} = 5$$

$$\text{Limiting error \%} = \frac{5}{150} \times 100\%$$

$$\in_r\% = 3.33\%$$

1.49 (a)
Gross errors includes human mistakes in reading instruments and recording and calculating measurement results.
These errors can be avoided by
(*i*) taking great care in reading and recording the data.
(*ii*) taking more readings of the quantity under measurement.
Hence, option (a) is correct.

1.50 (c)

$$\text{Relative error} = \frac{\text{Measured value} - \text{True value}}{\text{True value}} \times 100$$

$$= \frac{205.5 - 202.4}{202.4} \times 100 = 1.53\%$$

Hence, option (c) is correct.

1.51 (c)
Systematic errors are divided into three categories:
1. Instrumental errors:
   These errors occur due to
   (a) Instrument short comings in the instrument.
   (b) Misuse of the instruments.
   (c) Loading effects of instruments.
2. Environmental errors.
3. Observational errors.

Hence, option (c) is correct.

1.52 (a)
Significant figures convey actual information regarding the magnitude and the measurement precision of a quantity.
Hence, option (a) is correct.

1.53 (c)
Air friction does not require the use of permanent magnet whose introduction may lead to distortion of operating field. This damping is used where operating field is weak.
eg.: Moving iron, dynamometer, hot wire.
In fluid friction, oil which is required for damping can be used for insulation purpose. So it is suitable for high voltage instruments or electrostatic type.

1.54 (d)
PMMC ammeter is suitable for only dc applications and gives average value of current. Hence, option (d) is correct.

1.55 (a)
While selecting instruments, particular case should be taken as regards the range. The values to be measured should not lie in the lower third of the range. This is particularly important if the meter accuracy is specified in terms of the full scale deflection. Hence, option (a) is correct.

1.56 (b)
Magnitude of limiting error = 0.01 × 100 = 1 V

$$\% \text{ error} = \frac{1}{50} \times 100 = 2\%$$

Hence, option (b) is correct.

1.57 (d)
PMMC instruments are used only for measuring DC quantity.

1.58 (b)
Three types of forces are needed for the satisfactory operation of any indicating instrument. These are
1. Deflecting force
2. Controlling force
3. Damping force

1.59 (d)
Systematic errors are:
1. Instrument errors
2. Environmental errors
3. Observational errors

Instrumental errors arise due to
1. Inherent shortcomings in the instrument.
2. Misuse of the instruments.
3. Loading effects of instruments.

1.60 (c)

$$R = R_1 + R_2$$
$$= 100 + 300 = 400$$

Limiting errors $\frac{\delta R}{R} = \frac{R_1}{R}\cdot\frac{\delta R_1}{R_1} + \frac{R_2}{R}\cdot\frac{\delta R_2}{R_2}$

$$= \frac{100}{400}\cdot\frac{10}{100} + \frac{300}{400}\cdot\frac{5}{100} = \frac{25}{400}$$

$$\therefore \quad \delta R = \frac{25}{400}\times 400 = 25\ \Omega$$

Alternate solution:

$$R_1 = 100 \pm 10\% = (100 \pm 10)\ \Omega$$
$$R_2 = 300 \pm 5\% = (300 \pm 15)\ \Omega$$
$$R_S = R_1 + R_2 = 400 \pm 25$$
so, $\Delta R = 25\ \Omega$

1.61 (c)

The expansion in the wire is proportional to the heating effect of current and to the square of the rms value of the current. Therefore, the meter may be calibrated to read the rms value of the current.

1.62 (d)

Thermistors are highly sensitive, non linear elements i.e.

$$R_{T_1} = R_{T_2}\, e^{\beta\left(\frac{1}{T_1}-\frac{1}{T_2}\right)}$$

1.63 (b)

Readings are: 110.10 V
110.20 V
110.15 V
110.30 V
110.25 V

i.e. average reading is 110.20 V

range of error

$$= \pm\left|\text{extreme reading} - \text{average reading}\right| = \pm 0.10$$

Alternate solution:

$$\text{Range of error} = \frac{\text{Max. value} - \text{Min. value}}{2}$$
$$= \pm\frac{110.30 - 110.10}{2} = \pm 0.1$$

1.65 (c)

$$\%\frac{\Delta R_4}{R_4} = \%\frac{\Delta R_2}{R_2} + \%\frac{\Delta R_3}{R_3} + \%\frac{\Delta R_1}{R_1}$$
$$= 0.5\% + 0.5\% + 0.5\% = 1.5\%$$

i.e. option (c) is correct.

1.66 (b)

$$i = 5 + 10\sqrt{2}\sin(314t + 45°)$$

PMMC will measure: $i_{avg} = 5$

Hot-wire and moving iron instrument will measure:

$$i_{rms} = \sqrt{5^2 + \left(\frac{10\sqrt{2}}{\sqrt{2}}\right)^2} = \sqrt{125}$$

1.67 (d)

$$\text{Percentage accuracy} = \frac{15 - \frac{1}{2}}{15} = 96.67\%$$

1.68 (c)

At very high frequency, thermocouple type instruments are used.

1.70 (b)

Number of turns required

$$= \frac{1}{\text{resolution}}$$
$$= \frac{1}{\left(\frac{0.05}{100}\right)} = \frac{100\times 10^2}{5}$$
$$= 2000 \text{ turns}$$

1.71 (d)

Measured value should be nearer to full scale value of ammeters.

1.72 (b)

The reliability of an instrument refers to the degree to which repeatability continues to remain within specified limits.

1.75 (c)

Accuracy is improved by calibration.

1.76 (b)

Absolute error at full scale = Absolute error at reading value

$$= \frac{1\times 150}{100} = 1.5\ \text{V}$$

Limiting error at reading value

$$= \frac{\text{Absolute error}\times 100}{\text{Reading value}}$$
$$= \frac{1.5\times 100}{75} = 2\%$$

OR

Limiting error at reading value (75 V)

$$= \frac{\text{Full scale value} \times \text{Limiting error at full scale}}{\text{Reading value}}$$

$$\therefore \quad \% \in_r \text{ at } 75 \text{ V} = \frac{150 \times 1}{75} = 2\%$$

1.77 (d)

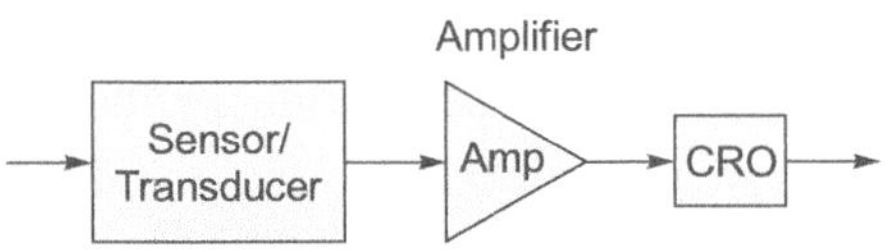

$S_1 = 0.4$ mV/° C
$S_2 = 5$ V/mV
$S_3 = 10$ mV/V

For cascade connection,

Overall sensitivity $= S_1 \times S_2 \times S_3$
$= 0.4 \times 5 \times 10 = 20$ mV/° C

1.78 (a)

For initial cost,

1 cubic meter= 450 brick

$2 \times 15 \times 1.2 = 30 \times 1.2 = 36$ cubic meter

36 cubic meter contains brick $= 450 \times 36 = 16200$

$$\text{Cost of 16200 bricks} = \frac{530}{1000} \times 16200$$
$$= 53 \times 162 = 8586$$

2nd volume $= 2 \times 15 \times 1.2 \times 1.01^3$
$= 37.08$ cubic meter

Extra volume = 1.08 Cubic meter

Bricks in extra area $= 450 \times 1.08 \approx 486$

$$\text{Extra cost} = \frac{530}{1000} \times 486 \approx 257.58$$

1.79 (c)

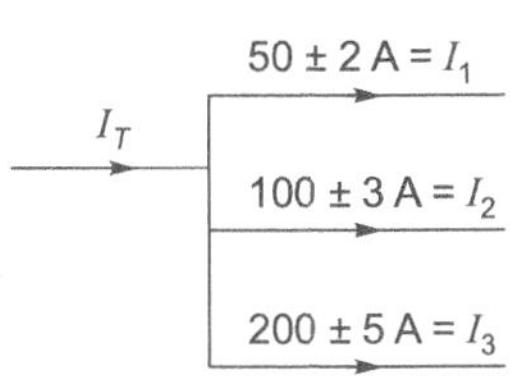

$$I_T = I_1 + I_2 + I_3$$

$$\frac{\delta I_T}{\delta I_1} = 1 = \frac{\delta I_T}{\delta I_2} = \frac{\delta I_T}{\delta I_3}$$

$$\sigma_{IT} = \sqrt{\left(\frac{\delta I_T}{\delta I_1}\right)^2 \sigma_1^2 + \left(\frac{\partial I_T}{\partial I_2}\right)^2 \sigma_2^2 + \left(\frac{\delta I_T}{\delta I_3}\right)^2 \sigma_3^2}$$

$$= \sqrt{1^2 \times 2^2 + 1^2 \times 3^2 + 1^2 \times 5^2}$$

$= 6.16$ A

$I_T = (50 + 100 + 200) \pm 6.16$
$= 350 \pm 6.16$ A

1.80 (b)

Given readings

$\Rightarrow$ 10.03 A, 10.10 A, 10.11 A, 10.08 A

$$I_{avg} = \frac{10.03 + 10.10 + 10.11 + 10.08}{4}$$

$$= \frac{40.32}{4} = 10.8 \text{ Amp}$$

$$\text{Range of error} = \frac{I_{max} - I_{min}}{2}$$

$$= \frac{10.11 - 10.03}{2} = \pm 0.04$$

1.81 (a)

Let, R = unknown resist

V.D. across low resistance

$\Rightarrow$ $E_1 = 0.4221$ volt

V.D. across standard resistance

$\Rightarrow$ $E' = 1.0235$ volt

Let, S = standard resistance

$\Rightarrow$ $S = 0.1\ \Omega$

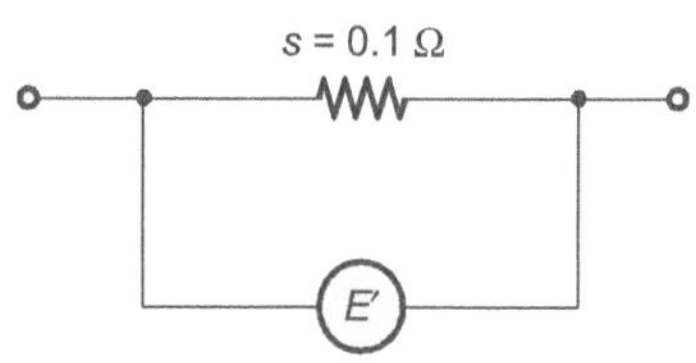

$$E_1 = \left(\frac{E'}{I'}\right) \times I_1$$

$$0.4221 = \frac{1.0235}{0.1} \times R$$

$\Rightarrow$ $R = 0.04124\ \Omega$

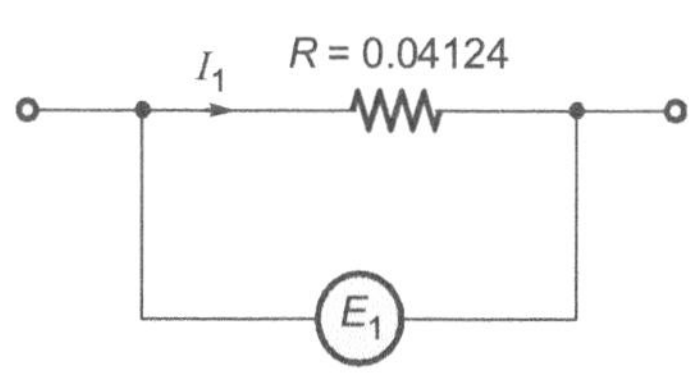

$$I_1 = \frac{E_1}{R} = \frac{0.4221}{0.04124} = 10.235 \text{ Amp}$$

■■■■

2 Measurement of Resistance and Potentiometers

2.1 Consider the following statements:
The value of earth resistance depends upon
1. shape of electrode
2. depth to which the electrode is driven into earth
3. specific resistance of soil
4. material of electrode

Which of the following statements are correct?
(a) 1,2,3 and 4 (b) 2,3 and 4
(c) 1 and 2 (d) 1,3 and 4 **[ESE-2001]**

2.2 **Assertion (A):** A resistance potentiometers used for measurement of displacement, sensitivity and linearity are conflicting requirements.
Reason (R): The voltmeter used for measurement of output voltage of the potentiometer has a finite resistance which causes loading effects.
(a) Both A and R are true and R is the correct explanation of A
(b) Both A and R are true but R is NOT the correct explanation of A
(c) A is true but R is false
(d) A is false but R is true **[ESE-2001]**

2.3 A dc circuit can be represented by an internal voltage source of 50 V with an output resistance of 100 kΩ. In order to achieve accuracy better than 99% for voltage measurement across its terminals, the voltage measuring device should have a resistance of at least
(a) 10 MΩ (b) 1 MΩ
(c) 10 kΩ (d) 1 kΩ **[ESE-2001]**

2.4 The accuracy of Kelvin's double bridge for the measurement of low resistance is high because the bridge
(a) uses two pairs of resistance arms.
(b) has medium value resistance in the ratio arms.
(c) uses a low resistance link between standard and test resistances.
(d) uses a null indicating galvanometer. **[ESE-2002]**

2.5 Which one of the following statements is NOT true?
(a) Potentiometric linear displacement can give high output signal.
(b) Linear variable differential transformer has low output impedance.
(c) Synchros and resolvers have low accuracy.
(d) Eddy current proximity transducers are non-contact type transducers. **[ESE-2002]**

2.6 Wire-wound resistors are unsuitable for use at high frequencies because
(a) They are likely to melt under excessive eddy current.
(b) They exhibit unwanted inductive and capacitive effects.
(c) They create more electrical noise.
(d) They consume more power. **[ESE-2003]**

2.7 A standard resistance is made 'Bifilar' type to eliminate
(a) Stray capacitance (b) Temperature effect
(c) Inductive effect (d) Skin effect **[ESE-2003]**

2.8 A resistor R is measured using the V-I method, with V read as 10.14 V and I as 5.07 mA. Which one of the following expresses the value of resistance?
(a) 2 kΩ (b) 2.00 kΩ
(c) 2000 Ω (d) 2.0 kΩ **[ESE-2004]**

2.9 Which one of the following is measured by the loss of charge method ?
(a) Low R (b) High R
(c) Low L (d) High L **[ESE-2004]**

2.10 Low resistance from few ohms down to one micro ohm is measured using which one of the following instruments?

(a) Ohmmeter
(b) A series type ohmmeter
(c) A shunt type ohmmeter
(d) A voltmeter and an ammeter [ESE-2004]

2.11 A single slide wire is used for the measurement of current in a circuit. The voltage drop across a standard resistance of 1.0 Ω is balanced at 70 cm. What is the magnitude of the current, if the standard cell having an e.m.f of 1.45 volts is balanced at 50 cm?
(a) 3.09 A (b) 2.65 A
(c) 2.03 A (d) 1.45 A [ESE-2006]

2.12 The materials to be used in the manufacture of a standard resistance should be of
(a) high resistivity and low temperature coefficient
(b) low resistivity
(c) high temperature coefficient
(d) low resistivity and high temperature coefficient [ESE-2006]

2.13 Which one of the following techniques reduces the residual inductance of standard resistance?
(a) Using high resistivity material
(b) Using material of low temperature coefficient of resistance
(c) Using proper shielding
(d) Making a bifilar winding on a card [ESE-2008]

2.14 For low resistance (from few micro ohms to one ohm) measurement, which bridge is used?
(a) Wheatstone bridge
(b) Kelvin bridge
(c) Guarded Wheatstone bridge
(d) Maxwell bridge [ESE-2008]

2.15 Which one of the following is not the criterion used to select potentiometer in a control system?
(a) Accuracy (b) Noise
(c) Time response (d) Frequency response [ESE-2009]

2.16 Which one of the following bridges will be used for the measurement of very low resistance?
(a) Kelvin bridge
(b) Maxwell's bridge
(c) Wheatstone bridge
(d) Hay's bridge [ESE-2009]

2.17 A RC potentiometer to measure ac voltage, it is desired that V_0/V_i should be independent of frequency. The value of C should be

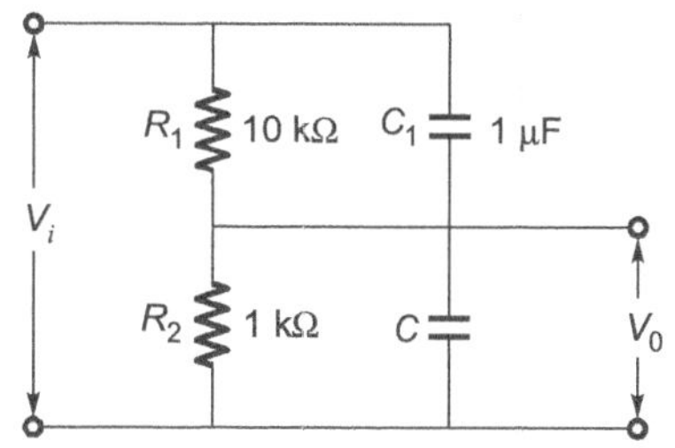

(a) 10 μF (b) 11 μF
(c) 0.1 μF (d) 0.09 μF [ESE-2011]

2.18 A resistance of 108 Ω is specified using significant figures as indicated below:
1. 108 Ω
2. 108.0 Ω
3. 0.000108 MΩ

Among these:
(a) 1 represents greater precision than 2 and 3
(b) 2 represents greater precision but 1 and 3 represents same precision
(c) 2 and 3 represent greater precision than 1
(d) 1, 2 and 3 represent the same precision [ESE-2011]

2.19 The preferred methods of measuring low resistance and the resistance of cable insulation are respectively
(a) *V/I* method and loss-of-charge method.
(b) Kelvin's double-bridge and Megger test.
(c) Wheatstone bridge and Kelvin's double bridge.
(d) potentiometer method and Wheatstone bridge. [ESE-2012]

2.20 In measuring resistance by voltmeter-ammeter method, the voltmeter can be connected either across supply or across the resistance. If the resistance is low, the voltmeter should be connected.
(a) across the supply
(b) across the resistance
(c) either across the supply or across the resistance
(d) neither across the supply nor across the resistance [ESE-2015]

Directions:

Each of the next items consists of two statements, one labelled as the **'Statement (I)'** and the other as

'Statement (II)'. Examine these two statements carefully and select the answers to these items using the codes given below:

Codes:

(a) Both Statement (I) and Statement (II) are individually true and Statement (II) is the correct explanation of Statement (I).

(b) Both Statement (I) and Statement (II) are individually true but Statement (II) is not the correct explanation of Statement (I).

(c) Statement (I) is true but Statement (II) is false.

(d) Statement (I) is false but Statement (II) is true.

2.21 **Statement (I) :** Ammeter and voltmeter method is used for measurement of low as well as medium resistances.

Statement (II) : Carey-Foster slide wire bridge is a modification of the Wheatstone bridge. [ESE-2015]

2.22 **Statement (I) :** In the Kelvin double-bridge method, provision has been made to eliminate the errors due to contact and lead resistances.

Statement (II) : The Schering bridge is used for measuring small capacitance at low voltages with very high precision. [ESE-2015]

2.23 The precision resistors are

(a) carbon composition resistors
(b) wire-wound resistors
(c) resistors with a negative temperature coefficient
(d) resistors with a positive temperature coefficient

[ESE-2016]

2.24 During the measurement of resistance by Carey Foster bridge, no error is introduced due to

1. contact resistance
2. connecting leads
3. thermoelectric e.m.f.

Which of the above are correct?

(a) 1 and 2 only (b) 1 and 3 only
(c) 2 and 3 only (d) 1, 2 and 3 [ESE-2016]

2.25 Which of the following methods is used for the measurement of high resistances?

(a) Carey-Foster bridge method
(b) Substitution method
(c) Loss of charge method
(d) Potentiometer method [ESE-2017]

2.26 In the measurement of a low resistance using a potentiometer, voltage drop across the low resistance under test is 0.4 V, voltage drop across 0.1 Ω standard resistance is 1 V. What is the power loss in unknown resistance?

(a) 0.04 W (b) 0.4 W
(c) 4 W (d) 40 W [ESE-2023]

2.27 Consider the following statements :

1. Potentiometer is the usual basis for calibration of voltmeters, ammeters and wattmeters.
2. The potentiometers may not be used for measurement of current, power and resistance.

What of the above statement(s) is/are correct?

(a) 1 only (b) 2 only
(c) Both 1 and 2 (d) Neither 1 nor 2

[ESE-2023]

2.28 A simple slide wire is used for measurement of current in a circuit. The voltage drop across a standard resistor of 0.1 Ω is balanced at 75 cm. What is the magnitude of current in the circuit if the standard cell emf of 1.45 V is balanced at 50 cm?

(a) 21.75 A (b) 10.87 A
(c) 9.66 A (d) 4.83 A [ESE-2023]

2.29 Which one of the following statements is correct regarding potentiometric recorders?

(a) A sensitivity of 4 VB/mm is attained with an error of less than ±0.25% with a bandwidth of 0.8 Hz.
(b) A sensitivity of 10 VB/mm is attained with an error of less than ±0.35% with a bandwidth of 0.9 Hz.
(c) A sensitivity of 5 VB/mm is attained with an error of less than ±0.25% with a bandwidth of 1 Hz.
(d) A sensitivity of 8 VB/mm is attained with an error of less than ±0.45% with a bandwidth of 0.8 Hz.

[ESE-2024]

■■■■

Answers Measurement of Resistance and Potentiometers

2.1 (a)	2.2 (a)	2.3 (a)	2.4 (a)	2.5 (d)	2.6 (b)	2.7 (c)	2.8 (b)	2.9 (b)
2.10 (d)	2.11 (c)	2.12 (a)	2.13 (d)	2.14 (b)	2.15 (b)	2.16 (a)	2.17 (a)	2.18 (b)
2.19 (b)	2.20 (b)	2.21 (b)	2.22 (c)	2.23 (b)	2.24 (d)	2.25 (c)	2.26 (c)	2.27 (a)
2.28 (a)	2.29 (a)							

Explanations Measurement of Resistance and Potentiometers

2.1 (a)

The resistance of earth electrode should be low to give good protection. An earth electrode will only be effective so long it has a low resistance to the earth and can carry large currents without deteriorating.

2.2 (a)

For a high sensitivity, the input voltage should be large and this calls for a high value of resistance of potentiometer R_P. On the other if we consider the linearity the resistance of potentiometer R_P should be as low as possible. The resistance of the potentiometer R_P cannot be made low because if we do so the power dissipation goes up with the result that we have to make the input voltage small to keep the power dissipation to the acceptable level. This results in lower sensitivity. Thus linearity and sensitivity are therefore two conflicting requirements. In fact, for good linearity the ratio R_M/R_P should be as large as possible where R_M is finite input resistance of the voltmeter.

2.3 (a)

$$\frac{E_L}{E_0} = \frac{49.5}{50} = \frac{1}{1+\dfrac{100\text{ k}\Omega}{Z_L}}$$

$$\Rightarrow \quad Z_L \simeq 10\text{ M}\Omega$$

2.4 (a)

Kelvin bridge overcomes the difficulties that arise in a Wheatstone bridge on account of the resistance of the leads and the contact resistances while measuring low valued resistors. The Kelvin double bridge incorporates the idea of a second set of ratio arms hence the name double bridge and the use of four terminal resistors for the low resistance arms. The second set of ratio arms eliminate the effect of connecting lead resistance.

2.7 (c)

A bifilar coil is an electromagnetic coil that has two closely spaced, parallel windings or wire which is made of two filaments or strengths to reduce inductive effect.

2.8 (b)

$$R = \frac{V}{I} = \frac{10.14}{5.07\times10^{-3}} = 2\text{ k}\Omega$$

Note: Option (b) is more significant figure than other reading.
Because given values are upto two significant figures after decimal.

2.11 (c)

Voltage drop per unit length
$= 1.45/50 = 0.029$ V/cm
Voltage drop across 70 cm length
$= 0.029 \times 70 = 2.03$ V

$$\therefore \text{ Current through resistor} = \frac{2.03}{1} = 2.03\text{ A}$$

2.13 (d)

Bifilar winding is used for the purpose.

2.17 (a)

$$\frac{V_o}{V_i} = \frac{R_2 || \dfrac{1}{j\omega C}}{R_1 || \dfrac{1}{j\omega C_1} + R_2 || \dfrac{1}{j\omega C}}$$

Simplifying, the above expression

$$\frac{V_o}{V_i} = \frac{R_2(1+j\omega C_1 R_1)}{(R_1+R_2)+j\omega R_1 R_2(C_1+C)}$$

$$= \frac{R_2(1+j\omega C_1 R_1)}{(R_1+R_2)+j\omega R_1 R_2(C_1+C)} \times \frac{(R_1+R_2)-j\omega R_1 R_2(C_1+C)}{(R_1+R_2)-j\omega R_1 R_2(C_1+C)}$$

Imaginary part of the expression

$$\frac{j\omega[C_1 R_1(R_1+R_2)-R_1 R_2(C_1+C)]R_2}{(R_1+R_2)^2+\omega^2 R_1^2 R_2^2(C_1+C)^2}$$

For V_o/V_i to be independent of frequency, imaginary part should be

$C_1R_1(R_1 + R_2) - R_1R_2(C_1 + C) = 0$

$$C_1 R_1^2 - R_1 R_2 C = 0$$

$$\Rightarrow \quad C = \frac{C_1 R_1}{R_2} = 1 \times \frac{10}{1}\ \mu F = 10\ \mu F$$

Hence, option (a) is correct.

2.18 (b)

1. 108 Ω has 3 significant figures.
2. 108.0 Ω has 4 significant figures.
3. 0.000108 MΩ can be written has 108 Ω.

So, it has 3 significant figures.

The more the significant figures, the greater the precision of measurement.

Hence, option (b) is correct.

2.19 (b)

Methods for measurement of high resistance.

- Direct deflection method
- Loss of charge method
- Mega ohm bridge
- Megger

and for low resistance

- Kelvin's double bridge method
- Ammeter voltmeter method
- Potentiometer method.

2.20 (b)

When the voltmeter connected across the resistance then,

$$\text{error} \approx \frac{R_{load}}{R_V}$$

and, the R_{load} is load, so this method is appropriate.

2.23 (b)

Wire wound resistors are used as precision resistors. They have high power rating and low resistance value.

2.25 (c)

Following methods are used for high resistance measurement:

1. Megger
2. Mega ohms bridge
3. Loss of charge method
4. Direct deflection method

2.26 (c)

Voltage across low resistance = 0.4 V

Let R (R_{AC}) low resistance

and S = Standard resistance

$= 0.1\ \Omega = R_{AB}$

Voltage difference across standard resistance

$= 1\ V = E$

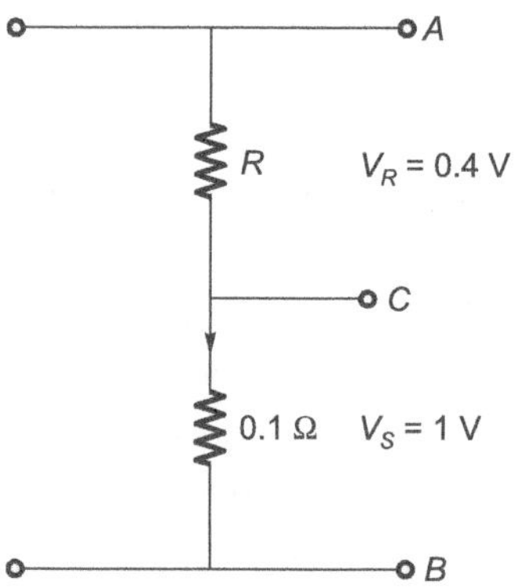

Length ratio = Voltage ratio = Resistance ratio

$$\frac{l_1}{l} = \frac{E_1}{E} = \frac{R_{AC}}{R_{AB}}$$

$$\frac{0.4}{1} = \frac{R_{AC}}{0.1\ \Omega}$$

$R_{AC} = 0.04\ \Omega$ (unknown resistance)

$$I = \frac{E_1}{R_{AC}} = \frac{0.4}{0.04} = 10\ A$$

Power loss in unknown resistance

$= I^2 R_{AC}$

$= (10)^2 \times 0.04$

$= 4\ W$

2.27 (a)

Potentiometers are not used for measurement of power.

2.28 (a)

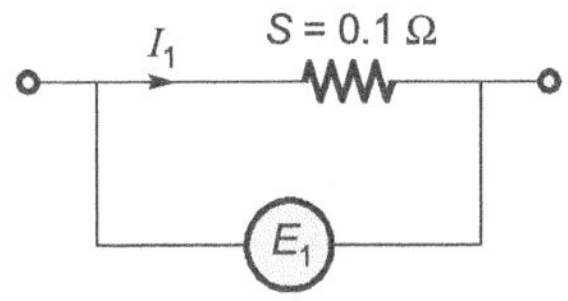

V.D. across standard resistor $\Rightarrow E_1$

Let $S \Rightarrow$ Standard resistance

$$S = 0.1\ \Omega \quad \Rightarrow \quad L_1 = 75\text{ cm}$$

Standard cell E.M.F.

$$\Rightarrow \quad E = 1.45\text{ Volt} \quad \Rightarrow \quad l = 50\text{ cm}$$

$$\frac{E_1}{E} = \frac{l_1}{l}$$

$$\Rightarrow \quad E_1 = \left(\frac{1.45\text{ Volt}}{50\text{ cm}}\right) \times 75\text{ cm}$$

$$= 2.175\text{ Volt} \Rightarrow E_1$$

$$I_1 = \frac{E_1}{S} = \frac{2.175\text{ Volt}}{0.1\ \Omega} = 21.75\text{ Amp}$$

■■■■

3 A.C. Bridges

3.1 Match **List-I** (Bridge) with **List-II** (Parameter to be measured) and select the correct answer:

List-I	List-II
A. Maxwell's bridge	**1.** Frequency
B. Hay's bridge	**2.** Inductance of medium Q-coils $(1 < Q < 10)$
C. Schering bridge	**3.** Inductance of high Q-coils $(Q > 10)$
D. Wien bridge	**4.** Capacitance

Codes:

	A	B	C	D
(a)	4	3	2	1
(b)	4	1	2	3
(c)	2	1	4	3
(d)	2	3	4	1

[ESE-2001]

3.2 Consider the following equations which can be derived from the ac bridges shown in the Fig. 1 and Fig. 2 by assuming $\Delta L/L = 0.1$ and $R = \omega L$:

1. $V_{o1} = V_{o2}$ 2. $V_{o1} = 0.05\ V_s$
3. $V_{o1} = 0.1\ V_s$ 4. $V_{o2} = 0.05\ V_s$
5. $V_{o2} = 0.1\ V_s$

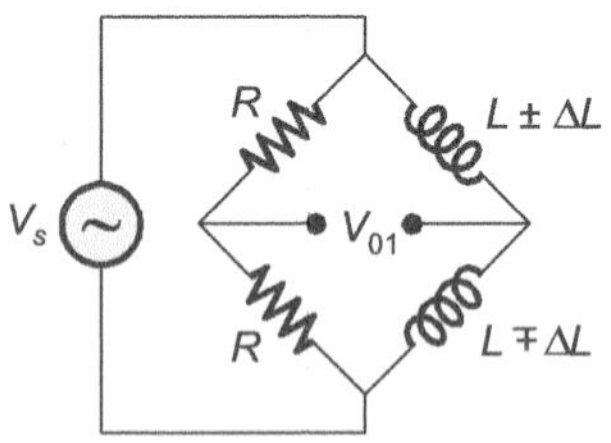

Fig. 1

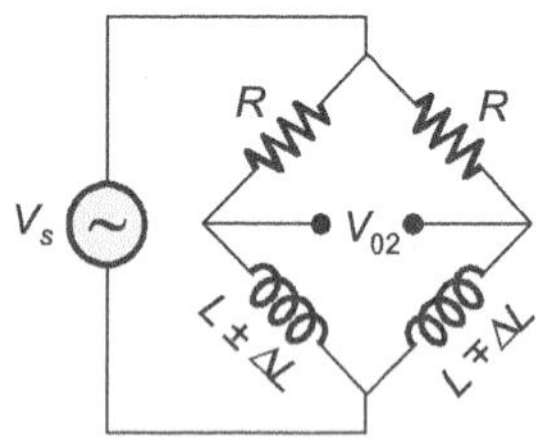

Fig. 2

The correct derived equations from these figures of ac bridges are

(a) 1, 2 and 4 (b) 2, 4 and 5
(c) 2 and 5 (d) 1, 3 and 5

[ESE-2001]

3.3 Match **List-I** (Device/network) with **List-II** (Application) and select the correct answer:

List-I

A. Twin-T network
B. Shielded decade capacitance box selective
C. Wagner earthing arrangement
D. Inter bridge transformer

List-II

1. For use in accurate a.c. bridge
2. For realising frequency amplifier
3. To match impedance and block d.c. noise in a.c. bridge
4. For minimising earth capacitance leakage

Codes:

	A	B	C	D
(a)	2	1	4	3
(b)	3	4	1	2
(c)	2	4	1	3
(d)	3	1	4	2

[ESE-2002]

3.4 For the bridge shown in the given figure,

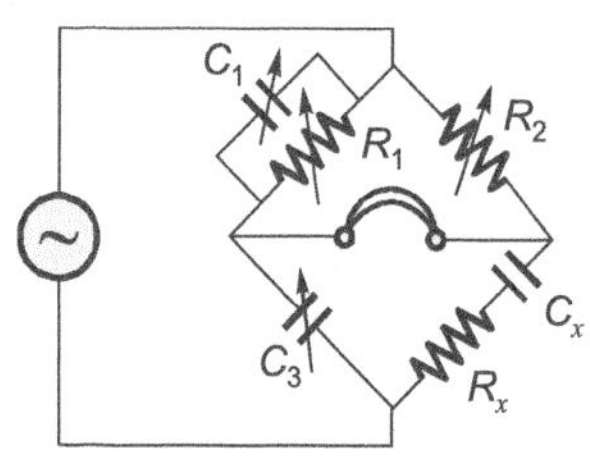

at balance the values of R_x, C_x and Q_x will be

(a) $R_x = C_1R_2/C_3$, $C_x = R_1C_3/R_2$, $Q_x = \omega C_1R_1$
(b) $R_x = C_1R_2/C_3$, $C_x = R_1C_3/R_2$, $Q_x = 1/\omega C_1R_1$
(c) $R_x = C_1R_1/C_3$, $C_x = R_2C_3/R_1$, $Q_x = 1/\omega C_1R_1$
(d) $R_x = C_1R_1/C_3$, $C_x = R_2C_3/R_1$, $Q_x = \omega C_1R_1$

[ESE-2002]

3.5 **Assertion (A):** A four-arm 'Wien' bridge network is sometimes used in feedback circuit of tuned amplifier.

Reason (R) : The balance equation of such a Wein bridge contains 'frequency' term along with arm parameters.

(a) Both A and R are true and R is the correct explanation of A
(b) Both A and R are true but R is NOT the correct explanation of A
(c) A is true but R is false
(d) A is false but R is true [ESE-2002]

3.6 **Assertion (A):** A Schering Bridge used for testing of a porcelain insulator is shielded by a metallic screen.
Reason (R): Earth's magnetic field is blocked by a metallic screen.
(a) Both A and R are true and R is the correct explanation of A
(b) Both A and R are true but R is NOT the correct explanation of A
(c) A is true but R is false
(d) A is false but R is true
[ESE-2003]

3.7 A Wien-bridge is used to measure the frequency of the input signal. However, the input signal has 10% third harmonic distortion. Specifically the signal is $2\sin 400\pi t + 0.2\sin 1200\pi t$ (with t in sec.). With this input the balance will
(a) Lead to a null indication and setting will correspond to a frequency of 200 Hz
(b) Lead to a null indication and setting will correspond to 260 Hz
(c) Lead to a null indication and setting will correspond to 400 Hz
(d) Not lead to null indication
[ESE-2003]

3.8 The capacitance and loss angle of a given capacitor specimen are best measured by
(a) Wheatstone bridge
(b) Maxwell bridge
(c) Anderson bridge
(d) Schering bridge
[ESE-2003]

3.9 Which of the following bridges can be used for inductance measurement?
1. Maxwell's bridge 2. Schering bridge
3. Wein bridge 4. Hay's bridge
5. Wheatstone bridge
Select the correct answer using the codes given below:
(a) 1 and 2 (b) 2 and 3
(c) 3, 4 and 5 (d) 1 and 4
[ESE-2005]

3.10 Which of the following factors decide the accuracy in a bridge measurement?
1. Accuracy of the null indicator
2. Accuracy of the bridge components
3. Sensitivity of the null indicator
4. Applied voltage to the bridge system
Select the correct answer using the codes given below:
(a) 1 and 2 (b) 2 and 3
(c) 3 and 4 (d) 1 and 3
[ESE-2005]

3.11 Wagner Earth devices in AC bridge circuits are used for
(a) Shielding all the bridge elements from external magnetic field
(b) Eliminating the effect of stray capacitance
(c) Minimizing the effect of inter-component capacitance
(d) Eliminating all the node to earth capacitances
[ESE-2005]

3.12 Which one of the following is represented by the circuit shown below

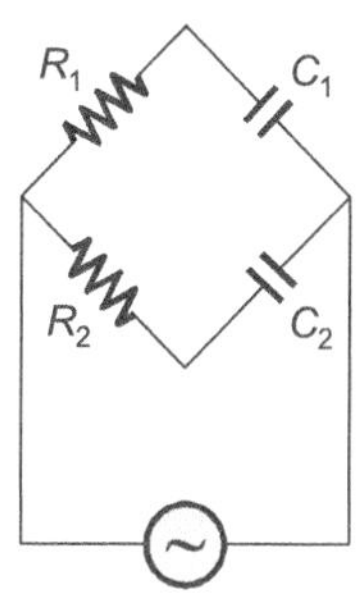

(a) De Sauty bridge
(b) Anderson bridge
(c) Heaviside-Campbell bridge
(d) Hay bridge [ESE-2006]

3.13 Maxwell's inductance-capacitance bridge is used for measurement of inductance of
(a) low Q coils only
(b) medium Q coils only
(c) high Q coils only
(d) low and medium Q coils
[ESE-2006]

3.14 Which of the following bridges can be used to construct a harmonic distortion analyzer?
1. Maxwell bridge 2. Hay bridge
3. Schering bridge 4. Wien bridge

Select the correct answer using the code given below:

(a) Only 1 and 2 (b) Only 2 and 3
(c) Only 3 and 4 (d) Only 4

[ESE-2006]

3.15 What should be the main characteristic(s) of the null detector in a bridge measurement?

1. Accuracy 2. Precision
3. Sensitivity 4. Resolution

Select the correct answer using the code given below:

(a) Only 1 and 2 (b) Only 2 and 3
(c) Only 3 and 4 (d) Only 3

[ESE-2006]

3.16 Match **List-I** (Parameter to be measured) with **List-II** (Instrument to be Used) and select the correct answer using the code given below the lists:

List-I

A. Average value of current
B. RMS value of current
C. Frequency of a wave
D. Strain gauge resistance

List-II

1. Self-balancing bridge
2. Wien bridge
3. PMMC ammeter
4. Moving-iron ammeter

Codes:

	A	B	C	D
(a)	3	4	2	1
(b)	2	1	3	4
(c)	3	1	2	4
(d)	2	4	3	1

[ESE-2006]

3.17 Hay's bridge is suitable for the measurement of which one of the following?

(a) Inductance with $Q < 10$
(b) Inductance with $Q > 10$
(c) Capacitance with high dissipating factors
(d) Capacitance with low dissipating factors

[ESE-2007]

3.18 Which one of the following bridges is used for measurement of dielectric loss and power factor of a capacitor?

(a) Maxwell's bridge (b) Anderson bridge
(c) De Sauty bridge (d) Schering bridge

[ESE-2007]

3.19 Schering bridge can be used to measure which one of the following?

(a) Q of a coil
(b) Inductance and its Q-value
(c) Very small resistance
(d) Capacitance and its power factor

[ESE-2008]

3.20 Inductance is measured by which one of the following?

(a) Wien bridge (b) Schering bridge
(c) Maxwell bridge (d) Owen bridge

[ESE-2008]

3.21 The dielectric loss of a capacitor can be measured by which one of the following?

(a) Wien bridge (b) Owen bridge
(c) Schering bridge (d) Maxwell bridge

[ESE-2008]

3.22 Which bridge is used to determine frequency?

(a) Anderson bridge (b) De Sauty bridge
(c) Wien bridge (d) Campbell bridge

[ESE-2008]

3.23 The imperfect capacitance which is shunted by a resistance can be measured by which one of the following?

(a) Carey Foster bridge
(b) Owen bridge
(c) Schering bridge
(d) Wien bridge [ESE-2008]

3.24 Which one of the following is a frequency sensitive bridge ?

(a) De-Sauty bridge (b) Schering bridge
(c) Wien's bridge (d) Maxwell's bridge

[ESE-2009]

3.25 **Assertion (A):** In a bridge type of measurement, it is required that the indicator used to show the balance condition of the bridge should have very high sensitivity.

Reason (R): The accuracy of the null-indicator does not play any role in a bridge measurement.

(a) Both A and R are true and R is the correct explanation of A
(b) Both A and R are true but R is NOT the correct explanation of A
(c) A is true but R is false
(d) A is false but R is true [ESE-2009]

3.26 Dissipation factor, tanδ, of a capacitor is measured by which bridge?
(a) Anderson bridge (b) Hay bridge
(c) Schering bridge (d) Wien bridge
[ESE-2009]

3.27 Match **List-I** with **List-II** and select the correct answer using the code given below the lists:

List-I	List-II
A. Hay bridge	1. Medium resistance
B. Wheatstone bridge	2. Frequency
C. Wien bridge	3. Capacitance
D. Schering bridge	4. High Q-inductance

Codes:

	A	B	C	D
(a)	4	2	1	3
(b)	3	2	1	4
(c)	4	1	2	3
(d)	3	1	2	4

[ESE-2011]

3.28 **Assertion (A):** AC bridge methods are the best and most usual methods for the precise measurement of self and mutual inductances and capacitances.
Reason (R): Wagner earthing device is used in AC bridge for eliminating the effect of the earth capacitance.
(a) Both A and R are true and R is the correct explanation of A
(b) Both A and R are true but R is NOT the correct explanation of A
(c) A is true but R is false
(d) A is false but R is true
[ESE-2011]

3.29 In the circuit shown, when the current through the branch *AD* is zero, the battery current I_B is

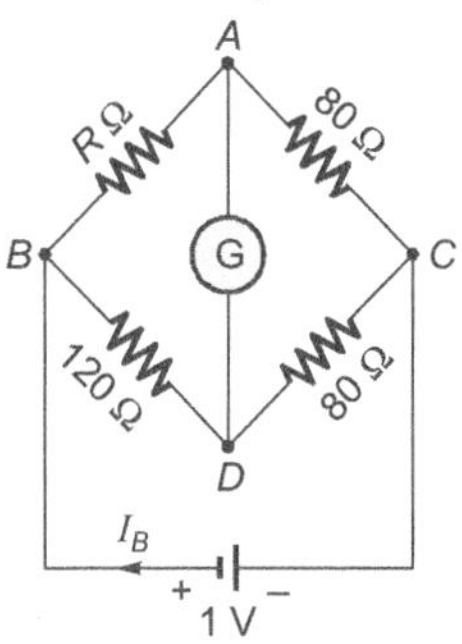

(a) 1 mA (b) 2 mA
(c) 10 mA (d) 20 mA
[ESE-2012]

3.30 The measurement of frequency can be carried out with
(a) Owen's bridge
(b) Wien's bridge
(c) Maxwell's inductance-capacitance bridge
(d) Schering's bridge
[ESE-2012]

3.31 Match **List-I** with **List-II** and select the correct answer using the code given below the lists:

List-I
A. Mutual inductance
B. High-Q inductance
C. Audio frequency
D. Dielectric loss

List-II
1. Wien-bridge
2. Schering-bridge
3. Hay bridge
4. Heaviside-Campbell bridge

Codes:

	A	B	C	D
(a)	4	1	3	2
(b)	2	3	1	4
(c)	4	3	1	2
(d)	2	1	3	4

[ESE-2012]

3.32 Wagner's earthing device is used in A.C. bridges for
(a) shielding the bridge elements
(b) eliminating the stray electrostatic field effects
(c) eliminating the effect of earth capacitance
(d) eliminating the effect of inter-component capacitances
[ESE-2013]

3.33 A bridge circuit works at a frequency of 2 kHz. The following can be used as detectors for detection of null conditions in the bridge
(a) Vibration galvanometers and Head-phones
(b) Headphones and Tunable amplifiers
(c) Vibration galvanometers and Tunable amplifiers
(d) Vibration galvanometers, Head-phones and Tunable amplifiers
[ESE-2013]

3.34 **Statement (I):** Bridge measurements are considered to be more accurate as compared to measurements done using indicating instruments.

Statement (II): In a bridge measurement, the accuracy of the components used in the different arms of the bridge along comes into picture.

[ESE-2014]

3.35 In De Sauty Bridge (unmodified form) it is possible to obtain balance.

(a) Even if both the capacitors are imperfect.
(b) If one of the capacitors is perfect.
(c) Only if both the capacitors are perfect.
(d) All of the above [ESE-2014]

3.36 With the help of which bridge are the capacitance and dielectric loss of a capacitor generally measured?

(a) De Sauty (b) Wien bridge
(c) Anderson (d) Schering [ESE-2014]

3.37 A bridge circuit works at a frequency of 2 kHz. Which of the following can be used as detectors for detection of null conditions in the bridge?

(a) Vibration galvanometers and headphones
(b) Headphones and tunable amplifiers
(c) Vibration galvanometers and tunable amplifiers
(d) Vibration galvanometers, headphones and tunable amplifiers

[ESE-2015]

3.38 A Wheatstone bridge has got three resistances taken in clockwise direction as 120 Ω, 150 Ω, and 150 Ω. The value of the fourth resistance for null balance would be

(a) 150 Ω (b) 120 Ω
(c) 300 Ω (d) 750 Ω [ESE-2015]

3.39 Schering bridge is a very versatile AC bridge and is used for capacitor testing in terms of

1. capacitance value (magnitude)
2. loss angle measurement
3. simple balance detector like PMMC instrument
4. providing safety to operators by incorporating Wagner earthing device

Which of the above are correct?

(a) 1 and 3 only (b) 3 and 4 only
(c) 1, 2 and 4 only (d) 1, 2, 3 and 4

[ESE-2016]

3.40 A Wien Bridge Oscillator is suitable for

1. Audio frequency applications
2. Radio frequency applications
3. Very low frequency applications

Which of the above frequency applications is/are correct?

(a) 1 only (b) 2 only
(c) 3 only (d) 1, 2 and 3

[ESE-2017]

3.41 The impedances of an AC bridge as shown in figure below are as follows :

Z_1 = 100 Ω with phase angle of 60°, Z_2 = 300 Ω with phase angle of 0°, and Z_3 = 50 Ω with phase angle of 30°. What is the nature of unknown impedance Z_4?

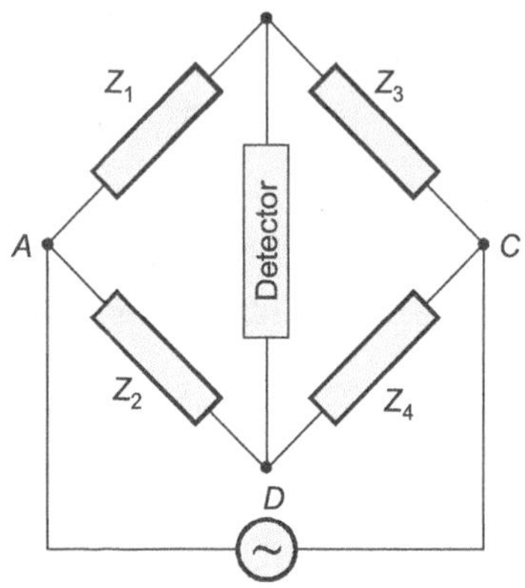

(a) Purely resistive circuit
(b) Series R-C circuit
(c) Series R-L circuit
(d) Parallel R-L circuit [ESE-2021]

3.42 The arms of a four-arm bridge abcd, supplied with sinusoidal voltage, have the following values:

Arm *ab* : A resistance of 200 Ω in parallel with a capacitance of 1 μF.
Arm *bc* : 400 Ω resistance
Arm *cd* : 1000 Ω resistance
Arm *da* : A resistance R_2 in series with a 2 μF capacitances.

What are the values of R_2 and the frequency respectively at which the bridge will balance?

(a) 200 Ω, 456 Hz (b) 400 Ω, 398 Hz
(c) 200 Ω, 398 Hz (d) 400 Ω, 456 Hz

[ESE-2022]

3.43 Which one of the following methods is **not** used in measurement of high resistance?

(a) Direct deflection method
(b) Mega Ohm bridge method
(c) Megger method
(d) Kelvin double bridge method

[ESE-2023]

3.44 Which one of the following is intended for the rapid measurement of the winding resistances of machines and transformers?
(a) Direct deflection
(b) Wheatstone bridge
(c) Megger
(d) Kelvin bridge ohmmeter

[ESE-2023]

3.45 A highly sensitive galvanometer can detect a current as low as 0.1 nA. This galvanometer is used in a Wheatstone bridge as a detector. Each arm of the bridge has a resistance of 1 kΩ. The input voltage applied to the bridge is 20 V. What is the smallest change in resistance which can be detected, if the resistance of the galvanometer can be neglected compared to the internal resistance of the bridge?
(a) 10 mΩ (b) 10 μΩ
(c) 20 mΩ (d) 20 μΩ

[ESE-2023]

3.46 Which one of the following methods is **not** used in measurement of low resistance?
(a) Direct deflection method
(b) Ammeter voltmeter method
(c) Potentiometer method
(d) Kelvin double bridge method

[ESE-2023]

3.47 A regular Wheatstone bridge is used to measure high resistance. The bridge has ratio arms of 10,000 Ω and 10 Ω. The adjustable arm has a maximum value of 10,000 Ω. A battery of 10 V emf and negligible resistance is connected from the junction of ratio arms to the opposite corner. What is the maximum resistance that can be measured by this arrangement?
(a) 10 kΩ (b) 10 MΩ
(c) 100 kΩ (d) 100 MΩ

[ESE-2023]

3.48 A resistance of approximately 3000 Ω is needed to balance a bridge. It is obtained on a 5 dial resistance box having steps of 1000 Ω, 100 Ω, 10 Ω, 1 Ω and 0.1 Ω. The measurement is to be guaranteed to 0.1 percent. For this accuracy, how many of these dials would it be worth to adjust?
(a) 2997 Ω to 3003 Ω (b) 3000 Ω to 3006 Ω
(c) 2994 Ω to 3000 Ω (d) 2996 Ω to 3002 Ω

[ESE-2023]

3.49 Consider the following statements :
1. Campbell's bridge measures an unknown mutual inductance in terms of a standard mutual inductance.
2. Campbell's bridge is used for measurement of self-inductance in terms of a standard mutual inductance.

Which of the above statement(s) is/are correct?
(a) 1 only (b) 2 only
(c) Both 1 and 2 (d) Neither 1 nor 2

[ESE-2023]

3.50 Consider the following statements :
1. Carey Foster bridge is used for measurement of a capacitance in terms of a standard mutual inductance.
2. Heydweiller bridge is used for measurement of mutual inductance in terms of a standard mutual inductance.

Which of the above statement(s) is/are correct?
(a) 1 only (b) 2 only
(c) Both 1 and 2 (d) Neither 1 nor 2

[ESE-2023]

■■■■

Answers A.C. Bridges

3.1 (d)	3.2 (a)	3.3 (a)	3.4 (a)	3.5 (a)	3.6 (c)	3.7 (d)	3.8 (d)	3.9 (d)
3.10 (b)	3.11 (d)	3.12 (a)	3.13 (b)	3.14 (d)	3.15 (d)	3.16 (a)	3.17 (b)	3.18 (d)
3.19 (d)	3.20 (c)	3.21 (c)	3.22 (c)	3.23 (c)	3.24 (c)	3.25 (b)	3.26 (c)	3.27 (c)
3.28 (b)	3.29 (c)	3.30 (b)	3.31 (c)	3.32 (c)	3.33 (b)	3.34 (a)	3.35 (c)	3.36 (d)
3.37 (b)	3.38 (b)	3.39 (c)	3.40 (a)	3.41 (b)	3.42 (b)	3.43 (d)	3.44 (d)	3.45 (d)
3.46 (a)	3.47 (b)	3.48 (a)	3.49 (a)	3.50 (a)				

Explanations A.C. Bridges

3.1 (d)

- The Wien's bridge is primarily known as a frequency determining bridge. It may be employed in a harmonic distortion analyzer, where it is used as notch filter, discriminating against one specific frequency. The Wien's bridge also finds applications in audio and HF oscillators as the frequency determining device.
- The Hay's bridge is suited for the measurement of high Q inductors, especially those inductors having a Q greater than 10. For inductors having Q values smaller than 10, the term $(1/Q)^2$ in the expression for inductance

$$L = \frac{R_2R_3C_4}{1+(1/Q)^2}$$

becomes rather important and thus cannot be neglected. Hence this bridge is not suited for measurement of coils having Q less than 10 and for these applications a Maxwell's bridge is more suited.

3.2 (a)

$$V_{01} = V_{02} = \frac{V_s}{2}\cdot\frac{\Delta L}{L} \text{ on solving bridge circuit.}$$

3.3 (a)

- Twin-T network is used in notch filter for realising frequency selection.
- Wagner earthing device removes all the earth capacitances from the bridge network.

3.4 (a)

Refer Schering Bridge

3.5 (a)

Wien bridge is a frequency determining network.

3.11 (d)

A special voltage divider circuit called a Wagner earth is used to maintain the null detector at ground potential without the need for a direct connection to the null detector (to ground). This is done to reduce stray capacitance errors.

3.14 (d)

Wien bridge can be tuned to frequencies.

3.15 (d)

In the null detector, the measured quantity is balanced out. This means the detector has to cover a small range around the balance (null) point and therefore can be made highly sensitive.

3.17 (b)

For high value of Q.

3.27 (c)

	Bridge	Measurement
(i)	Wheatstone bridge	Medium resistance
(ii)	Kelvin double bridge	Low resistance
(iii)	Maxwell's inductance capacitance bridge	Medium Q induc.
(iv)	Hay's bridge	High Q-inductance
(v)	Anderson's bridge	Low Q-inductance
(vi)	Owen's bridge	Low Q-inductance
(vii)	Schering bridge	Capacitance
(viii)	Wien bridge	Frequency

Hence, option (c) is correct.

3.28 (b)

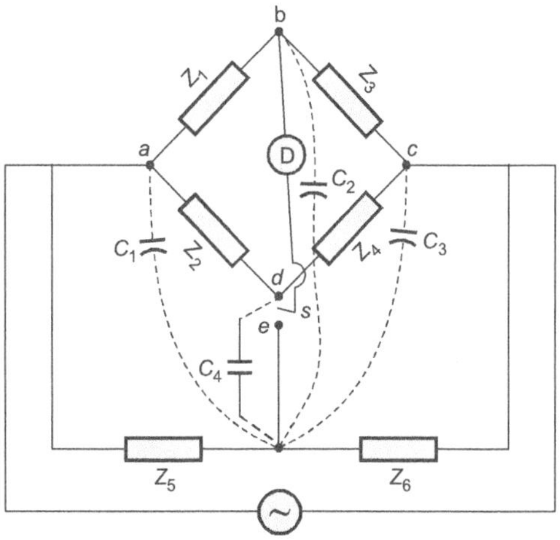

Wagner earthing device

All true points b, c and e are brought to earth potential by adjusting the bridge. Under these conditions no current flow in the earth capacitance C_2 and C_4 and since C_1 and C_3 shunt the wagner arms Z_5 and Z_6 these capacitances are eliminated from the bridge network.

3.29 (c)

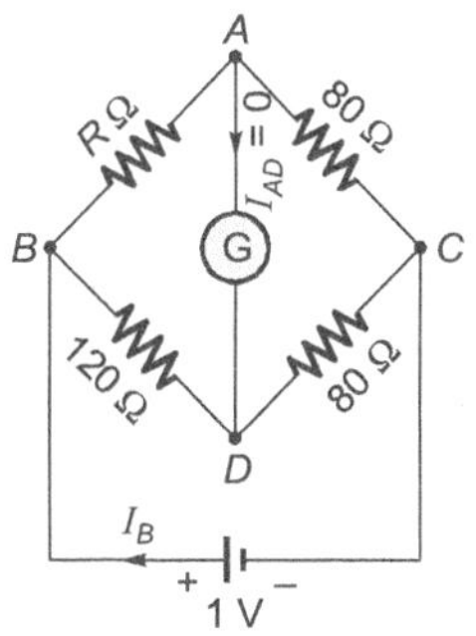

As $I_{AD} = 0$

⇒ Bridge is balanced

∴ $R = 120\ \Omega$

$$I_B = \frac{1}{(120+80)\|(120+80)}$$

$$= \frac{1}{100}\text{A} = 10\text{ mA}$$

3.30 (b)

Wien's bridge is used for frequency measurement.

3.31 (c)

- Heaviside-Campbell bridge is used for the measurement of mutual inductance.
- Hay bridge is used for coils of having $Q > 10$.
- Wien bridge is suitable for frequency measurement (100 Hz to 100 kHz).

3.32 (c)

Wagner's earthing device is used in AC bridges in order to eliminate the effect of earth capacitance.

3.33 (b)

0 Hz → D'Arsovnal Galvanometer
5 Hz - 200 Hz → Vibration Galvanometer
200 Hz - 1000 Hz → Telephone detector (Head phones)
>1000 Hz → Tunable detector

3.35 (c)

De-sauty bridge is used to obtain balance if both capacitor are perfect.

3.36 (d)

Schering bridge is used to determine capacitor and dielectric loss.

3.37 (b)

2 kHz in audible range

So, Headphones and tunable amplifiers is used.

3.38 (b)

$$\frac{x}{150} = \frac{120}{150} \Rightarrow x = 120\ \Omega$$

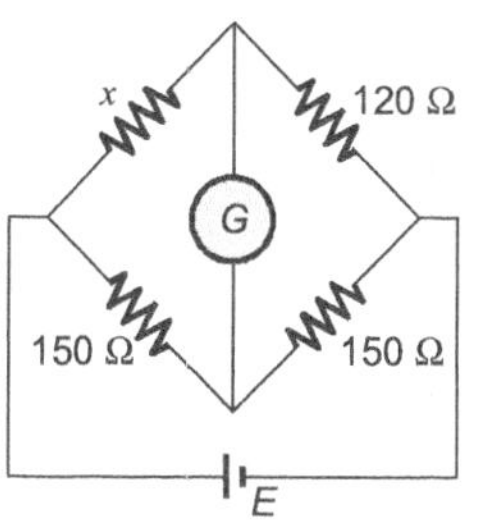

3.40 (a)

Wien bridge oscillator is suitable for audio frequency applications (20Hz – 20 kHz).

3.41 (b)

At balance,

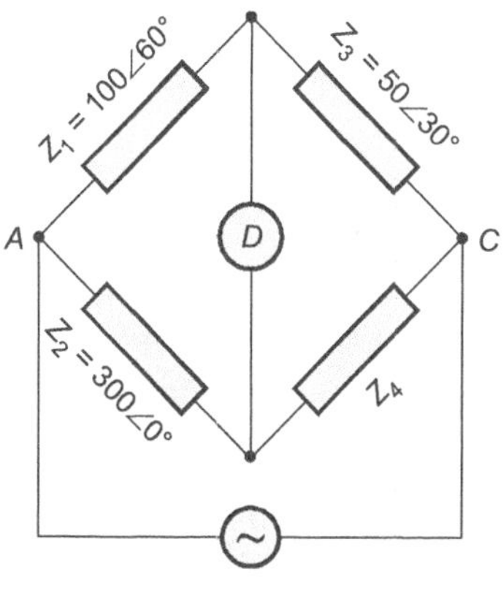

At bridge balance,

$$Z_1Z_4 = Z_2Z_3$$

$$Z_4 = \frac{Z_2Z_3}{Z_1} = \frac{300 \times 50\angle 30°}{100\angle 60°}$$

$$= 150\angle -30°$$

$$= 150 \cos 30° - j150 \sin 30°$$

$$= R - jX_C = (130 - j75)\ \Omega$$

Z_4 contains R and C.

3.42 (b)

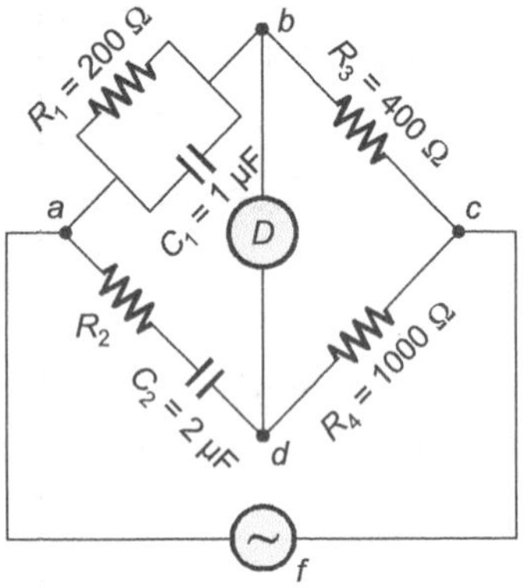

For Wien bridge under balance condition:

$$\frac{R_4}{R_3} = \frac{R_2}{R_1} + \frac{C_1}{C_2}$$

$$\frac{1000}{400} = \frac{R_2}{200} + \frac{1}{2}$$

$$R_2 = 400\ \Omega$$

$$f = \frac{1}{2\pi\sqrt{R_1R_2C_1C_2}}$$

$$= \frac{1}{2\pi\sqrt{200 \times 400 \times 1 \times 2 \times 10^{-12}}}$$

$$= 398\text{ Hz}$$

3.43 (d)

Kelvin double bridge is used for low resistance.

3.44 (d)

Kelvin double bridge is used for low resistance measurement.

3.45 (d)

Given that : I_g = 0.1 nAmp = 0.1×10^{-9} Amp

R = 1 kΩ, Supply = E = 20 Volt

It is a quarter bridge :

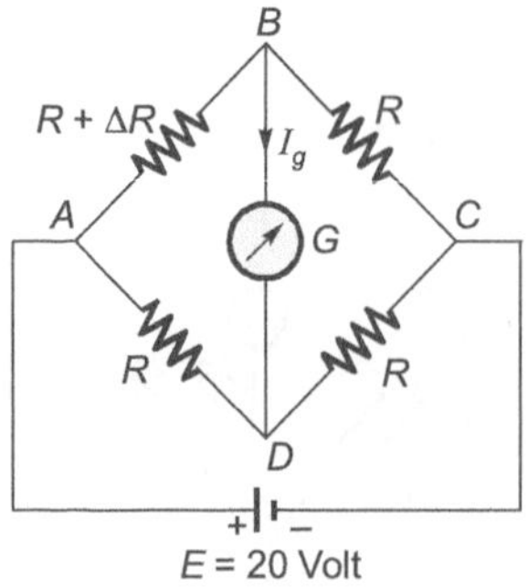

$$V_{Th} = \frac{E}{4} \times \frac{\Delta R}{R}$$

$$R_{Th} = R \parallel R + R \parallel R = R$$

$$= 1\ \text{k}\Omega$$

$$I_g = \frac{V_{Th}}{R_{Th}}$$

$$\Rightarrow \quad V_{Th} = I_g \times R_{Th}$$

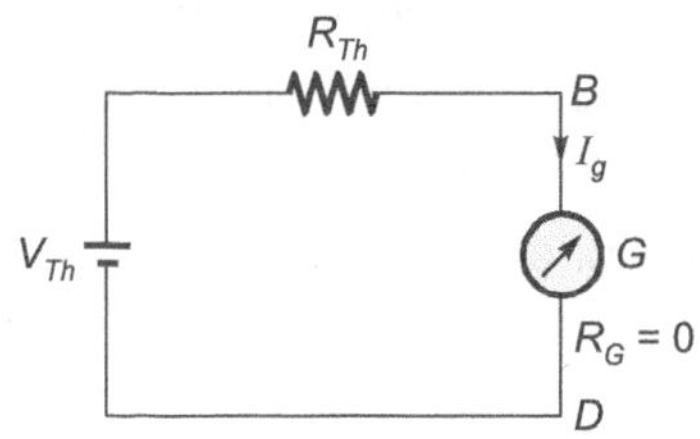

$$V_{Th} = 0.1 \times 10^{-9} \times 10^3$$

$$= 0.1 \times 10^{-6}\text{ Volt}$$

$$0.1 \times 10^{-6} = \frac{20}{4} \times \frac{\Delta R}{10^3}$$

$$\Rightarrow \quad \Delta R = \frac{0.1 \times 10^{-6} \times 10^3}{5} = \frac{0.1 \times 10^{-3}}{5}$$

$$\Delta R = 0.02 \times 10^{-3}$$

$$= 20 \times 10^{-6}$$

$$= 20\ \mu\Omega$$

3.46 (a)

Direct deflection method is used for high resistance measurement.

3.47 (b)

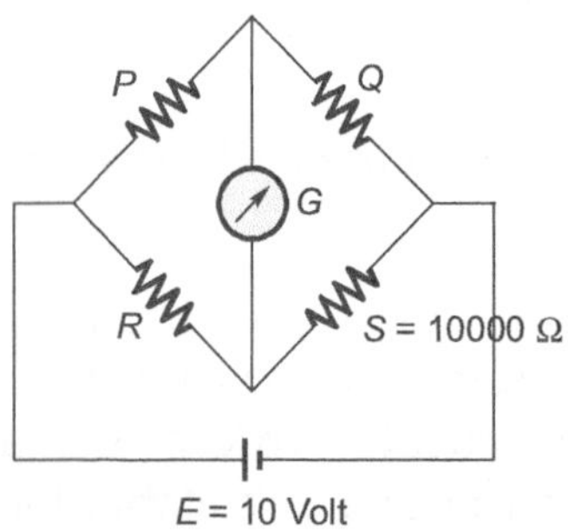

Given that : $P = 10000\ \Omega = 10\ k\Omega$

$E = 10$ Volt; $Q = 10\ \Omega$; $S = 10000\ \Omega = 10\ k\Omega$

$$R = \frac{P}{Q} \times S$$

$$R = \frac{10\ k\Omega}{10} \times 10\ k\Omega = 10\ M\Omega$$

3.48 (a)

$$3000 \pm 0.1\%$$

Limiting value of 3000 Ω

$$\Rightarrow \quad 3000 \pm 3000 \times 0.1\%$$

$$= 3000 \pm 3000 \times \frac{0.1}{100}$$

$\Rightarrow$ 2997 to 3003 Ω

3.49 (a)

Only 1 statement correct.

3.50 (a)

Only 1 statement is correct.

■■■■

4 Galvanometers, Voltmeters and Ammeters

4.1 Consider the following statements associated with moving iron instruments:

1. These can be used in d.c. as well a.c. circuits.
2. The scale is non-uniform.
3. The moving iron is placed in a field of a permanent magnet.

Which of these statements are correct ?

(a) 1, 2 and 3 (b) 1 and 2
(c) 2 and 3 (d) 1 and 3

[ESE-2002]

4.2 The reflecting mirror mounted on the moving coil of a vibration galvanometer is replaced by a bigger size mirror. This will result in

(a) lower frequency of resonance and lower amplitude of vibration.
(b) lower frequency of resonance but the amplitude of vibration is unchanged.
(c) higher frequency of resonance and lower amplitude of vibration.
(d) higher frequency of resonance but the amplitude of vibration is unchanged.

[ESE-2002]

4.3 A spring controlled moving iron voltmeter draws a current of 1 mA for full scale value of 100 V. If it draws a current of 0.5 mA, the meter reading is

(a) 25 V (b) 50 V
(c) 100 V (d) 200 V

[ESE-2002]

4.4 **Assertion (A):** A full-wave rectifier type a.c. voltmeter reads the true rms value of the input wave form.

Reason (R): The full wave rectifier type a.c. voltmeter has a rectifier unit first which feeds its output to the PMMC indicative instrument.

(a) Both A and R are true and R is the correct explanation of A
(b) Both A and R are true but R is NOT the correct explanation of A
(c) A is true but R is false
(d) A is false but R is true

[ESE-2002]

4.5 Vibration galvanometers, tuneable amplifiers and head phones are used in

(a) d.c. bridges
(b) a.c. bridges
(c) Both d.c. and a.c. bridges
(d) kelvin double bridge.

[ESE-2003]

4.6 Which one of the following multi-range voltmeters has high and constant input impedance ?

(a) Permanent magnet moving coil voltmeter
(b) Electronic voltmeter
(c) Moving iron voltmeter
(d) Dynamometer type voltmeter

[ESE-2003]

4.7 Three d.c. voltmeters are connected in series across a 120 V d.c. supply. The voltmeters are specified as follows :

Voltmeter *A* : 100 V, 5 mA
Voltmeter *B* : 100 V, 250 ohms/V
Voltmeter *C* : 10 mA, 15,000 ohms

The voltages read by the meters *A*, *B* and *C* are respectively

(a) 40, 50 and 30 V (b) 40, 40 and 40 V
(c) 60, 30 and 30 V (d) 30, 60 and 30 V

[ESE-2003]

4.8 Which one of the following statements is correct? Spiral springs are used in instruments to

(a) provide controlling torque
(b) provide damping torque
(c) lead the current to moving coil as well as to provide the controlling torque
(d) provide linear deflection

[ESE-2004]

4.9 Which one of the following types of instruments can be used to determine the r.m.s. value of a.c. voltage of high magnitude (10 kV) and of any wave shape?
(a) Moving iron instruments
(b) Dynamometer type instruments
(c) Induction instruments
(d) Electrostatic instruments

[ESE-2004]

4.10 Which one of the following statements is correct? The deflection of hot wire instrument depends on
(a) r.m.s. value of the a.c. current
(b) r.m.s. value of the a.c. voltage
(c) average value of the a.c. current
(d) average value of the a.c. voltage

[ESE-2004]

4.11 Which one of the following materials is used in the fabrication of swamping resistance of a PMMC instrument ?
(a) Copper (b) Aluminium
(c) Manganin (d) Tungsten

[ESE-2004]

4.12 A D'Arsonval movement with internal resistance $R = 100\ \Omega$ and full scale current of 1 mA is to be converted into (0 – 10) V range. What is the required resistance?
(a) 10 kΩ (b) 10100 Ω
(c) 9900 Ω (d) 12000 Ω

[ESE-2004]

4.13 **Assertion (A):** An electronic voltmeter measures the voltage across a high resistance more accurately as compared to an ordinary multimeter
Reason (R): The electronic voltmeter consists of a voltage amplifier which is not present in an ordinary multimeter.
(a) Both A and R are true and R is the correct explanation of A
(b) Both A and R are true but R is NOT the correct explanation of A
(c) A is true but R is false
(d) A is false but R is true

[ESE-2004]

4.14 Match **List-I** (Frequency) with **List-II** (Detector) and select the correct answer using the codes given below the lists:

List-I
A. Zero frequency
B. 50 Hz
C. 1200 Hz
D. 10 kHz

List-II
1. Head phone
2. D'Arsonval galvanometer
3. Cathode ray oscilloscope
4. Vibration galvanometer
5. Ballistic galvanometer

Codes:

	A	B	C	D
(a)	2	1	5	3
(b)	3	4	1	2
(c)	2	4	1	3
(d)	3	1	5	2

[ESE-2005]

4.15 **Assertion (A):** The flux densities used in PMMC instruments vary from 0.1 T to 1.0 T.
Reason (R): The power requirement for PMMC movement to give full scale deflection is small ranging from 25 μW to 200 μW.
(a) Both A and R are true and R is the correct explanation of A
(b) Both A and R are true but R is NOT the correct explanation of A
(c) A is true but R is false
(d) A is false but R is true

[ESE-2005]

4.16 **Assertion (A):** The PMMC type of indicating instruments are always critically damped.
Reason (R): A critically damped system directly moves to its steady state without oscillation.
(a) Both A and R are true and R is the correct explanation of A
(b) Both A and R are true but R is NOT the correct explanation of A
(c) A is true but R is false
(d) A is false but R is true

[ESE-2005]

4.17 Match **List-I** (Specifications of Voltages to be Measured) with **List-II** (Type of Most Suitable Instruments) and select the correct answer using the code given below the lists:

List-I

A. 0-10 mV from a source of internal resistance of 1 MΩ

B. Thermo-emf ranging up to 5 mV from a thermocouple

C. Supply voltage of 230 V, 50 Hz

D. R.M.S. value of a voltage containing dc and ripples of 50 Hz and harmonic

List-II

1. Thermal
2. Moving Iron
3. Permanent magnet moving coil
4. Electronic
5. Ballistic Galvanometer

Codes:

	A	B	C	D
(a)	2	3	5	1
(b)	4	1	2	3
(c)	2	1	5	3
(d)	4	3	2	1

[ESE-2005]

4.18 Which of the following factors limit the deflection of the pointer of a PMMC instrument to about 90°?

1. Its damping mechanism
2. Linearity of the magnetic field in which the coil moves
3. Controls spring arrangement
4. Shape of the pole shoe of the horseshoe magnet

Select the correct answer using the code given below:

(a) Only 1 and 3 (b) Only 2 and 4
(c) Only 2 and 3 (d) Only 1 and 4

[ESE-2006]

4.19 Which of the following indicating instruments has/have linear scale?

1. Moving-iron meter
2. Permanent magnet moving-coil meter
3. Thermocouple meter
4. Rectifier type meter

Select the correct answer using the code given below:

(a) Only 1 and 2 (b) Only 2 and 3
(c) Only 3 and 4 (d) Only 2

[ESE-2006]

4.20 A moving-coil instrument gives full-scale deflection for 1 mA and has a resistance of 5 Ω. If a resistance of 0.55 Ω is connected in parallel to the instrument, what is the maximum value of current it can measure?

(a) 5 mA (b) 10 mA
(c) 50 mA (d) 100 mA

[ESE-2006]

4.21 A sinusoidal voltage of 1 V rms value at 10 Hz is applied across the two terminals of a PMMC type of voltmeter. What is the deflection of the pointer?

(a) Zero volt
(b) 1 volt
(c) $\sqrt{2}$ volt
(d) The pointer oscillates around zero volt

[ESE-2006]

4.22 What is the series resistance required to extend the 0-100 V range of a 20000 Ω/V meter to 0-1000 V?

(a) 10 MΩ (b) 16 MΩ
(c) 18 MΩ (d) 20 MΩ

[ESE-2006]

4.23 **Assertion (A):** The accuracy of the null detector does not play any role in a bridge measurement.
Reason (R): Under balance condition of the bridge, no current flows through the null detector.

(a) Both A and R are true and R is the correct explanation of A
(b) Both A and R are true but R is NOT the correct explanation of A
(c) A is true but R is false
(d) A is false but R is true [ESE-2007]

4.24 Flux meter is a special type of ballistic galvanometer provided with which one of the following?

(a) Heavy electromagnetic damping and very small controlling torque
(b) Heavy electromagnetic damping and very large controlling torque
(c) Small electromagnetic damping and small controlling torque
(d) Large controlling torque and small electromagnetic damping

[ESE-2007]

4.25 Why is damping of a ballistic galvanometer kept small?
(a) To get minimum overshoot
(b) To make the system critically damped
(c) To make the system oscillatory
(d) To get first deflection large
[ESE-2007]

4.26 Which one of the following instruments is used for standardization of a Drysdale a.c. potentiometer?
(a) Rectifier type ammeter
(b) PMMC ammeter
(c) Precision type electrodynamometer ammeter
(d) Thermocouple ammeter
[ESE-2007]

4.27 What is clamp-on ammeter used for?
(a) Low a.c. current (b) High a.c. current
(c) Low d.c. current (d) High d.c. current
[ESE-2008]

4.28 What is the 'swamping' resistance which is connected in series with the working coil of a voltmeter to drastically reduce the error in measurement caused due to variation in temperature, made of?
(a) Constantan (b) Eureka
(c) Manganin (d) Nichrome
[ESE-2008]

4.29 How can a milli-ammeter be used as a voltmeter?
(a) By connecting a low resistance in parallel with the instrument
(b) By connecting a high resistance in parallel with the instrument
(c) By connecting a low resistance in series with the instrument
(d) By connecting a high resistance in series with the instrument
[ESE-2008]

4.30 **Assertion (A):** A PMMC instrument is used for reading both d.c. and a.c. signals.
Reason (R): The deflecting torque in a PMMC instrument is directly proportional to the current in the moving coil.
(a) Both A and R are true and R is the correct explanation of A
(b) Both A and R are true but R is NOT the correct explanation of A
(c) A is true but R is false
(d) A is false but R is true
[ESE-2008]

4.31 Due to which one of the following reasons bearings of PMMC Instrument are made of Jewel?
(a) To avoid wear and tear of the moving system
(b) To provide a small support
(c) It can be easily replaced
(d) To make the system robust
[ESE-2008]

4.32 In an electrometer, the movable plate is 11 cm in diameter. When 12 kV is applied between the movable plate and the fixed plate, the force is 0.006 N. The change in capacitance for 1.5 mm movement of the movable plate is
(a) 0.44×10^{-12} F (b) 0.37×10^{-6} F
(c) 0.125×10^{-12} F (d) 12.5×10^{-12} F
[ESE-2008]

4.33 Consider the following statements:
1. Amplifier gain and phase shift.
2. Filter transfer functions.
3. Two port network parameters.
4. Power gain in a two port circuit.

Which of the above quantities can be measured using a vector voltmeter?
(a) 1 and 3 only (b) 1, 2 and 4
(c) 1, 2 and 3 (d) 3 and 4
[ESE-2009]

4.34 A 100 kV, 50 Hz supply is fed to a rectifier ammeter (using a bridge rectifier) through a capacitor. The PMMC ammeter of the rectifier instrument reads 45×10^{-3} Amp. What is the value of the capacitor?
(a) 15.90×10^{-10} F (b) 15.90×10^{-12} F
(c) 17.66×10^{-9} F (d) 17.66×10^{-11} F
[ESE-2009]

4.35 **Assertion (A):** To increase the range of an ammeter to measure high currents, it is required to connect a high resistor in shunt across the ammeter.
Reason (R): The shunt resistor will divert the excess current and allow only the rated current to pass through the deflecting system of the ammeter.

(a) Both A and R are true and R is the correct explanation of A
(b) Both A and R are true but R is NOT the correct explanation of A
(c) A is true but R is false
(d) A is false but R is true

[ESE-2009]

4.36 **Assertion (A):** The sensitivity of a voltmeter is often expressed in terms of ohms-per-volt.
Reason (R): High sensitivity voltmeters use a basic D'Arsonval meter which has high sensitivity.
(a) Both A and R are true and R is the correct explanation of A
(b) Both A and R are true but R is NOT the correct explanation of A
(c) A is true but R is false
(d) A is false but R is true

[ESE-2009]

4.37 A D'Arsonval galvanometer, 1 mA, 50 ohm, is to be converted to a 5 Amp-ammeter. What is the value of the shunt resistor, R_{sh}?
(a) 10 ohm (b) 1 ohm
(c) 0.01 ohm (d) 100 ohm

[ESE-2009]

4.38 **Assertion (A) :** A thermocouple type of indicating instrument measures the true r.m.s. value of the current that passes through it.
Reason (R) : It uses a p.m.m.c type of indicating instrument to measure the current.
(a) Both A and R are true and R is the correct explanation of A
(b) Both A and R are true but R is NOT the correct explanation of A
(c) A is true but R is false
(d) A is false but R is true

[ESE-2010]

4.39 A moving coil ammeter having a resistance of 1 ohm gives full scale deflection when a current of 10 mA is passed through it. The instrument can be used for the measurement of voltage up to 10 V by
(a) connecting a resistance of 999 ohm in series with the instrument
(b) connecting a resistance of 999 ohm parallel to the ammeter
(c) connecting a resistance of 999 ohm parallel to the load
(d) connecting a resistance of 1000 ohm in series with the load

[ESE-2010]

4.40 The value of the multiplier resistance for a dc voltmeter, having 50 V range with 5 kΩ/V sensitivity, employing a 200 μA meter movement and having internal resistance of 100 Ω, is given by
(a) 249.9 kΩ (b) 200 kΩ
(c) 200 kΩ (d) 25 kΩ

[ESE-2010]

4.41 A basic D'Arsonval movement with a full scale deflection of 50 μA and internal resistance of 500 Ω is used as voltmeter. The value of the multiplier resistance needed to employ this meter to measure a voltage range of (0 – 10) V is given by
(a) 100 kΩ (b) 500 kΩ
(c) 199.5 kΩ (d) 2×10^5 kΩ

[ESE-2010]

4.42 The sensitivity of 200 μA meter movement when it is used as a dc voltmeter is given by
(a) 500 Ω/mV (b) 5 Ω/V
(c) 0.5 Ω/mV (d) 5 Ω/mV

[ESE-2010]

4.43 A 1 mA meter movement with an internal resistance of 100 Ω is to be converted into (0 – 100) mA. To achieve this, value of shunt resistance R_{sh} is given by
(a) 1 kΩ (b) 200 kΩ
(c) 1.01 Ω (d) 1.01 kΩ

[ESE-2010]

4.44 A (0 – 25) Amp ammeter has a guaranteed accuracy of 1 percent of full scale reading. The current measured by this ammeter is 10 Amp. The limiting error in percentage for this instrument is
(a) 2.5% (b) 0.5%
(c) 0.25% (d) 0.025%

[ESE-2010]

4.45 Consider the following statements regarding the controlling torque:

1. It is not present in power factor meter.
2. It opposes the deflecting torque.
3. It is provided by air friction or by fluid friction.

Which of these statements are correct?

(a) 1, 2 and 3 (b) 1 and 3 only
(c) 2 and 3 only (d) 1 and 2 only

[ESE-2011]

4.46 Consider the following statements in connection with deflection and null type instruments:

1. Null type instrument is more accurate than the deflection type one.
2. Null type of instrument can be highly sensitive as compared with deflection type instrument.
3. Under dynamic conditions, null type instrument is not preferred to deflection type instrument.
4. Response is faster in null type instrument as compared to deflection type instrument.

Which of these statements are correct?

(a) 1, 2 and 3 only (b) 1, 2 and 4 only
(c) 2, 3 and 4 only (d) 1, 2, 3 and 4

[ESE-2011]

4.47 The value of a shunt resistance required to convert an ammeter of 1 mA with 100 Ω internal resistance into 0-100 mA ammeter is

(a) 2.2 Ω (b) 1.01 Ω
(c) 1.2 Ω (d) 1.1 Ω

[ESE-2011]

4.48 An angular deflection type indicating PMMC meter is provided with spring control and

(a) damping by air dashpot.
(b) electromagnetic damping in the aluminium former only.
(c) electromagnetic damping in the aluminium former and the moving coil as well.
(d) no damping.

[ESE-2012]

4.49 An advantage of PMMC instrument is that it

(a) is free from friction error.
(b) has high torque-to-weight ratio of moving parts.
(c) has low torque-to-weight ratio.
(d) can be used on both AC and DC.

[ESE-2012]

4.50 When a steady current is passed through a ballistic galvanometer, then the deflection will be

(a) maximum
(b) zero
(c) twice the normal value as it depends on Hibbert magnetic
(d) None of the above

[ESE-2012]

4.51 By mistake, an ammeter is used as a voltmeter. In all probabilities, it will

(a) give much higher reading.
(b) give extremely low reading.
(c) indicate no reading at all.
(d) get damaged.

[ESE-2012]

4.52 A shunt resistance of 25 Ω is required for extending the range of an ammeter from 100 μA to 500 μA. The value of internal resistance of this ammeter will be

(a) 25 Ω (b) 50 Ω
(c) 100 Ω (d) 1000 Ω

[ESE-2012]

4.53 Guard circuits are used in insulation resistance measurements to

(a) increase the range of resistance values measured.
(b) reduce the effect of leakage current on measurement.
(c) protect against external electric fields.
(d) protect against external magnetic fields.

[ESE-2012]

4.54 A 50 μA meter with an internal resistance of 1 kΩ is to be used as a DC voltmeter of range 50 V. Then the voltage multiplying factor n is

(a) 100 (b) 10
(c) 1000 (d) 10000 [ESE-2012]

4.55 The value of resistance R_s to be added in series with an ammeter whose full scale deflection is of 0.1 mA and internal resistance is of 500 Ω, to make it suitable to measure (0 – 10) V is

(a) 0.02 kΩ (b) 99.5 kΩ
(c) 500.02 Ω (d) 499.98 Ω

[ESE-2012]

4.56 **Statement (I):** Vibration galvanometer is widely used as detector in bridge measurements.

Statement (II): Since the damping is very small, the deflection of the moving system is very much greater in neighbourhood of resonance than at any other frequency.

(a) Both Statement (I) and Statement (II) are individually true and Statement (II) is the correct explanation of Statement (I).
(b) Both Statement (I) and Statement (II) are individually true but Statement (II) is not the correct explanation of Statement (I).
(c) Statement (I) is true but Statement (II) is false.
(d) Statement (I) is false but Statement (II) is true.

[ESE-2012]

4.57 If one of the control springs of a permanent magnet coil ammeter is broken, then on being connected it will read

(a) zero
(b) half of the correct value
(c) twice of the correct value
(d) an finite value

[ESE-2013]

4.58 A 0.5 Ω resistance is required to be connected in parallel to a moving coil instrument whose full scale deflection is 1 mA; so that this instrument can measure 10 mA current. Internal resistance of this instrument is

(a) 5.0 Ω (b) 4.5 Ω
(c) 2.25 Ω (d) 0.45 Ω

[ESE-2013]

4.59 For a certain dynamometer ammeter the mutual inductance (M) varies with deflection θ° as $M = -6\cos(\theta + 30°)$ mH. Find the deflecting torque produced by a direct current of 50 mA corresponding to a deflection of 60°.

(a) 10 N-m (b) 20 N-m
(c) 15 μN-m (d) 1.5 μN-m

[ESE-2013]

4.60 An 1 mA, 50 Ω Galvanometer is required to measure 5 Amp (full scale). Find out the value of resistance to be added, across (shunt) the Galvanometer to accomplish this measurement.

(a) 10 Ω (b) 0.01 Ω
(c) 1.0 Ω (d) 0.001 Ω

[ESE-2013]

4.61 Volt-box is basically a device used for

(a) measuring the voltage
(b) extending the range of voltmeter
(c) extending the voltage range of the potentiometer
(d) measuring power

[ESE-2013]

4.62 To minimize voltmeter loading

(a) voltmeter operating current has to be very small.
(b) voltmeter operating current has to be very high.
(c) resistance connected in series with the coil should be low.
(d) resistance connected in parallel with the coil should be high.

[ESE-2013]

4.63 The galvanometer is protected during transportation by

(a) connecting critical damping resistance across the galvanometer terminals.
(b) shorting the galvanometer terminals.
(c) keeping the galvanometer terminals open-circuited.
(d) connecting a capacitor across the galvanometer terminals.

[ESE-2013]

Directions:

Each of the next items consists of two statements, one labelled as the **'Statement (I)'** and the other as **'Statement (II)'**. Examine these two statements carefully and select the answers to these items using the codes given below:

Codes:

(a) Both Statement (I) and Statement (II) are individually true and Statement (II) is the correct explanation of Statement (I).
(b) Both Statement (I) and Statement (II) are individually true but Statement (II) is not the correct explanation of Statement (I).
(c) Statement (I) is true but Statement (II) is false.
(d) Statement (I) is false but Statement (II) is true.

4.64 Statement (I): A permanent magnet moving coil instrument is always slightly under damped.

Statement (II): The pointer of the PMMC instrument should overshoot a little beyond the steady-state position to give the accurate reading.

[ESE-2014]

4.65 Statement (I): RF voltage is measured by rectifying the alternating voltage first and then amplifying the resulting dc output.

Statement (II): Amplification of RF signal itself is very difficult.

[ESE-2014]

4.66 A basic D'Arsonval movement showing full scale deflection for a current of 50 μA and having internal resistance of 500 Ω is used as a voltmeter. What is the value of multiplier resistance needed to measure a voltage range of 0 – 20 V?

(a) 398.5 kΩ (b) 399 kΩ
(c) 399.5 kΩ (d) 400 kΩ

[ESE-2014]

4.67 A galvanometer has a current sensitivity of 1 μA/mm and a critical damping resistance of 1 kΩ. The voltage sensitivity and the megaohm sensitivity respectively are

(a) 1 mV/mm and 1 MΩ
(b) 1 mV/mm and 2 MΩ
(c) 2 mV/mm and 2 MΩ
(d) 2 mV/mm and 1 MΩ [ESE-2014]

4.68 Electrostatic voltmeters are particularly suitable for measuring high voltages because the construction is simplified due to

(a) large electrostatic forces
(b) small electrostatic forces
(c) large value of current
(d) small value of current [ESE-2014]

4.69 A moving coil instrument of resistance 5 Ω requires a potential difference of 75 mV to give a full scale deflection of 75 mV to give a full scale deflection. The value of shunt resistance needed to give a full scale deflection at 30 A is

(a) 2.5 mΩ (b) 9.99 Ω
(c) 5 Ω (d) 9.95 Ω

[ESE-2014]

4.70 The deflection of a hot wire instrument depends on

(a) instantaneous value of alternating current
(b) average value of current
(c) rms value of alternating current
(d) voltage instead of current

[ESE-2014]

4.71 A dc voltmeter has a sensitivity of 1000 Ω/V. When it measures half full scale in 100 V range, the current through the voltmeter is

(a) 100 mA (b) 50 mA
(c) 1 mA (d) 0.5 mA

[ESE-2014]

4.72 In moving iron instruments, eddy current damping cannot be used as

(a) they have a strong operating magnetic field.
(b) they are not normally used in vertical position.
(c) they need a large damping force, which can only be provided by air friction.
(d) the introduction of a permanent magnet required for eddy current damping would distort the existing weak operating magnetic field.

[ESE-2014]

4.73 In the circuit shown below, the ammeter reads 0.1 A and the voltmeter reads 10 V. The internal resistance of the ammeter is 1 Ω and that of the voltmeter is 500 Ω. What is the value of R?

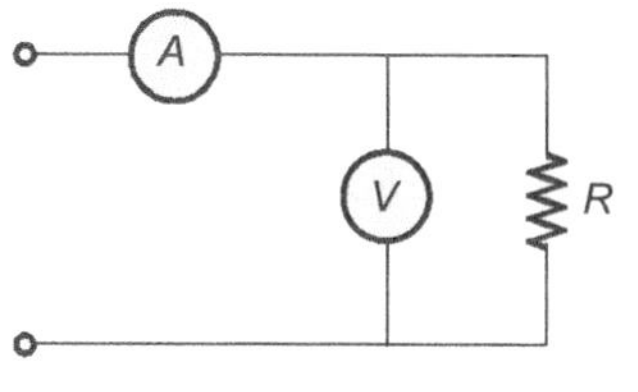

(a) 100 Ω (b) 125 Ω
(c) 90 Ω (d) 120 Ω

[ESE-2015]

4.74 A 1 mA galvanometer with internal resistance of 50 Ω is to be converted to measure 5 A (full-scale). What is the value of the shunt resistance required for this conversion?

(a) 1 Ω (b) 0.01 Ω
(c) 1 kΩ (d) 10 Ω

[ESE-2015]

4.75 A 50 μA basic d'Arsonval movement with an internal resistance of 500 Ω is to be used as a voltmeter. The value of the multiplier resistance required to measure a full-scale voltage range of 0 - 5 volts is

(a) 999.5 kΩ (b) 99.5 kΩ
(c) 9.99 kΩ (d) 0.99 kΩ

[ESE-2015]

Directions:

Each of the next items consists of two statements, one labelled as the **'Statement (I)'** and the other as **'Statement (II)'**. Examine these two statements carefully and select the answers to these items using the codes given below:

Codes:

(a) Both Statement (I) and Statement (II) are individually true and Statement (II) is the correct explanation of Statement (I).
(b) Both Statement (I) and Statement (II) are individually true but Statement (II) is not the correct explanation of Statement (I).
(c) Statement (I) is true but Statement (II) is false.
(d) Statement (I) is false but Statement (II) is true.

4.76 **Statement (I)** : The ammeter loading effect is due to the high resistance of the ammeter.
Statement (II) : Increasing the resistance of voltmeter will reduce the voltmeter loading effect.

[ESE-2015]

4.77 **Statement (I)** : In instruments where spring control is used for providing controlling torque, the scale is uniform, and where gravity control is used, the scale is non-uniform.
Statement (II) : In instruments where controlling torque is provided by spring control, the current is proportional to the deflection, and where the controlling torque is provided by gravity control, the current is proportional to sine of the deflection.

[ESE-2015]

4.78 A 3-turn 100 kΩ potentiometer with 1% linearity uses 30 V supply. What is the potentiometer constant?

(a) 0.1 V/turn (b) 10 V/turn
(c) 33.33 V/turn (d) 0.3 V/turn

[ESE-2015]

4.79 Time response of an indicating instrument is decided by which of the following systems?

(a) Mechanical system provided by pivot and jewel bearing
(b) Controlling system
(c) Deflecting system
(d) Damping system

[ESE-2015]

4.80 A certain fluxmeter has the following specifications:
Air gap flux density = 0.05 Wb/m^2
Number of turns on moving coil = 40
Area of moving coil = 750 mm^2
If the flux linking 10 turns of a search coil of 200 mm^2 area connected to the fluxmeter is reversed in a uniform field of 0.5 Wb/m^2, then the deflection of the fluxmeter will be

(a) 87.4° (b) 76.5°
(c) 65.6° (d) 54.7°

[ESE-2016]

4.81 A 100 μA ammeter has an internal resistance of 100 Ω. For extending its range to measure 500 μA, the required shunt resistance is

(a) 10 Ω (b) 15 Ω
(c) 20 Ω (d) 25 Ω

[ESE-2016]

4.82 A 200 V PMMC voltmeter is specified to be accurate within ±2% of full scale. The limiting error, when the instrument is used to measure a voltage of 100 V, is

(a) ±8% (b) ±4%
(c) ±2% (d) ±1%

[ESE-2016]

4.83 A moving-coil galvanometer can be used as a DC ammeter by connecting

(a) a high resistance in series with the meter
(b) a high resistance across the meter
(c) a low resistance across the meter
(d) a low resistance in series with the meter

[ESE-2016]

4.84 Consider the following types of damping:

1. Air-friction damping
2. Fluid-friction damping
3. Eddy-current damping

PMMC type instruments use which of the above?

(a) 1 only (b) 2 only
(c) 3 only (d) 1, 2 and 3

[ESE-2016]

4.85 Consider the following instruments:

1. MI instrument
2. Electrostatic instrument
3. Electrodynamometer instrument

Which of the above instruments is/are free from hysteresis and eddy-current losses?

(a) 1 only (b) 2 only
(c) 3 only (d) 1, 2 and 3

[ESE-2016]

4.86 A moving-coil instrument gives full-scale deflection of 10 mA, when a potential difference 10 mV is applied across its terminals. To measure currents up to 100 A, the same instrument can be used

(a) with shunt resistance of 0.0001 Ω
(b) with series resistance of 0.01 Ω
(c) with shunt resistance of 0.01 Ω
(d) with series resistance of 0.0001 Ω

[ESE-2016]

4.87 A current of $-4 + 3\sqrt{2}\sin(\omega t + 30°)$ A is passed through a centre zero PMMC meter and a moving-iron meter. The two meters will read respectively

(a) –4 A and –5 A (b) 4 A and –5 A
(c) –4 A and 5 A (d) 4 A and 5 A

[ESE-2016]

4.88 The values of ammeter and voltmeter resistances are 0.1 Ω and 2000 Ω respectively as shown in the figure below. The percentage error in the calculated value of R = 100 Ω (voltmeter reading 200 V/ammeter reading 2 A) is nearly

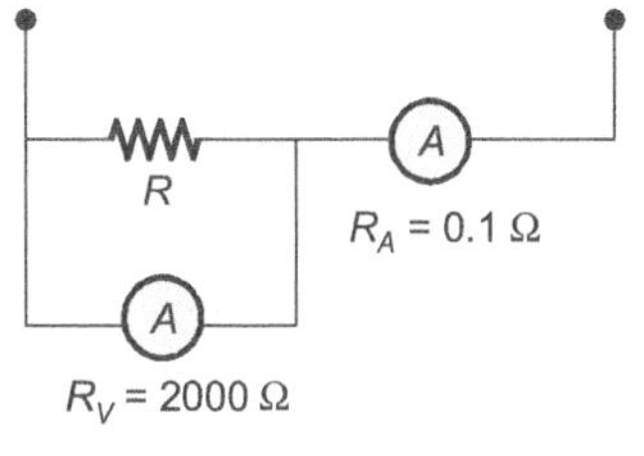

(a) –2% (b) –5%
(c) 2% (d) 5%

[ESE-2016]

4.89 What is the multiplying power of a shunt of 200 Ω resistance when used with a galvanometer of 1000 Ω resistance?

(a) 4 (b) 6
(c) 12 (d) 20

[ESE-2016]

4.90 A current of $(10 + 5\sin\omega t + 3\sin 2\omega t)$ is measured using a moving iron instrument. The reading would be

(a) 08.82 A (b) 10.00 A
(c) 10.82 A (d) 12.75 A

[ESE-2017]

Direction : Each of the next FOUR (4) items consist of two statements, one labelled as 'Statements (I) and the other as 'Statement (II)'. Examine these two statements carefully and select the answers to these items using the codes given below:

Codes:

(a) Both Statement (I) and Statement (II) are individually true and Statement (II) is the correct explanation of Statement (I).
(b) Both Statement (I) and Statement (II) are individually true but Statement (II) is not the correct explanation of Statement (I).
(c) Statement (I) is true but Statement (II) is false.
(d) Statement (I) is false but Statement (II) is true.

4.91 **Statement (I):** An instrument manufactured as an ammeter should not be used as a voltmeter.

Statement (II): The high resistance winding of an ammeter will suffer serious damage if connected across a high voltage source.

[ESE-2017]

4.92 **Statement (I):** Moving iron instruments are used in ac circuits only.

Statement (II): The deflecting torque in moving iron instruments depends on the square of the current.

[ESE-2017]

4.93 **Statement (I):** PMMC instruments are suitable in aircraft and air space applications.

Statement (II): PMMC instruments use a core magnet which possesses self-shielding property.

[ESE-2017]

4.94 **Statement (I):** A ballistic galvanometer is preferred as a detector in an AC bridge to measure inductance supplied by a source at power frequency.

Statement (II): An AC bridge to measure inductance is balanced at the fundamental component.

[ESE-2017]

4.95 Consider the following statements with regard to Moving Iron (MI) instruments:

1. These instruments posses high operating torque.
2. These instruments can be used in ac and dc circuits.
3. Power consumption in these instruments is lower for low voltage range.

Which of the above statements are correct?

(a) 1 and 2 only (b) 1 and 3 only
(c) 2 and 3 only (d) 1, 2 and 3

[ESE-2017]

4.96 Which one of the following material is used for the swamping resistance of moving coil instruments?

(a) Carbon (b) Manganin
(c) Silver (d) Brass

[ESE-2019]

4.97 In a PMMC instrument, the swamping resistor is used to

(a) increase the damping of the instrument.
(b) reduce the current within the limits.
(c) compensate for temperature variations.
(d) increase the full-scale sensitivity.

[ESE-2019]

4.98 A moving coil ammeter has a fixed shunt of 0.02 Ω. With a coil resistance of R = 1000 Ω and a potential difference of 500 mV across it, full scale deflection is obtained. The current through the moving coil to give full scale deflection will be

(a) 25 A (b) 0.5×10^{-2} A
(c) 0.25×10^{-3} A (d) 0.5×10^{-3} A

[ESE-2019]

4.99 A moving iron instrument has full scale current of 100 mA. It is converted into a 250 V voltmeter by using a series resistance made of a material having negligible resistance temperature coefficient. The meter has a resistance of 320 Ω at 20°C. After carrying a steady current of 100 mA for a long time, the resistance of the coil increases to 369 Ω due to self heating. When a voltage of 250 V is applied continuously, the error due to self-heating will be nearly.

(a) –1.1% (b) –1.9%
(c) –2.5% (d) –3.3%

[ESE-2019]

4.100 A 0 - 150 V voltmeter has a guaranteed accuracy of 1% full scale reading. The voltage measured by this instrument is 83 V. The limiting error will be nearly.

(a) 1.2% (b) 1.8%
(c) 2.4% (d) 3.2%

[ESE-2019]

4.101 The moving iron instruments when measuring voltages or currents,

(a) indicate the same values of the measurement for both ascending and descending values of current.
(b) indicate higher values of the measurement for ascending values of current.
(c) indicate higher values of the measurement for descending values of current.
(d) indicate lower values of the measurement for both ascending and descending values of current.

[ESE-2020]

4.102 True RMS-reading voltmeter

1. Measures the RMS value of voltage accurately.
2. Eliminates the error due to waveform.
3. Uses the thermocouple for heating.

Which of the above statements are correct?

(a) 1 and 2 only (b) 1 and 3 only
(c) 2 and 3 only (d) 1, 2 and 3

[ESE-2020]

4.103 What is the value of shunt resistance of an ammeter, if the range is extended from 1 mA to 10 mA and metre has a resistance of 27 Ω?

(a) 3 Ω (b) 4 Ω
(c) 2.5 Ω (d) 3.5 Ω

[ESE-2021]

4.104 Which one of the following instruments is the most sensitive of the moving iron mechanism and has the most linear scale?

(a) Moving coil
(b) Hot wire
(c) Electrodynamometer
(d) Radial vane repulsion

[ESE-2021]

4.105 A voltmeter has a resistance of 300 Ω and inductance of 0.12 H. This instrument reads correctly on DC. What is the reading on AC at 100 V when the frequency is 25 Hz?

(a) 99.80 V (b) 120.36 V
(c) 142.00 V (d) 151.00 V

[ESE-2021]

4.106 A voltmeter has a resistance of 20 kΩ and connected in series with an external resistance across a 230 V supply. If the instrument reads 160 V, then the value of external resistance is

(a) 6745 Ω (b) 7748 Ω
(c) 8750 Ω (d) 9752 Ω

[ESE-2021]

4.107 A moving coil instrument gives full-scale deflection with 15 mA and has a resistance of 5 Ω. The value of resistance to be connected in series to enable it to read up to 100 V is

(a) 6666.66 Ω (b) 6660.66 Ω
(c) 6661.60 Ω (d) 6662.60 Ω

[ESE-2021]

4.108 A moving coil meter of 50 Ω resistance reads up to 25 mA. What is the value of series resistance, so that it can be read upto 10 V?

(a) 300 Ω (b) 250 Ω
(c) 350 Ω (d) 400 Ω

[ESE-2021]

4.109 A half-wave rectifier type a.c. voltmeter is fed with a 20 V rms signal. What is the equivalent d.c. output voltage?

(a) 6.4 V (b) 9 V
(c) 12.8 V (d) 18 V

[ESE-2022]

4.110 A D'Arsonval movement has a resistance of 100 Ω and full deflection current of 2 mA and is used for construction of multirange ammeter as shown in the figure:

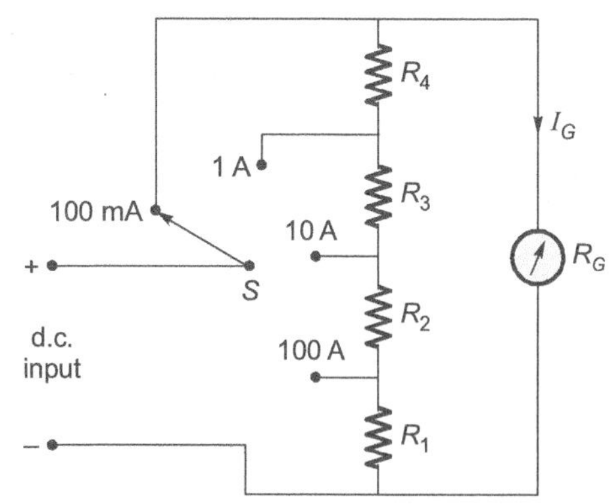

What are the values of resistances R_1, R_2, R_3 and R_4, if the shunt resistances are connected as shown in the figure?

(a) R_1 = 0.002 Ω, R_2 = 0.018 Ω, R_3 = 0.18 Ω and R_4 = 1.836 Ω.
(b) R_1 = 0.002 Ω, R_2 = 0.18 Ω, R_3 = 0.018 Ω and R_4 = 1.836 Ω.
(c) R_1 = 1.836 Ω, R_2 = 0.18 Ω, R_3 = 0.018 Ω and R_4 = 0.002 Ω.
(d) R_1 = 1.836 Ω, R_2 = 0.018 Ω, R_3 = 0.18 Ω and R_4 = 0.02 Ω.

[ESE-2022]

4.111 A d.c. galvanometer of 4 Ω resistance reads up to 50 mA. What is the value of the resistance in parallel to enable the instrument to read up to 1 A?

(a) 0.21053 Ω (b) 0.26316 Ω
(c) 0.31285 Ω (d) 0.37347 Ω

[ESE-2022]

4.112 Which one of the following statements is not correct for electrodynamometer type instrument?

(a) It can measure a range of currents and voltages up to 10 A and 600 V respectively.

(b) The deflecting torque is inversely proportional to the square of the current.
(c) It can be used for both a.c. and d.c. systems.
(d) It has the same calibration for d.c. instruments as well as a.c. measurements.

[ESE-2022]

4.113 The controlling torque of a gravity-controlled measuring instrument is directly proportional to

(a) θ
(b) $\cos\theta$
(c) θ^2
(d) $\sin\theta$

[ESE-2022]

4.114 **Statement (I) :** When the signal is of the form of current, then series input devices are used.

Statement (II) : An ammeter, which is a series device, thus should be designed with a low input impedance so that the current is correctly measured.

[ESE-2022]

4.115 Which one of the following is used to protect the standard cell and galvanometer against over currents?

(a) A pressure coil in series
(b) A current coil in parallel
(c) A resistance in series
(d) A resistance in parallel

[ESE-2023]

4.116 The coil of a moving-coil voltmeter is 40 mm long and 30 mm wide, and has 100 turns on it. The control spring exerts a torque of 240×10^{-6} N-m when the deflection is 100 divisions on full scale. If the flux density of the magnetic field in the air gap is 1.0 Wb/m^2, what is the value of the resistance that must be put in series with the coil to give one volt per division by neglecting the resistance of the voltmeter coil?

(a) 5 kΩ
(b) 15 kΩ
(c) 50 kΩ
(d) 75 kΩ

[ESE-2024]

4.117 Which one of the following statements is correct regarding moving-iron instrument when voltages or currents are measured?

(a) It indicates the same value of the measurement for both ascending and descending values.
(b) It indicates the higher value of the measurement for ascending values.
(c) It indicates the higher value of the measurement for descending values.
(d) It indicates the lower value of the measurement for descending values.

[ESE-2024]

4.118 A moving-coil instrument gives a full scale deflection of 10 mA, when the potential difference across its terminals is 100 mV. What is the value of shunt resistance for a full-scale deflection corresponding to 100 A?

(a) 0.0001 Ω
(b) 0.001 Ω
(c) 0.01 Ω
(d) 0.1 Ω

[ESE-2024]

■■■■

Answers Galvanometers, Voltmeters and Ammeters

4.1	(b)	4.2	(a)	4.3	(a)	4.4	(b)	4.5	(b)	4.6	(b)	4.7	(a)	4.8	(c)	4.9	(d)
4.10	(a)	4.11	(c)	4.12	(c)	4.13	(b)	4.14	(c)	4.15	(b)	4.16	(d)	4.17	(b)	4.18	(b)
4.19	(d)	4.20	(b)	4.21	(d)	4.22	(c)	4.23	(a)	4.24	(a)	4.25	(d)	4.26	(c)	4.27	(b)
4.28	(c)	4.29	(d)	4.30	(d)	4.31	(a)	4.32	(c)	4.33	(c)	4.34	(a)	4.35	(d)	4.36	(b)
4.37	(c)	4.38	(a)	4.39	(a)	4.40	(a)	4.41	(c)	4.42	(d)	4.43	(c)	4.44	(a)	4.45	(d)
4.46	(a)	4.47	(b)	4.48	(b)	4.49	(b)	4.50	(b)	4.51	(d)	4.52	(c)	4.53	(b)	4.54	(c)
4.55	(b)	4.56	(a)	4.57	(a)	4.58	(b)	4.59	(c)	4.60	(b)	4.61	(c)	4.62	(a)	4.63	(a)
4.64	(a)	4.65	(a)	4.66	(c)	4.67	(a)	4.68	(d)	4.69	(a)	4.70	(c)	4.71	(d)	4.72	(d)
4.73	(b)	4.74	(b)	4.75	(b)	4.76	(b)	4.77	(a)	4.78	(a)	4.79	(d)	4.80	(b)	4.81	(d)
4.82	(b)	4.83	(c)	4.84	(c)	4.85	(b)	4.86	(a)	4.87	(c)	4.88	(b)	4.89	(b)	4.90	(c)
4.91	(c)	4.92	(d)	4.93	(a)	4.94	(d)	4.95	(c)	4.96	(b)	4.97	(c)	4.98	(d)	4.99	(b)
4.100	(b)	4.101	(c)	4.102	(d)	4.103	(a)	4.104	(d)	4.105	(a)	4.106	(c)	4.107	(c)	4.108	(c)
4.109	(b)	4.110	(a)	4.111	(a)	4.112	(b)	4.113	(d)	4.114	(b)	4.115	(d)	4.116	(c)	4.117	(c)
4.118	(b)																

Explanations Galvanometers, Voltmeters and Ammeters

4.1 (b)

Moving iron is not placed in a field of a permanent magnet. Moving iron instrument employs an electromagnet and the iron vane moves in such a way so as to increase the flux of the electromagnet.

4.2 (a)

Bigger size mirror results in increased inertia constant J.

$$f_n = \frac{1}{2\pi}\sqrt{\frac{K}{J}}$$

and $$A = \frac{GI_m}{\sqrt{(D\omega)^2 + (K - J\omega^2)^2}}$$

as J increases both f_n and A decrease.

4.3 (a)

$$\theta \propto I^2$$

$$\Rightarrow \quad \frac{100}{\theta} = \left(\frac{1}{0.5}\right)^2 \Rightarrow \theta = 25 \text{ V}$$

4.4 (b)

PMMC instrument is calibrated to read true d.c. value.

4.6 (b)

In electronic volt meter, operational amplifier is used in between the input and the meter which is the reason for high and constant input impedance.

4.7 (a)

$$R_1 = \frac{100}{5\times10^{-3}} = 20000\ \Omega$$

$$R_2 = 250 \times 100 = 25000\ \Omega$$

$$R_3 = 15000\ \Omega$$

$$\therefore \quad I = \frac{120}{R_1 + R_2 + R_3} = 2 \text{ mA}$$

$$\therefore \quad V_1 = 2 \times 10^{-3} \times 20000 = 40 \text{ V}$$

$$\therefore \quad V_2 = 2 \times 10^{-3} \times 25000 = 50 \text{ V}$$

$$\therefore \quad V_3 = 2 \times 10^{-3} \times 15000 = 30 \text{ V}$$

4.10 (a)

The expansion (deflection) is proportional to the heating effect of the current and hence to the square of the rms value of the current.

Therefore, the meter may be calibrated to read the rms value of the a.c. current.

4.11 (c)

Manganin has a negligible temperature coefficient. A high swamping resistance made of manganin (20 to 30 times the coil resistance) is connected in series with the coil so that although the resistance of the coil may change considerably, the change in the total resistance of circuit is small.

4.12 (c)

$$R + 100 = \frac{10}{1\times10^{-3}} = 10{,}000$$

$$\Rightarrow \quad R = 9900\ \Omega$$

4.13 (b)
An electronic voltmeter measures the voltage across a high resistance more accurately due to its high input resistance.

4.15 (b)
PMMC is very sensitive instrument.

4.16 (d)
They are slightly underdamped.

4.18 (b)
No actual part of the coil is allowed to reach the extreme positions near the pole tips where, there is fringing field. Owing to fringing the flux density near the pole tips is smaller than that at the centre and also the field is not radial. Thus the angular span of scale is restricted to 90°.

4.19 (d)
Moving-iron meter and thermocouple meter have square-law response so do not have linear scale.

4.20 (b)

$$V = 5\times 1\text{ mA} = 5\text{ mV}$$

$$\text{New resistance} = \frac{5\times0.55}{5.55} = \frac{55}{111}\ \Omega$$

$$\therefore \quad I_{max} = \frac{5\times111}{55} \simeq 10\text{ mA}$$

4.21 (d)
In PMMC type meter torque is not unidirectional as it employs permanent magnet not electromagnet.

4.22 (c)
Initial resistance = 20000 × 100 = 2 MΩ
Final resistance = 20000 × 1000 = 20 MΩ
∴ Series resistance = 20 – 2 = 18 MΩ

4.23 (a)
Only presence is checked not magnitude so accuracy of detector does not play any role in bridge measurement.

4.24 (a)
Flux meter is a special type of ballistic galvanometer in which controlling torque is very small and electromagnetic damping is heavy. There are no controlling springs as controlling torque is very small. The high electromagnetic damping is obtain by having a low resistance of the circuit comprising the flux meter and search coil.

4.25 (d)
The damping of the ballistic galvanometer should be small in order such that amplitude of first swing is large.

4.26 (c)
Electrodynamometer type instruments are mainly used for calibration because of their accuracy.

4.27 (b)
Clamp on ammeter has current transformer.

4.28 (c)
Manganin is used as it has high resistance with minimum temperature coefficient.

4.29 (d)
So by connecting a high resistance in series milli ammeter is used as voltmeter.

4.30 (d)
PMMC instrument can be used only for DC.

4.32 (c)

$$\frac{1}{2}(\Delta C)V^2 = F\times\text{displacement}$$

$$\text{So,} \quad \Delta C = \frac{2\times\text{force}\times\text{displacement}}{V^2}$$

4.34 (a)

$$\text{Output voltage} = V_0 = \frac{2V_m}{\pi}$$

$$V_0 = I\times X_c$$

$$\frac{2V_m}{\pi} = \frac{I}{\omega C}$$

$$\frac{2\times100\times10^3\times\sqrt{2}}{\pi} = \frac{45\times10^{-3}}{2\times\pi\times50\times C}$$

$$\therefore \quad C = 15.90\times10^{-10}\text{ F}$$

4.35 (d)
To increase the range, it is required to connect a small resistor in shunt across the ammeter.

4.37 (c)

$$R_{sh} = \left(\frac{I_m}{I - I_m}\right) \times R_m = \frac{1 \times 10^{-3}}{5 - 10^{-3}} \times 50 = 0.01\ \Omega$$

4.39 (a)

$$R_m = 1\ \Omega$$
$$I_m = 10\text{ mA}$$
$$V = 10\text{ V}$$
$$I = \frac{10}{I} = 10\text{ A}$$
$$m = \frac{I}{I_m} = 1000$$
$$R_S = R_m(m-1) = 999\ \Omega - \text{in series}$$

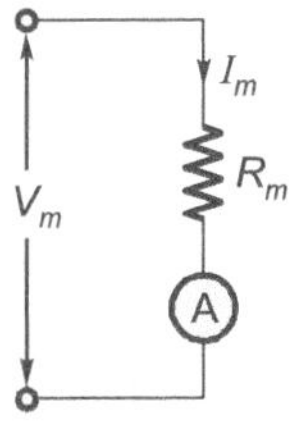

4.40 (a)

$$V = 50\text{ V}$$
$$S_V = 5\text{ k}\Omega/\text{V}$$
$$I_{FS} = 200\ \mu\text{A}$$
$$R_m = 100\ \Omega$$
$$V_m = I_{FS} \times R_m = 20\text{ mV}$$
$$m = \frac{V}{V_m} = 2500$$
$$R_s = R_m(m-1) = 249.9\text{ k}\Omega$$

4.41 (c)

$$I_m = 50\ \mu\text{A}$$
$$R_m = 500\ \Omega$$
$$V = 10\text{ V}$$
$$V_m = I_m R_m = 25\text{ mV}$$
$$m = \frac{V}{V_m} = \frac{10\text{ V}}{25\text{ mV}} = 400$$
$$R_S = R_m(m-1)$$
$$R_S = 199.5\text{ k}\Omega$$

4.42 (d)

$$S_V = \frac{1}{I_m} = \frac{1}{200 \times 10^{-6}} = 5\ \Omega/\text{mV}$$

4.43 (c)

$$I_m = 1\text{ mA}$$
$$R_m = 100\ \Omega$$
$$I = 100\text{ mA}$$
$$m = \frac{I}{I_m} = 100$$
$$R_{sh} = \frac{R_m}{m-1} = \frac{100}{99} = 1.01\ \Omega$$

4.44 (a)

$$\text{Error} = 25 \times \frac{1}{100} = \frac{1}{4}$$
$$\text{limiting error} = \frac{1/4}{10} \times 100\% = 2.5\%$$

4.45 (d)

Air friction and fluid friction provide damping torque.

Hence, option (d) is correct.

4.46 (a)

Null type instruments require many manipulations before null conditions are obtained and hence are apparently not suitable for dynamic measurements wherein the measured quantity changes with time. On the other hand, deflection type of instruments can follow the variations of the measured quantity more rapidly and hence are more suitable for dynamic measurements on account of their faster response.

Hence, option (a) is correct.

4.47 (b)

$$R_{sh} = \frac{R_m}{m-1}$$

where, $R_m = 100\ \Omega$ and $m = \frac{100}{1} = 100$

$$R_{sh} = \frac{100}{100-1} = 1.01\ \Omega$$

Hence, option (b) is correct.

4.48 (b)

In PMMC the damping torque is produced by movement of the aluminium former, moving in the magnetic field of the permanent magnet.

4.49 (b)

The PMMC instruments have high torque to weight ratio of moving parts.

4.50 (b)

As long as charge passes through the meter, there is no significant deflection of the moving system. The moving system deflects only after the charge has completely passed through the meter.

4.51 (d)

An ammeter has low internal resistance and is connected in series to the circuit, if we use it as a voltmeter then the high current will flow through the ammeter and then it will get damaged.

4.52 (c)

Multiplying power of shunt

$$m = \frac{I}{I_m} = \frac{500}{100} = 5$$

Internal resistance of ammeter

$$R_m = R_{sh}(m-1) = 25\,(5-1) = 100\ \Omega$$

4.53 (b)

By using guard circuit in the measurement of insulation resistance, leakage current passes through guard wire. Hence it reduces the effect of leakage current.

4.54 (c)

Multiplying factor (n)

$$= \frac{50}{50\times 10^{-6}\times 10^{3}} = 1000$$

4.55 (b)

$$\text{Multiplying factor } (m) = \frac{10}{0.1\times 10^{-3}\times 500}$$

$$m = 200$$

$$R_s = (m-1)\cdot R_m = 199\times 500 = 99500\ \Omega = 99.5\ \text{k}\Omega$$

4.56 (a)

The vibration galvanometer is used for detecting alternating currents in the frequency of its natural resonance. Most common application is as a null indicating instrument in AC bridge circuits and current comparators.

4.57 (a)

In PMMC type instruments control springs are connected in series with the moving coil. Therefore current through the coil will be zero and hence meter will read zero.

4.58 (b)

$$R_{sh} = \frac{R_m}{\frac{I_{fs}}{I_m}-1}$$

$$0.5 = \frac{R_m}{\frac{10}{1}-1} \Rightarrow R_m = 4.5\ \Omega$$

4.59 (c)

Given, $M = -6\cos(\theta + 30°)$ mH

$I_d = 50$ mA

$\therefore$ At $\theta = 60°$

$$\frac{dM}{d\theta} = +6\sin(60° + 30°) = 6$$

$$\therefore\quad T_d = I_d^2\frac{dM}{d\theta} = 50\times 50\times 6\times 10^{-9}\ \text{N-m} = 15\ \mu\text{N-m}$$

4.60 (b)

$$R_{sh} = \frac{R_m}{\frac{I_{fs}}{I_m}-1} = \frac{50}{\frac{5}{1\times 10^{-3}}-1} = \frac{50}{4999} \simeq 0.01\ \Omega$$

4.63 (a)

During transportation, if any emf induced due to stray fields etc. then there must be a critical damping resistance across the galvanometer terminals, in order to protect the galvanometer.

4.66 (c)

$$R_{se} = \frac{20}{50\times 10^{-6}} - 500 = 399.5\ \text{k}\Omega$$

4.67 (a)

Voltage sensitivity = Current sensitivity × Critical damping resistance

$= 1\ \mu$A/mm × 1 kΩ

$= 1$ mV/mm

$$\text{Resistance sensitivity} = \frac{1}{I_{fd}} = \frac{1}{1\,\mu A} = 1\,\text{M}\Omega$$

4.69 (a)

$$m = \frac{30}{\left(\frac{75\times 10^{-3}}{5}\right)} = 2000$$

$$R_{sn} = \frac{5}{(2000-1)} = 2.5\ \text{m}\Omega$$

4.71 (d)

$$I_{fsd} = \frac{1}{1000\ \Omega/\text{V}} = 1\text{mA}$$

$$I_{half\ fsd} = 0.5\ \text{mA}$$

4.73 (b)

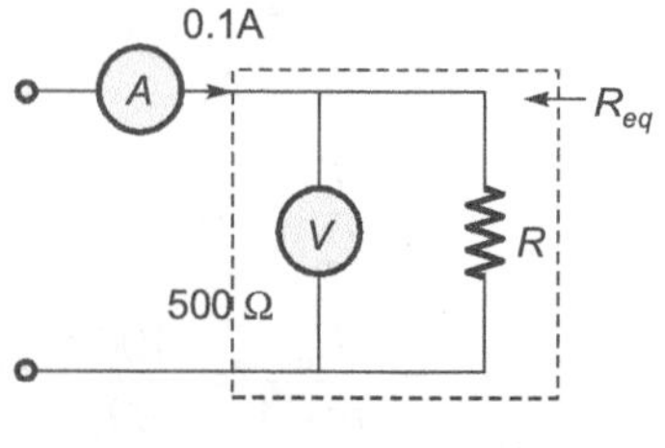

$$R_{eq} = \frac{500R}{500+R}$$

$$\Rightarrow \quad 10 = \frac{500R}{500+R} \times 0.1$$

$$\Rightarrow \quad 5000 + 10R = 50R \Rightarrow 40R = 5000$$

$$\Rightarrow \quad R = 125\ \Omega$$

4.74 (b)

$$m = \frac{5}{1\times10^{-3}} = 5000$$

$$R_{sh} = \frac{50}{(5000-1)} = \frac{R_m}{(m-1)}$$

$$R_{sh} = 0.01\ \Omega$$

4.75 (b)

$$V_m = 50\times10^{-6}\times500 = 25\text{ mV}$$

$$m = \frac{5}{25\times10^{-3}} = 200$$

$$R_{se} = R_m(m-1) = 500\,(200-1)$$

$$R_{se} = 99.5\text{ k}\Omega$$

4.79 (d)

Time response of an indicating instrument is decided by Damping system.

4.80 (b)

Constant of Fluxmeter

$$G = N_c\,B_c\,A_c = 40\times0.05\times750\times10^{-6} = 1500\times10^{-6}$$

Flux linkage with search coil

$$= 0.5\times200\times10^{-6} = 1\times10^{-4}\ wb$$

As the flux is reversed, the change in flux linking with search coil is

$$\Delta\phi = 2\times1\times10^{-4} = 2\times10^{-4}\text{ Wb}$$

$$\Delta\phi = \frac{G\theta}{N}$$

$$2\times10^{-4} = \frac{1500\times10^{-6}\times\theta}{10}$$

$$\theta = 1.33\text{ rad}$$

$$= 1.33\times\frac{180}{\pi} = 76.2°$$

4.81 (d)

$$R_{sh} = \frac{R_m}{\left(\frac{I}{I_m}-1\right)} = \frac{100}{\left(\frac{500}{100}-1\right)} = 25\ \Omega$$

4.82 (b)

% Error at any reading

$$= \frac{\text{Full scale error}\times\text{Full scale value}}{\text{Full scale reading value}}$$

$$= \frac{2\times200}{100} = 4\%$$

4.86 (a)

$$R_{sh} = \frac{R_m}{m-1} = \frac{V_m/I_m}{\left(\frac{I}{I_m}-1\right)}$$

$$= \frac{10/10}{\left(\frac{100}{10\times10^{-3}}-1\right)} = 0.0001\ \Omega$$

4.87 (c)

PMMC reads avg. value = –4 A

MI reads RMS value

$$= \sqrt{I_0^2+\frac{1}{2}I_1^2} = \sqrt{(-4)^2+\frac{1}{2}\left(3\sqrt{2}\right)^2} = 5\text{ A}$$

4.88 (b)

$$\%\,E_r = -\frac{V}{IR_v}\times100$$

$$= -\frac{200}{2\times2000}\times100 = -5\%$$

4.89 (b)

$$m = \frac{R_m}{R_{sh}}+1 = \frac{1000}{200}+1 = 6$$

4.90 (c)

MI reads rms current as given below,

$$I_{rms} = \left[I_0^2+\frac{1}{2}\left(I_{m_1}^2+I_{m_2}^2+\ldots\right)\right]^{1/2}$$

$$i(t) = 10 + 5\sin\omega t + 3\sin 2\omega t$$

$$I_{rms} = \left[10^2+\frac{1}{2}\left(5^2+3^2\right)\right]^{1/2} = 10.82\text{ A}$$

4.91 (c)

Ammeter has low resistance not high resistance.

4.92 (d)

MI is used for both AC and DC circuits.

4.93 (a)

PMMC has self shielding property due to permanent magnet so used for air space and aircraft applications.

4.94 (d)

Vibration Galvanometer is used as a detector in AC bridges.

4.95 (c)

MI instruments used for both AC and DC measurements and consumes less power.

4.98 (d)

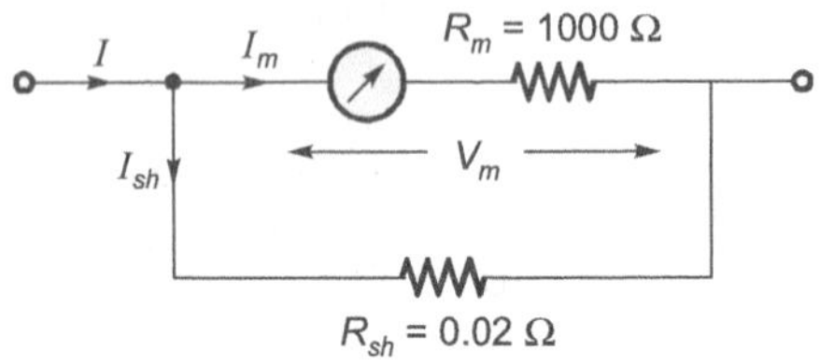

$$m = 1+\frac{R_m}{R_{sh}} = 1+\frac{1000}{0.02} = 50001$$

$$I_m = \frac{V_m}{R_{sh}} = \frac{500\times10^{-3}}{1000}$$

$$= 0.5 \times 10^{-3} \text{ A}$$

4.99 (b)

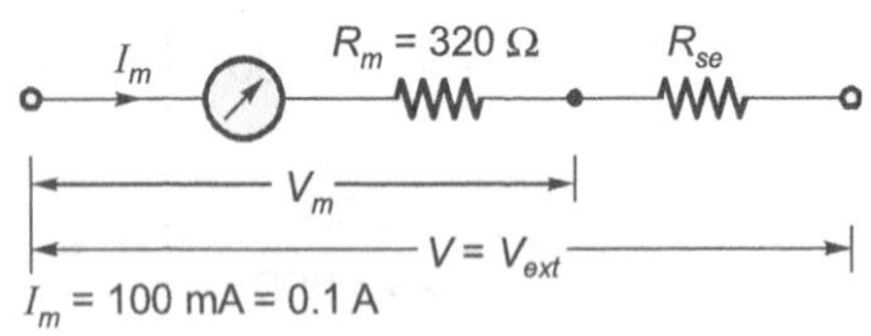

$$I_m = 100 \text{ mA} = 0.1 \text{ A}$$

$$V_m = I_m R_m = 320 \times 0.1 = 32 \text{ V}$$

$$R_{se} = R_m[m-1] = R_m\left[\frac{V}{V_m}-1\right]$$

$$R_{se} = 320\left[\frac{250}{32}-1\right] = 2180\ \Omega$$

With Temperature 20°C:

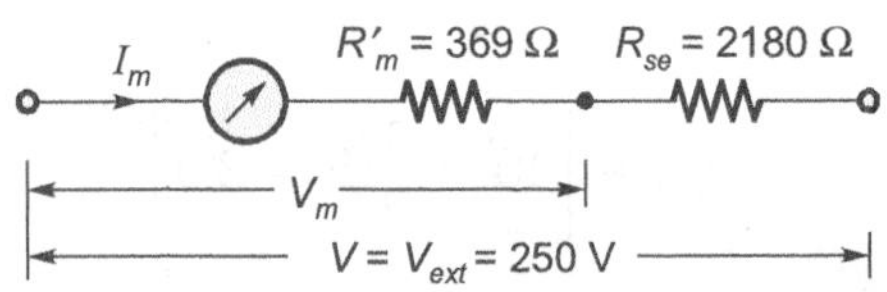

$$I'_m = \frac{250}{2180+369} = 0.098077$$

Calibrated as $\Rightarrow$ 0.1 A $\rightarrow$ 250 V

0.098 A $\rightarrow$?

$$\Rightarrow \frac{250}{0.1}\times0.098077 = 245.19 \text{ V}$$

True value = 250 volt

$$\% \text{ Error} = \frac{245.19-250}{250}\times100 = -1.924\%\%$$

4.100 (b)

$$\%\text{L.E.} = \frac{150}{83} = 1.81\%$$

4.103 (a)

Multiplication factor,

$$m = \frac{10 \text{ mA}}{1 \text{ mA}} = 10$$

$$\text{Value of shunt resistance} = \frac{R_m}{m-1}$$

$$R_{sh} = \frac{27}{10-1}$$

$$R_{sh} = 3\ \Omega$$

4.104 (d)

Repulsion instruments are more suitable for economical production in manufacture, and a nearly uniform scale is more easily obtained. Out of the other moving iron mechanism, radial vane repulsion is the most sensitive and has most linear scale.

4.105 (a)

For DC measured correctly

$$I_1 = I_{fs} = \frac{100\text{V}}{300\Omega} = \frac{1}{3} = 0.333\text{A}$$

For AC of 100 V, f = 25 Hz

$$I_2 = \frac{100}{\sqrt{(300)^2+(2\pi\times25\times0.12)^2}}$$

$$= \frac{100}{300.59} \Rightarrow 0.3326 \text{ A}$$

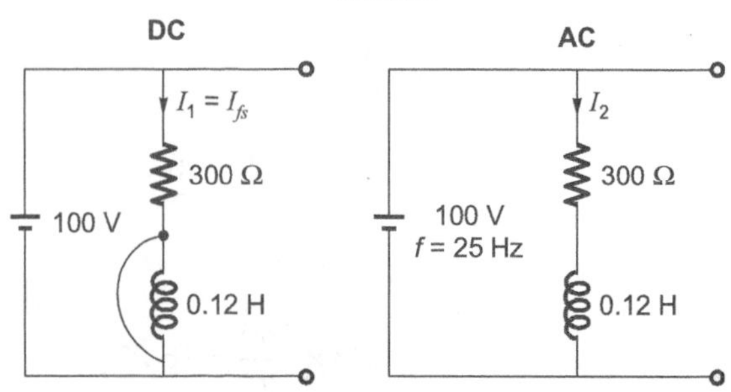

DC : 0.3333 A → 100 V

For AC current of 0.3326 A → ?

$$\text{Reading} = \frac{100}{0.3333} \times 0.3326 = 99.8 \text{ V}$$

4.106 (c)

$$V_m = 160 \text{ V}$$
$$R_m = 20 \text{ k}\Omega$$
$$V = 230 \text{ V}$$

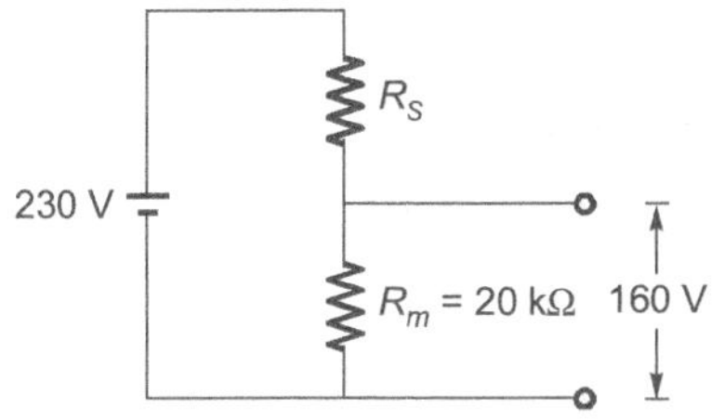

$$R_s = R_m(m-1)$$
$$= R_m\left(\frac{V}{V_m} - 1\right)$$
$$= 20 \times 10^3\left(\frac{230}{160} - 1\right)$$
$$= 8.75 \text{ k}\Omega = 8750 \ \Omega$$

4.107 (c)

$$V_m = R_m \times I_m = 15 \times 5 = 75 \text{ mV}$$
$$V = 100 \text{ V}$$

Resistance to be added in series

$$R_s = R_m\left(\frac{V}{V_m} - 1\right)$$
$$= 5\left(\frac{100}{75 \times 10^{-3}} - 1\right)$$
$$= 6661.66 \ \Omega$$

4.108 (c)

Full scale meter current

$$I_{fs} = 25 \text{ mA} = I_m$$
$$R_m = 50 \ \Omega$$
$$V_m = I_m \cdot R_m = 1.25 \text{ V}$$
$$R_S = R_m(m-1)$$

$$\text{Multiplication factor } (m) = \frac{10 \text{ V}}{1.25 \text{ V}} = 8$$

$$R_S = R_m\left(\frac{V}{V_m} - 1\right) = 50\left(\frac{10}{1.25} - 1\right)$$
$$= 350 \ \Omega$$

4.109 (b)

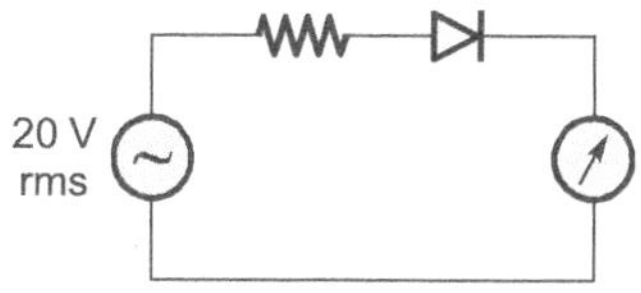

$$V_{DC} = 0.45 \ V_{rms}$$
$$= 0.45 \times 20$$
$$= 9 \text{ V}$$

4.110 (a)

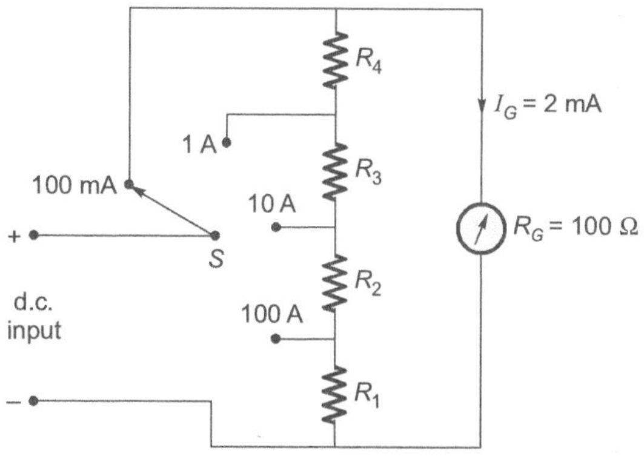

For 100 mA:

$$R_{sh} = R_1 + R_2 + R_3 + R_4$$
$$= \frac{R_m}{\left(\frac{I}{I_m} - 1\right)} = \frac{100}{\left(\frac{100}{2} - 1\right)}$$

$$R_1 + R_2 + R_3 + R_4 = 2.0408 \ \Omega \quad \text{...(i)}$$

For 1 A:

$$R_2 + R_3 + R_1 = \frac{100 + R_4}{\left(\frac{1}{2 \times 10^{-3}} - 1\right)} = \frac{100 + R_4}{499} \quad \text{...(ii)}$$

For 10 A:

$$R_2 + R_1 = \frac{100 + R_3 + R_4}{\left(\frac{10}{2 \times 10^{-3}} - 1\right)}$$
$$= \frac{100 + R_3 + R_4}{4999} \quad \text{...(iii)}$$

Switched at 100 A:

$$R_{sh} = R_1$$
$$= \frac{R_2 + R_3 + R_4 + 100}{\left(\frac{100}{2 \times 10^{-3}} - 1\right)}$$
$$= \frac{R_2 + R_3 + R_4 + 100}{49999} \quad \text{...(iv)}$$

By solving equation (i), (ii), (iii) and (iv)

$R_1 = 0.002\ \Omega$; $R_2 = 0.018\ \Omega$

$R_3 = 0.18\ \Omega$

and $R_4 = 1.836\ \Omega$

4.111 (a)

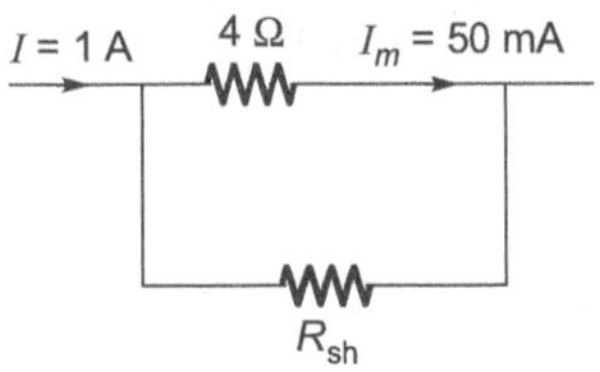

$$R_{sh} = \frac{R_m}{(m-1)},$$

$$m = \frac{I}{I_m} = \frac{4}{\left(\frac{1}{50\times10^{-3}} - 1\right)}$$

$$= 0.21053\ \Omega$$

4.112 (b)

In electrodynamometer type instrument deflection torque,

$$\tau_d \propto I^2$$

∴ (b) option is right choice.

4.113 (d)

$$T_d = (mgl)\sin\theta$$

⇒ For gravity control

4.115 (d)

A resistance in parallel.

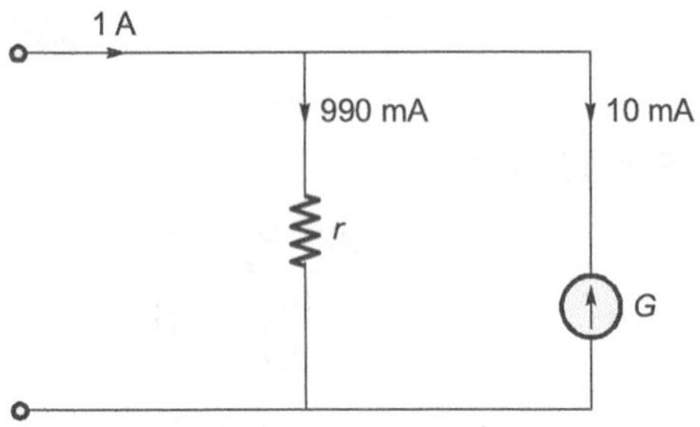

4.116 (c)

Moving coils – (V)

$$\Rightarrow \quad A = 40\text{ mm} \times 30\text{ mm}$$

$$= 40 \times 30 \times 10^{-6}\ \text{m}^2$$

Number of turns,

$N = 100,$

$\tau_c = 240 \times 10^{-6}$ N-m

Deflections = 100 division FSD,

$B = 1.0\ \text{Wb/m}^2$

at S.S $|\tau_d| = |\tau_c|$

$$\Rightarrow \quad B\,INA = \tau_c$$

$$B \times \frac{V}{R} \times N \times A = \tau_c$$

$$1 \times \frac{100}{R} \times 100 \times 40 \times 30 \times 10^{-6} = 240 \times 10^{-6}$$

$$\Rightarrow \quad R = 50000\ \Omega = 50\ \text{k}\Omega$$

4.118 (b)

$I_{FSD} = 10$ mA,

$V_{FSD} = 100$ mV

$$\Rightarrow \quad R_m = \frac{V_{FSD}}{I_{FSD}} = \frac{100\text{ mV}}{10\text{ mA}} = 10\ \Omega$$

$$\underset{I_m}{(0 - 10\text{ mA})} = \underset{I_{ext}}{(0 - 100\text{ A})}$$

$$m = \frac{I_{ext}}{I_m} = \frac{100}{10\times10^{-3}} = 10000$$

$$R_{sh} = \frac{R_m}{m-1} = \frac{10}{(10000-1)} = 0.001\ \Omega$$

■■■■

5 Measurement of Energy and Power

5.1 In a single-phase power factor meter, the controlling torque is
(a) provided by spring control
(b) provided by gravity control
(c) provided by stiffness of suspension
(d) not required

[ESE-2001]

5.2 In electrodynamometer type wattmeters, the inductance of pressure coil produces error. The error is
(a) constant irrespective of the power factor of the load
(b) higher at higher power factor loads
(c) higher at lower power factor loads
(d) highest at unity power factor loads

[ESE-2001]

5.3 If an induction type energy meter runs fast, it can be slowed down by
(a) lag adjustment
(b) light load adjustment
(c) adjusting the position of braking magnet and moving it closer from the centre of the disc
(d) adjusting the position of braking magnet and moving it away from the centre of the disc

[ESE-2001]

5.4 When two-Wattmeter method of measurement of power is used to measure power in a balanced three phase circuit; if the Wattmeter reading is zero, then
(a) power consumed in the circuit is zero
(b) power factor of the circuit is zero
(c) power factor is unity
(d) power factor is 0.5

[ESE-2002]

5.5 In the measurement of power on balanced load by two Wattmeter method in a 3-phase circuit, the readings of the Wattmeters are 3 kW and 1 kW respectively, the later being obtained after reversing the connections to the current coil. The power factor of the load is

(a) 0.277 (b) 0.554
(c) 0.625 (d) 0.866

[ESE-2002]

5.6 Two types of connection of Wattmeter pressure coil are shown in the figures.

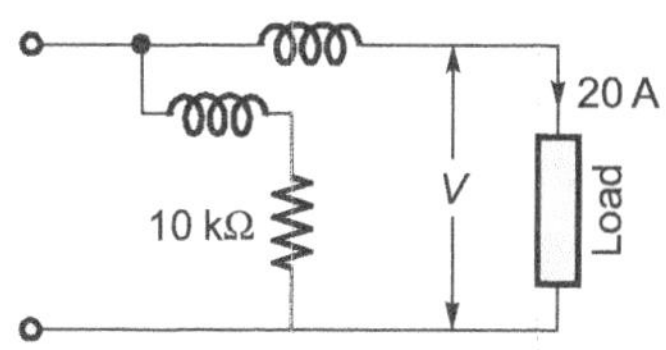

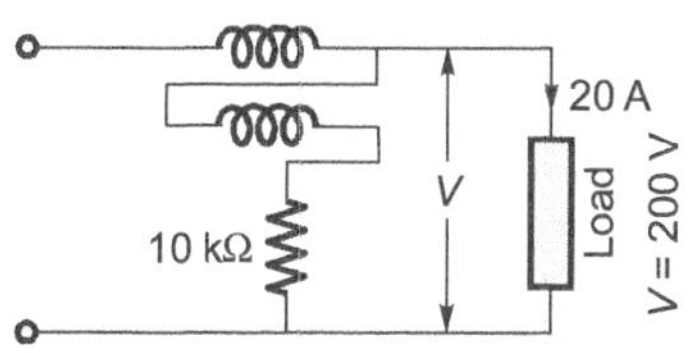

The value of the Wattmeter current coil resistance r, which makes the connection errors the same in the two cases is

(a) 0.05 Ω (b) 0.1 Ω
(c) 0.01 Ω (d) 0.125 Ω [ESE-2002]

5.7 In calibration of a dynamometer Wattmeter by potentiometer, phantom loading arrangement is used because
(a) the arrangement gives accurate results.
(b) the power consumed in calibration work is minimum.
(c) the method gives quick results.
(d) the onsite calibration is possible.

[ESE-2002]

5.8 An energy-meter having a meter constant of 1200 rev/kWh is found to make 5 revolutions in 75s. The load power is

(a) 500 W (b) 100 W
(c) 200 W (d) 1000 W [ESE-2002]

5.9 The energy capacity of a storage battery is rated in

(a) kWh (b) kW
(c) Ampere hour (d) Joules

[ESE-2003]

5.10 The pressure coil of an induction type energy meter is

(a) Highly resistive (b) Highly inductive
(c) Purely resistive (d) Purely inductive

[ESE-2003]

5.11 Which one of the following methods decreases the error due to connections in a dynamometer type Wattmeter?

(a) Using bifilar compensating winding in place of current coil.
(b) Using non-inductive pressure coil circuit.
(c) Using a capacitor across a part of high resistance of pressure coil circuit.
(d) Using a swamping resistance.

[ESE-2004]

5.12 The power of a three-phase, three-wire balanced system was measured by two Wattmeter method. The reading of one of the Wattmeters was found to be double that of the other. What is the power factor of the system?

(a) 1 (b) 0.866
(c) 0.707 (d) 0.5

[ESE-2004]

5.13 What is the effect of inductance in the pressure coil on performance of a dynamo-meter type wattmeter?

(a) It reads low on lagging power factor and high on leading power factor.
(b) It reads high on lagging power factor and low on leading power factor.
(c) Its reading is not affected at all.
(d) It always reads low. [ESE-2005]

5.14 Which one of the following defects is responsible for creeping in an induction type energy meter?

(a) Imperfect lag compensation
(b) Over friction compensation
(c) Imperfect overload compensation
(d) Misalignment of brake magnet.

[ESE-2005]

5.15 **Assertion (A):** The rotating disc in an energy meter is made up of a magnetic material.

Reason (R): Braking takes place due to eddy current generated by the braking magnet.

(a) Both A and R are true and R is the correct explanation of A
(b) Both A and R are true but R is NOT the correct explanation of A
(c) A is true but R is false
(d) A is false but R is true

[ESE-2006]

5.16 **Assertion (A):** General purpose dynamo-meter type Wattmeter cannot indicate the correct value of power at low power factors.

Reason (R): The presence of self-inductance in the pressure coil circuit introduces an error in the indicated value which increases appreciably with decrease in power factor.

(a) Both A and R are true and R is the correct explanation of A
(b) Both A and R are true but R is NOT the correct explanation of A
(c) A is true but R is false
(d) A is false but R is true

[ESE-2006]

5.17 The voltage coil of a single-phase house service energy meter

(a) is highly resistive
(b) is highly inductive
(c) is highly capacitive
(d) has a phase angle equal to load power factor angle

[ESE-2006]

5.18 Which one of the following is the correct statement?

Due to inductance in the pressure coil, the dynamometer wattmeter

(a) reads low on lagging p.f. and high on leading p.f.
(b) reads high on lagging p.f. and low on leading p.f.
(c) reading is not affected at all
(d) always reads low

[ESE-2007]

5.19 Which one of the following is the main cause of creeping in the induction type energy meters?
(a) Friction compensation
(b) Lag/Lead compensation
(c) Overload compensation
(d) Braking torque producing system
[ESE-2007]

5.20 Which of the following statements are correct in case of a power factor meter?
1. The deflection is proportional to the phase angle between field coil and crossed coil.
2. The restoring torque is provided by a controlling torque.
3. It consists of two coils mounted at right angles to each other.

Select the correct answer using the code given below:
(a) 1 and 2 (b) 2 and 3
(c) 1 and 3 (d) 1, 2 and 3
[ESE-2007]

5.21 In a low power factor wattmeter, why is a compensating coil employed?
(a) To neutralize the capacitive effect of pressure coil
(b) To compensate for inductance of pressure coil
(c) To compensate for the error caused by power loss in the pressure coil
(d) To compensate for the error caused by eddy currents
[ESE-2007]

5.22 Which one of the following is used for the measurement of 3-phase power factor?
(a) Power factor meter
(b) Crossed-coil power factor meter
(c) Phase-angle watt hour meter
(d) Polarised-vane power factor meter
[ESE-2008]

5.23 If the current flowing through a 20 Ω resistor is given as,
$i(t) = 4 + 5 \sin\omega t - 3 \cos 3\omega t$, then what is the power consumed by the resistor?
(a) 1000 W (b) 660 W
(c) 500 W (d) 180 W
[ESE-2009]

5.24 How can the power supplied to a high frequency heating system be measured?
(a) By dynamometer wattmeter
(b) By induction wattmeter
(c) By thermocouple type wattmeter
(d) By moving iron ammeter and voltmeter
[ESE-2009]

5.25 Consider the following statements regarding measurement of 3-phase power by two-wattmeter method; one of the wattmeter reads negative implying:
1. Power factor is less than 0.5.
2. Power flow is in the reverse direction.
3. Load power factor angle is greater than 60°.
4. Load is unbalanced.

Which of the above statements are correct?
(a) 1 and 2 only (b) 2 and 3 only
(c) 1 and 3 only (d) 1, 2, 3 and 4
[ESE-2010]

5.26 Match **List-I** with **List-II** and select the correct answer using the code given below the lists:

List-I
A. Electrostatic
B. Induction
C. Dynamometer
D. Thermal

List-II
1. Power at 50 Hz only
2. Power at frequencies ranging from dc to RF
3. Power at unity power factor at high values
4. Power at frequency ranging from dc to 150 Hz

Codes:

	A	B	C	D
(a)	3	1	4	2
(b)	2	1	4	3
(c)	3	4	1	2
(d)	2	4	1	3

[ESE-2010]

5.27 In the power measurement by ammeter-voltmeter method, if the voltmeter is connected across the load, then the value of the power will be
(a) the power consumed by the load
(b) the sum of power consumed by the load and ammeter
(c) the sum of power consumed by the load and voltmeter
(d) the sum of power consumed by the load, ammeter and voltmeter
[ESE-2011]

5.28 The current and potential coils of a dynamometer type wattmeter were accidentally interchanged while connecting. After energizing the circuit, it was observed that the wattmeter did not show the reading. This could be due to the
(a) Damage to potential coil
(b) Damage to current coil
(c) Damage to both the potential and current coil
(d) Loose contacts

[ESE-2011]

5.29 Consider the following statements associated with an energy meter:
1. It is an integrating type instrument.
2. It is an induction type instrument.
3. It uses a permanent magnet for rotation of aluminium disc.
4. It employs a high control torque.

Which of these statements are correct?
(a) 1, 2, 3 and 4 (b) 1 and 2 only
(c) 2 and 3 only (d) 3 and 4 only

[ESE-2011]

5.30 A capacitor is connected across a portion of resistance of the multiplier in order to make the pressure coil circuit of the wattmeter non-inductive. The value of this resistance is r while the total resistance and inductance of the pressure circuit are respectively R_p and L. The value of the capacitance C is
(a) $\frac{L}{R_p^2}$ (b) $\frac{0.41L}{r^2}$
(c) $\frac{L}{r^2}$ (d) $\frac{0.41L}{R_p^2}$

[ESE-2011]

5.31 The magnetic field responsible for the production of the deflecting torque in an accurate dynamometer type wattmeter, being very weak, the accuracy of the measurement can be increased by providing a
(a) Magnetic shield around the instrument
(b) Compensating winding along with the pressure coil
(c) Astatic arrangement to the moving system of the instrument
(d) Capacitance shunt across a portion of the pressure coil

[ESE-2011]

5.32 Due to the effect of inductance in the pressure coil, a dynamometer type wattmeter
(a) Reads low on lagging power factor and high on leading power factor
(b) Reads high on lagging power factor and low on leading power factor
(c) Reading is independent of the power factor
(d) Always reads lower than actual value

[ESE-2011]

5.33 **Assertion (A):** Electrodynamometer wattmeter is not suitable for low power factor power measurement.
Reason (R): Many wattmeters are compensated for errors caused by inductance of voltage coil by means of a capacitor connected in parallel with a portion of multiplier series resistance.
(a) Both A and R are true and R is the correct explanation of A
(b) Both A and R are true but R is NOT the correct explanation of A
(c) A is true but R is false
(d) A is false but R is true

[ESE-2011]

5.34 A single-phase energy meter having meter constant of 200 rev/kWh is operating on 230 V, 50 Hz supply with a load of 10 A, and at unity power factor for three hours continuously. The number of revolutions shown by the meter during this period is
(a) 13800 (b) 1380
(c) 276 (d) 138

[ESE-2012]

5.35 The meter constant of a single-phase energy meter is 500 rev/kWh. It is found that with a load of 5 kW, it makes 40 revolutions in 50 sec. The percentage error is
(a) 5.25% (b) 10.5%
(c) 15.25% (d) 20%

[ESE-2012]

5.36 In two-wattmeter method of measuring power in a balanced 3-phase circuit, the readings of the two wattmeters are in the ratio of 1 : 2, the circuit power factor is

(a) $\frac{1}{\sqrt{2}}$ (b) $\frac{1}{2}$

(c) $\frac{\sqrt{3}}{2}$ (d) 1 [ESE-2013]

5.37 Consider the following statements:
Adjustment is required in an induction type energy meter in the following manner so that it can be compensated for slowdown of speed on the specified load due to some unspecified reason:

1. Adjusting the position of braking magnet and moving it away from the centre of the disc.
2. Adjusting the position of braking magnet and moving it closer to the centre of the disc.
3. Adjusting the load.

Which of these statements are correct?
(a) 1, 2 and 3 (b) 1 only
(c) 2 only (d) 3 only

[ESE-2013]

5.38 A 3-phase moving coil type power factor meter has three fixed and symmetrically spaced current coils, inside of which are three other similarly placed moving potential coils. While in operation, rotating magnetic field is produced
(a) in the current coils but not in the potential coils.
(b) in the potential coils but not in the current coils.
(c) in both potential coils and the current coils.
(d) in neither the potential coils nor the current coils.

[ESE-2013]

5.39 In a low power factor wattmeter, some-times compensating coil is connected in order to
(a) neutralize the capacitive effect of pressure coil.
(b) compensate for inductance of pressure coil.
(c) compensate for power loss in the pressure coil.
(d) reduce the error caused by eddy current.

[ESE-2013]

5.40 The current and potential coils of a wattmeter were accidentally interchanged while connecting. After energizing the circuit, it was observed that the wattmeter did not show the reading. This would be due to
(a) damage done to the potential coil
(b) damage done to the current coil
(c) damage done to both potential and current coils
(d) loose contact

[ESE-2013]

5.41 Creep error may occur in induction type energy meter due to
(a) incorrect position of brake magnet
(b) incorrect adjustment of position of shading band
(c) overvoltage across voltage coil
(d) increase in temperature

[ESE-2013]

Directions:

Each of the next items consists of two statements, one labelled as the **'Statement (I)'** and the other as **'Statement (II)'**. Examine these two statements carefully and select the answers to these items using the codes given below:

Codes:
(a) Both Statement (I) and Statement (II) are individually true and Statement (II) is the correct explanation of Statement (I).
(b) Both Statement (I) and Statement (II) are individually true but Statement (II) is not the correct explanation of Statement (I).
(c) Statement (I) is true but Statement (II) is false.
(d) Statement (I) is false but Statement (II) is true.

5.42 **Statement (I):** In an ac dynamometer type wattmeter the instantaneous value of developed deflecting torque is proportional to the product of voltage and current in the respective coils at the corresponding instant, the constant of proportionality being the same as in the case of dc use.

Statement (II): The moving system of the meter is prevented by its inertia from following the variations in the deflecting torque which takes place during each cycle, and takes up a position corresponding to the average value of the torque.

[ESE-2013]

5.43 **Statement (I):** Electrostatic Wattmeter is not widely used commercially because of its inability to measure power of high value.

Statement (II): It is used mainly for very small power measurement at high voltages and low power factors.

[ESE-2013]

5.44 **Statement (I):** A Watt-hour meter must be calibrated at both full rated load as well as at 10% of rated load.

Statement (II): The source of error at full load is inaccurate damping and at light loads, the torque is not exactly proportional to load.

[ESE-2013]

5.45 A 3-phase load of 0.8 pf lag is supplied from a balanced 3-phase supply of phase sequence *RYB*. With V_{YB} as reference the current I_R will (Given $\cos^{-1}(0.8) = 36.76°$)

(a) In-phase with V_{YB}
(b) Lag V_{YB} by 36.76°
(c) Lead V_{YB} by 53.14°
(d) Lag V_{YB} by 53.14° [ESE-2013]

5.46 In the two-wattmeter method of measuring 3-ϕ power, the wattmeters indicate equal and opposite readings when load power factor is

(a) 90 leading (b) 90 lagging
(c) 30 leading (d) 30 lagging

[ESE-2013]

5.47 If 3-phase power is measured with the help of two-wattmeter method in a balanced load with the application of 3-phase balanced voltage, variation in readings of wattmeters will depend on

(a) load only
(b) power factor only
(c) load and power factor
(d) neither load nor power factor

[ESE-2013]

5.48 **Statement (I):** The simplest method of power measurement is by means of electrodynamic type wattmeters, having two fixed coils, and one moving coil.

Statement (II): Either of the fixed and the moving coils can be used as the current or the voltage coils.

(a) Both Statement (I) and Statement (II) are individually true and Statement (II) is the correct explanation of Statement (I).
(b) Both Statement (I) and Statement (II) are individually true but Statement (II) is not the correct explanation of Statement (I).
(c) Statement (I) is true but Statement (II) is false.
(d) Statement (I) is false but Statement (II) is true. [ESE-2014]

5.49 One single-phase energy meter operating on 230 V and 5 A for 5 hours makes 1940 revolutions. Meter constant is 400 rev/kWh. The power factor of the load is

(a) 1.0 (b) 0.8
(c) 0.7 (d) 0.6 [ESE-2014]

5.50 The current coil of a single-phase energy meter is wound on

(a) one limb of the laminated core
(b) both the limbs of the laminated core with same number of turns
(c) both the limbs of the laminated core with different number of turns
(d) the center of the limb on the laminated core

[ESE-2014]

5.51 For controlling the vibration of the disc of ac energy meter, damping torque produced by

(a) Eddy current
(b) Chemical effect
(c) Electrostatic effect
(d) magnetic effect [ESE-2014]

5.52 The meter constant of a single-phase 230 V induction watt hour meter is 400 rev/kWh. The speed of the meter disc for a current of 10 A of 0.9 pf lagging will be

(a) 13.80 rpm (b) 16.02 rpm
(c) 18.20 rpm (d) 21.10 rpm

[ESE-2014]

5.53 Two wattmeters are used to measure the power in a 3-phase balanced system. What is the power factor of the load when one wattmeter reads twice the other?

(a) 0 (b) 0.5
(c) 0.866 (d) 1

[ESE-2014]

5.54 In a balanced 3-phase 200 V circuit, the line current is 115.5 A. When the power is measured by two wattmeter method, one of the wattmeter reads 20 kW and the other one reads zero. What is the power factor of the load?

(a) 0.5 (b) 0.6
(c) 0.7 (d) 0.8

[ESE-2014]

5.55 A coil having an inductance of 0.5 H and a resistance of 60 Ω is connected in series with a capacitance of 10 μF. The coil is connected to 100 V ac supply. What is the source frequency and current flowing in the circuit under resonance condition?

(a) 7.121 Hz and 16.7 A
(b) 7.121 Hz and 1.67 A
(c) 71321 Hz and 16.7 A
(d) 71.21 Hz and 1.67 A

[ESE-2014]

5.56 In an induction type energy meter, the steady speed attained by the rotating disc is

1. Proportional to the deflecting torque.
2. Proportional to the resistance of the path of eddy currents.
3. Inversely proportional to the effective readings of disc from its axis.
4. Inversely proportional to the square of brake magnet flux.

Which of the above are correct?

(a) 1, 2 and 3 only (b) 1, 2 and 4 only
(c) 2, 3 and 4 only (d) 1, 2, 3 and 4

[ESE-2014]

5.57 Two wattmeter method is employed to measure power in a 3-phase balanced system with the current coil connected in the *A* and *C* lines. The phase sequence is *ABC*. If the wattmeter with its current coil in A phase line reads zero, then the power factor of the 3-phase load will be

(a) zero lagging (b) zero leading
(c) 0.5 lagging (d) 0.5 leading

[ESE-2014]

5.58 In an electrodynamometer wattmeter

(a) the fixed coils providing magnetic flux are connected across the power line.
(b) the compensated wattmeter improves its accuracy by using windings with opposite currents with respect to series windings
(c) if the full-scale power measured is 100 W, then the half-scale power will be 10 W
(d) It can measure a.c. power but is unsuitable for d.c. power

[ESE-2015]

5.59 In a two-wattmeter method of measuring power in a balanced 3-phase circuit, the ratio of the two wattmeter readings is 1:2. The circuit power factor is

(a) 0.707 (b) 0.5
(c) 0.866 (d) indeterminate

[ESE-2015]

5.60 The maximum power demand of a consumer is 2 kW and the corresponding daily energy consumption is 30 units. What is the corresponding load factor?

(a) 0.25 (b) 0.5
(c) 0.625 (d) 0.75 [ESE-2015]

5.61 The current $i(t)$ through a 10 Ω resistor in series with an inductance is given by

$i(t) = 3 + 4\sin(100t + 45°) + 4\sin(300t + 60°)$ A

The RMS value of the current and the power dissipated in the circuit are respectively

(a) 5 A and 150 W (b) 11 A and 250 W
(c) 5 A and 250 W (d) 11 A and 150 W

[ESE-2016]

5.62 Three star-connected loads of $3\angle 60°$ Ω each and three delta-connected loads of $9\angle 60°$ Ω each are connected in parallel and fed from a three-phase balanced source having line-to-neutral voltage of 120 V. The line currents drawn from the supply will be

(a) 10 A each (b) 20 A each
(c) 80 A each (d) 160 A each

[ESE-2016]

5.63 A wattmeter reads 10 kW, when its current coil is connected in *R* phase and the potential coil is connected across *R* and neutral of a balanced 400 V (RYB sequence) supply. The line current is 54 A. If the potential coil reconnected across B-Y phases with the current coil in *R* phase, the new reading of the wattmeter will be nearly

(a) 10 kW (b) 13 kW
(c) 16 kW (d) 19 kW

[ESE-2016]

5.64 Two-wattmeter method of power measurement in three-phase system is valid for
(a) balanced star-connected load only
(b) unbalanced star-connected load only
(c) balanced delta-connected load only
(d) balanced or unbalanced star- as well as delta-connected loads

[ESE-2016]

5.65 If the total powers consumed by three identical phase loads connected in delta and star configurations are W_1 and W_2 respectively, then W_1 is
(a) $3\,W_2$ (b) $\frac{W_2}{3}$
(c) $\sqrt{3}W_2$ (d) $\frac{W_2}{\sqrt{3}}$

[ESE-2016]

5.66 A wattmeter is measuring the power supplied to a circuit whose power factor is 0.7. The frequency of the supply is 50 c/s. The wattmeter has a potential coil circuit of resistance 1000 Ω. and inductance 0.5 H. The error in the meter reading is
(a) 4% (b) 8%
(c) 12% (d) 16%

[ESE-2016]

5.67 A 400 V, three-phase, rated frequency balanced source is supplying power to a balanced three-phase load carrying a line current of 5 A at an angle of 30° lagging. The readings of the two wattmeters W_1 and W_2, used for measuring the power drawn by the circuit, are respectively
(a) 2000 W and 1000 W
(b) 1500 W and 1500 W
(c) 2000 W and 1500 W
(d) 1500 W and 1000 W

[ESE-2016]

5.68 If an energy meter makes 5 revolutions in 100 seconds, when a load of 225 W is connected, the meter constant is
(a) 800 rev/kWh (b) 222 rev/kWh
(c) 147 rev/kWh (d) 13 rev/kWh

[ESE-2016]

5.69 Consider the following statements with regard to induction type wattmeter:
1. It can be used on both ac and dc systems.
2. Power consumption is relatively low.
3. It is accurate only at stated frequency and temperature.

Which of the above statements is/are correct?
(a) 1 only (b) 2 only
(c) 3 only (d) 1, 2 and 3

[ESE-2017]

5.70 **Statement (I):** To measure power consumed by the load, it is necessary to interchange the pressure coil terminals when the pointer of a wattmeter kicks back.

Statement (II): The pressure coil terminals are interchanged to get upscale reading in a wattmeter without affecting the continuity of power to the load.

[ESE-2017]

5.71 Consider the following statements regarding power measurement of three-phase circuits by two-wattmeter method:
1. Total power can be measured if the three-phase load is balanced and can be represented by an equivalent Y-connection only.
2. Total power can be measured for the three-phase load irrespective of, whether the load is balanced or not and connected in Y or D.
3. Power factor can be calculated only if the three-phase load is balanced.

Which of the above statements are correct?
(a) 1 and 2 only (b) 1 and 3 only
(c) 2 and 3 only (d) 1, 2 and 3

[ESE-2018]

5.72 An energy meter makes 100 revolutions of its disc per unit of energy. The number of revolutions made by the disc during one hour when connected across 210 V source and drawing a current of 20 A at 0.8 p.f. leading is
(a) 336 (b) 316
(c) 286 (d) 256

[ESE-2018]

5.73 **Statement (I) :** A general purpose dynamometer type wattmeter does not read accurately at low power factors.

Statement (II) : The presence of self-inductance of the pressure coil introduces an error.

[ESE-2018]

5.74 **Statement (I) :** A dynamometer type wattmeter has a linear scale while a dynamometer type voltmeter has a non-linear scale.

Statement (II) : Deflecting torque developed in a dynamometer type wattmeter is proportional to the power and that developed in a dynamometer type ammeter is proportional to the square of the current.

[ESE-2018]

5.75 **Statement (I) :** The rotating disc in an energy meter is made of a magnetic material.

Statement (II) : Braking takes place due to eddy current generated by the braking magnet.

[ESE-2018]

5.76 There will be serious errors if power factor of non-sinusoidal waveform is measured by electrodynamometer power factor meter. This is true for

(a) Single-phase meters alone
(b) 3-phase meters only
(c) Both Single-phase meters and 3-phase meters
(d) 3-phase meters with balanced loads

[ESE-2019]

5.77 An electrodynamometer instrument can be used as

1. Wattmeter and VAR meter.
2. Power factor meter and frequency meter.
3. Transfer instrument.

Which of the above statements are correct?

(a) 1 and 2 only (b) 1 and 3 only
(c) 2 and 3 only (d) 1, 2 and 3

[ESE-2020]

5.78 The power in a 3-phase circuit is measured with the help of 2-wattmeters; the readings of one of the wattmeters is positive and that of the other is negative. The magnitude of readings is different. It can be concluded that the power factor of the circuit will be

(a) Unity (b) Zero
(c) 0.5 (d) Less than 0.5

[ESE-2020]

5.79 In a dynamometer type wattmeter at low power factor, the inductance of the pressure coil introduces a serious error. This error can be minimized by

(a) connecting a high resistance in series with it and connecting a capacitor across a part of this resistance.
(b) connecting a high resistance in parallel with it and connecting a capacitor across a part of this resistance.
(c) connecting a low resistance in series with it and connecting a capacitor across a part of this resistance.
(d) connecting a low resistance in parallel with it and connecting a capacitor across a part of this resistance.

[ESE-2021]

5.80 The torque generated in the aluminium disc of induction type energy meter is maximum when the phase difference between the magnetic fields of shunt and series electromagnets is equal to

(a) 180° (b) 90°
(c) 45° (d) 0° **[ESE-2022]**

5.81 The two wattmeter method produces wattmeter readings P_1 = 1500 W and P_2 = 2500 W when connected to a delta-connected load. What is the per-phase reactive power if the line voltage is 220 V?

(a) 3117 VAR (b) 311.7 VAR
(c) 577.3 VAR (d) 5773 VAR

[ESE-2023]

5.82 A three-phase balanced star-connected motor draws a real power of 5.6 kW from the line voltage of 220 V and the line current of 18.2 A. What is the approximate power factor of this motor?

(a) 0.6075 (b) 0.8075
(c) 0.4075 (d) 0.2075 **[ESE-2023]**

5.83 The power in a three-phase circuit is measured with the help of 2 wattmeters. The reading of one of the wattmeters is positive and that of the other is negative. The magnitudes of readings are different. What is the value of the power factor of the circuit under this condition?

(a) 0.5 (lagging)
(b) Zero (lagging)
(c) Less than 0.5 (lagging)
(d) Unity **[ESE-2024]**

■■■■

Answers Measurement of Energy and Power

5.1 (d)	5.2 (c)	5.3 (d)	5.4 (d)	5.5 (a)	5.6 (c)	5.7 (b)	5.8 (c)	5.9 (c)
5.10 (b)	5.11 (a)	5.12 (b)	5.13 (b)	5.14 (b)	5.15 (d)	5.16 (a)	5.17 (b)	5.18 (b)
5.19 (a)	5.20 (c)	5.21 (b)	5.22 (d)	5.23 (b)	5.24 (c)	5.25 (c)	5.26 (a)	5.27 (c)
5.28 (b)	5.29 (b)	5.30 (c)	5.31 (c)	5.32 (b)	5.33 (b)	5.34 (b)	5.35 (c)	5.36 (c)
5.37 (c)	5.38 (c)	5.39 (c)	5.40 (b)	5.41 (c)	5.42 (a)	5.43 (b)	5.44 (a)	5.45 (c)
5.46 (b)	5.47 (c)	5.48 (c)	5.49 (b)	5.50 (b)	5.51 (a)	5.52 (a)	5.53 (c)	5.54 (a)
5.55 (b)	5.56 (c)	5.57 (c)	5.58 (b)	5.59 (c)	5.60 (c)	5.61 (c)	5.62 (c)	5.63 (b)
5.64 (d)	5.65 (a)	5.66 (d)	5.67 (a)	5.68 (a)	5.69 (c)	5.70 (a)	5.71 (c)	5.72 (a)
5.73 (a)	5.74 (a)	5.75 (d)	5.76 (a)	5.77 (d)	5.78 (d)	5.79 (a)	5.80 (b)	5.81 (c)
5.82 (b)	5.83 (c)							

Explanations Measurement of Energy and Power

5.1 (d)

There will be two deflecting torques, one acting on coil *A* and the other on coil *B*. The coil windings are so arranged that the torques due to the two coils are opposite in direction. So, there is no controlling torque.

5.2 (c)

The error caused by pressure coil inductance is $VI \sin\phi \tan\beta$. With low power factor, the value of ϕ is large and therefore, the error is correspondingly large.

5.3 (d)

Braking torque $T_D \propto R^2$ where *R* is the radial position of braking magnet with respect to the centre of the disc.

Note: $N \propto \frac{1}{R}$ (N = speed)

5.4 (d)

In two-wattmeter method

$$W_1 = V_{RY} I_R \cos(30^\circ + \phi)$$
$$W_2 = V_{BY} I_B \cos(30^\circ - \phi)$$

For p.f. = 0.5 or $\phi = 60^\circ$

W_1 reads zero.

5.5 (a)

When p.f. < 0.5 one of the wattmeter reads negative power; which is read after reversing the connections to the current coil or pressure coil. Also by formula:

$$\cos\phi = \cos\left(\tan^{-1}\sqrt{3}\,\frac{P_1 - P_2}{P_1 + P_2}\right) = 0.277$$

where $P_1 = 3$ kW and $P_2 = -1$ kW.

5.6 (c)

For type-1 error = $I^2 R_C$

For type-2 error = $\frac{V^2}{R_P}$

So, $$I^2 R_C = \frac{V^2}{R_P} = \frac{(200)^2}{10{,}000}$$

$$\Rightarrow \quad R_C = \frac{(200)^2}{10{,}000 \times (20)^2} = 0.01\,\Omega$$

5.7 (b)

Actual loading arrangements would involve a considerable waste of power. In order to avoid this 'Phantom' or 'Fictitious' loading is done.

5.8 (c)

Number of revolutions = $K \times P \times t$

$$\therefore\ 5 = 1200 \times P \times \frac{75}{3600} \Rightarrow P = 200\text{ W}$$

5.11 (a)

For compensation of error due to connection bifilar compensating winding is used.

5.12 (b)

$$\cos\phi = \cos\left(\tan^{-1}\sqrt{3}\,\frac{2P - P}{2P + P}\right) = 0.866$$

5.14 (b)
In some meters a slow but continuous rotation is observed even when there is no current flowing through the current coil and only pressure coil is energized. This is called creeping. The major cause for creeping is over-compensation for friction.

5.15 (d)
The rotating disc in an energy meter is made up of aluminium.

5.16 (a)
Error in dynamometer type Wattmeter is appreciably larger at lower p.f. due to pressure coil inductance.

5.19 (a)
Over compensation for friction is the main reason of creeping.

5.20 (c)
No controlling torque in p.f. meter, hence point 2 is incorrect.

5.22 (d)
Crossed coil is used for 1-ϕ supply. Polarized vane power factor meter needs rotating field so primarily used for three phase.

5.23 (b)

$$I_{rms} = \sqrt{(4)^2 + \left(\frac{5}{\sqrt{2}}\right)^2 + \left(\frac{3}{\sqrt{2}}\right)^2} = \sqrt{33}$$

and $P = I_{rms}^2 R$

$= 33 \times 20$

$= 660$ W

5.27 (c)
Ammeter not only indicates the current through the load but also additional current in the voltmeter.

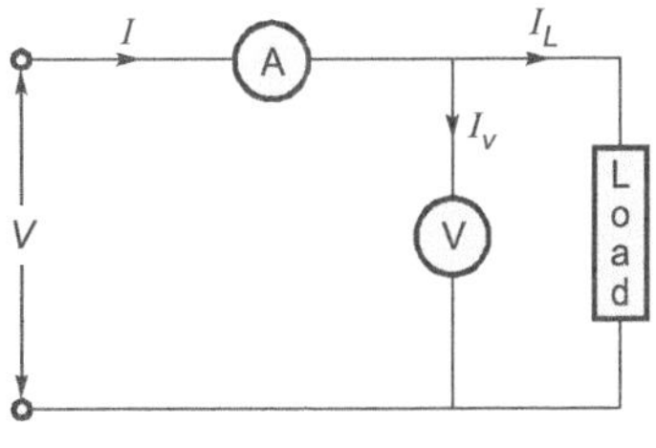

Current through the voltmeter,

$$I_v = \frac{V}{R_v}$$

where, R_v = Resistance of the voltmeter

Power consumed by load = VI_L

$$= V(I + I_v) = V\left(I + \frac{V}{R_v}\right) = VI + \frac{V^2}{R_v}$$

= Power indicated by the instruments + Power consumed by the voltmeter

Hence, power indicated by the instruments = Power consumed in load + Power consumed in voltmeter. Hence, option (c) is correct.

5.28 (b)
Current coil is connected in series with the load and carries the load current. If the coils are interchanged, current coils gets directly connected across the supply. Large current flows through the coil resulting into damage to current coil. Hence, option (b) is correct.

5.29 (b)
Energy meter is an integrating and induction type instrument and uses permanent magnet for producing braking torque.
Hence, option (b) is correct.

5.30 (c)

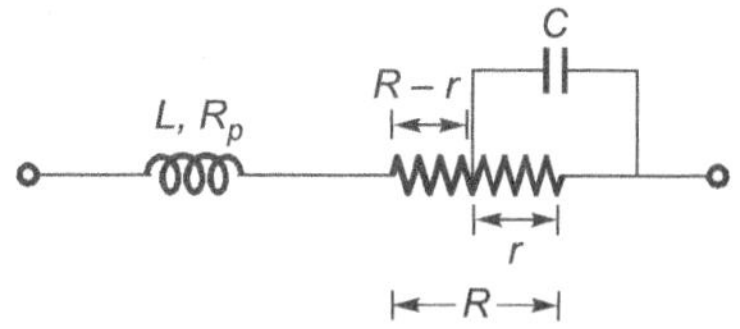

$$Z_p = R_p + (R-r) + j\omega L + r \,||\, \frac{1}{j\omega C}$$

$$= R_p + (R-r) + j\omega L + \frac{r}{1 + j\omega Cr}$$

$$= R_p + (R-r) + j\omega L + \frac{r(1 - j\omega Cr)}{1 + \omega^2 C^2 r^2}$$

Neglecting the term $\omega^2 C^2 r^2 << 1$

$$Z_p = R_p + (R-r) + j\omega L + r(1 - j\omega Cr)$$

$$= (R_p + R) + j(\omega L - \omega Cr^2)$$

For non-inductive pressure coil circuit,

$$\omega L = \omega Cr^2$$

$$C = \frac{L}{r^2}$$

Hence, option (c) is correct.

5.31 (c)

Astatic electrodynamometer instruments are constructed with two similar sets of fixed and moving coils are mounted on the same shaft. The two pairs of fixed coils are so connected that their magnetic fields are in opposition. Since the two fields are in opposition and therefore any uniform external field reduces the field of one coil and increases by an equal amount, the field of the other coil. Thus the net torque on account of the external magnetic field is zero.

Hence, option (c) is correct.

5.32 (b)

Due to the effect of inductance in the pressure coil, deflecting torque in a dynamometer type wattmeter.

(i) For lagging power factor,

$T_d \propto VI\cos(\phi-\beta)$ $\{\cos(\phi-\beta) > \cos\phi\}$

so wattmeter reads high.

(ii) For leading power factor,

$T_d \propto VI\cos(\phi+\beta)$ $\{\cos(\phi+\beta) < \cos\phi\}$

so wattmeter reads low.

Hence, option (b) is correct.

5.33 (b)

Electrodynamometer wattmeter is not suitable for low power factor power measurement, due to:

(i) Low value of deflecting torque.

(ii) Increased errors due to pressure coil inductance.

(iii) Power loss in the pressure coil current.

Hence, option (b) is correct.

5.34 (b)

Meter constant = 200 rev./ kWh

$$= \frac{\text{Number of revolutions}}{\text{kilowatt hour}}$$

$$200 = \frac{n}{VI\cos\phi \cdot t}$$

$$= \frac{n}{230\times10\times1\times3\times10^{-3}}$$

$$\Rightarrow \quad n = 1380$$

5.35 (c)

Meter constant according to given data

$$K = \frac{40\times60\times60}{5\times50} = 576 \text{ rev/kWh}$$

but it is given that meter constant is 500.

Therefore % error in meter constant

$$= \frac{576-500}{500}\times100 = 15.2\%$$

5.36 (c)

$$\tan\phi = \sqrt{3}\left(\frac{W_1-W_2}{W_1+W_2}\right) = \sqrt{3}\left(\frac{2-1}{2+1}\right)$$

$$\tan\phi = \frac{1}{\sqrt{3}} \Rightarrow \phi = 30°$$

$$\text{p.f.} = \cos\phi = \cos 30° = \frac{\sqrt{3}}{2}$$

5.37 (c)

To increase the speed of the disk (on the specified load), braking magnet is needed to adjust closer to the centre of the disc.

Note: $\uparrow N \propto \frac{1}{\downarrow d}$

5.40 (b)

Current coil has very low resistance, if it is connected in parallel, then due to high current passing through it, it will get damage.

5.41 (c)

Cause of creeping are over compensation for friction, excessive voltage across the potential coil, vibration and magnetic field.

5.43 (b)

Electrostatic wattmeter is not commercial instrument but a precision or laboratory instrument. Statement-II is true but not an explaination to statement-I.

5.45 (c)

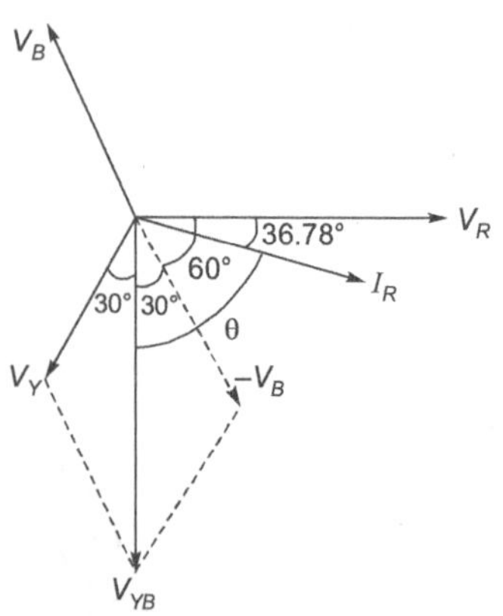

From the phasor diagram, I_R will lead V_{YB} by 30° + 60° – 36.78° = 53.14°.

5.46 (b)

$$\tan\phi = \sqrt{3}\left(\frac{W_1-(-W_2)}{W_1+(-W_2)}\right) = \infty$$

$\Rightarrow \quad \phi = 90°$ lagging

5.47 (c)

$$w_1 = \sqrt{3}VI\cos(30°+\phi)$$
$$w_2 = \sqrt{3}VI\cos(30°-\phi)$$

Hence readings of wattmeters depend on both loading as well as power factor.

5.48 (c)

Fixed coil carry current and moving coil carry voltage. This configuration is fixed not flexible.

5.49 (b)

$$\text{Energy consumed} = \left(\frac{1940}{400}\right) = 4.85 \text{ kWh}$$

$$\text{Power} = \left(\frac{4.85}{5}\right) \text{kWh} = 970 \text{ watts}$$

$$VI\cos\phi = 970 \text{ W}$$
$$230\times 5\cos\phi = 970 \text{ W}$$
$$\cos\phi = \left(\frac{970}{230\times 5}\right) = 0.843$$

5.52 (a)

$$\text{Power} = 230\times 10\times 0.9 = 2.07 \text{ kW}$$

Meter constant = 400 revolutions/kWh

$$\text{Revolutions in rpm} = \frac{400\times 2.07}{h} = \frac{400\times 2.07}{60 \text{ min.}} \text{revolution} = 13.80 \text{ rpm}$$

5.53 (c)

$W_1 \propto \cos(30+\phi)$ and $W_2 \propto \cos(30-\phi)$

if $\quad \phi = 30°$

$$W_1 = \frac{1}{2}; \qquad W_2 = 1$$

i.e. one is twice the other,

$$\text{p.f. of load} = \cos\phi = \cos 30° = \frac{\sqrt{3}}{2} = 0.866$$

5.54 (a)

$W \propto \cos(30+\phi)$ or $\cos(30-\phi)$

if $\quad \phi = 60°$

$$W = \cos(30°+60°) = 0$$

so, $\quad \text{p.f.} = \cos(60°) = \frac{1}{2}$

5.55 (b)

$$I_0 = \frac{100}{60} = 1.67 \text{ A}$$

$$f_0 = \frac{1}{2\pi\sqrt{0.5\times 10\times 10^{-6}}} = \frac{10^3}{2\pi\times 2.236} = 7.12 \text{ Hz}$$

5.57 (c)

$$W \propto \cos(30+\phi)$$

If $\quad \phi = 60°, \quad W \to 0$

$$\text{p.f.} = \cos 60° = \frac{1}{2} = 0.5 \text{ lagging}$$

Note: $w_1 = w_2$, unity pf
$w_1 = 2w_2$, 0.866 pf
$w_1 = 0$, 0.5 pf
$w_1 = -w_2$, 0 pf

5.58 (b)

In an electrodynamometer wattmeter the compensated wattmeter improves its accuracy by using windings with opposite currents with respect to series windings.

5.59 (c)

$$\phi = \tan^{-1}\left[\frac{\sqrt{3}(2-1)}{2+1}\right] = \tan^{-1}\left[\sqrt{3}\times\frac{1}{3}\right] = 30°$$

$$\text{p.f.} = \cos 30° = \frac{\sqrt{3}}{2} = 0.866$$

5.60 (c)

$$\text{Load factor} = \frac{P_{avg.}}{P_{max}} = \frac{30 \text{ kWhr}}{2 \text{ kW}\times 24 \text{ hr}} = 0.625$$

5.61 (c)

$$I_{rms} = \sqrt{I_{rms1}^2 + I_{rms2}^2 + \ldots\ldots}$$

$$= \sqrt{(3)^2 + \left(\frac{4}{\sqrt{2}}\right)^2\left(\frac{4}{\sqrt{2}}\right)^2}$$

$$= \sqrt{25} = 5 \text{ A}$$

$$P = I_{rms}^2 \cdot R = (5)^2\cdot(10) = 250 \text{ W}$$

5.62 (c)

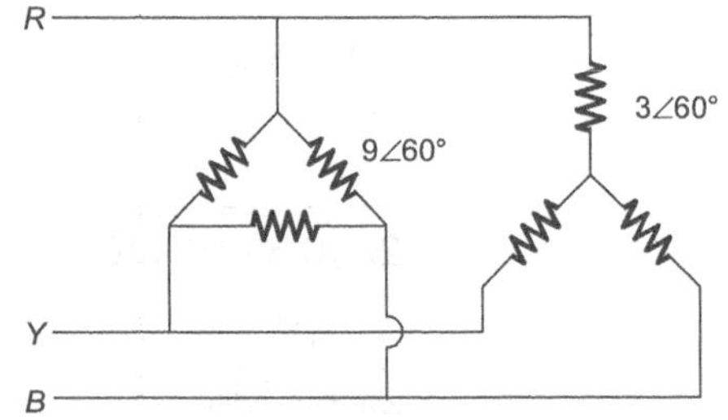

Converting Δ-Load into star

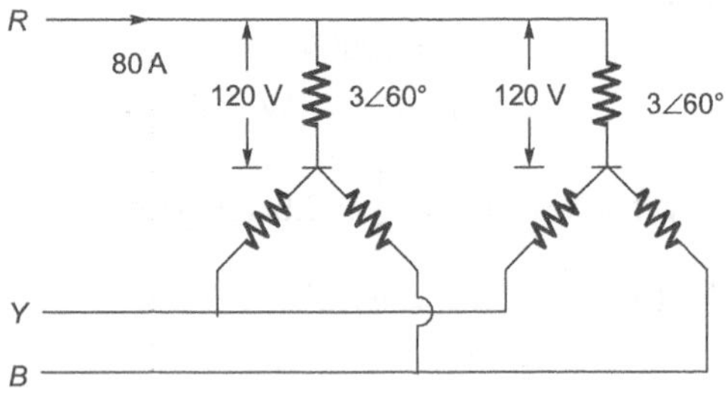

$$I = \frac{120}{3} = 40 \text{ A}$$

$$I_{total} = 40 + 40 = 80 \text{ A}$$

5.63 (b)

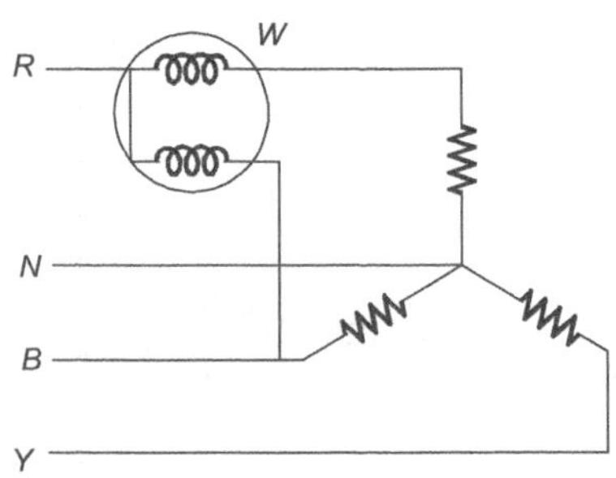

$$P = V_{ph} I_{ph} \cos\phi$$

$$10 \times 10^3 = \frac{400}{\sqrt{3}} \times 54 \times \cos\phi$$

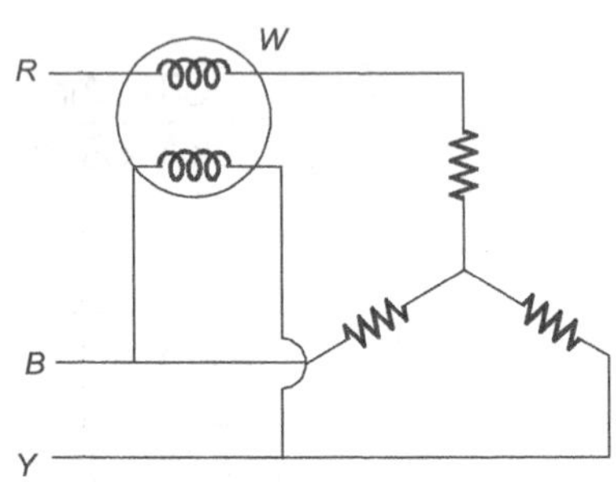

$$\cos\phi = 0.8$$

$$P = V_L I_L \sin\phi$$

$$= 400 \times 54 \times 0.6$$

$$= 12.46 \text{ kW} \simeq 13 \text{ kW}$$

5.65 (a)

$$P_\Delta = 3P_Y$$

$$W_1 = 3W_2$$

5.66 (d)

$$\%E_r = \frac{P_m - P_T}{P_T} \times 100$$

$$= \tan\phi \times \tan\beta \times 100$$

$$= \tan\phi \cdot \left(\frac{2\pi f L_P}{R_P}\right) \times 100$$

$$= \tan(\cos^{-1} 0.7) \cdot \left(\frac{2\pi \times 50 \times 0.5}{1000}\right) \times 100 = 16\%$$

5.67 (a)

$$W_1 = V_L I_L \cos(30 - \phi)$$

$$= 400 \times 50 \times \cos(30 - 30) = 2000 \text{ W}$$

$$W_2 = V_L I_L \cos(30 + \phi)$$

$$= 400 \times 50 \times \cos(30 + 30) = 1000 \text{ W}$$

5.68 (a)

$$\text{Meters constant, } k = \frac{\text{No. of revolutions}}{\text{kW hr}}$$

$$= \frac{5}{\frac{225}{1000} \times \frac{100}{3600}}$$

$$= 800 \text{ Rev/kWhr}$$

5.69 (c)

Induction wattmeter measures only AC. It is accurate at stated frequency and temperature.

5.70 (a)

The pressure coil interchange will not disturb the power flow to the load.

5.72 (a)

$$V = 210 \text{ V},$$

$$I = 20 \text{ A, p.f.} = 0.8,$$

$$t = 1 \text{ Hr}$$

$$\text{Energy consumed} = \frac{V.I.\cos\phi \times t}{1000} \text{ kWh}$$

$$= \frac{210 \times 20 \times 0.8 \times 1}{1000}$$

$$= 3.36 \text{ kWh}$$

For 1 kWH of energy → 100 revolution

For 3.36 kWH → ?

$$= 3.36 \times 100 = 336 \text{ rev.}$$

5.73 (a)

Due to inductance of potential coil of wattmeter error is more at low p.f. power measurements.

5.74 (a)

In dynamometer wattmeter,

$$T_d = \frac{P}{R_S} \cdot \frac{dM}{d\theta} = K\theta$$

$$\theta \; \alpha \; P = \text{linear scale}$$

Voltmeter, $$T_d = \frac{V^2}{R_s^2} \frac{dM}{d\theta} = K\theta$$

$$\theta \; \alpha \; V^2$$

$\Rightarrow$ Non-linear scale

5.75 (d)

Rotating disc of energy meter mode of Aluminium which is non-magnetic materials.

5.78 (d)

$$P_1 = \sqrt{3} V_0 I \cos(30 - \phi)$$

$$P_2 = \sqrt{3} V I \cos(30 + \phi)$$

If, $\phi > 60$ then p.f. < 0.5

5.79 (a)

LPF wattmeter :

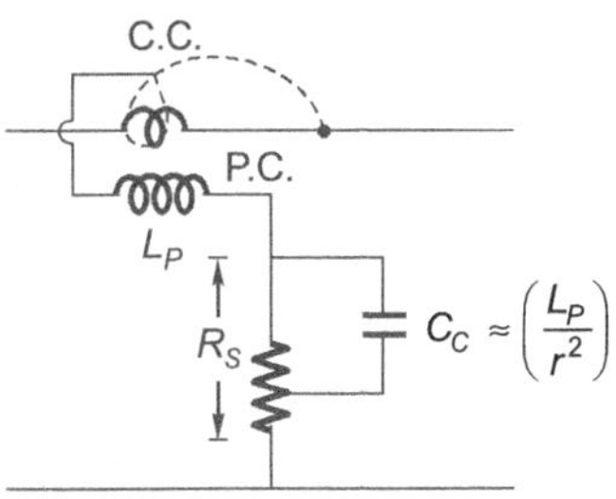

Connecting a capacitor (compensating capacitor) parallel to some part of high resistance (R_s) which is in series with potential coil.

Additionally, a special Bifilar winding is also connected to minimise the effect of mutual inductance caused by heavy CC current due to low pf.

5.80 (b)

For energy meter

$\tau_d \propto \phi_{sh} \, \phi_{se} \times \sin(\angle\phi_{sh} - \angle\phi_{se})$; τ_d is maximum when $\sin(\angle\phi_{sh} - \angle\phi_{se}) = 1$.

So angle between $\angle\phi_{sh}$ and $\angle\phi_{se}$ should be 90°.

$\therefore$ (b) option is correct.

5.81 (c)

Given : $P_1 = 1500$ Watt, $P_2 = 2500$ Watt

Reactive power, $Q = \sqrt{3}[P_1 - P_2]$

$|Q| = \sqrt{3}[1500 - 2500] = 1000\sqrt{3}$ (for 3-phase)

$$Q_{\text{per phase}} = \frac{Q}{3}$$

$$\Rightarrow Q_{\text{per phase}} = \frac{1000\sqrt{3}}{3} = \frac{1000}{\sqrt{3}}$$

$$= 577.3 \text{ VAR}$$

5.82 (b)

Given that :

Real power, P= 5.6 kW = 5600 Watt

$V_{line} = 220$ Volt, $I_{line} = 18.2$ Amp

$$P = \sqrt{3} V_L \cdot I_L \cos(\theta)$$

$$\Rightarrow \quad 5600 = \sqrt{3} \times 220 \times 18.2 \times \cos(\theta)$$

$$\text{P.f} = \cos(\theta) = 0.8075$$

5.83 (c)

$P_1 = +\text{ve}$

and $P_2 = -\text{ve}$

$\Rightarrow$ $0 < \text{p.f.} < 0.5$

■■■■

6 CRO and Frequency Measurement

6.1 Which one of the following statements correctly represents the post acceleration in a Cathode-Ray Tube ?
(a) It provides deflection of the beam.
(b) It increases the brightness of the trace if the signal frequency is higher than 10 MHz.
(c) It accelerates the beam before deflection.
(d) It increases the brightness of the trace of low frequency signal. [ESE-2001]

6.2 If the Q-factor of a coil is measured by varying the frequency, then
(a) The plot between *Q* and frequency is linear.
(b) The value of *Q* initially decreases with increase of frequency and afterwards it will increase with increase of frequency.
(c) The value of *Q* initially increases with increase of frequency and afterwards it will decrease with increase of frequency.
(d) The *Q*-factor remains constant irrespective of value of frequency.
[ESE-2001]

6.3 Which of the following electronic instruments (or equipment) can be used to measure correctly the fundamental frequency component of a waveform and its higher harmonics?
1. Cathode ray oscilloscope
2. Vacuum tube voltmeter
3. Spectrum analyzer
4. Distortion factor meter

Select the correct answer using the codes given below:
(a) 1 and 2 (b) 2 and 3
(c) 3 and 4 (d) 1 and 4 [ESE-2001]

6.4 Spectrum analyser is a combination of
(a) narrow band superheterodyne receiver and CRO
(b) signal generator and CRO
(c) oscillator and wave analyser
(d) VTVM and CRO [ESE-2001]

6.5 A screen pattern oscillogram, shown in the given figure is obtained when a sine-wave signal of unknown frequency is connected to the vertical input terminals, and at the same time, a 600 Hz sine-wave voltage is connected to the horizontal input terminals of an oscilloscope.

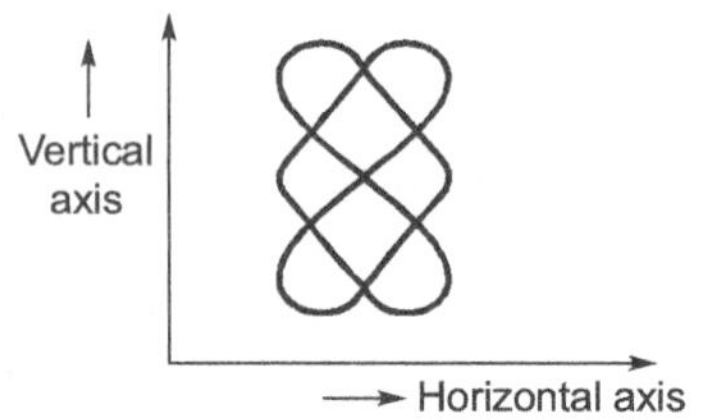

What is the value of unknown frequency?
(a) 300 Hz (b) 400 Hz
(c) 600 Hz (d) 900 Hz
[ESE-2002]

6.6 Which of the following statements is NOT correct for a storage type oscilloscope?
(a) The storage target is a conductive mesh covered area with magnesium fluoride
(b) Secondary emission electrons fetch a positively charged pattern
(c) The flood guns used for display, emit high velocity electrons
(d) The flood guns are placed between the deflection plates and storage target.
[ESE-2002]

6.7 In an oscilloscope, two Lissajous figure (X) and (Y) are observed. This indicates that ratio of vertical input signal frequency to that of horizontal input frequency are

(X)

(Y)

(a) 5/3 for X and 3/2 for Y
(b) 3/2 for X and 5/3 for Y
(c) 5/3 for X and 5/3 for Y
(d) 3/2 for X and 3/2 for Y [ESE-2002]

6.8 **Assertion (A):** A spectrum analyzer displays the frequency spectrum of a signal on the CRT screen.
Reason (R): It uses a sweep voltage generator to sweep the electron beam as well as the input signal.
(a) Both A and R are true and R is the correct explanation of A
(b) Both A and R are true but R is NOT the correct explanation of A
(c) A is true but R is false
(d) A is false but R is true
[ESE-2002]

6.9 A C.R.O. is operated with X and Y settings of 0.5 ms/cm and 100 mV/cm. The screen of the C.R.O. is 10 cm × 8 cm (X and Y). A sine wave of frequency 200 Hz and r.m.s. amplitude of 300 mV is applied to the Y-input. The screen will show
(a) one cycle of the undistorted sine wave.
(b) two cycles of the undistorted sine wave.
(c) one cycle of the sine wave with clipped amplitude.
(d) two cycles of the sine wave with clipped amplitude.
[ESE-2003]

6.10 Which one of the following truly represents the output on the screen of spectrum analyzer when an amplitude modulated wave is connected to it?
(a) Single vertical line on the screen
(b) Two vertical lines on the screen
(c) Three vertical lines with amplitude
(d) Three vertical lines out of which two have equal magnitude
[ESE-2003]

6.11 **Assertion (A):** Cathode ray oscilloscopes using CRT employing electrostatic deflection are used in laboratories for scientific measurements
Reason (R): CRT using electrostatic deflection systems have more deflection sensitivity as compared to CRT employing magnetic deflecting system.
(a) Both A and R are true and R is the correct explanation of A
(b) Both A and R are true but R is NOT the correct explanation of A
(c) A is true but R is false
(d) A is false but R is true [ESE-2004]

6.12 **Assertion (A):** Data acquisition systems are used in a large and ever increasing number of applications in a variety of industrial and scientific areas.
Reason (R): Transducer converts the analog signal to a form acceptable by the A/D convertor.
(a) Both A and R are true and R is the correct explanation of A
(b) Both A and R are true but R is NOT the correct explanation of A
(c) A is true but R is false
(d) A is false but R is true [ESE-2004]

6.13 A single channel digital storage oscilloscope uses a 12 bit, 10^8 samples/s ADC. For a 10 kHz sine wave input, what is the number of samples taken per cycle of input?
(a) 10^{12} (b) 10^8
(c) 10^4 (d) 10^2 [ESE-2005]

6.14 Which of the following measurements can be made using Lissajous figures ?
1. Frequency
2. Phase difference
3. Time interval between pulses
4. Pulse width
5. Fundamental and higher harmonic components

Select the correct answer using the code given below:
(a) 1 and 2 (b) 2 and 3
(c) 3 and 4 (d) 4 and 5 [ESE-2005]

6.15 Beam of electrons in a cathode ray tube emanates because of
(a) Second emission
(b) Thermionic emission
(c) Diffusion
(d) Post acceleration [ESE-2005]

6.16 Which one of the following frequency meter is suitable for measuring radio frequency?
(a) Vibrating reed frequency meter
(b) Weston frequency meter
(c) Electrical resonance frequency meter
(d) Heterodyne frequency meter
[ESE-2005]

6.17 One cycle of a square wave signal observed on an oscilloscope is found to occupy 6 cm at a scale setting of 30 μs/cm. What is the signal frequency?

(a) 1.8 kHz (b) 5.55 kHz
(c) 18 kHz (d) 55.5 kHz

[ESE-2006]

6.18 Match **List-I** with **List-II** and select the correct answer using the code given below the lists:

List-I

A. Digital Counter
B. Schering Bridge
C. Megger
D. Spectrum Analyzer

List-II

1. Measurement of harmonics
2. Measurement of frequency
3. Measurement of dielectric loss
4. Measurement of insulation resistance

Codes:

	A	B	C	D
(a)	1	3	4	2
(b)	2	4	3	1
(c)	1	4	3	2
(d)	2	3	4	1

[ESE-2006]

6.19 The measurement of very low and very high frequencies is invariably done using a frequency counter in which one of the following?

(a) Frequency measurement mode only
(b) Period measurement mode only
(c) Frequency and period measurement modes, respectively
(d) Period and frequency measurement modes, respectively

[ESE-2007]

6.20 What is the approximate input impedance of a CRO?

(a) Zero (b) 1 MΩ
(c) 10 Ω (d) 10 μΩ

[ESE-2007]

6.21 Match **List-I** (Name of Instrument) with **List-II** (Frequency Range) and select the correct answer using the code given below the lists:

List-I	**List-II**
A. Telephone detector	**1.** 200 Hz
B. Vibration galvanometer	**2.** 100 Hz
C. Tunable amplifier	**3.** 1 kHz
D. D' Arsonval galvanometer	**4.** dc

Codes:

	A	B	C	D
(a)	3	4	2	1
(b)	2	1	3	4
(c)	3	1	2	4
(d)	2	4	3	1

[ESE-2007]

6.22 In a CRO astigmatism is

(a) a source of generating fast electrons
(b) a medium for absorbing secondary emission electrons
(c) an additional focus control
(d) a time-delay control in the vertical deflection system

[ESE-2008]

6.23 If the bandwidth of an oscilloscope is given as direct current to 10 MHz, what is the fastest rise time a sine wave can have to be produced accurately by the oscilloscope?

(a) 35 nsec (b) 10 nsec
(c) 3.5 nsec (d) 0.035 nsec

[ESE-2009]

6.24 Assertion (A): Electron beam switch is used in a multitrace CRO.

Reason (R): Electron beam switch synchronizes the input signal and gives a steady waveform on the CRO screen.

(a) Both A and R are true and R is the correct explanation of A
(b) Both A and R are true but R is NOT the correct explanation of A
(c) A is true but R is false
(d) A is false but R is true

[ESE-2009]

6.25 The oscilloscope has an input capacitance of 50 pF and a resistance of 2 MΩ and the voltage divider ratio (k) of 10. What are the parameters of a high-impedance probe?

(a) C_1 = 5.55 pF and R_1 = 9 MΩ
(b) C_1 = 5.55 pF and R_1 = 18 MΩ
(c) C_1 = 3.33 pF and R_1 = 9 MΩ
(d) C_1 = 1.11 pF and R_1 = 18 MΩ

[ESE-2009]

6.26 In a vibrating reed type frequency meter, all the reeds
(a) are of identical dimensions and weight
(b) have different natural frequencies
(c) have the same natural frequencies
(d) are not placed closed to an electromagnet
[ESE-2011]

6.27 A frequency counter can be used for the measurement of
1. fundamental frequency of input signal.
2. time interval between two pulses.

Which of these is/are correct?
(a) 1 only (b) 2 only
(c) Neither 1 nor 2 (d) Both 1 and 2
[ESE-2012]

6.28 **Statement (I):** The measurement of voltage magnitude by a cathode-ray oscilloscope is very fast as compared to other methods of measurement.

Statement (II): Cathode-ray beam travels at the speed of light.
(a) Both Statement (I) and Statement (II) are individually true and Statement (II) is the correct explanation of Statement (I).
(b) Both Statement (I) and Statement (II) are individually true but Statement (II) is not the correct explanation of Statement (I).
(c) Statement (I) is true but Statement (II) is false.
(d) Statement (I) is false but Statement (II) is true.
[ESE-2012]

6.29 In a two-channel oscilloscope operating in x-y mode, two in phase 50 Hz sinusoidal waveforms of equal amplitude are fed to the two channel. What will be the resultant pattern on the screen?
(a) An ellipse
(b) A parabola
(c) Straight line inclined at 45° with respect to x-axis
(d) A circle
[ESE-2014]

6.30 The function of input attenuators in measuring instruments, like VTVM and CRO, is to
(a) increase the input impedance
(b) attenuate the frequency range
(c) attenuate the input signal amplitude without altering the frequency contents
(d) attenuate the input impedance
[ESE-2014]

6.31 Delay line is essential in a CRO, to ensure that
(a) vertical signal starts after the retrace period of sweep signal.
(b) the sweep reaches the horizontal plates before the desired signal under consideration.
(c) initial part of signal to be observed is not lost.
(d) all of the above
[ESE-2014]

6.32 A dual-beam CRO
(a) has one set of vertical deflection plates
(b) has two sets of horizontal deflection plates
(c) has two separate electron beams
(d) None of the above
[ESE-2015]

6.33 If the bandwidth of an oscilloscope is 10 MHz, what is the fastest rise time a square wave can have to be accurately reproduced by the instrument?
(a) 10 ns (b) 35 ns
(c) 28 ns (d) 100 ns
[ESE-2015]

6.34 If V_1 is the fundamental voltage, V_3 and V_5 are the amplitudes of the 3rd and 5th harmonic and

$$\frac{V_3}{V_1} = x\%, \quad \frac{V_5}{V_1} = y\%$$

then the total harmonic distortion of the system will be
(a) $\sqrt{x^2 + y^2}$ (b) $\frac{y}{x}$
(c) $x + y$ (d) $\frac{1}{\sqrt{x^2 + y^2}}$
[ESE-2015]

6.35 Frequency counter can be used to measure
1. fundamental frequency of input signal
2. fundamental and harmonic frequencies of input signal
3. time interval between two pulses
4. pulse width

Which of the above statements are correct?
(a) 1, 2, 3 and 4 (b) 1, 2 and 3 only
(c) 2, 3 and 4 only (d) 1, 3 and 4 only
[ESE-2015]

6.36 A CRO screen has 10 divisions on the horizontal scale. If a voltage signal 5 sin ($314t$ + 45°) is examined with a line base setting of 5 ms/div, the number of signals displayed on the screen will be

(a) 1.25 cycles (b) 2.5 cycles
(c) 5 cycles (d) 10 cycles

[ESE-2016]

6.37 Consider the following statements with regard to Lissajous pattern on a CRO:

1. It is a stationary pattern on the CRO.
2. It is used precise measurement of frequency of a voltage signal.
3. The ratio between frequencies of vertical and longitudinal voltage signals should be an integer to have a steady Lissajous pattern.

Which of the above statements is/are correct?

(a) 1 only (b) 2 only
(c) 3 only (d) 1, 2 and 3

[ESE-2017]

6.38 The self-capacitance of a coil is measured by the resonating capacitor. The measurement gives the value of tuning capacitor as C_1 = 460 pF at a frequency, f_1 = 2 MHz. The second measurement at f_2 = 4 MHz yields a new value of tuning capacitor, C_2 = 100 pf. The distributed capacitance C_d will be

(a) 10 pF (b) 20 pF
(c) 30 pF (d) 40 pF [ESE-2019]

6.39 Vertical delay line in CRO

(a) Gives proper time for thermionic emission of electrons.
(b) Delays the signal voltage by 200 ns.
(c) Allows the horizontal sweep to start prior to vertical deflection.
(d) Delays the generation of sweep voltage.

[ESE-2019]

6.40 A CRT has an anode voltage of 2000 V and parallel deflecting plates 2 cm long and 5 mm apart. The screen is 30 cm from the centre of the plates. If the input voltage is applied to the deflecting plates through amplifiers having an overall gain of 100, the input voltage required to deflect the beam through 3 cm will be

(a) 1 V (b) 3 V
(c) 5 V (d) 7 V

[ESE-2020]

6.41 An aquadag is used in a CRO to collect

(a) Primary electrons only
(b) Secondary emission electrons only
(c) Both primary electrons and secondary emission electrons
(d) Heat emission electrons

[ESE-2020]

6.42 A piezoelectric transducer has a capacitance of 1500 pF and leakage resistance of 10^5 MΩ. The oscilloscope used for read-out has a resistance of 1 MΩ in parallel with a capacitance of 500 pF. What is the time constant of the entire system?

(a) 0.002 s (b) 0.20 s
(c) 20 s (d) 200 s

[ESE-2021]

6.43 What is the range of frequency measured by the typical frequency meter?

(a) 1 MHz (b) 10 Hz
(c) 1 kHz (d) 1 GHz

[ESE-2024]

■■■■

Answers CRO and Frequency Measurement

6.1 (b)	6.2 (c)	6.3 (c)	6.4 (a)	6.5 (b)	6.6 (c)	6.7 (d)	6.8 (a)	6.9 (c)
6.10 (d)	6.11 (c)	6.12 (a)	6.13 (c)	6.14 (a)	6.15 (b)	6.16 (d)	6.17 (b)	6.18 (d)
6.19 (d)	6.20 (b)	6.21 (c)	6.22 (c)	6.23 (a)	6.24 (a)	6.25 (b)	6.26 (b)	6.27 (d)
6.28 (b)	6.29 (c)	6.30 (c)	6.31 (d)	6.32 (c)	6.33 (b)	6.34 (a)	6.35 (d)	6.36 (b)
6.37 (b)	6.38 (b)	6.39 (b)	6.40 (a)	6.41 (b)	6.42 (a)	6.43 (c)		

Explanations CRO and Frequency Measurement

6.1 (b)

In an electrostatic system of deflection, a low value of accelerating voltage (E_a) should be used to give high deflection sensitivity $\left(S \propto \frac{1}{E_a}\right)$. This voltage is usually kept below 4 kV. Although this voltage gives good sensitivity but reduces brightness, which can be seriously impaired at high frequencies. If, signals of frequencies higher than 10 MHz are to be displayed, post deflection acceleration is necessary to increase the brightness of the trace which otherwise would be dim.

6.2 (c)

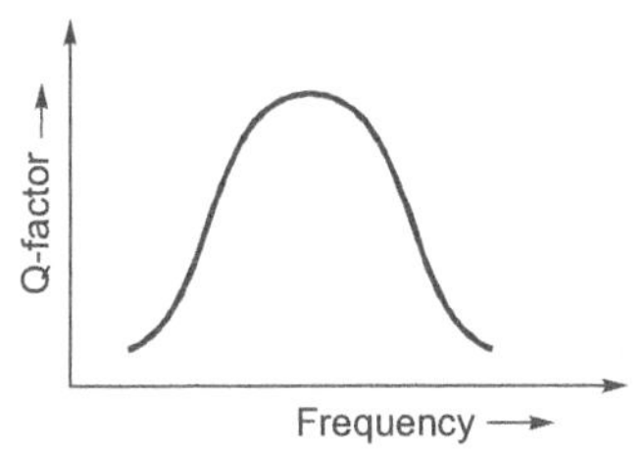

6.3 (c)

Both spectrum analyzer and distortion factor meter are signal analysers. A spectrum analyser sweeps the signal frequency band and displays a plot of amplitude versus frequency. Distortion factor meter tunes out the fundamental signal and gives an indication of the harmonics.

6.5 (b)

Refer Lissajous Pattern

$$\frac{f_y}{f_x} = \frac{\left(\begin{array}{l}\text{No. of intersections of the}\\ \text{horizontal line with the curve}\end{array}\right)}{\left(\begin{array}{l}\text{No. of intersections of the}\\ \text{vertical line with the curve}\end{array}\right)} = \frac{2}{3}$$

$$\therefore \quad f_y = \frac{2}{3} \times 600 = 400 \text{ Hz}$$

6.6 (c)

Refer Analog storage oscilloscope (variable persistence storage type).

6.7 (d)

For (X); $\frac{f_y}{f_x} = \frac{3}{2}$ and for (Y); $\frac{f_y}{f_x} = \frac{1+1/2}{1} = \frac{3}{2}$

6.9 (c)

$$\text{Time period of sine wave} = \frac{1}{200} = 5 \text{ ms}$$

$$V_{pp} = (300\sqrt{2}) \times 2 > 800 \text{ mV}$$

6.11 (c)

CRT employing electromagnetic deflection system has more sensitivity vis-a-vis CRT employing only electrostatic deflection system.

6.13 (c)

$$\text{Number of samples} = 10^8 \times \frac{1}{10 \times 10^3}$$

$$= 10^4 \text{ per cycle of input.}$$

6.17 (b)

$$T = 6 \times 30 \times 10^{-6} = 180 \text{ μs}$$

$$f = \frac{1}{180 \times 10^{-6}} = 5.55 \text{ kHz}$$

6.19 (d)

Low frequency measurement in frequency mode gives low-accuracy. So period measurement is used.

6.20 (b)

CRO has high input impedance.

6.21 (c)

D'Arsonval galvanometer is used for dc, Telephone detector for telephonic range of frequencies (kHz).

6.22 (c)

In modern oscilloscope, it is present.

6.23 (a)

$$t_r = \frac{0.35}{\text{B.W}} = \frac{0.35}{10 \times 10^6} = 35 \text{ nsec}$$

6.25 (b)

High impedance probe

$$R_1 = 2(k-1)$$

$$= 2(10-1) = 18 \text{ M}\Omega$$

$$C_1 = \frac{50 \text{ pf}}{k-1} = 5.55 \text{ pf}$$

6.26 (b)

Vibrating reed frequency meter indicates the supply frequency by means of individual reeds, when rated voltage is applied across the terminals of the meter, the particular reed whose natural frequency of vibration coincides with the supply frequency, vibrate with full amplitude.

6.27 (d)

Frequency counter can be used for measurement of time interval between two pulses.

6.32 (c)

Dual-beam *CRO* has two separate electron beams

6.33 (b)

$$B.W. \times t_r = 0.35$$

$$t_r = \frac{0.35}{10 \times 10^6} = 35 \text{ ns}$$

6.35 (d)

Frequency counter can be used to measure fundamental frequency of input signal, time interval between two pulses & pulse width but not harmonic frequencies of input signal.

6.36 (b)

$\because$ Time period of *I/P* signal 5 sin (314 t + 45°)

$$\frac{2\pi}{T} = 314$$

$$T = 20 \text{ m sec}$$

$\because$ CRO horizontal scale has 10 divisions with base setting 5 ms/div

$\therefore$ Total time period of horizontal scale

$$= 10 \times 5 = 50 \text{ msec}$$

$$\text{No. of cycle} = \frac{50 \text{ msec}}{20 \text{ msec}} = 2.5$$

6.37 (b)

For constant or equal frequencies the signal is stationary.

$$\frac{f_y}{f_x} = \frac{\text{No. of Horizontal tangencies}}{\text{No. of Vertical tangencies}}$$

(should be integer)

f_y = Frequency of vertical voltage signal

f_x = Freqency of Horizontal voltage signal

6.38 (b)

$f_2 = 2f_1 \Rightarrow$ double frequency

$\Rightarrow$ $n = 2$

$$C_d = \frac{C_1 - n^2C_2}{n^2-1} = \frac{C_1 - 4C_2}{3} = \frac{460 - 4 \times 100}{3} = 20 \text{ pF}$$

6.40 (a)

$V_a = 2000$ V, $l_d = 2$ cm,

$d = 5$ mm $= 0.5$ cm,

$L = 30$ cm $D = 3$ cm,

$V_d = ?$

$$D = \frac{Ll_d \cdot V_d}{2V_a \cdot d}$$

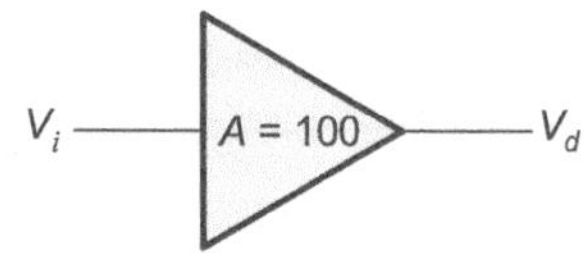

$$\Rightarrow V_d = \frac{2V_a \cdot d \cdot D}{L \cdot l_d}$$

$$V_d = \frac{2 \times 2000 \times 0.5 \times 3}{30 \times 2} = 100 \text{ Volt}$$

$$V_i = \text{Input} = \frac{V_d}{\text{gain}} = \frac{100 \text{ V}}{100} = 1 \text{ V}$$

6.41 (b)

The bombarding electrons, striking the screen, release secondary emission electrons. These secondary electrons are collected by an aqueous solution of graphite called aquadag which is connected to the second anode, collection of secondary electrons is necessary to keep the CRT in a state at electrical equilibrium.

6.42 (a)

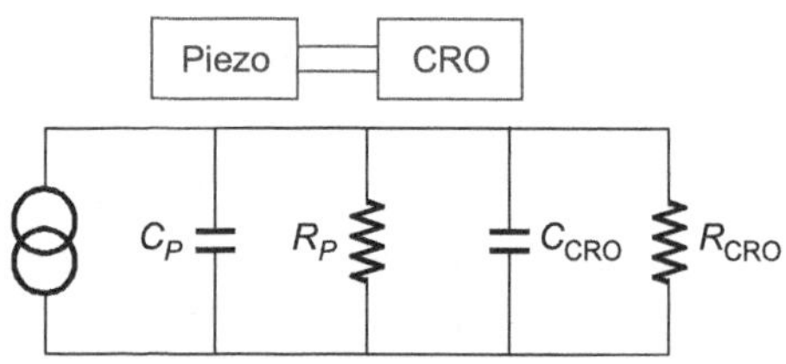

$C_p = 1500$ pF

$C_{CRO} = 500$ pF

$R_p = 10^5$ MΩ

$R_{CRO} = 1$ MΩ

$C_T = C_P + C_{CRO} = 1500 + 500$

$= 2000$ pF

$R_T = R_p \,||\, R_{CRO} = 1 \text{ M}\Omega \,||\, 10^5 \text{ M}\Omega$

$\simeq 1$ MΩ

Time constant $= R_T \times C_T$

$= 2000 \times 10^{-12} \times 1 \times 10^6$

$= 2$ msec $= 0.002$ sec

6.43 (c)

Typical frequency meter can measure the frequency from few Hz to second GHz.

7 Transducers and Sensors

7.1 Consider the following statements in connection with measurement of temperature:

1. A thermistor is highly sensitive as compared with platinum resistance thermometer
2. The resistance of a thermistor is solely a function of its absolute temperature whether the source of heat is external, internal or both.
3. A thermistor has linear resistance temperature characteristics
4. Most thermistors exhibit negative resistance temperature coefficient.

Which of these statements are correct ?

(a) 1, 2 and 3 (b) 1, 2 and 4
(c) 2, 3 and 4 (d) 1, 3 and 4

[ESE-2001]

7.2 **Assertion (A):** In digital transducers used for measurement of displacement, it is quite common to use Gray Code instead of natural binary code.
Reason (R): Gray Code is used for overcoming the inherent disadvantages of a natural binary code. These include errors that are caused on account of even slight misalignment which may cause change of decimal number thereby changing the number of bits. Gray Code does not suffer from this disadvantage.

(a) Both A and R are true and R is the correct explanation of A
(b) Both A and R are true but R is NOT the correct explanation of A
(c) A is true but R is false
(d) A is false but R is true

[ESE-2001]

7.3 Which one of the following transducers can be used for measurement of pressures as high as 100,000 atmosphere?

(a) Mcleod gauge (b) Pirani gauge
(c) Bridgman gauge (d) Knudsen gauge

[ESE-2001]

7.4 Which of the following transducers is classified as an active transducer ?

(a) Metallic strain gauge
(b) Capacitive microphone
(c) LVDT
(d) Piezoelectric transducer

[ESE-2002]

7.5 An imperfect capacitor is represented by a capacitance *C* in parallel with a resistance *R*. The value of its dissipation factor tan δ is

(a) ωCR (b) $\omega^2 CR$
(c) $1/\omega^2 CR$ (d) $1/\omega CR$

[ESE-2002]

7.6 Match **List-I** (Transducer) with **List-II** (Range of temperature) and select the correct answer:

List-I

A. Mercury in glass thermometer
B. Platinum resistance thermometer
C. Thermocouple with lead compensation
D. Optical radiation pyrometer

List-II

1. can be used upto 300°C in normal conditions
2. can be used upto 900°C with sufficient accuracy
3. can be used upto 1400°C
4. can be used for temperatures upto several thousand degrees.

Codes:

	A	B	C	D
(a)	3	2	1	4
(b)	1	4	3	2
(c)	3	4	1	2
(d)	1	2	3	4

[ESE-2002]

7.7 Match **List-I** (Instrument) with **List-II** (Measurand) and select the correct answer:

List-I	List-II
A. McLeod gauge	1. Temperature
B. Turbine meter	2. Pressure
C. Pyrometer	3. Flow
D. Synchros	4. Displacement

Codes:

	A	B	C	D
(a)	1	4	2	3
(b)	2	3	1	4
(c)	1	3	2	4
(d)	2	4	1	3

[ESE-2002]

7.8 In a stroboscopic method of rotational speed measurement of a machine shaft,
N = the machine shaft speed of rotation of the shaft in revolutions/min
n = No. of points on the circuit pattern
F = No. of flash per min.
The speed of rotation N will be
(a) $N = F + n$ (b) $N = F - n$
(c) $N = F/n$ (d) $N = F \cdot n$

[ESE-2002]

7.9 Which curve in the given figure represents resistance temperature characteristics of a thermistor?

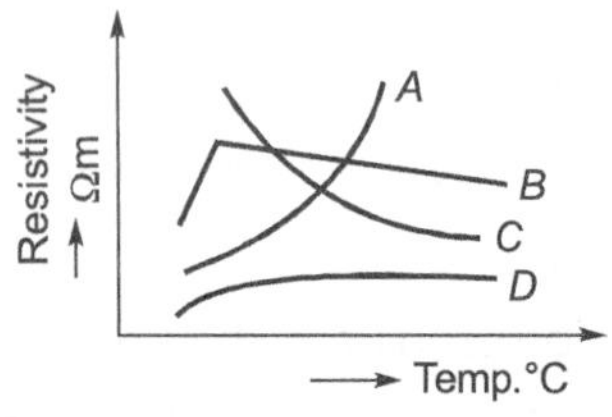

(a) Curve A (b) Curve B
(c) Curve C (d) Curve D

[ESE-2002]

7.10 A 1 cm piezoelectric transducer having a g-coefficient of 58 V/kg/m^2 is subjected to a constant pressure of 10^{-3} kg/m^2 for about 15 minutes. The Piezo voltage developed by the transducer will be
(a) 116 mV (b) 58 mV
(c) 29 mV (d) 0 mV

[ESE-2002]

7.11 Match **List-I** (Parameter) with **List-II** (Transducer) and select the correct answer:

List-I
A. Pressure
B. Temperature
C. Displacement
D. Stress

List-II
1. Thermistor
2. Piezoelectric crystal
3. Capacitance transducer
4. Resistance strain gauge
5. Ultrasonic waves

Codes:

	A	B	C	D
(a)	1	2	5	3
(b)	2	1	4	3
(c)	1	2	5	4
(d)	2	1	3	4

[ESE-2003]

7.12 Measurement of flow, thermal conductivity and liquid level using thermistors make use of
(a) Resistance decrease with temperature
(b) Resistance increase with temperature
(c) Self-heating phenomenon
(d) Change of resistivity

[ESE-2003]

7.13 A strain gauge with a resistance of 250 ohms undergoes a change of 0.150 ohm during a test. The strain is 1.5×10^{-4}. Then the gauge factor is
(a) 2.0 (b) 3.0
(c) 4.0 (d) 100

[ESE-2003]

7.14 Which one of the following gives Gauge factor of a strain gauge?
(a) $\frac{\Delta L}{L} \Big/ \frac{\Delta R}{R}$ (b) $\frac{\Delta R}{R} \Big/ \frac{\Delta L}{L}$
(c) $\frac{\Delta R}{R} \Big/ \frac{\Delta D}{D}$ (d) $\frac{\Delta R}{R} \Big/ \frac{\Delta \rho}{\rho}$

[ESE-2004]

7.15 Pair of active transducers is
(a) Thermistor, Solar cell
(b) Thermocouple, Thermistor
(c) Thermocouple, Solar cell
(d) Solar cell, LVDT

[ESE-2003]

7.16 Sensitivity of LVDT is mainly due to
(a) Magnetic shielding of the core
(b) Permeability of the core
(c) Exact cancellation of secondary voltages
(d) Insulation used in the winding

[ESE-2003]

7.17 A seismic transducer using a spring-mass-damper system as shown below will have an output displacement of zero when the input x_i is a

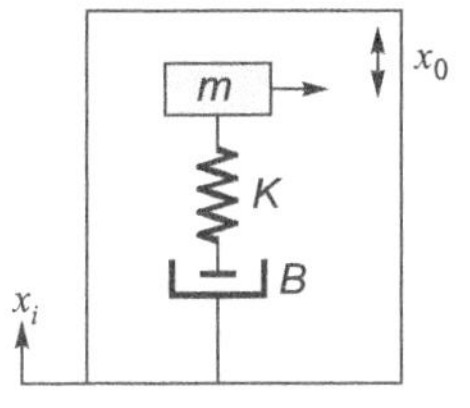

(a) constant displacement
(b) constant velocity
(c) constant acceleration
(d) sinusoidal displacement [ESE-2003]

7.18 Match **List-I** (Transducer) with **List-II** and select the correct answer using the codes given below:

List-I
A. Piezo-electric crystal
B. Manometer
C. Resistance bridge
D. Resistance strain gauge

List-II
1. Static pressure measurement
2. Dynamic force measurement
3. Study of vibration of a cantilever
4. Vibration in position of slider by an external force changes the resistance

Codes:

	A	B	C	D
(a)	4	1	2	3
(b)	4	3	2	1
(c)	2	1	4	3
(d)	2	3	4	1

[ESE-2004]

7.19 Which one of the following thermocouples has the highest temperature measuring range?
(a) Copper-Constantan
(b) Iron-Constantan
(c) Alumel-Chromel
(d) Platinum Rhodium-Platinum

[ESE-2004]

7.20 Thermistors are essentially semiconductors
(a) Well suited to precision measurement of temperature
(b) Widely used in the lower temperature range of –100°C to 300°C
(c) Which behave as resistors with a high negative temperature coefficient of resistance
(d) All of the above [ESE-2005]

7.21 Why are dummy strain gauges employed?
(a) For calibration of strain gauges
(b) For increasing the sensitivity of the bridge
(c) For compensation of temperature variations
(d) For neutralizing the influence of bridge voltage supply variations.

[ESE-2005]

7.22 Match **List-I** with **List-II** and select the correct answer using the code given below the lists:

List-I
A. Bolometer
B. Hot-wire Anemometer
C. C-type Bourdon Tube
D. Optical Pyrometer

List-II
1. Measurement of temperature of a furnace
2. Measurement of high pressure
3. Measurement of flow of air around an aeroplane
4. Measurement of power at 500 MHz
5. Measurement of angular position

Codes:

	A	B	C	D
(a)	2	3	5	1
(b)	4	1	2	3
(c)	2	1	5	3
(d)	4	3	2	1

[ESE-2006]

7.23 Which of the following can be used/modified for measurement of angular speed?
1. LVDT 2. Magnetic pick-up
3. Tacho-generator 4. Strain gauge

Select the correct answer using the code given below:
(a) Only 1 and 2 (b) Only 2 and 3
(c) Only 3 (d) Only 2, 3 and 4

[ESE-2006]

7.24 Piezo-electric crystal is generally employed for the measurement of which one of the following?
(a) Flow (b) Velocity
(c) Acceleration (d) Temperature
[ESE-2006]

7.25 Which one of the following transducers is an active transducer?
(a) Piezo-electric pressure transducer
(b) Metallic strain gauge
(c) Semi-conductor strain gauge
(d) Platinum resistance thermometer
[ESE-2007]

7.26 Which one of the following pressure transducers is suitable for measurement of high pressure?
(a) Alphatron (b) McLeod Gauge
(c) Pirani Gauge (d) Bourdon Gauge
[ESE-2007]

7.27 In strain gauge torque transducers, the strain gauges are mounted at which one of the following?
(a) 0° to the shaft axis
(b) 45° to the shaft axis
(c) 90° to the shaft axis
(d) at any angle with the shaft axis
[ESE-2007]

7.28 Why is strain gauge bridge sometimes excited with ac?
(a) It has a stable performance with ac
(b) Its sensitivity is more with ac
(c) Power frequency pick-up can be avoided with ac
(d) ac output can be easily amplified
[ESE-2007]

7.29 A thermo-couple arrangement is to be used to measure temperature in the range of 700-800°C. Which one of the following pairs would be most suitable?
(a) Copper-constantan
(b) Iron-constantan
(c) Chromel-alumel
(d) Platinum-platinum rhodium
[ESE-2007]

7.30 **Assertion (A):** Piezoelectric transducers can be used for measurement of both static and dynamic phenomena.
Reason (R): Piezoelectric transducers have very good high frequency response.
(a) Both A and R are true and R is the correct explanation of A
(b) Both A and R are true but R is NOT the correct explanation of A
(c) A is true but R is false
(d) A is false but R is true
[ESE-2008]

7.31 **Assertion (A):** Capacitive transducers can be used for measurement of both static and dynamic phenomena.
Reason (R): Capacitive transducers are extremely sensitive.
(a) Both A and R are true and R is the correct explanation of A
(b) Both A and R are true but R is NOT the correct explanation of A
(c) A is true but R is false
(d) A is false but R is true
[ESE-2008]

7.32 Which displacement transducer is used for accurate and linear measurement?
(a) LVDT
(b) Strain gauge
(c) Potentiometer
(d) Capacitive displacement transducer
[ESE-2008]

7.33 Match **List-I** with **List-II** and select the correct answer using the code given below the lists:
List-I
A. Variable capacitance device
B. Orifice meter
C. Thermistors
List-II
1. Flow measurement
2. Temperature measurement
3. Pressure transducer
4. Force and torque measurement
Codes:

	A	B	C
(a)	3	1	2
(b)	2	4	3
(c)	2	1	3
(d)	3	4	2

[ESE-2008]

7.34 Which of the following primary detector type transducers are employed for displacement measurement?

1. Thermistor 2. Diaphragm
3. Thermocouple 4. Pivot Torque

Select the correct answer using the code given below:

(a) 1 and 2 (b) 2 and 4
(c) 3 and 4 (d) 1 and 4

[ESE-2008]

7.35 The principle of Hall effect is made use of in the construction of which one of the following?

(a) Ammeter (b) Voltmeter
(c) Gaussmeter (d) Galvanometer

[ESE-2008]

7.36 What is a differential transformer?

(a) Constant pressure transducer
(b) Variable pressure transducer
(c) Constant displacement transducer
(d) Variable inductance transducer

[ESE-2008]

7.37 For a certain thermistor, the material constant (β) is 3000 kelvin and its resistance at 27°C is 1050 ohm. What is the temperature coefficient of resistances for this thermistor?

(a) 0.033×10^{-3} ohm/ohm/°C
(b) –0.033 ohm/ohm/°C
(c) –3.33 ohm/ohm/°C
(d) –3.0 ohm/ohm/°C

[ESE-2009]

7.38 The dead zone in a pyrometer is 0.12 percent of the span. The instrument is calibrated from 500°C to 2000°C. What temperature change must occur before it can be detected in degree Centigrade?

(a) 187.5 (b) 1.875
(c) 18.75 (d) 0.1875

[ESE-2009]

7.39 The strain gauge with a resistance of 250 ohm undergoes a change of 0.15 ohm. During a test the strain is 1.5×10^{-4}. What is the gauge factor?

(a) 4.7 (b) 4.0
(c) 3.5 (d) 2.0

[ESE-2009]

7.40 A strain gauge having a resistance of 500 ohm and a gauge factor 3.0 is bonded on a member of structure undergoing tensile stress. If the change in resistance of the gauge is accurately measured as 1.5 ohm, what is the value of strain suffered by the member?

(a) 0.01 (b) 0.001
(c) 0.1 (d) 0.003

[ESE-2009]

7.41 Match **List-I** with **List-II** and select the correct answer using the code given below the lists:

List-I	**List-II**
A. Hall effect	1. Varistor
B. Light energy	2. Photodiodes
C. Electric field	3. Measuring low magnetic field
D. Applied voltage	4. Liquid crystal display

Codes:

	A	B	C	D
(a)	1	2	4	3
(b)	3	2	4	1
(c)	1	4	2	3
(d)	3	4	2	1

[ESE-2010]

7.42 Consider the following statements:

1. Electromagnetic flowmeter is independent of liquid density.
2. Electromagnetic flowmeter cannot be employed for measuring flow of non-conducting fluids.

Which of these statements is/are correct?

(a) 1 only (b) Both 1 and 2
(c) 2 only (d) Neither 1 nor 2

[ESE-2010]

7.43 A piezoelectric crystal has a thickness of 2.5 mm and a voltage sensitivity of 0.05 V_m/N. The piezoelectric crystal is subjected to an external pressure of 1.6×10^6 N/m^2, then the corresponding output voltage is

(a) 200 volts
(b) 3.2×10^9 volts/m of thickness
(c) 0.07×10^9 V/(m^3/Newton)
(d) 200 m volts

[ESE-2010]

7.44 Consider the following units for the measurement of pressure directly:

1. Rota meter 2. Bourdon tube
3. Plani meter 4. Vanes

Of these, the pressure can be measured by

(a) 1 and 2 only (b) 3 and 4 only
(c) 2 only (d) 1, 2, 3 and 4

[ESE-2010]

7.45 Consider the following statements:

1. The main shortcomings of diaphragms are that they are prone to shock vibrations.
2. Diaphragms have the advantages of high accuracy and good dynamic response.
3. Selection of material for diaphragms mainly, depends upon temperature range and chemical nature of fluid coming in contact with diaphragm during pressure measurement.

Which of these statements is/are correct?

(a) 1, 2 and 3 (b) 2 and 3 only
(c) 1 only (d) 1 and 2 only

[ESE-2010]

7.46 A resistance strain gauge with gauge factor (S_f) of 2 is bonded to a steel member, which is subjected to a strain of 1×10^{-6}. The original resistance value of this strain gauge is 120. The change in resistance due to the applied strain is

(a) 60 Ω (b) 240×10^{-6} Ω
(c) 240 Ω (d) 60×10^{-6} Ω

[ESE-2010]

7.47 **Assertion (A) :** Piezoelectric transducers can be used for measurement of both static and dynamic phenomena.

Reason (R) : Piezoelectric transducers have very good high frequency response.

(a) Both A and R are true and R is the correct explanation of A
(b) Both A and R are true but R is NOT the correct explanation of A
(c) A is true but R is false
(d) A is false but R is true

[ESE-2010]

7.48 **Assertion (A):** Bellows are quite suitable for dynamic pressure measurements.

Reason (R) : Bellows are rugged, simple and rugged in construction and are capable of providing large force and wide pressure range of measurements.

(a) Both A and R are true and R is the correct explanation of A
(b) Both A and R are true but R is NOT the correct explanation of A
(c) A is true but R is false
(d) A is false but R is true

[ESE-2010]

7.49 The commercial thermopiles are formed by

(a) Series of Si-Al thermocouples in an IC by doping Ai layers on p-type Si on n-type Si epitaxial layers
(b) Series of Cu-W thermocouples strips
(c) Piezoelectric material strips piled together
(d) Series of bismuth-telluride couples

[ESE-2011]

7.50 Which one of the following is not a self-generating type transducer?

(a) Thermocouple and thermopile
(b) Piezoelectric pick-up
(c) Photovoltaic cell
(d) Magnetostriction gauge

[ESE-2011]

7.51 A coil 250 turns is connected to a 50 V DC source. If the coil resistance is 10 Ω, the mmf (magnetomotive force) developed in AT would be

(a) 500 (b) 1250
(c) 2500 (d) 250

[ESE-2012]

7.52 The pressure in a tank varies from 20 psi to 100 psi. The pressure in the tank is desired to be kept at 50 psi. The full-scale error when the pressure inside the tank is 30 psi will be

(a) 35% (b) 25%
(c) 40% (d) 15%

[ESE-2012]

7.53 Match **List-I** with **List-II** and select the correct answer using the code given below the lists:

List-I

A. Thermocouple
B. Thermistor
C. Strain gauge
D. LVDT

List-II
1. Modulated output
2. Resistance changes with pressure
3. Negative temperature coefficient
4. Constant temperature at one end

Codes:

	A	B	C	D
(a)	1	3	2	4
(b)	4	3	2	1
(c)	1	2	3	4
(d)	4	2	3	1

[ESE-2012]

7.54 An LVDT is used to measure 1 mm displacement for which a voltmeter of range 0 to 2 V through an amplifier having a gain of 500 is connected at the output of the LVDT. If the output of the LVDT is 2 mV, then the sensitivity of the instrument will be
(a) 1 V/mm (b) 0.5 V/mm
(c) 0.1 V/mm (d) 0.05 V/mm

[ESE-2012]

7.55 The expansion for the acronym LVDT, a transducer used for displacement measurement, is
(a) low voltage displacement transducer.
(b) light vision displacement transducer.
(c) linear variable displacement transducer.
(d) linear variable differential transformer.

[ESE-2012]

7.56 Match **List-I** with **List-II** and select the correct answer using the code given below the lists:

List-I
A. Electric field
B. Variable applied voltage
C. Light energy
D. Magnetic field

List-II
1. Photodiodes
2. Liquid crystal display
3. Hall effect sensor
4. Varistor

Codes:

	A	B	C	D
(a)	2	1	4	3
(b)	3	1	4	2
(c)	2	4	1	3
(d)	3	4	1	2

[ESE-2012]

7.57 A linear displacement digital transducer uses
(a) BCD code (b) Gray code
(c) Hexadecimal code (d) Binary code

[ESE-2012]

7.58 When compared with other transducers measuring temperature, a four-lead platinum RTD
1. has better linearity over a wide operating range.
2. has better accuracy and precision.
3. has better stability at high temperature.
4. is inexpensive.

Which of these are correct?
(a) 1, 2 and 3 (b) 1, 2 and 4
(c) 1, 3 and 4 (d) 2, 3 and 4

[ESE-2012]

7.59 **Statement (I):** Force and pressure can be measured by using capacitive transducer.

Statement (II): Capacitive transducer can be used to measure both static and dynamic phenomena.
(a) Both Statement (I) and Statement (II) are individually true and Statement (II) is the correct explanation of Statement (I).
(b) Both Statement (I) and Statement (II) are individually true but Statement (II) is not the correct explanation of Statement (I).
(c) Statement (I) is true but Statement (II) is false.
(d) Statement (I) is false but Statement (II) is true.

[ESE-2013]

7.60 **Statement (I):** Platinum resistance thermometers are widely used for temperature measurements for variety of industrial applications.

Statement (II): Platinum resistance thermometers provide the highest temperature sensitivity as compared to all known temperature transducers.
(a) Both Statement (I) and Statement (II) are individually true and Statement (II) is the correct explanation of Statement (I).
(b) Both Statement (I) and Statement (II) are individually true but Statement (II) is not the correct explanation of Statement (I).
(c) Statement (I) is true but Statement (II) is false.
(d) Statement (I) is false but Statement (II) is true.

[ESE-2014]

7.61 The principle of operation used in capacitive transducers to measure level of liquid is change of
(a) Area of plates
(b) Dielectric strength
(c) Distance between plates
(d) Shape of plates [ESE-2014]

7.62 A resistance strain gauge with a gauge factor of 2.0 is fastened to a steel member subjected to a stress of 100 N/mm^2. The modulus of elasticity of steel is approximately 2×10^5 N/mm^2. The percentage change in resistance is
(a) 1.50 (b) 1.00
(c) 0.15 (d) 0.10 [ESE-2014]

7.63 The resistance of 125 Ω strain gauge changes by 1 Ω for 4000 micro-strain. The gauge factor for strain gauge is
(a) 1.5 (b) 2.0
(c) 2.5 (d) 3.0 [ESE-2014]

7.64 Consider the following transducers:
1. LVDT 2. Piezoelectric
3. Thermocouple 4. Photovoltaic cell
Which of the above are active transducers?
(a) 1, 2 and 3 (b) 1, 2 and 4
(c) 2 and 4 only (d) 2, 3 and 4
[ESE-2014]

7.65 A capacitive transducer consists of two parallel plates of diameter 2 cm each and separated by an air gap of 0.25 mm. What is the displacement sensitivity?
(a) +200 pF/cm (b) –300 pF/cm
(c) –444 pF/cm (d) +44.4 pF/cm
[ESE-2015]

Directions:
Each of the next items consists of two statements, one labelled as the **'Statement (I)'** and the other as **'Statement (II)'**. Examine these two statements carefully and select the answers to these items using the codes given below:

Codes:
(a) Both Statement (I) and Statement (II) are individually true and Statement (II) is the correct explanation of Statement (I).
(b) Both Statement (I) and Statement (II) are individually true but Statement (II) is not the correct explanation of Statement (I).
(c) Statement (I) is true but Statement (II) is false.
(d) Statement (I) is false but Statement (II) is true.

7.66 **Statement (I) :** A 'strain gauge' is an example of a 'Transducer' or an electromechanical transformer.
Statement (II) : In the 'strain gauge', displacement is used to vary the resistance of a circuit component and the 'strain' is measured in terms of the change in the resistance.
[ESE-2015]

7.67 **Statement (I) :** The principle of the resistance strain gauge is that if gauge factor is known, the measurement of $\frac{dR}{R}$ allows the measurement of strain $\frac{dL}{L} = \varepsilon$.
Statement (II) : The output of an LVDT is of the order of millivolt.
[ESE-2015]

7.68 **Statement (I):** The gauge factor of a strain gauge is the ratio of strain to per unit change in resistance.
Statement (II): Poisson's effect is defined as producing less strain with opposite sign on the plane perpendicular to the applied load.
[ESE-2016]

7.69 **Statement (I):** A constant temperature type hot-wire anemometer is suitable for turbulent flow measurements.
Statement (II): When the resistance of the hot wire is kept constant by incorporating current feedback, the bandwidth is increased.
[ESE-2016]

7.70 **Statement (I):** Optical pyrometers are used as transducers for the measurement of flame temperature in a boiler.
Statement (II): Non-invasive methods are suitable for flame temperature measurement in a boiler.
[ESE-2016]

7.71 **Statement (I):** The null voltage of an LVDT cannot be reduced to an insignificant value.
Statement (II): Hall effect transducers are primarily used to measure flux density.
[ESE-2016]

7.72 The Hall voltage, V_H, for a thin copper plate of 0.1 mm carrying a current of 100 A with the flux density in the z-direction, $B_z = 1$ Wb/m^2 and the Hall coefficient, $R_H = 7.4 \times 10^{-11}$ m^3/C, is
(a) 148 μV (b) 111 μV
(c) 74 μV (d) 37 μV
[ESE-2016]

7.73 A piezoelectric crystal has a coupling coefficient K of 0.32. How much electrical energy must be applied to produce output energy of 7.06×10^{-3} J?
(a) 25.38 mJ (b) 22.19 mJ
(c) 4.80 mJ (d) 2.26 mJ

[ESE-2016]

7.74 Dummy strain gauges are used for
(a) compensation of temperature changes
(b) increasing the sensitivity of bridge
(c) compensating for different expansions
(d) calibration of strain gauge

[ESE-2016]

7.75 A structural member is compressed to produce a strain of 5 μm/m. The nickel wire strain gauge has a gauge factor of –12.1. The pre-stress resistance of the gauge is 120 Ω. The change in resistance due to compressive strain will
(a) increase the resistance by 7.26 mΩ
(b) decrease the resistance by 7.26 mΩ
(c) increase the resistance by 49.6 mΩ
(d) decrease the resistance by 49.6 mΩ

[ESE-2016]

7.76 **Statement (I) :** Electromagnetic flow meter is preferred for flow velocity measurement of slurries in pipes as long as the slurry has adequate electrical conductivity.

Statement (II) : Electromagnetic flow meter does not insert any instrument parts into the body of the fluid flow to cause obstruction as in most of other flow-meters.

[ESE-2018]

7.77 The variations in the measured quantity due to sensitivity of transducer to any plane other than the required plane is
(a) Cross sensitivity
(b) Sensitivity
(c) Interference
(d) Distributed sensitivity [ESE-2019]

7.78 A resistance strain gauge with a gauge factor of 2 is fastened to a steel member subjected to a stress of 1050 kg/cm^2. The modulus of elasticity of steel is 2.1×10^6 kg/cm^2. The change in resistance ΔR of the strain gauge element due to applied stress will be
(a) 0.1% (b) 0.2%
(c) 0.3% (d) 0.4%

[ESE-2019]

7.79 The ramp type digital voltmeter can measure accurately with
(a) a positive going ramp voltage only.
(b) a negative or positive going linear ramp voltage.
(c) a negative going ramp voltage only.
(d) an asymptotic ramp voltage only.

[ESE-2019]

7.80 A resistance wire strain gauge with a gauge factor of 2 is bonded to a steel structural member subjected to a stress of 100 MN/m^2. The modulus of elasticity of steel is 200 GN/m^2. The percentage change in the value of the gauge resistance due to the applied stress will be
(a) 0.1% (b) 0.3%
(c) 0.5% (d) 0.7%

[ESE-2020]

7.81 Capacitive transducers can be used for the measurement of liquid level. The principle of operation used in this case is the change of capacitance with change of
(a) Distance between plates
(b) Area of plates
(c) Dielectric
(d) Resonance

[ESE-2020]

7.82 **Statement I :** A strain gauge pressure sensor converts the physical quantity "pressure" into an electrical signal and electrical resistance of the strain gauge changes with the tensile strain.

Statement II : Bourdon tube pressure sensors work on the principle of change of any form of deformation, the cross-sectional tubing tends to recover its circular form under the action of pressure, which is used as mechanical pressure measuring instrument.
(a) Both Statement (I) and Statement (II) is the correct explanation of Statement (I).

(b) Both Statement (I) and Statement (II) are individually true, but Statement (II) is not the correct explanation of Statement (I).
(c) Statement (I) is true, but Statement (II) is false.
(d) Statement (I) is false, but Statement (II) is true.

[ESE-2021]

7.83 Which one of the following is not a self-generating type of transducer?
(a) Bourdon gauge for the measurement of pressure
(b) Pitot tube for the measurement of fluid flow velocity
(c) Thermistor for the measurement of temperature
(d) Photovoltaic cell

[ESE-2022]

7.84 The approximate range of gauge factor for a semiconductor strain gauge is
(a) 2 - 3 (b) 50 - 90
(c) 10 - 20 (d) 100 - 200

[ESE-2022]

7.85 The approximate pressure range for ionization gauge measuring device is
(a) 10^{-8} to 10^{-3} torr (b) 10^{-3} to 10^{-2} torr
(c) 10^{-10} to 10^{-8} torr (d) 10^{-2} to 10^{-1} torr

[ESE-2022]

7.86 What is the key factor that must be considered while selecting a transducer for a particular application?
(a) Only the input characteristics should be considered
(b) Only the output characteristics should be considered
(c) Only the transfer characteristics should be considered
(d) Input, output and transfer characteristics should be considered

[ESE-2024]

7.87 If a transducer has an output impedance of 1 Ω and a load resistance of 1 kΩ, it behaves as
(a) a constant current source
(b) a constant voltage source
(c) a constant power source
(d) a constant energy source

[ESE-2024]

■■■■

Answers Transducers and Sensors

7.1 (b)	7.2 (a)	7.3 (c)	7.4 (d)	7.5 (d)	7.6 (d)	7.7 (b)	7.8 (c)	7.9 (c)
7.10 (d)	7.11 (d)	7.12 (a)	7.13 (c)	7.14 (b)	7.15 (c)	7.16 (d)	7.17 (a, b)	7.18 (c)
7.19 (c)	7.20 (d)	7.21 (c)	7.22 (d)	7.23 (b)	7.24 (c)	7.25 (a)	7.26 (b)	7.27 (b)
7.28 (c)	7.29 (*)	7.30 (d)	7.31 (b)	7.32 (a)	7.33 (a)	7.34 (b)	7.35 (c)	7.36 (d)
7.37 (b)	7.38 (b)	7.39 (b)	7.40 (b)	7.41 (b)	7.42 (c)	7.43 (a)	7.44 (c)	7.45 (a)
7.46 (b)	7.47 (d)	7.48 (a)	7.49 (a)	7.50 (d)	7.51 (b)	7.52 (b)	7.53 (b)	7.54 (a)
7.55 (d)	7.56 (c)	7.57 (b)	7.58 (a)	7.59 (a)	7.60 (c)	7.61 (b)	7.62 (d)	7.63 (b)
7.64 (d)	7.65 (c)	7.66 (a)	7.67 (c)	7.68 (d)	7.69 (a)	7.70 (a)	7.71 (b)	7.72 (c)
7.73 (b)	7.74 (a)	7.75 (b)	7.76 (b)	7.77 (a)	7.78 (a)	7.79 (b)	7.80 (a)	7.81 (c)
7.82 (b)	7.83 (c)	7.84 (d)	7.85 (a)	7.86 (d)	7.87 (b)			

Explanations Transducers and Sensors

7.1 (b)
Thermistor is a semiconductor device which is non-linear and highly sensitive as well as exhibit negative resistance-temperature coefficient. Thermistors having positive resistance-temperature coefficient are not useful as temperature transducers.

7.2 (a)
Gray code is used for overcoming the inherent disadvantages of a natural binary code as explained in the Reason (R).

7.4 (d)
Piezoelectric transducer is a active transducer.

7.5 (d)

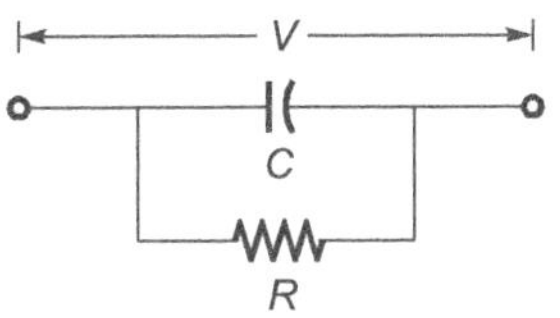

$$Q = \frac{\text{Energy stored}}{\text{Energy dissipated}} = \frac{V^2 / X_c}{V^2 / R} = \omega CR$$

$\because$ Dissipation factor $= \frac{1}{Q}$

$$\Rightarrow \quad \tan\delta = \frac{1}{\omega CR}$$

7.6 (d)
Optical radiation pyrometer used for temperatures upto several thousand degrees and mercury thermometer upto 300°C.

7.7 (b)
Pressure: McLeod gauge (very high pressure), Pirani gauge, (low pressure), Bourdon (medium), Bellows, Diaphragm, Manometer
Flow: Turbine meter, orifice plate, venturi meter, pitot tube
Displacement: Synchrose, LVDT, Strain gauge

7.9 (c)
Thermistor has a very high negative temperature coefficient of resistance.

7.10 (d)
It can't measure static pressure.

7.12 (a)
Thermistors have negative resistance-temperature coefficient.

7.13 (c)

$$\text{G.F.} = \frac{\Delta R / R}{\Delta l / l} = \frac{0.150}{250 \times 1.5 \times 10^{-4}} = 4$$

7.21 (c)
Dummy strain gauges are not part of measurement setup but they are employed just to compensate for temperature variations. A dummy strain gauge forms one arm of the Wheatstone bridge adjacent to the active strain gauge.

7.23 (b)
LVDT and strain gauge measure linear displacement.

7.25 (a)
Piezo-electric pressure transducer itself work as source for power.

7.26 (b)
Pirani Gauge is used for low pressure and Bourdon for medium range.

7.29 (*)
Iron-constantan – 200 to + 850 °C
Chromel-Alumuel – 200 to +1100 °C
Platinum – Platinum Rhodium – 0 to +1400 °C

7.30 (d)
Voltage developed is not held under static conditions so cannot be used for static.
But this type transducer have good frequency response.

7.31 (b)
Capacitive transducer can be used for force and pressure hence can be used for both static and dynamic.

7.34 (b)
Thermistor and Thermocouple are used for temperature.

7.37 (b)

$$R_{T_1} = R_{T_2} \exp\left[\beta\left(\frac{1}{T_1} - \frac{1}{T_2}\right)\right]$$

R_{T_1} = Resistance at T_1^0 K

R_{T_2} = Resistance at T_2^0 K

β = Material constant

7.38 (b)
Span = 2000 – 500 = 1500°C

$$\therefore \text{Temperature change} = \frac{0.12}{100} \times 1500 = 1.875°\text{C}$$

7.39 (b)

$$\text{Gauge factor} = \frac{\frac{\Delta R}{R}}{\frac{\Delta l}{l}}$$

$$\text{Gauge factor} = \frac{\frac{0.15}{250}}{1.5 \times 10^{-4}} = 4$$

7.40 (b)

$$\text{Gauge factor} = \frac{\frac{\Delta R}{R}}{\frac{\Delta l}{l}}$$

$$3 = \frac{\frac{1.5}{500}}{\frac{\Delta l}{l}}$$

$$\therefore \boxed{\frac{\Delta l}{l} = 0.001}$$

7.43 (a)
Output voltage of piezoelectric crystal.

$e_0 = g \cdot p \cdot t$
$g = 0.05\ V_m/N$
$p = 1.6 \times 10^6\ N/m^2$
$t = 2.5\ mm = 2.5 \times 10^{-3}\ m$
$e_0 = 0.05 \times 1.6 \times 10^6 \times 2.5 \times 10^{-3}$
$e_0 = 200\ V$

7.44 (c)
- Bourdon tube is used to measure the pressure.
- Rolta meter is used to measure the flow.

7.46 (b)
Gauge factor,

$S_F = 2$

Strain, $\frac{\Delta L}{L} = 1 \times 10^{-6}$

$R = 120\ \Omega$

Gauge factor, $G_F = \frac{\Delta R / R}{\Delta L / L}$

$$\Delta R = S_F \times \frac{\Delta L}{L} \times R$$
$$= 2 \times 10^{-6} \times 120$$
$$\Delta R = 240 \times 10^{-6}\ \Omega$$

7.49 (a)
A thermopile consists of a number of thermocouples connected in series. The thermopiles have a slow response and the sensitive to changes in ambient temperature. They are used in radiation pyrometers and as infrared detectors. Hence, option (a) is correct.

7.50 (d)

Active transducer or self-generating transducers are devices that do not require any power supply for their operation.

eg.: Photovoltaic cell, thermocouple and thermopile, piezoelectric crystals.

7.51 (b)

$$\text{Current in coil} = \frac{50}{10} = 5$$

Magnetomotive force (mmf) = NI

$$= 250 \times 5 = 1250 \text{ AT}$$

7.52 (b)

$$\% \text{ full scale error} = \frac{50-30}{100-20} \times 100$$

$$= \frac{20}{80} \times 100 = 25\%$$

7.53 (b)

- Thermo-electric emf depends upon the difference in temperature between the hot junction and the reference junction i.e. at one end temperature is constant.
- Thermistors are generally composed of semi-conductor materials (negative temperature coefficient).

7.54 (a)

Sensitivity of instrument-amplification factor sensitivity of LVDT

$$= 500 \times \frac{2 \times 10^{-3}}{1} = 1 \text{ V/mm}$$

7.55 (d)

The most widely used inductive transducer to translate the linear motion into electrical signals is the linear variable differential transformer (LVDT).

7.56 (c)

Hall effect sensor is associated with magnetic field, photo diodes with light energy and varistor with variable applied voltage.

7.57 (b)

Gray code is used in a linear displacement digital transducer.

7.58 (a)

<u>RTD:</u>

Advantages:

- Most stable
- Most accurate
- More linear than thermocouple

Disadvantages:

- Expensive
- Self-heating
- Current-source required.

7.62 (d)

$$\text{G.F.} = \frac{\Delta R/R}{\Delta l/l} \quad \ldots(i)$$

$$\text{Modulus of elasticity} = \frac{\text{stress}}{\text{strain}}$$

$$2 \times 10^5 = \frac{100}{(\Delta l/l)}$$

$$\frac{\Delta l}{l} = \frac{100}{2 \times 10^5} = 50 \times 10^{-5}$$

From equation (i),

$$2 = \frac{\Delta R/R}{50 \times 10^{-5}}$$

$$\Rightarrow \quad \frac{\Delta R}{R} = 50 \times 2 \times 10^{-5} = 10^{-3}$$

$$\% \text{ change} = 10^{-3} \times 100 = 0.1$$

7.63 (b)

$$\text{G.F.} = \frac{\Delta R/R}{\Delta l/l} = \frac{\frac{1}{125}}{4000 \times 10^{-6}} = 2$$

7.68 (d)

$$\left(\because G_f = \frac{\Delta R/R}{\Delta l/l} \right)$$

7.72 (c)

$$V_H = ?$$
$$t = 0.1 \text{ mm}$$
$$I = 100 \text{ A}$$
$$B = 1 \text{ Wb/m}^2$$
$$R_H = 7.4 \times 10^{-11} \text{ m}^3\text{/c}$$

$$V_H = R_H \frac{BI}{t}$$

$$= \frac{7.4 \times 10^{-11} \times 1 \times 100}{0.1 \times 10^{-3}}$$

$$= 7.4 \times 10^{-5} \text{ V}$$

$$\Rightarrow \quad V_H = 74 \ \mu \text{ volt}$$

7.75 (b)

$$\text{Gauge factor} = G_f = \frac{\Delta R/R}{(\Delta l/l)}$$

$$\Delta_R = G_f \cdot R \cdot \left(\frac{\Delta l}{l}\right)$$

$$= -12.1 \times 120 \times 5 \times 10^{-6}$$

$$= -7.26 \text{ m}\Omega$$

–ve indicates decrease of resistance.

7.76 (b)

Electromagnetic flow meter is used to measure voltage proportional to velocity of liquid.
When compared to mechanical flow transducers electromagnetic flow meter is non-obstruction type device.

7.78 (a)

$$G_f = 2$$

$$\text{Stress} = 1050 \text{ kg/cm}^2$$

$$Y = 2.1 \times 10^6 \text{ kg/cm}^2$$

$$\text{Strain} = \in = \frac{\text{Stress}}{Y} = \frac{1050}{2.1 \times 10^6}$$

$$= 500 \times 10^{-6} = \in$$

$$\Delta R = G_f \in R$$

$$\% \frac{\Delta R}{R} = G_f \in \times 100$$

$$\% \frac{\Delta R}{R} = 2 \times 500 \times 10^{-6} \times 100$$

$$= 0.1\%$$

7.80 (a)

$$G_f = 2$$

$$\text{Stress} = 100 \times 10^6 \text{ N/m}^2,$$

$$y = 200 \times 10^9 \text{ N/m}^2$$

$$G_f = \frac{\Delta R/R}{\in}$$

$$\Rightarrow \quad \frac{\Delta R}{R} = G_f \cdot \in$$

$$\% \frac{\Delta R}{R} = G_f \cdot \in \times 100$$

$$y = \frac{\text{stress}}{\text{strain}}$$

$$\Rightarrow \quad \in = \frac{100 \times 10^6}{200 \times 10^9} = 0.5 \times 10^{-3}$$

$$\% \frac{\Delta R}{R} = 2 \times 0.5 \times 10^{-3} \times 100 = 0.1\%$$

7.81 (c)

$$C = \frac{\in A}{d} = \frac{\in_0 \in_r A}{d}$$

Since distance and area are constant.

∴ By changing the dielectric constant of liquid, we can measure the change in capacitance.

■■■■

8 Instrument Transformers

8.1 Which one of the following instruments is commonly used to measure primary current of a transformer connected to mains?

(a) Electrostatic meter
(b) Current transformer
(c) Moving coil type meter
(d) Moving iron meter

[ESE-2008]

8.2 A current transformer has a phase error of +3°. The phase angle between the primary and secondary current is

(a) 3° (b) 177°
(c) 180° (d) 183°

[ESE-2013]

8.3 In using instrument transformers, care should be taken not to open circuit the

(a) primary of a voltage transformer when the secondary is connected to the rated load.
(b) secondary of a voltage transformer when the primary is energized with the rated voltage.
(c) primary of a current transformer when the secondary is connected to the rated load.
(d) secondary of a current transformer when the primary is carrying the rated current.

[ESE-2016]

8.4 In potential transformers, the secondary turns are increased slightly and the primary and secondary windings are wound as closely as possible to compensate for

(a) phase angle and ratio error, respectively.
(b) ratio and phase angle error, respectively.
(c) any eddy current loss and hysteresis loss, respectively.
(d) the hysteresis loss and eddy current loss, respectively.

[ESE-2017]

8.5 Instrument transformers are

(a) Used to extend the range of the AC measuring instruments only.
(b) Used to isolate the measuring instruments from the high voltage only.
(c) Used to extend the range and isolate the measuring instruments.
(d) Not used at generating stations and transformer stations.

[ESE-2020]

■■■■

Answers Instrument Transformers

8.1 (d) 8.2 (b) 8.3 (d) 8.4 (b) 8.5 (c)

Explanations Instrument Transformers

8.1 (d)

Primary current is AC.

8.2 (b)

Phase angle between the primary and secondary current,

$$= 180° - \text{phase error}$$
$$= 180° - 3° = 177°$$

8.4 (b)

By increasing secondary turns (called auxiliary secondary turns) ratio error is reduced.

By closely placing primary and secondary windings no load current I_0 is minimized due to reduction in leakage flux hence phase angle and ratio error are reduced.

8.5 (c)

Use of instrument transformer is the extension of instrument range. So that current, voltage, power and energy can be measured with instruments or meters of moderate size is of very great importance in commercial metering.

■■■■

9 Digital Meter, Electronic Instruments

9.1 The circuit generally used in digital instruments to convert sine waves into rectangular pulses is a
(a) sawtooth generator
(b) differential amplifier
(c) sample and hold circuit
(d) Schmitt trigger [ESE-2001]

9.2 Which of the following are data representation elements in a generalized measurement system?
1. Analog indicator 2. Amplifier
3. A/D converter 4. Digital display
Select the correct answer using the codes given below:
(a) 1 and 2 (b) 1 and 4
(c) 2 and 4 (d) 3 and 4 [ESE-2003]

9.3 A multimeter is used for the measurement of the following:
1. Both a.c. and d.c. voltage
2. Both a.c. and d.c. current
3. Resistance
4. Frequency
5. Power
Select the correct answer using the codes given below:
(a) 1, 2 and 4 (b) 1, 2 and 5
(c) 1, 3 and 5 (d) 1, 2 and 3
[ESE-2003]

9.4 Which one of the following statements is correct? An electronic voltmeter is more reliable as compared to multimeter for measuring voltage across low impedance because
(a) its sensitivity is high
(b) it offers high input impedance
(c) it does not alter the measured voltage
(d) its sensitivity and input impedance are high and do not alter the measured value
[ESE-2004]

9.5 Which one of the following decides the precision of integrating digital voltmeter?
(a) Reference voltage of analog comparator
(b) Slope of the generated ramp
(c) Width of the generated pulses
(d) Electronic counter
[ESE-2004]

9.6 Which one of the following digital voltmeters is most suitable to eliminate the effect of period noise?
(a) Ramp type digital voltmeter
(b) Integrating type digital voltmeter
(c) Successive approximation type digital voltmeter
(d) Servo type digital voltmeter
[ESE-2005]

9.7 Consider the following statements:
1. Buffer 2. Differentiator
3. Integrator 4. Comparator
Which of the above is/are components in a dual slope integrating type voltmeter?
(a) 1, 3 and 4 (b) 1 and 2
(c) 3 and 4 (d) 2 only
[ESE-2005]

9.8 The reference voltage and the input voltage are sequentially connected to the integrator with the help of a switch in a
(a) Successive approximation A/D converter
(b) Dual slope integration A/D converter
(c) Voltage to time converter
(d) Voltage to frequency converter
[ESE-2005]

9.9 Match **List-I** (Type of DVM) with **List-II** (Sub-component in ADC) and select the correct answer using the codes given below the lists:
List-I
A. Ramp type
B. Dual-slope
C. Servo-type
D. Successive approximation

List-II
1. DAC
2. Voltage to time converter
3. Pulse-generator
4. Potentiometer
5. Capacitor

Codes:

	A	B	C	D
(a)	2	1	4	5
(b)	4	5	3	1
(c)	2	5	4	1
(d)	4	1	3	5

[ESE-2005]

9.10 In modern electronic multimeter a FET or MOSFET is preferred over BJT because
(a) Its input resistance is low.
(b) Its input resistance is high.
(c) Its input resistance is high and does not vary with the change of range.
(d) It is cheaper. [ESE-2005]

9.11 What is the range for a $3\frac{1}{2}$ digital meter?
(a) 0 to 1999 (b) 0 to 1500
(c) 0 to 999 (d) 0 to 19999
[ESE-2006]

9.12 A successive approximation A/D converter has a resolution of 20 mV. What is its digital output for an analog input of 2.17 V?
(a) 01101100 (b) 01101101
(c) 01101011 (d) 01110100
[ESE-2006]

9.13 A doctor is using a digital clinical thermometer, which employs an A/D converter. The converter provides for both +ve and –ve reference inputs in place of only +ve and ground potential reference inputs. $V_{ref}^{+} = 2.16$ V & $V_{ref}^{-} = 1.85$ V. The amplifier used for the converter input generates signal of 20 mV per °F of body temperature. The converter output is between 00000 and 11111. Which one of the following is correct?
(a) The thermometer range is 92.5°F to 108°F
(b) The thermometer range is 46°F to 216°F
(c) The thermometer range is 0°F to 108°F
(d) The thermometer cannot by used to measure body temperature 98.4°F
[ESE-2006]

9.14 Assertion (A): Dual slope ADC is the most preferred digital conversion approach in digital multimeters.
Reason (R): Dual slope ADC provides high accuracy in A/D conversion which at the same time suppressing the hum effect on the input signal.
(a) Both A and R are true and R is the correct explanation of A
(b) Both A and R are true but R is NOT the correct explanation of A
(c) A is true but R is false
(d) A is false but R is true [ESE-2007]

9.15 Match **List-I** (Type of Digital Voltmeter) with **List-II** (Its construction features) and select the correct answer using the code given below the lists:
List-I
A. Basic ramp type DVM
B. Staircase-ramp type DVM
C. Integrating type DVM
D. Continuous balance type DVM

List-II
1. Uses D/A converter
2. True average value of input voltage using voltage frequency converter
3. Servo-driven potentiometer is commonly used
4. Does not use D/A converter

Codes:

	A	B	C	D
(a)	1	2	3	4
(b)	4	1	2	3
(c)	1	2	4	3
(d)	4	1	3	2

[ESE-2007]

9.16 Match **List-I** (Type of Electronic Voltmeter) with **List-II** (Leading Characteristics) and select the correct answer using the code given below the lists:
List-I
A. Amplifier Rectifier
B. Rectifier Amplifier
C. True RMS
D. Logarithmic

List-II
1. Wide input signal, dynamic range
2. High resistivity, limited bandwidth
3. Limited resistivity, large bandwidth
4. Capability to read non-sinusoidal AC

Codes:

	A	B	C	D
(a)	1	2	3	4
(b)	2	3	4	1
(c)	1	2	4	3
(d)	2	3	1	4

[ESE-2007]

9.17 The successive approximation A/D output for a 4-bit converter to a 8.217 volt input (if the reference is 5 V) will be

(a) 1101 (b) 0110
(c) 1010 (d) 1001 [ESE-2008]

9.18 Which amplifier is used in an electronic multimeter?

(a) Power amplifier
(b) Buffer amplifier
(c) Differential amplifier
(d) Wideband amplifier [ESE-2008]

9.19 In a digital voltmeter, the oscillator frequency is 400 kHz. The ramp voltage falls from 8 V to 0 V in 20 ms. What is the number of pulses counted by the counter?

(a) 8000 (b) 4000
(c) 3200 (d) 1600

[ESE-2009]

9.20 **Assertion (A):** An electronic millivoltmeter used to read very low a.c. voltages at high frequencies is an amplifier-rectifier type of meter.

Reason (R): The diodes cannot rectify low a.c. voltages of millivolt order.

(a) Both A and R are true and R is the correct explanation of A
(b) Both A and R are true but R is NOT the correct explanation of A
(c) A is true but R is false
(d) A is false but R is true [ESE-2009]

9.21 Consider the following statements. The *A* to *D* converter used in a digital instrument could be

1. Successive approximation converter type.
2. Flash converter type.
3. Dual slope converter type.

The correct sequence in the increasing order of the conversion time taken by these types are

(a) 1, 2 and 3 (b) 2, 1 and 3
(c) 1,3 and 2 (d) 2, 3 and 1

[ESE-2010]

9.22 Consider the following statements:

1. The main drawback of digital system is that the real world is mainly analog.
2. The major advantage of digital instruments over analog instruments is higher accuracy and better resolution.
3. Digital instruments are ordinarily used for the measurement of both analog and digital quantities.

Which of these statements is/are correct?

(a) 1, 2 and 3 (b) 1 and 3 only
(c) 2 only (d) 1 and 2 only

[ESE-2010]

9.23 A 4-digit DVM (digital voltmeter) with a 100 mV lowest full-scale range would have a sensitivity of how much value while resolution of this DVM is 0.0001 ?

(a) 0.1 mV (b) 10 mV
(c) 0.1 mV (d) 0.01 mV

[ESE-2010]

9.24 The precision of a ramp type digital voltmeter depends on

(a) frequency of the generator and slope of the ramp
(b) frequency of the generator
(c) slope of the ramp
(d) switching time of the gate

[ESE-2010]

9.25 An electronic voltmeter gives more accurate readings in high-resistance circuits as compared to a non-electronic voltmeter because of its

(a) low meter resistance
(b) high kΩ/V rating
(c) high V/kΩ rating
(d) high resolution [ESE-2012]

9.26 Modern electronic multimeters measure resistance by

(a) taking advantage of an electronic bridge compensator for nulling.
(b) forcing a constant current and measuring the voltage across unknown resistance.
(c) using a bridge circuit.
(d) applying a constant voltage across the unknown resistance and measuring the current through it.

[ESE-2012]

9.27 **Statement (I):** An electronic voltmeter measures the voltage across a high-value resistor more accurately as compared with an ordinary multimeter.

Statement (II): The input impedance of the electronic voltmeter is many orders of magnitude higher than that of an ordinary multimeter.

(a) Both Statement (I) and Statement (II) are individually true and Statement (II) is the correct explanation of Statement (I).
(b) Both Statement (I) and Statement (II) are individually true but Statement (II) is not the correct explanation of Statement (I).
(c) Statement (I) is true but Statement (II) is false.
(d) Statement (I) is false but Statement (II) is true.

[ESE-2012]

9.28 An 8-bit successive approximation DVM of 5 V range is used to measure 1.2 V. The contents of the SAR after 5 clock pulses is

(a) 01010000 (b) 00111100
(c) 00111000 (d) 00110111

[ESE-2013]

9.29 In a digital voltmeter, the oscillator frequency is 400 kHz. A ramp voltage to be measured by this voltmeter falls from 8 V to 0 V in 20 ms. The number of pulses counted by the counter is

(a) 8000 (b) 4000
(c) 3200 (d) 1600 [ESE-2013]

9.30 **Statement (I):** Flash analog-to-digital conversion is the fastest but an expensive method for designing ADCs.

Statement (II): Flash analog-to-digital converters are employed on very high speed digital acquisition systems.

(a) Both Statement (I) and Statement (II) are individually true and Statement (II) is the correct explanation of Statement (I).
(b) Both Statement (I) and Statement (II) are individually true but Statement (II) is not the correct explanation of Statement (I).
(c) Statement (I) is true but Statement (II) is false.
(d) Statement (I) is false but Statement (II) is true.

[ESE-2014]

9.31 A DVM uses 10 MHz clock and has a voltage controlled generator which provides a width of 5 μs/volt of unit signal. 10 V input signal would correspond to a pulse count of

(a) 500 (b) 750
(c) 250 (d) 1000

[ESE-2014]

9.32 Which of the following ADC has highest accuracy?

(a) Successive is approximation type
(b) Flash or parallel type
(c) Single slope integration type
(d) Dual slope integration type [ESE-2014]

9.33 An integrator type DVM (digital voltmeter) contains a 100 kΩ and 1 μF capacitor. If the voltage applied to the integrator input is 1 volt, what voltage will be present at the output of the integrator after 1 second.

(a) 1.1 V (b) 1 V
(c) 10 V (d) 100 V

[ESE-2015]

9.34 Analog-to-digital converter with the minimum number of bits that will convert analog input signals in the range of 0-5 V to an accuracy of 10 mV is

(a) 6 (b) 9
(c) 12 (d) 15 [ESE-2016]

9.35 An 8-bit successive approximation A-to-D converter is driven by a 2 MHz clock. Its conversion time is

(a) 18 μs (b) 16 μs
(c) 8 μs (d) 4.5 μs

[ESE-2016]

9.36 In a Q-meter, distributed capacitance of a coil is measured by changing the capacitance of the tuning capacitor. The values of tuning capacitor are C_1 and C_2 for resonant frequencies f_1 and $2f_1$ respectively. The value of distributed capacitance will be

(a) $\frac{C_1 - C_2}{2}$ (b) $\frac{C_1 - 2C_2}{3}$

(c) $\frac{C_1 - 4C_2}{3}$ (d) $\frac{C_1 - 3C_2}{2}$

[ESE-2020]

9.37 In a digital voltmeter, during start of conversion, zero indication is displayed and is called auto zeroing. This is achieved by

(a) Using a positive reference voltage
(b) Using a negative reference voltage
(c) Properly charging the differentiator circuit capacitance to ground
(d) Properly discharging the integrator circuit capacitance to ground

[ESE-2020]

9.38 A $3\frac{1}{2}$ digit DVM has an accuracy specification of ±0.5 percent of reading ±1 digit. What is the possible error in volt, when the instrument is reading 5.00 V on the 10 V range?

(a) 0.015 V (b) 0.025 V
(c) 0.035 V (d) 0.045 V

[ESE-2024]

9.39 If the bandwidth of an oscilloscope is given as direct current to 10 MHz, what is the fastest rise time a sine wave can have to be accurately reproduced by the instrument?

(a) 35 ns (b) 17.5 ns
(c) 0.175 μs (d) 35 μs

[ESE-2024]

■■■■

Answers Digital Meter, Electronic Instruments

9.1 (d)	9.2 (b)	9.3 (d)	9.4 (d)	9.5 (a)	9.6 (d)	9.7 (c)	9.8 (b)	9.9 (c)
9.10 (c)	9.11 (a)	9.12 (b)	9.13 (a)	9.14 (a)	9.15 (b)	9.16 (b)	9.17 (a)	9.18 (c)
9.19 (a)	9.20 (a)	9.21 (b)	9.22 (a)	9.23 (d)	9.24 (a)	9.25 (b)	9.26 (b)	9.27 (a)
9.28 (b)	9.29 (a)	9.30 (a)	9.31 (a)	9.32 (d)	9.33 (c)	9.34 (b)	9.35 (d)	9.36 (c)
9.37 (d)	9.38 (c)	9.39 (a)						

Explanations Digital Meter, Electronic Instruments

9.2 (b)

Amplifier and A/D converter are signal conditioners.

9.6 (d)

Refer potentiometric type digital voltmeter.

9.10 (c)

The electronic multi meter (EMM) can have input resistance ranging from 10 MΩ to 100 MΩ with the input resistance remaining constant over all ranges instead of being different at different ranges. The EMM gives far less loading effects.

9.11 (a)

A three digit display on a digital voltmeter for (0-1) V range will be able to indicate values from zero to 999 mV, with smallest increment or resolution of 1 mV. In practice a fourth digit usually capable of indicating either 0 or 1 only is placed to the left of active digits. This permits going above 999 to 1999 to give overlap between ranges for convenience. This is called over ranging. This type of display is known as a $3\frac{1}{2}$ digit display.

9.12 (b)

Full scale voltage comes out to be

$$= 20 \times 10^{-3} \times (2^8 - 1) = 5.1 \text{ V}$$

$5.1 \rightarrow 255\,(11111111) = 108.5 \simeq 109\,(01101101)$

So, accordingly digital output can be predicted to get 2.17 V.

9.13 (a)

Here, $(11111)_b = 2^5 - 1 = 31$

Voltage corresponding to $31 \times 20 \times 10^{-3}$

$$= 0.62 \text{ volt}$$

0.62 V equivalent to 31 °F

2.16 equivalent to $\frac{2.16}{0.62} \times 31\,°F = 108\,°F$

Similarly, 1.85 V equivalent to

$$\frac{1.85}{0.62} \times 31\,°F = 92.5\,°F$$

9.15 (b)

Integrating type DVM measure true average using voltage to frequency conversion.

9.16 (b)

Amplifier Rectifier so limited B.W. True RMS has capability to read non sinusoidal AC.

9.17 (a)

MSB set for more than reference voltage i.e. 5 volt and lower then MSB will set for half of the reference i.e. 2.5. But input is more than (5 + 2.5) so first two MSB will be set

$$8.217 \text{ volt} = 5 + 2.5 + 0 + \left(\frac{1.25}{2}\right)$$

↓ ↓ ↓ ↓

1 1 0 1

MSB LSB

9.19 (a)

$$\text{No. of pulses} = 400 \times 10^3 \times 20 \times 10^{-3} = 8000$$

9.21 (b)

Type	Conversion time
1. SAR	nT_{CLK}
2. Flash converter	T_{CLK}
3. Dual slope type	$2^{n+1}\,T_{CLK}$

9.23 (d)

Sensitivity = Resolution × (minimum full scale value)

= Resolution × Range

= $0.0001 \times 100 \times 10^{-3}$ V

= 0.01 mV

9.25 (b)

Electronic voltmeters has low V/kΩ rating i.e. high kΩ/V rating and produce less loading effects especially in electronic and communication circuits.

9.26 (b)

In electronic multimeters, when the function switch is placed in the OHM position, a resistor is connected in series with an internal battery and the meter simply measures the voltage drop across the unknown resistor.

9.27 (a)

The electronic voltmeters, have high kΩ/V rating due to high input impedance in comparison of ordinary multimeter.

9.28 (b)

$$5\text{ V} \rightarrow 11111111 : 255$$

$$1.2\text{ V} \rightarrow \frac{255}{5} \times 1.2 \;:\; 61.2 \simeq 60$$

0 0 1 1 1 1 0 0

9.29 (a)

$$\text{Number of pulses} = \frac{\text{Pulse width}}{\text{Oscillator time period}}$$

$$= \frac{20 \times 10^{-3}}{\frac{1}{400} \times 10^{-3}} = 8000$$

9.31 (a)

$$\text{Time for one count} = \frac{1}{10 \times 10^6} = 10^{-7} \text{ sec}$$

$$\text{Total time} = 5 \times 10^{-6} \times 10 = 50 \times 10^{-6} \text{ sec}$$

$$\text{Count} = \frac{50 \times 10^{-6}}{10^{-7}} = 500$$

9.32 (d)

Dual slope integrating type ADC has highest accuracy and flash or parallel type is fastest.

9.33 (c)

$$V = \frac{1}{RC}\int_0^1 V dt = \frac{1}{100\,\text{k} \times 1\,\mu} \times 1 = 10\text{ V}$$

9.34 (b)

∵ In *A* to *D* converter

Accuracy = Resolution = Error

$$\therefore \quad \frac{V_{FS}}{2^n - 1} \le 10 \text{ mV}$$

$$\frac{5\text{ V}}{2^n - 1} \le 10 \text{ mV}$$

$$\frac{1}{2^n - 1} \le 2 \times 10^{-3}$$

$$2^n - 1 \ge 500$$

$$2^n \geq 501$$
$$n \approx 9$$

9.35 (d)

Conversion time = Clock period × (No. of bits being converted)

$$= \frac{1}{2 \times 10^6} \times 8 = 4 \text{ ms}$$

9.36 (c)

$$C_d = \frac{C_1 - nC_2}{(n^2 - 1)}$$
$$n = 2$$
$$\Rightarrow \quad C_d = \frac{C_1 - 4C_2}{3}$$

9.37 (d)

The zero indication is displayed when capacitor present in integrator circuit is discharge to zero via ground by a process of auto-zeroing of capacitor.

9.38 (c)

$3\frac{1}{2}$ DVM, reading = 5 volt

FSD = 10 volt

Accuracy = ±0.5% reading + 1 digit

$\Downarrow \quad \Downarrow$

Total error = Error - (1) + Error - (2)

$$\text{Error-1} = \pm\frac{0.5}{100} \times 5 = 0.025$$

Error-2 = digit

$$1 \text{ digit} = \frac{V_{FSD}}{10^N} = \frac{10}{10^3} = 0.01$$

Total error = 0.025 + 0.01 = ±0.035

9.39 (a)

Given, B.W = 10 MHz

$$t_r = \frac{0.35}{\text{B.W.}} = \frac{0.35}{10 \times 10^6} = 0.35 \times 10^{-7} = 35 \text{ nsec}$$

■■■■

10 Data Acquisition System, Recorders

10.1 Assertion (A): In good quality magnetic cassette tape recorders, ac bias is provided to record head for better recording and reproduction.

Reason (R): The use of ac bias avoids the non linear characteristic of the BH curve of the recording head.

(a) Both A and R are true and R is the correct explanation of A
(b) Both A and R are true but R is NOT the correct explanation of A
(c) A is true but R is false
(d) A is false but R is true

[ESE-2001]

10.2 Digital data acquisition systems are used

1. Only when the output of the transducers is in digital form
2. When physical process being monitored is slowly varying (narrow bandwidth)
3. When low accuracy can be tolerated
4. When high accuracy and low per channel cost is required

Which of these statements are correct ?

(a) 1, 2 and 3 (b) 1, 3 and 4
(b) 1 and 3 (d) 2 and 4

[ESE-2001]

10.3 Assertion (A): Multiplexing of signals is invariably used in telemetry systems.

Reason (R): Multiplexing improves signal to noise ratio over the communication channel and at destination.

(a) Both A and R are true and R is the correct explanation of A
(b) Both A and R are true but R is NOT the correct explanation of A
(c) A is true but R is false
(d) A is false but R is true

[ESE-2003]

10.4 Assertion (A) : Magnetic tape is not used for digital data recording directly.

Reason (R): Digital data requires high speed recording.

(a) Both A and R are true and R is the correct explanation of A
(b) Both A and R are true but R is NOT the correct explanation of A
(c) A is true but R is false
(d) A is false but R is true

[ESE-2003]

10.5 The correct sequence of the blocks is an analog data acquisition unit starting from the input is

(a) Transducer-Recorder-Signal Conditioner
(b) Transducer-Signal Conditioner-Recorder
(c) Signal Conditioner-Transducer-Recorder
(d) Signal Conditioner-Recorder-Transducer

[ESE-2003]

10.6 Which one of the following definition correctly represents a data acquisition system (DAS)?

(a) DAS is a group of electronic devices that are connected to perform the measurement and quantization of electrical signals for digital processing
(b) DAS is a group of devices that are connected to store different signals
(c) DAS is a system to control a process
(d) DAS is a signal conditioner

[ESE-2004]

10.7 Consider the following elements:

1. Low pass filter
2. Signal transmission medium
3. Amplifier
4. Digital to analog convertor
5. Analog to digital convertor

Which one of the following sequences is the correct sequence for effective signal reconstruction in a data acquisition and processing scheme?

(a) 1-3-5-2-4 (b) 3-1-5-2-4
(c) 1-5-3-2-4 (d) 3-5-1-4-2

[ESE-2004]

10.8 **Assertion (A):** A dual trace CRO can display two input signals simultaneously

Reason (R): A dual trace CRO uses a CRT having two electron guns to generate two electron beams simultaneously.

(a) Both A and R are true and R is the correct explanation of A
(b) Both A and R are true but R is NOT the correct explanation of A
(c) A is true but R is false
(d) A is false but R is true [ESE-2004]

10.9 Consider the signal, $V_m \sin 100t + 2\,V_m \sin 200t$ to be sampled and stored in a data acquisition system. The same is to be extracted off-line later on. In order to extract the signal effectively, the original sampling frequency has to be

(a) 100 rad/s (b) 200 rad/s
(c) 210 rad/s (d) $\sqrt{100^2 + 200^2}$ rad/s

[ESE-2005]

10.10 Which one of the following transmission systems for telemetry has largest bandwidth?

(a) FM/FM radio transmission system
(b) Co-axial copper cables transmission system
(c) Fibre-optic data transmission system
(d) Synchro-position repeater system

[ESE-2005]

10.11 Pulse code modulation is commonly used in telemetry because

(a) It ensures immunity from noise during transmission.
(b) The bandwidth requirement of the channel is reduced.
(c) It removes quantization error.
(d) It permits lower rate of sampling than what is normally required under Shanon's theorem.

[ESE-2005]

10.12 In an analog data acquisition unit, what is correct sequence of the blocks starting from the input?

(a) Transducer – Recorder – Signal conditioner
(b) Transducer – signal conditioner – Recorder
(c) Signal conditioner – Transducer – Recorder
(d) Signal conditioner – Recorder – Transducer

[ESE-2005]

10.13 In microwave telemetry, repeater stations are required at every

(a) 2 km (b) 5 km
(c) 40 km (d) 100 km

[ESE-2006]

10.14 Data acquisition systems are usually of

(a) analog type (b) digital type
(c) integrating type (d) hybrid type

[ESE-2006]

10.15 **Assertion (A):** Usually the magnetic recording head of a tape recorder is provided with a.c. bias for its operation.

Reason (R): It is more convenient to provide a.c. bias to the magnetic recording head as compared to the d.c. bias.

(a) Both A and R are true and R is the correct explanation of A
(b) Both A and R are true but R is NOT the correct explanation of A
(c) A is true but R is false
(d) A is false but R is true

[ESE-2007]

10.16 In an analog Data Acquisition System (DAS), what is the correct sequence of the blocks (therein) starting from the input?

(a) Transducer – recorder – filter – signal conditioner
(b) Transducer – signal conditioner – recorder
(c) Signal conditioner – transducer – recorder
(d) Signal conditioner – filter – transducer – recorder

[ESE-2007]

10.17 Continuous recording of a signal is not possible in a

(a) magnetic tape recorder
(b) strip chart recorder
(c) X-Y recorder
(d) Galvanometric recorder [ESE-2008]

10.18 Which of the following are the objectives of a data acquisition system?
1. It must acquire necessary data at correct speed and time.
2. It must collect and store data.
3. There should be provision for real-time data display.
4. There should be provision for stored data display on request.

(a) 1, 2 and 3 only (b) 1, 3 and 4 only
(c) 1, 2, 3 and 4 (d) 2, 3 and 4 only

[ESE-2012]

10.19 Analog data acquisition systems are used when
(a) wide bandwidth and low accuracy are required/sufficient.
(b) narrow bandwidth and low accuracy are required sufficient.
(c) wide bandwidth and high accuracy are required.
(d) narrow bandwidth and high accuracy are required.

[ESE-2012]

10.20 In a digital data acquisition system a scanner-multiplexer
(a) scans the printed diagram and converts it into digital data.
(b) accepts multiple digital inputs and output any one of them with select lines.
(c) accepts multiple analog inputs and sequentially connects them to an ADC.
(d) checks the correct functioning of the modules one by one.

[ESE-2013]

10.21 In data acquisition system, analog data acquisition system is used
(a) for narrow frequency width, while digital data acquisition system is used when wide frequency width is to be monitored.
(b) for wide frequency width, while digital data acquisition system is used when narrow frequency width is to be monitored.
(c) when quantity to be monitored varies slowly, while its counterpart is preferred if the quantity to be monitored varies very fast.
(d) when quantity to be monitored is time-variant, while digital data acquisition system is preferred when quantity is time-invariant.

[ESE-2016]

■■■■

Answers Data Acquisition System, Recorders

10.1 (a)	10.2 (d)	10.3 (c)	10.4 (b)	10.5 (b)	10.6 (a)	10.7 (b)
10.8 (a)	10.9 (b)	10.10 (c)	10.11 (a)	10.12 (b)	10.13 (c)	10.14 (b)
10.15 (c)	10.16 (b)	10.17 (d)	10.18 (c)	10.19 (a)	10.20 (c)	10.21 (b)

Explanations Data Acquisition System, Recorders

10.1 (a)

The Magnetic pattern impressed upon the tape is directly proportional to the flux density produced at the air gap. The relationship between current in the winding and the flux density in the gap is non-linear. This non-linearity produces a distortion in the signal applied to the tape. In order to avoid distortion, a high frequency bias of constant amplitude is applied with the signal input to the recorder. Both the amplitude and frequency of the bias is larger than the maximum amplitude and highest frequency contained in the input signal. The bias frequency is usually four times the highest signal frequency. The bias current is usually between 5 to 30 times the signal current and the exact value depends on the tape and head characteristics.

10.2 (d)

- Digital data acquisition systems can be used with analog output as well as digital output of the transducers. Analog data can be converted into digital data through A/D converter.

- Refer TDM; when telemetering slow-changing, low bandwidth data, it would be extremely wasteful if these measurements occupied a full RF communication link. To increase efficiency, the communication link should be used for other measurements during the "dead" interval between samples. We see that efficient use of the communication system may be achieved by time-sharing the communication channel among a number of measured parameters each being transmitted at a rate compatible with the rate of change of the parameters.
- Further, Digital data acquisition systems are highly accurate due to better noise immunity.

10.3 (c)

Multiplexing makes low cost of transmission per channel. It will improve SNR only when digital data signals are transmitted.

10.8 (a)

A dual trace CRO uses two electron beam to display two input signals, simultaneously whose readings can be taken alternating one at a time.

10.9 (b)

Using band pass sampling theorem rather than baseband (i.e. low pass) sampling theorem

$$f_u = \frac{100}{\pi}\text{ Hz};\ f_B = \frac{100}{\pi} - \frac{50}{\pi} = \frac{50}{\pi}\text{ Hz}$$

$$\therefore \quad \frac{f_u}{f_B} = 2 \rightarrow k = 2$$

$$\therefore \quad f_s = \frac{2f_u}{k} = \frac{2}{2} \times \frac{100}{\pi}$$

$$\Rightarrow \quad \omega_s = 2\pi f_s = 200 \text{ rad/s}$$

The minimum sampling rate, $2f_M$ samples per second is called the Nyquist rate. Which applies to baseband or lowpass signals. The bandpass sampling theorem states that if a bandpass signal $m(t)$ has a spectrum of bandwidth $\omega_B = 2\pi f_B$ and an upper frequency limit $\omega_u = 2\pi f_u$, then $m(t)$ can be recovered from $m_s(t)$ by bandpass filtering if $f_s = \frac{2f_u}{k}$,

where k is the largest integer not exceeding $\frac{f_u}{f_B}$.

All higher sampling rates are not necessarily usable unless they exceed $2f_u$.

10.10 (c)

Optical fibre has no inductance so it has largest bandwidth.

10.11 (a)

The stream of pulses and non-pulses streams of 1's and 0's are not easily affected by interference and noise. When in the presence of noise, the presence or absence of a pulse can be easily determined. Hence used in telemetry.

10.16 (b)

Transducer senses the input quantity first. After that signal conditioner for manipulation of signal.

10.18 (c)

Objectives of DAS:

- There should be provision for real-time data display and for stored data display on request.
- It must acquire necessary data at correct speed and time.
- It must collect and store data.

10.19 (a)

- The analog data acquisition system is used where wide bandwidth (BW) is required and where the accuracy has not much significance.
- The digital data acquisition system is used where narrow bandwidth is required and where a high accuracy is needed.

■■■■

11 Miscellaneous

11.1 The modern standard of time is

(a) a second defined as 1/86400 of a mean solar day.

(b) a second defined as time constant of an RC series circuit having $R = 2\ M\Omega$, $C = 500$ pF.

(c) A second which is duration of 9192631770 periods of radiation corresponding to the transition between the two hyperfine levels of the fundamental state of the atom cesium 133.

(d) a second defined as 1/31556925.9747 of the time required by the earth to orbit the sun in the year 1900.

[ESE-2001]

11.2 In a Q-meter measurement to determine the self-capacitance of a coil, the first resonance occurred at f_1 with $C_1 = 300$ pF. The second resonance occurred at $f_2 = 2f_1$ with $C_2 = 60$ pF. The self-capacitance of coil works out to be

(a) 240 pF (b) 60 pF
(c) 360 pF (d) 20 pF [ESE-2003]

11.3 Standard cell

(a) will have precise and accurate constant voltage when current drawn from it is few microamperes only.

(b) will have precise and accurate constant voltage when few milliamperes are drawn from it.

(c) will continue to have constant voltage irrespective of loading conditions.

(d) can supply voltages up to 10 V.

[ESE-2003]

11.4 The battery cells in an electronic multimeter are required to measure which one of the following?

(a) Resistance (b) Voltage
(c) Current (d) Power

[ESE-2004]

11.5 **Assertion (A)** : The Q-meter measures the Q-factor of a coil when the circuit is in resonance.
Reason (R) : The Q-factor of a coil depends only on its inductance and not on its resistance.

(a) Both A and R are true and R is the correct explanation of A

(b) Both A and R are true but R is NOT the correct explanation of A

(c) A is true but R is false

(d) A is false but R is true

[ESE-2010]

11.6 Which of the following moving particles cannot be deflected by magnetic fields?

(a) α-Particles (b) Neutrons
(c) Protons (d) Electrons

[ESE-2012]

11.7 The voltage sensitivities of Barium Titanate and Quartz are respectively 12×10^{-3} Vm/N and 50×10^{-3} Vm/N. Their respective permittivities are 12.5×10^{-9} F/m and 40.6×10^{-12} F/m. What are their charge sensitivities?

(a) 1.04×10^6 C/N and 1.23×10^9 C/N

(b) 150 pC/N and 2 pC/N

(c) 24.5 pC/N and 90.6 pC/N

(d) 0.9×10^6 C/N and 1.23×10^9 C/N

[ESE-2013]

11.8 A semiconductor device made out of a material having very high temperature coefficient of resistance is

(a) Transistor (b) Varistor
(c) Thyristor (d) Thermistor

[ESE-2014]

■■■■

Answers Miscellaneous

11.1 (c) 11.2 (d) 11.3 (a) 11.4 (a) 11.5 (c) 11.6 (b) 11.7 (c) 11.8 (d)

Explanations Miscellaneous

11.2 (d)

Self-capacitance, $C_d = \frac{C_1 - 4C_2}{3}$

11.3 (a)

The makers specify the maximum value as 100 μA. This means that the current drawn from the cell should be less than 100 μA and this current should flow momentarily.

11.5 (c)

$Q = \frac{\omega L}{R}$ Reason is false.

11.6 (b)

Magnetic field can deflect only charged particles

$$\vec{F} = q(\vec{u} \times \vec{B})$$

Hence option (b) is correct.

11.7 (c)

Charge sensitivity
= Permittivity × Voltage sensitivity

11.8 (d)

The thermistor is, as the name implies, a temperature sensitive resistor, that is its terminal resistance is related to its body temperature. The compound employed determines whether the device has a positive or a negative temperature coefficients.

Thermistor

The change in temperature can be due to internal effects such as caused by the current through the thermistor or due to external effects of heating or cooling.
Hence, this device is considered as having a very high temperature coefficient of resistance.

■■■■

UNIT

V

Basic Electronics Engineering

Syllabus

Basics of Semiconductor diodes, transistors and their characteristics, Junction and field effect transistors (BJT, FET and MOSFETS), different types of transistor amplifiers, equivalent circuits and frequency response; oscillators and other circuits, feedback amplifiers.

Contents

OOOO

1 Semiconductor Physics

1.1 A specimen of intrinsic germanium with the density of charge carriers of $2.5 \times 10^{13}/cm^3$, is doped with donor impurity atoms such that there is one donor impurity atom for every 10^6 germanium atoms. The density of germanium atoms is $4.4 \times 10^{22}/cm^3$. The hole density would be
(a) $4.4 \times 10^{16}/cm^3$ (b) $1.4 \times 10^{10}/cm^3$
(c) $4.4 \times 10^{10}/cm^3$ (d) $1.4 \times 10^{16}/cm^3$
[ESE-2001]

1.2 If for intrinsic Silicon at 27°C, the charge concentration and mobilities of free-electrons and holes are 1.5×10^{16} per m^3, 0.13 $m^2/(Vs)$ and 0.05 $m^2/(Vs)$ respectively, its conductivity will be
(a) 2.4×10^{-3} $(\Omega\text{-m})^{-1}$
(b) 3.15×10^{-3} $(\Omega\text{-m})^{-1}$
(c) 5×10^{-4} $(\Omega\text{-m})^{-1}$
(d) 4.32×10^{-4} $(\Omega\text{-m})^{-1}$ [ESE-2001]

1.3 **Assertion (A):** The intrinsic carrier concentration of Si at room temperature is more than that of GaAs.
Reason (R): Si is an indirect bandgap semiconductor while GaAs is a direct bandgap semiconductor.
(a) Both A and R are true and R is the correct explanation of A
(b) Both A and R are true but R is NOT the correct explanation of A
(c) A is true but R is false
(d) A is false but R is true
[ESE-2002]

1.4 The bonding forces in compound semiconductors, such as GaAs, arise from
(a) ionic bonding
(b) metallic bonding
(c) covalent bonding
(d) combination of ionic and covalent bonding
[ESE-2002]

1.5 In a p-type silicon sample, the hole concentration is 2.25×10^{15}/cc. If the intrinsic carrier concentration is 1.5×10^{10}/cc, what is the electron concentration in the p-type silicon sample
(a) zero (b) 10^{10}/cc
(c) 10^5/cc (d) 1.5×10^{25}/cc
[ESE-2005]

1.6 Match **List-I (Parameter)** with **List-II (Variation)** and select the correct answer using the code given below the lists:
List-I
A. Electron mobility around room temperature
B. Energy gap
C. Intrinsic carrier concentration
D. Mole density (gm/mole)
List-II
1. Increases with temperature
2. Decreases with temperature
3. Remains constant as temperature is varied
Codes:

	A	B	C	D
(a)	2	1	1	1
(b)	1	2	1	3
(c)	2	2	1	3
(d)	2	2	1	1

[ESE-2006]

1.7 Match **List-I** (Metal-semiconductor band diagram under equilibrium) with **List-II** (Type of contact) and select the correct answer using the code given below the lists:
List-I **List-II**

A. 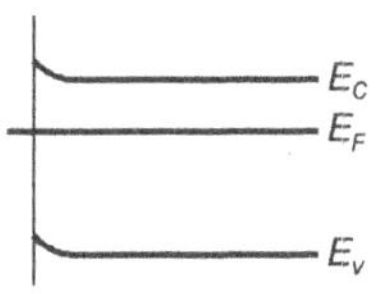

1. Rectifier to n-type

B. 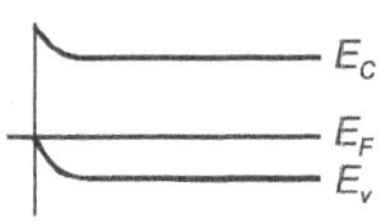

2. Ohmic to n-type

C.  3. Rectifier to p-type

D. 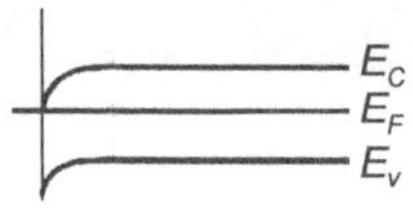 4. Ohmic to p-type

Codes:

	A	B	C	D
(a)	1	4	2	3
(b)	1	4	3	2
(c)	4	1	3	2
(d)	4	1	2	3

[ESE-2006]

1.8 Which one of the following equations represents the energy gap (E_G) variation of silicon with temperature (T)?

(a) $E_G(T) = 2.11 - 3.60 \times 10^{-4}\ T$

(b) $E_G(T) = 1.21 - 3.60 \times 10^{-4}\ T$

(c) $E_G(T) = 1.41 - 2.23 \times 10^{-4}\ T$

(d) $E_G(T) = 0.785 - 2.23 \times 10^{-4}\ T$

[ESE-2006]

1.9 In a Hall effect experiment, a p-type semiconductor sample with hole concen-tration p_1 is used. The measured value of the Hall voltage is V_{H1}. If the p-type sample is now replaced by another p-type sample with hole concentration p_2 where $p_2 = 2p_1$, what is the new Hall voltage V_{H2}?

(a) $2\ V_{H1}$ (b) $4\ V_{H1}$

(c) $(1/2)\ V_{H1}$ (d) $(1/4)\ V_{H1}$

[ESE-2006]

1.10 For an n-type semiconductor having any doping level, which of the following hold(s) good:

1. $p_n N_D = n_i^2$ 2. $p_p N_D = n_i^2$

3. $n_n N_D = n_i^2$ 4. $p_n n_n = n_i^2$

Select the correct answer using the code given below:

(a) 1 and 4 (b) 2 and 4

(c) 3 and 4 (d) Only 4

[ESE-2006]

1.11 Match **List-I** (Semiconductor Property) with **List-II** (Corresponding Unit) and select the correct answer using the code given below the lists:

List-I	List-II
A. Carrier mobility	1. eV(electron volt)
B. Diffusion length	2. m^2/V-sec
C. Diffusion coefficient	3. m
D. Energy gap	4. m^2/s

Codes:

	A	B	C	D
(a)	4	2	3	1
(b)	2	3	4	1
(c)	2	3	1	4
(d)	4	2	1	3

[ESE-2007]

1.12 **Assertion** (A): The semiconductor material used in making an optical source should be a direct bandgap material.

Reason (R): Carrier recombination time is shorter in a direct bandgap semiconductor.

(a) Both A and R are true and R is the correct explanation of A

(b) Both A and R are true but R is NOT the correct explanation of A

(c) A is true but R is false

(d) A is false but R is true

[ESE-2009]

1.13 **Assertion** (A): In a p-type semiconductor, with increase in the level of doping at a fixed temperature, the hole concentration increases but the electron concentration falls.

Reason (R): Law of mass action holds good in case of semiconductor.

(a) Both A and R are true and R is the correct explanation of A

(b) Both A and R are true but R is NOT the correct explanation of A

(c) A is true but R is false

(d) A is false but R is true

[ESE-2010]

1.14 **Assertion** (A): At 0°K, a semiconductor behaves as an insulator.

Reason (R): No free carries are available in the semiconductor at 0°K.

(a) Both A and R are true and R is the correct explanation of A

(b) Both A and R are true but R is NOT the correct explanation of A

(c) A is true but R is false

(d) A is false but R is true

[ESE-2010]

1.15 **Assertion** (A): The energy band diagram of a real semiconductor can not be represented by flat conduction and valence bands.
Reason (R): The presence of energy states in the forbidden energy gap results in carrier trapping, giving rise to finite curvature in the bands.
(a) Both A and R are true and R is the correct explanation of A
(b) Both A and R are true but R is NOT the correct explanation of A
(c) A is true but R is false
(d) A is false but R is true [ESE-2010]

1.16 The concentration of minority carriers in an extrinsic semiconductor under equilibrium is
(a) directly proportional to the doping concentration.
(b) inversely proportional to the doping concentration.
(c) directly proportional to the intrinsic concentration.
(d) inversely proportional to the intrinsic concentration.
[ESE-2012]

1.17 If n is the number of electrons per unit volume of the semiconductor and v_d is the drift velocity of the electrons, then the current flowing through a semiconductor is given by
(a) $i = \frac{n}{v_d}$ (b) $i = nv_d$
(c) $i = \frac{v_d}{n}$ (d) $i = n v_d^{1/2}$
[ESE-2012]

1.18 Consider the following statements:
The intrinsic carrier concentration of a semiconductor:
1. Depends on doping
2. Increases exponentially with decrease of band gap of the semiconductor
3. Increases non-linearity with increase of temperature
4. Increases linearly with increase of temperature
Which of the above statements are correct?
(a) 1, 2 and 3 (b) 1 and 2 only
(c) 2 and 3 only (d) 2 and 4 only
[ESE-2014]

1.19 The concentration of minority carriers in an extrinsic semiconductor under equilibrium is
(a) directly proportional to doping concentration.
(b) directly proportional to intrinsic concentration.
(c) inversely proportional to doping concentration.
(d) inversely proportional to intrinsic concentration.
[ESE-2014]

1.20 A semiconductor differs from a conductor in that it has
(a) only one path for the free electrons in the valence band.
(b) only one path for holes in the conductance band.
(c) two paths followed by free electrons and holes, one an ordinary path in the conduction band and the other one an extraordinary path in the valence band, respectively.
(d) two paths followed by free electrons and holes, one an extraordinary path in the conduction band and the other one an ordinary path in valence band, respectively.
[ESE-2016]

1.21 What is the type of breakdown that occurs in a Zener diode having breakdown voltage (6 V)?
(a) Avalanche breakdown only
(b) Zener breakdown only
(c) Avalanche breakdown where breakdown voltage is below 6 V and Zener breakdown otherwise
(d) Zener breakdown where breakdown voltage is below 6 V and Avalanche breakdown otherwise
[ESE-2016]

1.22 A Zener regulator has an input voltage varying between 20 V and 30 V. The desired regulated voltage is 12 V, while the load varies between 140 Ω and 10 kΩ. The maximum resistance in series with the unregulated source and Zener diode would be
(a) 3.3 Ω (b) 6.6 Ω
(c) 36.6 Ω (d) 93.3 Ω
[ESE-2016]

1.23 If V_m is the peak value of an applied voltage in a half wave rectifier with a large capacitor across the load, then the peak inverse voltage will be
(a) 0.5 V_m (b) V_m
(c) 1.5 V_m (d) 2.0 V_m **[ESE-2016]**

1.24 The thermal voltage V_T of a semiconductor diode at 27°C temperature is nearly
(a) 17 mV (b) 20 mV
(c) 23 mV (d) 26 mV **[ESE-2019]**

1.25 A two–step procedure in a typical diffusion apparatus to obtain the complementary-error-function Gaussian distribution involves the first step and second step respectively as
(a) Predeposition and Drive–in
(b) Predeposition and Drive–out
(c) Drive–in and Postdeposition
(d) Drive–out and Postdeposition
[ESE-2020]

1.26 An *n*-type semiconductor specimen has a Hall coefficient of 300 cm^3/C and its resistivity is 0.1 Ω-cm. Its electron mobility is
(a) 300 cm^2/V-sec (b) 30 cm^2/V-sec
(c) 3000 cm^2/V-sec (d) 3 cm^2/V-sec
[ESE-2022]

1.27 Consider the following statements regarding biasing of p-n junction :
1. The depletion layer width gets narrowed down on application of forward voltage.
2. A forward-biased p-n junction offers very high resistance to current flow.
3. The barrier potential increases when the junction is reverse-biased.
4. Majority charge carrier's current is established in a forward-biased p-n junction.

(a) 1, 2, 3 and 4 (b) 2, 3 and 4 only
(c) 1, 3 and 4 only (d) 3 and 4 only
[ESE-2023]

1.28 What is an optoisolator?
(a) LED is paired with a photodiode or phototransistor in the same package and it allows DC coupling.
(b) LED is paired with a silicon diode in the same package and it allows AC coupling.
(c) LED is paired with a photodiode or phototransistor in the same package and it does not allow DC coupling.
(d) LED is paired with a photodiode or phototransistor in the same package and it does not allow AC coupling.
[ESE-2024]

1.29 The mobilities of free electrons and holes in pure silicon are 1300 cm^2/V-s and 500 cm^2/V-s respectively. What is the value of intrinsic conductivity for silicon? (Assume $n_i = 1.5 \times 10^{10}$ cm^{-3} for silicon at room temperature)
(a) 2.69×10^{13} S/cm (b) 4.32×10^{-6} S/cm
(c) 4.32×10^{13} S/cm (d) 2.69×10^{-6} S/cm
[ESE-2024]

1.30 In an *n*-type semiconductor, the Fermi level is 0.3 eV below the conduction level at a room temperature of 300 K. If the temperature is increased to 360 K, what is the new position of the Fermi level?
(a) Remain unchanged
(b) 0.26 eV above the conduction level
(c) 0.36 eV above the conduction level
(d) 0.36 eV below the conduction level
[ESE-2024]

■■■■

Answers Semiconductor Physics

1.1 (b)	1.2 (d)	1.3 (b)	1.4 (c)	1.5 (c)	1.6 (c)	1.7 (a)	1.8 (b)	1.9 (c)
1.10 (a)	1.11 (b)	1.12 (b)	1.13 (a)	1.14 (a)	1.15 (a)	1.16 (b)	1.17 (b)	1.18 (c)
1.19 (c)	1.20 (c)	1.21 (d)	1.22 (d)	1.23 (d)	1.24 (d)	1.25 (a)	1.26 (c)	1.27 (c)
1.28 (a)	1.29 (b)	1.30 (d)						

Explanations Semiconductor Physics

1.1 (b)

In an n-type material $n \simeq N_D$,

$$n_i = p_i = 2.5 \times 10^{13}/\text{cm}^3$$

$$= \frac{4.4\times10^{22}}{10^6}/\text{cm}^3$$

$$n\cdot p = n_i^2$$

$$p = \frac{(2.5\times10^{13})^2}{4.4\times10^{16}} = 1.4 \times 10^{10}/\text{cm}^3$$

1.2 (d)

$$\sigma = (n\mu_n + p\mu_p)q; n = p = n_i$$

$$= (0.13+0.05)\times1.5\times10^{16}\times1.6\times10^{-19}$$

$$= 4.32\times10^{-4}(\Omega\text{-m})^{-1}$$

1.3 (b)

Both the statements are true but R is not the correct explanation of A because there is no relation in intrinsic concentration with direct band gap or indirect band gap.

1.5 (c)

Given that,

$$p = 2.25 \times 10^{15}/\text{cc}$$

and $n_i = 1.5 \times 10^{10}/\text{cc}$

According to mass action low

$$n\cdot p = n_i^2$$

$$n = \frac{n_i^2}{p} = \frac{(1.5\times10^{10})^2}{2.25\times10^{15}} = 10^5/\text{cc}$$

1.6 (c)

Mobility $\mu \propto T^{-m}$, m is a constant depending on crystal and type of doping.

So decreases with temperature.

Energy gap

$$E_g = E_{go} - \beta(T - T_0)$$

β – is a constant.

T is temperature at which E_g is to be found

E_{go} – energy gap at 0 K

$T_0 \to 0$ K

E_g decreases with rise in temperature

Intrinsic carrier concentration $\propto T^{3/2}$ increases with temperature.

1.7 (a)

In case of rectifier gap between E_c to E_v will be large in case of ohmic contact gap between E_c to E_v will be small and in n-type case E_f will be close to E_c and in p-type E_F will be close to E_v.

1.8 (b)

Energy gap of silicon at a temperature T

$$E_g = E_{go} - \beta T$$

where, E_{go} = 1.2 eV at 0 K

$$E_g = 1.21 - 3.60 \times 10^{-4} \times T$$

$$\beta = 3.60 \times 10^{-4} \text{ eV/K for Si}$$

1.9 (c)

$$V_H \propto \frac{1}{\text{Concentration}}$$

$$\therefore \frac{V_{H2}}{V_{H1}} = \frac{p_1}{2p_1} \Rightarrow V_{H2} = \frac{1}{2}V_{H1}$$

1.10 (a)

$$n_n = N_D \text{ so, } p_n\cdot n_n = n_i^2$$

$$\Rightarrow \quad p_n\cdot N_D = n_i^2$$

1.11 (b)

Diffusion length – meter

Energy gap – eV

1.12 (b)

Both statements are correct but reason for A is that in a direct bandgap material during recombination energy is emitted in the form of light (99%).

1.16 (b)

For an n-type semi conductor

$$p_n = \frac{n_i^2}{N_D}$$

where,

p_n is the concentration of holes.

N_D is the concentration of donor atoms.

and n_i is the intrinsic concentration.

1.17 (b)

Current, $i = neA\,v_d$

1.18 (c)

The intrinsic carrier concentration of a semiconductor increases non linearly with increase of temperature.

$$n_i^2 = A_0 T^3 e^{-E_{G0}/kT}$$

[Where E_{GO} = energy gap at 0°K in electron volts] and intrinsic carrier concentration of a semiconductor increases exponentially with decrease of band gap of the semiconductor.

1.19 (c)

$$np = n_i^2$$

In *n*-type semiconductor, *n* is the majority carrier and *p* is the minority carrier. The minority carrier is inversely related to majority carrier concentration.

1.22 (d)

$$I_{min} = I_{z\,min} + I_{z\,max}$$

$$\frac{0 + V_z}{R_{L\,min}} = \frac{12}{140}$$

$$R = \frac{V_{min} - V_z}{I_{min}}$$

$$= \frac{20 - 12}{12/140} = 93.3\ \Omega$$

1.23 (d)

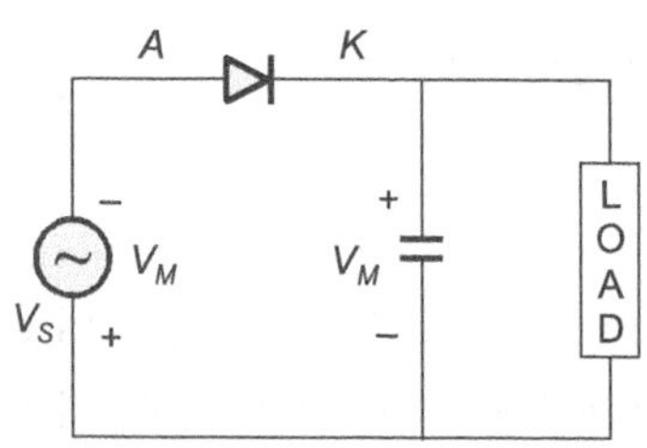

The steady state voltage across load $\simeq V_M$ [due to large capacitor]

$$\therefore \quad PIV = 2\,V_M$$

1.24 (d)

$$V_T = \frac{T}{11600} = \frac{300}{11600} = 26\text{ mV}$$

1.25 (a)

Impurity diffusion is of two types:

(i) Predeposition which results in complementary error function profile.

(ii) Drive in diffusion which results in Gaussian distribution profile.

1.26 (c)

$$R_H = \frac{1}{ne}\ \text{(Hall coefficient)}$$

$$\sigma = n \times e \times \mu = \frac{1}{\rho}$$

$$\sigma = \frac{\mu}{R_H} = \frac{1}{\rho}$$

$$\mu = \frac{R_H}{\rho} = \frac{300}{0.1} = 3000\ \text{cm}^2/\text{V-sec}$$

1.27 (c)

Statement 1 : In forward bias depletion layer width will become narrower (True).

Statement 3 : In reverse bias depletion layer width will be wider (True).

Statement 4 : In forward bias majority charge carriers current is established (True).

1.29 (b

Intrinsic conductivity,

$$\sigma_i = n_i(\mu_n + \mu_p)q$$

$$= 1.5 \times 10^{10}(1300 + 500)1.6 \times 10^{-19}$$

$$= 4.32 \times 10^{-6}\ \text{s/cm}$$

1.30 (d)

The fermi level is 0.3 eV below the conduction level at 300 K in n-type. As temperature increases to 360 K then the gap between conduction level and fermi level increases.

■■■■

2 Semiconductor Diode, Characteristics & Circuits

2.1 **Assertion (A):** In avalanche breakdown, the reverse current sharply increases with voltage due to a field emission.

Reason (R): The field emission requires lightly doped '*p*' and '*n*' regions.

(a) Both A and R are true and R is the correct explanation of A
(b) Both A and R are true but R is NOT the correct explanation of A
(c) A is true but R is false
(d) A is false but R is true

[ESE-2001]

2.2 In a forward biased photo diode, an increase in incident light intensity causes the diode current to

(a) increase
(b) remain constant
(c) decrease
(d) remain constant while the voltage drop across the diode increases

[ESE-2001]

2.3 Bridge rectifiers are preferred because

(a) they require small transformer
(b) they have less peak inverse voltage
(c) they need small transformer and also have less peak inverse voltage
(d) they have low ripple factor

[ESE-2001]

2.4 Consider the following statements in connection with the biasing of semi-conductor diodes:

1. LEDs are used under forward-bias condition
2. Photodiodes are used under forward-bias condition
3. Zener diodes are used under reverse-bias condition
4. Variable capacitance diodes are used under reverse-bias condition

Which of these statements are correct?

(a) 1, 2 and 3 (b) 1, 2 and 4
(c) 2, 3 and 4 (d) 1, 3 and 4

[ESE-2002]

2.5 The junction capacitance of a linearly graded junction varies with the applied reverse bias V_r as

(a) V_r^{-1} (b) $V_r^{-1/2}$
(c) $V_r^{-1/3}$ (d) $V_r^{1/2}$ [ESE-2002]

2.6 The diffusion capacitance of a forward biased p^+-n junction diode with a steady current I depends on

(a) width of the depletion region
(b) mean lifetime of the holes
(c) mean lifetime of the electrons
(d) junction area

[ESE-2002]

2.7 In the circuit shown below, the average value of $V_0(t)$ will be

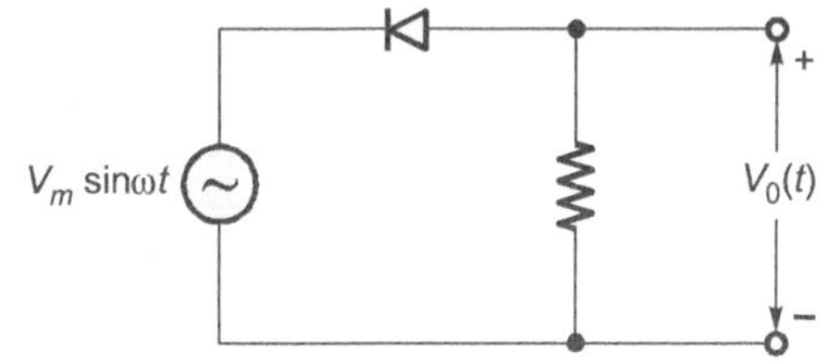

(a) 0 (b) $-V_m/\pi$
(c) $-V_m/\sqrt{2}$ (d) $-V_m$ [ESE-2002]

2.8 In the circuit shown below, the average value of $V_0(t)$ will be

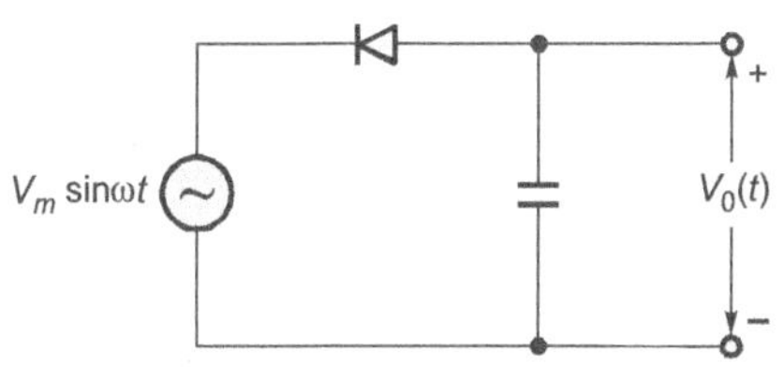

(a) 0 (b) $-V_m/\pi$
(c) $-V_m/\sqrt{2}$ (d) $-V_m$

[ESE-2002]

2.9 **Assertion** (A): A tunnel diode can be used as an oscillator.
Reason (R): Voltage controlled negative resistance is exhibited by a tunnel diode.
(a) Both A and R are true and R is the correct explanation of A
(b) Both A and R are true but R is NOT the correct explanation of A
(c) A is true but R is false
(d) A is false but R is true

[ESE-2002]

2.10 In a p-n junction, to make the depletion region extent predominantly into p-region, the concentration of impurities in the p-region must be
(a) Much less than the concentration of impurities in n-region
(b) Much higher than the concentration of impurities in n-region
(c) Equal to the concentration of impurities in n-region
(d) zero

[ESE-2003]

2.11 Match **List-I** (Circuits) with **List-II** (Characteristics/Applications) and select the correct answer:

List-I	List-II
A. High-pass RC circuit	**1.** Comparator
B. Low-pass RC circuit	**2.** DC Restorer
C. Clamping circuit	**3.** Integrator
D. Clipping circuit	**4.** Differentiator
	5. Compensated Attenuator

Codes:

	A	B	C	D
(a)	5	4	2	1
(b)	4	3	1	2
(c)	5	4	1	2
(d)	4	3	2	1

[ESE-2003]

2.12 Which one of the following statements is correct? A photodiode works on the principle of
(a) photo-voltaic effect
(b) photo-conductive effect
(c) photo-electric effect
(d) photo-thermal effect

[ESE-2004]

2.13 Two p-n junction diodes are connected back to back to make a transistor. Which one of the following is correct?
(a) The current gain of such a transistor will be high
(b) The current gain of such a transistor will be moderate
(c) It cannot be used as a transistor due to large base width
(d) It can be used only for pnp transistor

[ESE-2005]

2.14 What is the reverse recovery time of a diode when switched from forward bias V_F to reverse bias V_R?
(a) Time taken to remove the stored minority carriers.
(b) Time taken by the diode voltage to attain zero value.
(c) Time to remove stored minority carriers plus the time to bring the diode voltage to reverse bias V_R.
(d) Time taken by the diode current to reverse.

[ESE-2005]

2.15 Find the break region (voltage range) over which the dynamic resistance of a diode is multiplied by a factor of 1000. Let this region be contained between v_1 and v_2, then is $|v_1 - v_2|$ given by
(a) $\log_C(1000V_T)$
(b) $1000\ V_T$
(c) $(\log_e 10^3)\ V_T$
(d) The value cannot be computed with the given data

[ESE-2005]

2.16 Match **List-I** (Type of Device) with **List-II** (Characteristics/Application) and select the correct answer using the codes given below the lists:

List-I	List-II
A. Zener diode	**1.** Display panel
B. Tunnel diode	**2.** Voltage reference
C. Schottky diode	**3.** Light detection
D. Photo diode	**4.** Negative resistance
	5. High frequency switching

Codes:

	A	B	C	D
(a)	3	4	5	2
(b)	2	5	1	3
(c)	3	5	1	2
(d)	2	4	5	3

[ESE-2005]

2.17 Match **List-I** (Device) with **List-II** (Application) and select the correct answer using the code given below the lists:

List-I
A. p-n junction diode
B. tunnel diode
C. JFET
D. Schottky barrier diode

List-II
1. Microwave generator
2. Low frequency rectifier
3. High frequency rectifier
4. Voltage variable resistor

Codes:

	A	B	C	D
(a)	2	3	4	1
(b)	4	1	2	3
(c)	2	1	4	3
(d)	4	3	2	1

[ESE-2005]

2.18 In the rectifier circuit shown above, what should be minimum peak-inverse-voltage (PIV) rating of the diode?

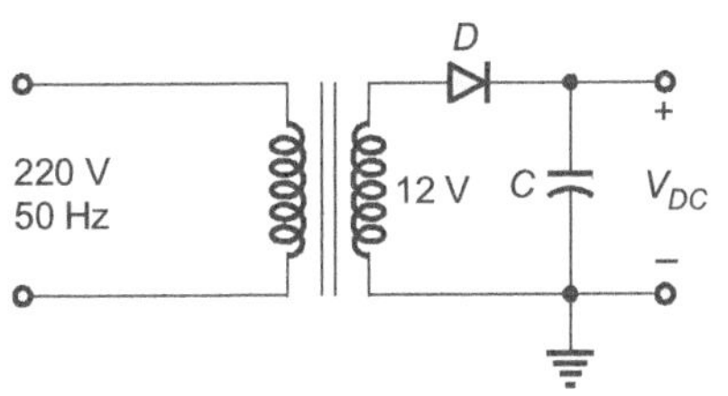

(a) 12 V (b) $12\sqrt{2}$ V
(c) 24 V (d) $24\sqrt{2}$ V

[ESE-2005]

2.19 In a centre tap full wave rectifier, 100 V is the peak voltage between the centre tap and one end of the secondary. What is the maximum voltage across the reverse biased diode?
(a) 200 V (b) 141 V
(c) 100 V (d) 86 V [ESE-2005]

2.20 Match **List-I** (Device) with **List-II** (Application) and select the correct answer using the code given below the lists:

List-I	List-II
A. Diode	1. Amplifier
B. Transistor	2. Oscillator
C. Tunnel Diode	3. Rectifier
D. Zener Diode	4. Voltage Regulator

Codes:

	A	B	C	D
(a)	4	1	2	3
(b)	3	2	1	4
(c)	4	2	1	3
(d)	3	1	2	4

[ESE-2006]

2.21 Match **List-I** (Type of Diode) with **List-II** (Characteristics/Applications) and select the correct answer using the code given below the lists:

List-I
A. Tunnel diode
B. Zener diode
C. Photo diode
D. Schottky diode

List-II
1. Reverse current varies directly with the amount of light
2. Exhibits negative resistance region in its I-C characteristic
3. Uses only majority carriers and is intended for high frequency operations
4. Silicon p-n junction diode that is designed for limiting the voltage across the terminals in reverse bias

Codes:

	A	B	C	D
(a)	2	3	1	4
(b)	1	4	2	3
(c)	2	4	1	3
(d)	1	3	2	4

[ESE-2006]

2.22 The circuit shown below is the Thevenin's equivalent circuit of a centre-tapped full wave rectifier with diode forward resistance $R_f = 100\ \Omega$, transformer secondary coil resistance $R_S = 30\ \Omega$, peak input voltage $V_m = 10$ V. What are the values of V and R_0, respectively?

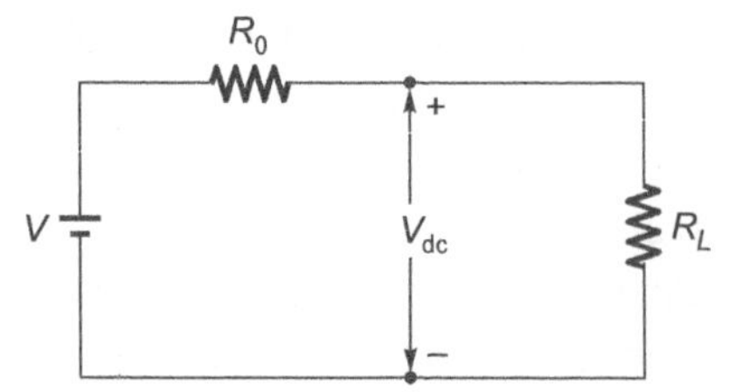

(a) 10 V, 100 Ω (b) 6.36 V, 130 Ω
(c) 6.36 V, 115 Ω (d) 4.54 V, 130 Ω

[ESE-2006]

2.23 Consider the circuit given below where R_f is the diode forward resistance and R_L the load resistance.
What is the average rectified current equal to?

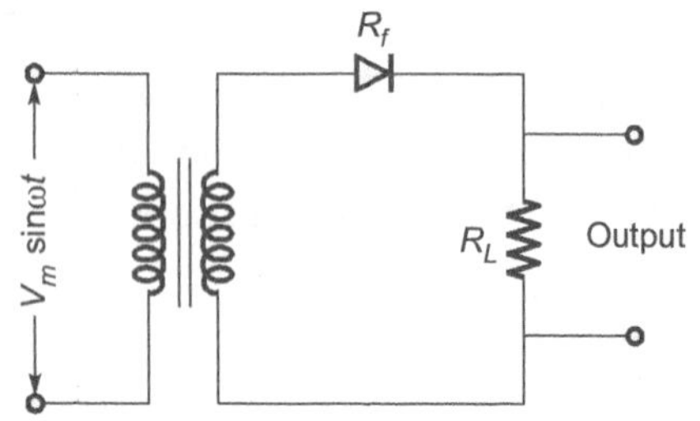

(a) $V_m/(R_L + R_f)$ (b) $V_m/\{\pi(R_f + R_L)\}$
(c) $2\,V_m/\pi$ (d) $V_m/\{\sqrt{2}(R_L + R_f)\}$

[ESE-2006]

2.24 The reverse saturation current of a Si-based p-n junction diode increases 32 times due to a rise in ambient temperature. If the original temperature was 40°C, what is the final temperature?
(a) 90°C (b) 72°C
(c) 45°C (d) 50°C [ESE-2006]

2.25 Match the I-V characteristics given below with Gunn diode, Photo diode and Tunnel diode.

(A)
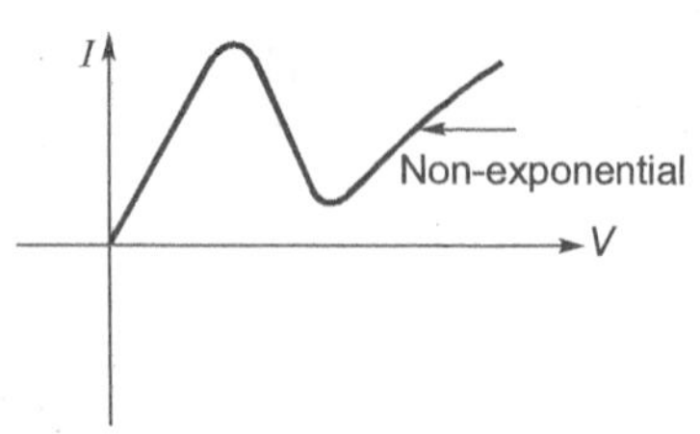

(B)
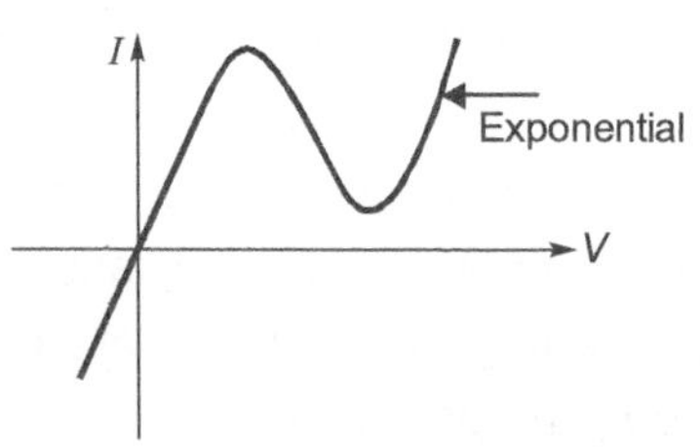

(C)
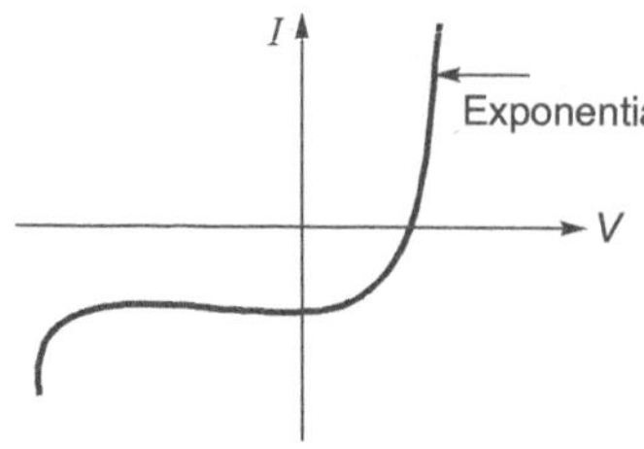

Select the correct answer using the code given below:

	Gunn diode	Photo diode	Tunnel diode
(a)	A	B	C
(b)	B	A	C
(c)	A	C	B
(d)	B	C	A

[ESE-2006]

2.26 Consider the following statements:
An applied bias voltage in a *p-n* junction diode (*n* region positive with respect to *p* region) results in
1. increase in potential barrier
2. reduction in space charge layer width
3. increase in space charge layer width
4. increase in magnitude of electric field

Which of these statements are correct?
(a) 1 and 2 (b) 1 and 3
(c) 1 and 4 (d) 1, 3 and 4

[ESE-2006]

2.27 In a biased step-graded p-n junction, what is the correct expression for the equilibrium contact potential (V_0)?

(a) $V_0 = V_T ln\left(N_A/N_D\, n_i^2\right)$

(b) $V_0 = V_T ln\left(N_D/N_A\, n_i^2\right)$

(c) $V_0 = V_T ln\left(N_A N_D/n_i^2\right)$

(d) $V_0 = V_T ln\left(n_i^2/N_A\, N_D\right)$

where $V_T \propto T/q$, T being the temperature, q the electronic charge, N_A and N_D are the doping levels of the p and n regions, respectively, and n_i is the intrinsic carrier concentration.

[ESE-2006]

2.28 In an unbiased p-n junction, the junction current at equilibrium is
(a) due to diffusion of majority carriers only
(b) due to diffusion of minority carriers only

(c) zero, because equal and opposite drift and diffusion currents for electrons and holes cross the junction
(d) zero, because no charges cross the junction

[ESE-2006]

2.29 In a bridge a.c. to d.c. converter using *p-n* diodes, if the input voltage is $V\sin\omega t$, what is the peak inverse voltage across any diode?

(a) V (b) 2 V
(c) $V/2$ (d) $V/\sqrt{2}$

[ESE-2006]

2.30 A Zener diode regulator shown in the figure given below is to be designed to meet the following specifications:

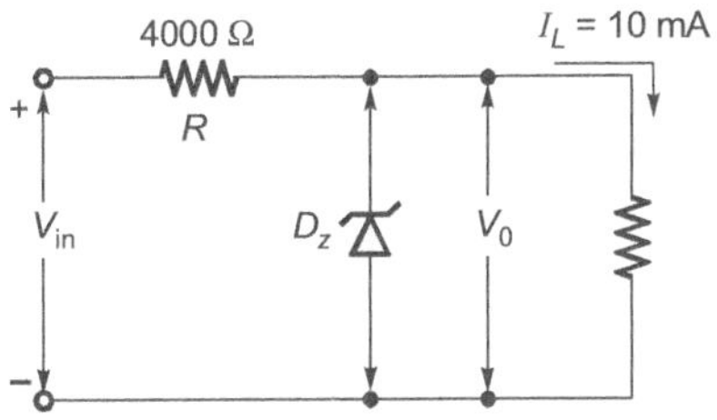

$I_L = 10$ mA, $V_0 = 10$ V, V_{in} varies from 30 V to 50 V. The Zener diode has $V_z = 10$ V and I_{zk} (knee current) = 1 mA. For satisfactory operation, which one of the following is correct?

(a) $R \le 1800\ \Omega$
(b) $2000\ \Omega \le R \le 2200\ \Omega$
(c) $3700\ \Omega \le R \le 4000\ \Omega$
(d) $R > 4000\ \Omega$

[ESE-2007]

2.31 For a half-wave rectifier, what is the output dc voltage? (with peak voltage = V_m, dc current = I_{DC} and forward resistance of diode = R_f)

(a) $V_{DC} = \frac{V_m}{\pi} + I_{DC} R_f$

(b) $V_{DC} = \frac{V_m}{\pi}$

(c) $V_{DC} = \frac{V_m}{\pi} - I_{DC} R_f$

(d) $V_{DC} = 0.707 \frac{V_m}{\pi} - I_{DC} R_f$

[ESE-2007]

2.32 Intermediate (I) layer of PIN-diode imparts which one of the following features to a *p-n* junction diode?

(a) High reverse blocking capability
(b) High forward current rating
(c) Inverting capability
(d) Poor turn off performance

[ESE-2007]

2.33 In a single phase full wave controlled bridge rectifier, minimum output voltage and maximum output voltage are obtained at which conduction angles?

(a) 0°, 180° respectively
(b) 180°, 0° respectively
(c) 0°, 0° respectively
(d) 180°, 180° respectively [ESE-2007]

2.34 In the circuit given below, D_1 and D_2 are ideal. Which one of the following represents the transfer characteristics of the circuit?

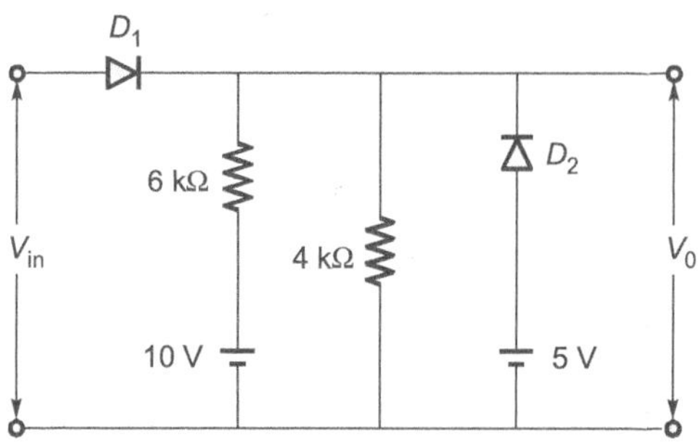

(a)
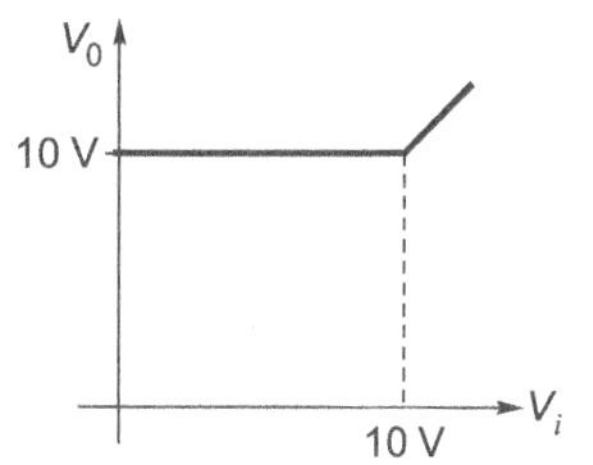

(b)
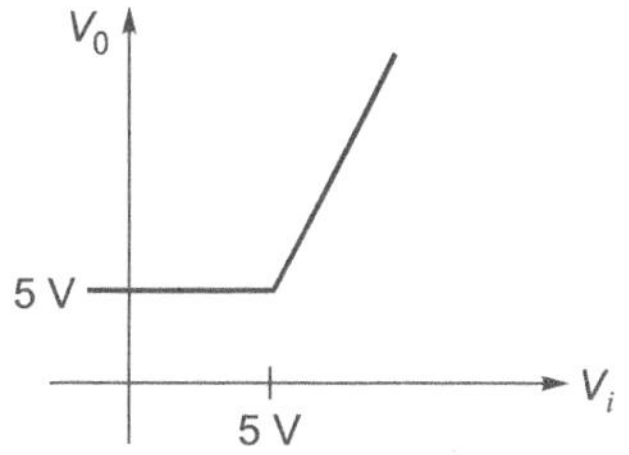

(c)
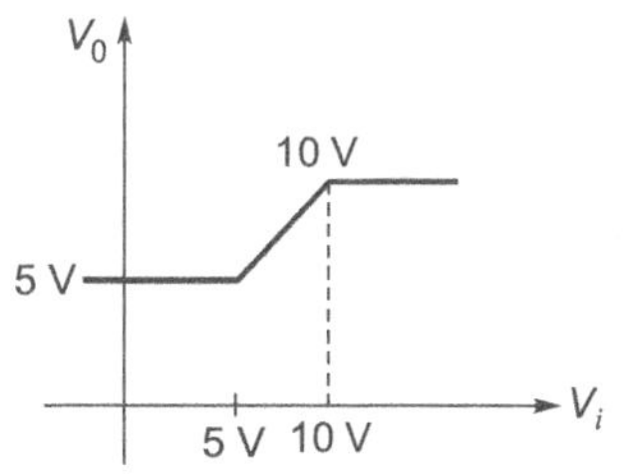

(d)
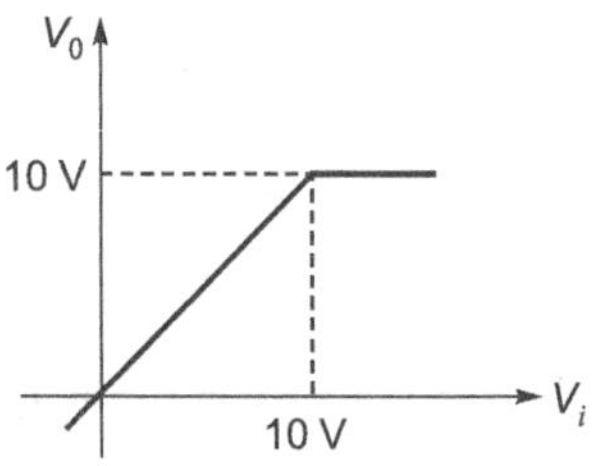

[ESE-2007]

2.35 **Assertion (A) :** A clamping network produces an output signal which is a replica of the input signal.

Reason (R) : A clamping circuit removes a part of the input signal.

(a) Both A and R are true and R is the correct explanation of A
(b) Both A and R are true but R is NOT the correct explanation of A
(c) A is true but R is false
(d) A is false but R is true [ESE-2007]

2.36 For the circuit shown in Figure-A the *V-I* (voltage-current) characteristic of the circuit using ideal components is given by curve a in Figure B.

Figure-A

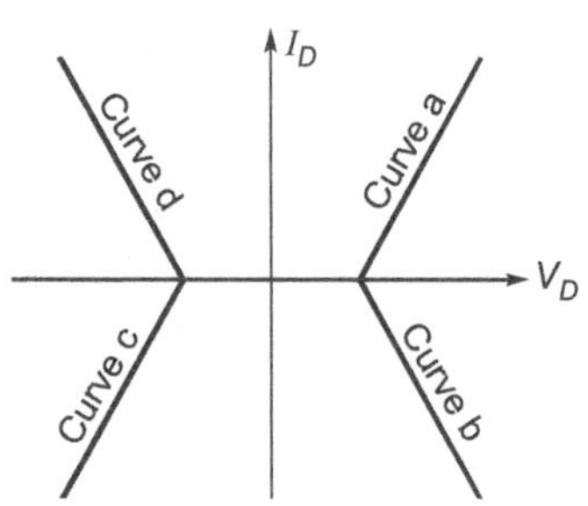

Figure-B

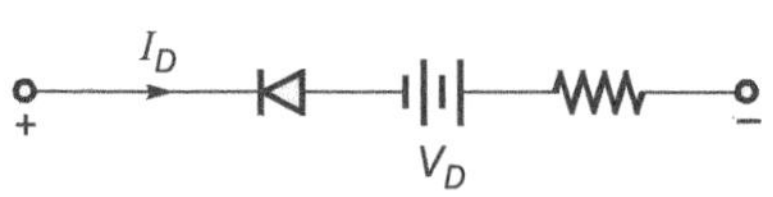

Figure-C

Which curve in Figure-B represents the *V-I* characteristic for the circuit shown in Figure-C?

(a) Curve *a* (b) Curve *b*
(c) Curve *c* (d) Curve *d*

[ESE-2008]

2.37 What is the peak current through the resistor in the circuit given below assuming the diode to be ideal?

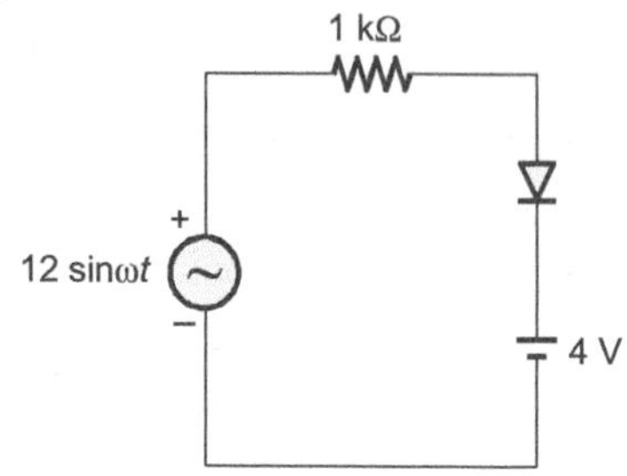

(a) 4 mA (b) 8 mA
(c) 12 mA (d) 16 mA

[ESE-2008]

2.38 In the circuit given below the Zener diode D_1 has a reverse breakdown voltage of 100 V and reverse saturation current of 25 μA. The corresponding values for D_2 are 50 V and 50 μA. What is the current in the circuit?

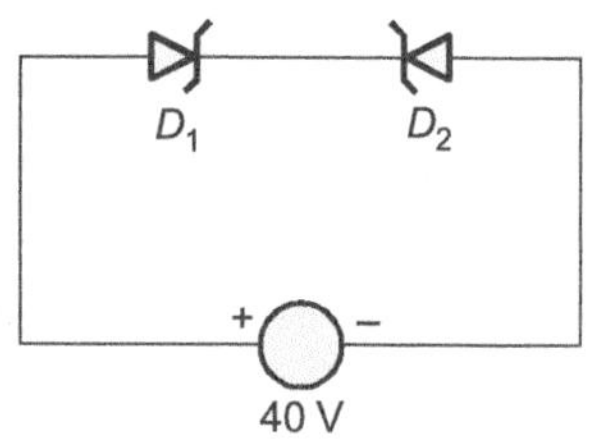

(a) 25 μA anticlockwise
(b) 25 μA clockwise
(c) 50 μA anticlockwise
(d) 50 μA clockwise

[ESE-2008]

2.39 What are the states of the three ideal diodes in the circuit as shown below?

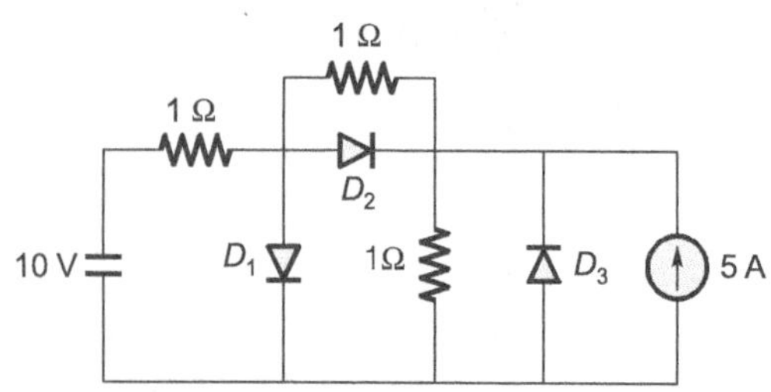

	D_1	D_2	D_3
(a)	On	Off	Off
(b)	Off	On	Off
(c)	On	Off	On
(d)	Off	On	On

[ESE-2009]

2.40 The cut-in voltage of Zener diode D_z and diode D shown in the figure below is 0.7 V. Breakdown voltage of D_z is 3.3 V and reverse breakdown voltage of D is 50 V. The other parameters can be assumed to be the same as those of an ideal diode. Then what are the values of the peak output voltage V_0?

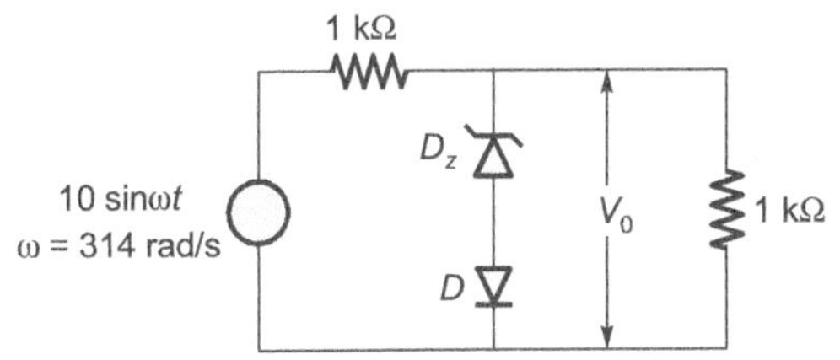

	Positive Half cycle in V	Negative Half cycle in V
(a)	3.3	1.4
(b)	4	5
(c)	3.3	3.3
(d)	4	4

[ESE-2009]

2.41 Consider the following statements about a Tunnel diode:

1. Tunneling takes place at a speed decided by junction temperature.
2. Concentration of impurities is of the order of 1 part in 10^3.
3. Both tunneling current and normal pn junction injection current exist.
4. Tunnel diode exhibits current controlled negative resistance, characteristic, only.

Which of these statements is/are correct?

(a) 1 only (b) 1 and 2
(c) 2 and 3 (d) 3 and 4 [ESE-2009]

2.42 The forward resistance of the diode shown in the below circuit is 5 ohms, and the other parameters are same as those of an ideal diode. Then what is the d.c. component of the source current?

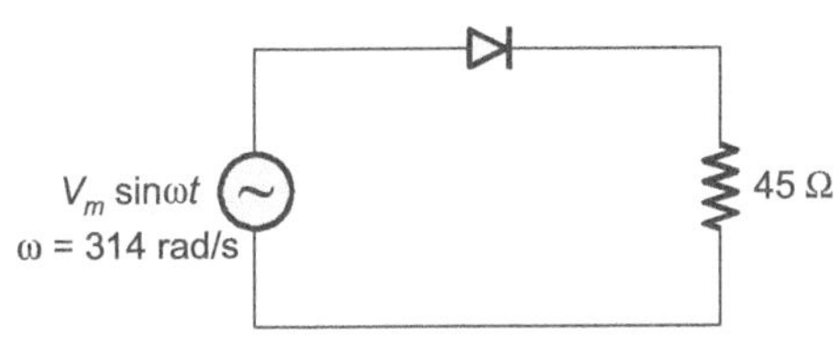

(a) $\frac{V_m}{50\pi}$ (b) $\frac{V_m}{50\pi\sqrt{2}}$

(c) $\frac{V_m}{100\pi\sqrt{2}}$ (d) $\frac{2V_m}{50\pi}$ [ESE-2009]

2.43 The 6 V Zener diode as shown in the circuit below, has zero Zener resistance and a knee current of 5 mA. Then what is the minimum value of R so that the voltage across it does not fall below 6 V?

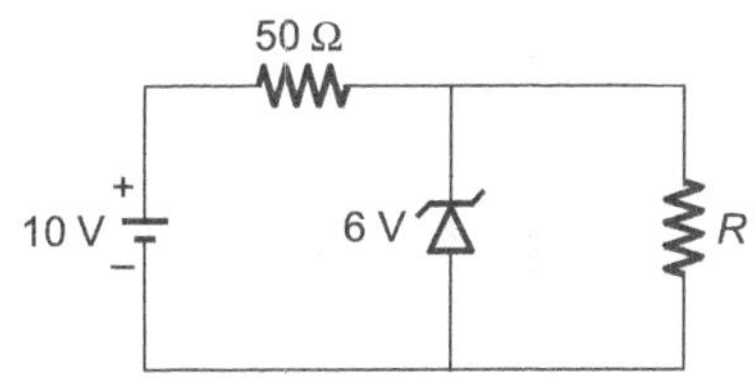

(a) 1200 ohms (b) 80 ohms
(c) 50 ohms (d) 40 ohms

[ESE-2009]

2.44 For a sinusoidal input of peak value V_p, the output waveform V_0 will be

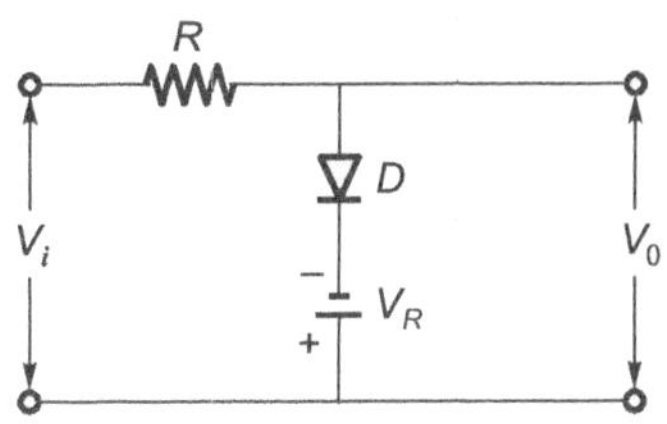

(a)

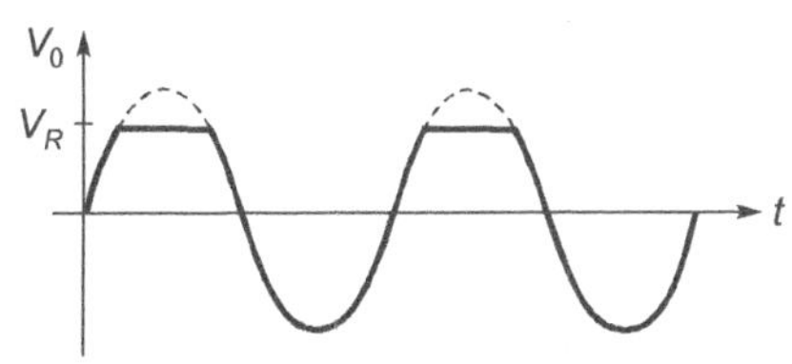

(b)

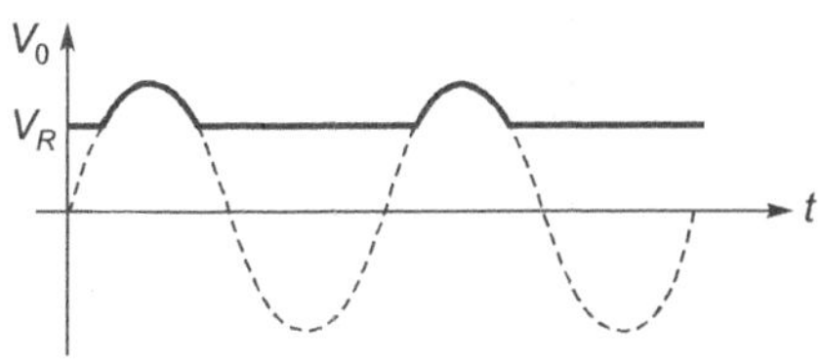

(c)

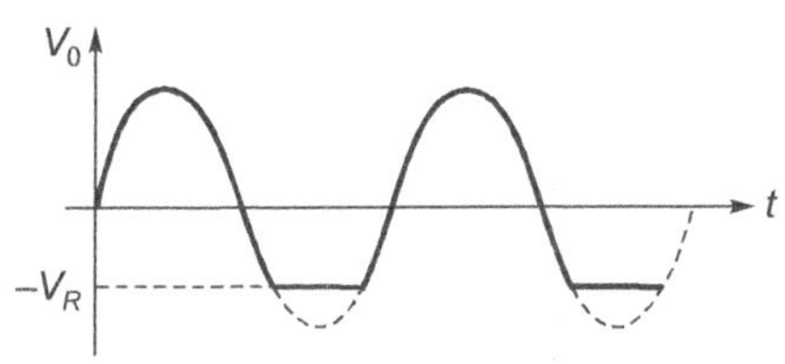

(d)

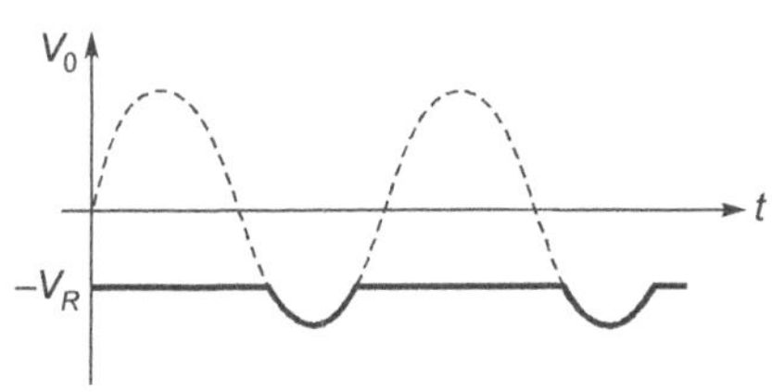

[ESE-2010]

2.45 Assertion (A): In a full-wave rectifier, it is absolutely necessary to use identical diodes having identical characteristics.

Reason (R): In practice, it is never possible to have two diodes having identical characteristics due to technological limitations.

(a) Both A and R are true and R is the correct explanation of A
(b) Both A and R are true but R is NOT the correct explanation of A
(c) A is true but R is false
(d) A is false but R is true

[ESE-2010]

2.46 Assertion (A): In the work function model of Schottky barrier diode contacts to n-type semiconductor, the metal work function will be higher than the semiconductor work function.

Reason (R): To equalise the Fermi levels on the Ferro sides of the contact, electrons from semiconductor surface move to metal side forming a space charge region at the semiconductor surfaces.

(a) Both A and R are true and R is the correct explanation of A
(b) Both A and R are true but R is NOT the correct explanation of A
(c) A is true but R is false
(d) A is false but R is true

[ESE-2010]

2.47 Dark current in a semiconductor photo-diode is
(a) the forward bias current.
(b) the forward saturation current.
(c) the reverse saturation current.
(d) the transient current.

[ESE-2011]

2.48 Consider the following statements:
1. A Zener diode degrades the input signal at high frequencies due its transition capacitance.
2. The Zener voltage V_z does not vary with temperature.
3. Regulation the Zener diode is adversely affected at the knee current $I_{Z(min)}$ due to limited power dissipation capacity.
4. In a simple Zener diode regulated circuit, amplification is not possible.

Which of these statements are correct?
(a) 1, 2, 3 and 4 (b) 3 and 4
(c) 2 and 4 (d) 1 and 4

[ESE-2011]

2.49 In an open circuited p-n junction diode space charge density at the junction is
(a) Maximum (b) Zero
(c) Positive (d) Negative

[ESE-2012]

2.50 Material used for fabrication of Tunnel diode is
(a) Ge or GaAs (b) Si and GaAs
(c) Si and InSb (d) Ge and InSb

[ESE-2012]

2.51 The 6 V Zener diode shown in figure has zero Zener resistance and a knee current of 5 mA. The minimum value of R, so that the voltage across it does not fall below 6 V is

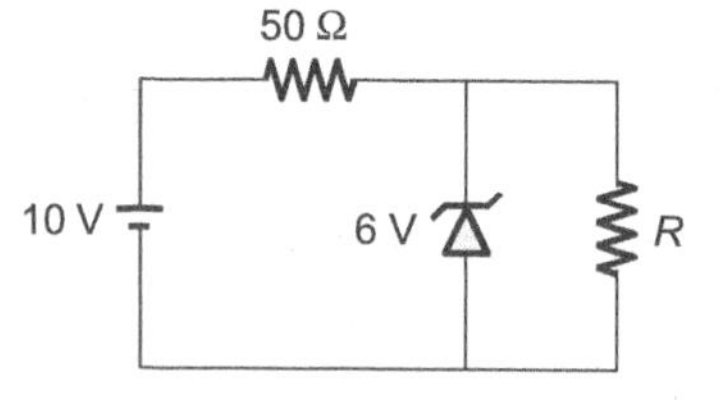

(a) 1.2 kΩ (b) 80 Ω
(c) 50 Ω (d) Zero

[ESE-2013]

2.52 A full-wave rectifier using centre tapped transformer and a bridge rectifier used similar diodes and have equal no load output voltage. Under equal load conditions
(a) output of bridge rectifier will be less than that of the other.
(b) output of bridge rectifier will be more than that of the other.
(c) output voltage of both will be exactly equal.
(d) output voltage of anyone may be more that of the other.

[ESE-2013]

2.53 For forward biased diode
(a) transition capacitance is larger than diffusion capacitance
(b) diffusion capacitance is larger than transition capacitance.
(c) both capacitances are having same value.
(d) cannot predict with certainty.

[ESE-2014]

2.54 In a *p-n* junction diode under reverse bias, the magnitude of electric field is maximum at
(a) the edge of the depletion region on the *p* side.
(b) the edge of the depletion region on the *n* side.
(c) the *p-n* junction.
(d) the center of the depletion region on the side.
[ESE-2014]

2.55 As compared to an LED, an LCD has the distinct advantage of
(a) extremely low power consumption
(b) providing a silver display
(c) being extremely thin
(d) giving two types of display [ESE-2014]

2.56 The increase in value of β of transistor can cause the fixed bias circuit to
(a) shift from saturation region to active region.
(b) shift the operation from active mode to saturation mode.
(c) shift the operation from saturation mode to cut-off mode.
(d) shift the operation from cut-off mode to active mode.
[ESE-2014]

2.57 In a *P-N* junction diode under reverse bias, the magnitude of electric field is maximum at
(a) the edge of the depletion region on the *P* side
(b) the edge of the depletion region on the *N* side
(c) the centre of the depletion region on the *N* side
(d) the *P-N* junction [ESE-2015]

2.58 A silicon diode is preferred to a germanium diode because of its
(a) higher reverse current
(b) lower reverse current and higher reverse break down voltage
(c) higher reverse current and lower reverse break down voltage
(d) None of the above [ESE-2015]

2.59 Compared to an ordinary semiconductor diode, a Schottky diode has
(a) higher reverse saturation current and zero cut-in voltage
(b) higher reverse saturation current and higher cut-in voltage
(c) higher reverse saturation current and lower cut-in voltage
(d) lower reverse saturation current and lower cut-in voltage
[ESE-2015]

2.60 Which of the following is called 'hot carrier diode'?
(a) PIN diode (b) LED
(c) Photo diode (d) Schottky diode
[ESE-2015]

2.61 Ripple-rejection ratio of voltage regulator is the ratio of
(a) output voltage to input ripple voltage
(b) output power to input power of regulator
(c) input power to output power of regulator
(d) input ripple voltage to output ripple voltage
[ESE-2015]

2.62 A Schottky diode is
(a) a majority carrier device
(b) a minority carrier device
(c) a fast recovery diode
(d) both majority and minority carrier diode
[ESE-2015]

2.63 In an *L*-section filter, a bleeder resistance connected across the load
(a) provides good regulation for all values of load
(b) ensures lower PIV of the diodes
(c) ensures lower values of capacitance in the filter
(d) reduces ripple content
[ESE-2015]

2.64 Consider the following statements:
A power supply uses bridge rectifier with a capacitor input filter. If one of the diodes is defective, then
1. The dc load voltage will be lower than its expected value.
2. Ripple frequency will be lower than its expected value.
3. The surge current will increase considerably.

Which of the above statements are correct?
(a) 1 and 2 only (b) 1 and 3 only
(c) 2 and 3 only (d) 1, 2 and 3
[ESE-2016]

2.65 A clamper circuit
1. Adds or subtracts a dc voltage to or from a waveform.
2. Does not change the shape of the waveform.

Which of the above statements is/are correct?
(a) 1 only (b) 2 only
(c) Both 1 and 2 (d) Neither 1 nor 2
[ESE-2016]

2.66 The total characteristic of a stabilizer is
(a) constant output voltage with low internal resistance.
(b) constant output current with low internal resistance.
(c) constant output voltage with high internal resistance.
(d) constant internal resistance with variable output voltage.
[ESE-2016]

2.67 Consider the following statements in the relevant context:
1. The two types of currents that flow in semiconductor diodes and transistors are drift and diffusion currents.
2. The junction region is called depletion region or space-charge region.
3. When currents flow through the diode in forward bias, the depletion region current is mostly of 'diffusion' type.

Which of the above statements are correct?
(a) 1, 2 and 3 (b) 1 and 2 only
(c) 1 and 3 only (d) 2 and 3 only
[ESE-2018]

2.68 What is Shockley's equation of a semiconductor diode in the forward bias regions?
(a) $I_D = I_S(e^{V_D^2/nV_T} - 1)$
(b) $I_D = I_S(e^{V_D/nV_T} - 1)$
(c) $I_D = I_S(e^{nV_D/nV_T} - 1)$
(d) $I_D = I_S(e^{V_T/nV_D} - 1)$
where,
I_S is reverse saturation current
V_D is applied forward-bias voltage across the diode
V_T is thermal voltage
n is an ideality factor [ESE-2019]

2.69 In emergency lighting system, the component used for maintaining the charge on the battery is
(a) LED (b) Shockley diode
(c) Thermistor (d) SCR [ESE-2019]

2.70 In a tunnel diode, the width of the junction barrier is
(a) directly proportional as the square root of impurity concentration
(b) inversely proportional as the square root of impurity concentration
(c) directly proportional as square of impurity concentration
(d) inversely proportional as square of impurity concentration [ESE-2020]

2.71 For an ideal power supply, the output voltage is independent of the load and the percentage regulation is
(a) equal to zero.
(b) equal to unity.
(c) greater than unity.
(d) less than unity but not zero. [ESE-2021]

2.72 In P-N junction diode, the reverse saturation current increases by 7.2% by a degree rise in junction temperature (in Celsius) and gets
(a) halved for every 10° C rise in temperature
(b) doubled for every 10° C rise in temperature
(c) halved for every 20° C rise in temperature
(d) doubled for every 20° C rise in temperature
[ESE-2022]

2.73 Which one of the following is the correct representation of Boltzmann diode equation with standard notations? (where I is the diode current, V is the bias voltage and V_T is the thermal voltage)
(a) $I = I_0\left(e^{\frac{V}{\eta V_T}} - 1\right)$ (b) $I = I_0\left(e^{\frac{V}{\eta V_T}} + 1\right)$
(c) $I = I_0\left(e^{-\frac{V}{\eta V_T}} - 1\right)$ (d) $I = I_0\left(e^{-\frac{V}{\eta V_T}} + 1\right)$
[ESE-2023]

2.74 A silicon diode made half-wave rectifier produces a maximum value of load current as 20 mA through a resistance of 1 kΩ. What is the PIV of the diode?
(a) 14.7 V (b) 16.7 V
(c) 18.7 V (d) 20.7 V
[ESE-2023]

2.75 Ripple factor for a half-wave and a full-wave rectifier circuits respectively are
(a) 0.48 and 1.21 (b) 0.37 and 2.12
(c) 1.21 and 0.48 (d) 2.12 and 0.37
[ESE-2023]

2.76 Statement (I): Diodes designed expressly to operate in the forward biased region are called zener diodes.

Statement (II): The principal operating region for a zener diode is negative in terms of both voltage and current.

[ESE-2023]

2.77 What are the nominal values of drop in potential across conducting silicon diodes, Schottky diodes and light- emitting diodes respectively?

(a) 0.6 to 0.7 volt, 0.2 V and 1.4 V
(b) 0.6 to 0.7 volt, 1.2 V and 0.2 V
(c) 0.6 to 0.7 volt, 2.2 V and 1.4 V
(d) 0.6 to 0.7 volt, 0.2 V and 5.4 V

[ESE-2024]

2.78 Consider the following statements regarding semiconductor diode:

1. In the non-conducting region (when the p-n junction is reverse biased), the diode current is exactly zero.
2. The diode requires a small positive voltage to be applied before it enters the conducting region.
3. For large input voltages and/or currents, the diode enters breakdown regions in the forward direction.

Which of the above statements is/are correct?

(a) 1 and 2 (b) 2 only
(c) 1 only (d) 2 and 3

[ESE-2024]

2.79 What happens to the depletion region and the depletion region capacitance of a p-n junction diode with increased reverse-biased voltage?

(a) The depletion region shrinks and the depletion region capacitance decreases
(b) The depletion region shrinks and the depletion region capacitance increases
(c) The depletion region widens and the depletion region capacitance decreases
(d) The depletion region widens and the depletion region capacitance increases

[ESE-2024]

■■■■

Answers Semiconductor Diode, Characteristics & Circuits

2.1 (d)	2.2 (b)	2.3 (c)	2.4 (d)	2.5 (c)	2.6 (b)	2.7 (b)	2.8 (d)	2.9 (a)
2.10 (a)	2.11 (d)	2.12 (b)	2.13 (c)	2.14 (c)	2.15 (c)	2.16 (d)	2.17 (c)	2.18 (d)
2.19 (a)	2.20. (d)	2.21 (c)	2.22 (c)	2.23 (b)	2.24 (a)	2.25 (c)	2.26 (d)	2.27 (c)
2.28 (d)	2.29 (a)	2.30 (a)	2.31 (c)	2.32 (a)	2.33 (b)	2.34 (b)	2.35 (c)	2.36 (c)
2.37 (b)	2.38 (d)	2.39 (a)	2.40 (b)	2.41 (c)	2.42 (a)	2.43 (b)	2.44 (d)	2.45 (d)
2.46 (a)	2.47 (c)	2.48 (d)	2.49 (b)	2.50 (a)	2.51 (b)	2.52 (b)	2.53 (b)	2.54 (c)
2.55 (a)	2.56 (b)	2.57 (d)	2.58 (b)	2.59 (c)	2.60 (d)	2.61 (d)	2.62 (a)	2.63 (a)
2.64 (d)	2.65 (c)	2.66 (a)	2.67 (a)	2.68 (b)	2.69 (a)	2.70 (b)	2.71 (a)	2.72 (b)
2.73 (a)	2.74 (d)	2.75 (c)	2.76 (d)	2.77 (a)	2.78 (b)	2.79 (c)		

Explanations Semiconductor Diode, Characteristics & Circuits

2.1 (d)

A is wrong because avalanche breakdown takes place due to a large no. of breaking of covalent bonds across the junction due to impact ionization due to highly energetic e^{-s} collide with negative ions and excite e^- from negative ion.

2.2 (b)

In case of F.B. photo diode it is insensitive to incident light and behave as a normal diode so diode current remains same irrespective of intensity of incident light.

2.3 (c)

Bridge rectifier requires comparatively small transformers as required for centre tap full wave rectifier and this is due to no requirement of centre tapping at all.

Also PIV for diodes in bridge rectifier is V_m while in centre tap full wave rectifier it is 2 V_m, so less peak inverse voltage.

2.4 (d)

Photodiodes are used under reverse-bias condition not forward-bias condition.

2.5 (c)

For step-graded junction $C_T \propto V^{-1/2}$ and for linearly graded junction $C_T \propto V^{-1/3}$

2.6 (b)

$C_D = \dfrac{\tau}{r_d} = \dfrac{\tau I}{\eta V_T}$; as p-side is heavily doped (p^+) than the n-side, so that the current at the junction is entirely due to holes.

2.7 (b)

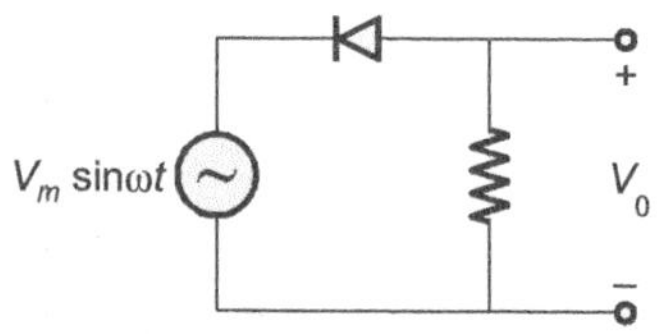

For positive half cycle of input diode D will be off and output will be zero. and for negative half of input diode D will be on and we will get the output equal to input as

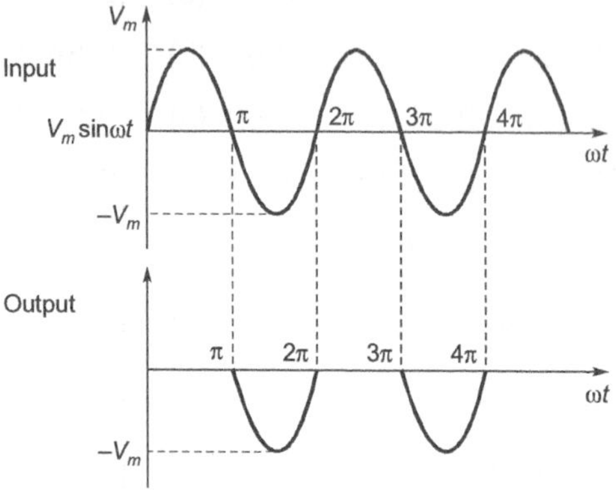

Just same as half wave rectifier circuit but we are getting the output in negative half of input.

So, average value of output = $-V_m/\pi$.

2.8 (d)

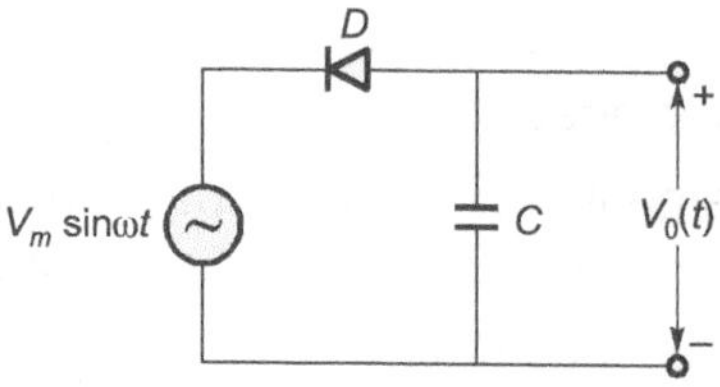

Diode D will be on in –ve half of input and capacitor C will be charged to $-V_m$ instantly because on resistance of diode D is zero so time constant $RC = 0$. Once C will charged it will never discharged because when diode D is off it's resistance $R \to \infty$ discharging time constant will be ∞. So output, $V_0 = -V_m$

2.9 (a)

A device can be used as microwave oscillator if it exhibited –ve resistance property and tunnel diode exhibited this property so, A and R both are true and R is correct explanation.

2.10 (a)

∵ depletion region predominantly in p-region that means diode is of type p-n^+.

i.e. lesser the doping higher will depletion region on that side.

∵ depletion width $\propto \dfrac{1}{\text{Doping concentration}}$

∴ holes on p-side are lesser compare to electrons on n-side.

2.11 (d)

High pass RC circuit → act as differentiator for low RC.

Low pass RC circuit → act as integrator for high RC (time constant)

Clamping circuit act as DC inserter or DC restorer.

Clipping circuit act as comparator → as it compares input w.r.t. certain DC voltage and clips off the portion accordingly.

2.13 (c)

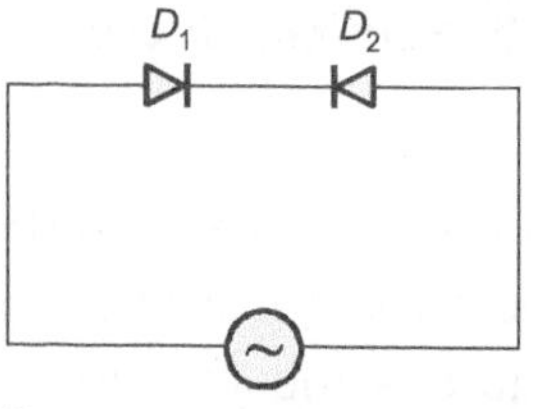

This configuration will not work as a transistor because base width will be too large. So that carrier injected from forward bias diode (equivalent to emitter-base junction) will not reach to collector.

2.14 (c)

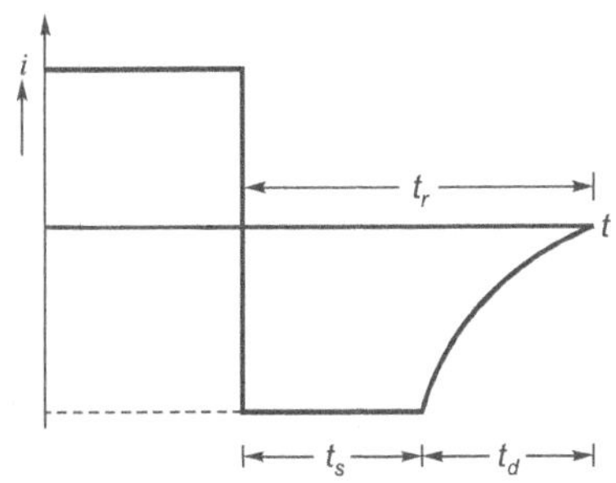

Recovery time, $t_r = t_s + t_d$

t_s = storage time (to remove stored minority carriers)

t_d = to bring the diode in reverse bias condition i.e. $V_D = V_R$) and current $i = 0$.

2.15 (c)

$$r_d = \frac{\eta V_T}{I_0 e^{v.\eta/V_T}} \Rightarrow \frac{r_{d,2}}{r_{d,1}} = \frac{e^{v_1.\eta/V_T}}{e^{v_2.\eta/V_T}}$$

$$\therefore \quad \frac{1000 \,.\, r_{d,1}}{r_{d,1}} = e^{|v_1 - v_2| \cdot \frac{\eta}{V_T}}$$

$$\therefore \quad \log_e 10^3 = |v_1 - v_2| \cdot \frac{\eta}{V_T}$$

$$\Rightarrow \quad |v_1 - v_2| = V_T(\log_e 10^3) \text{ for } \eta = 1$$

2.16 (d)

Zener diode are used as voltage reference in voltage regular circuit.

Tunnel diodes are –ve resistance devices and used as high speed switching in microwave oscillators.

Photodiodes respond to light falls on it as photo current changes due to change in intensity of light.

Schottky diode are used for high speed switching.

2.17 (c)

p-n junction diode is used for rectification at low frequency since switching time is large.

Schottky barrier diode is used for rectification at high frequency since switching time is small for this diode.

Tunnel diode is used in microwave generator, oscillator circuit.

JFET is used as voltage variable resistor in active region.

2.18 (d)

During positive half cycle, capacitor would be charged to 12 V rms.

So, PIV = $(12 + 12)\sqrt{2} = 24\sqrt{2}$ V

2.19 (a)

In case of full wave rectifier PIV = 2 V_m

where V_m maximum voltage across centre tap and one end of primary or secondary.

So, PIV = 2 × 100 = 200 V.

2.20 (d)

Diode → used for rectification.

Transistor → used as an amplifier.

Tunnel diode → used as a switch in microwave oscillator circuit.

Zener diode → used in voltage regulator circuit.

2.21 (c)

Photodiode: Current varies in accordance with intensity of light falls as electron-hole pair generation depends on light intensity.

Tunnel diode: Exhibit –ve resistance property due to which used in microwave oscillator circuit.

Zener diode: Used for voltage reference, i.e. maintain the constant voltage across the load terminals (independent of load) i.e. limiting the voltage across the terminals in reverse bias.

Schottky diode: Uses only majority carrier (because it is metal-semiconductor junction) due to this there will be no storage time required for storing minority carriers in case of making of the switch so lesser switching time and therefore high frequency of operation.

2.22 (c)

Thevenin's equivalent of full wave rectifier (centre tapped) is

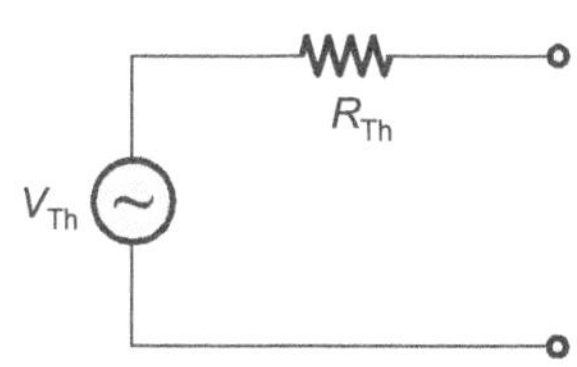

Where,

$$V_{th} = \frac{2V_m}{\pi}$$

$$R_{th} = R_f + \frac{R_s}{2}$$

From the data given

$$R_{th} = 100 + \frac{30}{2} = 115\,\Omega$$

$$V_{th} = \frac{2 \times 10}{\pi} = 6.36\text{ V}$$

2.23 (b)

This is a half wave rectifier circuit so average value of current

$$I_{dc} = I_m/\pi$$

Where I_m is maximum current

$$= \frac{V_m}{(R_f + R_L)}$$

$$\therefore \quad I_{dc} = \frac{V_m}{\pi(R_f + R_L)}$$

2.24 (a)

We know that,

$$I_0' = I_0 2^{(T_2 - T_1)/10}$$

I_0 – reverse saturation current at temp T_1

I_0' – reverse saturation current at temp T_2

$$\therefore \quad 32 I_0 = I_0 2^{(T_2 - 40)/10}$$

$$\frac{T_2 - T_1}{10} = 5$$

$$T_2 = T_1 + 50 = 90°\text{C}$$

2.25 (c)

These all are the standard I-V characteristic of Gun diode, tunnel diode, and photo diode respectively.

2.26 (d)

Potential barrier now becomes $(V_0 + V_r)\uparrow$

where $V_r \rightarrow$ applied reverse voltage

Space charge layer width $w \propto \sqrt{V_0 + V_r}$

So, $\uparrow$ electric field

$$E \propto x_{no} N_D \text{ or } E \propto x_{po} N_A$$

where,

x_{no} – part of depletion width on n-side increases as w increases with V_r

x_{po} – part of depletion width on p-side increases as w increases with V_r.

So, electric field $E\uparrow$.

2.28 (d)

When diode is unbiased $V = 0$

$$\therefore \quad i = I_0[e^{V/\eta V_T} - 1]$$

$$i = 0$$

This is due to no charge carriers are crossing the junction.

2.29 (a)

We know that peak inverse voltage across any diode in bridge rectifier is = V_m (V_m is maximum value of input voltage) this is due to sharing of voltages by the two diodes in reverse biased condition.

2.30 (a)

$$\frac{V_{in} - V_z}{R} = I_z + I_L$$

$$\frac{V_{in_{min}} - V_z}{R} = I_{zk} + I_L, \quad I_{zk} = \text{knee current}$$

$$R \le \frac{V_{in_{min}} - V_z}{I_{zk} + I_L} \Rightarrow R \le \frac{30 - 10}{(1\text{ mA} + 10\text{ mA})}$$

$$R \le 1818\,\Omega$$

2.31 (c)

For half wave rectifier,

$$I_m = \frac{V_m}{(R_f + R)}$$

If R is load resistance

$$\therefore \quad I_{DC} = \frac{I_m}{\pi} = \frac{V_m}{\pi(R_f + R)}$$

$$\therefore \quad V_{DC} = I_{DC} R = \frac{V_m \cdot R}{\pi(R_f + R)}$$

$$= \frac{V_m[R + R_f - R_f]}{\pi(R_f + R)}$$

$$V_{DC} = \frac{V_m}{\pi}\left[1 - \frac{R_f}{R_f + R}\right]$$

$$V_{DC} = \frac{V_m}{\pi} - \frac{V_m \cdot R_f}{\pi(R_f + R)}$$

$$V_{DC} = \frac{V_m}{\pi} - I_{DC} \cdot R_f$$

$$\because \quad I_{DC} = \frac{V_m}{\pi(R_f + R)}$$

2.32 (a)

In PIN diode intrinsic layer decides the reverse blocking capacity.

2.33 (b)

Output of a bridge rectifier is

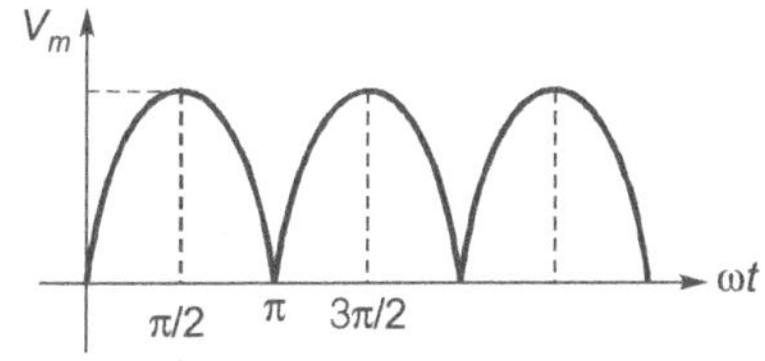

So maximum is obtained at $\frac{\pi}{2}, \frac{3\pi}{2} ...$

and minimum is obtained 0°, 180° ...

2.34 (b)

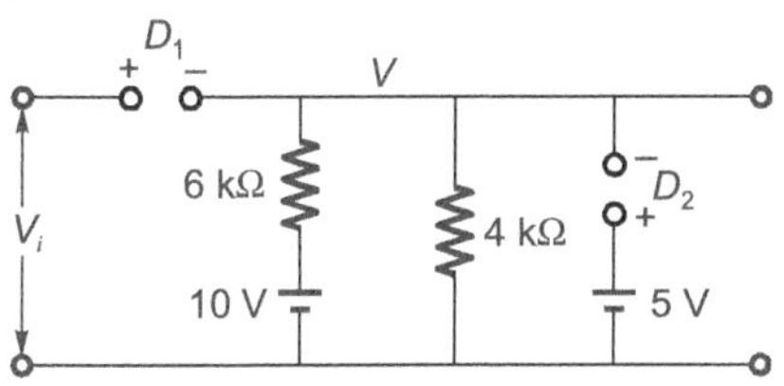

Now get V_{D_1} and V_{D_2} to know about forward or reverse biasing condition of diodes.

$$V\left(\frac{1}{6}+\frac{1}{4}\right) = \frac{10}{6} \Rightarrow V = 4 \text{ volt}$$

$$V_{D_1} = V_i - 4$$

(for $V_i < 4$ V it is off initially)

$$V_{D_2} = 1 \text{ volt, hence}$$

$$V_{D_2} = \text{forward biased so, } V_0 = 5 \text{ V}$$

For $V_i > 5$ V, $V_0 = V_i$ ($\because D_1$ gets on)

2.36 (c)

As shown in figure '*a*' I_D will flow in the direction shown on application of V_D hence both will be positive.

But in figure '*c*' current I_D will only flow opposite to the direction shown on application of opposite voltage because only then diode will be forward biased hence curve '*c*' means both current and voltage opposite to curve '*a*'.

2.37 (b)

$$\text{Peak current} = \frac{\text{Peak voltage across the circuit}}{\text{circuit resistance}}$$

$$= \frac{12-4}{1\text{K}} = 8 \text{ mA}$$

2.38 (d)

$\because$ Due to given connection D_1 will be in forward biased and D_2 will be in reverse biased.

Since, 40 V < 50 V (reverse break down voltage of D_2).

So it will be in reverse bias condition so allows to flow only reverse saturation (not in breakdown condition) current i.e. 50 μA and it flows in clockwise direction as source supports it.

2.39 (a)

Analyzing the circuit, we can see that D_1 and D_2 are in forward bias and D_3 is in reverse bias. But no current flow through D_2, because current gets shortest path through D_1.

2.40 (b)

In positive half cycle when

$$V_i > 4 \text{ volt}$$

Diode D_Z will be breakdown and diode D will be ON

So, output voltage

$$= V_Z + V_{\text{cut in}} \text{ of } D$$
$$= 3.3 + 0.7 = 4 \text{ V}$$

and for $V_i < 4$ V

D_Z will be open circuited so

$$\text{output} \quad V_0 = V_i \sin\omega t \times \frac{1}{1+1}$$

$$= \frac{1}{2} V_i \sin\omega t$$

So maximum output in positive half cycle = 4 V

in negative half cycle diode D will be open circuited as it is in reverse biased condition and reverse voltage is < 50 V.

$$\text{So,} \quad \text{output} = \frac{1}{1+1} \times V_i \sin\omega t = \frac{V_i}{2} \sin\omega t$$

So maximum (magnitude) in negative half

$$= \frac{10}{2} = 5 \text{ V}$$

As $V_i = 10$ V

(since maximum of $V_i \sin\omega t = 10$)

2.41 (c)

Tunneling takes place when charge carriers cross the junction at a speed at light, it is not depend on junction temperature.

Doping levels are very high i.e. 1 : 10^3
Both tunneling and normal *p-n* junction current flows.
Tunnel diode exhibit three properties.

1. Works as linear resistance for smaller voltage.
2. Works as negative resistance in moderate voltage ranges.
3. Works as normal diode (exponential characteristic) at high voltage.

2.42 (a)

This is a half wave rectifier circuit.
So, maximum value of current

$$I_m = \frac{V_m}{R_f + R_L} = \frac{V_m}{5+45} = \frac{V_m}{50}\text{ A}$$

output is as

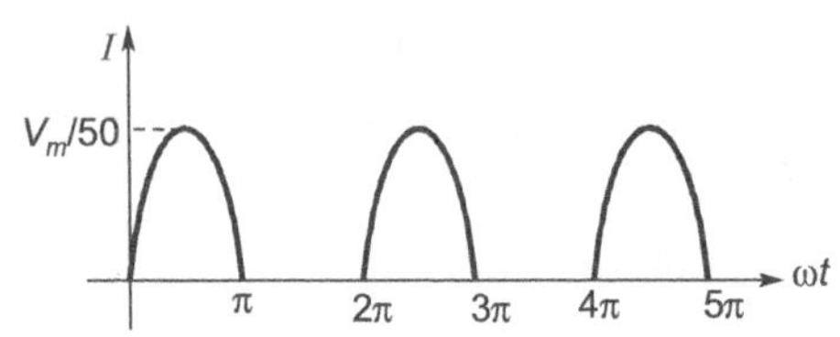

So average value of o/p current

$$I_{DC} = \frac{I_m}{\pi} = \frac{V_m}{50\pi}$$

2.43 (b)

$$R_1 = \frac{V_{i(\text{min.})} - V_z}{I_{z(\text{min})} + I_{L\max}}$$

$$50 = \frac{10-6}{5\times 10^{-3} + I_{L(\max)}}$$

$\therefore \quad I_{L(\max)} = 0.075$ A

and $\quad V = I_{L(\max)} \times R$

$\therefore \quad R = \dfrac{6}{0.075} = 80\ \Omega$

2.44 (d)

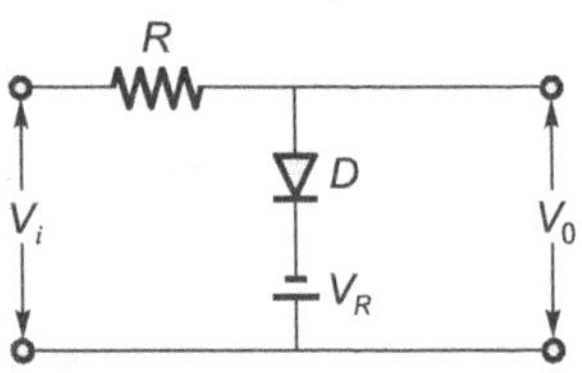

where $V_i > V_R$
Diode *D* is forward biased so act as short circuit.

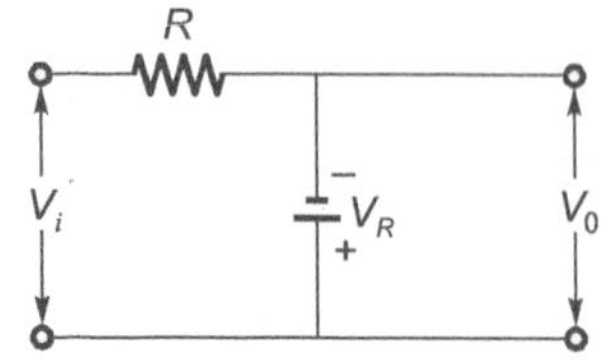

So, $\quad V_0 = -V_R$
when $\quad V_i = V_R$
then diode *D* is reverse biased and act as open circuit.
So, $\quad V_0 = V_i$

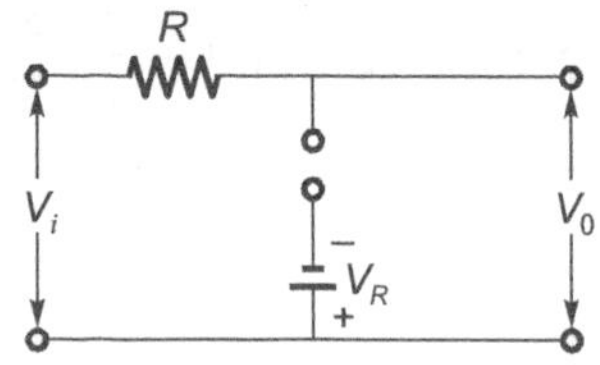

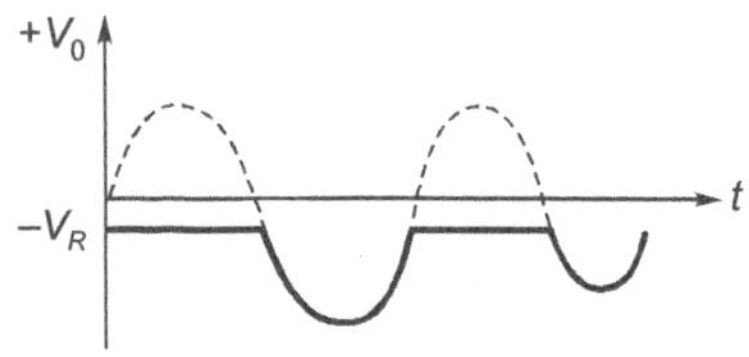

2.47 (c)

Photodiode is always operated in reverse bias and when light is off current flowing through this diode is reverse saturation current as working as a normal diode when light is off.

2.48 (d)

In Zener diode regulated circuit, amplification is not possible because, it is not an active element which can amplify the signal. So, statement 4 is correct.
Breakdown voltage varies with temperature so, statement 2 is wrong.
Regulation of Zener diode is adversely affected at $I_{z\max}$ due to limited power dissipation capacity $P_{z\max}$ so, statement 3 is wrong.

2.49 (b)

- In an open circuited p-n junction diode there is no space charge at the junction i.e. space charge density is zero.
- In the absence of an applied bias across a semi conductor diode, the net flow of charge in one direction is zero or the current under no-bias conditions is zero.

2.50 (a)
The tunnel diode is fabricated by doping the semiconductor materials that will form the p-n junction at a level one hundred to several thousand times that of a typical semiconductor diode. It is usually made from Ge (germanium) but can also be made in gallium arsenide and silicon materials.

2.51 (b)

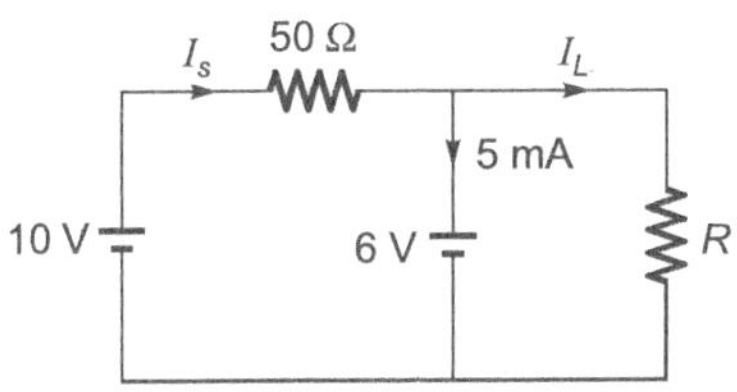

$$I_s = \frac{10-6}{50} = 80 \text{ mA}$$

$$I_L = I_s - 5$$
$$= 80 - 5 = 75 \text{ mA}$$

$$\therefore \quad R = \frac{6}{75 \times 10^{-3}} = 80\ \Omega$$

2.52 (b)
In bridge rectifier voltage drop across two diodes and in midpoint, voltage drop across single diode takes place. (or)
In reality, during each half cycle current flows through two diodes instead of just one. So the amplitude of the output voltage is two voltage drop ($2 \times 0.7 = 1.4$ V) less than the input voltage amplitude.

2.53 (b)
The diffusion, or storage capacitance is proportional to the current I. So, in forward biased diode the current I increases. Hence for a forward current C_D (diffusion capacitance) is usually much larger than C_T (transition capacitance).

2.54 (c)

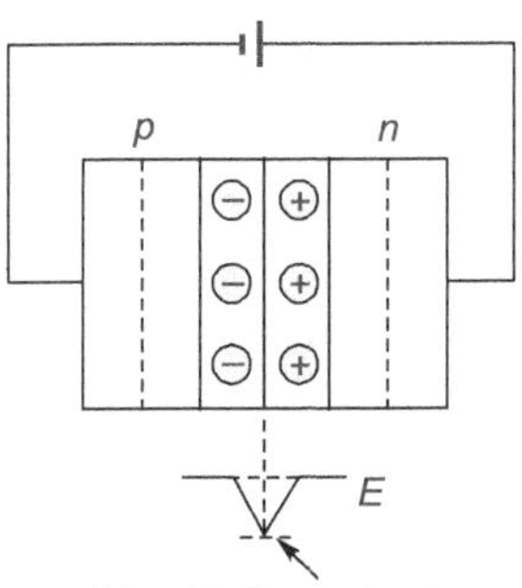

Magnitude maximum

2.55 (a)
The liquid crystal display (LCD) has the distinct advantage of having a lower power requirement than the LED, typically on the order of microwatts for the display, compared to the order of milliwatts for LEDs.

2.56 (b)
The increase in value of β of transistor can cause the fixed bias circuit to shift the operation from active mode to saturation mode.

2.57 (d)
The electric field is always maximum at the PN-junction.

2.58 (b)
Silicon diode has reverse current in range of nano ampere but Germanium diode has reverse current in range of micro ampere.

2.59 (c)
Schottky diode (metal-semiconductor junction diode) has higher reverse saturation current and lower cut in voltage with respect to ordinary diode.

2.60 (d)
In Schottky diode current flow because of thermionic emission of e^- and thus it is called "hot carrier diode".

2.61 (d)
Actually it is power supply rejection ratio (PSRR)

$$\text{and PSRR} = \frac{\text{Ripple}_{\text{input}}}{\text{Ripple}_{\text{output}}}$$

2.62 (a)
Schottky diode is a metal-semiconductor junction diode in which current flow due to majority carrier only.

2.63 (a)
The bleeder resistor help in discharging the capacitor when load is not attached so it help in providing good regulation for all value of loads.

2.67 (a)
1. The two type of current that flow in semi conductor diodes and transistors are drift and diffusion current.

In forward bias → diffusion current

In reverse bias → drift current

2. The junction region is called depletion region or space charge region.
3. In forward bias diffusion current will flow.

2.68 (b)

$$I_D = I_S\left[e^{V/\eta V_T} - 1\right]$$

2.69 (a)

LED is used in emergency lighting system.

2.70 (b)

$$W = \left(\frac{2\epsilon_0\epsilon_r V_j}{q} \times \frac{1}{\text{doping}}\right)^{1/2}$$

$$W \propto \frac{1}{\sqrt{\text{doping}}}$$

2.71 (a)

Output voltage is independent of load

% VR will be zero.

2.72 (b)

It double for every 10° C rise in temperature.

2.73 (a)

The Boltzmann diode equation is

$$I = I_o\left[e^{V/\eta VT} - 1\right]$$

2.74 (d)

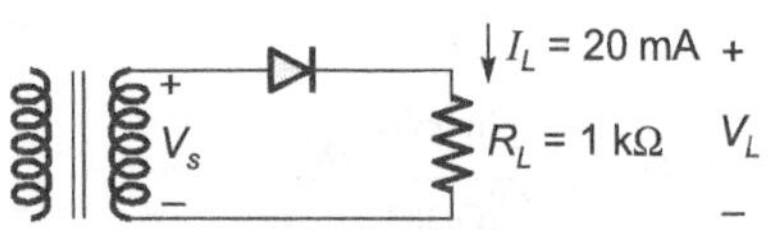

$$V_L = I_L \times R_L = 20 \text{ mA} \times 1 \text{ K}$$
$$= 20 \text{ V}$$

P.I.V. = V_L = 20 V (ideal)

P.I.V. = $V_{D\,on} + V_L$ = 20.7 V (practical)

2.75 (c)

Ripple factor for HWR is 1.21.

Ripple factor for FWR is 0.48.

2.76 (d)

- Zener didoe will operate in reverse bias region not in forward bias.
- Zener diode is negative in terms of voltage and current because the region will be in 3rd quadrant.

2.77 (a)

Silicon diodes,

V_{drop} = 0.6 to 0.7 V

Schottky diodes,

V_{drop} = 0.2 V

LED, V_{drop} = 1.4 V

2.78 (b)

Regarding semiconductor diode,

The diode requires a small positive voltage to be applied before it enters the conducting region.

2.79 (c)

As reverse bias voltage increases the depletion region widens and the depletion region capacitance decreases.

■■■■

3 BJT-Characteristics, Biasing and Thermal Stabilization

3.1 A circuit using the BJT is shown in the below figure, the value of β is

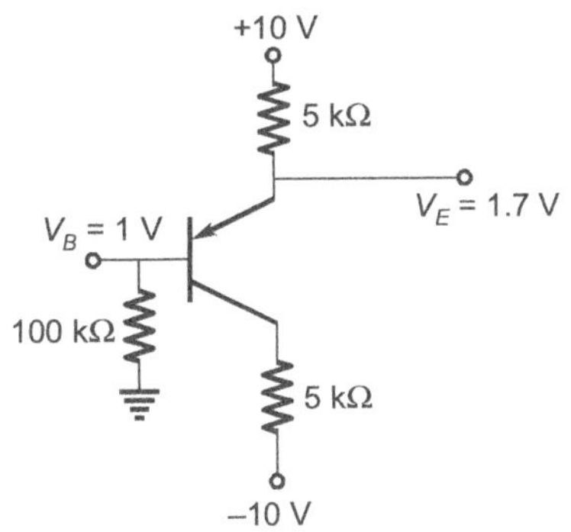

(a) 120 (b) 150
(c) 165 (d) 166

[ESE-2001]

3.2 The Darlington pair has a current gain of approximately β^2, the voltage gain A_v, the input resistance R_i and the output resistance R_0. When the Darlington pair is used in the emitter-follower configuration, A_v, R_i and R_0 are respectively
(a) very large, very large and very small
(b) unity, very large and very small
(c) unity, very small and very large
(d) very large, very small and very large

[ESE-2001]

3.3 For the circuit shown in the below figure, assuming β = 100 for the transistor, the transistor will be in

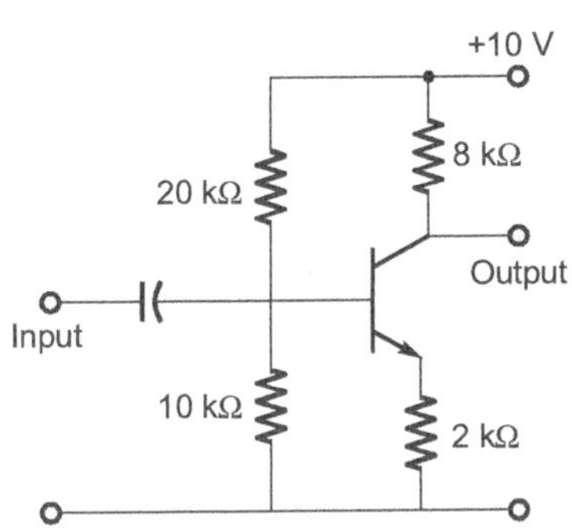

(a) cut off region
(b) inverse active region
(c) active region
(d) saturation region

[ESE-2001]

3.4 For the circuit shown in the below figure, by assuming β = 200 and V_{BE} = 0.7 V, the best approximation for the collector current I_c in the active region is

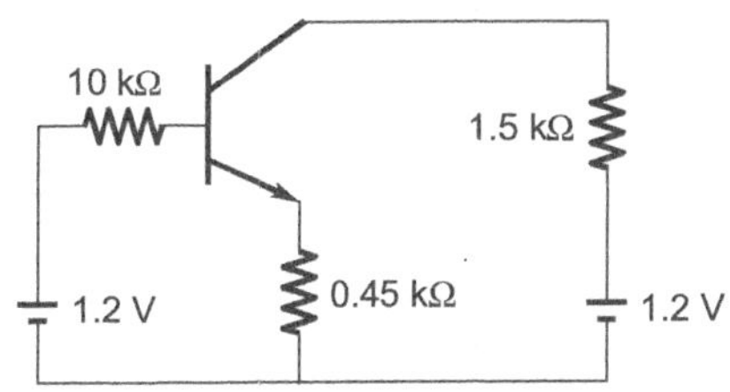

(a) 1 mA (b) 2.4 mA
(c) 3 mA (d) 9.6 mA

[ESE-2001]

3.5 A switch circuit using the transistor is shown in the figure below. Assume $h_{FE(min)}$ = 20 and f_T = 100 MHz. The most dominant speed limitation is brought by

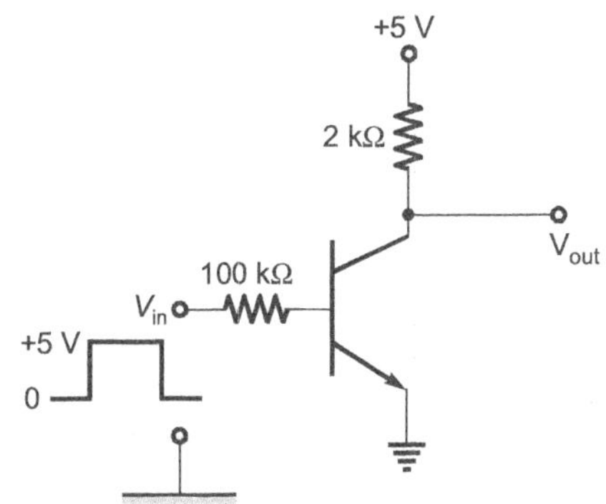

(a) rise time (b) fall time
(c) storage time (d) delay time

[ESE-2001]

3.6 Early effect in BJT refers to
(a) avalanche breakdown
(b) thermal runaway
(c) base narrowing
(d) zener breakdown

[ESE-2002]

3.7 The Darlington pair is mainly used for
(a) impedance matching
(b) wideband voltage amplification
(c) power amplification
(d) reducing distortion

[ESE-2002]

3.8 Consider a silicon transistor connected as a common emitter amplifier as shown below. The quiescent collector voltage of the circuit is approximately

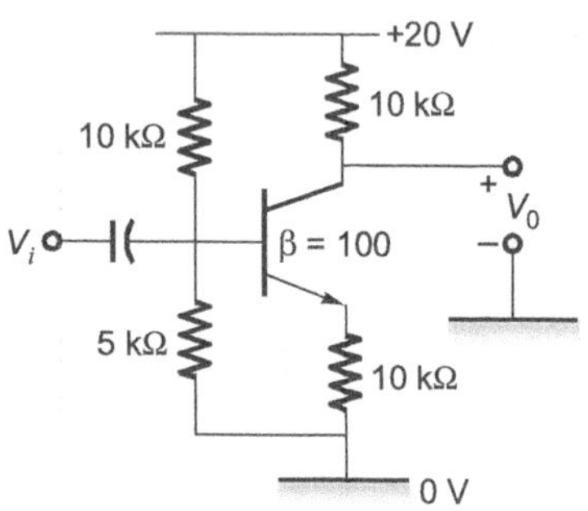

(a) 6.67 (b) 10 V
(c) 14 (d) 20 V

[ESE-2003]

3.9 In the transistor circuit as shown below, the collector to ground voltage is +20 V. The possible condition is

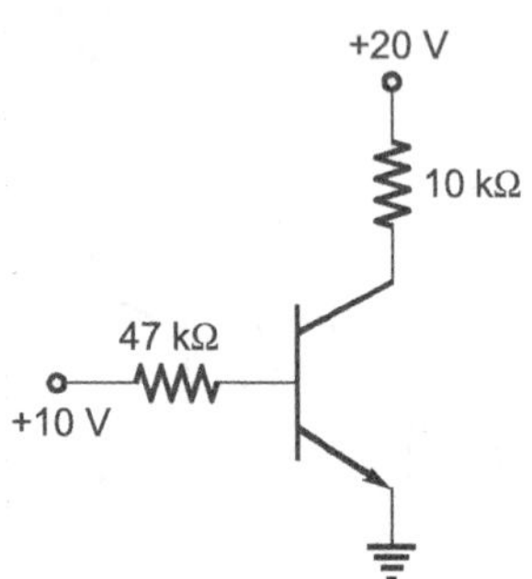

(a) Collector-emitter terminals shorted
(b) Emitter to ground connection open
(c) 10 kilo-ohms resistor open
(d) Collector-base terminals shorted

[ESE-2003]

3.10 Consider the following with reference to a CE transistor amplifier:
1. The use of negative feedback
2. The conversion of d.c. power to a.c.
3. High voltage and current gains
4. The use of a step-up transformer

The power gain is due to
(a) 1 and 2 (b) 2 and 3
(c) 1 and 3 (d) 1 and 4

[ESE-2003]

3.11 In the circuit as shown below, the ratio of V_0 to $(V_2 - V_1)$ would approximately (neglecting constant due to V_{CC}) be

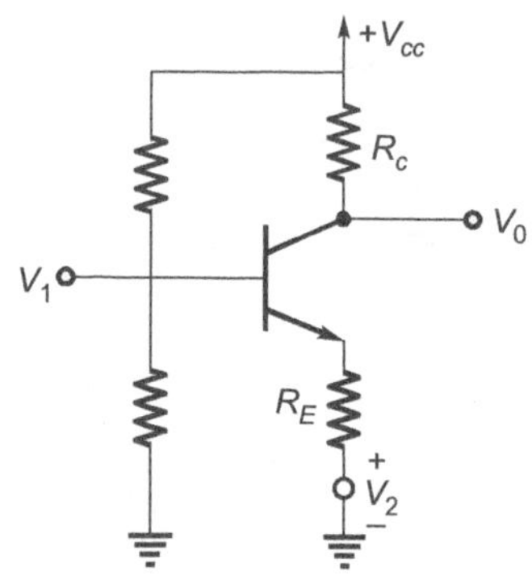

(a) R_C/R_E (b) R_E/R_C
(c) $-R_C/R_E$ (d) $-R_E/R_C$

[ESE-2003]

3.12 A silicon transistor with

$$V_{BE_{sat}} = 0.8\text{ V}, \beta_{dc} = 100$$

and $V_{CE_{sat}} = 0.2$ V

is used in the circuit shown below:

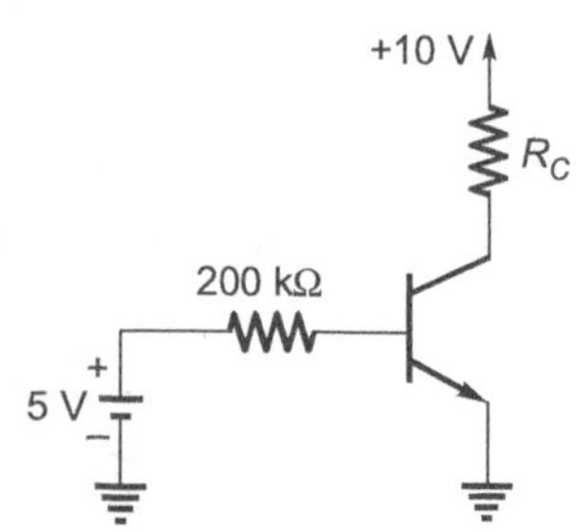

What is the minimum value of R_C for which transistor is in saturation?
(a) 4286 Ω (b) 4667 Ω
(c) 5000 Ω (d) 1000 Ω

[ESE-2004]

3.13 For the circuit shown in figure given below, assume $\beta = h_{FE} = 100$. The transistor is in

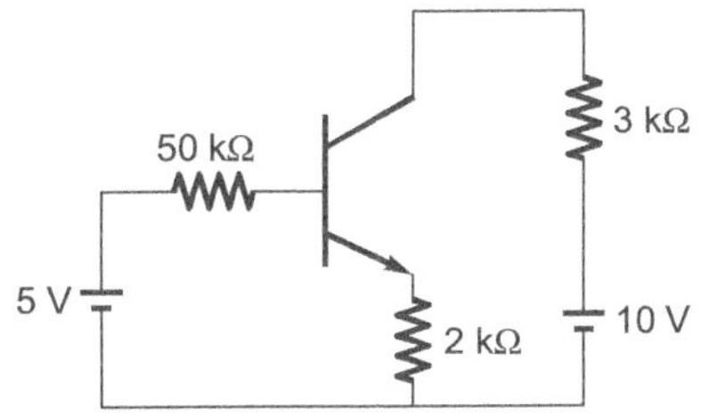

(a) Active region and $V_{CE} = 5$ V
(b) Saturation region
(c) Active region and $V_{CE} = 1.42$ V
(d) Cut-off region

[ESE-2005]

3.14 Which one of the following statement is correct? In a transistor,
(a) I_{CBO} is greater than I_{CEO}, and does not depend upon temperature.
(b) I_{CBO} is greater than I_{CO}, and doubles for every ten degrees rise in temperature.
(c) I_{CBO} is equal to I_{CO} and double for every ten degrees rise in temperature.
(d) I_{CEO} is equal to I_{CO} and doubles for every ten degrees rise in temperature.

[ESE-2005]

3.15 The dynamic transfer characteristics of a transistor is represented by

$$i_C = A_1 i_b + A_2 i_b^2$$

Where A_1 and A_2 are constants. If input signal $i_b = I_1 \cos \omega_1 t + I_2 \cos\omega_2 t$ the output will contain
(a) $\omega_1, \omega_2, 2\omega_1, 2\omega_2$
(b) dc term, $\omega_1, \omega_2, \omega_1 + \omega_2, \omega_1 - \omega_2$
(c) dc term, $\omega_1, \omega_2, 2\omega_1, 2\omega_2, 2\omega_1 + 2\omega_2, 2\omega_1 - 2\omega_2$
(d) dc term, $\omega_1, \omega_2, 2\omega_1, 2\omega_2, \omega_1 + \omega_2, \omega_1 - \omega_2$

[ESE-2005]

3.16 An emitter in a bipolar junction transistor is doped much more heavily than the base as it increases the
(a) Emitter efficiency
(b) Base transport factor
(c) Forward current gain
(d) All the three given above

[ESE-2005]

3.17 For common emitter configuration which one of the following statement is correct?
(a) large current gain and high input resistance
(b) large voltage gain and low output resistance
(c) small voltage gain and low input resistance
(d) small current gain and high output resistance

[ESE-2005]

3.18 Match **List-I** (Type of Amplifier/Configuration) with **List-II** (Characteristic Property) and select the correct answer using the code given below the lists:

List-I
A. Common emitter amplifier
B. Emitter follower
C. Common base amplifier
D. Darlington pair

List-II
1. Very low output resistance
2. Current gain $\simeq 1$
3. Beta multiplication
4. Very high power gain

Codes:

	A	B	C	D
(a)	4	1	2	3
(b)	2	3	4	1
(c)	4	3	2	1
(d)	2	1	4	3

[ESE-2006]

3.19 What is the thermal runaway in a bipolar junction transistor biased in the active region due to?
(a) Heating of the transistor emitter region
(b) Changes in 'β' which increases with temperature
(c) Base emitter voltage V_{BE} which decreases with rise in temperature
(d) Increases in reverse collector-base saturation current due to rise in internal device temperature

[ESE-2006]

3.20 In the circuit given below, if the output is taken from point *E* instead of node *C*, what will be the result?

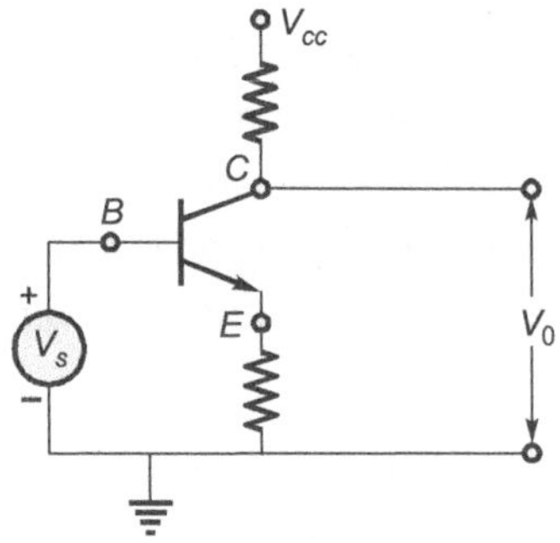

(a) An increase in the output impedance
(b) A reduction in the output impedance
(c) An increase in the input impedance
(d) A reduction in the input impedance

[ESE-2006]

3.21 Which one of the following statements is correct in respect of BJT?

(a) Avalanche multiplication starts when the reverse biased collector-base voltage V_{CB} equals the avalanche breakdown voltage BV_{CBO}.
(b) The early effect starts as soon as punch-through occurs in a transistor.
(c) The small signal current gain h_{fe} = large signal current gain h_{FE} when $\partial h_{FE} / \partial I_C = 0$.
(d) In the CE mode, a transistor can be cut off by reducing I_B to zero.

[ESE-2006]

3.22 If α_F and α_R denote the forward and inverted mode current gains of a BJT, which one of the following is correct?

(a) $\alpha_F = \alpha_R$ (b) $\alpha_F < \alpha_R$
(c) $\alpha_F \geq \alpha_R$ (d) $\alpha_F >> \alpha_R$

[ESE-2007]

3.23 Fixed biasing of CE configuration is shown in the figure given below.

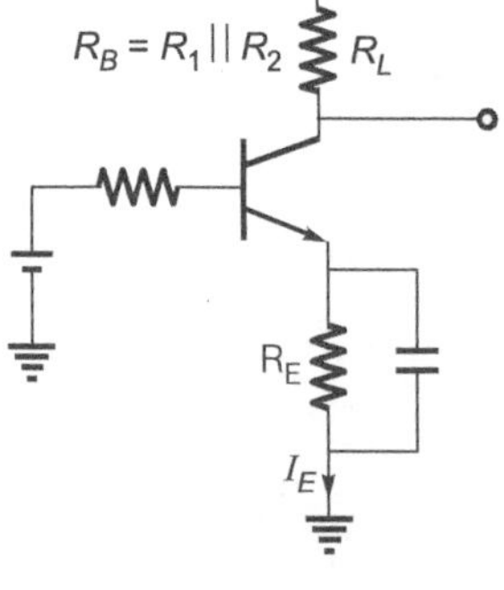

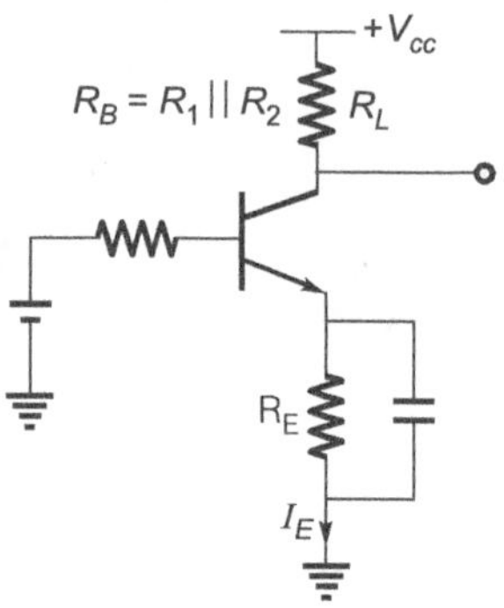

The current stabilization factor for $R_B << \beta R_E$ is $S_i = 1 + \frac{R_B}{R_E}$.

For $R_B << R_E$, what is the voltage stabilization factor S_v?

(a) $= \frac{R_E}{R_B + R_E}$ (b) $= \frac{R_B}{R_E}$
(c) $\approx -\frac{1}{R_E}$ (d) $\approx -\frac{R_E}{R_B + R_E}$

[ESE-2007]

3.24 The maximum junction-temperature of a transistor is 150°C and the ambient temperature is 25°C. If the total thermal impedance is 1°C/W, what is the maximum power dissipation?

(a) 1/175 W (b) 175 W
(c) 125 W (d) 1/125 W

[ESE-2008]

3.25 What is the transistor combination shown in the figure given below?

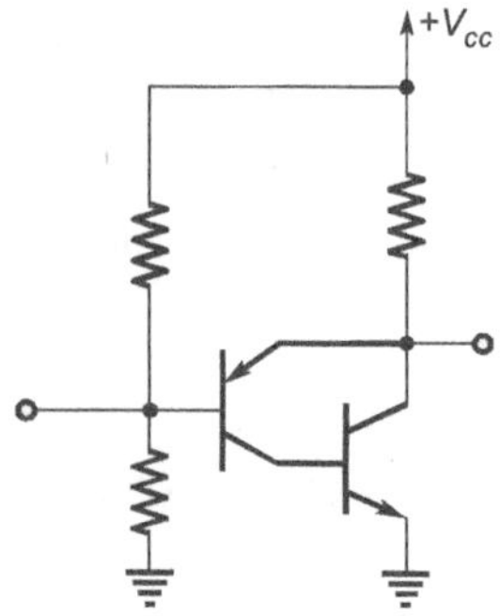

(a) A Darlington pair
(b) A complementary pair
(c) It effectively acts as a single p-n-p transistor
(d) It effectively acts as a single n-p-n transistor

[ESE-2008]

3.26 A transistor has a maximum power dissipation limit of 300 mW for ambient temperature up to 25°C. If the maximum allowable junction temperature is 175°C, then what is the limit of the device in an ambient temperature of 55°C?
(a) 120 mW (b) 240 mW
(c) 300 mW (d) 360 mW [ESE-2009]

3.27 β of a BJT varies from 15 to 65. R_L = 10 Ω, V_{CC} = 120 V and V_{BB} = 8 V. If $V_{CE(Sat)}$ = 1.5 V and $V_{BE(Sat)}$ = 1.75 V then what is the value of R_B that will result in saturation with an overdrive factor of 10?

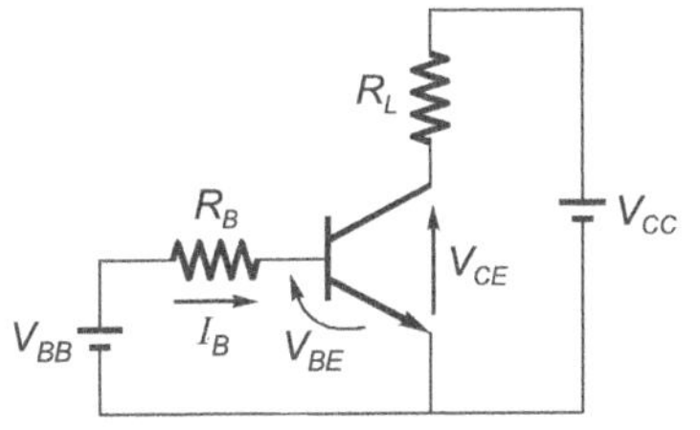

(a) 7.9 Ω (b) 0.79 Ω
(c) 79 Ω (d) 7.9 kΩ [ESE-2009]

3.28 In a certain self biased Si npn transistor the d.c. base voltage is 3.2 V then what is the d.c. emitter voltage? (Assume the transistor is in linear-active mode)
(a) 0.7 V (b) 2.5 V
(c) 3.2 V (d) 3.9 V [ESE-2009]

3.29 When a transistor is used in switching mode then what is the turn-on time?
(a) Sum of delay time and rise time
(b) Sum of rise time and storage time
(c) Sum of delay time and storage time
(d) Sum of rise time and fall time

[ESE-2009]

3.30 The Si transistor as shown in the circuit below has β = 50 and negligible leakage current. If V_{CC} = 18 V, V_{EE} = 4 V, R_E = 200 Ω, R_C = 4 kΩ, R_B = 72 kΩ, what is the value of the quiescent collector current I_{CQ}?

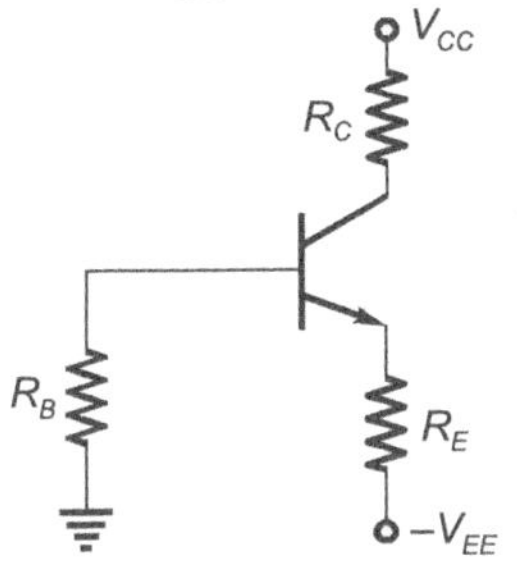

(a) 1.1 mA (b) 2 mA
(c) 5 mA (d) 3.6 mA

[ESE-2009]

3.31 In the below circuit as shown β = 99, V_{BE} = 0.6 V, then what are the values of V_C and I_C corresponding to the operating point?

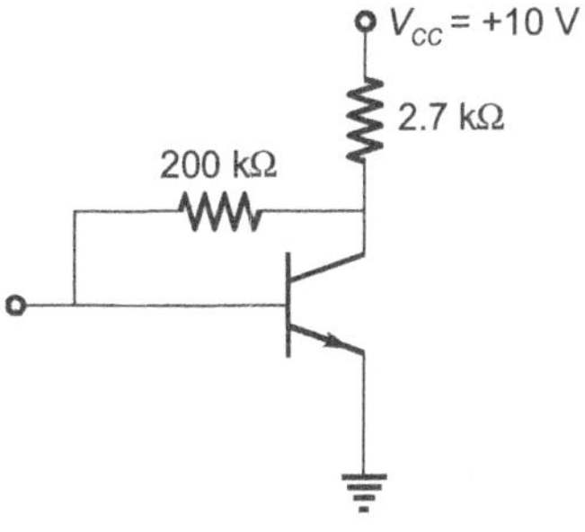

(a) 4.6 V and 1.98 mA
(b) 4.7 V and 2.00 mA
(c) 5.4 V and 1.56 mA
(d) 4.2 V and 2.1 mA

[ESE-2009]

3.32 If the amplification of a single stage is not sufficient or the input or output impedance is not of the correct magnitude for the intended application how may two stages be connected to achieve desired result?
(a) Cascode connection
(b) Complementary symmetry connection
(c) Cascade connection
(d) Totem pole connection

[ESE-2009]

3.33 Consider the following statements:
When Darlington circuit and normal single stage emitter followers are compared
1. Both have near unity voltage gain.
2. Both have equal current gain.
3. Darlington circuit has higher output resistance.
4. Emitter follower has lower input resistance.

Which of these statements is/are correct?
(a) 1 only (b) 1 and 4 only
(c) 2 and 4 only (d) 1, 2, 3 and 4

[ESE-2010]

3.34 In the difference amplifier as shown below, the differential output

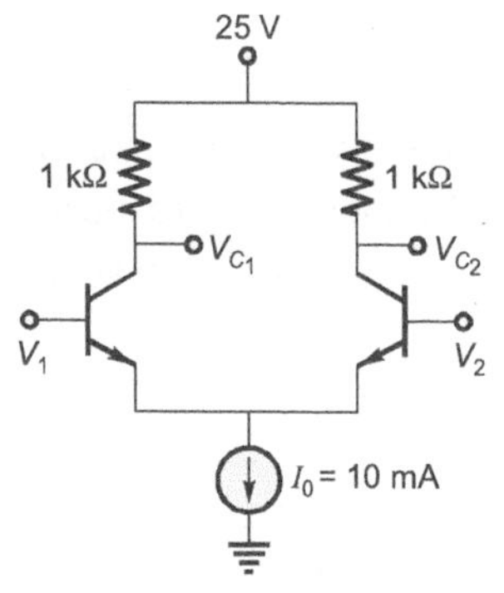

$(V_{C_2} - V_{C_1})$ for $V_1 = +5$ V, $V_2 = 0$ V is
(a) 0 V (b) 5 V
(c) 10 V (d) 15 V [ESE-2010]

3.35 An emitter follower regulator has the following disadvantage:
(a) It does not provide high gain.
(b) No provision exists for varying the output voltage.
(c) It output resistance is high.
(d) It cannot withstand high load current.
[ESE-2011]

3.36 For a transistor, turn-off time is
(a) Sum of storage time and fall time
(b) Maximum value of storage time
(c) Maximum value of fall time
(d) Sum of rise time and fall time
[ESE-2011]

3.37 In a common collector amplifier the voltage gain is
(a) Constant
(b) Less than 1
(c) Varies with input voltage
(d) Varies with load impedance [ESE-2011]

3.38 **Assertion (A):** The collector current in a BJT in common base mode increases slightly with increase in collector base voltage.
Reason (R): Increased collector-base bias causes avalanche breakdown to increase the current.
(a) Both A and R are true and R is the correct explanation of A
(b) Both A and R are true but R is NOT the correct explanation of A
(c) A is true but R is false
(d) A is false but R is true [ESE-2011]

3.39 **Assertion** (A): The operating point in transistor amplifier shift with changes in temperature.
Reason (R): Reverse saturation current approximate doubles for every 10°C increase in temperature.
(a) Both A and R are true and R is the correct explanation of A
(b) Both A and R are true but R is NOT the correct explanation of A
(c) A is true but R is false
(d) A is false but R is true
[ESE-2011]

3.40 For a Silicon n-p-n transistor, the base to emitter voltage (V_{BE}) is 0.7 V and the collector to base voltage (V_{CB}) is 0.2 V. Then the transistor is operating in the
(a) Normal active mode
(b) Saturation mode
(c) Inverse active mode
(d) Cut-off mode
[ESE-2012]

3.41 Early effect is the modulation of effective base width by
(a) Emitter voltage (b) Emitter current
(c) Collector voltage (d) Junction temperature
[ESE-2012]

3.42 A Bipolar Junction Transistor (BJT) works in three regions:
1. Saturation
2. Active
3. Cut-off

If BJT is to be used in amplifier circuit, the region it works in is/are
(a) 1, 2 and 3 (b) 1 and 2 only
(c) 2 only (d) 1 only
[ESE-2012]

3.43 **Statement (I):** In a transistor switching circuit, it is desirable that the transistor should not be driven into hard saturation for fast switching applications.
Statement (II): When a transistor is under saturation state, both its emitter-base and collector-base junctions remain under forward bias.

(a) Both Statement (I) and Statement (II) are individually true and Statement (II) is the correct explanation of Statement (I).
(b) Both Statement (I) and Statement (II) are individually true but Statement (II) is not the correct explanation of Statement (I).
(c) Statement (I) is true but Statement (II) is false.
(d) Statement (I) is false but Statement (II) is true.
[ESE-2012]

3.44 In a BJT, $I_{CO} = I_{CBO} = 2\ \mu A$. Given $\alpha = 0.99$, the value of I_{CEO} is
(a) 2 μA (b) 99 μA
(c) 198 μA (d) 200 μA
[ESE-2013]

3.45 The CE short circuit current gain, β of a transistor
(a) is a monotonically increasing function of the collector current I_C.
(b) is a monotonically decreasing function of I_C.
(c) for low values of I_C, it increases and reaches a maximum and then decreases with further increase in I_C.
(d) is not a function of I_C.
[ESE-2014]

3.46 Consider the following steps in the fabrication of a monolithic bipolar junction transistor:
1. Emitter diffusion
2. Buried diffusion
3. Buried layer formation
4. Epi-layer formation

What is the correct sequence of these steps?
(a) 1, 2, 3 and 4 (b) 4, 2, 3 and 1
(c) 1, 3, 2 and 4 (d) 4, 3, 2 and 1
[ESE-2014]

3.47 When a transistor is connected in common emitter mode it will have
(a) negligible input resistance and high output resistance.
(b) high input resistance and low output resistance.
(c) medium input resistance and high output resistance.
(d) low input resistance as well as output resistance.
[ESE-2014]

3.48 In case of amplifiers, which coupling gives the highest gain?
(a) Transformer coupling
(b) Resistance coupling
(c) Impedance coupling
(d) Capacitance coupling [ESE-2014]

3.49 When a PNP transistor is properly biased, the holes from the emitter.
(a) diffuse through the base into the collector region.
(b) recombine with the electrons in the base region.
(c) recombine with the electrons in the emitter region
(d) diffuse through the emitter to collector
[ESE-2015]

3.50 Variation in β in a BJT can cause a fixed bias circuit to go
(a) into active mode of operation from saturation mode
(b) out of active mode
(c) out of saturation
(d) into cutoff mode from active mode of operation
[ESE-2015]

3.51 If an *npn* silicon transistor is operated at $V_{CE} = 5$ V and $I_C = 100\ \mu A$ and has a current gain of 100 in the *CE* connection, then the input resistance of this circuit will be
(a) 250 kΩ (b) 450 kΩ
(c) 4500 kΩ (d) 2500 kΩ
[ESE-2015]

3.52 For a BJT; I_C = 5 mA, I_B = 50 μA and $I_{CBO} = 0.5\ \mu A$, then the value of β is
(a) 99 (b) 91
(c) 79 (d) 61
[ESE-2016]

3.53 Which of the following conditions must be satisfied for a transistor to be in saturation?
1. Its collector to base junction should be under forward bias.
2. Its collector to base junction should be under reverse bias.
3. Its emitter to base junction should be under reverse bias.
4. Its emitter to base junction should be under forward bias.

Which of the above conditions are correct?

(a) 1 and 3 (b) 2 and 3
(c) 2 and 4 (d) 1 and 4

[ESE-2016]

3.54 A switched-capacitor network is/are

1. Time variant sample data network
2. Non linear network
3. Linear time invariant network

(a) 1 only (b) 2 only
(c) 3 only (d) 1 and 2

[ESE-2016]

3.55 A bipolar transistor has $\alpha = 0.98$, $I_{CO} = 10\ \mu A$. If the base current is 100 μA, then collector current would be

(a) 2.91 mA (b) 3.49 mA
(c) 4.91 mA (d) 5.49 mA

[ESE-2018]

3.56 The important fact about the collector current is:

(a) It is greater than emitter current.
(b) It equals the base current divided by the current gain.
(c) It is small.
(d) It approximately equals the emitter current.

[ESE-2019]

3.57 In a grounded-emitter transistor, when emitter current becomes zero in cut-off region the emitter potential is called

(a) Floating Emitter Potential
(b) Breaking Emitter Potential
(c) Cascading Emitter Potential
(d) Cut-off Emitter Potential

[ESE-2020]

3.58 When maximum reverse-biasing voltage is applied between the collector and base terminals of the transistor and emitter is open circuited, breakdown occurs due to

(a) Avalanche breakdown
(b) Avalanche multiplication
(c) Punch–through
(d) Reach–through

[ESE-2020]

3.59 In a CB configuration, the current amplification factor is 0.97. If the emitter current is 1 mA, the value of base current is

(a) 0.97 mA (b) 1.0 mA
(c) 0.03 mA (d) 1.03 mA

[ESE-2021]

3.60 In a common base configuration, the alpha of the transistor is 0.99, its collector current is 1 mA and the collector to base current with emitter open is 1 μA. The value of base current is

(a) 29 μA (b) 19 μA
(c) 9 μA (d) 39 μA

[ESE-2021]

3.61 A transistor has $\beta = 105$ and $I_C = 840\ \mu A$. What is the value of I_B?

(a) 0.008 μA (b) 0.08 μA
(c) 0.8 μA (d) 8 μA

[ESE-2022]

3.62 Which one of the following statements is not correct for a transistor?

(a) The region at the centre is always base region.
(b) Emitter region is more heavily doped and based region is very lightly doped.
(c) Collector region is very lightly doped compared to base region.
(d) Low power input can be converted to a large power output with the half of a small piece of semi conductor without any hassles of preheating and handling of large heat dissipation.

[ESE-2022]

3.63 For certain of the reverse voltage in a transistor, the effective base width may reduce to zero resulting into the voltage breakdown. This phenomenon is called

(a) early effect (b) avalanche multiplication
(c) punch through (d) zones breakdown

[ESE-2022]

3.64 If a transistor has the value of common base current gain 0.98, then what is the value of common emitter current gain?

(a) 48 (b) 49
(c) 50 (d) 51

[ESE-2023]

3.65 **Statement (I):** A small-signal equivalent circuit of a BJT that applies to both *n-p-n* and *p-n-p* transistors is valid at lower frequencies by ignoring capacitance effects.

Statement (II): Despite the structural similarities, a p-n-p BJT has higher current gain than a comparable *n-p-n* BJT because holes are less mobile than electrons.

[ESE-2023]

■■■■

Answers BJT-Characteristics, Biasing and Thermal Stabilization

3.1 (c)	3.2 (b)	3.3 (d)	3.4 (a)	3.5 (b)	3.6 (c)	3.7 (a)	3.8 (c)	3.9 (b)
3.10 (c)	3.11 (c)	3.12 (b)	3.13 (c)	3.14 (b)	3.15 (d)	3.16 (d)	3.17 (a, b)	3.18 (a)
3.19 (d)	3.20 (b)	3.21 (a)	3.22 (d)	3.23 (c)	3.24 (c)	3.25 (b)	3.26 (b)	3.27 (c)
3.28 (b)	3.29 (a)	3.30 (b)	3.31 (a)	3.32 (c)	3.33 (b)	3.34 (d)	3.35 (b)	3.36 (a)
3.37 (b)	3.38 (b)	3.39 (a)	3.40 (a)	3.41 (c)	3.42 (c)	3.43 (b)	3.44 (d)	3.45 (c)
3.46 (d)	3.47 (c)	3.48 (a)	3.49 (a)	3.50 (b)	3.51 (*)	3.52 (a)	3.53 (d)	3.54 (d)
3.55 (d)	3.56 (d)	3.57 (a)	3.58 (b)	3.59 (c)	3.60 (c)	3.61 (d)	3.62 (c)	3.63 (c)
3.64 (b)	3.65 (c)							

Explanations BJT-Characteristics, Biasing and Thermal Stabilization

3.1 (c)

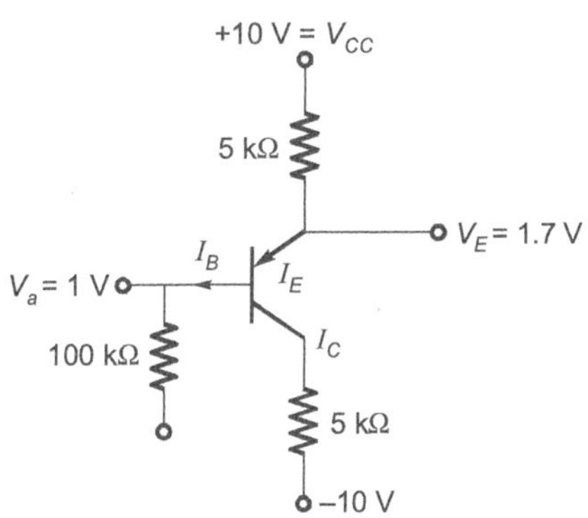

Current, $I_E = \dfrac{V_{CC} - V_E}{5\,\text{k}\Omega} = \dfrac{10 - 1.7}{5\,\text{k}\Omega}$

$= 1.66$ mA

Current, $I_B = \dfrac{V_a}{100\,\text{k}\Omega} = \dfrac{1}{100\,\text{k}\Omega}$

$= 0.01$ mA

we know that,

$$I_E = I_C + I_B$$

But, $I_C = \beta I_B$

where β is current gain of CE amplifier.

$\therefore \quad I_E = (\beta + 1)I_B$

$$(\beta + 1) = \frac{I_E}{I_B} = \frac{1.66}{0.01} = 166$$

$\Rightarrow \quad \beta = 165$

3.2 (b)

We know that voltage gain of CC (Common collector) is around unity. When we will cascade two CC (i.e. Darlington pair) voltage gain will be

$$AV = AV_1 \cdot AV_2 \cong 1$$

$\because \quad AV_1 \simeq 1$ and $AV_2 \simeq 1$

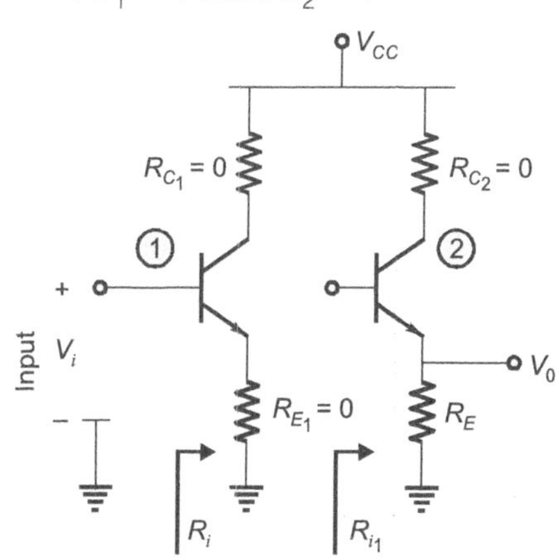

we know that for CC,

(Large)$R_{i_1} = h_{ie} + (1 + h_{fe})\, R_E$ ...(i)

this will work as R_E for amplifier (1)

so, over all input impedance

$$R_i = h_{ie} + (1 + h_{fe})\, R_{i_1}$$

$$R_i = h_{ie} + (1 + h_{fe})\, [h_{ie} + (1 + h_{fe})R_E]$$

so, R_i further increases, so very large.

Similarly output impedance further decreases and becomes very small.

3.3 (d)

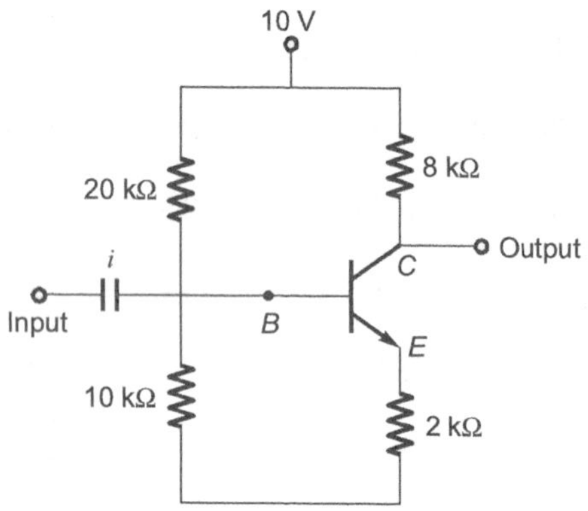

$\beta = 100$

Let us find Thevenin equivalent w.r.t. to terminal B.

$$V_{th} = V_{OC} = \frac{10}{10+20} \times 10$$

$$= \frac{10}{3}\text{ V}$$

$$R_{th} = 10\text{ k} \,||\, 20\text{ k} = 6.67\text{ k}\Omega$$

∴ Equivalent circuit,

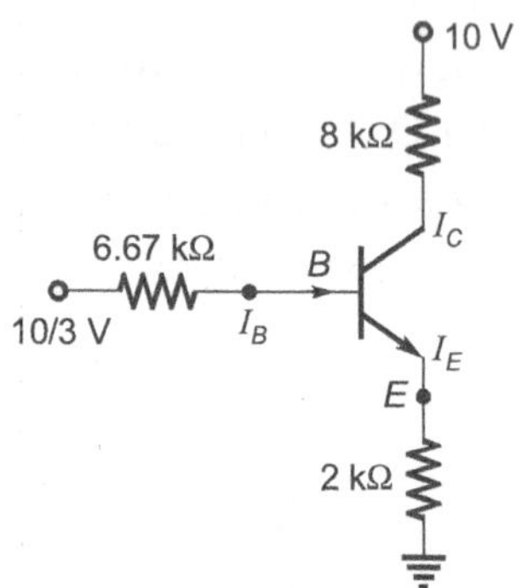

$I_E = (\beta + 1)\, I_B$ in active region.

Since for transistor to be on i.e. (either in active or saturation) emitter base junction should be forward based for this open circuit voltage across emitter base junction should be greater that, $V_{BE} = 0.7$ V

Now check for this is

$$V_{OC} > (V_B - V_E) \text{ when } I_B = I_E = 0$$

$$= \left(\frac{10}{3} - 0\right) = \frac{10}{3}\text{ V}$$

∵ $V_{OC} > V_{BE}(0.7\text{ V})$

∴ Transistor is on.

So it is either in active or in saturation.

Let it is in active region

So first in base-emitter junction loop according to KVL →

$$\frac{10}{3} - 6.67 I_B - V_{BE} - 2 \times (\beta + 1) I_B = 0$$

putting the values we get

$$I_B = 0.0126\text{ mA}$$

∴ Now find,

$$V_{CE} = 10 - 8 \times I_C - 2 \times I_E$$
$$= 10 - 8 \times \beta I_B - 2 \times (\beta + 1) I_B$$
$$= 10 - 8 \times 1.26 - 2 \times 101 \times 0.0126$$
$$= -2.6252\text{ V}$$

Since, V_{CE} is not greater that 0.2 V so our assumption that this is in active region is wrong so the transistor is in saturation region.

3.4 (a)

Taking $I_E \simeq I_C$ for common-base configuration

$10 \times 1000 \times I_B + 0.45 \times 1000 \times I_E = 1.2 - 0.7$

$$10000 \times \frac{I_C}{\beta} + 450 I_C = 0.5 \text{ as } I_C = \beta I_B$$

∴ $$I_C = \frac{0.5}{500} = 1\text{ mA}$$

3.5 (b)

For saturation,

$$I_C \cong \frac{V_{CC}}{2} = \frac{5}{2} = 2.5\text{ mA}$$

$$I_B \cong \frac{5}{10} = 0.5\text{ mA}$$

so for saturation,

$$h_{fc} \times I_B \geq I_{csat}$$

$$h_{fe} \times 0.5 \geq 2.5 \Rightarrow h_{fe} \geq 5$$

So minimum value of h_{fe} for saturation is 5.

∵ given $h_{fe} = 20$

∴ $h_{fe} > 5$

∴ transistor will go in saturation and due to this storage time will limit speed of operation.

3.6 (c)
The decrease in effective base width with increasing reverse collector voltage is known as the early effect or base-width modulation.

3.7 (a)
Darlington's pair has a very high input impedance and a very low output impedance so it is used as a buffer between high impedance source and low impedance load i.e. in impedance matching.

3.8 (c)

$$V'_{BB} = \left(\frac{5}{5+10}\right)20 = \frac{20}{3}$$

Due to large β $\boxed{I_B \simeq 0}$

$$\therefore \quad \frac{20}{3} = 0.7 + 10 \times 10^3 I_C$$

$$\boxed{I_C = 0.6 \text{ mA}}$$

$$V_{CQ} = 20 - 10 \times 10^3 I_C$$
$$= 20 - 6 = 14 \text{ V}$$

3.9 (b)

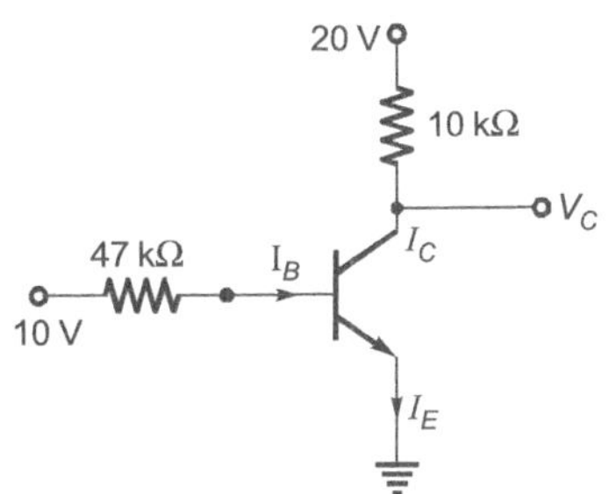

Here, $V_C = 20 - 10 I_C$

$\because$ $V_C = 20$

$\therefore$ $I_C = 0$

This is possible if $I_B = I_E = 0$ and this is possible if emitter is open.

3.10 (c)
Power gain is due to high voltage gain > 1, and high current gain > 1. CE transistor amplifier does not use step-up transformer. Also there is no conversion of dc power to ac.

3.11 (c)
$V_0 = V_{CC} - I_C R_C = -I_C R_C$ neglecting constant due to V_{CC}

$V_1 - V_2 = I_E R_E = I_C R_E$ as $I_E \cong I_C$ neglecting R_{Th} and V_{Th} to neglect constant due to V_{CC}

$$\therefore \quad \frac{V_0}{V_1 - V_2} = -\frac{R_C}{R_E}$$

3.12 (b)
Let transistor is in saturation

So, $V_{BEsat} = 0.8$ V

$V_{CEsat} = 0.2$ V

$$\therefore \quad I_B = \frac{5-0.8}{200\text{k}} = 0.021 \text{ mA}$$

$$\left(\because I_B = \frac{V_{BB} - V_{BE\,sat}}{200\text{k}}\right)$$

$$I_{Csat} = \frac{V_{CC} - V_{CE\,sat}}{R_C} = \frac{10-0.2}{R_C} = \frac{9.8}{R_C}$$

For transistor to be in saturation

$$\beta I_B \geq I_{Csat}$$

$$10^{-3} \times 100 \times 0.021 \geq \frac{9.8}{R_C}$$

$$R_C = \frac{9.8}{10^{-1} \times 0.021}$$

$$R_C \geq 4666.67$$

$$\therefore \quad R_{C\min} = 4667\ \Omega$$

3.13 (c)
Let the transistor is in active region.

So we can write the KVL in base emitter junction loop.

$5 - 50 \times I_B - 0.7 - 2 \times I_E = 0$

$5 - 50 I_B - 0.7 - 2 \times 101\, I_B = 0$

$\because I_E = (\beta + 1) I_B = 101 I_B$

$$\therefore \quad I_B = \frac{4.3}{(50+202)} \text{mA} = 0.01706 \text{ mA}$$

$\therefore$ $I_C = 1.706$ mA

$I_E = 0.01706 \times 101 = 1.72306$ mA

$\therefore$ $V_{CE} = V_{CC} - I_C \times 3 - I_E \times 2$

$= 10 - 1.706 \times 3 - 2 \times 1.72306$

$V_{CE} = 1.4358$

$\because$ $V_{CE} > 0.2$ V so transistor is in active region, our assumption is correct.

3.14 (b)
We know that $I_{CEO} (\beta + 1) I_{CO}$ and I_{CBO} are slightly greater than I_{CO} due to some current induced and flow at the surface and it also doubles for every ten degree rise in temperature.

3.15 (d)

$$i_c = A_1 i_b + A_2 i_b^2$$

and $i_b = I_1 \cos\omega_1 t + I_2 \cos\omega_2 t$

$\therefore$ $i_c = A_1[I_1\cos\omega_1 t + I_2\cos\omega_2 t] + A_2[I_1\cos\omega_1 t + I_2\cos\omega_2 t]^2$

$i_c = A_1[I_1\cos\omega_1 t + I_2\cos\omega_2 t] + A_2[I_1^2\cos^2\omega_1 t + I_2^2\cos^2\omega_2 t + 2I_1 I_2\cos\omega_1 t\cos\omega_2 t]$

$i_c = A_1[I_1\cos\omega_1 t + I_2\cos\omega_2 t] + A_2[1/2 I_1^2\{1+\cos 2\omega_1 t\} + 1/2\, I_2^2\{1+\cos 2\omega_2 t\} + I_1 I_2 \cos(\omega_1+\omega_2)t + \cos(\omega_1-\omega_2)t]$

So frequency component at output

$\omega_1, \omega_2, 2\omega_1, 2\omega_2, (\omega_1+\omega_2), (\omega_1-\omega_2)$

3.16 (d)

Emitter efficiency, $\gamma = \dfrac{i_{Ep}}{i_{Ep} + i_{En}}$

(for pnp transistor) $i_{Ep} \to$ emitter current due to hole.

If emitter is heavily doped compare to base then $i_{Ep} >> i_{En}$ and $\gamma \to 1$ increases.

Base transport factor $\beta = \dfrac{i_c}{i_{Ep}}$

i_c is collector current and is due to holes which are not recombined in base as i_{Ep} increases the recombination in base is neglected so $i_c \simeq i_{Ep}$ so, B increases and tends to 1.

And forward current gain increases as recombination goes lesser and lesser compare to input current.

3.17 (a, b)

Chara.	CB	CE	CC
I/p resi	Very low (40Ω)	High (50 kΩ)	Very high (750 kΩ)
O/p resi	Very low (40Ω)	High (10 kΩ)	Low (50 Ω)
Current gain	Less than unity (0.98)	High (100)	High (100)
Voltage gain	Small (150)	High (500)	Less than unity

3.18 (a)

For common emitter amplifier voltage gain and current gain both are greater than unity and having moderate value so power gain = $A_V A_I$ very high.

Common base has high voltage gain but current gain $I_C/I_E \simeq$ unity.

Emitter follower has highest i/p impedance and lowest output impedance.

Darlington pair is used for high current gain in which two common collector (having current gain $(\beta + 1) = I_E/I_B$ are cascaded so overall current gain

$$A_I = A_{I_1} A_{I_2}$$
$$= (\beta + 1)(\beta + 1) \text{ normally } \beta >> 1 \cong \beta \cdot \beta$$
$$= \beta^2 \to \beta \text{ multiplier}]$$

3.19 (d)

Due to increase in temperature reverse collector base saturation current gets double for every 10°C rise in temperature and so increases the power dissipation across collector base junction which further increases the temperature which further increases the reverse saturation current again this process continues and thermal runaway occurs.

3.20 (b)

When output will be taken from E then this circuit become emitter follower where there is a voltage series feedback mechanism which decreases the output impedance.

3.21 (a)

The maximum reverse-biasing voltage which may be applied before breakdown between the collector and base terminals of the transistor, under the condition that the emitter lead be open-circuited is represented by the symbol βV_{CBO}.

This breakdown voltage is a characteristic of the transistor alone. Breakdown may occur because of avalanche multiplication of the current I_{CO} that crosses the collector junction.

3.22 (d)

In case of inverted mode, injection efficiency will be lower as collector is less heavily doped compare to emitter so

$$\alpha_R = \alpha_R \beta << \alpha_F$$
$$\alpha_R << \alpha_F$$

α_R = injection efficiency in case of inverted mode

α_F = injection efficiency in case of forward mode.

3.23 (c)

$$S = \frac{\partial I_C}{\partial V_{BE}} = \frac{-\beta}{R_B + R_E(1+\beta)} = \frac{-\frac{\beta}{R_E}}{1+\beta+\frac{R_B}{R_E}}$$

$$1+\beta \cong \beta$$

$$S = -\frac{1}{R_E} \qquad \left(\because \frac{R_B}{R_E} << 1+\beta\right)$$

3.24 (c)

Power dissipation

$$= \frac{\text{temperature difference}}{\text{thermal resistance}}$$

$$= \frac{150° - 25°}{1°C/W} = \frac{125}{1} = 125\,W$$

3.25 (b)

It is not Darlington because in Darlington both transistor are connected as common collector. Here one is NPN where as other is PNP so it is complementary pair.

3.26 (b)

Given that,

$$P_d = 300\text{ mW}$$
$$T_A = 25°C$$

So we know that,

$$(T_J - T_A) = R \times P_d$$

where, R – thermal resistance

$$R = \frac{150}{300\text{mW}} = \frac{1}{2}°C/\text{mW}$$

Now, $T_A = 55°C$

$$\therefore\ P_d = \frac{T_j - T_A}{R} = \frac{175-55}{(1/2)} = \frac{120}{1/2} = 240\,\text{mW}$$

3.27 (c)

Given, $V_{CC} = 120$ V, $V_{CE\,sat} = 1.5$ V

$V_{BE\,sat} = 1.75$ V, $R_L = 10\ \Omega$

$$I_{Csat} = \frac{V_{CC} - V_{CEsat}}{R_L}$$

$$= \frac{120-1.5}{10} = 11.85\,A$$

$$\therefore\quad I_{Bsat} = \frac{V_{BB} - V_{BEsat}}{R_B}$$

$$= \frac{8-1.75}{R_B} = \frac{6.25}{R_B}\,A$$

$\because$ overdrive factor is 10 i.e. current flows is 10 times as required for saturation

i.e. $$I'_B = 10 I_B = \frac{62.5}{R_B}$$

Now we know,

$$\beta I'_B \geq I_{Csat} \text{ for saturation}$$

$$\beta\frac{62.5}{R_B} \geq 11.85$$

$$R_B \leq \frac{62.5}{11.85} \times \beta \leq 5.27\,\beta$$

For $\beta = 15$

$$R_B \leq 79\ \Omega$$
$$(R_B)_{max} = 79\ \Omega$$

3.28 (b)

$$V_{BB} = V_{BE} + V_{EE}$$

In active mode

$$V_{BE} = 0.7\text{ V}$$
$$\therefore\quad 3.2 = 0.7 + V_{EE}$$
$$\therefore\quad V_{EE} = 2.5\text{ V}$$

3.29 (a)

It is equal to delay time + current to rise above a threshold current i.e. rise time.

3.30 (b)

Applying KVL,

$$0 - I_B R_B - V_{BE} - (I_B + I_C) R_E + V_{EE} = 0$$

$$-\frac{I_C}{\beta} \times R_B - V_{BE} - \left(\frac{I_C}{\beta} + I_C\right) R_E + V_{EE} = 0$$

$$\left(-\frac{I_C}{50} \times 72000\right) - 0.7 - \left(\frac{I_C}{50} + I_C\right) 200 + 4 = 0$$

$$\therefore\quad 3.3 = 1440\, I_C + 204\, I_C$$

$$\therefore\quad I_C = 2 \times 10^{-3} = 2\text{ mA}$$

3.31 (a)

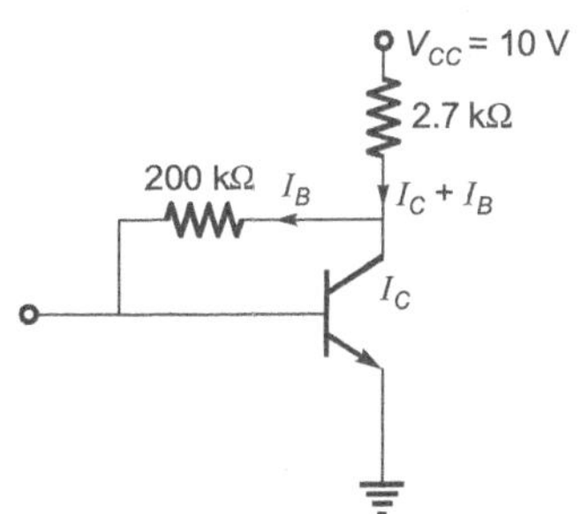

So we can write,

$$10 = (2.7)(I_C + I_B) + 200\text{ K}\, I_B + V_{BE}$$
$$10 = (2.7\text{ K})(I_C + I_B) + (200\text{ K}\, I_B) + 0.7$$

$$9.3 = 2.7\,\text{K}\,I_C + 202.7\,\text{K}\,I_B$$
$$= 2.7\text{K}I_C + \frac{202.7\,\text{K}I_C}{\beta}$$
$$I_C = \frac{9.3}{\left(2.7 + \frac{202.7}{99}\right)} = 1.9589 \text{ mA}$$
$$\therefore \quad V_C = V_{CC} - 2.7\,(I_C + I_B)$$
$$V_C = 10 - 2.7\left(I_C + \frac{I_C}{\beta}\right) = 4.657 \text{ V}$$

3.32 (c)

Due to cascading we have sufficient gain because $A = A_1 A_2$... can manage the input impedance and output impedance as per the requirement, normally high input impedance and low output impedance.

3.33 (b)

Darlington circuit and emitter follower does have same current gain.
Current gain of Darlington circuit = $(1 + \beta)^2 \approx \beta^2$
Darlington circuit has higher input resistance.
Input resistance = $(1 + \beta)^2 R_i$

3.34 (d)

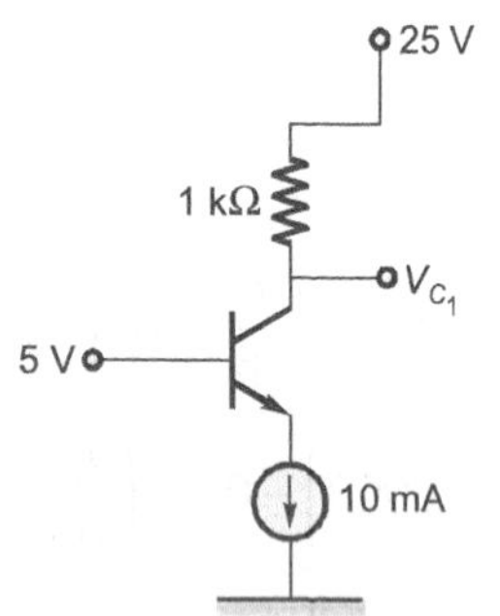

Since, $V_2 = 0$
Transistor is in cut off.

So, $V_{C_2} = 0$

$$25 - (10 \times 1) = V_{C_1} = 15 \text{ V}$$

3.35 (b)

For emitter follower regulator, there is no provision for varying the output voltage.

3.36 (a)

Turn-off time = storage time (for charge carriers to reach in their original condition.) + fall time (time taken for current to reduce to zero.)

3.37 (b)

Common collector is also emitters follower where input follows the output and voltage gain ≅ 1 (normally less than 1) but it is not constant it depends on R_E emitter resistance.

3.38 (b)

A is correct as i_c increases due to increase in collector base voltage due to base width modulation.

R is also correct but not always the reason as due to increased voltage, effective base width decreases so recombination decreases so i_c increases.

3.39 (a)

The operating point is always depends on transistor parameters and transistor parameters varies with temperature so as operating point, so A is correct.

Main effect is due to reverse saturation current which double for every 10°C rise in temperature.

3.40 (a)

According to the question $V_{CB} = 0.2$ V i.e. the collector to base junction is reverse bias and $V_{BE} = 0.7$ V i.e. base to emitter junction is forward bias. This condition meet in active region.

3.41 (c)

Collector voltage has some effect on collector current it increases slightly with increase in voltage. This is called the early effect and is modeled as a linear increase in total current with increases in V_{CE}.

$$i_c = I_s e^{\frac{v_{BE}}{V_T}}\left(1 + \frac{v_{CE}}{V_A}\right)$$

V_A is called the early voltage.

3.42 (c)

In active region I_c is almost constant on increasing V_{CE}. In this region transistor acts like an amplifier.

3.43 (b)

In hard saturation base current is 2 to 10 times larger than minimum base current $I_{B,\text{min}}$. So, for fast switching application, the transistor should not be driven into hard saturation.

In saturation state *EB* and *CB* junction are under forward bias condition.

The both the statements are true but statement II is not the correct explanation of statement I.

3.44 (d)

$$I_{CEO} = \frac{I_{CBO}}{1-\alpha} = \frac{2\,\mu A}{1-0.99} = 200\,\mu A$$

3.45 (c)

The small signal β or h_{fe} is defined and measured at a constant V_{CE} that is, with a zero signal component between collector and emitter it is known as the short circuit common emitter current gain.

The value of β depends on the current at which the transistor is operating, and the relationship takes the form shown below.

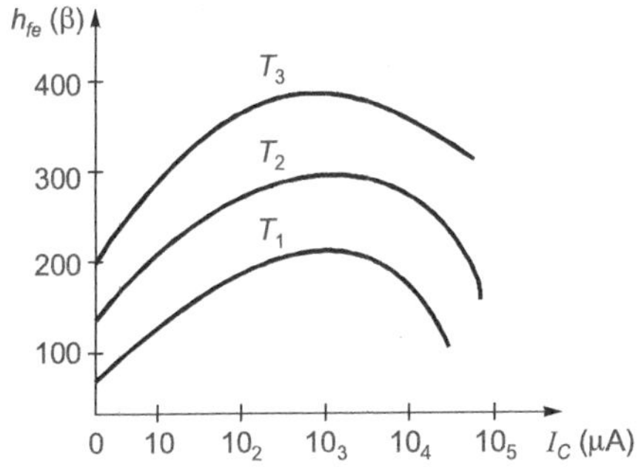

$T_3 > T_2 > T_1$

Where T = temperature

β of a transistor for low values of I_C, it increases and reaches a maximum and then decreases with further increases in I_C.

3.46 (d)

The following steps are followed in the fabrication of monolithic bipolar junction transistor.

1. Epi-layer formation
2. Buried layer formation
3. Base diffusion
4. Emitter diffusion

3.47 (c)

When a transistor is connected in common emitter mode it will have medium input resistance and high output resistance.

3.49 (a)

When PNP transistor is forward biased then holes from emitter diffuses through base into collector.

3.50 (b)

Since we bias our device in active regions thus variation in β will take the device out of active region.

3.51 (*)

Since, $I_C = 100$ mA and $\beta = 100$

so, $I_B = 1$ mA

and input resistance is

$$\frac{V_T}{I_B} = \frac{25\,\text{mV}}{1\,\mu\text{A}} = 25\,\text{k}\Omega$$

3.52 (a)

$$I_C = \beta I_B + \beta I_{CBO}$$

$$\beta = \frac{I_C}{I_B + I_{CBO}} = \frac{5\times10^{-3}}{50.5\times10^{-6}} = 99$$

3.55 (d)

$$I_C = \beta I_B + (1+\beta)I_{CO}$$

where, $$\beta = \frac{\alpha}{1-\alpha} = \frac{0.98}{1-0.98} = 49$$

$$I_C = 49 \times 100\,\mu A + (50) \times 10\,\mu A$$
$$= 10\,\mu A\,[490 + 50]$$
$$= 5.4\text{ mA}$$

3.56 (d)

$$I_E = I_B + I_C$$

I_B is in μA, $I_E \approx I_C$

3.58 (b)

$$BV_{CEO} = \frac{BV_{CBO}}{\sqrt[n]{\beta}}$$

$$\Rightarrow \quad BV_{CEO} = \left(\sqrt[n]{\frac{1}{\beta}}\right)BV_{CBO}$$

where, n = Avalanche multiplication factor

3.59 (c)

$$\alpha = 0.97$$
$$I_E = 1\text{ mA}$$
$$I_B = ?$$
$$I_E = (1+\beta)I_B$$
$$= \frac{1}{1-\alpha}\times I_B$$
$$I_B = (1-\alpha)I_E$$
$$= 0.03 \times 1\text{ mA} = 0.03\text{ mA}$$

3.60 (c)

$$\alpha = 0.99$$
$$I_C = 1 \text{ mA}$$
$$I_{CBO} = 1 \text{ μA}$$
$$I_B = ?$$
$$I_C = \beta I_B + (1+\beta)I_{CBO}$$
$$I_C = \frac{\alpha}{1-\alpha}I_B + \frac{1}{1-\alpha}I_{CBO}$$
$$1 \text{ mA} = \frac{0.99}{1-0.99}I_B + \frac{1}{1-0.99} \times 1 \text{ μA}$$
$$1 \text{ mA} = 99I_B + 100 \times 1 \text{ μA}$$
$$= 99I_B + 0.1 \text{ mA}$$
$$1 \text{ mA} - 0.1 \text{ mA} = 99I_B$$
$$99I_B = 0.9 \text{ mA}$$
$$I_B = \frac{0.9 \text{ mA}}{99}$$
$$= 9.09 \text{ μA}$$

3.61 (d)

Given, $\beta = 105$

$$I_C = 840 \text{ μA}$$
$$I_B = ?$$
$$I_C = \beta I_B$$
$$I_B = \frac{I_C}{\beta} = \frac{840 \text{ μA}}{105} = 8 \text{ μA}$$

3.62 (c)

Collector region is very lightly doped compared to base region is wrong statement.

Theoretical transistor has

Emitter - heavily doped,

Collector - moderate

Base - lightly-doped

3.63 (c)

For certain of reverse voltage in transistor, the effective base width may reduce to zero. This phenomena is known as punch through.

3.64 (b)

Common base current gain

$$\alpha = 0.98$$

Common emitter current gain

$$\beta = \frac{\alpha}{1-\alpha} = \frac{0.98}{1-0.98} = 49$$

3.65 (c)

Statement (I): Small signal equivalent model for BJT will not have junction capacitances.

Statement (II): *p-n-p* current gain will be less than *n-p-n* because $\mu_n > \mu_p$.

■■■■

4 Field Effect Transistors

4.1 The modified work function of an n-channel MOSFET is –0.85 V. If the interface charge is 3×10^{-4} C/m^2 and the oxide capacitance is 300 μF/m^2, the flat band voltage is

(a) –1.85 V (b) –0.15 V
(c) +0.15 V (d) +1.85 V

[ESE-2002]

4.2 Consider following statements:

1. BJT is a current controlled device with high input impedance and high gain bandwidth
2. FET is a voltage controlled device with high input impedance and low gain bandwidth
3. UJT is a negative resistance device and can be used as an oscillator
4. BJT, FET and UJT can all be used for amplification.

Which of these statements are correct?

(a) 1 and 2 (b) 2 and 3
(c) 3 and 4 (d) 1 and 4

[ESE-2005]

4.3 Match **List-I (Circuit Symbol)** with **List-II (Device)** and select the correct answer using the code given below the lists:

List-I List-II

A. 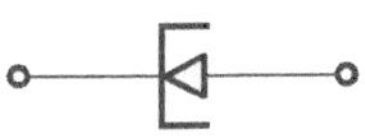 1. N-channel FET

B.  2. Varactor

C. 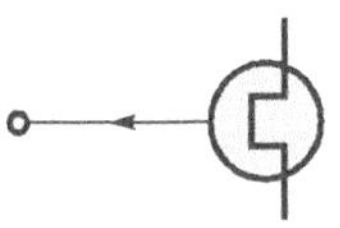 3. Tunnel diode

D.  4. P-channel MOSFET

Codes:

	A	B	C	D
(a)	3	2	1	4
(b)	1	4	3	2
(c)	3	4	1	2
(d)	1	2	3	4

[ESE-2005]

4.4 What is the main difference between MOSFETs and BJTs in terms of their I-V characteristics?

(a) Current is quadratic with V_{GS} for MOSFETs and linear with V_{BE} for BJTs
(b) Current is linear with V_{GS} for MOSFETs and exponential with V_{BE} for BJTs
(c) Current is exponential with V_{GS}/V_{BE} in both these devices, but rise is faster in MOSFETs
(d) Current is quadratic with V_{GS} for MOSFETs and exponential with V_{BE} for BJTs

[ESE-2006]

4.5 **Assertion (A):** In JFET, the phenomenon of thermal runaway is not observed around room temperature.

Reason (R): The heat dissipation in the semiconductor increases its temperature which increases the carrier mobility with temperature.

(a) Both A and R are true and R is the correct explanation of A
(b) Both A and R are true but R is NOT the correct explanation of A
(c) A is true but R is false
(d) A is false but R is true

[ESE-2006]

4.6 In a source follower, consider the following statements:

1. The voltage gain of a source follower is always less than one.
2. It has some current gain and power gain.
3. Its output resistance can be made low.

Which of these statements are correct?

(a) 1 and 2 only (b) 2 and 3 only

(c) 1 and 3 only (d) 1, 2 and 3

[ESE-2007]

4.7 **Assertion (A):** MOSFETs perform very well in parallel operation.

Reason (R): MOSFET has smaller turn-off time.

(a) Both A and R are true and R is the correct explanation of A

(b) Both A and R are true but R is not the correct explanation of A

(c) A is true but R is false

(d) A is false but R is true

[ESE-2007]

4.8 The lower turn off time of MOSFET when compared to BJT can be attributed to which one of the following?

(a) Input impedance

(b) Positive temperature coefficient

(c) Absence of minority carriers

(d) On-state resistance

[ESE-2007]

4.9 Two MOSFETS M_1 and M_2 are connected in parallel to carry a total current of 20 A. The drain to source voltage of M_1 is 2.5 V and that of M_2 is 3 V. What are the drain currents of M_1 and M_2 when the current sharing series resistances are each of 0.5 Ω?

(a) 10.5 A and 9.5 A

(b) 9.5 A and 10.5 A

(c) 10.5 A and 10.5 A

(d) 9.5 A and 9.5 A

[ESE-2009]

4.10 The JFET in the below, circuit has an I_{DSS} = 10 mA, V_P = 5 V. The value of the resistance R_S for a drain current of I_{DS} = 6.4 mA is

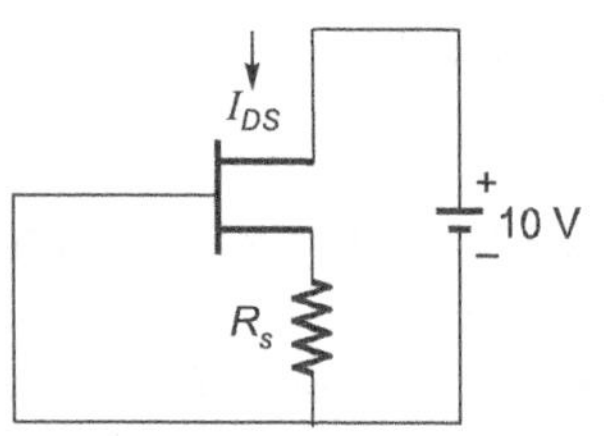

(a) 1.06 kΩ (b) 560 Ω

(c) 470 Ω (d) 156 Ω

[ESE-2010]

4.11 A JFET is set up as a follower, with μ = 200, r_d = 100 kΩ and source load resistor R_L = 1 kΩ. The output resistance R_0 is

(a) 1000 Ω (b) 500 Ω

(c) 333 Ω (d) 666 Ω

[ESE-2010]

4.12 Thermal runaway is not encountered in FETs because

(a) I_{DS} has a zero temperature coefficient.

(b) I_{DS} has a negative temperature coefficient.

(c) I_{DS} has a positive temperature coefficient

(d) The mobility of the carriers increases with increase in temperature.

[ESE-2010]

4.13 The FET shown in the figure below is a

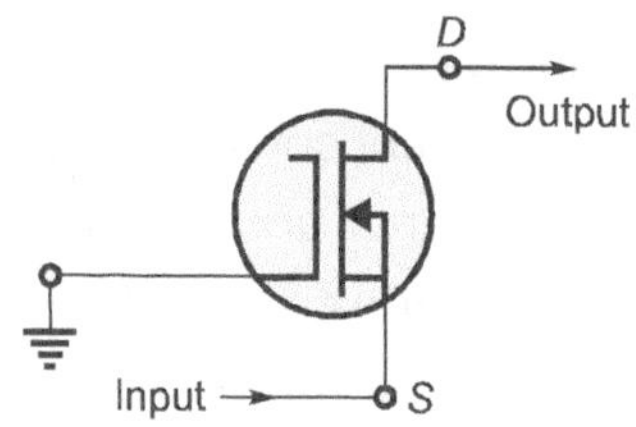

(a) Common drain

(b) Common gate

(c) Common source

(d) Common source follower

[ESE-2010]

4.14 Compared to the bipolar junction transistor, a JFET:

1. Has a larger gain bandwidth produce
2. Is less noisy
3. Has less input resistance
4. Has current flow due to only majority carriers

(a) 1, 2, 3 and 4 are correct

(b) 1 and 2 are correct

(c) 2 and 4 are correct

(d) 3 and 4 are correct

[ESE-2011]

4.15 Which one of the following statement is not correct for a MOSFET?

(a) Are easy to parallel for higher current

(b) Leakage current is relatively high

(c) Have more linear characteristic

(d) Overload and peak current handling capability are high

[ESE-2011]

4.16 Match **List-I** with **List-II** and select the correct answer using the code given below the lists:

List-I	List-II
A. BJT	1. Population inversion
B. MOS capacitor	2. Pinch-off voltage
C. LASER diode	3. Early effect
D. JFET	4. Flat-band voltage

Codes:

	A	B	C	D
(a)	3	1	4	2
(c)	2	1	4	3
(c)	3	4	1	2
(d)	2	4	1	3

[ESE-2012]

4.17 **Statement (I):** Most JFETs are designed to work in depletion mode.

Statement (II): Depletion mode takes advantage of very high input resistance or reverse biased state.

(a) Both Statement (I) and Statement (II) are individually true and Statement (II) is the correct explanation of Statement (I).
(b) Both Statement (I) and Statement (II) are individually true but Statement (II) is not the correct explanation of Statement (I).
(c) Statement (I) is true but Statement (II) is false.
(d) Statement (I) is false but Statement (II) is true.

[ESE-2012]

4.18 The regions of operation of a MOSFET to work as a linear resistor and linear amplifier are

(a) cut-off and saturation respectively
(b) triode and cut-off respectively
(c) triode and saturation respectively
(d) saturation and triode respectively

[ESE-2013]

4.19 The following statements refer to an n channel FET operated in the active region:

1. The gate voltage V_{GS} reverse biases the junction.
2. The drain voltage V_{DD} is negative with respect to the source.
3. The current in the n channel is due to electrons.
4. Increasing the reverse bias V_{GS} increases the cross section for conduction.

Which of these statements are correct?

(a) 1 and 2 (b) 1 and 3
(c) 2 and 3 (d) 3 and 4

[ESE-2013]

4.20 The value of the capacity reactance obtainable from a reactance FET whose g_m is 12 ms when the gate-to-source resistance is 1/9 of the reactance of the gate-to-drain capacitor at frequency 5 MHz is

(a) 650 Ω (b) 750 Ω
(c) 775 Ω (d) 800 Ω

[ESE-2013]

4.21 **Statement (I):** MOSFET's are intrinsically faster than bipolar devices.

Statement (II): MOSFETs have excess minority carrier.

(a) Both Statement (I) and Statement (II) are individually true and Statement (II) is the correct explanation of Statement (I).
(b) Both Statement (I) and Statement (II) are individually true but Statement (II) is not the correct explanation of Statement (I).
(c) Statement (I) is true but Statement (II) is false.
(d) Statement (I) is false but Statement (II) is true.

[ESE-2013]

4.22 Which of the following transistor is symmetrical in the same that emitter and collector or source and drain terminals can be interchanged?

(a) JFET (b) MOSFET
(c) NPN transistor (d) PNP transistor

[ESE-2014]

4.23 When the drain voltage in an *n*-MOSFET is negative, it is operating in

(a) active region (b) inactive region
(c) ohmic region (d) reactive region

[ESE-2015]

4.24 Which is the important factor in the steady state characteristics of a MOSFET?

(a) Current gain
(b) Transconductance
(c) Output resistance
(d) Drain-source voltage

[ESE-2015]

4.25 In a 2 input CMOS logic gate, one input is left floating i.e. connected neither to ground nor to a signal. What will be the state of that input?
(a) 1
(b) 0
(c) same as that of the other input
(d) indeterminate (neither 1 nor 0)
[ESE-2016]

4.26 The disadvantage of a typical MOSFET as compared to BJT is
(a) Increased power-handling levels
(b) Reduced power-handling levels
(c) Increased voltage-handling levels
(d) Reduced voltage-handling levels
[ESE-2019]

4.27 In a Field Effect Transistor (FET) the maximum voltage that can be applied between any two terminals is given by
(a) Low $|V_{DS}|$ causing avalanche breakdown
(b) Low $|V_{GS}|$ causing avalanche breakdown
(c) $|V_{DS}| = 0$ when gate is reverse–biased
(d) $|V_{GS}| = 0$ when gate is reverse–biased
[ESE-2020]

4.28 A depletion–type MOSFET can be operated in an enhancement mode where negative charges are induced into n-type channel by applying
(a) Positive Gate Voltage
(b) Negative Gate Voltage
(c) Positive Drain Voltage
(d) Negative Drain Voltage
[ESE-2020]

4.29 The double base diode which is operated with the emitter forward biased and a smaller emitter junction is called
(a) Field Effect Transistor (FET)
(b) Uni-Junction Transistor (UJT)
(c) Bipolar Junction Transistor (BJT)
(d) Metal Oxide Semiconductor Field Effect Transistor (MOSFET)
[ESE-2020]

4.30 For depletion type MOSFET : I_D = 4.5 mA at V_{GS} = –2 V. What is the value of I_{DSS} if V_P = –5 V?
(a) 5.5 mA (b) 12.5 mA
(c) 5.0 mA (d) 15 mA
[ESE-2021]

4.31 Which one of the following consists of a layer of metal at the top named gate, a silicon dioxide layer below it, and a semiconductor substrate as the bottom-most layer?
(a) JFET (b) BJT
(c) MOSFET (d) DMOSFET
[ESE-2021]

4.32 The MOSFET is in the cut-off state, when the gate source voltage is
(a) less than gate to source voltage
(b) greater than gate to source voltage
(c) less than the threshold voltage
(d) greater than the threshold voltage
[ESE-2021]

4.33 For an n-channel silicon FET with $a = 3 \times 10^{-4}$ cm and $N_D = 10^{15}$ electron/cm^3, what is the pinch-off voltage if the dielectric constant of silicon is $\varepsilon = 12\varepsilon_0$ and $\varepsilon_0 = \frac{1}{36\pi} \times 10^{-9}$?
(a) 6.8 V (b) 5.2 V
(c) 8.8 V (d) 9.2 V [ESE-2021]

4.34 The common emitter configuration is widely used due to its
(a) low voltage and low power gain
(b) low voltage and high power gain
(c) high voltage and high power gain
(d) high voltage and low power gain
[ESE-2023]

4.35 Consider the following statements regarding voltage-divider bias in a transistor :
1. The voltage-divider bias configuration uses two DC bias sources to provide forward-reverse bias to the transistor.
2. The voltage-divider bias provides a very small base current to the transistor compared to the bias current.
3. Two resistors R_1 and R_2 form a voltage divider that provides the base bias voltage to the transistor.

Which of the above statements is/are correct?
(a) 1 and 2 only (b) 2 and 3 only
(c) 2 only (d) 1,2 and 3
[ESE-2024]

■■■■

Answers Field Effect Transistors

4.1 (d)	4.2 (b)	4.3 (a)	4.4 (d)	4.5 (c)	4.6 (c)	4.7 (b)	4.8 (c)	4.9 (a)
4.10 (d)	4.11 (c)	4.12 (b)	4.13 (b)	4.14 (c)	4.15 (c)	4.16 (c)	4.17 (a)	4.18 (c)
4.19 (b)	4.20 (*)	4.21 (c)	4.22 (a)	4.23 (b)	4.24 (b)	4.25 (d)	4.26 (b)	4.27 (a)
4.28 (a)	4.29 (b)	4.30 (b)	4.31 (c)	4.32 (c)	4.33 (a)	4.34 (c)	4.35 (b)	

Explanations Field Effect Transistors

4.1 (d)

Given that, $W_C = -0.85$

interface charge $q = 3 \times 10^{-4}$ c/m^2

oxide capacitance,

$$C = 300 \times \mu F/m^2 = 300 \times 10^{-6} F/m^2$$

we know that flat band voltage

$$V = \frac{q}{C}(1 - W_C)$$

$$V = \frac{3 \times 10^{-4}}{300 \times 10^{-6}}[1-(-0.85)] = 1.85 \text{ V}$$

4.2 (b)

BJT is current controlled current source having low i/p impedance and high gain bandwidth used as amplifier as well as oscillator FET is voltage controlled device having high input impedance and low gain bandwidth used as amplifier as well as oscillator. UJT is –ve resistance device and used as oscillator it can't be used as an amplifier because there is no controlled action.

4.4 (d)

For MOSFET,

$$I_{DS} = I_{DSS}\left(1 - \frac{V_{GS}}{V_P}\right)^2$$

and for BJT, $I = I_0 (e^{V/\eta V_T} - 1)$

4.5 (c)

FET devices have positive temp-resistance coefficient. So thermal runaway is not observed and mobility ↓ with T↑.

4.6 (c)

No power gain in source follower because $A_V < 1$.

4.7 (b)

MOSFET is a high frequency device hence both *A* and *R* are correct but reason of A is positive resistance temperature coefficient.

4.8 (c)

During turnoff in case of BJT storage time required to sweptout the minority carriers from n and p regions and this is not required in case of MOSFET because it is a majority carrier device so lesser turnoff time.

4.9 (a)

Given as:

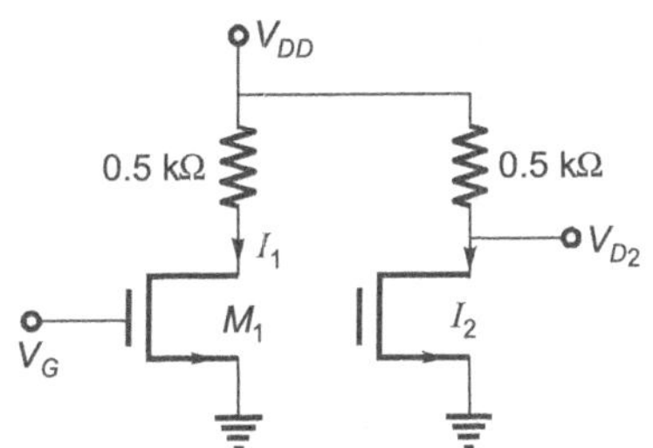

Given that,

$$I_1 + I_2 = 20 \text{ mA} \quad \ldots(i)$$

$$(V_{DS}) = 2.5 \text{ V}$$

$$V_{D1} = V_{S1} = 2.5 \text{ V}$$

$$V_{D1} = 2.5 \text{ V} \quad (\because V_{S1} = 0)$$

Similarly, $V_{D2} = 3$ V

∴ From equation (*i*),

$$\frac{V_{DD} - V_{D1}}{0.5 \times 10^3} + \frac{V_{DD} - V_{D2}}{0.5 \times 10^3} = 20 \text{ mA} = 20 \times 10^{-3} \text{ A}$$

$$2V_{DD} - V_{D1} - V_{D2} = 10 \times 10^{-3} \text{ A} \times 10^3$$

$$2V_{DD} - 2.5 - 3 = 10 \times 10^{-3} \text{ A} \times 10^3$$

$$V_{DD} = \frac{5.5}{2} + 5 \times 10^{-3} \times 10^3 = \frac{15.5}{2} \text{V}$$

Now, $$I_1 = \frac{V_{DD} - V_{D1}}{0.5 \text{ k}\Omega} = 2[V_{DD} - V_{D1}] \text{ mA}$$

$$= 2\left[\frac{5.5}{2} + 5 \times 10^{-3} \times 10^{-3} - 2.5\right] \text{mA}$$

$$= (5.5 + 10 - 5) \text{ mA} = 10.5 \text{ mA}$$

$$I_2 = \frac{V_{DD} - V_{D2}}{0.5 \text{ k}\Omega} = \frac{\frac{15.5}{2} - 3}{0.5 \text{k}\Omega}$$

$$= 15.5 - 6 = 9.5 \text{ mA}$$

4.10 (d)

$$i_{DSS} = 10 \text{ mA}$$
$$V_P = 5 \text{ V}$$
$$i_{DS} = 6.4 \text{ mA}$$

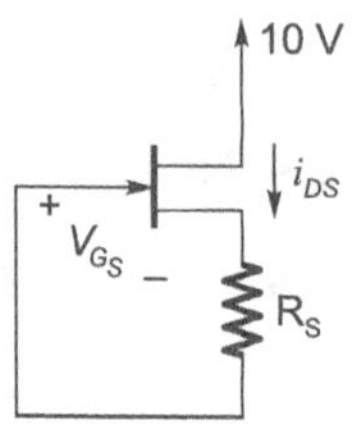

$$i_{DS} = I_{DSS}\left(1 - \frac{V_{GS}}{V_P}\right)^2$$

$$6.4 = 10\left(1 - \frac{V_{GS}}{V_P}\right)^2$$

$$V_{GS} = 1 \text{ V}$$

By applying KVL,

$$V_{GS} + R_S i_{DS} = 0$$
$$1 - 6.4 R_S = 0$$
$$R_S = 156.25\ \Omega$$

4.11 (c)

$$\mu = g_m r_d$$
$$200 = g_m \times 100 \text{ k}$$
$$g_m = 2 \text{ ms}$$

$$R_0 = \left(\frac{1}{g_m}\right) \| R_L = \frac{1}{2} \| 1$$

$$R_0 = \frac{1/2}{3/2} \text{k} = 333\ \Omega$$

4.12 (b)

Thermal runaway is not encountered in FET because as temperature increases, i_{DS} decreases.

4.13 (b)

Since gate is grounded so configuration is common gate.

4.14 (c)

Compared to bipolar junction transistor JFET has lower gain bandwidth product.

Current flow in JFET due to majority carrier, and there is no role of minority carriers and noise is due to minority carriers mainly so less noisy.

4.15 (c)

Some of the feature of MOSFET:

(*i*) MOSFET is a unipolar device.

(*ii*) It has high input impedance.

(*iii*) It has lower switching losses but its on-resistance and conduction losses are more.

(*iv*) It is voltage controlled device.

(*v*) It has positive temperature coefficient for resistance. This makes parallel operation of MOSFETs easy.

(*vi*) Secondary breakdown does not occur, because it has positive temperature coefficient.

Hence, option (c) is correct.

4.16 (c)

Pinch-off voltage is associated with JFET, early effect with BJT and population inversion with LASER diode hence option (c) is correct.

4.17 (a)

JFETs are typically depletion mode. Since the gate junction would be forward bias if the gate were taken more than a little from source toward drain voltage and in this mode the input resistance of reverse biased state is very high.

4.21 (c)

MOSFET's are unipolar device therefore they are faster than bipolar device.

4.22 (a)

For JFET the source and drain terminals can be interchanged because the structure of JFET is as shown below,

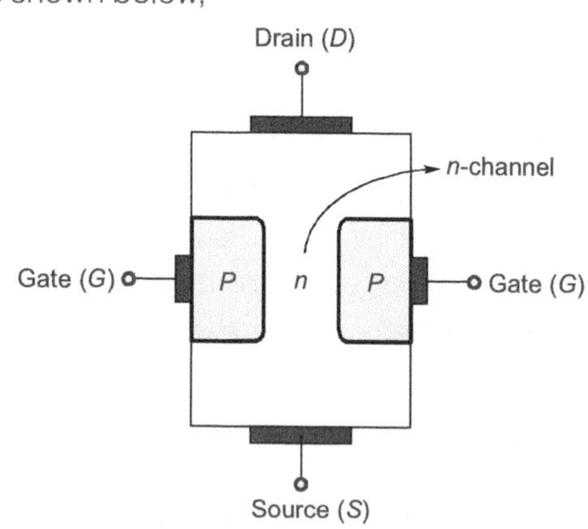

To the either sides of the *n*-channel drain and source are their, so they can be inter-changed.

4.23 (b)

When drain voltage is negative then no current flow in device and thus it is in inactive region.

4.26 (b)

The disadvantage of a typical MOSFET compared to BJT is reduced power handling levels.

4.27 (a)

The maximum voltage (V_{DS}) that can be applied, is the lowest voltage that will cause avalanche breakdown across the gate junction.

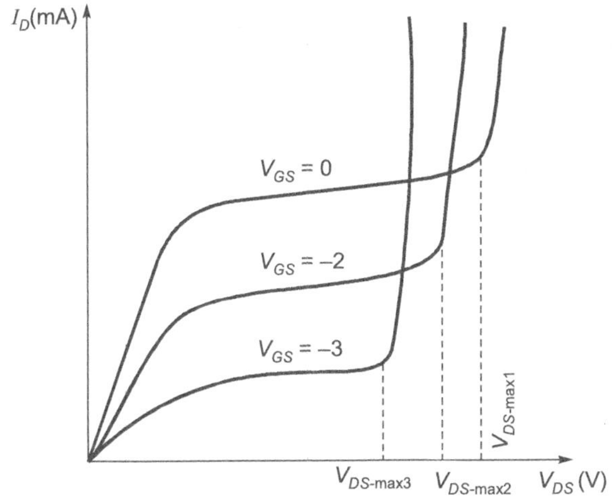

As it can be seen for above figure ($I_D - V_{DS}$) characteristics that for V_{GS} = reverse bias, less voltage is required.

4.28 (a)

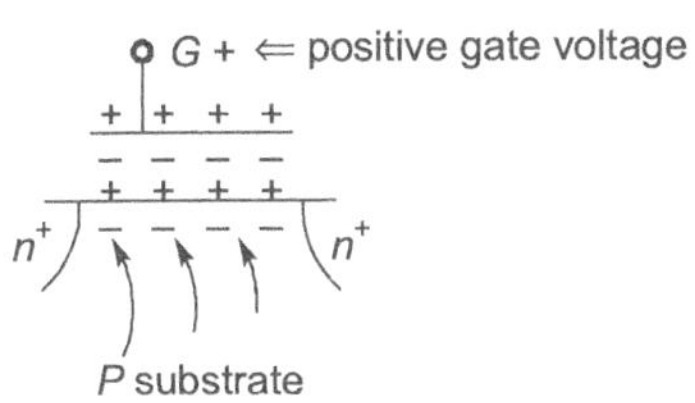

4.29 (b)

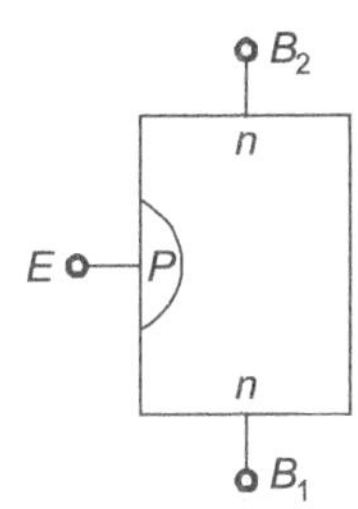

Uni junction transistor (UJT)

4.30 (b)

Depletion MOSFET :

$$I_D = 4.5 \text{ mA at } V_{GS} = -2 \text{ V}$$
$$I_{DSS} = ?$$
$$V_P = -5 \text{ V}$$

$$I_D = I_{DSS}\left[1-\frac{V_{GS}}{V_P}\right]^2$$

$$4.5 \text{ mA} = I_{DSS}\left[1-\frac{-2}{-5}\right]^2$$

$$= I_{DSS}\left[1-\frac{2}{5}\right]^2$$

$$4.5 \text{ mA} = I_{DSS}\left[\frac{5-2}{5}\right]^2$$

$$4.5 \text{ mA} = I_{DSS}\left[\frac{3}{5}\right]^2$$

$$I_{DSS} = \frac{4.5 \text{ mA} \times 25}{9} = 12.5 \text{ mA}$$

4.31 (c)

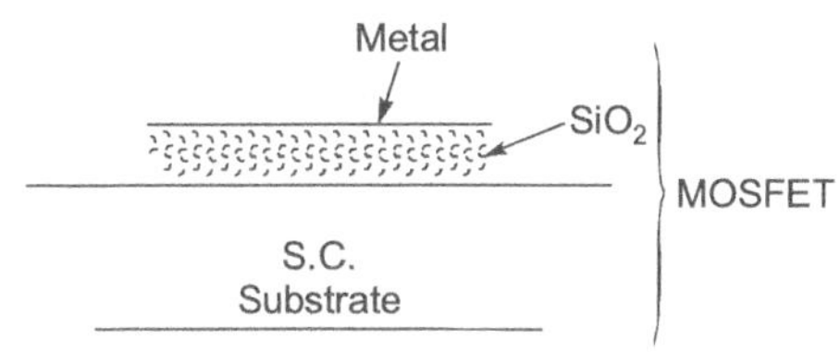

4.32 (c)

MOSFET is in cut-off state.

$$V_{GS} < V_t$$

4.33 (a)

n-channel Si FET

$$a = 3 \times 10^{-4} \text{ cm}$$
$$N_D = 10^{15} \text{ electron/cm}^3$$
$$V_p = ?$$
$$\varepsilon = 12\varepsilon_0$$

where $\varepsilon_0 = \frac{1}{36\pi}\times 10^{-9}$ F/m

$= 8.85 \times 10^{-12}$ F/m or 8.85×10^{-14} F/cm

$$V_p = \frac{a^2 q N_D}{2\epsilon}$$

$$V_p = \frac{(3\times10^{-4})^2 \times 1.6\times10^{-19}\times10^{15}}{2\times12\times8.85\times10^{-14}}$$

$= 6.8$ V

4.34 (c)

Common emitter configuration is widely used due to high voltage and high power gain.

4.35 (b)

For voltage divider bias 2 and 3 statements are correct.

■■■■

5 BJT and FET Amplifier

5.1 In an RC coupled amplifier, the gain decreases in the frequency response due to the
(a) coupling capacitor at low frequency and bypass capacitor at high frequency.
(b) coupling capacitor at high frequency and bypass capacitor at low frequency.
(c) coupling junction capacitance at low frequency and coupling capacitor at high frequency.
(d) device junction capacitor at high frequency and coupling capacitor at low frequency.
[ESE-2001]

5.2 Match **List-I (Hybrid parameter)** with **List-II (Units/definitions)** and select the correct answer:

List-I	List-II
A. h_{ie}	1. Forward Current transfer ratio
B. h_{fe}	2. Ohms
C. h_{re}	3. Siemens
D. h_{oe}	4. Reverse Voltage transfer ratio

Codes:

	A	B	C	D
(a)	2	1	3	4
(b)	1	2	4	3
(c)	1	2	3	4
(d)	2	1	4	3

[ESE-2001]

5.3 In an RC coupled common emitter amplifier
(a) Coupling capacitance affects the h.f. response and bypass capacitance affects the L.f. response.
(b) Both coupling and bypass capacitances affect the h.f. response.
(c) Both coupling and bypass capacitances affect the L.f. response only.
(d) Coupling capacitance affects the L.f. response and the bypass capacitance affects the h.f. response.
[ESE-2003]

5.4 **Assertion** (A): Transistor h-parameter equivalent circuit can be used for the analysis irrespective of the configuration (*CE*, *CB* or *CC*) of the transistor used.
Reason (R): The values of h-parameter remain unchanged with transistor configuration.
(a) Both A and R are true and R is the correct explanation of A
(b) Both A and R are true but R is NOT the correct explanation of A
(c) A is true but R is false
(d) A is false but R is true
[ESE-2003]

5.5 What is the purpose of impedance matching between the output of previous stage and input of next stage in a cascaded amplifier?
(a) To achieve high efficiency
(b) To achieve maximum power transfer
(c) To achieve reduced distortion
(d) To achieve reduced noise
[ESE-2006]

5.6 Which of the following are true for h-parameters of transistors?
1. They are real numbers at audio frequencies.
2. They are easy to measure.
3. They vary widely with temperature.
Select the correct answer using the code given below:
(a) 1 and 2 only (b) 2 and 3 only
(c) 1 and 3 only (d) 1, 2 and 3
[ESE-2007]

5.7 The input resistance R_i and output resistance R_0 of an ideal current amplifier, in ohms, are
(a) 0 and 0 (b) 0 and ∞
(c) ∞ and 0 (d) ∞ and ∞
[ESE-2010]

5.8 **Assertion** (A): If the output voltage waveform in a transistor amplifier is a true replica of the input waveform, then it is said to be operated in class A mode. The Q point cannot be fixed near $I_C = 0$.
Reason (R): The positive half cycle of the input will be clipped off in the output due to transistor being drawn into cut-off from active region.
(a) Both A and R are true and R is the correct explanation of A
(b) Both A and R are true but R is NOT the correct explanation of A
(c) A is true but R is false
(d) A is false but R is true [ESE-2011]

5.9 Consider the following statements.
The h-parameters of a BJT:
1. Represent the input resistance, output conductance, current gain etc.
2. Can be easily measured in the laboratory.
3. Can be used in circuit analysis over a wide range of frequencies.
4. Constitute a simple small signal model for easy circuit analysis at low frequencies where junction capacitances can be neglected.

Which of these statements are correct?
(a) 1, 2, 3 and 4 (b) 1, 2 and 4
(c) 1, 3 and 4 (d) 2, 3 and 4
[ESE-2011]

5.10 **Statement (I):** A good amplifier should not only amplify but also should faithfully reproduce the input signal.
Statement (II): Distortion takes place in amplifiers due to non-linearity of the devices.
(a) Both Statement (I) and Statement (II) are individually true and Statement (II) is the correct explanation of Statement (I).
(b) Both Statement (I) and Statement (II) are individually true but Statement (II) is not the correct explanation of Statement (I).
(c) Statement (I) is true but Statement (II) is false.
(d) Statement (I) is false but Statement (II) is true.
[ESE-2012]

5.11 **Statement (I) :** Both Coupling capacitance and emitter bypass capacitance affect the low frequency response of an R-C-coupled amplifier.
Statement (II) : Both stray capacitances and emitter-to-base diffusion capacitance have a profound effect on the low frequency response of an R-C-coupled amplifier.
(a) Both Statement (I) and Statement (II) are individually true and Statement (II) is the correct explanation of Statement (I).
(b) Both Statement (I) and Statement (II) are individually true but Statement (II) is not the correct explanation of Statement (I).
(c) Statement (I) is true but Statement (II) is false.
(d) Statement (I) is false but Statement (II) is true.
[ESE-2018]

5.12 Which of the following conditions will be satisfied for an impedance matched system?
(a) The decibel power gain is equal to twice the decibel voltage gain.
(b) The decibel power gain is equal to the decibel voltage gain.
(c) The decibel power gain is half the decibel voltage gain.
(d) The decibel power gain is equal to the thrice the decibel voltage gain.
[ESE-2019]

5.13 The dB gain of cascaded system is simply
(a) The square of the dB gain of each stage.
(b) The sum of the dB gains of each stage.
(c) The multiplication of the dB gains of each stage.
(d) The division of the dB gains of each stage.
[ESE-2019]

5.14 The Miller effect input capacitance C_{M_i} is
(a) $(1 - A_V^2)C_f$ (b) $(1 - A_V)C_f$
(c) $(1 - C_f)A_V$ (d) $(1 - C_f^2)A_V$
Where,
C_f = feedback capacitance
$A_V = \frac{V_0}{V_i}$ [ESE-2019]

5.15 What is the power gain of transistor amplifier, if its current gain is 40 and voltage gain is 25?
(a) 100 (b) 1200
(c) 1000 (d) 950 [ESE-2022]

5.16 Which one of the following statements is not correct regarding common-base amplifier?
(a) The output is in the same phase as the input alternating signal.

(b) It cannot operate at higher frequency as compared to CE amplifier.
(c) The current gain is less than unity.
(d) Impedance matching is needed when cascading because there is very large difference in the input and output impedance, the input impedance is low and the output high. [ESE-2022]

5.17 Which one of the following is a disadvantages of CE amplifier?
(a) It provides good current as well as voltage gain.
(b) It provides the maximum power gain of the three configuration.
(c) It has medium both input and output impedances.
(d) Its frequency response bandwidth is lower than the amplifiers of the other two configurations. [ESE-2022]

5.18 **Statement (I)** : When negative feedback is applied to the ideal amplifier, the differential input voltage is zero.
Statement (II) : There is no current flow into either input terminal of the ideal op-amp.
[ESE-2022]

5.19 The junction FET is a three terminal
(a) voltage controlled voltage device
(b) current controlled voltage device
(c) voltage controlled current device
(d) current controlled current device
[ESE-2023]

5.20 Consider the basic MOSFET circuit as shown in the figure with variable gate voltage.

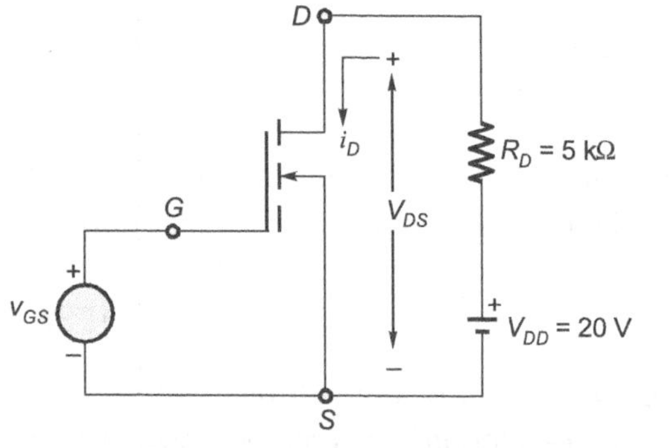

The MOSFET is given to have very large V_A, V_T = 4 V, and I_{DSS} = 8 mA. What is the value of i_D for V_{GS} = 5 V? (Where V_A is a constant in the rantge of 30 V to 200 V)
(a) 0.2 mA (b) 0.5 mA
(c) 1.0 mA (d) 2.0 mA
[ESE-2023]

5.21 Which one of the following is the correct circuit symbol representation of p-channel JFET?

(a)
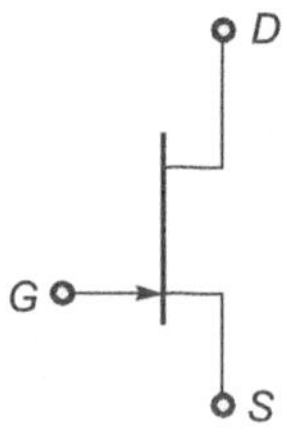
(b)
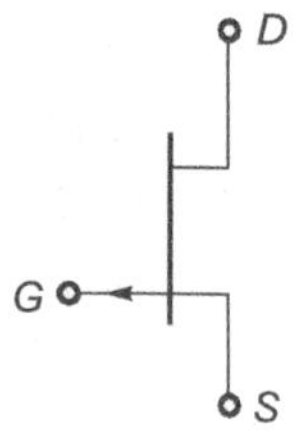
(c)
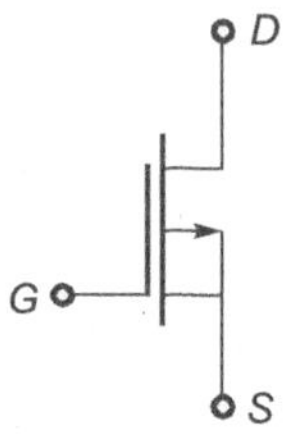
(d)
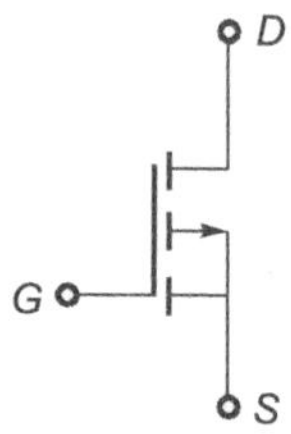

[ESE-2023]

5.22 Consider the following statements regarding FET amplifier configurations :
1. For common-source and common-drain configurations, $R_{in} = R_1 \parallel R_2$, which can be selected to be large during bias design.
2. For the common-drain configuration, the voltage gain is generally less than unity or near unity.
3. For the common-gate configuration, the current gain cannot be larger than unity.

Which of the above statement(s) is/are correct?
(a) 1 only (b) 1 and 2 only
(c) 2 and 3 only (d) 1, 2 and 3
[ESE-2023]

5.23 **Statement (I):** MOSFETs are preferred over JFETs for digital integrated circuits, either p-channel metal-oxide semiconductor or n-channel metal-oxide semiconductor logic circuits can be constructed.

Statement (II): The usage of a p-channel MOSFET as the active load for an n-channel MOSFET leads to a logic family known as complementary-symmetry MOS or 7 CMOS.
[ESE-2023]

5.24 Consider the following statements regarding common-base transistor amplifier:

1. In the active region, the collector-base junction is forward biased while the base-emitter junction is reverse biased.
2. In the cutoff region, the collector-base and base-emitter junctions of a transistor are both reverse biased.
3. In the saturation region, the collector-base and base-emitter junctions are both forward biased.

Which of the above statements is/are correct?

(a) 1 and 2 only (b) 2 and 3 only
(c) 2 only (d) 1,2 and 3

[ESE-2024]

5.25 What are the biasing states of collector-base junction and base-emitter junction in the active region of a common-emitter transistor amplifier?

(a) Both collector-base junction and base-emitter junction are forward biased
(b) Both collector-base junction and base-emitter junction are reverse biased
(c) The collector-base junction is forward biased and the base-emitter junction is reverse biased
(d) The collector-base junction is reverse biased and the base-emitter junction is forward biased

[ESE-2024]

5.26 Match the following lists:

List-I :
(n-p-n BJT operating regions)

P. Cutoff region
Q. Saturation region
R. Forward active region

List-II :
(n-p-n BJT characteristics)

1. $I_C = \beta_F I_B$, $I_B > 0$, $V_{BE} > V_{BE(on)}$, $V_{CE} > V_{CE(sat)}$,
2. $I_C = I_B = I_E = 0$, $V_{BE} < V_{BE(on)}$, $V_{BC} < V_{BC(on)}$
3. $I_B > 0$, $I_C > 0$, $I_C < \beta_F I_B$, $V_{BE} > V_{BE(on)}$

Select the correct answer using the code given below,

Codes:

	P	Q	R
(a)	1	2	3
(b)	2	3	1
(c)	1	3	2
(d)	2	1	3

[ESE-2024]

5.27 Consider the following statements regarding negative feedback in amplifier circuits:

1. It has reduction in overall voltage gain.
2. It has enhanced frequency response.
3. It has higher output impedance.

Which of the above statements is/are correct?

(a) 1 and 2 only (b) 2 and 3 only
(c) 2 only (d) 1,2 and 3

[ESE-2024]

■■■■

Answers BJT and FET Amplifier

5.1 (d)	5.2 (d)	5.3 (c)	5.4 (c)	5.5 (b)	5.6 (d)	5.7 (b)	5.8 (a)	5.9 (b)
5.10 (b)	5.11 (c)	5.12 (b)	5.13 (b)	5.14 (b)	5.15 (c)	5.16 (b)	5.17 (d)	5.18 (b)
5.19 (c)	5.20 (b)	5.21 (b)	5.22 (d)	5.23 (b)	5.24 (b)	5.25 (d)	5.26 (b)	5.27 (a)

Explanations BJT and FET Amplifier

5.1 (d)

Coupling capacitors are used to block the dc and pass the ac in amplifiers as

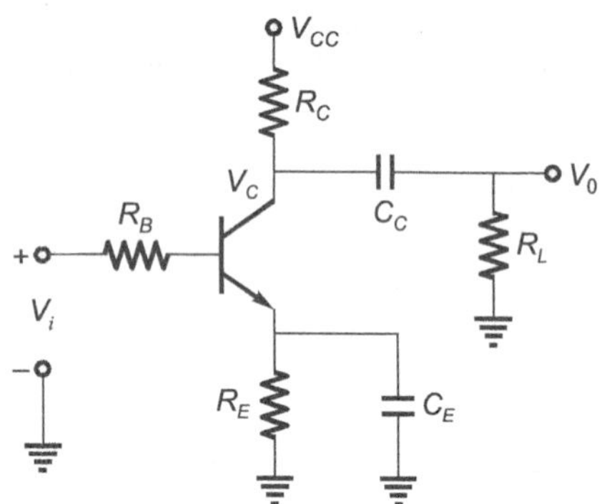

C_C = Coupling capacitor

C_E = Bypass capacitor

This is a circuit of a BJT amplifier.

C_C = block the dc and pass the AC

We take the value of C_C very high.

So that, $1/j\omega C_c \to 0$ so that we get output across R_L and No drop across C_c.

But at low frequency $1/j\omega C_c \to$ finite value because C_c very high makes it towards 0 but ω very low make it very high so overall $1/j\omega C_c \to$ finite, so a portion of V_c is drop across C_c, so less o/p will appear across R_L so gain decreases similarly bypass capacitor C_E for dc $1/j\omega C_E \to \infty$ $\to$ O.C. and for A.C. $1/j\omega\, C_E \to 0$ as $C_E \to \infty$ so R_E will be present for dc so stabilize operating point and but for A.C. R_E will not be present so no feedback so a higher gain but at low frequency $1/j\omega C_E \to$ finite because $\omega \to 0$ (very low) makes it high and $C_E \to \infty$ makes it low so overall it is finite small value.

So, $Z_E = R_E \,||\, Z_C$(finite) $\to$ so there will be feedback (–ve) due to Z_E so gain will decreases due to this –ve feedback at low frequencies.

Due to junction capacitances (active at high frequency) gain will decreases at high frequencies.

5.2 (d)

We know for h parameters

$$\begin{bmatrix} V_1 \\ I_2 \end{bmatrix} = \begin{bmatrix} h_{ie} & h_{re} \\ h_{fe} & h_{oe} \end{bmatrix} \begin{bmatrix} I_1 \\ V_2 \end{bmatrix}$$

$$\therefore \quad V_1 = h_{ie}I_1 + h_{re}V_2$$

$$I_2 = h_{fe}I_1 + h_{oe}V_2$$

$$\therefore \quad h_{ie} = \left.\frac{V_1}{I_1}\right|_{\text{at } V_2=0} \quad \text{i.e. (ohms)}$$

$$h_{re} = \left.\frac{V_1}{V_2}\right|_{\text{at } I_1=0}$$

$\Rightarrow$ Reverse voltage transfer ratio.

$$h_{fe} = \left.\frac{I_2}{I_1}\right|_{\text{at } V_2=0} \quad \text{forward current ratio}$$

$$h_{oe} = \left.\frac{I_2}{V_2}\right|_{\text{at } I_1=0} \left(\frac{1}{\text{ohm}}\right)$$

$\Rightarrow$ Siemens

5.4 (c)

The value of h-parameters change with change in configuration so R is false.

5.5 (b)

Impedance matching is required to achieve maximum power transfer, because if there is no impedance matching then power transfer from one stage to other will be poor.

5.6 (d)

h-parameters have real values $\to$ at small frequencies (audio)

They are easy to measure, they make transistor

in amplification mode analysis easy.
They vary with temperature.

5.7 (b)
For current amplifier

$R_{if} = 0$

$R_{of} = \infty$

5.8 (a)
A is correct as for class A, operating point is in mid of load line so for whole input cycle transistor is in active region so output is $V_0 = AV_i$ so output is exact replica of input and R is also correct and reason for A because if Q point can be fixed near $I_C = 0$ i.e. $V_{CE} = V_{CC}$, for –ve half cycle transistor will be in cut-off region clipping off –ve half of input so not operated in class A region.

5.9 (b)
h-parameters of BJT,
$h_i \rightarrow$ corresponds to input resistance
$h_f \rightarrow$ corresponds current gain
$h_0 \rightarrow$ corresponds to output conductance
h-parameters are useful for circuit analysis in low frequency region and low signal condition.
Statement 3 is wrong.
So, (b) is correct.

5.10 (b)
An ideal amplifier would be a totally linear device, but real amplifier are only linear within limits.
When the signal drive to the amplifier is increased, the output also increases until saturation and cannot produce any more output after saturation and result in distortion therefore can not reproduce the input signal.

5.11 (c)
Statement-I is correct, statement-II is not correct.

5.12 (b)

$$\text{Voltage gain}, A_v = \frac{V_o}{V_i}$$

$$\text{Power gain} = \frac{P_o}{P_i} = \frac{V_o^2 / R_o}{V_i^2 / R_i} = A_V^2 \times \frac{R_i}{R_o}$$

$$\text{Power gain} = 10\log\left(A_V^2 \frac{R_i}{R_o}\right)$$

$$= 2 \times 10\log A_v + 10\log\left(\frac{R_i}{R_o}\right)$$

where, $R_i = R_o$

$$|\text{Power gain}|_{\text{dB}} = 2 \times 10\log A_v + 10\log 1$$

$$= 20\log A_v = |\text{Voltage gain}|_{\text{dB}}$$

5.13 (b)

$$A_V = A_{V1} \cdot A_{V2} \ldots A_{Vn}$$

$$20 \log A_V = 20\log(A_{V1} \cdot A_{V2} \ldots A_{Vn})$$

$$= 20\log A_{V1} + 20\log A_{V2} + \ldots 20\log A_{Vn}$$

5.14 (b)
Miller's input capacitance,

$$C_{Mi} = (1 - A_V)C_M$$

5.15 (c)
Given data, $A_I = 40$

$A_V = 25$

$P_{\text{gain}} = A_V \cdot A_I$

$= 40 \times 25 = 1000$

5.16 (b)
In common base amplifier,
"It can not operate at higher frequency as compared to CE amplifier" is wrong statement.
∴ (b) option is right choice.

5.17 (d)
The disadvantages of CE amplifier is its frequency response bandwidth is lower.

5.18 (b)
Statement I : When negative feedback is applied to ideal amp, the differential i/p voltage is zero (virtual short) (true).
Statement II : There is no current flow into either input terminal or ideal op amp :
$Z_i = \infty$ (True)

5.19 (c)
JFET is a VCCS (voltage controlled current device)

$$I_D \propto V_{GS}$$

5.20 (b)

$$I_D = I_{DSS}\left[1-\frac{V_{GS}}{V_T}\right]^2$$

KVL in output loop

$$V_{DD} = I_D r_D + V_{DS}$$

$$\frac{V_{DD}-V_{DS}}{R_D} = I_{DSS}\left[1-\frac{V_{GS}}{V_T}\right]^2$$

$$\frac{20-V_{DS}}{5K} = 8\text{ mA}\left[1-\frac{5}{4}\right]^2$$

$$V_{DS} = 17.5\text{ V}$$

$$V_{DD} = I_D R_D + V_{DS}$$

$$I_D = \frac{V_{DD}-V_{DS}}{R_D}$$

$$= \frac{20-17.5}{5K} = \frac{2.5}{5K}$$

$$= \frac{1}{2}\text{ mA}$$

$$= 0.5\text{ mA}$$

5.21 (b)

The symbol representation of p-channel JFET is

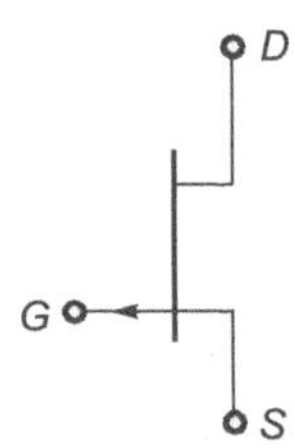

5.22 (d)

Statement 1 : For common source & common drain

$$R_{in} = R_1 \parallel R_2 \text{ (True)}$$

Statement 2 : For common drain configuration, the voltage gain is nearly unity (True).

Statement 3 : For common gate configuration, current gain is nearly unity (True).

5.23 (b)

Statement (I): MOSFETs are preferred over JFET for digital integrated circuits.

Statement (II): CMOS is constructed by NMOS and PMOS where PMOS will behave as a active load.

5.24 (b)

Common base amplifier, (2) and (3) are correct.

5.25 (d)

In common emitter transistor amplifier,

V_{BE} forward biased

V_{CB} reverse biased

5.27 (a)

In negative feedback amplifier

1.It has reduction in overall voltage gain.

$$\left[A_F = \frac{A}{1+A\beta}\right]$$

2.It has enhanced frequency response.

$[BW_F = (1 + A\beta) \times BW]$

■■■■

6 Frequency Response

6.1 An amplifier of gain A is bridged by a capacitance C as shown below.

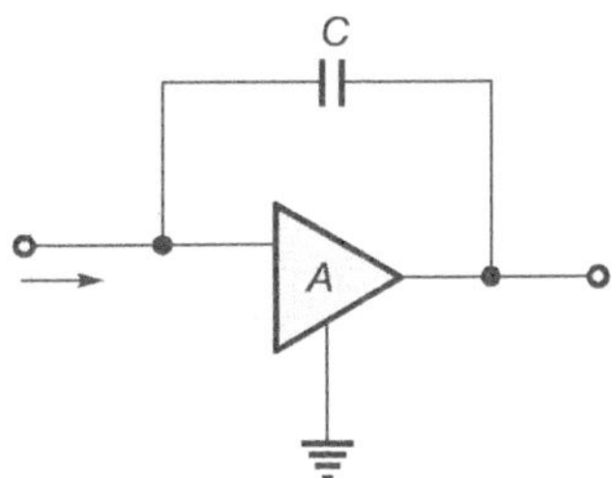

The effective input capacitance is

(a) C/A (b) $C(1 - A)$

(c) $C(1 + A)$ (d) CA

[ESE-2002]

6.2 Two identical RC coupled amplifiers, each having an upper cut-off frequency f_u, are cascaded with negligible loading. What is the upper cut-off frequency of the overall amplifier?

(a) $\dfrac{f_u}{\sqrt{\sqrt{2}-1}}$ (b) $f_u\sqrt{\sqrt{2}-1}$

(c) $f_u/2$ (d) $2\,f_u$

[ESE-2004]

6.3 Two identical RC coupled amplifiers, each having a lower cut-off frequency f_l, are cascaded with negligible loading. What is the lower cut-off frequency of the overall amplifier?

(a) $\dfrac{f_l}{\sqrt{\sqrt{2}-1}}$ (b) $f_l\sqrt{\sqrt{2}-1}$

(c) $f_l/2$ (d) $2\,f_l$

[ESE-2004]

6.4 If the coupling capacitors of a CE transistor amplifier is shorted, which one of the following graphs will represent the frequency response curve of the amplifier?
(A_V = voltage gain, f = frequency in Hertz, A_{max} = maximum value of A_V)

(a)

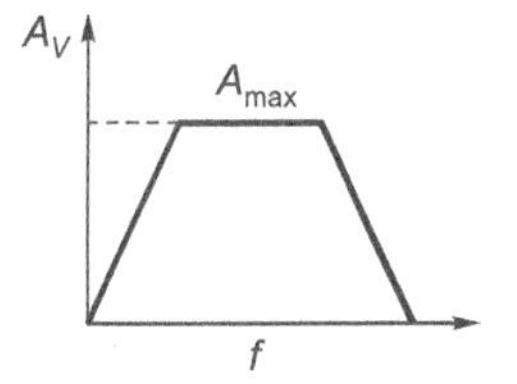

(b)

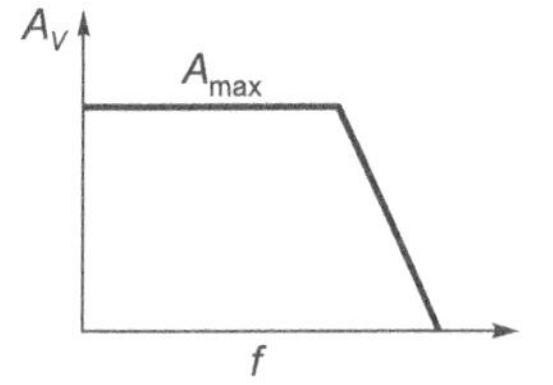

(c)

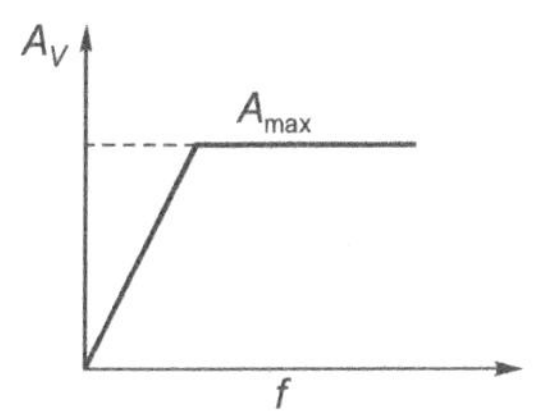

(d)

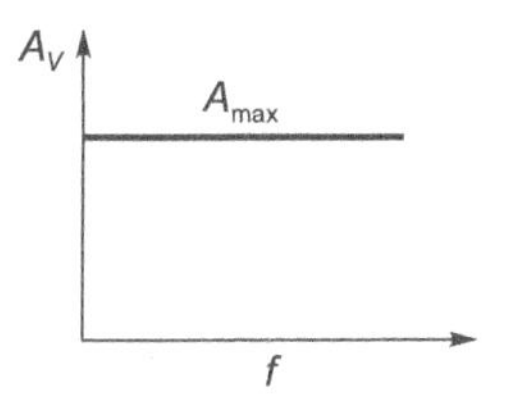

[ESE-2004]

6.5 In a single stage RC-coupled amplifier stage, what are the phase shifts introduced at lower and upper 3-dB frequencies, respectively?

(a) 45°, 225° (b) 45°, 135°

(c) 90°, 180° (d) 45°, 180°

[ESE-2005]

6.6 The gain of a bipolar transistor drops at high frequencies. This is because of the

(a) Coupling and bypass capacitors
(b) Early effect
(c) Inter-electrode transistor capacitances
(d) Coupling and bypass capacitors, and inter-electrode transistor capacitances

[ESE-2005]

6.7 Consider the following statements:
In frequency response of n-stage amplifiers,

1. upper cut-off frequency is $\sqrt{(2^{1/n}-1)}$ times that of a single stage
2. lower cut-off frequency is $\sqrt{(2^{1/n}-1)}$ times that of a single stage
3. lower cut-off frequency is $\sqrt{(2^{1/n}-1)^{-1}}$ times that of a single stage
4. upper cut-off frequency is $\sqrt{(2^{1/n}-1)^{-1}}$ times that of a single stage

Which of these statements are correct?
(a) 1 and 2 (b) 1 and 3
(c) 2 and 4 (d) 3 and 4

[ESE-2006]

6.8 Following transistor parameters were measured at I_c = 5 mA, V_{CE} = 10 V, h_{fe} = 100, $[A_1]$ = 10 at 10 M rad/s. What is the value for $(C_e + C_c)$, if the unity-gain frequency $\omega_T = \dfrac{g_m}{C_e + C_c}$
(a) 20 nF (b) 2 nF
(c) 5 pF (d) 5 nF

[ESE-2007]

6.9 A resistance R_f is connected across the collector and base of a BJT amplifier of gain $-A(A > 0)$. The input impedance of the amplifier will consist of transistor internal resistance $r_{b'e}$ shunted by which one of the following?
(a) $R_f(1 + A)$ (b) $R_f(1 - A)$
(c) $R_f/(1 + A)$ (d) $R_f/(1 - A)$

[ESE-2008]

6.10 What is the effect of cascading the amplifier stages?
(a) To increase the voltage gain and increase the bandwidth
(b) To increase the voltage gain and reduce the bandwidth
(c) To decrease the voltage gain and increase the bandwidth
(d) To decrease the voltage gain and reduce the bandwidth

[ESE-2008]

6.11 **Assertion** (A): In small-signal R-C coupled amplifiers the gain falls in the high frequency range.
Reason (R): The decline in gain with frequency is due to the reactance of the coupling capacitor.
(a) Both A and R are true and R is the correct explanation of A
(b) Both A and R are true but R is NOT the correct explanation of A
(c) A is true but R is false
(d) A is false but R is true

[ESE-2009]

6.12 **Assertion** (A): In a typical RC coupled amplifier, the gain falls at high frequencies.
Reason (R): The amplifier has to use coupling capacitors in the input and output side for isolating dc biasing circuit and ac signal.
(a) Both A and R are true and R is the correct explanation of A
(b) Both A and R are true but R is NOT the correct explanation of A
(c) A is true but R is false
(d) A is false but R is true

[ESE-2009]

6.13 Which one of the following is not true of direct coupled amplifiers?
(a) Low cost
(b) Can amplify high frequency signals
(c) Can amplify low frequency signals
(d) Operating point shifts with temperature variations

[ESE-2011]

6.14 Generally, the gain of transistor falls at high frequencies due to the
(a) internal capacitances of the device
(b) coupling capacitor at the input
(c) skin effect
(d) coupling capacitor at the output

[ESE-2012]

6.15 The bandwidth of an RC-coupled amplifier is limited by
(a) coupling capacitors at the low frequency end and bypass capacitors at the high frequency end.
(b) coupling capacitors at the high frequency end and bypass capacitors at tie low frequency end.

(c) bypass and coupling capacitors at the low frequency end and device shunt capacitors at the high frequency end.

(d) device shunt capacitors at the low frequency end and bypass as well as coupling capacitors at the high frequency end.

[ESE-2014]

6.16 Consider the following components in a multistage R-C coupled amplifier:

1. Parasitic capacitance of transistor
2. Coupling capacitance
3. Stray capacitance
4. Wiring capacitance

Which of the above components effectively control high frequencies?

(a) 1, 2 and 3 (b) 1, 2 and 4
(c) 1, 3 and 4 (d) 2, 3 and 4

[ESE-2017]

6.17 For most FET configurations and for common-gate configurations, the input impedances are respectively

(a) High and high (b) High and low
(c) Low and low (d) Low and high

[ESE-2019]

6.18 During a low frequency response of an amplifier which is invariably of RC–coupled type, there is a range of frequency characteristics over which the amplification is constant and delay is also constant, called

(a) Low band frequency
(b) Mid band frequency
(c) High band frequency
(d) Hyper band frequency

[ESE-2020]

6.19 Consider the following statements regarding R-C coupled amplifiers :

1. The cost of R-C coupled amplifier is low because of low cost of coupling capacitors and resistors.
2. They occupy less space because of small size of resistors and capacitors.
3. They have better frequency response.

Which of the above statements are correct?

(a) 1 and 2 only (b) 2 and 3 only
(c) 1, 2 and 3 only (d) 1 and 3 only

[ESE-2021]

■■■■

Answers Frequency Response

6.1 (b)	6.2 (b)	6.3 (a)	6.4 (b)	6.5 (a)	6.6 (c)	6.7 (b)	6.8 (c)	6.9 (c)
6.10 (b)	6.11 (c)	6.12 (b)	6.13 (b)	6.14 (a)	6.15 (c)	6.16 (c)	6.17 (b)	6.18 (b)
6.19 (c)								

Explanations Frequency Response

6.1 (b)

Assuming voltage Amplifier with zero input current

$$q_i = C(V_i - AV_i) \Rightarrow \frac{q_i}{V_i} = C(1-A)$$

6.2 (b)

$$f'_u = f_u\sqrt{2^{1/n}-1}$$

Here, $n = 2$ (for two amplifiers.)

6.3 (a)

$$f'_e = \frac{f_e}{\sqrt{2^{1/n}-1}}$$

and $n = 2$ for two amplifiers.

6.4 (b)

We know that due to coupling capacitor gain is reduced at low frequency, at high frequency it has no effect on gain so if coupling capacitor is shorted, gain will not be reduced at low frequency.

6.6 (c)

At high frequency junction capacitances offer finite impedance where as at low frequency they are open circuit.

So a part of current flows through it so actual input current decreases so o/p decreases so gain decreases.

6.7 (b)

As gain increases ⇒ BW decreases

Upper cutoff frequency decreases due to cascading as

$$f'_H = f_h\left[\sqrt{2^{1/n}-1}\right]$$

Lower cutoff frequency increases due to cascading

$$f'_l = \frac{f_l}{\sqrt{(2^{1/n}-1)}}$$

6.8 (c)

$$\text{Gain} \times \text{B.W.} = \frac{g_m}{C};\ g_m = \frac{I_C}{V_{CE}}$$

$$10 \times 10 \times 10^6 = \frac{5\text{ mA}}{10 \times (C_c + C_e)}$$

$$C_c + C_e = 5 \times 10^{-12}\text{ F}$$

6.9 (c)

As per Miller theorem, $R_i = \dfrac{R_f}{1 - A_v}$

Here, $A_v = -A$

So, $R_i = \dfrac{R_f}{1 + A}$

6.10 (b)

Due to cascading voltage gain

$$A_V = A_{V1} \cdot A_{V2}$$

A_{V1} – Voltage gain of 1st stage

A_{V2} – Voltage gain of 2nd stage

So it increases.

While upper cutoff frequency

$$f^*_H = f_h\sqrt{(2^{1/n}-1)} \text{ decreases}$$

While lower cutoff frequency

$$f^*_V = \frac{f_l}{\sqrt{(2^{1/n}-1)}}$$

increases compare to single state

So, B.W. = $(f^*_H - f^*_V)$ decreases

as f^*_H decreases and f^*_V increases.

6.11 (c)

Gain at high frequency falls due to internal junction capacitances so, R is false because due to coupling capacitor gain decreases at low frequencies not at high frequencies.

6.12 (b)

Both the statements are correct but R is not a explanation for A.

∵ Gain decreases at high frequency due to internal junction capacitances this is the reason of A.

6.13 (b)

Direct coupled amplifier are designed for amplification of dc and low frequency signal.

6.14 (a)

The gain of a transistor falls at high frequencies due to the various parasitic capacitances of the devices.

6.16 (c)

At high frequencies, internal capacitors like parasitic, stray, wiring capacitances effect. The frequency response. But coupling capacitor effect the frequency response at low frequency.

6.17 (b)

For CS and CD the input impedance Z_i is high, for CG the input impedance Z_i is low.

6.18 (b)

RC coupled amplifier frequency response

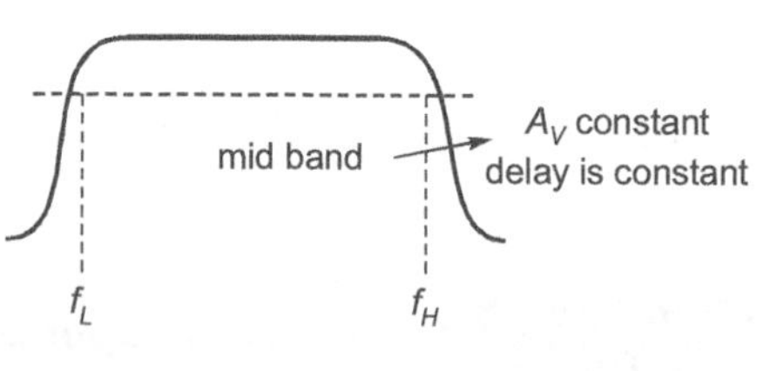

6.19 (c)

RC Coupled Amplifier :

1. The cost of RC coupled amplifier is low because of low cost of coupling capacitors and resistors.
2. They are not used in IC design, because they occupy large area on IC.
3. They have better frequency response.

■■■■

7 Feedback Amplifiers

7.1 **Assertion** (A): In small signal class 'A' amplifier, the output is a magnified replica of the input without any change in frequency.

Reason (R): The dc operating point is fixed in class 'A' position.

(a) Both A and R are true and R is the correct explanation of A
(b) Both A and R are true but R is NOT the correct explanation of A
(c) A is true but R is false
(d) A is false but R is true

[ESE-2001]

7.2 The open-loop voltage gain of an amplifier is 240. The noise level in the output without feedback is 100 mV. If a negative feedback with $\beta = 1/60$ is used, the noise level in the output will be

(a) 1.66 mV (b) 2.4 mV
(c) 4.0 mV (d) 20 mV

[ESE-2003]

7.3 An FET oscillator uses the given phase shift network as shown below. The minimum gain required for oscillation is

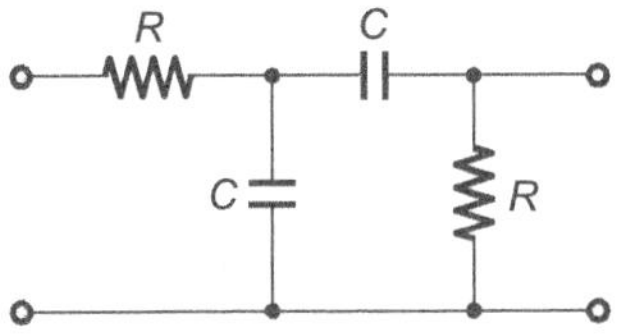

(a) –29 (b) 1
(c) 3 (d) 29

[ESE-2003]

7.4 Negative feedback in an amplifier leads to which one of the following?

(a) Decrease in bandwidth
(b) Increase in current gain
(c) Increase in voltage gain
(d) Decrease in voltage gain

[ESE-2007]

7.5 A negative feedback amplifier with open-loop gain

$$\frac{-A_0}{1+j\dfrac{\omega}{\omega_0}}.$$

$A_0 > 0$ and feedback factor $\beta (> 0)$ will have a 3 dB cut-off at what frequency?

(a) $\omega_0 A_0 \beta$ (b) $\omega_0(1 + A_0\beta)$
(c) $\omega_0/(1 + A_0\beta)$ (d) $w_0(1 - A_0\beta)$

[ESE-2008]

7.6 An amplifier without feedback, when fed with a 1 V, 50 Hz input signal gives an output of 30 V, 50 Hz with a 5% 2nd order distortion. When 10% of the output is feedback what is the 2nd order distortion?

(a) 0.375 (b) 1.3 V
(c) 0.75 V (d) 3 V

[ESE-2009]

7.7 For a transconductance amplifier, input and output resistances are respectively

(a) ∞ and 0 (b) 0 and ∞
(c) 0 and 0 (d) ∞ and ∞

[ESE-2011]

7.8 Three identical amplifiers, each having a gain of $\frac{A_0}{2}\angle 60°$ are connected in cascade. The positive feedback loop has a gain of 0.008. The value of A_o that will render the cascaded system oscillatory is

(a) 10 (b) –10
(c) 250/3 (d) 250

[ESE-2012]

7.9 **Statement (I):** In a negative feedback amplifier, the noise voltage due to noise generated within the amplifier is reduced.

Statement (II): Negative feedback reduces the amplifier gain.

(a) Both Statement (I) and Statement (II) are individually true and Statement (II) is the correct explanation of Statement (I).
(b) Both Statement (I) and Statement (II) are individually true but Statement (II) is not the correct explanation of Statement (I).
(c) Statement (I) is true but Statement (II) is false.
(d) Statement (I) is false but Statement (II) is true. [ESE-2012]

7.10 A feedback amplifier is designed with an amplifier gain of –1000 and feedback of $\beta = -0.1$. If the amplifier had a gain change of 20% due to temperature, the change in gain of the feedback amplifier is
(a) 10% (b) 5%
(c) 0.2% (d) 0.01% [ESE-2013]

7.11 A signal $V_m \sin(\omega t + \phi)$ is applied to an amplifier whose gain A is independent of frequency. The amplifier will preserve the form of the input signal (though with a delay) if the phase shift ϕ is
(a) constant
(b) inversely proportional to frequency
(c) proportional to frequency
(d) proportional to the square of the frequency
[ESE-2013]

7.12 In any function where microphone, amplifier and speakers are used often one would hear a humming sound, which increases in volume gradually. This is due to
(a) positive feedback between micro-phone and speaker
(b) negative feedback between micro-phone and speaker
(c) inadequate frequency response of amplifier
(d) noise pickup (50 Hz) from power supply
[ESE-2013]

7.13 The effect of current shunt feedback in an amplifier is to
(a) increase the input resistance and decrease the output resistance
(b) increase both input and output resistance
(c) decrease both input and output resistance
(d) decrease the input resistance and increase the output resistance
[ESE-2013]

7.14 An amplifier circuit has an overall current gain of –100 and an input resistance of 10 kΩ with a load resistance of 1 kΩ. The overall voltage gain of the amplifier is
(a) 5 dB (b) 10 dB
(c) 20 dB (d) 40 dB [ESE-2013]

Directions:
Each of the next items consists of two statements, one labelled as the **'Statement (I)'** and the other as **'Statement (II)'**. Examine these two statements carefully and select the answers to these items using the codes given below:

Codes:
(a) Both Statement (I) and Statement (II) are individually true and Statement (II) is the correct explanation of Statement (I).
(b) Both Statement (I) and Statement (II) are individually true but Statement (II) is not the correct explanation of Statement (I).
(c) Statement (I) is true but Statement (II) is false.
(d) Statement (I) is false but Statement (II) is true.

7.15 **Statement (I)** : A large negative feedback is deliberately introduced in an amplifier to make its gain independent of variation in parameters of active device and other components.
Statement (II) : A large negative feedback results in a high value of return difference compared to unity which makes the feedback gain inversely proportional to feedback factor.
[ESE-2015]

7.16 **Statement (I)** : An astable multivibrator is also called a free-running oscillator.
Statement (II) : An astable multivibrator circuit forces an operational amplifier to operate in the saturation region.
[ESE-2015]

7.17 In a voltage-series-feedback amplifier with open loop gain A_v and the feedback factor β, the input resistance becomes
(a) $\frac{R_i}{(1+\beta A_v)}$ (b) $R_i\sqrt{(1+\beta A_v)}$
(c) $R_i(1 + \beta A_v)$ (d) $\frac{R_i}{\sqrt{(1+\beta A_v)}}$
[ESE-2015]

7.18 In an amplifier with a gain of 1000 without feedback and cut-off frequencies at 2 kHz and 20 kHz, negative feedback of 1% is employed. The cut-off frequencies with feedback would be
(a) 220 Hz and 22 kHz
(b) 182 Hz and 220 kHz
(c) 220 kHz and 220 kHz
(d) 182 Hz and 22 kHz [ESE-2016]

7.19 Consider the following circuits:
1. Oscillator
2. Emitter follower
3. Power amplifier
Which of the above circuits employ feedback?
(a) 1 and 2 only (b) 2 and 3 only
(c) 1 and 3 only (d) 1, 2 and 3
[ESE-2016]

7.20 Applications of negative feedback to a certain amplifier reduced its gain from 200 to 100. If the gain with the same feedback is to be raised to 150, in the case of another such appliance, the gain of the amplifier without feedback must have been
(a) 400 (b) 450
(c) 500 (d) 600 [ESE-2018]

7.21 A negative feedback amplifier where an input current controls an output voltage is called
(a) Current amplifier
(b) Transconductance amplifier
(c) Transresistance amplifier
(d) Voltage amplifier [ESE-2019]

7.22 Which one of the following is **not** a distortion type that exists either separately or simultaneously in amplifiers?
(a) Linear distortion
(b) Non-linear distortion
(c) Frequency distortion
(d) Delay distortion [ESE-2020]

7.23 Transistor noise caused by the recombination and generation of carriers on the surface of the crystal is called
(a) Thermal noise (b) Excess noise
(c) White noise (d) Shot noise
[ESE-2020]

7.24 In a feedback amplifier, which configuration increases bandwidth, decreases non-linear distortion and improves transconductance with negative feedback ?
(a) Voltage-series (b) Current-series
(c) Voltage-shunt (d) Current-shunt
[ESE-2020]

7.25 In a paraphase amplifier, where two amplifiers are connected in cascade, the output from second stage
(a) Equals signal input without change of sign
(b) Equals signal input with change of sign
(c) Does not equal to signal input and has no sign change
(d) Does not equal to signal input but has sign change
[ESE-2020]

7.26 The power amplifiers in which the operating point is so adjusted that the collector current flows only during the positive half-cycle of the input signal are known as
(a) Class-B amplifiers
(b) Class-A amplifiers
(c) Class-AB amplifiers
(d) Class-C amplifiers [ESE-2021]

7.27 A multistage amplifier employs four stages, each of which has a voltage gain of 40. The overall gain of the amplifier is approximately
(a) 32 dB (b) 64 dB
(c) 96 dB (d) 128 dB [ESE-2021]

7.28 When the quiescent point of an amplifier is biased just at the cut-off axis, so that only the positive half of the signal input is amplified and the negative half of the signal is cut off, it is referred as
(a) class AB amplification
(b) class A amplification
(c) class B amplification
(d) class C amplification [ESE-2022]

7.29 **Statement (I)** : If the gain margin is negative, this gives the decibel rise in open-loop gain, which is theoretically permissible without oscillation.
Statement (II) : For a multistage amplifier, if the open-loop gain $|\beta A|$ is unity when the phase shift is 180°, then the closed-loop amplifier will oscillate.
[ESE-2022]

7.30 An amplifier has a bandwidth of 500 kHz and voltage gain of 100. What is the amount of negative feedback if the bandwidth is extended to 5 MHz?

(a) 9.0 (b) 0.9
(c) 0.09 (d) 0.009

[ESE-2023]

7.31 An amplifier has an open loop voltage gain of 1000. If 10% negative voltage series feedback is used, what is the closed loop voltage gain?

(a) 0.19 (b) 1.9
(c) 9.9 (d) 0.99

[ESE-2023]

7.32 What is the value of Q factor if the band pass filter is having frequencies from 800 Hz to 1200 Hz

(a) 1.30 (b) 2.45
(c) 3.40 (d) 4.35

[ESE-2023]

Answers Feedback Amplifiers

7.1 (b)	7.2 (d)	7.3 (c)	7.4 (d)	7.5 (b)	7.6 (a)	7.7 (d)	7.8 (b)	7.9 (a)
7.10 (c)	7.11 (c)	7.12 (a)	7.13 (d)	7.14 (c)	7.15 (a)	7.16 (b)	7.17 (c)	7.18 (b)
7.19 (a)	7.20 (d)	7.21 (c)	7.22 (a)	7.23 (b)	7.24 (b)	7.25 (a)	7.26 (a)	7.27 (d)
7.28 (c)	7.29 (c)	7.30 (c)	7.31 (c)	7.32 (b)				

Explanations Feedback Amplifiers

7.1 (b)

Both are correct since operating point for class A amplifier is at the middle of dc load line so as for whole cycle of input it work in linear range of transistor so as to work as amplifier so having characteristic like $y = kx$ where y is output and x is input so at the output frequencies at input will be present. But R is not the correct explanation of A.

7.2 (d)

We know that due to negative feedback effect of noise reduces and it is reduces by a factor $(1 + AB)$

where

$A \rightarrow$ gain without feedback

$B \rightarrow$ feedback factor

$$\therefore \text{ Noise level} = \frac{100 \text{ mV}}{(1+AB)}$$

$$= \frac{100 \text{ mV}}{1+240\times\frac{1}{60}} = \frac{100 \text{ mV}}{5}$$

$$= 20 \text{ mV}$$

7.3 (c)

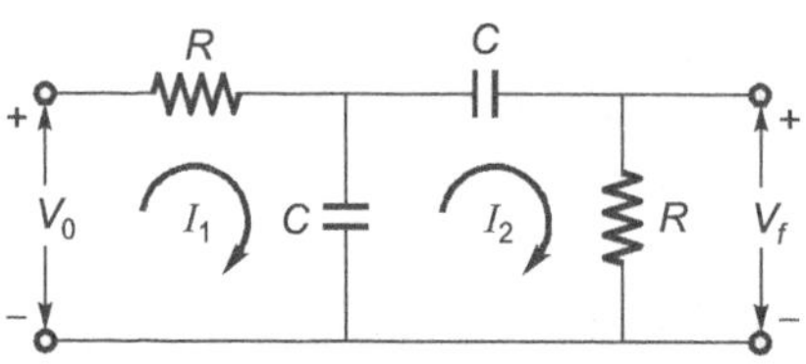

Apply KVL in first loop

$$V_0 = I_1\left(R-\frac{j}{\omega C}\right) + I_2\left(\frac{j}{\omega C}\right) \quad ...(i)$$

Apply KVL in second loop

$$0 = I_2\left(R-\frac{2j}{\omega C}\right) + I_1\left(\frac{j}{\omega C}\right)$$

$$I_2\left(R-\frac{2j}{\omega C}\right) = -I_1\frac{j}{\omega C} \quad ...(ii)$$

From equation (i) and (ii),

$$\frac{V_0}{I_2 R} = \frac{V_0}{V_f} = 3+\frac{j(1-x^2)}{x} \quad \text{where } x = \frac{1}{\omega RC}$$

For phase shift to be 180° (for oscillations)

$$\tan^{-1}\left(\frac{1-x^2}{3x}\right) = \pi \Rightarrow \frac{1-x^2}{3x} = 0$$

$$\Rightarrow \quad x = 1 \text{ i.e. } \omega = \frac{1}{RC}$$

$$\therefore \quad \frac{V_0}{V_f} = 3 \text{ at } \omega = \frac{1}{RC}$$

$$\therefore \quad \beta = \frac{V_f}{V_0} = \frac{1}{3}$$

Loop gain $\beta A \geq 1$ (for oscillations)

$$\therefore \quad A_{minimum} = \frac{1}{\beta} = 3$$

7.4 (d)

Negative feedback increases the BW by a factor $(1 + A\beta)$. Where as it reduces the gain.

7.5 (b)

Here cut-off frequency is asked but not mentioned that it is lower or upper. Generally we take upper cut off which increases due to negative feed back by a factor of $(1 + A\beta)$.

7.6 (a)

Given that without feedback input is

$$V_i = 1 \text{ V, 50 Hz}$$

and output is $V_0 = 30$ V, 50 Hz

So open loop gain

$$A_{OL} = \frac{V_0}{V_i} = \frac{30}{1} = 30$$

% feedback = 10%

So, $\beta = 0.1$

$\therefore$ gain with –ve feedback

$$= \frac{A_{OL}}{1 + A_{OL}\beta} = \frac{30}{1 + 30 \times 0.1} = 7.5$$

$\therefore$ 2nd harmonic distortion is reduced by a factor of $(1 + \beta A_{OL})$

$$= \frac{5}{(1 + A_{OL}\beta)} = \frac{5}{1 + 30 \times 0.1} = 1.25\%$$

So second harmonic distortion after feedback

$$= \text{distortion} \times V_0$$

$$= \frac{1.25}{100} \times 30 = 0.375 \text{ V}$$

7.7 (d)

For a transconductance amplifier input is voltage and output is current. So, at output there is series connection also at input there is series mixing, and for series connection impedance is high so input impedance and output impedance both are ∞, another explanation for this is; input is voltage so to avoid voltage drop due to source resistance input impedance of transconductance amplifier should be as high as possible ideally ∞.

$\because$ output is current, so in order to avoid bypass output current to output resistance, output resistance must be ∞ so that whole current flows through load R_L.

7.8 (b)

Gain of cascaded system = $A_1 \cdot A_2 \cdot A_3$

$$A = \left(\frac{A_0}{2}\angle 60^\circ\right)^3 = \frac{A_0^3}{8}\angle 180^\circ$$

From Barkhausen criteria.

$$|A\beta| = 1 \text{ and } \angle A\beta = \pm 180^\circ$$

$$\Rightarrow \quad \angle A = \pm 2n\pi$$

$$-\frac{A_0^3}{8} \times 0.008 = 1 \Rightarrow A_0 = -10$$

7.9 (a)

The negative feedback reduces the amplifier gain but it has a numerous advantages such as gain stability, reduction in non-linear distortion, reduction in noise, increase in bandwidth etc.

7.10 (c)

$$A_f = \frac{A}{1 + A\beta} \quad \ldots(i)$$

$$\frac{\partial A_f}{\partial T} = \frac{1}{(1 + A\beta)^2} \cdot \frac{\partial A}{\partial T} \quad \ldots(ii)$$

Equations $(i) \div (ii)$, we get

$$\frac{1}{A_f} \cdot \frac{\partial A_f}{\partial T} = \frac{1}{(1 + A\beta)} \times \left(\frac{1}{A} \cdot \frac{\partial A}{\partial T}\right)$$

Given,

$$\left(\frac{1}{\partial T} \times \frac{\partial A}{A}\right) = \text{gain change due to temperature}$$

So change in gain of feedback amplifier is

$$\frac{1}{\partial T} \frac{\partial A_f}{A_f} = \frac{1}{1 + (-1000 \times -0.1)} \times 20\%$$

$$= \frac{20\%}{1 + 100} = \frac{20\%}{101} \approx 0.2\%$$

7.11 (c)

Gain A is independent of frequency. So in order to preserve the form (shape) of input signal, the phase should be linear function of frequency.

Because for distortion less filter

$\alpha \neq$ function of frequency and $\beta \propto \omega$

where α is magnitude and β are represent the phase.

7.12 (a)

Input to microphone (mic) is voice, which is amplified by amplifier and this amplified voice is coming out from speaker, which again work as input to mic, this amplified signal again amplified by amplifier and coming out from speaker and again go to mic, this is cumulative process and example of positive feedback between microphone and speaker.

7.14 (c)

$$|\text{Voltage gain}| = \left|\text{Current gain} \times \frac{R_L}{R_i}\right|$$

$$= \left|-100 \times \frac{1}{10}\right| = 10$$

$\therefore$ Gain in dB = 20 log 10 = 20 dB

7.16 (b)

Both the statements are correct and statement-II is not reason of statement-I.

7.17 (c)

In voltage series feedback the input impedance increases by factor of $(1 + \beta A_v)$, so input impedance will be $R_i(1 + \beta A_v)$.

7.18 (b)

$$f_L^1 = \frac{f_L}{1 + A\beta} = \frac{2 \times 10^3}{1 + 1000 \times 0.01}$$

$$= 182 \text{ Hz}$$

$$f_H^1 = f_H(1 + A\beta)$$

$$= 20 \times 10^3 \times 11 = 220 \text{ kHz}$$

7.20 (d)

Open loop gain, $A = 200$

Closed loop gain,

$$A_f = 100$$

$$A_f = \frac{A}{1 + A\beta}$$

$$100 = \frac{200}{1 + 200\beta}$$

$$100 + 20000\,\beta = 200$$

$$20000\,\beta = 100$$

$$\beta = \frac{1}{200} = 0.005$$

$$A_f' = 150$$

$$A' = ?$$

$$A_f' = \frac{A'}{1 + A'\beta}$$

$$150 = \frac{A'}{1 + A' \times 0.005}$$

$$150 + A' \times 0.75 = A'$$

$$150 = A'(1 - 0.75) = A'(0.25)$$

$$A' = \frac{150}{0.25} = 600 \text{ (Open loop)}$$

7.21 (c)

CCVS, $A = \frac{V_0}{I_i}$ (Transfer f_n)

= Transresistance amplifier

7.22 (a)

Linear distortion is not a distortion type in Amplifiers.

7.23 (b)

The noise which occurs at low frequencies varies approximately as $1/f$, and is called excess or flicker noise. The source of this noise is recombination and generation of carries an the surface of crystal.

7.24 (b)

To improve transconductance current series is used.

7.25 (a)

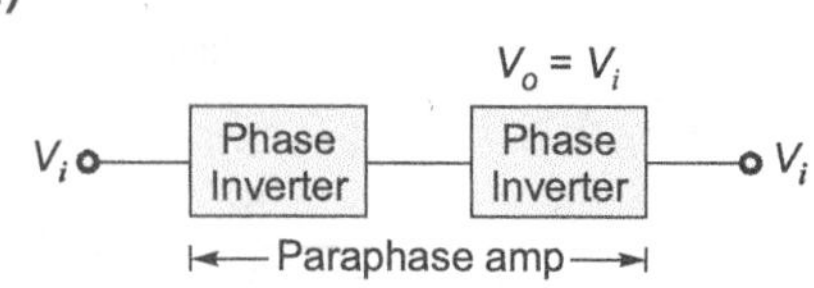

7.26 (a)

Power amplifier in which operating point is so adjusted that collector current flows only during positive half cycle is class B amplifiers.

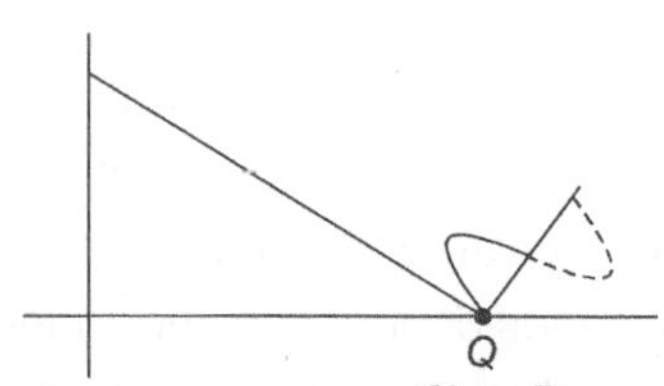

Class B

7.27 (d)

Given multistage amplifier has four stages. Each has voltage gain 40.

$$\text{Overall gain} = 40 \times 40 \times 40 \times 40 = 256 \times 10^4$$

$$A_V|_{dB} = 20\log(256 \times 10^4) = 20\log 256 + 20\log 10^4 = 20\log 256 + 80 = 48.16 + 80 \simeq 128 \text{ dB}$$

7.28 (c)

When the operating or quiescent point of an amplifier is biased at cut-off

Then it is class B amplification.

7.29 (c)

Gain margin = –[Open loop gain] dB

Statement I : $A\beta \geq 1$ or 0 dB (True)

Gain margin < 1 or 0 dB

Statement II : Multistage amp. if $A\beta = 1$ and phase shift is 180° closed loop amp. will oscillate. (False)

7.30 (c)

An amplifier,

Bandwidth, $BW = 500$ kHZ

$$A_V = 100$$

$$(BW)_F = 5 \text{ MHz}$$

$$A_F(BW)_F = A_V BW$$

$$A_F = \frac{A_V \cdot BW}{(BW)_F} = \frac{100 \times 500 \text{ kHz}}{5 \text{ MHz}} = 10$$

$$A_F = \frac{A_V}{1 + A_V\beta}$$

$$10 = \frac{100}{1 + 100\beta}$$

$$1 + 100\beta = 10$$

$$100\beta = 9$$

$$\beta = \frac{9}{100} = 0.09$$

7.31 (c)

Open loop voltage gain, $A = 1000$

$$\beta = \frac{10}{100} = 0.1$$

$$A_F = \frac{A}{1 + A\beta} = \frac{1000}{1 + 1000 \times 0.1} = \frac{1000}{1 + 100} = \frac{1000}{101} = 9.9$$

7.32 (b)

$$Q = \frac{\text{Centre frequency}}{\text{BW}} = \frac{\sqrt{f_{c1} \cdot f_{c2}}}{\text{BW}} = \frac{\sqrt{f_{c1} \cdot f_{c2}}}{f_{c1} - f_{c2}} = \frac{\sqrt{800 \times 1200}}{400} = 2.45$$

Alternate Solution:

$$f_L = 800 \text{ Hz}$$

$$f_H = 1200 \text{ Hz}$$

$$\text{BW} = f_H - f_L = 1200 - 800 = 400 \text{ Hz}$$

$$\frac{f_H}{f_L} \geq 1.1$$

$$f_C = \sqrt{f_L f_H} = \sqrt{1200 \times 800} = 979 \text{ Hz}$$

$$\text{Quality factor, } Q = \frac{f_C}{\text{BW}} = \frac{979\text{Hz}}{400\text{Hz}} = 2.45$$

■■■■

8 Oscillators

8.1 A sinusoidal waveform can be converted to a square waveform by using a
(a) two stage transistorized overdriven amplifier
(b) two stage diode detector circuit
(c) voltage comparator based on op-amp
(d) regenerative voltage comparator circuit
[ESE-2001]

8.2 Consider the following bistable multi-vibrator circuit?

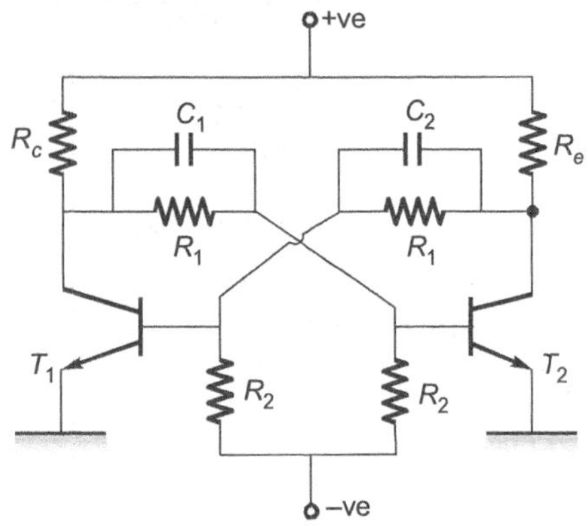

Which one of the following statements is correct? In the above circuit, condensers C_1 and C_2 are used
(a) for passing the noise
(b) to compensate the collector-emitter capacitance
(c) to speed up the switching action
(d) to provide negative feedback path
[ESE-2004]

8.3 Match **List-I** with **List-II (Application)** and select the correct answer using the codes given below:

List-I
A. Monostable multivibrator
B. Bistable multivibrator translator
C. Clamping circuit
D. Schmitt trigger

List-II
1. Comparator
2. D.C. level
3. Delay
4. Voltage controlled oscillator
5. Counter

Codes:

	A	B	C	D
(a)	4	3	1	2
(b)	3	5	2	1
(c)	4	5	2	1
(d)	5	4	1	2

[ESE-2004]

8.4 Consider the following statements:
1. A Hartley oscillator circuit uses a tapped inductor for inductive feedback.
2. Oscillator circuit can be operated in class A condition for better wave shape.
3. Frequency stabilization is obtained by use of automatic biasing.

Which of these statements are correct?
(a) 1, 2 and 3
(b) 1 and 2
(c) 2 and 3
(d) 1 and 3
[ESE-2005]

8.5 **Assertion** (A): Emitter coupled oscillator is capable of high frequency operation
Reason (R): It consists of saturating NPN BJTs.
(a) Both A and R are true and R is the correct explanation of A
(b) Both A and R are true but R is NOT the correct explanation of A
(c) A is true but R is false
(d) A is false but R is true
[ESE-2005]

8.6 **Assertion** (A): An FET operated crystal oscillator operates on the concept of feedback.
Reason (R): The feedback is provided by the drain-to-gate capacitance C_{dg}.

(a) Both A and R are true and R is the correct explanation of A
(b) Both A and R are true but R is NOT the correct explanation of A
(c) A is true but R is false
(d) A is false but R is true

[ESE-2005]

8.7 An amplifier without feedback has a gain of 1000. What is the gain with a negative feedback of 0.009?
(a) 900 (b) 125
(c) 100 (d) 10

[ESE-2007]

8.8 Consider the following statements in respect of the Wien bridge oscillator shown in the figure below:

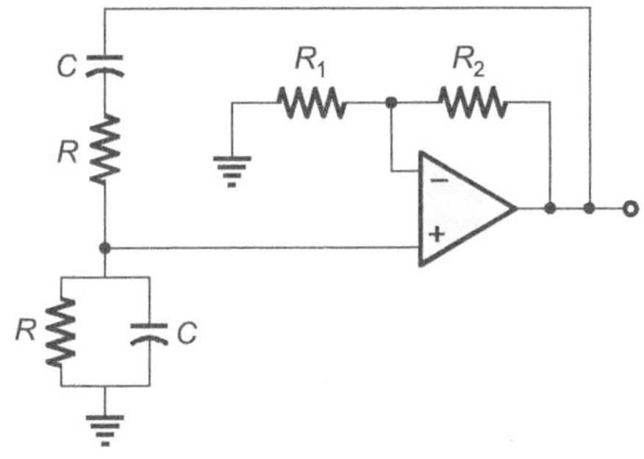

1. For $R = 1$ kilo-ohm

$$C = \left(\frac{1}{2\pi}\right) \mu\text{F}, f = 1\text{kHz}$$

2. For $R = 3$ kilo-ohms

$$C = \left(\frac{1}{18\pi}\right) \mu\text{F}, f = 3\text{kHz}$$

Which of these statements is/are correct?
(a) 1 only (b) 2 only
(c) Both 1 and 2 (d) Neither 1 nor 2

[ESE-2008]

8.9 Consider the following statements:
1. Wein bridge oscillator is suitable for generating 1 kHz.
2. Colpitts oscillator is suitable for generating 1 MHz.

Which of these statements is/are correct?
(a) 1 only (b) 2 only
(c) Both 1 and 2 (d) Neither 1 nor 2

[ESE-2008]

8.10 The highest frequency stability is achieved by using an oscillator of the type
(a) Colpitts (b) Crystal controlled
(c) Hartley (d) RC oscillator

[ESE-2011]

8.11 One shot multi-vibrator, with a pulse input gives an output:
(a) A single triangular pulse
(b) A square pulse
(c) An impulse
(d) A single sinusoid pulse

[ESE-2011]

Directions:
Each of the next items consists of two statements, one labelled as the **'Statement (I)'** and the other as **'Statement (II)'**. Examine these two statements carefully and select the answers to these items using the codes given below:

Codes:
(a) Both Statement (I) and Statement (II) are individually true and Statement (II) is the correct explanation of Statement (I).
(b) Both Statement (I) and Statement (II) are individually true but Statement (II) is not the correct explanation of Statement (I).
(c) Statement (I) is true but Statement (II) is false.
(d) Statement (I) is false but Statement (II) is true.

8.12 **Statement (I):** In tuned amplifier, a high gain is achieved at a narrow band of frequencies.
Statement (II): A high-Q parallel resonant circuit is used as load in the amplifier.

[ESE-2012]

8.13 **Statement (I):** A feedback amplifier can oscillate under certain conditions.
Statement (II): Loop gain is infinity and phase shift is zero during oscillation.

[ESE-2012]

8.14 **Statement (I):** An RC phase shift oscillator satisfies Barkhausen criteria of oscillations.
Statement (II): An RC phase shift oscillator must use 3 equal valued resistors and 3 equal valued capacitors.

[ESE-2012]

8.15 The pulse width out of a one slot multivibrator increases when the

(a) supply voltage increases
(b) timing resistor decreases
(c) UTP decreases
(d) timing capacitance increases **[ESE-2014]**

8.16 In order to generate a square waveform a sinusoidal input signal, one can use

1. Schmitt trigger circuit
2. Clippers and amplifiers
3. Monostable multivibrator

Which of the above statements are correct?

(a) 1, 2 and 3 (b) 1 and 2 only
(c) 1 and 3 only (d) 2 and 3 only

[ESE-2015]

8.17 Which of the following circuits is used for converting a sine wave into a square wave?

(a) Monostable multivibrator
(b) Bistable multivibrator
(c) Schmitt trigger circuit
(d) Darlington complementary pair **[ESE-2016]**

8.18 In an R-C phase shift oscillator using FET and 3-section R-C phase shift network, the condition for sustained oscillation is

When, $n = \frac{R_d}{R}$

(a) $\beta > 6n$
(b) $\beta > 29$
(c) $\beta > 4n + 23 + \frac{29}{n}$
(d) $\beta > 23 + \frac{29}{n}$ **[ESE-2017]**

8.19 A tuned-collector oscillator has a fixed inductor of 100 μH and has to be tunable over the frequency band of 500 kHz to 1500 kHz. What is the range of variable capacitor to be used?

(a) 115 - 1021 pF (b) 113 - 1015 pF
(c) 93 - 1015 μF (d) 110 - 1021 μF

[ESE-2017]

8.20 For RC phase shift oscillator using FET, the gain of the amplifier stage must be practically somewhat greater than

(a) 27 (b) 28
(c) 29 (d) 30 **[ESE-2019]**

8.21 In a crystal oscillator, especially when piezoelectric crystal like quartz is applied, then the inductor L, capacitor C and resistor R are the analogs of the mechanical system as

(a) Mass, compliance, viscous-damping factor
(b) Mass, spring constant, viscous-damping
(c) Mass, momentum, viscous-damping factor
(d) Mass, displacement, viscous-damping factor

[ESE-2020]

8.22 In a phase shift oscillator using an FET, at a certain frequency if the phase shift introduced by the RC network is 180°, then the total phase shift from gate around the circuit and back to the gate will be

(a) 0° (b) 90°
(c) 180° (d) 270°

[ESE-2020]

8.23 Which one of the following statements is not correct related to oscillators?

(a) The frequency of a sinusoidal oscillator is determined by the condition that the loop-gain phase shift is zero.
(b) In every practical oscillator, the loop gain is slightly larger than unity and the amplitude of the oscillations is limited by onset of non-linearity.
(c) The condition of unity loop gain $-A\beta = 1$ is called the Barkhausen criterion.
(d) Oscillations will be sustained if, at the oscillator frequency, the magnitude of the product of the transfer gain of the amplifier and the magnitude of the feedback factor the feedback network are less than unity.

[ESE-2022]

8.24 Which one of the following is correct in the feedback circuit, having three resistance and capacitance elements with conditions as $R_1 = R_2 = R_3 = \bar{R}$ and $C_1 = C_2 = C_3 = \bar{C}$ for phase-shift oscillator?

(a) $f = \frac{0.065}{\bar{R}\bar{C}}$ (b) $f = \frac{0.078}{\bar{R}\bar{C}}$
(c) $f = \frac{0.045}{\bar{R}\bar{C}}$ (d) $f = \frac{0.038}{\bar{R}\bar{C}}$

[ESE-2022]

■■■■

Answers Oscillators

8.1 (d)	8.2 (c)	8.3 (b)	8.4 (b)	8.5 (c)	8.6 (a)	8.7 (c)	8.8 (c)	8.9 (c)
8.10 (b)	8.11 (b)	8.12 (a)	8.13 (c)	8.14 (c)	8.15 (d)	8.16 (c)	8.17 (c)	8.18 (*)
8.19 (b)	8.20 (c)	8.21 (b)	8.22 (a)	8.23 (d)	8.24 (a)			

Explanations Oscillators

8.1 (d)
Using Schmitt trigger circuit (i.e. regenerative voltage comparator) we can convert sinusoidal waveform into square wave form.

8.2 (c)
These are the commutating capacitors and assist the binary in making abrupt transition b/w the states so speed up the switching action so also known as speed up capacitors.

8.3 (b)
Monostable multi-vibrator is used to introduce delay.
Bistable multi-vibrator translator is nothing but a counter.
Clk + translator = counter
Clamping circuit → dc restorer
Schimit trigger → covert any i/p waveform in square waveform of duty cycle 50% or other than 50% and it is basically regenerative comparator circuit.

8.4 (b)
For frequency stabilization biasing is done such that operating point lies in linear range of transistor characteristic i.e. near to middle of DC load line not by automatic biasing.

8.5 (c)
Emitter coupled oscillator is capable of high frequency operation because it consists of non-saturating NPN BJTs.

8.6 (a)
Any oscillator is combination of amplifier with –ve feedback without any external input.
An in case of FET operated crystal oscillator feedback is given by C_{dg} i.e. drain to gate capacitance.

8.7 (c)

$$A_{if} = \frac{A}{1+A\beta}; \qquad \beta = \text{feedback factor}$$

A = gain without feedback
A_{if} = gain with feedback

8.8 (c)

$$f = \frac{1}{2\pi RC}$$

For $C = \frac{1}{2\pi}\mu F$, $R = 1\ k\Omega$, $f = 1$ kHz

For $C = \frac{1}{18\pi}\mu F$, $R = 3\ k\Omega$;

$$f = \frac{1}{2\pi \times 3 \times \frac{1}{18\pi}} = 3\text{ kHz}$$

Hence both statement correct.

8.9 (c)
Wein bridge oscillator is audio frequency oscillator used for frequency ≤ 20 kHz.
While Colpitts oscillators are used at RF frequencies ≥ 20 kHz.

8.10 (b)
Stability factor of crystal controlled oscillator is highest.

8.11 (b)
One shot multivibrator is monostable multivibrator with pulse input it will generate a square pulse of finite width.

8.12 (a)
Tuned amplifiers are amplitude inverting a resonant circuit, and are intended for selective amplification within a narrow band of frequency. Radio and TV amplifiers employ tuned amplifiers to select one broadcast channel from among the many.

8.13 (c)

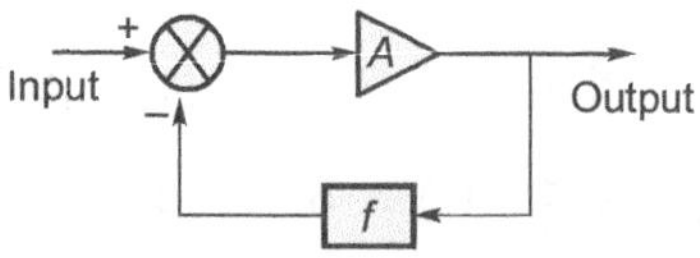

If the signal, after making a round trip through A and f, has a gain of 1 and a phase shift of exactly 360°, then it oscillates oscillation criteria.

1. $|A_f| = 1$ 2. $\angle A_f = \pm 180°$

But, in the negative feedback system, there is already a 180° phase shift. Therefore, the phase shift caused by A and f together will only need to be 180° to cause oscillation.

8.14 (c)

RC phase shift oscillator satisfies Barkhausen criteria.

$|A\beta| = 1$ and $\angle A\beta = \pm 180°$

By cascading three or even four RC stages together ($4 \times 45°$), the stability of the oscillator can be greatly improved.

8.19 (b)

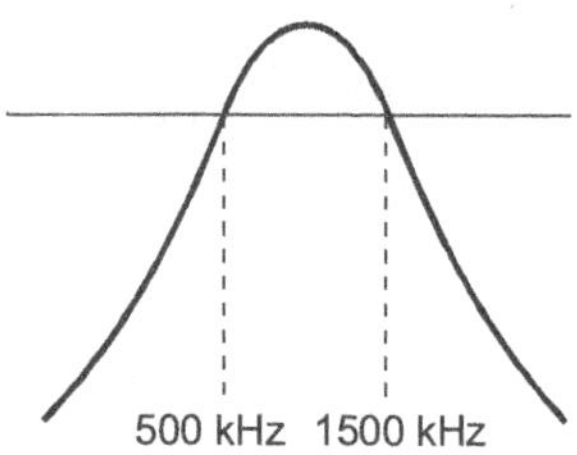

$$f_L = \frac{1}{2\pi\sqrt{LC_H}}$$

$$C_H = 1014.23 \text{ pF}$$

$$f_H = \frac{1}{2\pi\sqrt{LC_L}}$$

$$C_L = 112.6 \text{ pF}$$

8.20 (c)

For RC phase shift oscillator,

$$|A_V| = 29$$

8.21 (b)

Mass, spring constant, viscous-damping factor.

8.22 (a)

0° (zero degree), overall phase shift will be 0° or 2π

8.23 (d)

In relation to oscillators

$A\beta < 1$ is not correct

∴ (d) option is right choice.

8.24 (a)

$$R_1 = R_2 = R_3 = \bar{R}$$

$$C_1 = C_2 = C_3 = \bar{C}$$

RC phase shift oscillator

$$f = \frac{1}{2\pi RC\sqrt{6}}$$

$$= \frac{1}{2 \times 3.14 \times 2.44 \times \bar{R}\bar{C}}$$

$$= \frac{1}{15.32\bar{R}\bar{C}} = \frac{0.065}{\bar{R}\bar{C}}$$

■■■■

UNIT

VI

Computer Fundamentals

Syllabus

Number systems, Boolean algebra, arithmetic functions, Basic Architecture, Central Processing Unit, I/O and Memory Organisation; peripheral devices, data represenation and programming, basics of Operating system and networking, virtual memory, file systems; Elements of programming languages, typical examples.

Contents

OOOO

1 Data Representation

1.1 What is the octal equivalent of $(5621.125)_{10}$?
(a) 11774.010 (b) 12765.100
(c) 16572.100 (d) 17652.010 **[ESE-2017]**

1.2 What is the hexadecimal representation of $(657)_8$?
(a) 1 AF (b) D 78
(c) D 71 (d) 32 F **[ESE-2017]**

1.3 The decimal value 0.5 in IEEE single precision floating point representation has fraction bits of
(a) 000...000 and exponent value of 0
(b) 000...000 and exponent value of –1
(c) 100...000 and exponent value of 0
(d) 100...000 and exponent value of –1
[ESE-2018]

1.4 The decimal equivalent of binary number 1001.101 is
(a) 9.750 (b) 9.625
(c) 10.750 (d) 10.925 **[ESE-2019]**

1.5 Convert decimal 41.6875 into equivalent binary:
(a) 100101.1011 (b) 100101.1101
(c) 101001.1011 (d) 101001.1101
[ESE-2019]

1.6 The decimal value of signed binary number 11101000 expressed in 1's complement is
(a) –223 (b) –184
(c) –104 (d) –23 **[ESE-2019]**

1.7 **Statement-I:** Sign magnitude representation is generally used in implementing the integer portion of the ALU.
Statement-II: In sign magnitude representation there are two representation of 0.
(a) Both Statement (I) and Statement (II) are individually true and Statement (II) is the correct explanation of Statement (I)
(b) Both Statement (I) and Statement (II) are individually true; but Statement (II) is **not** the correct explanation of Statement (I)
(c) Statement (I) is true; but Statement (II) is false
(d) Statement (I) is false; but Statement (II) is true
[ESE-2019]

1.8 The hexadecimal of the binary number $(11010011)_2$ is
(a) $D3_{16}$ (b) $D4_{16}$
(c) $C3_{16}$ (d) $C4_{16}$ **[ESE-2020]**

1.9 Which of the following factors are to be considered while selecting number representations to be used in a computer?
1. Number types to be represented
2. Range of values to be encountered
3. Cost of the hardware to store and process the numbers
4. Positional notation with fixed weight

(a) 1,2 and 4 only (b) 1,3 and 4 only
(c) 2, 3 and 4 only (d) 1,2 and 3 only
[ESE-2020]

1.10 If a negative binary number is to be represented by n–bits, then the standard format will be
(a) Sign bit '0' on left and magnitude right
(b) Sign bit '1' on left and magnitude on right
(c) Sign bit '0' on right and magnitude on left
(d) Sign bit '1' on right and magnitude on left
[ESE-2020]

1.11 What is the exponent length in bits for IEEE double-precision floating point format?
(a) 32 bits (b) 8 bits
(c) 16 bits (d) 11 bits **[ESE-2023]**

Answers Data Representation

1.1 (b) 1.2 (a) 1.3 (b) 1.4 (b) 1.5 (c) 1.6 (d) 1.7 (d) 1.8 (a) 1.9 **(a)**
1.10 (b) 1.11 (d)

Explanations Data Representation

1.1 (b)

8	5621	
8	702	5
8	87	6
8	10	7
8	0	2
		1

$(5621.125)_{10} = (?)_8$

Conversion of Integer part

$\Rightarrow \quad (5621)_{10} = (12765)_8$

Conversion of fractional part

$0.125 \times 8 = 1.000$

$\therefore \quad (5621.125)_{100} = (12765.100)_8$

1.2 (a)

$(657)_8 = (?)_{16}$

$(657)_8 = (110\ 101\ 111)_2$

$S = (1\ AF)_{16}$

$\left(\underbrace{0001}_{1}\ \underbrace{1010}_{A}\ \underbrace{1111}_{F}\right)_2 = (1\ AF)_{16}$

Alternative method:

$(657)_8$ = 1 AF H as 657 oct

= 11010111 in binary

So take group to 4 bits then it would be 1 AF in hex.

1.3 (b)

Raw data:

$(0.5)_{10}$
↓
Binary data
↓
(0.1)
↓
Normalized data (1.*bbb*...)
↓
Align to left upto 1 time
↓
$(1.0) \times 2^{-1}$

So, fraction (*M*) is all 0's and actual exponent is –1.

1.4 (b)

$1001.101 = 1 \times 2^3 + 1 \times 2^0 + 1 \times 2^{-1} + 1 \times 2^{-3}$

$= 9.625$

1.5 (c)

41.6875

41 ⇒ 32 16 8 4 2 1
     1 0 1 0 0 1

$0.6875 \times 2 = 1.375$

$0.375 \times 2 = 0.75$

$0.75 \times 2 = 1.50$

$0.50 \times 2 = 1.00$

.1011

41.6875 ⇒ 101001.1011

1.6 (d)

11101000 (–23)

11101000 ⟶ 00010111 (23)

Option (d) is correct.

1.7 (d)

Statement-II is correct and statement-I is false.

1.8 (a)

To convert a binary number into hexadecimal, we start from the LSB binary digit and from groups of 4 binary digits. Next, we write the hexadecimal equivalent of each group of 4 bits.

$\underbrace{1101}_{D}\ \underbrace{0011}_{3}$

1.10 (b)

In signed binary number representation, the MSB represents the sign of the number and remaining bits represent the magnitude of the number.

Sign bit	Magnitude bits

Moreover, if the number is positive the sign bit = 0; and if the number is negative, the sign bit = 1.

1.11 (d)

Double precision (64 bit) format

64 bit

S	BE	M
1 bit	11 bit	52 bit

■■■■

2 Computer Architecture

2.1 Consider the following statements regarding Computer Architecture:

1. The advantage with dedicated bus is decrease in size and cost.
2. In synchronous timing, the occurrence of events on the bus is determined by the clock.
3. Data bus width decides the number of bits transferred at one time.

Which of the above statements are correct?

(a) 1 and 2 only (b) 1 and 3 only
(c) 2 and 3 only (d) 1, 2 and 3 **[ESE-2018]**

2.2 Which of the following design levels of a computer are widely used in computer design?

1. Gate level 2. Processor level
3. Register level 4. User level

(a) 1 and 3 only (b) 2 and 4 only
(c) 3 and 4 only (d) 1, 2 and 3 only **[ESE-2020]**

2.3 Consider the following statements for a simple assembler :

1. It scans the entire assembly program twice, where each scan is called a pass.
2. It generates a table that includes all symbols and their binary values.
3. It will use the symbol table and other tables to generate the object program and output some information that will be needed by the linker.

Which of the above statements are correct?

(a) 1 and 2 only (b) 1 and 3 only
(c) 2 and 3 only (d) 1, 2 and 3 **[ESE-2021]**

2.4 Which one of the following controllers is to check the status of each device and inform the centre processing unit of the status of each?

(a) Programmable I/O interrupt controller
(b) DMA controller
(c) Disk controller
(d) Pipeline controller **[ESE-2022]**

2.5 Which one of the following bus architectures is used to maximize throughput of video graphics memory?

(a) EISA bus (b) VESA bus
(c) PCI bus (d) MCA bus **[ESE-2022]**

2.6 Which one of the following bus architectures is having maximum data rate (MB/s)?

(a) VESA bus (b) PCI-64 bus
(c) MCA bus (d) EISA bus **[ESE-2022]**

2.7 Which one of the following layers in PCI express protocol architecture is used for compatibility with PCI, initialization and enumeration of the devices connected to the PCI express?

(a) PCI express physical layer
(b) Software layer
(c) Data link layer
(d) Hardware layer **[ESE-2022]**

2.8 Which of the following statements are correct regarding multiprocessing architecture?

1. It can cause a system to change its memory access model from uniform memory access to non-uniform memory access.
2. There are two types of systems such a asymmetric multiprocessing and symmetric multiprocessing.
3. It adds CPUs to decrease computing power.

Select the correct answer using the code given below:

(a) 1 and 3 only (b) 1 and 2 only
(c) 2 and 3 only (d) 1, 2 and 3 **[ESE-2022]**

2.9 A benchmark program is run on a 200 MHz processor. The executed program consists of 1 million instruction executions, with the following instruction mix and clock cycle count :

Instruction Type	Instruction Count (Million)	Cycles per Instruction
Arithmetic and logic	8	1
Load and store	4	3
Branch	2	4
Others	4	3

What is the effective MIPS rate for the given program?

(a) 90 (b) 70
(c) 110 (d) 80 [ESE-2023]

2.10 Which one of the following statement(s) is/are correct for machine parallelism?

(a) It is a measure of the ability of the processor to take disadvantage of instruction-level parallelism.
(b) It is determined by the frequency of true data dependencies and procedural dependencies in the code.
(c) It is determined by the number of instructions that can be fetched and executed at the same time and by the speed and sophistication of the mechanisms that the processor uses to find independent instructions.
(d) It is determined by the time until the result of an instruction is available for use as an operand in a subsequent instruction.

[ESE-2023]

2.11 Which one of the following methods is used when the processor issues an I/O command, continues to execute other instructions, and is interrupted by the I/O module when the latter has completed its work?

(a) Interrupt-driven I/O
(b) I/O device controller
(c) I/O channel processor controller
(d) Programmed I/O [ESE-2023]

2.12 Consider the following characteristics of Complex Instruction Set Computer (CISC) processor :

1. It is having a small number of instructions.
2. It is having less addressing modes.
3. Most instructions can manipulate operands in the memory.
4. Control unit is microprogrammed.

Which of the above characteristics is/are **not** correct?

(a) 1 only (b) 1 and 2
(c) 1 and 4 (d) 3 and 4 [ESE-2024]

2.13 In which of the following processors, the designer can add easily new instruction without changing the architecture of the processor?

(a) CISC processor only
(b) RISC processor only
(c) Both CISC and RISC processors
(d) Neither CISC processor nor RISC processor

[ESE-2024]

2.14 Which one of the following is used to increase the performance of CPU, that means executing more instructions in less time?

(a) Sequencing (b) Pipelining
(c) Scheduling (d) Spooling [ESE-2024]

2.15 In Double Data Rate SDRAM, if 100 MHz clock rate and 64 bits data bus transfers data, then what is the approximate transfer rate for DDR3?

(a) 1600 MB/s (b) 6400 MB/s
(c) 3200 MB/s (d) 800 MB/s [ESE-2024]

2.16 What is the approximate rotational delay if the disk drive has 8 surfaces, each surface has 1024 tracks, each track has 64 sectors, each sector can hold 512 bytes and rotation speed of 6000 r.p.m.? (It is assumed that the sector is away from head half of the track)

(a) 0.005 s (b) 0.050 s
(c) 0.500 s (d) 0.505 s [ESE-2024]

2.17 Consider the following set of processes that arrive at time 0 with the length of the CPU burst given in milliseconds:

Process	Burst time
P_1	24
P_2	3
P_3	3

What is the average waiting time under the Round-Robin Scheduling, if we use a time quantum of 4 milliseconds?

(a) 5.66 milliseconds (b) 3.50 milliseconds
(c) 7.00 milliseconds (d) 6.55 milliseconds

[ESE-2024]

2.18 Consider the following set of processes, assumed to have arrived at time 0 in the order P_1, P_2, ..., P_5, with the length of the CPU burst given in milliseconds:

Process	Burst time	Priority
P_1	10	3
P_2	1	1
P_3	2	4
P_4	1	5
P_5	5	2

Assuming that low numbers represent high priority, what is the average waiting time under the priority scheduling?

(a) 5.66 milliseconds (b) 4.50 milliseconds
(c) 8.20 milliseconds (d) 6.55 milliseconds

[ESE-2024]

■■■■

Answers Computer Architecture

2.1 (c)	2.2 (d)	2.3 (d)	2.4 (a)	2.5 (c)	2.6 (b)	2.7 (a)	2.8 (b)	2.9 (a)
2.10 (c)	2.11 (a)	2.12 (b)	2.13 (a)	2.14 (b)	2.15 (a)	2.16 (a)	2.17 (a)	2.18 (c)

Explanations Computer Architecture

2.1 (c)

- Separate buses increases the cost and complexity.
- Clock based transmission is called as synchronous transmission.
- Data bus width is the word length of a CPU.

2.3 (d)

All are true.

2.4 (a)

Programmable I/O interrupt controller monitors the devices and checks which device has interrupted the CPU.

2.5 (c)

Peripheral component interconnect bus replaced other slow speed slots and also VESA bus which was under ISA (Industry Standard Association).

It has more data rate and can be used for general applications including video graphic applications.

2.6 (b)

PCI-64 bus.

Remaining were slow as compared to PCI-64 bus in data rate.

2.7 (a)

Peripheral component inter connect, initialization are done in physical layer.

2.8 (b)

Main advantages of multiprocessor systems is to change its memory access model from uniform memory access to non uniform memory access.

There two types of systems:

1. Symmetric multiprocessing
2. Asymmetric multiprocessing

2.9 (a)

$$\text{Cycle time} = \frac{1}{200\text{ MHz}}\text{sec}$$

$$= 0.005\ \mu\text{sec}$$

$$\text{Average instruction ET} = \left(\frac{\text{Time req/prog}}{\text{Instruction/prog}}\right)$$

$$= \frac{[(8*1)+(4*3)+(2*4)+(4*3)]0.005\ \mu\text{sec}}{8+4+2+4}$$

$$= 0.01111\ \mu\text{sec}$$

$$1\text{ Instruction} = 0.01111\ \mu\text{sec}$$

$$\text{Instructions in 1 sec} = \frac{1\text{ Instruction}}{0.01111\ \mu\text{sec}}$$

$$= 90\text{ MIPS}$$

2.10 (c)

Machine parallelism is a measure of the ability of the processor to take advantage of instruction-level parallelism. Machine parallelism is determined by the number of instructions that can be fetched and executed at the same time (the number of parallel pipelines) and by the speed and sophistication of the mechanisms that the processor uses to find independent instructions.

2.11 (a)

In interrupt driven IO, CPU initializes the IO module for IO job and gets busy with other task. Meanwhile, when interrupt occurs, then CPU services the interrupt by suspending the normal flow of execution.

2.12 (b)

CISC cpu contain more addressing modes and larger instruction set.

2.13 (a)

CISC cpu supports microprogrammed control unit. So, this control unit is flexible to implement new instructions.

2.14 (b)

Pipelining allows overlapping execution to improve the performance.

2.15 (a)

"DDR SDRAM → Double data rate synchronous data RAM

Given frequency = 100 MHz,

So, $T = \frac{1}{f} = 0.01$ μsec

In one clock 2 bits are transferred in double data RAM since in both rising and falling edge of clock the data transfer takes place. Hence, through 64 bits bus, 16 bytes are transferred.

$$\therefore \text{Transfer rate} = \frac{16\text{B}}{0.01\,\mu\text{sec}} = 1600 \text{ MB/sec}$$

2.16 (a)

Rotational latency:

6000 revolutions – 1 min (60 sec)

⇒ 1 revolution – ?

$$\frac{60}{6000}\text{sec} = 0.01 \text{ sec}$$

Average rotational latency

$$= \frac{1}{2}\text{revolution time}$$

$$= \frac{1}{2} \times 0.01\,\text{sec} = 0.005 \text{ sec}$$

2.17 (a)

P_{id}	Burst Time	
P_1	~~24~~	20
P_2	~~3~~	0
P_3	~~3~~	0

$TQ = 4$

Ready queue :

P_1	P_2	P_3	P_1

Chart :

P_1	P_2	P_3	P_1

0 4 7 10 30

	CT	TAT (CT – AT)	WT (TAT – BT)
P_1	30	30	6
P_2	7	7	4
P_3	10	10	7

$$\text{Avg. WT} = \frac{6+4+7}{3} = 5.66$$

2.18 (c)

P_{id}	Burst Time	Priority
P_1	10	3
P_2	1	1 (High)
P_3	2	4
P_4	1	5
P_5	5	2

Chart:

P_2	P_5	P_1	P_3	P_4

0 1 6 16 18 19

	CT	TAT (CT – AT)	WT (TAT – BT)
P_1	16	16	6
P_2	1	1	0
P_3	18	18	16
P_4	19	19	18
P_5	6	6	1

AT = 0

$$\text{Average WT} = \frac{(6+0+16+18+1)}{5} = 8.2$$

■■■■

3 Memory Organization

3.1 A computer system has a cache with a cache access time $T_C = 10$ ns, a hit ratio of 80% and an average memory access time $T_M = 20$ ns. What is the access time for physical memory T_P?

(a) 90 ns (b) 80 ns
(c) 60 ns (d) 20 ns **[ESE-2017]**

3.2 **Statement (I) :** Direct access method is based on a disk model of a file.

Statement (II) : Disks allow random access to any file block. **[ESE-2018]**

3.3 Consider the following statements:

1. Better memory utilization is possible with non-contiguous allocation using fixed size pages.
2. Associated memory is used for providing fast access to data stored in cache memory.
3. Direct mapping of cache memory is hard to implement.

Which of the above statements are correct?

(a) 1 and 2 only (b) 1 and 3 only
(c) 2 and 3 only (d) 1, 2 and 3 **[ESE-2018]**

3.4 What is the effective access time, if the average page-fault service time is 25 ms, memory access time is 100 ns and page-fault rate is *P*?

(a) 100 + 24,999,900 × P ns
(b) 100 + 25,000,000 × P ns
(c) 100 + 25,000 × P ns
(d) 25,000,000 + 100 P ns **[ESE-2018]**

3.5 The Central Processing Unit (CPU) consists of

(a) ALU and control unit only
(b) ALU, control unit and Resisters only
(c) ALU, Control unit and System bus only
(d) ALU, Control unit, Registers and Internal bus **[ESE-2019]**

3.6 When enough total memory space exists to satisfy a request, but it is not contiguous, then this problem is known as

(a) Internal Fragmentation
(b) External Fragmentation
(c) Overlays
(d) Partitioning **[ESE-2019]**

3.7 If a cache has 64-byte cache lines, how long does it take to fetch a cache line if the main memory takes 20 cycles to respond to each memory request and returns 2 bytes of data in response to each request?

(a) 980 cycles (b) 640 cycles
(c) 320 cycles (d) 160 cycles **[ESE-2019]**

3.8 Which of the following statements are correct about SRAM?

1. It provides faster access as compared to DRAM.
2. It is cheaper than DRAM.
3. It is more expensive than DRAM.
4. It has higher bit density than DRAM.

(a) 1 and 4 only (b) 1 and 3 only
(c) 1, 3 and 4 only (d) 2 and 4 only
[ESE-2019]

3.9 Features of solid state drives (SSDs) are:

1. High-performance in input/output operations per second.
2. More power consumption than comparable size HDDS.
3. Lower access times and latency rates.
4. More susceptible to physical shock and vibration.

(a) 2 and 3 only (b) 2 and 4 only
(c) 1 and 3 only (d) 1 and 4 only
[ESE-2019]

3.10 The memory management function of virtual memory includes

1. Space allocation
2. Program relocation
3. Program execution

4. Code sharing
(a) 1, 2 and 3 only (b) 1, 2 and 4 only
(c) 1, 3 and 4 only (d) 2, 3 and 4 only
[ESE-2019]

3.11 The physical address translation in virtual memory address with Memory Management Unit (MMU) is done by which one of the following mechanisms?
(a) Multiply virtual, address by some constant
(b) Translation lookaside buffer (TLB)
(c) Encryption key
(d) Using general purpose register in CPU
[ESE-2020]

3.12 **Statement (I) :** A cache is a memory unit placed between the CPU and main memory M and is used to store instructions, data or both.
Statement (II) : The cache's effect is to increase the average time required to access an instruction or data word, typically to just a single-clock cycle.
(a) Both Statement (I) and Statement (II) are individually true and Statement (II) is the correct explanation of Statement (I)
(b) Both Statement (I) and Statement (II) are individually true; but Statement (II) is **not** the correct explanation of Statement (I)
(c) Statement (I) is true; but Statement (II) is false
(d) Statement (I) is false; but Statement (II) is true [ESE-2020]

3.13 **Statement (I) :** A buffer is not an area in RAM or on the hard drive designated to hold input and output on their way in or out of the system.
Statement (II) : The process of placing items in a buffer so they can be retrieved by the appropriate device when needed is called spooling.
(a) Both Statement (I) and Statement (II) are individually true and Statement (II) is the correct explanation of Statement (I)
(b) Both Statement (I) and Statement (II) are individually true; but Statement (II) is **not** the correct explanation of Statement (I)
(c) Statement (I) is true; but Statement (II) is false
(d) Statement (I) is false; but Statement (II) is true
[ESE-2020]

3.14 Daisy-chaining method is used for
(a) establishing priority
(b) data transfer
(c) initiating input and output
(d) direct memory access [ESE-2021]

3.15 The performance of cache memory is measured in terms of
(a) Read ratio (b) Reference ratio
(c) Hit ratio (d) Locality ratio
[ESE-2021]

3.16 Which of the following issues are addressed by redundant arrays of independent disks?
(a) Performance and reliability
(b) Performance and stability
(c) Performance and process
(d) Performance and storage space
[ESE-2022]

3.17 Stack-oriented machine
(a) contains any accumulator or general purpose registers.
(b) does not contain only a stack pointer which points to the stack top.
(c) requires any operand address for arithmetic, logical and comparison instructions
(d) does not contain any accumulator or general purpose registers.
[ESE-2022]

3.18 Virtual memory implements the translation of a program's address space to
(a) virtual addresses
(b) physical addresses
(c) mapping addresses
(d) page addresses [ESE-2022]

3.19 Consider a main memory consisting of 4 K blocks, a cache memory consisting of 64 blcoks, and a block size of 32 words. Compute the number of bits in the main memory address using direct-mapped cache techniques.
(a) 16 (b) 17
(c) 19 (d) 14 [ESE-2023]

3.20 How many pages are in the disk, if the capacity of a virtual disk is 2 MB and each page is 2 kB in a byte-addressable system?
(a) 2048 (b) 1024
(c) 100 (d) 500
[ESE-2024]

■■■■

Answers Memory Organization

3.1 (c) 3.2 (a) 3.3 (a) 3.4 (a) 3.5 (d) 3.6 (b) 3.7 (b) 3.8 (b) 3.9 (c)
3.10 (b) 3.11 (b) 3.12 (c) 3.13 (b) 3.14 (a) 3.15 (c) 3.16 (a) 3.17 (a) 3.18 (b)
3.19 (b) 3.20 (b)

Explanations Memory Organization

3.1 (c)

$$T_C = 10 \text{ ns}$$
$$H_C = 0.8$$
$$T_{avg} = 20 \text{ ns}$$
$$T_m = ?$$
$$T_{avg} = H_C T_C + (1 - H_C)(T_m + T_C);$$
(Hierarchical memory org.)
$$20 = (0.8 \times 10) + (1 - 0.8)(x + 10)$$
$$20 = 8 + 0.2x + 2$$
$$x = 50 \text{ ns (Option not matching)}$$
$$T_{avg} = H_C T_C + (1 - H_C)\, T_m;$$
(Simultaneous memory org.)
$$20 = (0.8 \times 10) + (1 - 0.8)\, T_m$$
$$T_m = 60 \text{ ns}$$

3.3 (a)

Direct mapping is easy to implement.

3.4 (a)

$$\text{EMT} = (P \times S) + (1 - P)\, T_m$$
$$= (P \times 25 \text{ ms}) + (1 - P)\, 100 \text{ ns}$$
$$= (25000000 \text{ ns} \times P) + 100 \text{ ns} - 100P$$
$$= 100 \text{ ns} + 24{,}999{,}900 \times P$$

3.5 (d)

CPU contains ALU, CU and registers and internal buses.

3.7 (b)

CM line size = 64 B

Memory access time = 20 cycles/2 byte transfer

Total time required to fetch the cache line is,

⇒ (Memory referred number of times) × (Memory access time)

$$\Rightarrow \frac{64}{2} \times 20 \text{ cycles}$$

⇒ 640 cycles

3.8 (b)

SRAM is faster than DRAM and more expensive.

3.9 (c)

SSD's uses software instructions to go directly to the location where data is stored to access, therefore time required to access the data is low. Where as in the hard disk drive mechanically move its read/write head across a spinning platter to locate data to access.

3.10 (b)

Program execution is a CPU function. Virtual memory concept is used to increase the address space.

3.12 (c)

Statement-I is true. Statement-II is false.

3.13 (b)

Both statements are individually true.

3.14 (a)

Daisy-chaining method is used for establishing priority. It consists of a serial connection of all devices that request an interrupt. The device with the high priority is placed in the first position followed by lower priority devices.

3.15 (c)

- The performance of cache memory is frequently measured in terms of a quantity called Hit ratio. We can improve Cache performance using higher cache block size, higher associativity, reduce miss rate, reduce miss penalty, and reduce the time to hit in the cache.
- $\text{Hit Ratio (H)} = \dfrac{\text{\# Hits}}{\text{Total \# Accesses}}$

3.16 (a)

The main purpose of maintaining RAID is for performance and reliability.

3.17 (a)

In stack oriented machine CPU, data must be pushed onto stack first, then operated but

ACC and other registers also required.

Stack pointer is must.

∴ b, c, d are incorrect.

ALU instructions are operated on data at top of stack.

3.18 (b)

Virtual memory implements the translation of a program's address space to physical addresses.

CPU always generates virtual address but actual programs resides in physical address, so it should translate to physical address in order to access any instruction of the program.

3.19 (b)

Blocks in main memory = 4 K

Lines in cache memory = 64

Block size = 32 words

∴ Main memory size = 4 K × 32 W

$= 2^{12} \times 2^5$ W

$= 2^{17}$ W

So, MM addr = $\log_2 2^{17} = 17$ bit

3.20 (b)

$$\text{Pages in SM} = \frac{\text{SM size}}{\text{Page size}}$$

$$= \frac{2\text{ MB}}{2\text{ kB}} = \frac{2^{20}}{2^{10}} = 2^{10} = 1024$$

■■■■

4 Networking Fundamentals

4.1 Consider the following statements regarding Peer-to-Peer computing environment:
1. In this system, clients and servers are not distinguished from one another.
2. All nodes distributed throughout the system (within) are considered Peers and each may act as either a client or a server.
3. Peer-to-Peer system assuredly offers certain advantages over the traditional client-server system.
4. Peer-to-Peer system is just a replica of the file-server system.

Which of the above statements are correct?
(a) 1, 2, 3 and 4 (b) 1, 2 and 3 only
(c) 1 and 4 only (d) 2, 3 and 4 only
[ESE-2017]

4.2 The length of cable required for transmitting a data at the rate of 500 Mbps in an Ethernet LAN with frames of size 10,000 bits and for signal speed 2,00,000 km/s is
(a) 2.5 km (b) 2.0 km
(c) 1.5 km (d) 1.0 km **[ESE-2018]**

4.3 In which one of the following classes of computers, is the relationship between architecture and organization very close?
(a) Microcomputers
(b) Mini computers
(c) Mainframe computer
(d) Super computers **[ESE-2019]**

4.4 The total average read of write time T_{total} is
(a) $T_s + \frac{1}{2r} + \frac{b}{N}$ (b) $T_s + \frac{1}{2r} + \frac{b}{rN}$
(c) $\frac{T_s}{rN} + \frac{b}{N}$ (d) $T_s + 2r + \frac{b}{rN}$

where,
T_s = Average seek time
b = Number of bytes to be transferred
N = Number of bytes on a track
r = Rotation speed, in revolution per second
[ESE-2019]

4.5 Which one of the following is a powerful web platform for web applications and web services, built-in virtualization technologies, variety of new security tools, enhancements and streamlined configuration and management tools?
(a) Internet Explorer
(b) Internet Information Services
(c) Web Matrix
(d) Visual Web Developer **[ESE-2020]**

4.6 Which one of the following provides an interface to which a client can send a request to perform an action, in response, the server executes the action and sends back results to the client?
(a) File server system
(b) Open-source system
(c) Compute server system
(d) Peer to peer system **[ESE-2022]**

4.7 Which one of the following functions is not performed by the USB host controller?
(a) Configure the scheduling algorithms
(b) Packet generation
(c) Serializer/Deserializer
(d) Process request from device and host
[ESE-2022]

4.8 Consider the following functions of Root Hub :
1. It performs power distribution to the devices.
2. It enables and disables the ports.
3. It reports status of each port to the user.

Which of the above functions is/are correct?
(a) 1 only (b) 2 and 3 only
(c) 1 and 2 only (d) 1,2 and 3 **[ESE-2024]**

4.9 Which one of the following is used to keep the track of program statistics that may be a valuable tool for system administrators who wish to reconfigure the system to improve computing services?
(a) Programming table (b) Spooling
(c) Logging (d) Making file
[ESE-2024]

■■■■

Answers Networking Fundamentals

4.1 (b) 4.2 (b) 4.3 (a) 4.4 (b) 4.5 (d) 4.6 (c) 4.7 (a) 4.8 (c) 4.9 (c)

Explanations Networking Fundamentals

4.2 (b)

Transmission time (T.T.) ≥ 2 × Propagation Time (PT)

$$TT = \frac{\text{Frame size}}{\text{Band width}} = \frac{10,000}{500 \times 10^6}$$

$$PT = \frac{\text{Distance}}{\text{Speed}}$$

$$= \frac{x}{200000 \times 10^3 \text{ m/sec}}$$

$$TT \geq 2\ PT$$

$$\frac{10000}{500 \times 10^6} \geq 2 \times \frac{x}{200000 \times 10^3}$$

$$x = \frac{10^4 \times 2 \times 10^8}{2 \times 5 \times 10^8}$$

$$\Rightarrow \quad x = 2000 \text{ meter} \Rightarrow 2 \text{ km}$$

4.3 (a)

Microcomputer is a personal computer. So, architecture and organization is very close.

4.4 (b)

Seek time = T_s

"r" revolution ⇒ 1 sec

1 revolution ⇒ $\frac{1}{r}$sec

∴ Rotational latency = $\frac{1}{2}$ revolution time

$$= \frac{1}{2} * \frac{1}{r}\text{sec} = \frac{1}{2r}\text{sec}$$

Transfer time:

1 revolution time ⇒ 1 track data (N-bytes)

$$\text{time required for } b \text{ bytes} = \frac{\frac{1}{r}\text{sec} * \text{b}}{N} = \frac{b}{rN}$$

Total time = Seek time + Average rotational latency + Transfer time

$$= T_s + \frac{1}{2r} + \frac{b}{rN}$$

4.6 (c)

What ever the request sent by the client will be sent to server system and request will be processed in the several system and processed result will be sent back to client system. This is generally known as client server system and it is also known as computer server system.

4.7 (a)

Configuration the scheduling algorithms not performed by the USB host controller but other operations are performed in the USB host controller.

4.8 (c)

Root hub has the capability to enable and disable USB ports. This functionality can be useful for trouble shooting H/w issues, conserving power and controlling H/W issues, conserving power and controlling access to USB devices.

■■■■

5 Programming Elements

5.1 If n has the value 3, then C language statements:
$a[++ n] = n ++$; assigns

(a) 3 to $a[5]$ (b) 4 to $a[5]$
(c) 4 to $a[4]$ (d) 5 to $a[5]$

[ESE-2017]

5.2 The minimum number of arithmetic operations required to evaluate the polynomial $P(X) = X^5 + 8X^3 + X$ for a given value of X using only one temporary variable is

(a) 8 (b) 7
(c) 6 (d) 5

[ESE-2017]

5.3 What does the following program print?

```
void f(int*p, int*q)
{
    p = q;
    *p=2;
}
int i = 0, j = 1;
int main( )
{
    f(&i, &j);
    printf("%d %d \n", i,j);
    getchar ( );
    return 0;
}
```

(a) 2 2 (b) 2 1
(c) 0 1 (d) 0 2 [ESE-2018]

5.4 Consider the function fun1 shown below:

```
    int fun1 (int num)
{
    int count=0;
    while (num)
    { count ++;
      num>>=1;
    }
    return (count);
}
```

The value returned by fun1 (435) is

(a) 10 (b) 9
(c) 8 (d) 7

[ESE-2018]

5.5 **Statement (I) :** Variables that are defined inside subprograms are local variables.

Statement (II) : Their scope is in the body of the subprogram in which they are defined.

(a) Both Statement (I) and Statement (II) are individually true and Statement (II) is the correct explanation of Statement (I)
(b) Both Statement (I) and Statement (II) are individually true; but Statement (II) is **not** the correct explanation of Statement (I)
(c) Statement (I) is true; but Statement (II) is false
(d) Statement (I) is false; but Statement (II) is true

[ESE-2018]

5.6 **Statement (I):** High-level programming languages preferred by the scientific community as they are user friendly.

Statement (II): High level programming languages provide ways of detailing instruction for problem-solving that are translated into low-level language via compilers and interpreters before being executed by the computer.

(a) Both Statement (I) and Statement (II) are individually true and Statement (II) is the correct explanation of Statement (I)
(b) Both Statement (I) and Statement (II) are individually true; but Statement (II) is **not** the correct explanation of Statement (I)
(c) Statement (I) is true; but Statement (II) is false
(d) Statement (I) is false; but Statement (II) is true

[ESE-2018]

5.7 **Statement-I:** One can formulate problems more efficiently in a high level language and need not have a precise knowledge of the architecture of the computer.

Statement-II: High level languages permit programmers to describe tasks in a form which is problem oriented than computer oriented.

(a) Both Statement (I) and Statement (II) are individually true and Statement (II) is the correct explanation of Statement (I)
(b) Both Statement (I) and Statement (II) are individually true; but Statement (II) is **not** the correct explanation of Statement (I)
(c) Statement (I) is true; but Statement (II) is false
(d) Statement (I) is false; but Statement (II) is true

[ESE-2019]

5.8 A physical implementation of the type declaration in high-level programming languages where major information types should be assigned formats for identification is called

(a) Storage order (b) Tag
(c) Error correction (d) Error detection

[ESE-2020]

5.9 Which one of the following instruction formats is designed to be a large-scale time-shared system, with an emphasis on making the system easy to program, even if additional hardware expense was involved?

(a) PDP-8 (b) PDP-10
(c) PDP-6 (d) PDP-12

[ESE-2023]

5.10 When a process creates a new process using the fork() operation, which one of the following states is shared between the parent process and the child process?

(a) Stack
(b) Heap
(c) Linked list
(d) Shared memory segments

[ESE-2023]

■■■■

Answers Programming Elements

5.1 (c) 5.2 (d) 5.3 (d) 5.4 (b) 5.5 (a) 5.6 (a) 5.7 (a) 5.8 (b) 5.9 (b)
5.10 (d)

Explanations Programming Elements

5.1 (c)

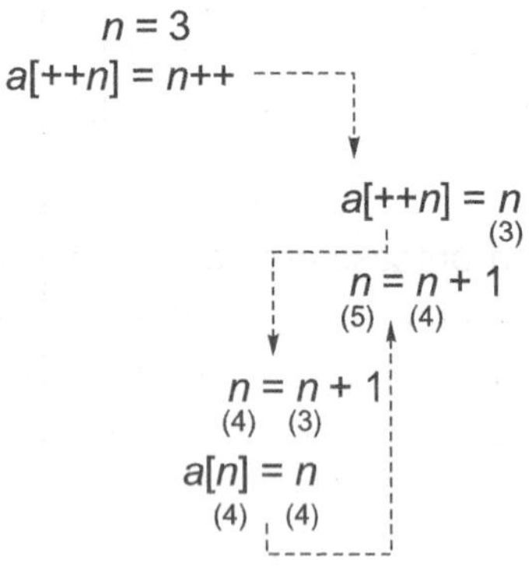

5.2 (d)

$$P(x) = x^5 + 8x^3 + x = x[x^4 + 8x^2 + 1]$$
$$= x[x^2 (x^2 + 8) + 1]$$

Here, $t = x \times x$ {i.e. $t = x^2$} ①

So, $P(x) = x[t[t + 8] + 1]$ (operations ②, ③, ④, ⑤)

∴ Number of arithmetic operations are '5'.

5.3 (d)

i: [0] (100) j: [1] (200)

$f(100, 200)$

$f(*p, *q)$

$*p$: [100] (2000) $*q$: [200] (3000)

$p = q$; p: [200] (2000)

$*p = 2$; [[p]] ← 2
200
j: [2] (200)

return function
printf (i, j)
i.e. (0, 2)

5.4 (b)

Fun1(435)

$$(435)_{10} = (1\ 1\ 0\ 1\ 1\ 0\ 0\ 1\ 1)_2$$

So, count value 9. i.e., number of bits in binary code.

5.7 (a)

High level languages are machine independent whereas low level languages are machine dependent.

A H.L.L. programmer requires knowledge of logic implementation for the problem.

5.9 (b)

In large scale time shared system, PDP-10 instruction formats are used.

5.10 (d)

Shared memory segments are shared between the parent and child processes.

■■■■

6 Operating System Concepts

6.1 Consider the following set of processes with data thereof as given here:

Process	Arrival time	CPU Burst time
P_1	0 ms	12 ms
P_2	2 ms	4 ms
P_3	3 ms	6 ms
P_4	8 ms	5 ms

An operating system uses shortest remaining time first scheduling algorithm for pre-emptive scheduling of processes. The average waiting time of the processes is

(a) 7.5 ms (b) 6.5 ms
(c) 5.5 ms (d) 4.5 ms [ESE-2018]

6.2 Consider the following statements:

1. System calls provide the interface between a process and the operating system.
2. PERL implementations include direct system call access.
3. System calls occur in different ways depending on the computer in use.

Which of the above statements are correct?

(a) 1 and 2 only (b) 1 and 3 only
(c) 2 and 3 only (d) 1, 2 and 3 only

[ESE-2018]

6.3 A multiprocessing technology which enables software to treat a single processor as two processors to utilize the processing power in the chip that would otherwise go unused and lets the chip operate more efficiently resulting in faster processing is called

(a) Systematic multiprocessing
(b) Massively parallel processing
(c) Co-processing
(d) Hyper threading

[ESE-2020]

6.4 According to the Belady's anomaly, the page-fault rate may

(a) increase as the number of allocated frames increases.
(b) increase as the number of allocated frames decreases.
(c) not change as the number of allocated frames increases.
(d) not change as the number of allocated frames decreases.

[ESE-2021]

6.5 Which one of the following tools is used to protect critical sections and prevent race conditions?

(a) Mutex lock (b) Semaphore lock
(c) Spooling lock (d) Livelock

[ESE-2022]

6.6 To arise a deadlock situation, which of the following conditions hold simultaneously in a system?

1. Mutual exclusion
2. Hold and wait
3. Preemption

Select the correct answer using the code given below:

(a) 1 and 3 only (b) 1 and 2 only
(c) 2 and 3 only (d) 1, 2 and 3

[ESE-2022]

6.7 Which one of the following statements is correct for atomic data?

(a) Aggregation of atomic and composite data into a set with defined relationships are validated.
(b) Heterogeneous combination of data into a single structure with an identified key is validated.
(c) Homogeneous combination of data into a single structure with an identified key is validated.

(d) It is a data that consists of a single piece of information; that is, they cannot be divided into other meaningful pieces of data.

[ESE-2023]

6.8 Consider the four processes $P_1 = 6$, $P_2 = 8$, $P_3 = 7$ and $P_4 = 3$ with the length of the CPU burst time given in milliseconds and all four processes arrive at time 0. What is the average waiting time for these processes with the shortest job first algorithm?

(a) 5 ms (b) 7 ms
(c) 9 ms (d) 11 ms

[ESE-2023]

6.9 The processes that are residing in main memory and are ready and waiting to execute are kept on a list called the ready queue. Then, this queue is generally stored as

(a) a double ended queue
(b) a priority queue
(c) a linked list
(d) a double ended list

[ESE-2023]

6.10 Consider the following statements regarding thread pool :

1. It is to create a number of threads at process startup and place them into a pool, where they sit and wait for work.
2. A thread pool limits the number of threads that exist at any one point and this is particularly important on systems that cannot support a large number of concurrent threads.
3. Once the summation thread is created, the parent must wait for it to complete before outputtting the value of sum, as the value is set by the summation thread.
4. It allocates memory and initializes a new thread in the JVM.

Which of the above statements are correct?

(a) 1 and 2 only (b) 3 and 4 only
(c) 1 and 3 only (d) 1, 2, 3 and 4

[ESE-2023]

6.11 If communicating processes reside in a temporary queue and the queue has a maximum length of zero, then the link cannot have any messages waiting in it. The sender must block until the recipient receives the message. In what way can such queues be implemented?

(a) Bounded capacity (b) Unbounded capacity
(c) Non-zero capacity (d) Zero capacity

[ESE-2024]

■■■■

Answers Operating System Concepts

6.1 (c) 6.2 (d) 6.3 (d) 6.4 (a) 6.5 (a) 6.6 (b) 6.7 (d) 6.8 (b) 6.9 (b)
6.10 (d) 6.11 (d)

Explanations Operating System Concepts

6.1 (c)

Gantt Chart:

P_1	P_1	P_2	P_2	P_2	P_2	P_3	P_3	P_3	P_4	P_1

0 1 2 3 4 5 6 7 8 12 17 27

	CT	TAT (CT − AT)	WT (TAT − BT)
P_1 :	27	27	15
P_2 :	6	4	0
P_3 :	12	9	3
P_4 :	17	9	4

$$\text{Average waiting time} = \frac{(15+0+3+4)}{4} = 5.5 \text{ ms}$$

6.4 (a)

Belady's anomaly states that page fault rate may increase as the number of allocated frames increase.

6.5 (a)

Critical section means shared resources, this will be protected by mutex lock.

6.6 (b)

To arise the deadlock four different conditions are necessary

mutual exclusion

hold and wait

no pre-emption

circular wait

Only two conditions are matching from given points: Mutual exclusion and hold and wait.

6.7 (d)

Atomic data is a primitive data which does not have the further decomposition. So, atomic data is a final meaning of the data.

6.8 (b)

Pid	BT
P1	6
P2	8
P3	7
P4	3

AT = 0

Chart :

P4	P1	P3	P2

0 3 9 16 24

	CT	TAT	WT
P_1	9	9	3
P_2	24	24	16
P_3	16	16	9
P_4	3	3	0

$$\therefore \text{Average weigth} = \frac{(3+16+9+0)}{4} = 7$$

6.9 (b)

Processes are inserted into a queue based on the arrival time.

6.10 (d)

A thread pool reuses previously created threads to execute current tasks and offers a solution to the problem of thread cycle overhead and resource thrashing.

Thread pool is a pool of worker threads. Worker thread is thread which can accept the task then complete it and come back again to thread pool to accept another task.

6.11 (d)

Zero capacity queue is referred to as a message system with no buffering. Zero capacity queue has maximum capacity of zero. So message queue doesn't have any waiting message in it.

■■■■